# INTRODUCTION TO THE CALCULUS OF VARIATIONS AND ITS APPLICATIONS

# INTRODUCTION TO THE CALCULUS OF VARIATIONS AND ITS APPLICATIONS

**Frederick Y.M. Wan**

*University of California, Irvine*

CRC Press
Taylor & Francis Group
Boca Raton London New York

CRC Press is an imprint of the
Taylor & Francis Group, an **informa** business
A CHAPMAN & HALL BOOK

First published 1995 by Chapman & Hall

Published 2019 by CRC Press
Taylor & Francis Group
6000 Broken Sound Parkway NW, Suite 300
Boca Raton, FL 33487-2742



First issued in paperback 2019



ISBN-13: 978-0-367-44924-7 (pbk)
ISBN-13: 978-0-412-05141-8 (hbk)



**Library of Congress Cataloging-in-Publication Data**

**Wan, Frederic Y.M.**
**Introduction to the calculus of variations and its applications/Frederic Y.M. Wan.**
**p. cm.**
**Includes bibliographical references.**
**ISBN 0-412-05141-9**
**1. Calculus of variations. I. Title.**
**QA315.W34 1993**
**515'.64—dc20 93-30157**
**CIP**

**Cover Design: Andrea Meyer, EmDash Inc.**

To my teacher, Eric Reissner,
who made it all possible.

# Contents

# Preface

It is human nature to optimize, and the *calculus of variations* has been developed for more than two centuries to quantify the optimization process for a certain class of optimization problems. Many new applications in science and engineering since the early twentieth century have led to the theory of *optimal control* as a variance of this classical subject. Yet, not many mathematics departments in our universities have courses on variational calculus or its modern derivative.

I became aware of the important role of the calculus of variations in continuum mechanics from courses in *elasticity theory* taught by Eric Reissner, my eventual Ph.D. thesis advisor. I was particularly impressed by Eric's ingenious use of the *semidirect method* to obtain from the theory of elasticity results for beams, plates, and shells. Later, I learned from the book by Bryson and Ho (1969) techniques in optimal control needed for my collaborated research in resource economics with Robert M. Solow. These important applications in mechanics and economics have already made their way into advanced texts in the two disciplines. However, it would be too narrowly focused for the calculus of variations or optimal control to be taught as a course in aeronautical engineering or management science. My own preference has always been to have a sequence of mathematics courses on these two topics which includes a wide range of applications. Through such courses, we may (re)kindle an interest on variational methods among students in mathematics along with other students from applications areas.

In 1974, I left Massachusetts Institute of Technology (MIT) for the University of British Columbia. Among the new courses I worked to introduce for its applied mathematics program was an upper division, two-semester (honors) course sequence entitled Calculus of Variations and Optimal Control. The contents and flavor of that course were pretty much set by its first instructor, Frank H. Clarke, who wrote the course description and prepared a set of lecture notes for the course on material similar to what we find in Courant and Hilbert

(1953), Gelfand and Fomin (1963), and Caratheodory (1935). Others who have taught the course include Colin W. Clark, Ulrich Haussmann, and myself; we all made use of Clarke's notes to different extents.

Shortly after I moved to the University of Washington in 1983, a three-quarter sequence on the calculus of variations and optimal control was organized for the existing graduate degree programs in our Department of Applied Mathematics and Department of Mathematics. The first instructor for the sequence was R. Tyrrell Rockafellar, who just happened to be the thesis advisor of Frank Clarke! Other instructors for the courses in this sequence have been Carl E. Pearson and myself. Neither Rockafellar nor Pearson followed any existing text or produced any lecture notes for the students. This volume is an edited version of the course note I wrote for my students when I taught the first two quarters of the sequence in 1988–90. It is organized as a text for an upper division undergraduate or beginning graduate course in mathematics with broad applications in many different areas of science and engineering.

## The Basic Problem

Pedagogically, it seems reasonable to begin such a volume by developing a calculus of variations for *smooth solutions* of the *basic problem* of the theory. (For kinematics and dynamics, for example, we expect force and acceleration to be well-defined and piecewise continuous.) At the end of Chapter 1, however, an example shows that even a very simple basic problem may have only a *piecewise smooth* (PWS) solution. This leads to a discussion of the integro-differential equation form of the Euler–Lagrange necessary condition for PWS *extremals* in Chapter 2.

A highlight of Chapter 2 is a discussion of the *Erdmann–Weierstrass corner conditions*. There are very few techniques for solving integro-differential equations. It is, therefore, desirable to work with the differentiated form of the Euler–Lagrange equation (called the *Euler differential equation* in this volume) applicable over intervals of the solution domain where the PWS solution (known as an *extremal*) is smooth. The Erdmann conditions help us find the possible location of the kinks of the PWS solution.

Most texts do not make this point clear; nor do they show how these conditions can be applied in a systematic way for locating points of discontinuity of the derivative of the solution. Chapter 2 contains many examples to illustrate how the Erdmann conditions are to be used. The simplest way to deduce the second corner condition is by way of the parametric form of the basic problem. The vector form of the basic problem is needed for this purpose and, therefore, is treated in that chapter.

Several modifications of the basic problem are treated in Chapter 3. They involve principally various relaxations of the prescribed end conditions, including the free-boundary problem. The extension to Lagrangians containing higher derivatives than the first is also discussed. The chapter ends with an example which shows that there are physically meaningful problems for which the solution lies outside the class of PWS functions.

Chapters 4 and 5 are concerned with sufficient conditions for a *weak* and *strong minimum* (or maximum). The Legendre, Jacobi, and Weierstrass necessary conditions are deduced and then strengthened for sufficiency theorems. Convexity as a sufficient condition is also analyzed. A specific example is produced to show that a weak minimum may not be the true minimum. It justifies the time and effort we choose to devote to a study of the theory for a strong minimum.

## Methods of Solution

The most efficient way to find the solution of the basic problem or its modified form is to find the extremals. In other words, we should know how to solve the second-order (scalar or vector) Euler differential equation. Chapter 6 offers a different-perspective to the solution process by casting the Euler differential equation as a first-order system, called the *Hamiltonian system*. The *Hamilton–Jacobi theory* is then developed for the solution of such a system by way of a generating function. An alternative solution process is to find the first integral of the Hamiltonian system. This leads to a discussion of Noether's theorem and its applications in Chapter 7, preceded by some material on analytical mechanics.

All the special techniques discussion in Chapters 1, 2, 6, and 7 notwithstanding, it is still a fact that exact solutions of Euler differential equations are rare events. Only adequate approximate solutions can be expected in most cases. Three chapters are devoted to methods for an approximate solution. Chapter 8 describes the *direct method* of the Rayleigh–Ritz type for constructing an approximate minimum point from a finite dimensional subspace of the original comparison function space. The spline or one-dimensional finite element method is introduced to show the advantages of local basis (or coordinate) functions. A *dual* problem is formulated to provide a lower bound of the approximate solution. Because any approximate solution of the *primal* problem is an upper bound, we have a squeeze on the approximate solution for assessing its accuracy. The concept and method of a *weak solution* are also discussed in this chapter. Chapter 9 introduces the technique of *dynamic programming* and shows how it can be used to solve problems in the calculus of variations. A third approach is to solve the two-point boundary-value problem defined by the Euler differential equation(s) and the auxiliary boundary conditions numerically. This approach involves no optimization and is, therefore, relegated to the appendix of the book.

### Constraints

Most problems in the calculus of variations have some restrictions on their solution. Five chapters are devoted to variational problems with additional constraints. Chapter 10 discusses problems with *isoperimetric constraints* and applies the theory developed to mechanical vibration, optimal forest harvesting schedule, and to the general Sturm–Liouville problem in ordinary differential equations. The *Rayleigh quotient* is deduced for the lowest eigenvalue; techniques for obtaining higher eigenvalues are also formulated. Pointwise *holonomic* equality and inequality constraints are treated in Chapter 11. Pointwise *nonholonomic* equality and inequality constraints involving derivatives of the unknowns are treated in Chapter 12. The possibility of a singular solution is indicated and the Most Rapid Approach technique is described for problems with a singular solution.

Chapters 13 and 14 are devoted to constraints in an optimal control setting. Only problems with linear dynamics and a linear Lagrangian are treated in Chapter 13. For these problems, a complete discussion of *controllability* is possible. More general optimal control problems and *sufficient conditions* for optimality are treated in Chapter 14. Many illustrative examples from the social sciences are solved in both chapters.

For all types of constraints, an effort is made through specific examples to show that *abnormal problems* exist for which the constraints determine the solution and the Lagrangian plays no role in the optimization process. The various *multiplier methods* are formulated to allow for abnormality whenever appropriate.

### Applications in Higher Dimensions

Variational problems in several independent variables are studied in Chapter 15. A variational formulation is shown to be possible for Maxwell's equations of electromagnetics. Various types of modifications of the basic problems similar to those in one dimensional problems are also considered. The finite element method is now more important for approximate solutions and is discussed in more detail. The technique is illustrated through the torsion problem of elastic bars. With several independent variables, we have now the option of using the Rayleigh–Ritz method with coordinate functions of known dependence in some variable(s) and not in the others. This so-called *semidirect method* of the calculus of variations is introduced by a simple membrane problem and used extensively later in Chapter 17 for the derivation of various plate theories.

The calculus of variations finds many applications in solid and fluid mechanics. Chapter 16 introduces the basic elements of elasticity theory and some variational formulations of the relevant boundary-value problems. Aside from the classical principle of minimum potential energy and the principle of minimum complementary energy, we also establish their relations to Reissner's variational principle and the Hu–Washizu principle which are less restrictive and, therefore, have wider applications. The theory of Saint–Venant torsion is deduced by an application of the semidirect method in conjunction with the theorem of minimum potential energy.

In Chapter 17, we see how the variational approach to continuum mechanics was used to resolve a long-standing paradox concerning the appropriate stress boundary conditions for the classical *thin plate theory*. It is relatively straightforward to obtain the two correct (contracted) conditions for this theory once we have a variational formulation of the problem. Reissner's ingenious application of *semidirect method* provides us for the first time with a plate theory which allows for the satisfaction of three physically expected edge stress conditions. This theory is known to be adequate for relatively thick plates. A similar semidirect method also enables us to derive the *Kirchhoff–Karman finite deflection (small strain) plate theory*.

I learned all the material in Chapters 16 and 17 when I was a student in Eric Reissner's courses on Elasticity Theory and have been living with it since that time. In contrast, I have worked much less with the material on *fluid mechanics* in Chapter 18 and have contended to limit the discussion there to some basic variational principles. A highlight of the chapter is a variational principle for *water waves* with a nonuniform bottom and its application to waves with slowly varying amplitude and phase. The chapter ends with a description of the variational formulation of Stokes' flow and Oseen's improvement.

## Selection of Course Material

There is obviously more material than we can teach in an introductory course. For a first quarter course on the classical subject, I normally cover most of the topics in chapters 1-5, 10-11 or 1-4, 10-11 and 15, omitting the material on inequality constraints in chapter 11. For a second quarter which is to include some optimal control theory, a combination of chapters 6-9 and 11-14 would be appropriate. If applications to continuum mechanics should be the theme, we should choose chapters 6-8, 15-18.

## Acknowledgments

In putting together this volume, I have obviously been influenced by Clarke's lecture notes and the book by Gelfand and Fomin in my choice of material for the basic problem. It is equally obvious that I have departed from their approaches (and others) by making extensive use of the parametric formulation. Some of the examples from management science and economics have previously appeared in Kamien and Schwarz (1981); but they are no less attractive or pedagogically appropriate for this volume. Many exercises were cribbed from all of the above as well as texts referenced in the Bibliography at the end of this book and are so acknowledged. I am also very much indebted to my colleague, Carl E. Pearson, who read chapters 1-12, 15, and 18 critically. Some of these chapters were rewritten completely because of his invaluable suggestions.

As a student interested in the applications of the calculus of variations and optimal control (and not just in the mathematics itself), I have always wanted to see examples which illustrate the applications of the mathematical techniques or an explanation on why we should know about a particular mathematical result. This was how the present book was conceived and put together. Several classes of students at the University of Washington have used and liked earlier versions of the course notes for the book. I am especially indebted to Tom Milac who proof-read most chapters at least once, and to Chonghua Gu and Dantong Zhu who generated some of the graphs and figures.

The first few drafts of the manuscript were prepared by Lilly Harper who retired in June, 1991. Frances Chen picked up where Lilly left off and did not miss a beat. I often marvel at their infinite patience in putting up with my seemingly endless changes and yet they still came up with a beautiful final product. As I have indicated to them before, I am most grateful for their prompt and helpful response to my requests.

My wife, Julia, has always had her own demanding professional career. But she has somehow managed to find time from her hectic schedule to get me to care for my own dahlias, to take part in civic and community activities, to relate to my nieces and nephews, and not to be totally consumed by the preparation of this volume. I would like her to know how much I have appreciated her effort to keep me from my tendency to be narrowly focused.

Frederic Y.M. Wan
Bainbridge Island, WA

# 1

# The Basic Problem

## 1. Introduction

It is human nature to optimize. Typically, we try to maximize profit, minimize cost, travel to a destination in the quickest time, and put out the least effort to get our work done. Nature, too, is seen to strive for efficiency. By Fermat's principle, light travels a route of the least time between two points, and by Hamilton's principle, mass particles move with the "least action" (giving Newton's second law of motion as a consequence). Our natural propensity to optimize has led to a long-standing effort to systematically determine the optimal realization of a variety of activities in science and engineering. This continuing effort has created a body of mathematical methods called (the mathematics of) *optimization*. We are concerned in these pages with a class of problems in optimization which involves minimizing or maximizing the value of an integral. In the simplest setting, we seek a function $y(x)$ on the interval $a \leq x \leq b$ which maximizes or minimizes the definite integral

$$J[y] \equiv \int_a^b F(x, y(x), y'(x))\, dx, \qquad (\ )' \equiv \frac{d(\ )}{dx}. \tag{1.1}$$

There are countless problems of this type in science and engineering and many others which can be recast into the same form. We will see a few of them in the next section and many more later, including some with $x$ and $y$ being vector quantities. The mathematical theory for this type of optimization problems is called the *calculus of variations*. Various kinds of auxiliary constraint conditions will also be incorporated in later chapters, leading eventually to the study of *optimal control theory*.

In the first few chapters, we will focus mainly on the scalar case (1.1) to present the main ideas of the subject. For many problems, the unknown $y(x)$ in (1.1) is required to satisfy the end conditions $y(a) = A$ and $y(b) = B$, where $A$ and $B$ are given constants. Whenever the problem of optimizing $J[y]$ is subject to these end constraints, we have a *basic problem* in the calculus of variations. For definiteness, we will focus on minimization problems unless the context of a specific problem requires that we do otherwise. (Note that a maximization problem for $J[y]$ may be restated as a minimization problem for $-J[y]$ instead.) In that case, the basic problem is stated more succinctly as

$$\min_{y \in S} \{J[y] \mid y(a) = A, \, y(b) = B\}, \tag{1.2}$$

where $S$ is the collection of eligible *comparison functions* to be specified by the problem. In general, the minimum value of $J$ will be different depending on the eligible comparison functions specified. For example, we may limit the eligible comparison functions $y$ to $n$ times continuously differentiable functions, denoted by $C^n$, or continuous and piecewise smooth functions, denoted by PWS. Given that $y'$ appears in the integrand $F(x, y, y')$ of $J[y]$, we will begin the theoretical development rather naturally with $C^1$ comparison functions, also known as *smooth comparison functions*. The choice of comparison functions will be modified in different ways later. A comparison function is called an *admissible comparison function* if it also satisfies all the prescribed constraints on $y(x)$ such as the end conditions.

The basic techniques for problems of the calculus of variations and optimal control theory will be developed in subsequent chapters. In Section 2 of this chapter, we formulate a few specific examples in these two related problem areas. They will be used later for illustrating the different solution techniques of the theory. We then work out the solution of a specific problem in Section 3. In the process, we will be led to the *Euler differential equation* (Euler DE) for the solution $\hat{y}(x)$ of the basic problem for a general $F(x,y,y')$. This equation forms the cornerstone of the classical theory of the calculus of variations; its smooth solution is called a smooth *extremal* for the problem. The integrand $F(x,y,y')$ of $J[y]$ is called the *Lagrangian* of the problem. The integral $J[y]$ itself is called the *performance index* or the *penalty integral*. Methods of solution for smooth extremals will be developed for special cases of Euler's differential equation in Section 4.

In courses on mathematical methods for applications, the treatment of the calculus of variations is invariably based on the Euler DE. This basic result is then generalized to allow for vector unknowns and higher dimensions. It is also extended to allow for various types of equality constraints. However, a solution of the optimization problem by way of the Euler DE presumes the solution curve to be at least smooth or, as Euler's differential equation is often used in its "ultradifferentiated" form, twice continuously differentiable. The smoothness requirement is not always met by the actual solutions of many problems. It is well known that the solution of many "time optimal problems" (TOP) in aerospace engineering and the social sciences involves a "bang-bang control" which corresponds to a PWS $y(x)$ in our basic problem. Bang-bang control problems will be discussed extensively in Chapter 13. In Section 6 of the present chapter, we will show the need to allow for PWS comparison functions even for a simple basic problem. This need will motivate the development of a more general variational calculus based on the Euler–Lagrange integrodifferential equation to be introduced in Chapter 2.

## 2. Some Examples

In this section, we formulate a few mathematical models of natural and social phenomena which pose problems in the calculus of variations. As indicated in Section 1, these problems will be used throughout this volume for illustrating the application of the different variational techniques to be established. The sample models are chosen in part because they are easily described and formulated; a few are chosen also for their historical importance.

### a. Minimal Surface of Revolution

As a typical example of the calculus of variations, consider a curve $y(x)$ which lies above a segment of the $x$-axis $a \leq x \leq b$ (Fig. 1.1). The surface area of the surface of revolution generated by rotating this curve about the $x$-axis is given by

$$J[y] = 2\pi \int_a^b y(x)\sqrt{1 + [y'(x)]^2}\, dx, \tag{1.3}$$

where $(\ )' = d(\ )/dx$. Let $y(a) = A$ and $y(b) = B$ be fixed. Finding the curve $\hat{y}(x)$ with these fixed end points which minimizes the total surface area has been an optimization problem of long-standing interest. It is in the form of the basic problem of the calculus of variations as defined by (1.1) and (1.2) with a Lagrangian

$$F(x, y, y') = 2\pi y\sqrt{1 + (y')^2}\,.$$

### b. The General Geodesic Problem

The minimum surface area problem (1.3) is a special case of the general geodesic problem. This is a basic problem for the integral

$$J[y] = k \int_a^b y^n \sqrt{1 + (y')^2}\, dx. \tag{1.4}$$

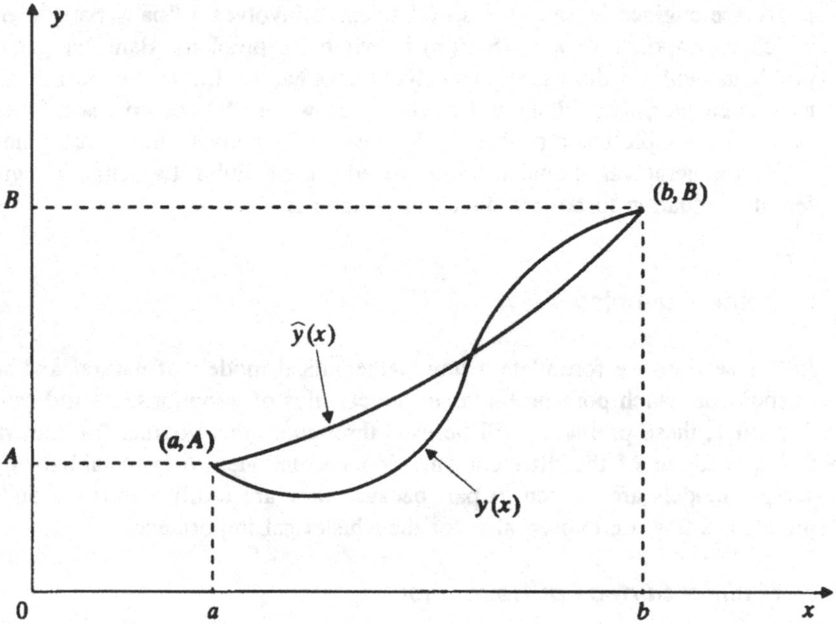


**Figure 1.1**

The $n = 0$ and $k = 1$ case corresponds to the planar geodesic problem, i.e., the problem of finding the shortest distance between two points in the plane. The $n = 1$ case corresponds to (1.3) if we take $k = 2\pi$.

The classical *brachistochrone* (meaning "shortest time" in Greek) problem corresponds to $n = -1/2$. The problem was originally posed by Johann Bernoulli in 1696 as a challenge to his brother Jacob (Anglicized as James in some references) in particular, and to the mathematical world in general. Let the plane curve $y(x)$ characterize the shape of a frictionless wire connectin two given points in space. The brachistochrone problem of Bernoulli seeks the curve $\widehat{y}(x)$ which minimizes the time of descent under gravity of a bead strung on the wire from the higher point $(a, A)$ to the lower point $(b, B)$. The formulation and the solution of this problem will be discussed in Section 5. Both Johann and Jacob solved the problem independently and eventually recognized the solution curve to be a piece of a *cycloid*. Sir Isaac Newton learned about the problem six months later and how it had baffled mathematicians in Europe. He solved it that same evening and sent in his solution anonymously the next day. On seeing the solution, Bernoulli at once exclaimed, "*Ah! I recognize the paw of the lion.*"

Two other special cases of the general geodesic problem are also of interest. For $n = -1$, the performance index (1.4) again gives the elapsed time for the bead to get from $(a,A)$ to $(b,B)$ as in the brachistochrone problem, but now, instead of the usual gravity field, the bead is in the force field of a quadratic potential. For $n = 1/2$, (1.4) gives the action integral for the free motion of a mass particle in a uniform field.

### c. Isoperimetric Problems

Often the class of admissible comparison functions for a basic problem in the calculus of variations is constrained by some additional requirements beyond those specified at the end points. One of the oldest problems of this type is Queen Dido's problem. Here we are to find the curve of a given length which connects two points $(a,0)$ and $(b,0)$ and, together with the $x$-axis, encloses the largest area possible.

Legend has it that this problem was solved intuitively by Queen Dido of Carthage in about 850 BC after she persuaded a North African chieftain to give her as much land as she could enclose within the hide of a bull. According to the legend, she cut the bull's hide into many thin strips and laid them out along a circular arc, using the Mediterranean coast as a supplementary boundary. In this way, she got the largest possible area for the state of Carthage.

The mathematical problem here can be stated by anyone who has had a course in calculus. The area to be maximized is given by the integral

$$J[y] = \int_a^b y(x)\,dx. \tag{1.5}$$

The function $y(x)$ is constrained by $y(a) = y(b) = 0$ and by the fact that the curve must have a given arc length $L$ between the two end points:

$$\int_a^b \sqrt{1+(y')^2}\,dx = L. \tag{1.6}$$

Although this mathematical problem had been known for some time, it was first solved by Jacob Bernoulli at the end of the 17th century.

In general, the minimization of the performance index (1.1) may be constrained by one or more integral conditions of the form

$$C[y] \equiv \int_a^b G(x,y,y')\,dx = L. \tag{1.7}$$

Because of the meaning of $C[y] = L$ for Queen Dido's problem, condition (1.7) is called an *isoperimetric constraint*. A variational problem with one or more isoperimetric constraints is called an *isoperimetric problem* in the calculus of variations. The solution of such a problem by the method of *Lagrange multiplier* will be discussed in Chapter 10.

### d. Minimum Cost Production Schedule

A firm has an order for a volume $B$ of its product to be delivered by time $T$. It would like to schedule production to meet the order at minimum cost. Let $y(t)$ be the total volume produced and accumulated by time $t$. The total cost for filling the order comes from two sources: (1) a production cost per unit item $c_p$ which increases linearly with the production rate, so that $c_p = ky'$ for some positive constant $k$ with $(\ )'$ now indicating $d(\ )/dt$ and (2) a constant holding cost per unit item per unit time $c_h$ to store inventory until delivery time. Suppose that it is possible for production to begin immediately at $t = 0$, and there is no accumulated inventory initially so that $y(0) = 0$. To meet the delivery schedule, we must have $y(T) = B$.

The total production cost over the time interval $(t, t + dt)$ for this producer is $c_p dy + c_h y(t)dt = k[y']^2 dt + c_h y\, dt$. It follows that the total cost over the period $[0, T]$ is given by

$$J[y] = \int_0^T \{k[y'(t)]^2 + c_h y(t)\}\, dt. \tag{1.8}$$

We want to choose a production strategy so that the corresponding $\hat{y}(t)$ minimizes $J[y]$ subject to the end constraint $y(0) = 0$ and $y(T) = B$. One simple production strategy is to produce at a uniform rate which meets the delivery requirement, namely, $y = Bt/T$. However, we will soon see that this strategy does not minimize $J[y]$.

For the present production cost problem, the admissible comparison functions must also satisfy the inequality constraints $y(t) \geq 0$ and $y'(t) \geq 0$ (which are met by the admissible comparison function $y = Bt/T$) if the products are not perishable and cannot be destroyed. Variational problems with equality and inequality constraints imposed on the derivative(s) of the unknown(s) are often more conveniently treated by *optimal control theory*, a modern derivative of the calculus of variations. This is especially the case if we are to actually choose $y'$, and not $y$, as in the present problem; what we can physically regulate here is the production rate. The perspective and approach of optimal control theory are sufficiently different from those of the calculus of variations to deserve a separate treatment. The basic theory of optimal control will be developed in Chapters 13 and 14.

### e. Optimal Harvesting Strategy for a Fishery

When left alone, a fish population will grow within the limit of the carrying capacity of the body of water and its nutrients. Typically, the growth process is modeled by the first-order ODE

$$y' = f(t,y),$$

where $y(t)$ is the total tonnage of fish biomass at time $t$. For $y(t)$ small compared to the region's *carrying capacity* $y_c$, $f(t,y)$ may be taken to be linear in $y$, i.e., $f(t,y) = ky + f_0(t)$; the undisturbed fish population growth is exponential. A growth (rate) function adequate for a much wider range of $y(t)$ is the *logistic growth* (rate)

$$f(t,y) = ky\left(1 - \frac{y}{y_c}\right).$$

Other more complex growth functions have been proposed and, given the initial population $y(0) = A$, the corresponding initial-value problems (IVP) have been analyzed to learn about the resulting growth behavior of fish populations.

A commercially valuable fish population is not likely to be left alone to grow naturally. More often, it is harvested continually at a rate $z(t)$. [For a seasonal fishery, $z(t)$ may be a periodic function which vanishes over regular intervals.] The actual growth rate of the fish population is then

$$y' = f(t,y) - z, \qquad y(0) = A. \tag{1.9}$$

The owner of a fishery with exclusive right to the fishing ground would naturally want to choose a harvesting strategy $\hat{z}(t)$ which maximizes the firm's profit.

Suppose the fishery is small so that its sale volume has no effect on the unit fish price $p$. (Similar situations arise often enough that the owner of such an economic activity is called a *price taker* in the economics literature). On the other hand, the cost of harvesting a unit of fish biomass, $c(z,y)$, is expected to vary with the harvest rate $z(t)$ and the fish population $y(t)$ (given that capital investment in boats and equipment have already been made and that it is more difficult to catch a unit fish biomass when there are fewer fish). The profit over a time increment $dt$ is $[p - c(z,y)]z\,dt$. The total profit over the planning period would be the integral of this quantity from 0 to $T$. It is possible to choose $z(t)$ to maximize this total profit. However, a little reflection suggests that we do something different.

Most of us know that a dollar of income now ($t = 0$) can (and should) multiply with time. If nothing else, we can always put it in the bank or buy bonds to earn interest. If the interest is compounded continuously at a rate $r$, we know that the principal plus interest of a current dollar will be $e^{rt}$ dollars at a later time $t$. Looking at it another way, a dollar of income at some later time $t$ is equivalent to $e^{-rt}$ dollars now. In particular, the profit stream of the fishery over an incremental time $dt$ at some later time $t$ has only a current worth (or, in the jargon of economics, a *present value*) of $e^{-rt}[p - c(z,y)]z\,dt$.

Correspondingly, the present value of the entire profit stream over the planning period is

$$P = \int_0^T e^{-rt}[p - c(z,y)]z\,dt. \tag{1.10a}$$

A shrewd fishery owner would want to choose $z$ to maximize the present value of the profit over the planning period (and not the undiscounted profit itself). The maximization is subject to the constraint of the initial-value problem (1.9) which characterizes the growth of the fish population.

The above optimization problem can be recast as a problem of the calculus of variations with one end constraint, $y(0) = A$. This is accomplished by using (1.9) in the form of $z = y' - f(t, y)$ to eliminate $z$ from the integrand of (1.10a). While we have not stipulated a terminal fish population [such as $y(T) = B$] to be left for the next planning period (or a salvage value for the unknown terminal population), it is not always the case that the optimal strategy should have $y(T) = 0$. The cost of harvesting the last few fish may be more than the income to be had from them.

With the harvesting rate being the physical quantity we can regulate, it is not always appropriate or attractive to use the differential equation in (1.9) to eliminate $z$ from the integrand. Unless the unit cost $c$ is independent of $z$, it turns out to be advantageous to incorporate the constraining ODE in (1.9) by the method of (Lagrange) multipliers and obtain the solution by the ***maximum principle*** of optimal control theory to be discussed in Chapters 13 and 14. Result obtained by this method are generally easier to interpret in the context of the modeled phenomenon.

If, instead of private ownership, the fishery is run by the federal government through some kind of central planning board, the objective of the board may no longer be profit maximization. A more likely goal may be to maximize the welfare of the peoples as measured by the aggregate of their satisfaction over time derived from consuming the fish harvested. In welfare economics, the satisfaction of the constituents over the incremental time interval $dt$ is taken to be $U(z)dt$, where $U(z)$ is a monotone increasing concave ***utility function.*** There is, generally, a social preference for immediate and near-future gratification. ("Let us eat, drink and be merry for tomorrow we shall die.") In that case, the social welfare index over a time period $[0, T]$ is taken to be the sum of the discounted satisfaction over that period:

$$J[y] = \int_0^T e^{-rt}U(z)\,dt, \tag{1.10b}$$

where $r$ is now the social preference rate. The maximization of $J$ is again subject to the equation for the growth dynamics and the initial condition in (1.9).

The examples of variational problems listed above involve mainly geometric and simple economic phenomena. They were chosen because the actual problem in each case can be described (relatively) simply. The formulation of the relevant mathematical models does not require a great deal of technical background knowledge. This is not usually the case for phenomena in the physical sciences and more classical engineering areas. For example, many variational problems in the mechanics of elastic bodies involve the *strain energy* in the body of interest. The concept of strain energy, however, can only be discussed with a proper understanding of stresses and strains in a deformed elastic body.

We do not intend to avoid elasticity theory or the mechanics of deformable media in general, for these subjects have been important to the development of the calculus of variations (especially in higher dimensions). But, we will introduce variational problems in these areas gradually and only after suitable background preparation. Only simple models, such as the brachistochrone and simple spring-mass systems, will be discussed in the early chapters. A more general treatment of particle dynamics, including a discussion of Hamilton's principle, will be presented in Chapters 6 and 7, in conjunction with the theory of Hamiltonian systems. The deformation and vibration of strings and membranes will be used to illustrate the material in Chapters 10, 12 and 15. The basic elements of the general theory of elasticity will be presented in Chapter 16, culminating in the establishment of several variational principles which are fundamental in elasticity theory and its applications. More variational developments in the mechanics of deformable bodies will be found in Chapters 17 and 18. In particular, the semidirect method of the calculus of variations will be exploited to establish various theories for elastic flat plate in Chapter 17.

Variational methods are now known to be useful in many areas other than mechanics. An effort has been made in this volume to use illustrative examples which demonstrate its wide applicability. An historical account of the development of the calculus of variations can be found in Goldstine (1980).

## 3. The Euler Differential Equation

The variational problems in Section 2 are rather rich in content and their solutions are not always straightforward. To develop a systematic method of solution for such problems, consider the minimum cost production problem (now using $x$ instead of $t$ as the independent variable):

$$J[y] = \int_0^T \{k[y'(x)]^2 + c_k y(x)\}\, dx, \tag{1.11a}$$

with

$$y(0) = 0, \tag{1.11b}$$

$$y(T) = B. \tag{1.11c}$$

Suppose that we have found some $C^1$ function $\hat{y}(x)$ which renders $J[y]$ a minimum value, so that $J[y] \geq J[\hat{y}]$ for all admissible comparison functions. Let $y(x) = \hat{y}(x) + \varepsilon h(x)$, where $\varepsilon$ is a (small) parameter and $h(x)$ is a smooth function in $(0, T)$ with $h(0) = h(T) = 0$. The stipulation that $h$ vanishes at the end points is necessary for $y(x)$ to satisfy the end contraints (1.11b) and (1.11c). With $y(x)$ written in this way, the value of $J[y]$ now depends on the value of the parameter $\varepsilon$. For a given $h(x)$, we indicate this dependence by writing $J[y] \equiv J(\varepsilon)$:

$$J[y] = \int_0^T \{k[\hat{y}' + \varepsilon h']^2 + c_h[\hat{y} + \varepsilon h]\}\, dx \equiv J(\varepsilon)$$

Evidently, we have $J(\varepsilon) \geq J(0)$. For $\varepsilon = 0$ to be a local minimum point of the ordinary function $J(\varepsilon)$, we know from elementary calculus that $J^{\cdot}(\varepsilon) \equiv dJ/d\varepsilon$ must vanish at $\varepsilon = 0$. For $J[y]$ given by (1.11a), we have

$$\begin{aligned} \left.\frac{dJ}{d\varepsilon}\right|_{\varepsilon=0} &= \int_0^T \{2k[\hat{y}'h' + c_h h]\}\, dx \\ &= [2k\hat{y}'h]_0^T + \int_0^T \{-(2k\hat{y}')' + c_h\} h\, dx \qquad (1.12) \\ &= \int_0^T \{-2k\hat{y}'' + c_h\} h\, dx = 0, \end{aligned}$$

where we have used the end conditions $h(0) = h(\mathrm{T}) = 0$ to eliminate the terms outside the integral sign. Condition (1.12) is of the form

$$\int_a^b g(x) h(x)\, dx = 0; \tag{1.13}$$

it is to hold for all smooth functions $h(x)$ with $h(a) = h(b) = 0$ and

$$g(x) = -2k\hat{y}'' + c_h$$

keeping in mind that $\hat{y}$ is a known function of $x$. If $g(x)$ is continuous in $(a, b)$, we can conclude that $g(x) \equiv 0$ for all $x$ in $[a, b]$. Why is this so?

To deduce $g(x) \equiv 0$, suppose $g(x)$ is not identically zero but is, say, positive for some point $\bar{x}$ in $(a, b)$. By the continuity of $g(x)$, we have $g(x) > 0$ for all $x$ in some interval $(c, d)$ with $a \leq c < \bar{x} < d \leq b$. Now, (1.13) must hold for all admissibile $h(x)$.

Consider the function

$$h(x) = \begin{cases} (x-c)^2(d-x)^2 & (c \le x \le d) \\ 0 & \text{(otherwise)}, \end{cases} \tag{1.14}$$

which is $C^1$ with $h(a) = h(b) = 0$; it is an admissible comparison function. For this particular $h(x)$, the integral on the left-hand side of (1.13) is positive. Hence, $g(x)$ cannot be positive in $[a, b]$. By a similar argument, $g(x)$ also cannot be negative. Together, they effectively prove the following *basic lemma* in the calculus of variations:

**[I.1]** ***If*** $\mathbf{g(x)}$ ***is continuous in*** $\mathbf{[a,b]}$ ***and (1.13) is satisfied for every choice of the (continuous and) smooth function*** $\mathbf{h(x)}$ ***with*** $\mathbf{h(a) = h(b) = 0}$***, then*** $\mathbf{g(x) \equiv 0}$ ***in*** $\mathbf{[a,b]}$.

It is not difficult to see that this conclusion remains valid if $h$ is $C^n$ for $n \ge 2$; it is only necesary to choose a similar but smoother comparison function $h(x)$ in (1.14).

Applied to the special case (1.11a), [I.1] now requires

$$-2k\hat{y}'' + c_h = 0, \qquad (0 < x < T).$$

This second-order ordinary differential equation and the fixed end conditions (1.11b) and (1.11c) define a two-point boundary-value problem (BVP) for the determination of $\hat{y}(x)$. The equation itself is called the *Euler differential equation* for the Lagrangian of the basic problem (1.11). For the present minimum cost production problem, the Euler DE can be integrated twice to give

$$\hat{y}(x) = \frac{c_h}{4k}x^2 + c_1 x + c_0,$$

where $c_0$ and $c_1$ are two constants of integration, as a smooth extremal of the problem. The end condition $\hat{y}(0) = 0$ requires $c_0 = 0$. The other end condition $\hat{y}(T) = B$ determines $c_1$ to be $(B - c_h T^2/4k)T^{-1}$ so that we have

$$\hat{y}(x) = \left[\frac{c_h T^2}{4k}\left(\frac{x}{T} - 1\right) + B\right]\frac{x}{T}.$$

The smooth extremal $\hat{y}(x)$ obtained above is useful for the minimum production cost provided $B \ge c_h T^2/4k$. Otherwise, we have $\hat{y}'(0) < 0$ so that $\hat{y}(x) < 0$ for (a range of) $x > 0$, which is unacceptable. The method of solution for the minimum cost problem with $B < C_h T^2/4k$ will be discussed in Chapter 14. It is important to keep in mind at this time that the (smooth) extremal corresponds to a stationary point of $J(\varepsilon)$; it is not necessarily a minimum point. Sufficient conditions for a minimum will be discussed in Chapters 4 and 5 (see also Exercise 13 of this chapter).

Still, what we have accomplished with the help of the general mathematical result [I.1] is already rather spectacular. In its original form (1.2), the solution of the minimization problem requires that we compare the values of $J[y]$ for all admissible comparison functions. This is an impossible task because there are generally infinitely many admissible comparison functions. With the help of [I.1], we have now reduced the search for the (smooth) minimum points $\hat{y}$ to a feasible process. All we have to do is to determine the solutions of the Euler differential equation which satisfy the end conditions, and we know how to solve boundary-value problems in ODE, numerically if necessary. If the basic problem is known to have a minimum value and if the BVP for the Euler DE has only one solution, then the unique solution is the minimum point of the basic problem.

The reduction of the basic problem to solving an Euler differential equation is also possible for a general Lagrangian $F(x, y, y')$. To see this, we again set $y(x) = \hat{y}(x) + \varepsilon h(x)$ in (1.1) so that $F(x, y, y') = F(x, \hat{y} + \varepsilon h, \hat{y}' + \varepsilon h')$ and

$$J[y] = \int_a^b F(x, \hat{y} + \varepsilon h, \hat{y}' + \varepsilon h')\, dx \equiv J(\varepsilon).$$

We assume in the following calculations (as we will throughout this volume) that all partial derivatives of $F$ with respect to its three arguments are continuous. With $F_{,\xi}$ indicating a partial derivative of $F$ with respect to an argument $\xi$, we have

$$\begin{aligned} J^{\bullet}(0) \equiv \left.\frac{dJ}{d\varepsilon}\right|_{\varepsilon=0} &= \int_a^b [F_{,y}(x,\hat{y},\hat{y}')h + F_{,y'}(x,\hat{y},\hat{y}')h']\, dx \\ &= [F_{,y'}(x,\hat{y},\hat{y}')h]_a^b + \int_a^b \left\{F_{,y}(x,\hat{y},\hat{y}') - \frac{d}{dx}[F_{,y'}(x,\hat{y},\hat{y}')]\right\} h\, dx \\ &= \int_a^b \left\{\widehat{F}_{,y} - \frac{d}{dx}[\widehat{F}_{,y'}]\right\} h\, dx \equiv \int_a^b G(x,\hat{y},\hat{y}')\, h\, dx = 0, \end{aligned} \tag{1.15}$$

where $\hat{f}(x) \equiv f(x, \hat{y}(x), \hat{y}'(x))$ for any function $f(x, y, v)$, and

$$F_{,y}(x) \equiv F_{,u}(x, u, \hat{y}'(x))\big|_{u=\hat{y}(x)}$$

$$\widehat{F}_{,y'}(x) \equiv F_{,y'}(x,\hat{y},\hat{y}') = F_{,v}(x,\hat{y},v)\big|_{v=\hat{y}'(x)}.$$

The vanishing of $J^{\bullet}(0)$ must hold for all (continuous and) smooth $h(x)$ with $h(a) = h(b) = 0$. Because $\widehat{F}_{,y} - [\widehat{F}_{,y'}]' \equiv G(x,\hat{y},\hat{y}') \equiv \widehat{G}(x)$ is continuous in $(a, b)$, the basic lemma [I.1] applies and we obtain the following fundamental result of the calculus of variations:

**[I.2]** ***The Euler differential equation of the performance index*** **J[y]** ***given by (1.1) is***

$$(F_{,y'})' = F_{,y}. \tag{1.16}$$

***The*** **C**$^1$ ***solution*** $\hat{\mathbf{y}}$ ***which minimizes*** **J** ***must be a solution of this second-order ODE.***

As in the minimum cost production problem, a $C^1$ solution $\hat{y}$ of the Euler DE (1.16) is called a *smooth extremal* of the given Lagrangian. An extremal which also satisfies the prescribed end conditions in (1.2) is called an *admissible extremal.* We emphasize that an admissible extremal $\hat{y}$ may not minimize $J[y]$, just as stationary point $\bar{x}$ of a function $f(x)$ may not minimize $f(x)$; it may be a maximum point or an inflection point of $f(x)$. In general, it is necessary to examine the second derivative of $J(\varepsilon)$ with respect to $\varepsilon$ for additional information about the desired minimum; we will do so in a later chapter.

It may be instructive to note that the Euler DE (1.16) can also be deduced as a consequence of the familiar stationarity condition for the extrema of an ordinary function in elementary calculus. Set $\Delta x = (b - a)/(N + 1)$ for some positive integer $N$ and $x_k = a + k\Delta x$, $k = 0, 1, 2, \dots, (N + 1)$ so that $x_0 = a$ and $x_{N+1} =$ b. Consider now the integral $J[y]$ as the limit, as $N \to \infty$, of the lower Riemann sum

$$S_N = \sum_{n=0}^{N} F(x_n, y_n, y'_n)\,\Delta x,$$

where $y_n = y(x_n)$ and $y'_n = y'(x_n)$ with $y_0 = A$ and $y_{N+1} = B$ by the prescribed boundary conditions of the problem. We further approximate $y'_n$ by the difference quotient $(y_{n+1} - y_n)/\Delta x$ so that

$$J[y] \approx \sum_{n=0}^{N} F_n \Delta x \equiv j(\mathbf{y})$$

where $\mathbf{y} = (y_1, y_2, \dots, y_N)^T$ and

$$F_n = F\left(x_n, y_n, \frac{y_{n+1} - y_n}{\Delta x}\right) \equiv f_n(\mathbf{y}).$$

In terms of this discrete analogue of the basic problem, the determination of $\hat{y}(x)$ to minimize $J[y]$ becomes a problem of finding a combination of the unknown independent variables $\{y_1, \dots, y_n\}$ wich minimizes $j(\mathbf{y})$.

A necessary condition for $\hat{\mathbf{y}} \equiv (\hat{y}_1, \hat{y}_2, \dots, \hat{y}_N)^T$ to be a local minimum point of $j(\mathbf{y})$ is that it makes $j(\mathbf{y})$ stationary, i.e., $j_{,y_k}(\hat{\mathbf{y}}) = 0$, $k = 1, 2, \dots, N$. For each $k$, $y_k$ appears in both $f_k$ and $f_{k-1}$. Hence, the stationarity condition with respect to $y_k$ takes the form

$$\left.\frac{\partial j}{\partial y_k}\right|_{y=\hat{y}} = \Delta x \left.\left\{\frac{\partial f_k}{\partial y_k} + \frac{\partial f_{k-1}}{\partial y_k}\right\}\right|_{y=\hat{y}} = 0$$

or

$$\frac{1}{\Delta x}[(\widehat{F}_{,y'})_k - (\widehat{F}_{,y'})_{k-1}] = (\widehat{F}_{,y})_k, \tag{*}$$

where

$$(\widehat{F}_{,z})_n = F_{,z}\left(x_n, \widehat{y}_n, \frac{\widehat{y}_{n+1} - \widehat{y}_n}{\Delta x}\right).$$

The necessary condition (*) for $j(\hat{\mathbf{y}})$ to be a local minimum holds for $k = 1, 2, \ldots, N$. We have then $N$ conditions for the $N$ unknowns $\{\widehat{y}_1, \ldots, \widehat{y}_N\}$. As $N \to \infty$ so that $\Delta x \to 0$, the condition (*) formally tends to the Euler DE for every value of $x$ in $(a, b)$. It remains only to show that the limiting process and the optimization process can be interchanged to complete the derivation.

## 4. Integration of the Euler Differential Equation

The unique admissible smooth extremal for the basic problem (1.11) was obtained by a straightforward calculation. This is often not the case for other problems because the Euler differential equation $(F_{,y'})' = F_{,y}$ is generally a nonlinear second-order ODE. An exact solution of such an equation in terms of elementary or special functions is not always possible or easily recognized when it is possible. It is, therefore, helpful to know that useful information about extremals can be extracted from the Euler DE for the following six subclasses of problems.

### a. $F(x, y, y') = F(x)$

This is not a problem in the calculus of variations. $J[y]$ cannot be minimized by a proper choice of $y(x)$ as the Lagrangian $F$ does not depend on $y$ (or its derivatives).

### b. $F(x, y, y') = F(y)$

The Euler DE degenerates to $F_{,y}(\widehat{y}) = 0$ so that $\widehat{y}$ is a constant. There are not enough free parameters to satisfy the end conditions. Hence, there is no solution for the basic problem except for special cases.

### c. $F(x, y, y') = F(y')$

There are many problems in science and engineering with this type of Lagrangian. The minimum cost production problem in Section 2 becomes a problem of this type if there is no cost for holding the inventory ($c_h = 0$). Another example is the geodesic problem for a plane curve $y$ with $J[y]$ given by (1.4) for $n = 0$. The Euler DE for this type of Lagrangian takes the form $[F_{,y'}(y')]' = 0$ or $F_{,y'}(y') = c_0$. The latter can be solved for $y'$ to get $y' = c_1$, where $c_1$ is an arbitrary constant. Upon integration, we obtain $y = c_1 x + c_2$. The two constants $c_1$ and $c_2$ may be chosen to satisfy the two end conditions. For the typical basic problem with $y(a) = A$ and $y(b) = B$, we find $c_1 = (B - A)/(b - a)$ and $c_2 = (bA - aB)/(b - a)$ so that $y(x) = A + (B - A)(x - a)/(b - a)$. Hence, a straight line connecting $(a, A)$ and $(b, B)$ is the only smooth candidate for the plane geodesics between these points. In general, it is useful to remember that

**[I.3]** ***When the Lagrangian* F *in (1.1) depends only on* y′, *the smooth extremals are straight lines.***

For many problems, it is straightforward to show that the straight admissible extremal, in fact, minimizes $J[y]$ (see Exercises). The corresponding sufficient conditions for a general $F(y')$ will be discussed in Chapter 4.

### d. $F(x, y, y') = F(x, y)$

The Euler DE takes the form $F_{,y}(x, \hat{y}) = 0$ which may be solved for $\hat{y}$ as a function of $x$. Hence, $\hat{y}$ is completely determined without any free parameters and, in general, will not satisfy the given end conditions. There is usually no admissible extremal for this type of basic problems. On the other hand, variational problems with a Lagrangian which does not depend on $y'$ sometimes occur naturally without specified end conditions. A solution of the Euler equation alone may provide the minimum (or maximum) point for these problems as illustrated by the following two examples.

**EXAMPLE 1:** A firm mass-produces a product and sells the total amount $y$ produced at time $t$ at a unit price $p(t, y)$ which is a monotone decreasing, concave function of $y$ and has a seasonal fluctuation. Suppose the profit earned from the sale at time $t$ is given by

$$F(t, y) = \frac{1}{2} p_0 (3 - \cos t) y \ln\left(\frac{y_0}{y}\right) - \alpha - \beta y,$$

where $p_0$, $y_0$, $\alpha$, and $\beta$ are know positive costants. Find the optimal production strategy $y(t)$ which maximizes the total profit over the planning period $0 \leq t \leq T$.

Suppose that the total profit over the short time interval interval $[0, T]$ is adequately given by the undiscounted sum

$$J[y] = \int_0^T F(t, y)\, dt.$$

By [I.2], the optimal strategy $\widehat{y}(t)$ must be a solution of the Euler DE for the given Lagrangian $F(t, y)$. Because $y'$ does not appear in $F$, the Euler DE simplifies to

$$F_{,y}(t, y) = \frac{1}{2} p_0 (3 - \cos t)\left[-1 + \ln\left(\frac{y_0}{y}\right)\right] - \beta = 0.$$

This equation can be solved to give

$$\widehat{y} = y_0 \exp\left(-1 - \frac{2\beta}{p_0(3 - \cos t)}\right).$$

Observe that nothing is (or can be) said about the end values $y(0)$ and $y(T)$ until we have solved the problem. They, too, are to be determined by the optimization process. Hence, there are no constraints to be met at the end points. At the same time, the conditions $h(0) = h(T) = 0$ are not needed in the derivation of the Euler DE; integration by parts is not required as there is no term involving $h'$ in $dJ/d\varepsilon$.

The fact that the Lagrangian $F$ depends only on $t$ and $y$ reduces the optimization problem to one of optimizing $F$ at each instance $t$. As such, we really have just a *classical* parameter optimization problem in elementary calculus. For the present profit maximization problem, $F(t, \widehat{y}(t))$ is in fact, a maximum value for each $t$ because

$$F_{,yy}(t, \widehat{y}(t)) = -\frac{p_0(3 - \cos t)}{2\widehat{y}} < 0$$

given the unique extremal $\widehat{y}(t)$ is positive for all $t \geq 0$. It follows that $J[\widehat{y}]$ is the maximum value.

**EXAMPLE 2:** A thin-walled cylindrical container of radius $R$ is partially filled with a uniform fluid. The container is rotating with a constant angular velocity $\omega$ about its own central axis (taken to be the $z$-axis in a cylindrical coordinate system). Determine the free surface height $z = Z(r)$ of the fluid by minimizing the potential energy of the fluid (see Fig. 1.2)

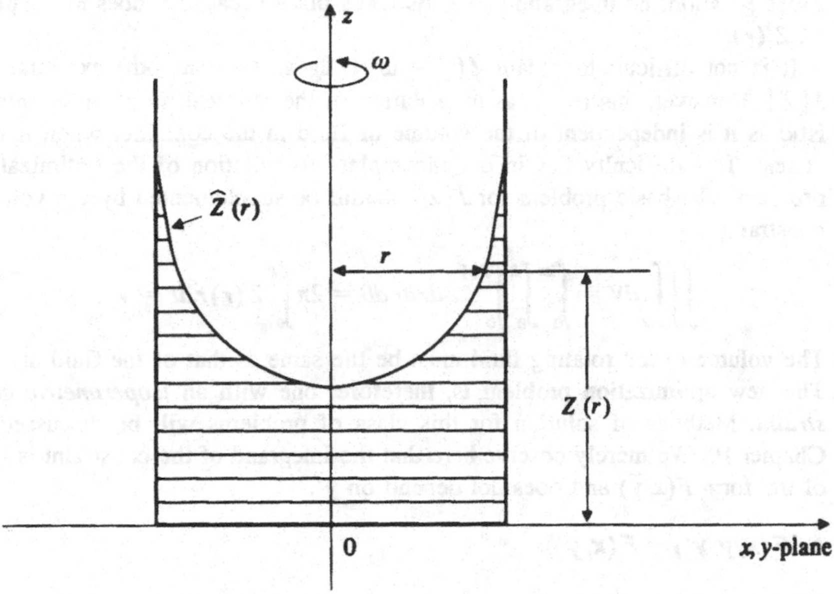


**Figure 1.2**

To solve this problem, we note that the vertical force $F_z$ acting on an elemental fluid volume $dV$ at a point with cylindrical coordinates $(r, \theta, z)$ is $-\rho g dV$ and the radial force $F_r$ acting on the same elemental volume is $\rho\omega^2 r dV$, where $\rho$ is the uniform mass density of the fluid and $\rho dV = dm$ is the mass of the elemental volume. The corresponding potential energy in the elemental fluid volume is $dE = \left(\rho g z - \frac{1}{2}\rho\omega^2 r^2\right) dV = \mathcal{E} dV$. Note that we have $F_z = -\mathcal{E}_{,z}$ and $F_r = -\mathcal{E}_{,r}$ as they should be. The total potential energy in the entire fluid is then given by

$$J[Z] = \iiint_V dE = \int_0^{2\pi}\int_0^R\int_0^{Z(r)}\left[\rho g z - \frac{1}{2}\rho\omega^2 r^2\right] r\, dz\, dr\, d\theta$$

$$= \pi\rho\int_0^R (gZ^2 - \omega^2 r^2 Z) r\, dr,$$

where $Z(r)$ is the free surface (of revolution) at the fluid top.

Observe that we do not know in advance the fluid height at the central axis $(r = 0)$ or at the container inner wall $(r = R)$ in the steadily rotating configuration. Hence, no end conditions can be prescribed for the variational problem. Again, the conditions $h(0) = h(R) = 0$ are not needed in the derivation of the

Euler equation; no integration by parts takes place because $F$ does not depend on $Z'(r)$.

It is not difficult to obtain $Z(r) = \omega^2 r^2/2g$ as the (smooth) extremal for $J[Z]$. However, this result, as the solution of the physical problem, is unrealistic as it is independent of the volume of fluid in the container when it was at rest. This difficulty lies in our incomplete formulation of the optimization problem. The basic problem for $J[Z]$ should be supplemented by the volume constraint

$$\iiint_V dV = \int_0^{2\pi}\int_0^R\int_0^{Z(r)} r\,dz\,dr\,d\theta = 2\pi\int_0^R Z(r)\,r\,dr = V_0.$$

The volume of the rotating fluid must be the same as that of the fluid at rest. The new optimization problem is, therefore, one with an *isoperimetric constraint*. Methods of solution for this class of problems will be discussed in Chapter 10. We merely observe here that the integrand of the constraint is also of the form $F(x, y)$ and does not depend on $y'$.

### e. $F(x, y, y') = F(x, y')$

With $F_{,y} \equiv 0$, the Euler DE can be integrated immediately to give $F_{,y'}(x, y') = c_1$. This equation may be solved for $y'$ to get $y' = f(x; c_1)$. Upon integration, we obtain

$$y(x) = A + \int_a^x f(t;\, c_1)\, dt \tag{1.17}$$

with the constant $c_1$ determined by

$$B = A + \int_a^b f(t;\, c_1)\, dt. \tag{1.18}$$

Hence, we conclude that

**[I.4]** ***When the Lagrangian*** **F** ***of (1.1) does not depend on*** **y** ***explicitly, all admissible extremals are obtained in the form of a quadrature as in (1.17) with the constant*** $c_1$ ***determined by (1.18).***

**EXAMPLE 3:** A solid of revolution moves in a perfect incompressible fluid with constant velocity $v_0$ in the direction of its axis of revolution (see Fig. 1.3). If the frictional force at any point on the surface is proportional to the square of the normal component of the velocity, find the shape of the surface which minimizes the total pressure drag resistance on the body.

This problem was solved by Newton in his *Principia* (1687) and, with some modification, continues to be of engineering interest today. Let $\psi$ be the angle

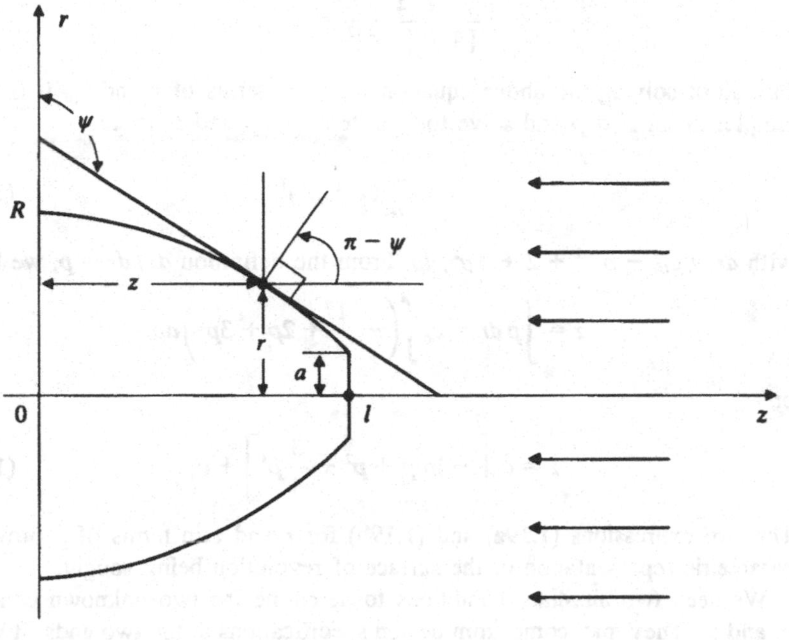


Figure 1.3

made by the meridional tangent at a point on the surface of the solid (at a radial distance $r$ from the $z$-axis) and the $x, y$-plane. The component of velocity normal to the surface at that point is then $v_0 \cos(\pi - \psi) = -v_0 \cos\psi$ and the frictional force is $c_D v_0^2 \cos^2\psi$, where $c_D$ is a known constant. The pressure drag resistance over a ring band of arclength $ds$ is $(2\pi C_D v_0^2 \cos^2\psi r\, ds)\cos\psi$. With $ds = dr/\cos\psi$, the total drag resistance over the whole body is

$$J = 2\pi c_D v_0^2 \int_0^R (\cos^2\psi)\, r\, dr.$$

Let the surface of revolution be given by $z(r)$. Then we have $\tan\psi = dz/dr \equiv z'$ and $\cos\psi = [1 + (z')^2]^{-1/2}$ so that

$$J[z] = 2\pi c_D v_0^2 \int_0^R \frac{r\, dr}{1 + (z')^2}.$$

The problem of minimizing $J[z]$ is evidently one with a Lagrangian of the form $F(x, y')$.

By [I.4], the Euler DE for $J[z]$ can be integrated once to give

$$\frac{2rz'}{[1+(z')^2]^2} = 2c_0.$$

Instead of solving the above equation for $z'$ in terms of $r$ and $c_0$, it is much simpler to set $z' \equiv p$ and solve for $r$ in terms of $p$ and $c_0$ to get

$$r = \frac{c_0}{p}(1+p^2)^2 \tag{1.19a}$$

with $dr = c_0(-p^{-2} + 2 + 3p^2)\,dp$. From the definition $dz/dr = p$, we have

$$z = \int p\,dr = c_0 \int \left(-\frac{1}{p} + 2p + 3p^3\right) dp$$

or

$$z = c_0\left[-\ln p + p^2 + \frac{3}{4}p^4\right] + c_1. \tag{1.19b}$$

The two expressions (1.19a) and (1.19b) for $r$ and $z$ in terms of $p$ provide a parametric representation of the surface of revolution being sought.

We need two auxiliary conditions to detemine the two unknown constants $c_1$ and $c_0$. They may come from design specifications at the two ends. It would seem natural to take $r = a$ for some $a \geq 0$ (with $0 < a < R$ for a blunt-nose body) at $z = 1$ and $r = R$ at the other end for some prescribed radius $R$. We will not determine the unknowns $c_1$ and $c_0$ here because the optimal shape for the solid of revolution obtained by the present frictional force assumption is know to agree poorly with that found by experiments. A more complete solution of this problem allowing for a blunt-nose and a piecewise smooth surface can be found in Bryson and Ho (1969).

It is more difficult to decide whether an admissible extremal $\hat{z}(r)$ actually minimizes $J[z]$ for this problem than it was for Example 1. We will deal with the general question of sufficient conditions for a minimum in Chapters 4 and 5.

### f. $F(x, y, y') = F(y, y')$

Multiply both sides of the Euler DE by $y'$ and rewrite the result as

$$[y'F_{,y'}]' - y''F_{,y'} = [F]' - y''F_{,y'}$$

or

$$[F - y'F_{,y'}]' = 0.$$

It follows that

**[I.5]** ***When the Lagrangian*** $\mathbf{F}$ ***of (1.1) does not depend on*** $\mathbf{x}$ ***explicitly, any nontrivial smooth extremal satisfies the following first integral of the Euler DE***

$$F - y'F_{,y'} = C_0 \tag{1.20}$$

***where*** $C_0$ ***is a constant of integration.***

In principle, relation (1.20) can be solved to obtain $y' = f(y; C_0)$ which is separable and can, therefore, be integrated to give $y$ as a function of $x$ and two arbitrary constants $C_0$ and $C_1$.

EXAMPLE 4: $J[y] = \int_a^b y^2(1-y')^2\,dx$, with $y(2) = 1$, $y(3) = 3$.

The Lagrangian $y^2(1-y')^2$ does not depend on $x$ explicitly. Its first integral (1.20) takes the form

$$y^2(1-y')^2 + 2y'y^2(1-y') = y^2[1-(y')^2] = C_0 \tag{1.21}$$

or

$$\frac{y\,dy}{\sqrt{y^2 - C_0}} = \pm\, dx.$$

Upon integration, we obtain

$$\sqrt{y^2 - C_0} = \pm C_1 \pm x \qquad \text{or} \qquad y^2 = C_0 + (C_1 + x)^2. \tag{1.22}$$

The end condition $y(2) = 1$ requires $C_0 = 1 - (C_1 + 2)^2$. This relation can be used to eliminate $C_0$ from the end condition $y(3) = 3$ to get $C_1 = -3/4$ and therewith $C_0 = 3/4$. Hence, the admissible smooth extremal for our basic problem is a segment of the hyperbola

$$y^2 - \left(x - \frac{3}{2}\right)^2 = \frac{3}{4} \qquad (2 \le x \le 3) \tag{1.23}$$

in the first quadrant (see Fig. 1.4).

Note that it is far from obvious how we may deduce (1.23) from the Euler DE without [I.5]. Does the extremal (1.23), in fact, minimize $J[y]$? We will consider this question in Exercise 20 of Chapter 4.

It may be said that only two of the six classes of Lagrangians merit special attention. Class C is a special case of class E whereas classes A, B and D do not lead to problems in the calculus of variations. However, the separate treatments of the six classes were designed to facilitate the recall of their

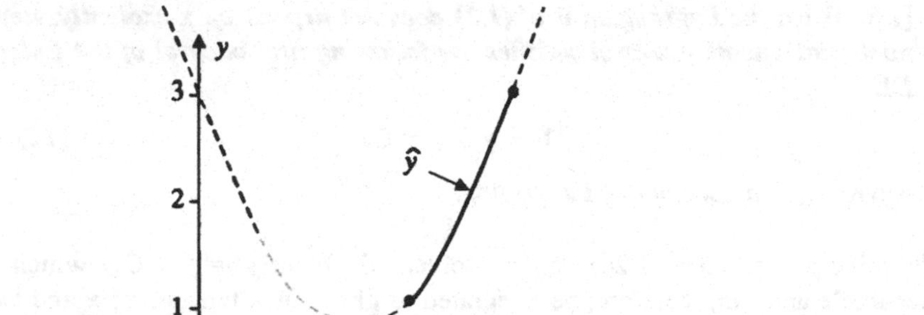

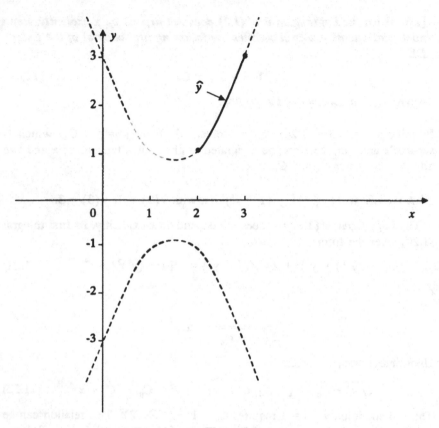

Figure 1.4

methods of solution and to allow the inclusion of more illustrative examples. Singling out class C also permits us to emphasize that extremals must be straight (or, as we shall see later in Chapter 2, polygonal) lines in this case.

With the experience gained from the few relatively simple examples in the last two sections, we should be in a position now to apply the same techniques to more elaborate problems. The historically important brachistochrone problem will be solved in the next section as an illustration.

## 5. The Brachistochrone Problem

A bead of mass $m$ with no initial velocity slides frictionlessly down along a curved wire under the influence of gravity from a point $(a, A)$ to a point $(b, B)$ (Fig. 1.5) What shape should the curved wire have for the time of descent of

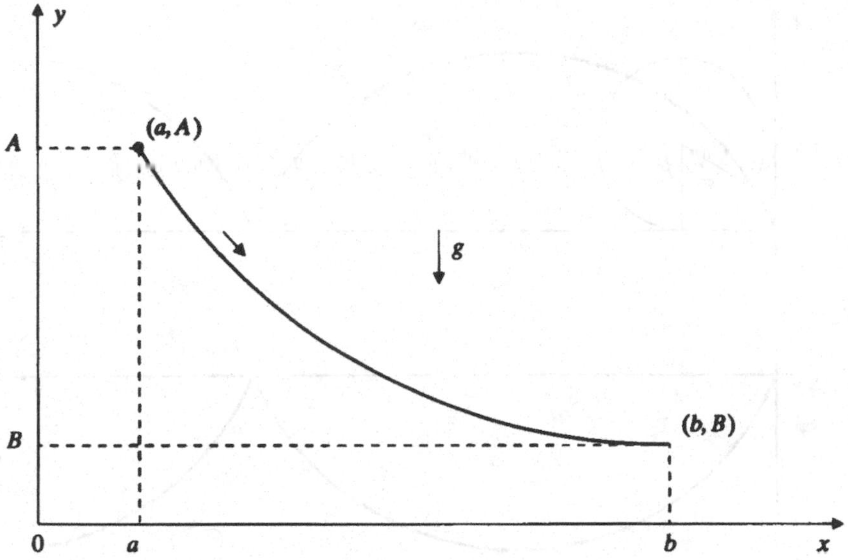


**Figure 1.5**

the bead to be a minimum? This problem was proposed by Johann Bernoulli in 1696 to the mathematical community, particularly to his brother Jacob with whom the contentious Johann had an ongoing feud. Several leading mathematicians at that time, including Leibnitz and Newton, found the correct solution. The solution curve for the wire is an arc of a cycloid, which is defined to be a curve traced out by a point on the rim of a rolling wheel (Fig. 1.6). The problem itself is now known as the *brachistochrone* (meaning "shortest time" in Greek). The brachistochrone is important in the history of mathematics because the method of Euler and Lagrange for the solution of this problem provided the framework for solving a variety of related optimization problems. It marks the inception of an area of mathematics known today as the *calculus of variations*, and variational mathematics has closely intertwined with the development in science and engineering ever since.

Given its prominent historical role in the development of the calculus of variations, the brachistochrone will be discussed in some detail in this section. As with most problems in applied mathematics, we begin with the modeling phase by formulating a mathematical model for the physical problem. From the relation between the velocity of the bead and the total (arc) length traveled, we have $v = ds/dt$ or $dt = ds/v$ so that the total time of descent $T$ is given by

$$T = \int_{x=a}^{x=b} \frac{ds}{v} = \int_a^b \frac{1}{v}\sqrt{1+(y')^2}\, dx, \tag{1.24}$$

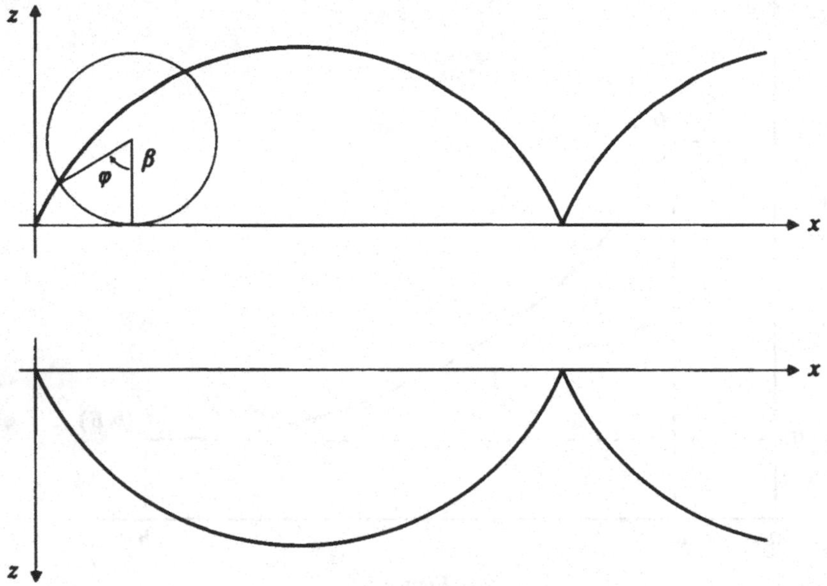


Cycloid: $x = \beta(\varphi - \sin\varphi)$, $z = \beta(1 - \cos\varphi)$

**Figure 1.6**

where $y(x)$, $a \leq x \leq b$, is the curve connecting the two points $(a, A)$ and $(b, B)$,

$$y(a) = A, \qquad y(b) = B \qquad (B < A). \tag{1.25}$$

We still need to express $v$ in terms of $y$ and possibly $x$. To do this, we note that without friction or any other dissipation, the energy of the bead-wire system is conserved. Let $P(t)$ and $K(t)$ denote the *potential* and *kinetic energy* of the system at time $t$. Conservation of energy requires

$$P(t) + K(t) = P(0) + K(0). \tag{1.26}$$

The known expressions for $P$ and $K$ for a particle in a gravity field enable us to write condition (1.26) as

$$mgy + \frac{1}{2}mv^2 = mgA + \frac{1}{2}mv_0^2 \equiv E, \tag{1.27}$$

where $v_0 = 0$ in our case because the bead is initially at rest. We solve (1.27) for $v$ to get

$$v = \sqrt{2g(\gamma - y)}, \qquad \gamma = \frac{E}{mg} \tag{1.28}$$

and then use (1.28) to write (1.24) as

$$T = \int_a^b \sqrt{\frac{1+(y')^2}{2g(\gamma - y)}}\,dx, \tag{1.29}$$

where $y(x)$ should not be greater than $\gamma$ for all $x$ in $[a, b]$. The problem of choosing a function $y(x)$ which satisfies (1.25) and minimizes $T$ is a basic problem of the calculus of variations. For a bead initially at rest, we have $v_0 = 0$ so that $\gamma = A$. In that case, the integrand of (1.29) has a integrable singularity at the end point at $x = a$. Such a singularity would not be present fo $v_0 > 0$.

By setting $z(x) = \gamma - y$ with $z \geq 0$ for $a \leq x \leq b$, expression (1.29) becomes

$$T = \frac{1}{\sqrt{2g}} \int_a^b \sqrt{\frac{1+(z')^2}{z}}\,dx. \tag{1.30}$$

The integral on the right of (1.30) is of the form (1.4) with $k = 1/\sqrt{2g}$ and $n = -1/2$. Its integrand does not depend on $x$ explicitly. We may, therefore, use [I.5] to obtain the following first integral of the Euler DE for the smooth extremal(s) of the problem:

$$F - z'F_{,z'} = \frac{1}{\sqrt{z[1+(z')^2]}} = c_0. \tag{1.31}$$

The second half of (1.31) is solved for $z'$ (expected to be non-negative) to give

$$\frac{dz}{dx} = \sqrt{\frac{1-c_0^2 z}{c_0^2 z}} \quad \text{or} \quad c_1 + \int_0^u \sqrt{\frac{u}{1-u}}\,du = c_0^2 x, \tag{1.32}$$

where $u = c_0^2 z$. The integral of $\sqrt{u/(1-u)}$ can be evaluated by setting $u = \sin^2\theta$ with $du = 2\cos\theta\sin\theta\,d\theta$. This substitution allows us to obtain from (1.32)

$$c_1 + \int_0^\theta 2\sin^2\theta\,d\theta = c_1 + \int_0^\theta (1-\cos 2\theta)\,d\theta = c_0^2 x$$

or

$$x = \alpha + \beta(2\theta - \sin 2\theta), \tag{1.33}$$

where $\beta = 1/(2c_0^2)$ and $\alpha = c_1/c_0^2$. But $\theta$ is related to $z$ by $z = u/c_0^2 = (\sin^2\theta)/c_0^2$ or

$$z = \beta(1 - \cos 2\theta). \tag{1.34}$$

If the coordinate system is chosen so that $\alpha = 0$, (1.33) and (1.34) give the cycloid of Figure 1.6 (with $\varphi = 2\theta$ and $\beta$ being the radius of rolling circle).

Equations (1.33) and (1.34) provide a parametric representation of the extremal $z(x)$. For a nontrivial solution which passes through the point $(a, \gamma - A) = (a, 0)$ (because $v_0 = 0$ for our case), we take $0 \leq \theta \leq \theta_0$ with

$$x(0) = \alpha = a. \tag{1.35}$$

Correspondingly, we have $z(0) = A - y(a) = 0$ which is automatically satisfied. For $z(x)$ to be admissible, $\beta$ must be chosen to satisfy the other end condition at $x = b$, namely, $z(\theta = \theta_0) = \gamma - B = A - B$ with $x(\theta = \theta_0) = b$. This requires $\beta$ and $\theta_0$ to satisfy

$$\beta(2\theta_0 - \sin 2\theta_0) = b - a \tag{1.36}$$

$$\beta(1 - \cos 2\theta_0) = A - B > 0. \tag{1.37}$$

Relations (1.36) and (1.37) can be rearranged to read

$$\frac{b-a}{B-A} - 2\theta_0 = \frac{b-a}{B-A}\cos 2\theta_0 - \sin 2\theta_0, \tag{1.38a}$$

$$\beta = \frac{A-B}{1-\cos 2\theta_0}. \tag{1.38b}$$

The right-hand side of (1.38a) is a periodic function of $2\theta_0$, whereas the left-hand side is linear (with a downward unit slope). A graph of both would show that they intersect exactly once to give a unique extremal for a given pair of $(a, A)$ and $(b, B)$ with $B < A$. Hence, there is a unique admissible extremal for our problem.

We see from the development above that it is convenient sometimes (and necessary at other times) to obtain the solution curve $y(x)$ in parametric form. In that case, we need to have a theory appropriate for a parametric representation of the solution. This will be discussed in Chapter 2. In the next section, we will return to the (seemingly) simpler problem in Example 4. In particular, we will show that a small change in that problem requires that we allow comparison functions to be piecewise smooth.

## 6. Piecewise-Smooth Extremals

Consider the same perfomance index $J[y]$ as in Example 4 but now over the interval $(0, 2)$ with $y(0) = 0$ and $y(2) = 1$. The condition $y(0) = 0$ requires $C_0 = -C_1^2$ in the solution (1.22). The condition $y(2) = 1$ then gives $\pm C_1 = -3/4$ and therewith $C_0 = -9/16$. Hence, the admissible smooth extremal is

$$\left(x - \frac{3}{4}\right)^2 - y^2 = \frac{9}{16} \qquad (0 \leq x \leq 2) \tag{1.39}$$

A graph of (1.39) shows that no part of the smooth extremal found lies in the strip $0 < x < 3/2$ (see Fig. 1.7). One branch of the hyperbola is located to the right of $x = 3/2$ and the other to the left of $x = 0$. As such, expression (1.39) for $y$ is at best incomplete (as it does not define $y$ in $0 < x < 3/2$) and may be incorrect as the solution for the new problem.

It is possible to patch up this expression by defining $y \equiv 0$ in the interval $0 \le x \le 3/2$. We see from (1.21) that $y \equiv 0$ is an extremal with $C_0 = 0$. The composite function (see Fig. 1.7)

$$y_m(x) = \begin{cases} 0, & \left(0 \le x \le \dfrac{3}{2}\right) \\ \left[\left(x - \dfrac{3}{4}\right)^2 - \dfrac{9}{16}\right]^{1/2}, & \left(\dfrac{3}{2} \le x \le 2\right), \end{cases} \tag{1.40}$$

satisfies the Euler DE in $0 < x < 3/2$ and $3/2 < x < 2$; moreover, it is continuous at $x = 3/2$ and satisfies the prescribed end conditions. We might be tempted to conclude that $J[y]$ would be minimized by $y_m$ in this case. Given

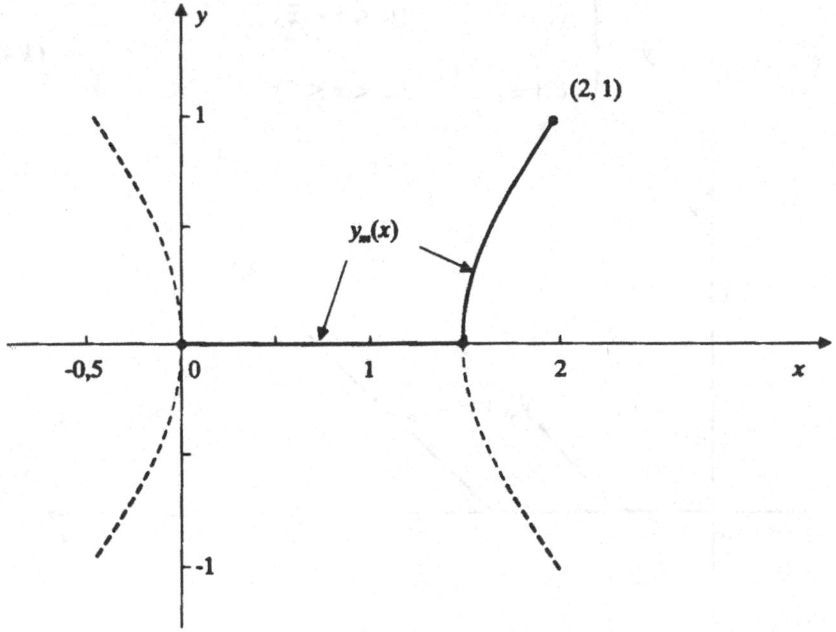


Figure 1.7

that the integrand of $J[y]$ is non-negative, the contribution from the portion of $J[y_m]$ over the interval $0 \leq x \leq 3/2$ simply cannot be smaller.

Unfortunately, the composite function (1.40) cannot be accepted as the solution of our problem as we have formulated it. The function $y_m(x)$ is only piecewise smooth (with $y_m'$ having a jump discontinuity at $x = 3/2$). The Euler DE from which we obtain (1.40) is deduced on the assumption that the quantity $\widehat{G}(x) = G(x, \widehat{y}(x), \widehat{y}'(x))$ in (1.15) is a continuous function. This assumption is violated by the comparison function (1.40) at $x = 3/2$; moreover, the limit of $y_m'(3/2)$ is unbounded. Hence, we must reconsider the whole solution process before we can accept (1.40) as the solution of our problem. In particular, we need to examine the possibility of allowing for PWS comparison functions in our derivation of necessary conditions for an optimum. We will do this in the next chapter.

Lest the readers get the impression that (1.40) gives the correct solution for our basic problem (in a theory for PWS extremals), *it does not*. The integrand $F = y^2(1 - y')^2$ is non-negative; the smallest value possible for $J[y]$ is zero. This value is attained by having $\widehat{y} \equiv 0$ or $\widehat{y}' = 1$ (corresponding to $\widehat{y} = c + x$). Both are extremals of the problem correspondig to $C_0 = 0$ in (1.21). (A third possible extremal corresponding to $\widehat{y}' = -1$ is of no interest here because of the prescribed end conditions). Consider the function

$$\widehat{y} = \begin{cases} 0, & (0 \leq x < \bar{x}) \\ c + x, & (\bar{x} < x \leq 2), \end{cases} \tag{1.41}$$

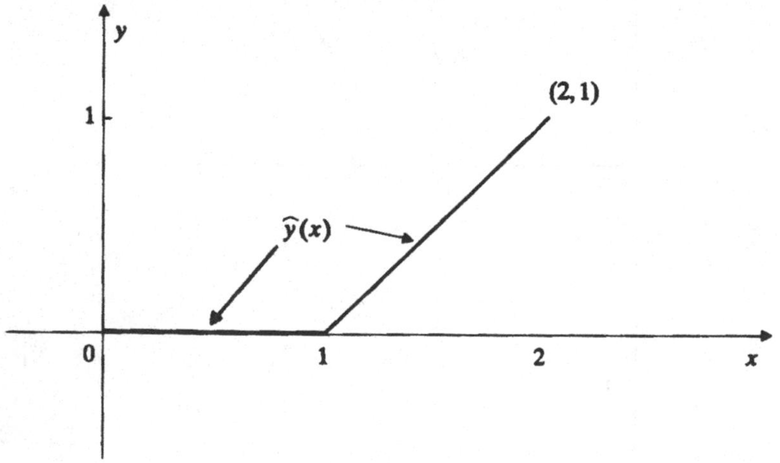


Figure 1.8

for some $\bar{x}$ in $(0,2)$. It satisfies the end condition $\hat{y}(0) = 0$. The other end condition $y(2) = 1$ is satisfied by setting $c = -1$. The function can be made continuous (and PWS) by taking $\hat{x} = 1$. In that case, we have $J[\hat{y}] = 0$ which is a global minimum because $J[y] \geq 0$ for any admissible $y(x)$. Thus, the actual solution of our basic problem is not a $C^1$ function but only PWS (see Fig. (1.8)). For other problems, it may not be possible to spot their PWS solution so easily and we must have a theory for PWS admissible comparison functions to obtain the correct solution of a given problem. Note that a similar PWS solution is not available for the basic problem in Example 4 because $\hat{x} = 1$ is not inside the solution domain $(2,3)$.

## 7. Exercises

1. Find all admissible $C^2$ extremals (if any) of the following basic problems and evaluate $J[\hat{y}]$. (It is not a part of the problem but you may wish to try to determine whether $J[y]$ is a minimum in each case.)

   (a) $F = [1-(y')^2]^2$, $y(0) = y(1) = 0$;

   (b) $F = [1-(y')^2]^2$, $y(0) = 0$, $y(1) = 1$;

   (c) $F = [1-(y')^2]^2$, $y(0) = 1$, $y(1) = 0$;

   (d) $F = (y')^2 - y^2$, $y(0) = 0$, $y(\pi/2) = 1$;

   (e) $F = (y')^2 - y^2$, $y(0) = 0$, $y(\pi) = 0$;

   (f) $F = (y')^2 - y^2$, $y(0) = 0$, $y(\pi) = 1$;

   (g) $F = x^{-1}\sqrt{1+(y')^2}$, $y(1) = 0$, $y(2) = 1$.

2. Let $\{x_0 = a, x_1, x_2, \ldots, x_{N-1}, x_N = b\}$ be $N+1$ equally spaced points in the interval $[a,b]$ with equal spacing $\Delta x = (b-a)/N$ so that $x_k = k\Delta x + a$. Write $y_k \equiv y(x_k)$.

   (a) By approximating $y'(x_k)$ by $(y_{k+1} - y_{k-1})/2\Delta x$ and the integral $J[y]$ by its lower Riemann sum obtain

   $$J[y] \simeq \sum_{n=1}^{N-1} F\left(x_n, y_n, \frac{y_{n+1}-y_{n-1}}{2\Delta x}\right)\Delta x \equiv J(y_1, y_2, \ldots, y_{N-1}),$$

   where $y_0 = A$ and $y_N = B$ are known from the end conditions.

   (b) To minimize the approximate expression $J(y_1, y_2, \ldots, y_{N-1})$, we seek a stationary point $(\hat{y}_1, \ldots, \hat{y}_{N-1})$ of $J$. Show that at the stationary point, we have

$$\Delta x \hat{F}_{,y}(x_n) = \frac{1}{2}[\hat{F}_{,y'}(x_{n+1}) - \hat{F}_{,y'}(x_{n-1})] \qquad (n = 1, 2, ..., N-1),$$

where

$$\hat{f}(x_n) \equiv f\left(x_n, \hat{y}_n, \frac{\hat{y}_{n+1} - \hat{y}_{n-1}}{2\Delta x}\right).$$

(c) In the limit as $\Delta x \to 0$, show that the stationarity condition of (b) tends to Euler DE (1.16). [The results provides a heuristic argument for (1.16) to be a necessary condition for a minimum $J[\hat{y}]$].

3. The rate $v$ (units of goods per unit time) at which a product can be sold depends not only on the unit price $p(t)$ but also on its rate of change $p'(t)$ so that $v = v(p, p')$. A monopolist who can set the price at will wants to choose $p(t)$ to maximize

$$J[p] = \int_0^T [p - c(v)]v \, dt,$$

where $c$ is the cost to produce one unit of the goods. (This cost varies with the rate of production and hence depends on $v$).

(a) Interpret the action of the monopolist.

(b) Suppose

$$v = v_0\left(1 - \frac{p}{p_0} - \frac{p'}{q_0}\right), \qquad c = c_0 v^{-1} + c_1 + c_2 v,$$

where $v_0$, $p_0$, $q_0$, $c_0$, $c_1$, and $c_2$ are known constants. Obtain the Euler DE for $J[p]$.

(c) Obtain directly from [I.5] a first integral of the Euler DE for $J[p]$.

4. Consider the basic problem with $F = y[1 - (y')^2]^{1/2}$, $y(0) = 0$, $y(b) = B$.

(a) Find all admissible smooth extremals if $B \neq 0$.

(b) Find all admissible smooth extremals if $B = 0$.

(c) Compare $J[\hat{y}]$ for part (b) with $J[y = 0]$. Note that $y = 0$ is an admissible comparison function for the problem with $B = 0$. Is it an extremal?

5. The flight path of an aircraft is required to be between two points of distance $D$ apart on a level desert. Assume for simplicity that the cost of flying the aircraft a unit (arc length) distance of the path at a height $y$ is

$\exp(-y/H)$ for some positive constant $H$. Find the possible smooth flight paths for minimum total cost.

6. Find the $C^2$ extremals of the basic problem with $F = (x^2 + y^2)^{1/2} \cdot [1 + (y')^2]^{1/2}$.

7. Consider the basic problem with $F = y^2(y')^2$, $y(0) = 0$ and $y(1) = 1$.
   (a) Show that the smooth extremals are parabolas.
   (b) Determine the unique admissible smooth extremal $\hat{y}$.
   (c) Compare $J[\hat{y}]$ with $J[\bar{y}]$, where $\bar{y}$ is the line joining the two ends.
   (d) Is $\bar{y}$ an extremal?

8. A shrewd monopolist would not maximize $J[p]$ in Exercise 3.
   (a) What would you maximize instead for that problem if the interest rate is fixed at $r$? [Hint: See Example e of section 2.]
   (b) Obtain the Euler DE for your Lagrangian.
   (c) What simplification occurs if $c^2 = 0$?

9. Given $J[y]$ with $y$ being a function of $x$, there are occasions when it is advantageous to treat $x$ as a function of $y$ instead. In that case, we have

$$J[y(x)] = \int_a^b F(x, y(x), y'(x))\, dx$$

$$= \int_A^B F(x(y), y, 1/x^\bullet) x^\bullet\, dy \equiv I[x(y)],$$

   where $(\ )^\bullet = d(\ )/dy$. Obtain the Euler DE for $x(y)$.

10. (a) Find all $C^2$ admissible extremals for the basic problem with $F = (x - y)^2$, $y(0) = 0$ and $y(1) = 1$.
    (b) For what values of $A$ and $B$ do we have a unique smooth extremal for the basic problem with $F = xy^2 + x^2 y$, $y(0) = A$ and $y(1) = B$?

11. Obtain the Euler DE of a basic problem with the following Lagrangian. Do not solve the Euler DE.
    (a) $F = (y')^2 + k^2 \cos y$ ($k$ is a constant);
    (b) $F = x(y')^2 - yy' + y$;
    (c) $F = (y')^2 + yy' + y^2$;

(d) $F = a(x)(y')^2 - b(x)y^2$;

(e) $F = y^2 + (y')^2 + 2e^x y$.

12. **Fermat's Principle of "Least Time" for Light Propagation.** (Actually, there are some restrictions on Fermat's principle for it to be a minimum principle. See Chapters 4 and 5). A light ray travels through an inhomogeneous (transparent) medium from a fixed point $A = (A_1, A_2, A_3)$ to another fixed point $B = (B_1, B_2, B_3)$. The total time it takes to traverse the curve path form A to B is given by

$$T = \int_0^T dt = \int_0^{s_B} \frac{ds}{v},$$

where $s$ is the arclength measured from the point A with $s_B = s(t = T)$ and $v$ is the light speed (equal to $ds/dt$).

(a) If $v$ is a given scalar function of position, say $v(y_1, y_2, y_3)$, obtain an expression for $T$ in terms of (an integral involving) the Cartesian coordinates $\{y_k\}$ (and their derivatives with respect to time).

(b) As a medium for the travelling light ray, the Earth's atmosphere is known to be inhomogeneous, with $v$ being faster at higher altitude (where the air density is lower). Assuming $v$ is uniform in planes parallel to the surface of a *flat* Earth, we may introduce a Cartesian coordinate system so that a ray of sunlight lies in the $y_1, y_3$ plane and set $y_3 = y$ and $y_1 = x$. We further let the light ray path be given by $y(x)$. Show that the expression $T[y_1, y_2, y_3]$ simplifies to

$$T[y] = \int_a^b \frac{1}{v(y)} \sqrt{1 + (y')^2}\, dx.$$

(c) With $v(y)$ being an increasing function of $y$, show that $y''(x) < 0$. What does this result say about sunset?

13. For the minimum production cost problem with $c_h = 0$, show that $J[\hat{y} + \varepsilon h] > J[\hat{y}]$ for $\varepsilon \neq 0$.

14. When there are many producers of a commodity, each alone has little or no influence on the unit price $p_0$ of the commodity. These producers are, therefore, *price takers* (just the opposite of a monopolist). Suppose the commodity is produced with a stock of capital $k$ (machines, factory, etc.) at a rate $f(k)$. The capital stock itself depreciates with time at a rate proportional to $k$, say, $\mu k$. The producer will have to invest to replenish the depreciated capital (to stay in business) and possibly to increase the existing capital stock. Suppose the producer infuses new capital at a rate $I(t)$.

(a) There is naturally a cost to the new capital infusion. Denote this cost by $c(I)$. The price taker producer wants to maximize.

$$J[k] = \int_0^T [p_0 f(k) - c(k' + \mu k)]\, dt.$$

Interpret the action of the price taker.

(b) Without specifying $f(\cdot)$ and $c(\cdot)$, obtain an appropriate form of the Euler DE.

# 2

# Piecewise-Smooth Extremals

## 1. Piecewise-Smooth Solutions for the Basic Problem

The informal discussion of Section 6 of Chapter 1 suggests that it is sometimes necessary to extend the class of admissible solutions to include piecewise-smooth (PWS) functions. We begin our development of this more general theory by defining some terms and notations.

A function $f(x)$ defined on the interval $[a,b]$ is said to be *piecewise-continuous* (PWC) if it is continuous over a finite number of subintervals $(a_i, a_{i+1})$ of $[a,b]$ with $f$ having finite limits at $a_i$ (from the right) and at $a_{i+1}$ (from the left) for $i = 0, \ldots, n$, where $a_0 = a$ and $a_{n+1} = b$ (see Fig. 2.1.).

A function $y(x)$ is said to be *piecewise-smooth* (PWS) on $[a,b]$ (see Fig. 2.2.) if it is

(i) continuous on $[a,b]$,
(ii) differentiable on $[a,b]$ except possibly for a finite number of points,
(iii) the derivative $y'$ is PWC on $[a,b]$.

(A PWS function on $[a,b]$ is sometimes called an *arc* in the literature). At any point $t$ where $y'$ fails to exist, we denote the limit of $y'(x)$ as $x$ approaches $t$ from the left by $y'_-(t)$ (or simply $y'_-$) and as $x$ approaches $t$ from the right by $y'_+(t)$ (or simply $y'_+$). If $y'$ exists and is continuous on $[a,b]$, then $y$ is said to be *smooth* or a $C^1$ function on $[a,b]$ (see Fig. 2.3.).

A scalar function of $m$ variables $F(x_1, \ldots, x_m)$ defined in a given region of the $m$-dimensional Euclidean space $R^m$ is said to be $n$ times continuously differentiable and is denoted by $C^n$ if $F$ and each of its mixed partial derivatives of order $\leq n$ are defined and continuous throughout the domain of $F$. For example, a function $F(x,y)$ is $C^2$ provided $F, F_{,x}, F_{,y}, F_{,xx}, F_{,xy}$, and $F_{,yy}$ are all continuous.

Unless specifically stated otherwise, the Lagrangian $F(x, y, y')$ in expression (1.1) for $J[y]$ is assumed to be $C^n$ (in all three arguments), where $n$ is

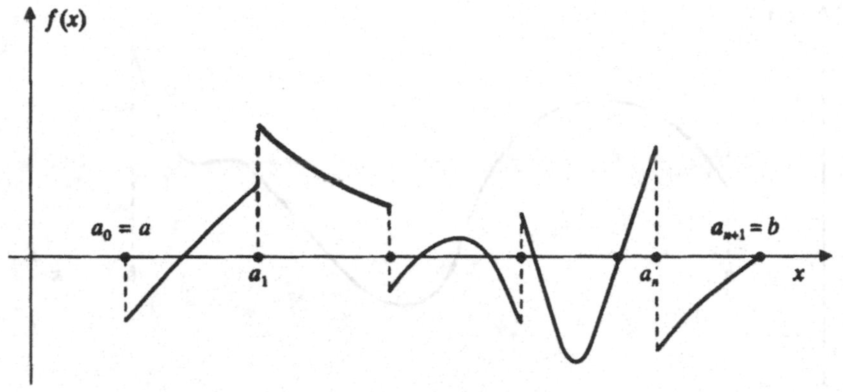


**Figure 2.1**

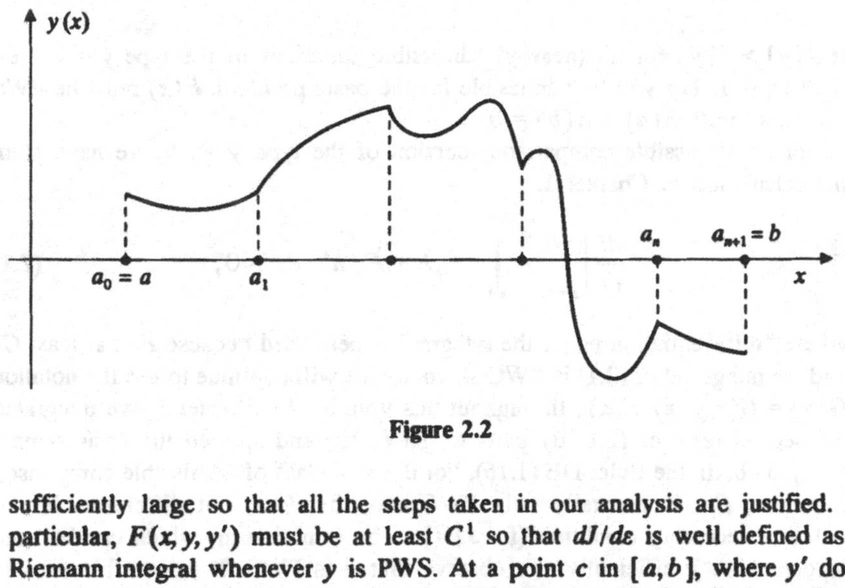


**Figure 2.2**

sufficiently large so that all the steps taken in our analysis are justified. In particular, $F(x, y, y')$ must be at least $C^1$ so that $dJ/d\varepsilon$ is well defined as a Riemann integral whenever $y$ is PWS. At a point $c$ in $[a, b]$, where $y'$ does not exist, it is replaced by $y'_-$ and $y'_+$, and the integral $J[y]$ is the sum of two integrals, one from $a$ to $c$ and the other from $c$ to $b$.

An *admissible comparison function* $y(x)$ is henceforth a PWS which satisfies the end constraints of the problem:

$$y(a) = A, \qquad y(b) = B.$$

(It is important to keep in mind that we will work with the larger class of PWS comparison functions from now on, not just $C^1$ functions). An admissible comparison function $\widehat{y}(x)$ is a *weak local minimum point* for the basic problem

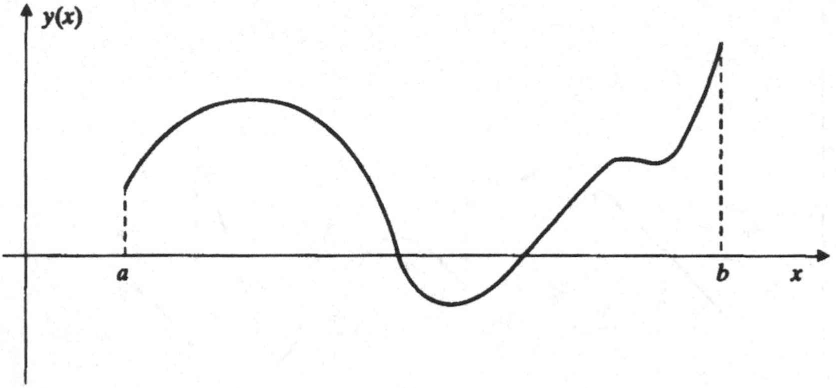


Figure 2.3

if $J[y] \geq J[\hat{y}]$ for all (nearby) admissible functions of the type $y = \hat{y} + \varepsilon h$ with $|\varepsilon| \ll 1$. For $y$ to be admissible for the basic problem, $h(x)$ must be PWS (at worst) with $h(a) = h(b) = 0$.

For an admissible comparison function of the type $\hat{y} + \varepsilon h$, we have from the calculation of Chapter 1.

$$\left.\frac{dJ}{d\varepsilon}\right|_{\varepsilon=0} = \int_a^b \{\hat{F}_{,y} h + \hat{F}_{,y'} h'\}\, dx = 0, \tag{2.1}$$

where "differentiation under the integral" is permitted because $F$ is at least $C^1$ and the integrand of (2.1) is PWC at worst. We will continue to use the notation $\hat{G}(x) = G(x, \hat{y}(x), \hat{y}'(x))$ throughout this volume. In Chapter 1, we integrated the second term of (2.1) by parts to get (1.15) and applied the basic lemma [I.1] to obtain the Euler DE (1.16). For the new class of admissible comparison functions, $\hat{F}_{,y'}$ is generally only PWC and, therefore, not differentiable. To obtain a necessary condition (for $J[\hat{y}]$ to be a minimum) which can be used to determine $\hat{y}$ efficiently, we observe that $h$ is PWS. It is possible then to integrate the first term of (2.1) by parts instead. We will exploit this observation in the next section.

Because $h(x)$ is bounded in $[a, b]$, we have $|y - \hat{y}| = |\varepsilon||h| < c_0\varepsilon = \varepsilon_0$ for some positive constant $c_0$ and for all $x$ in $[a, b]$. The minimum $J[\hat{y}]$ obtained by comparing the group of admissible $\hat{y} + \varepsilon h$ for sufficiently small $\varepsilon$ is, therefore, only a *local minimum*; only admissible comparison functions close to $\hat{y}$ are included in the competition. The situation is analogous to the minimum points of an ordinary function $f(x)$. If there are several stationary points $\{x_1, x_2, \ldots, x_n\}$ of $f$ for which $f''(x_i) > 0$, we can only say that each $f(x_i)$ is a local minimum. A global minimum is found by comparing these local minima with each other.

For comparison functions of the form $y = \hat{y} + \varepsilon h$, the derivative $y'$ is also close to $\hat{y}$ because $|y' - \hat{y}'| = |\varepsilon||h'| < c_1 \varepsilon = \varepsilon_1$ for some positive constant $c_1$ and all $x$ in $[a,b]$. We call the resulting (local) minimum a *weak (local) minimum*, as a smaller value for $J[y]$ may be obtained if we remove the restriction that $y'$ also be close to $\hat{y}'$. This restriction will be removed in Chapter 5 and the corresponding minimum will be called a *strong (local) minimum.*

## 2. The Euler–Lagrange Equation

Condition (2.1) is of the form

$$\int_a^b \{M(x)h(x) + N(x)h'(x)\}\, dx = 0, \tag{2.2}$$

where $M$ and $N$ are PWC functions. We want to show that for this condition to hold for all PWS $h(x)$ with $h(a) = h(b) = 0$, there has to be a certain relationship between $M(x)$ and $N(x)$.

For the basic problem with a PWS solution, let

$$\varphi(x) = \int_a^x M(t)\, dt. \tag{2.3}$$

We integrate the first term of (2.2) by parts to get

$$\int_a^b \{N(x) - \varphi(x)\}h'(x)\, dx = 0, \tag{2.4}$$

where we have used $h(a) = h(b) = 0$ to eliminate the integrated term $[\varphi(b)h(b) - \varphi(a)h(a)]$. Now let

$$c = \frac{1}{b-a}\int_a^b [N(x) - \varphi(x)]\, dx. \tag{2.5}$$

With the identity

$$c\int_a^b h'(x)\, dx = [ch(x)]^{b}_{a} = 0,$$

we can write (2.4) as

$$\int_a^b [N(x) - \varphi(x) - c]h'(x)\, \mathrm{dx} = 0. \tag{2.6}$$

This condition holds for all admissible functions $h(x)$, including

$$h(x) = \int_a^x [N(t) - \varphi(t) - c]\, dt.$$

This particular choice of $h(x)$ evidently satisfies the end condition $h(a) = 0$ and, by (2.5), also $h(b) = 0$. Substituting it into (2.6) transforms that condition into

$$\int_a^b [N(x) - \varphi(x) - c]^2\, dx = 0.$$

Because the integrand is (PWC and) non-negative, we must have

$$N(x) = c + \varphi(x) = c + \int_a^x M(t)\, dt \tag{2.7}$$

for all $x$ in $(a, b)$ except possibly for a finite number of points where $N(x)$ has a finite jump discontinuity. However, (2.7) also holds at these points in the sense that it holds for the left-sided limit and the right-sided limit at any point where $N(x)$ is not continuous. Thus, we have the following *DuBois – Reymond lemma* first published in 1879:

**[II.1]** *If for the PWC functions* M *and* N, *(2.2) holds for all PWS* h(x) *with* h(a) = h(b) = 0, *then* M *and* N *satisfy (2.7) for some constant* c *and for all* x *in* (a, b).

We apply this result to (2.1) and obtain immediately that

**[II.2]** *In order for the admissible (PWS) comparison function* $\hat{y}$ *of the basic problem (1.1) and (1.2) to render the integral* J[y] *a weak local minimum, it is necessary that*

$$\hat{F}_{,y'}(x) = c + \int_a^x \hat{F}_{,y}(t)\, dt \tag{2.8}$$

*for some constant* c.

Again, we adopt the convention that equations such as (2.8) are understood to hold with $y'$ replaced by $y'_-$ and $y'_+$ when $y'$ does not exist. The result [II.2] is a consequence of $J^\bullet(\varepsilon = 0) = 0$ with $J^\bullet \equiv dJ/d\varepsilon$; hence, the necessary condition (2.8) holds also for a (weak local) maximum or just a "stationary point." Lemma [II.1] was established about a century after the fundamental work of Euler (1707-1783) and Lagrange (1736-1813). Nevertheless, we will call (2.8) the *Euler–Lagrange equation* in honor of these two mathematicians for their pioneering contributions. It is a more general result than [I.2], which applies only to $C^1$ comparison functions.

The Euler–Lagrange equation is an integrodifferential equation for the unknown function $\widehat{y}(x)$. There are very few analytical methods available for determining its exact or approximate solution(s). Except for some special cases, a useful general method is to work with the differentiated form of this equation. If we compute the derivative of both sides of (2.8) with respect to $x$, we get [I.2] again as a corollary (of [II.2]) but now in the form

**[I.2′]** ***At any point*** $\mathbf{x}$ ***where*** $\widehat{\mathbf{y}}'$ ***is continuous, the extremal*** $\widehat{\mathbf{y}}$ ***satisfies the Euler DE*** **(1.16).**

For PWS comparison functions, the Euler DE does not hold for the entire interval $(a,b)$. We now formally define an *extremal* of the basic problem to be any PWS $\widehat{y}$ which satisfies (2.8). It is also a solution of the corresponding Euler DE (1.16) in subintervals of $(a,b)$ where $\widehat{y}'$ is continuous. For this reason, a solution of the Euler DE which is $C^1$ in the entire interval $(a,b)$ is called a *smooth extremal*. It is more effective (and convenient) to work with the Euler DE and we should do so whenever possible. Hence, we want to know (without solving the Euler DE) whether the extremals for a basic problem are smooth and where are the discontinuities of their derivative when they are not smooth. These discontinuities are known as *corners* of the extremals. Knowing the locations of the corners allows us to use the Euler DE in the intervals between two discontinuities. Information on discontinuities can be obtained from the so-called *Erdmann-Weierstrass corner conditions* which are to be derived and applied in a later section of this chapter. For a simple derivation of these corner conditions, we first consider necessary conditions similar to (2.8) for several unknowns and for the basic problem (1.1) and (1.2) in a parametric formulation.

## 3. Several Unknowns

Many problems arise naturally in science and engineering in the form of basic problems with several unknown functions. Obvious examples include the motion of a single particle in space and the one-dimensional motion of several masses connected by springs. In the first case, we often describe the motion of the point mass by the time-varying Cartesian coordinate functions $\{x(t),y(t),z(t)\}$. Hamilton's principle (see Section 2 of Chapter 6) requires that the difference between the kinetic and potential energy of the point mass

$$J[x,y,z] = \int_{t_1}^{t_2} \left\{ \frac{m}{2}[(x')^2 + (y')^2 + (z')^2] - V(x,y,z) \right\} dt, \qquad (2.9)$$

be rendered a minimum value by its actual motion, $\{\hat{x}(t), \hat{y}(t), \hat{z}(t)\}$, between prescribed initial and, end positions: $\{\hat{x}(t_i), \hat{y}(t_i), \hat{z}(t_i)\} = \{X_i, Y_i, Z_i\}, i = 1,2.$ In (2.9), $m$ is the mass of the particle and $V(x, y, z)$ is the potential which characterizes the force acting on the point mass with

$$f_x = -\frac{\partial V}{\partial x}, \quad f_y = -\frac{\partial V}{\partial y}, \quad f_z = -\frac{\partial V}{\partial z}, \tag{2.10}$$

being the components of force in the $x$, $y$, and $z$ directions, respectively.

For the one-dimensional motion of the spring-mass system shown in Figure (2.4), Hamilton's principle requires that we minimize

$$J[y_1(t), y_2(t)] = \int_{t_1}^{t_2} \left\{ \frac{1}{2}[m_1(y_1')^2 + m_2(y_2')^2 - V(y_1, y_2)\right\} dt, \tag{2.11}$$

where $y_1(t)$ and $y_2(t)$ are the displacement from equilibrium of the masses $m_1$ and $m_2$, respectively. In terms of the spring constants, $k_1$, $k_2$, and $k_3$, the potential $V$ is given by

$$V = \frac{1}{2}k_{11}y_1^2 + \frac{1}{2}k_{22}y_2^2 + k_{12}y_1y_2 - y_1f_1 - y_2f_2 \tag{2.12}$$

with $k_{11} = k_1 + k_2, k_{22} = k_2 + k_3$ and $k_{12} = -k_2$. The minimization is subject to the fixed end conditions $y_j(t_k) = Y_{jk}$ for prescribed values of $Y_{jk}$, $j, k = 1,2$.

For these and other many-body problems, it is often less cumbersome to work with a vector unknown $\mathbf{y}(x) = (y_1(x), y_2(x), \ldots, y_n(x))^T$. In terms of $\mathbf{y}(x)$, we may write the basic problem as

$$J[\mathbf{y}] = \int_a^b F(x, \mathbf{y}, \mathbf{y}')\, dx, \quad \mathbf{y}(a) = \mathbf{A}, \quad \mathbf{y}(b) = \mathbf{B}, \tag{2.13}$$

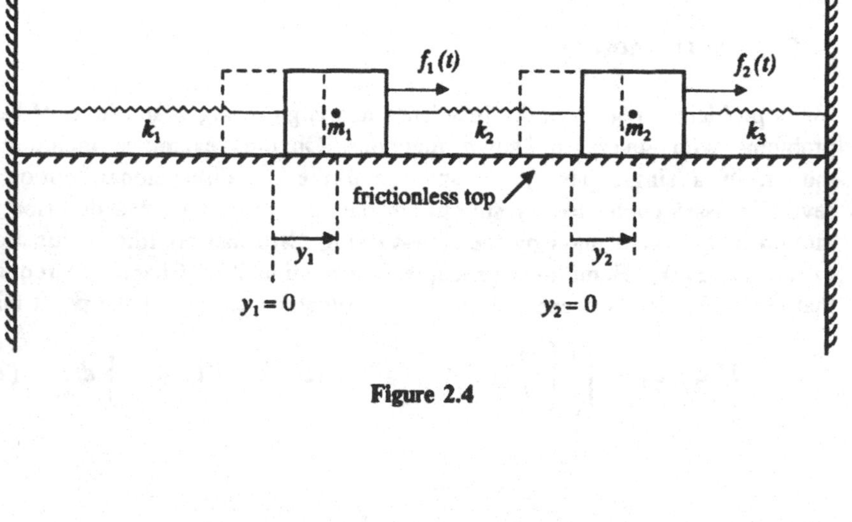


Figure 2.4

where **A** and **B** are two constant $n$-vectors. An admissible ($n$-vector) $\mathbf{y}(x)$ is a vector whose components $y_1(x), \ldots, y_n(x)$ are all admissible comparison functions, i.e., they are PWS and satisfy the end conditions. We would like to find an admissible (vector function) $\widehat{\mathbf{y}}$ which minimizes $J$.

Suppose $\widehat{\mathbf{y}}(x)$ minimizes $J$ and let $\mathbf{y}(x) = \widehat{\mathbf{y}}(x) + \varepsilon\mathbf{h}(x), \mathbf{h} = (h_1, \ldots, h_n)^T$ and $\mathbf{h}(a) = \mathbf{h}(b) = \mathbf{0}$. Then $J[\mathbf{y}]$ becomes an ordinary function of $\varepsilon$ after integration:

$$J[\mathbf{y}] = \int_a^b F(x, \widehat{\mathbf{y}} + \varepsilon\mathbf{h}, \widehat{\mathbf{y}}' + \varepsilon\mathbf{h}')\, dx = J(\varepsilon).$$

For $J[\widehat{\mathbf{y}}]$ to be a minimum, we need $J^{\bullet}(0) = 0$ so that

$$\int_a^b [\widehat{F}_{,\mathbf{y}} \cdot \mathbf{h} + \widehat{F}_{,\mathbf{y}'} \cdot \mathbf{h}']\, dx = \int_a^b \sum_{i=1}^n [\widehat{F}_{,y_i} h_i + \widehat{F}_{,y_i'} h_i']\, dx$$

$$= \sum_{i=1}^n \left\{ [\varphi_i(x) h_i]_a^b + \int_a^b [\widehat{F}_{,y_i'}(x) - \varphi_i(x)] h_i'\, dx \right\} = 0,$$

where

$$\varphi_i(x) = \int_a^x \widehat{F}_{,y_i}(t)\, dt \quad (i = 1, 2, \ldots, n).$$

With $h_i(a) = h_i(b) = 0$, we conclude again, by the *DuBois–Reymond* lemma, that

**[II.3]** *If* $J[\widehat{\mathbf{y}}]$ *is a (weak local) minimum for the basic problem* (2.13), $\widehat{\mathbf{y}}(x)$ *must satisfy the Euler–Lagrange equations*

$$\widehat{F}_{,y_i'} = c_i + \int_a^x \widehat{F}_{,y_i}(t)\, dt \qquad (i = 1, 2, \ldots, n) \tag{2.14}$$

*which may be written as a vector equation*

$$\widehat{F}_{,\mathbf{y}'}(x) = \mathbf{c} + \int_a^x \widehat{F}_{,\mathbf{y}}(t)\, dt \tag{2.15}$$

*with* $F_{,\mathbf{z}} = (F_{,z_1}, F_{,z_2}, \ldots, F_{,z_n})^T$ *and* $\mathbf{c} = (c_1, \ldots, c_n)^T$.

For all $x$ between corners, we may differentiate (2.15) to obtain

**[II.3]** *The Euler DE for* (2.13) *is the vector DE*

$$[F_{,\mathbf{y}'}]' = F_{,\mathbf{y}} \tag{2.16}$$

***which is equivalent to the system of* n *simultaneous (scalar) differential equations***

$$[F_{,y_i'}]' = F_{,y_i} \qquad (i = 1, 2, \ldots, n) \tag{2.17}$$

For the spring-mass system (2.11) and (2.12), we obtain from (2.17)

$$\begin{aligned} m_1 y_1'' + (k_1 + k_2)y_1 - k_2 y_2 &= f_1(t) \\ m_2 y_2'' + (k_2 + k_3)y_2 - k_2 y_1 &= f_2(t) \end{aligned} \tag{2.18}$$

which are the equations of motion of the system.

## 4. Parametric Form

Up to now, we have limited ourselves to extremals which do not occupy more than two (upper) quadrants of the $(x, y)$ plane so that $y$ is a well-defined function of $x$. We know that many curves (including an entire circle) cannot be so described. For this reason alone, it is often necessary to work with the parametric representation of extremals. In other cases, the parametric form of space curves arises naturally; curves traced by the motion of particles (including the cycloid) are familiar examples.

Suppose that the extremal, previously characterized by $y(x)$, is now taken in parametric form

$$x = \varphi(t), \qquad y = \psi(t), \qquad (\alpha \le t \le \beta) \tag{2.19}$$

with the end points given by $\{\varphi(\alpha), \psi(\alpha)\} = (a, A)$ and $\{\varphi(\beta), \psi(\beta)\} = (b, B)$. To avoid confusion, we write $(\ )^{\bullet} \equiv d(\ )/dt$. We stipulate throughout this volume that $\varphi^{\bullet}(t)$ and $\psi^{\bullet}(t)$ do not vanish simultaneously so that $(x^{\bullet})^2 + (y^{\bullet})^2 > 0$. Consider now the integral $J[y]$ for the basic problem rewritten in terms of the parametric representation (2.19):

$$\begin{aligned} J[y] &= \int_a^b F(x, y, y')\,dx = \int_\alpha^\beta F\left(x, y, \frac{y^{\bullet}}{x^{\bullet}}\right) x^{\bullet}\,dt \\ &\equiv \int_\alpha^\beta f(x, y, x^{\bullet}, y^{\bullet})\,dt \equiv \bar{J}[x, y]. \end{aligned} \tag{2.20}$$

All the arguments of $f$ in (2.20) are functions of $t$. When the end points of the admissible comparison functions are fixed, we call minimization of $\bar{J}[x, y]$ the ***parametric form*** of the basic problem in the calculus of variations.

Consider a change of variable $t = t(\tau)$ with $dt/d\tau > 0$ which sends $[\alpha_\tau, \beta_\tau]$ one-one onto $[\alpha, \beta]$. The parametric form of the basic problem is the same in either parameter because

$$\bar{J} = \int_{\alpha_\tau}^{\beta_\tau} f\left(x, y, \frac{dx}{d\tau}, \frac{dy}{d\tau}\right) d\tau = \int_{\alpha_\tau}^{\beta_\tau} f\left(x, y, \frac{dx}{dt}\frac{dt}{d\tau}, \frac{dy}{dt}\frac{dt}{d\tau}\right) d\tau$$

$$= \int_{\alpha_\tau}^{\beta_\tau} f\left(x, y, \frac{dx}{dt}, \frac{dy}{dt}\right)\frac{dt}{d\tau}\, d\tau = \int_{\alpha}^{\beta} f(x, y, x^\bullet, y^\bullet)\, dt,$$

where we have made use of the relation

$$f(x, y, \lambda x^\bullet, \lambda y^\bullet) = \lambda x^\bullet F\left(x, y, \frac{\lambda y^\bullet}{\lambda x^\bullet}\right) = \lambda f(x, y, x^\bullet, y^\bullet).$$

Hence, $\bar{J}[x, y]$ *does not depend on the particular parametric representation of the curve* $y(x)$.

To find $\hat{x}(t)$ and $\hat{y}(t)$ which minimize $\bar{J}[x, y]$, we apply [II.3] to the parametric form (2.20) of the basic problem. The two resulting *Euler–Lagrange* equations are

$$\hat{f}_{,x^\bullet}(t) = c_1 + \int_{\alpha}^{t} \hat{f}_{,x}(\xi)\, d\xi, \tag{2.21a}$$

$$\hat{f}_{,y^\bullet}(t) = c_2 + \int_{\alpha}^{t} \hat{f}_{,y}(\xi)\, d\xi. \tag{2.21b}$$

With $f(x, y, x^\bullet, y^\bullet) = x^\bullet F(x, y, y^\bullet/x^\bullet)$, we have

$$f_{,y^\bullet} = [x^\bullet F(x, y, y^\bullet/x^\bullet)]_{,y^\bullet} = F_{,y'}(x, y, y^\bullet/x^\bullet) = F_{,y'}(x, y, y'), \tag{2.22a}$$

$$\begin{aligned} f_{,x^\bullet} &= [x^\bullet F(x, y, y^\bullet/x^\bullet)]_{,x^\bullet} = F(x, y, y^\bullet/x^\bullet) - \frac{y^\bullet}{x^\bullet} F_{,y'}(x, y, y^\bullet/x^\bullet) \\ &= F(x, y, y') - y' F_{,y'}(x, y, y') \end{aligned} \tag{2.22b}$$

It follows from (2.22a) and $f_{,y} = x^\bullet F_{,y}(x, y, y')$ that (2.21b) is identical to the Euler–Lagrange equation (2.8). From (2.22b), (2.21a), and $f_{,x} = x^\bullet F_{,x}(x, y, y')$, we obtain the following new result:

**[II.4]** *If* $\hat{y}(x)$ *is an extremal of the scalar basic problem (1.1) and (1.2), then* $\hat{y}(x)$ *satisfies*

$$\hat{F}(x) - \hat{y}'\hat{F}_{,y'}(x) = c_1 + \int_a^x \hat{F}_{,x}(t)\, dt \tag{2.23}$$

*for all* $x$ *in* $[a, b]$ *and some constant* $c_1$. *If* $F$ *does not depend explicitly on* $x$, *then (2.23) simplifies to*

$$F(\hat{y}, \hat{y}') - \hat{y}' F_{,y'}(\hat{y}, \hat{y}') = c_1. \tag{2.23'}$$

Relation (2.23′) is in agreement with (1.20), now without the requirement that the extremal be smooth.

Given the identity

$$x^{\bullet}f_{,x^{\bullet}}+y^{\bullet}f_{,y^{\bullet}}=x^{\bullet}[x^{\bullet}F(x,y,y^{\bullet}/x^{\bullet})]_{,x^{\bullet}}+y^{\bullet}[x^{\bullet}F(x,y,y^{\bullet}/x^{\bullet})]_{,y^{\bullet}}$$
$$=x^{\bullet}F(x,y,y^{\bullet}/x^{\bullet})-y^{\bullet}F_{,y'}(x,y,y^{\bullet}/x^{\bullet})+y^{\bullet}F_{,y'}(x,y,y^{\bullet}/x^{\bullet})$$
$$=f(x,y,x^{\bullet},y^{\bullet}),$$

we differentiate both sides with respect to $t$ to obtain

$$\frac{d}{dt}[x^{\bullet}f_{,x^{\bullet}}+y^{\bullet}f_{,y^{\bullet}}]=x^{\bullet\bullet}f_{,x^{\bullet}}+x^{\bullet}\frac{d}{dt}(f_{,x^{\bullet}})+y^{\bullet\bullet}f_{,y^{\bullet}}+y^{\bullet}\frac{d}{dt}(f_{,y^{\bullet}})$$
$$\frac{df}{dt}=f_{,x}x^{\bullet}+f_{,y}y^{\bullet}+f_{,x^{\bullet}}x^{\bullet\bullet}+f_{,y^{\bullet}}y^{\bullet\bullet}.$$

The equality of these two expressions gives

$$x^{\bullet}\left[\frac{d}{dt}(f_{,x^{\bullet}})-f_{,x}\right]+y^{\bullet}\left[\frac{d}{dt}(f_{,y^{\bullet}})-f_{,y}\right]=0.$$

It is evident from this result that *if one of the Euler DE corresponding to* (2.21) *is satisfied, the other is satisfied automatically by the same choice of* $\widehat{y}(t)$ *and* $\widehat{x}(t)$ (or the extremal is straight).

**EXAMPLE 1:** The parametric form of the geodesic problem in the plane.

For this problem, we have $f=\sqrt{(x^{\bullet})^2+(y^{\bullet})^2}$ and, hence, the following Euler differential equations:

$$f_{,x^{\bullet}}=\frac{x^{\bullet}}{\sqrt{(x^{\bullet})^2+(y^{\bullet})^2}}=c_1,\qquad f_{,y^{\bullet}}=\frac{y^{\bullet}}{\sqrt{(x^{\bullet})^2+(y^{\bullet})^2}}=c_2$$

These may be combined to give

$$\frac{y^{\bullet}}{x^{\bullet}}=\frac{c_2}{c_1}\equiv c\qquad\text{or}\qquad\frac{dy}{dx}=c$$

so that $y=cx+d$. It is the straight line we expected. In view of the preceding result, it is not surprising that we cannot proceed further to obtain $x$ and $y$ as functions of $t$. We may, of course, set $x$ as any well-defined PWS function of $t$, including $x=t$ to complete the parametric representation of the extremal.

**EXAMPLE 2:** The minimal surface of revolution.

For this problem, we have $f=y\sqrt{(x^{\bullet})^2+(y^{\bullet})^2}$ and

$$f_{,x^{\bullet}} = \frac{x^{\bullet}y}{\sqrt{(x^{\bullet})^2 + (y^{\bullet})^2}} = \frac{y}{\sqrt{1+(y')^2}} = c_1 .$$

For $c_1 \neq 0$, $y(x)$ will be shown in Section 6 to be a *catenary*. On the other hand, if we use the other *Euler–Lagrange* equation,

$$f_{,y^{\bullet}} = [y\sqrt{(x^{\bullet})^2 + (y^{\bullet})^2}\,]_{,y^{\bullet}} = c_2 + \int_a^t \sqrt{(x^{\bullet})^2 + (y^{\bullet})^2}\, dt,$$

the solution process is more difficult. For a smooth extremal, we may write the differential form of $f_{,y^{\bullet}}$ as

$$\left[\frac{yy'}{\sqrt{1+(y')^2}}\right]' = \sqrt{1+(y')^2}\,.$$

With $y/\sqrt{1+(y')^2} = c_1$ (from the equation for $f_{,x^{\bullet}}$), we obtain

$$c_1 y'' = \sqrt{1+(y')^2}$$

or

$$\frac{c_1 y' y''}{\sqrt{1+(y')^2}} = y'.$$

The last relation can be integrated to give $y = c_1\sqrt{1+(y')^2}$, which is a known result from $f_{,x^{\bullet}} = c_1$. Again, nothing new came from working with the second *Euler–Lagrange* equation, only what we had already found from the first.

## 5. Erdmann's Corner Conditions

A *corner* (point) of an extremal $\hat{y}(x)$ is a point $x$ where $\hat{y}'(x)$ is discontinuous, i.e., the left derivative and right derivative of $\hat{y}$ exist at $x$ (and are denoted by $y'_-$ and $y'_+$ respectively) but they are not equal. Now, the right-hand side of the Euler–Lagrange equation (2.8) is a continuous function of $x$. It follows immediately that the left-hand side must also be continuous. We have then

**[II.5]** $\hat{F}_{,y'}(x)$ ***is continuous even at a corner of*** $\hat{y}(x)$, ***i.e.,*** $F_{,y'}(x, \hat{y}(x), \hat{y}'_+) = F_{,y'}(x, \hat{y}(x), \hat{y}'_-)$ ***at a corner point.***

Sometimes, this *first corner condition* of Erdmann (or Erdmann–Weierstrass) alone requires the extremals of a particular basic problem to be smooth.

**EXAMPLE 3:** $J[y] = \frac{1}{2}\int_a^b [(y')^2 - y^2]\,dx.$

In this example, we have $F_{,y'} = y'$ so that the continuity of $\hat{F}_{,y'}$ at a corner of an extremal requires $\hat{y}'_+ = \hat{y}'_-$ there. In other words, $\hat{y}'$ is continuous even at a possible corner location and $\hat{y}(x)$ must, therefore, be smooth.

**EXAMPLE 4:** $J[y] = \int_a^b \sqrt{1 + (y')^2}\,dx.$

As a more interesting example, we have for the integral above

$$F_{,y'} = \frac{y'}{\sqrt{1 + (y')^2}}$$

To simplify notations, it is customary to write $p = \hat{y}'_+$ and $q = \hat{y}'_-$ so that the continuity of $\hat{F}_{,y'}$ at a corner requires

$$\frac{p}{\sqrt{1 + p^2}} = \frac{q}{\sqrt{1 + q^2}} \qquad \text{or} \qquad p^2(1 + q^2) = q^2(1 + p^2).$$

From the last equation, we deduce $p^2 = q^2$ which can then be used in the first equation to get $p = q$. Again, the extremals do not have corners and must, therefore, be smooth.

More often, however, Erdmann's first corner condition is not sufficient to determine whether or not $\hat{y}'$ is continuous. A second condition is needed to fix both $p$ and $q$. This is provided by *Erdmann's second corner condition*:

**[II.6]** ***The quantity*** $(\hat{F} - \hat{y}'\hat{F}_{,y'})$ ***is continuous at a corner of*** $\hat{y}$.

This second corner condition is a consequence of [II.4] given that the right-hand side of (2.23) is continuous.

Let us see how [II.6] can be used in conjunction with [II.5].

**EXAMPLE 5:** $F(x, y, y') = (y')^2 + (y')^3.$

The two Erdmann conditions for this Lagrangian are

$$(p - q)[2 + 3(p + q)] = 0,$$
$$-(p - q)[(p + q) + 2(p^2 + pq + q^2)] = 0.$$

We have $p \neq q$ at a corner so that the two Erdmann corner conditions may be taken in the form

$$2 + 3(p + q) = 0, \qquad (p + q) + 2[(p + q)^2 - qp] = 0.$$

The first corner condition [II.5] gives $p + q = -2/3$ and by itself would allow for the possibility of a jump in $\hat{y}'$. But by using this result in the second condition, we obtain $pq = 1/9$ or $q = 1/(9p)$. The first condition can, in turn, be reduced to $9p^2 + 6p + 1 = (3p + 1)^2 = 0$ or $p = -1/3$ and therewith $q = 1/(9p) = -1/3 = p$. Hence, the extremals do not have corners.

**EXAMPLE 6:** $F(x, y, y') = y^2(1 - y')^2$.

This is the integrand from our illustrative Example 4 of Chapter 1. The first Erdmann corner condition [II.5] requires that $\widehat{y}_+^2(1 - p) = \widehat{y}_-^2(1 - q)$. But $\widehat{y}$ is continuous so that we have either $p = q$ or $\widehat{y} = 0$ at the corner point; therefore, the extremals of this problem must be smooth unless we have $\widehat{y} = 0$ at the corner(s). The second Erdmann corner condition [II.6] requires $\widehat{y}^2(1 - \widehat{y}')(1 + \widehat{y}')$ be continuous at a corner. This condition is satisfied for the interesting case $\widehat{y}_\pm = 0$ without giving any additional information on $p$ or $q$. Hence, it says nothing about whether or not $\widehat{y}$ is smooth at the zeros of $\widehat{y}(x)$.

Nevertheless, the two Erdmann conditions have provided us with useful information on the corner points of the extremals for this example. If they exist, such corner points must occur at locations where $\widehat{y}$ vanishes. To make use of this information for the determination of appropriate admissible extremals for our problem, we note that (2.23) reduces to

$$\widehat{y}^2[1 - (\widehat{y}')^2] = C_0 \tag{2.24}$$

for all $x$ in $(a, b)$ for the present problem. We have already investigated the case $C_0 \neq 0$ in Chapter 1 and found the result inadequate for the end conditions $y(0) = 0$ and $y(2) = 1$. We now consider the other option $C_0 = 0$. For this case, we have one of three possibilities: $\widehat{y} = 0$, $\widehat{y}' = 1$, and $\widehat{y}' = -1$. The condition $\widehat{y}' = -1$ gives rise to a positive integrand for $J[y]$ of our problem and should therefore, be discarded. It is then a matter of repeating the analysis of Chapter 1, Section 6 to find that $\widehat{y}$ is given by (1.41) with two unknown parameters: the constant of integration $c$ and the location of the corner point $\bar{x}$. These two unknowns are fixed by the end condition $y(2) = 1$ and the continuity of $\widehat{y}$ at $\bar{x}$. For the resulting admissible extremal $\widehat{y}$, we have $J[\widehat{y}] = 0$ which is the smallest value for $J[y]$ possible.

If the end condition $y(2) = 1$ is replaced by $y(1) = 2$, the constant $c$ in the solution (1.41) must be set equal to 1 so that $c + 1 = 2$. In that case, we have $\widehat{y} = 1 + x$ for $\bar{x} \leq x \leq 1$. The corner point $\bar{x}$ should be located at a point where $\widehat{y}$ is continuous with $\widehat{y}(\bar{x}_+) = \widehat{y}(\bar{x}_-) = 0$. This requires $\bar{x} = -1$ which is outside the solution domain. We conclude that the extremal must be smooth for this case and condition (2.24) holds for $C_0 \neq 0$. It is a straightforward calculation to verify that the exact solution for the new problem is given by

$$\widehat{y} = \sqrt{\left(x + \frac{3}{2}\right)^2 - \frac{9}{4}}\,.$$

## 6. The Ultradifferentiated Form

The solution of the basic problem in the calculus of variations must be an admissible extremal, i.e., a solution of the Euler–Lagrange equation which satisfies the prescribed end conditions. To find such a solution, it is preferable to work with the differentiated form of the Euler–Lagrange equation, i.e., the Euler DE. In the actual solution process, we usually need to carry out the differentiation of the term $(F_{,y'})'$ in (1.16) and work with

$$(F_{,y'y'})y'' + (F_{,y'y})y' + (F_{,y'x} - F_{,y}) = 0 \tag{2.25}$$

Equation (2.25) is called the ***ultradifferentiated*** form of the Euler–Lagrange equation. It is valid for all $x$ in $(a, b)$ where $y$ is at least $C^2$. This is not always the case; $y$ is certainly not $C^2$ at a corner point. What about noncorners? When can we guarantee that $y$ has a continuous second derivative so that we can use (2.25)? In a sense, the answer is suggested by (2.25) itself. At a noncorner point, all terms in (2.25) except $y''$ are well defined and are known to be continuous. (Recall our earlier stipulation on $F$). With $(F_{,y'y'})y'' = (F_{,y}) - (F_{,y'x}) - (F_{,y'y})y'$, the second derivative $\hat{y}''$ would also be defined and continuous if $\hat{F}_{,y'y'}$ does not vanish at that point. We summarize these observations as the following theorem of Hilbert:

**[II.7]** ***Suppose that $\hat{y}$ is an extremal, $x_0$ is not a corner point, and $\hat{F}_{,y'y'}(x_0) \equiv F_{,y'y'}(x_0, \hat{y}(x_0), \hat{y}'(x_0)) \neq 0$. Then $\hat{y}$ is $C^2$ and (2.25) holds in some neighborhood of $x_0$.***

A rigorous proof of Hilbert's theorem can be found in Courant and Hilbert, (1953, p. 202). A few examples show how we can use this theorem to our advantage.

**EXAMPLE 7:** $F = \frac{1}{2}[(y')^2 - y^2]$.

$F_{,y'y'} = 1$ for this case; the Erdmann conditions and Hilbert's theorem combine to assure us that an extremal is at least $C^2$. From the ultradifferentiated form of the Euler–Lagrange equation $y'' = -y$, we see that $\hat{y}$ has continuous derivatives of all order (and is, in fact, analytic) in $(a, b)$.

**EXAMPLE 8:** $F = y^2(2x - y')^2$ with $y(-1) = 0, y(1) = 1$.

For this example, we have $F_{,y'} = -2y^2(2x - y')$. By Erdmann's first corner condition, $\hat{y}$ has no corner except possibly where it vanishes. With $F_{,y'y'} = 2y^2$, $y$ is $C^2$ everywhere except possibly at the zeros of $y$. To pursue the possibility of a corner point, we note that the Euler DE for this problem is

$$[-2y^2(2x - y')]' = 2y(2x - y')^2.$$

Two solutions of this equation are readily seen to be $y_1 = 0$ and $y_2' = 2x$ (and there may be others). Hence, two extremals for the problem are $y_1 = 0$ and $y_2 = x^2 + c$. The former satisfied the end condition $y(-1) = 0$ but not $y(1) = 1$. The latter can be made to satisfy the end constraint $y_2(1) = 1$ by choosing $c = 0$ so that $y_2 = x^2$. This extremal does *not* satisfy the other end condition $y(-1) = 0$. Consider then the possibility of a nonsmooth admissible extremal of the form:

$$\hat{y} = \begin{cases} y_1 = 0 & (-1 \leq x < \bar{x}) \\ y_2 = x^2 & (\bar{x} < x \leq 1). \end{cases}$$

The requirement that $\hat{y}$ be continuous at $\bar{x}$ determines $\bar{x}$ to be $\bar{x} = 0$. In that case, $\hat{y}'_+ = \hat{y}'_-$ at $\bar{x}$ so that $\hat{y}$ is smooth there! On the other hand, we have

$$\hat{y}'' = \begin{cases} 0 & (-1 \leq x < 0) \\ 2 & (0 < x \leq 1) \end{cases}$$

so that $\hat{y}$ is not $C^2$ at $\bar{x} = 0$. This is not surprising given $\hat{F}_{,y'y'} = 0$ at $x = 0$ so that Hilbert's theorem is not applicable. It is evident from this example that there are smooth extremals which are not $C^2$ in the solution domain.

## 7. Minimal Surface of Revolution

In this section, the extremals for the problem of a minimal surface of revolution, formulated in Chapter 1, will be derived systematically to illustrate the proper solution process. The Lagrangian for the integral $J[y]$ is $F = y\sqrt{1 + (y')^2}$ in this case where we have omitted the multiplicative constant $2\pi$ without affecting the solution.

The first task in the solution process is to determine whether the extremals can have corners. Erdmann's corner conditions require that the two quantities

$$F_{,y'} = \frac{yy'}{\sqrt{1 + (y')^2}} \tag{2.26a}$$

and

$$F - y'F_{,y'} = \frac{y}{\sqrt{1 + (y')^2}} \tag{2.26b}$$

be continuous. By observing the second condition, we get from the first that $\hat{y}'$ must be continuous (and, therefore, that a corner point is not possible) except at locations where we have $y = 0$.

Next, we check

$$F_{,y'y'} = \frac{y}{[1 + (y')^2]^{3/2}}$$

for the zeros of $\widehat{F}_{,y'y'}(x)$. Evidently, this quantity does not vanish, except where $\widehat{y}$ does. It follows from Hilbert's theorem that an extremal, $\widehat{y}(x)$, must be at least $C^2$ except possibly where it vanishes. If we wish, we can now work with the ultradifferentiated form of the Euler–Lagrange equation to obtain the extremals, keeping an eye on points in $(a, b)$ where $\widehat{y}$ vanishes. However, the integrand $F$ for the present problem does not depend explicitly on $x$. For this class of $F$, the Euler–Lagrange equation simplifies to a first-order ODE by [II.4], and it is simpler to work with (2.23′).

From (2.26b), the first-order ODE (2.23′) for the present problem is

$$F - y'F_{,y'} = \frac{y}{\sqrt{1 + (y')^2}} = C_0. \tag{2.27}$$

One extremal corresponding to the special case $C_0 = 0$ is immediately seen to be $y(x) = 0$. We put aside this particular extremal for the moment as it usually does not satisfy the end conditions.

For $C_0 \neq 0$, (2.27) may be rewritten as

$$y' = \left(\frac{y^2}{C_0^2} - 1\right)^{1/2}.$$

The solution of this first-order separable ODE is

$$y = C_0 \cosh\left(\frac{x}{C_0} + C_1\right).$$

Thus, the required minimal surface of revolution (if it exists) is obtained by revolving an appropriate catenary about the $x$-axis. An appropriate catenary is one which passes through the prescribed end points. This is accomplished by choosing the two constants of integration $C_0$ and $C_1$ so that

$$C_0 \cosh\left(\frac{a}{C_0} + C_1\right) = \mathrm{A}, \quad C_0 \cosh\left(\frac{b}{C_0} + C_1\right) = \mathrm{B}.$$

for positive A and B.

The system of two transcendental equations for $C_0$ and $C_1$ turns out to have two, one, or no solutions, depending on the magnitude of $A$ and $B$. We limit the discussion here to the case $a = 0$ and $A = B$. In that case, we have from the two end conditions $\cosh(C_1) = \cosh(C_1 + b/C_0)$. This relation requires $C_1 + b/C_0 = -C_1$ or

$$C_1 = -\frac{b}{2C_0}$$

so that $y(x) = C_0 \cosh([2x - b]/2C_0)$. Let $\gamma = b/2C_0$ and write $y(x)$ as

$$\frac{y}{A} = \frac{1}{\sigma\gamma}\cosh\left(\gamma\left[\frac{2x}{b} - 1\right]\right), \tag{2.28a}$$

$$\sigma = \frac{2A}{b}. \tag{2.28b}$$

Then, $y(b) = A$ becomes

$$\cosh(\gamma) = \sigma\gamma. \tag{2.29}$$

For given $A$ and $b$, $\sigma = 2A/b$ is a known value. Hence, the only unknown $\gamma(= b/(2C_0))$ is required to be root of (2.29). With $y(0) = A\cosh(-\gamma)/\sigma\gamma = A\cosh(\gamma)/\sigma\gamma = A$, the other end condition is also satisfied.

For a sufficiently large $\sigma$, the graph of $\cosh(\gamma)$ intersects the straight line $\sigma\gamma$ at two points. It lies above (and therefore does not intersect) the straight line if $\sigma$ is sufficiently small. For one special value of $\sigma$, denoted by $\sigma_c$, the line is tangent to the curve $\cosh(\gamma)$. Thus, when $b$ is sufficiently large or $A$ is sufficiently small, there cannot be a minimal surface of revolution in the form of a catenary.

The critical value $\sigma_c$ is determined by the condition of tangency

$$\sinh(\gamma_c) = \sigma_c, \tag{2.30a}$$

where $\gamma_c$ is the value of $\gamma$ corresponding to $\sigma_c$:

$$\cosh(\gamma_c) = \sigma_c\gamma_c. \tag{2.30b}$$

We can eliminate $\sigma_c$ from the two equations above to get

$$\tanh(\gamma_c) = \frac{1}{\gamma_c} \tag{2.30c}$$

or $\gamma_c = 1.199679\ldots$. The value $\sigma_c$ is computed from (2.30a) to be $\sigma_c = 1.508880\ldots$. The single catenary corresponding to this pair of values of $\sigma_c$ and $\gamma_c$ is shown in Figure 2.5. The pairs of catenaries for $\sigma = 2$ and for $\sigma = 20$ are also shown in the same figure. As $\sigma$ increases, the catenary for the larger value of $\gamma$ drops more and more sharply from $A$ toward the $x$-axis. This suggests the possibility of a limiting extremal, called the *Goldschmidt curve*, which drops vertically from $y = A$ to $y = 0$ along the $y$-axis, runs horizontally from $x = 0$ to $x = b$ along the $x$-axis and then rises vertically at $x = b$ from $y = 0$ to $y = A$. From condition (2.27), we see that $\hat{y}(x) = 0$ is a possible extremal corresponding to $C_0 = 0$. However, the two vertical segments of the

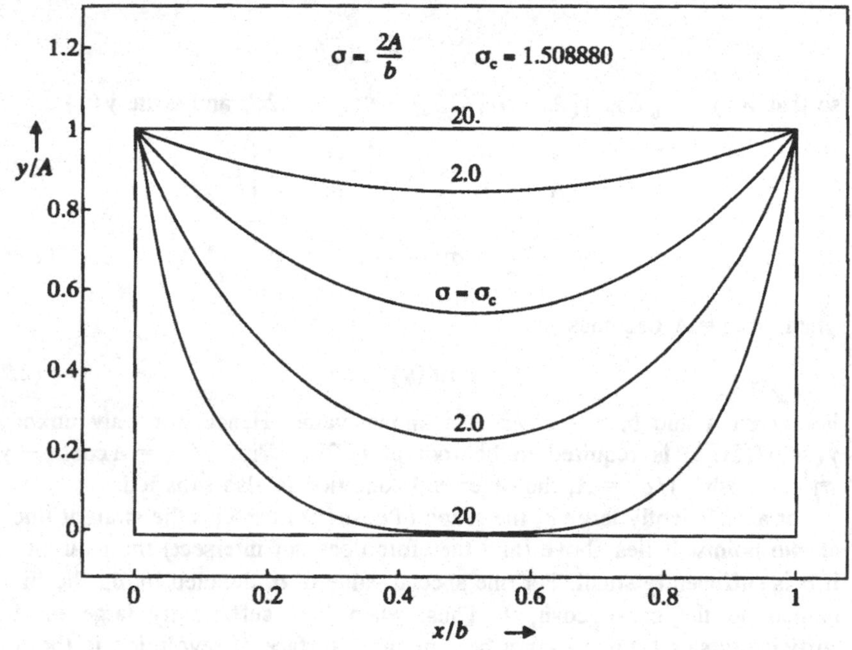


**Figure 2.5**

Goldschmidt curve cannot be represented by taking $y$ as a function of $x$; a parametric formulation of this scalar basic problem is needed to complete the solution process (see Exercise 15).

A physical setup which gives rise to the problem of minimal surface of revolution is the familiar soap film. It consists of two circular wire hoops of radius $A$ and $B$ perpendicular to the $x$-axis with centers on the $x$-axis. The hoops have been dipped in a soap solution and are gradually pulled apart (but kept in parallel) with a soap film connecting them. In a gravity-free space, the shape of the soap film when the hoop centers are located at $(a, 0)$ and $(b, 0)$ normally corresponds to a surface of revolution obtained by revolving a catenary about the $x$-axis. An exceptional case arises when the soap film breaks into two flat discs, each then spanning a hoop in its plane, possibly connected by a line. This exceptional case corresponds to the limiting Goldschmidt extremal (which is PWS).

It is not always possible to extract useful information about corner point locations or the existence and continuity of the second derivative of an extremal $\widehat{y}(x)$ from the results of Erdmann and Hilbert. The lack of information about the corners and flat points of $\widehat{y}$ should not deter anyone from making use of the ultradifferentiated form of the Euler–Lagrange equation to determine the extremals, exactly or approximately (by numerical methods for the Euler DE or direct methods for $J[y]$ to be discussed in later chapters). Experience and insight to the relevant technical phenomenon modeled by the variational problem usually allow us to construct the correct solution for our problem from the extremals obtained by such formal calculation. In fact, the ultradifferentiated form is often used for a formal solution of the problem with the possibility of nonsmooth (or flat) extremals investigated afterward for completeness.

On the other hand, the correct determination of an admissible extremal should not be mistaken for having the correct solution. For $\sigma > \sigma_c$, the problem of a minimal surface of revolution has two admissible $C^2$ extremals. Which one, if any, minimizes $J[y]$? This question will be answered in Chapter 4. Even if there is only one admissible extremal, $J[\hat{y}]$ could be a maximum instead of a minimum. The Euler–Lagrange equation is a consequence of the stationarity condition $dJ/d\varepsilon = 0$ at $\varepsilon = 0$ so that $\hat{y}$ is known only to give a stationary value of $J[y]$. Sufficient conditions for (various types of) minimum values of the general basic problem will be discussed in Chapters 4 and 5.

## 8. Maximum Rocket Height

Let $m_i$ be the initial mass of a rocket with a full tank of fuel moving vertically upward. Let $m_r$ be the terminal mass of the rocket after all the fuel has been burnt off. Suppose $m(t)$ is the total mass of the rocket and the remaining fuel at time $t$ and $v(t)$ is the upward velocity of the rocket at that time. Evidently, $m(t)$ is a decreasing function of time with $m_r \le m(t) \le m_i$. Over the time interval $(t, t+dt)$, an amount $-dm(>0)$ of fuel is ejected downward out of the rocket at a speed $c$ relative to the rocket. By $t + dt$, the rocket will be traveling at a new upward speed $v + dv$. Conservation of momentum requires

$$(m + dm)(v + dv) - (v - c)\,dm + mg\,dt + kv^2 dt = mv$$

where the term $mg\,dt$ corresponds to the effect of gravity and the term $kv^2dt$ is the momentum associated with air drag. After some cancellations, we are left with

$$\frac{dv}{dt} = -\frac{c}{m}\frac{dm}{dt} - g - \frac{k}{m}v^2. \tag{2.31}$$

We limit our discussion here to the case of no air resistance so that $k = 0$ (which is the situation for a rocket well above the Earth's atmosphere). In this case, we can integrate (2.31) to obtain

$$v(t) = c\ln\left(\frac{m_i}{m}\right) - gt \tag{2.32}$$

for a rocket initially at rest. After all the fuel is used up and the rocket reaches a maximum height $H$ at time $t = T$, we can determine $T$ by the condition

$$v(T) = c \ln\left(\frac{m_i}{m_r}\right) - gT = 0$$

or

$$T = \frac{c}{g} \ln\left(\frac{m_i}{m_r}\right) \tag{2.33}$$

The maximum height $H$ reached by the rocket of our simple model is then obtained by integrating $v(t)$ from 0 to $T$,

$$H = \int_0^T c \ln\left(\frac{m_i}{m}\right) dt - \frac{1}{2} gT^2. \tag{2.34}$$

The maximum rocket height $H$ obviously varies with the way in which $m_i$ is reduced to $m_r$, i.e., on the nonincreasing function $m(t)$. The problem of finding $\widehat{m}(t)$ to maximize $H$ with $m(0) = m_i$ and $m(T) = m_r$ is a basic problem in the calculus of variations with a Lagrangian of the form

$$F(t, m, m^\bullet) = F(m) = c \ln\left(\frac{m_i}{m}\right). \tag{2.35}$$

The Euler DE for this Lagrangian requires the quantity

$$\widehat{F}_{,m} = -\frac{c}{\widehat{m}(t)} \tag{2.36}$$

to vanish for all $0 \le t \le T$. This is not possible because $m(t)$ is restricted to be in the range $[m_r, m_i]$.

On the other hand, $H$ is clearly maximized if the integrand is maximized pointwise for all $0 \le t \le T$. For a fixed $t$, $\ln(m_i/m(t))$ is maximized if $m(t)$ is a minimum. Hence, $H$ is maximized by the function $\widehat{m}(t)$ which is $m_r$ for all $0 < t \le T$ with $\widehat{m}(0) = m_i$ as required by the end constraint:

$$\widehat{m}(t) = \begin{cases} m_i & (t = 0) \\ m_r & (0 < t \le T). \end{cases} \tag{2.37}$$

Thus, the "optimal" strategy is physically realized by burning up all the fuel immediately at take-off. Note that $\widehat{m}(t)$ satisfies the other end condition $m(T) = m_r$.

Mathematically, the maximizing function $\widehat{m}(t)$ is a PWC function and the calculus of variations for PWS functions developed in this chapter is not applicable. [Note that we arrive at the optimal solution $\widehat{m}(t)$ by a more or less intuitive argument starting with the expression for $v(t)$, and the determination of the vertical speed does not involve the calculus of variations]. A more general theory of the calculus of variations for PWC (or measurable) functions can be developed to deal with the present (and any other similar) problem. However, a theory for PWS functions is known to suffice for most applications. We limit ourselves in this volume to such a theory.

## 9. Exercises

1. (a) With $F = (y')^2 - y^2$, find all smooth admissible extremals for the basic problem with end conditions $y(0) = 0$ and $y(n\pi) = 0$, where $n$ is any positive integer greater than 1.

   (b) Show that $J[\widehat{y}]$ is not a minimum for any of the extremals of part (a).

2. (a) Find a PWS function $\widehat{y}$ which minimizes $J[y]$ of the basic problem with $F = [1 - (y')^2]^2$ and $y(0) = y(1) = 0$.

   (b) Show that $\widehat{F}_{,y'y'} > 0$ and, given the result of part (a), explain why this does not contradict Hilbert's theorem.

3. For the basic problem with $F = -(y')^2$, $y(0) = 0$, and $y(1) = 1$, show that:

   (a) $\widehat{y} = x$ is the unique $C^2$ admissible extremal.

   (b) All extremals cannot have corners.

   (c) The unique $C^2$ extremal of part (a) does not minimize $J[y]$. [*Hint*: Compare $J[y]$ for the admissible comparison function $y = x + \varepsilon h(x)$, where $h(x) = -x + x^2$.]

4. (a) Does $F = (y')^3$ admit extremals with corners?

   (b) Does the minimum cost production problem of Section 2, Chapter 1 admit extremals with corners?

   (c) Discuss the existence of extremals with corners for $F = (y')^4 - 6(y')^2$. What are the possible combinations of $p = \widehat{y}'_+$ and $q = \widehat{y}'_-$?

5. Consider the basic problem with $F = y[1 - (y')^2]^{1/2}$, $y(0) = 0$, and $y(b) = B$.

   (a) Is it possible to have PWS extremals?

   (b) Is it possible to have admissible PWS extremals for $B \neq 0$?

(c) Find an admissible extremal for $B = 0$.

(d) An admissible $y(x)$ for part $(c)$ is $\hat{y}_0 = 0$. Is it an extremal? Compare $J[\hat{y}_0]$ with $J[\hat{y}_c]$ where $\hat{y}_c$ is the extremal of part $(c)$.

6. Find the extremal(s) with only one corner for the basic problem with $F(x, y, y') = (y' - 1)^2(y' + 1)^4$, $y(0) = 0$, and $y(4) = 2$.

7. For [II.3] to be true, we only need to show that $F_{,y'}(x, \hat{y}(x), \hat{y}'(x)) \equiv \hat{F}_{,y'}(x)$ is differentiable. If $\hat{y}'$ is continuous, show that $(\hat{F}_{,y'})'$ is continuous.

8. Obtain the extremals(s) of the following Lagrangians and discuss the applicability of Euler DE and Euler–Lagrange (integrodifferential) equation in each case.

   (a) $F(x, y, y') = \sqrt{y'}$.

   (b) $F(x, y, y') = \sqrt{y[1 - (y')^2]}$.

   [*Note*: It is often prudent to apply the Euler DE by temporarily ignoring its inapplicability at (hopefully) isolated points in $[a, b]$ and then examine the behavior of the extremals at these points to check the usefulness of the solution obtained.]

9. For some problems which appear to be intractable, changing from Cartesian to polar coordinates often makes them more tractable. Transform the problem with $F(x, y, y') = \sqrt{x^2 + y^2}\,\sqrt{1 + (y')^2}$ by setting $x = r\sin\theta$ and $y = r\cos\theta$ and taking $r$ as the new independent variable.

10. [Queen Dido's Problem] Let the curve $y(x)$ with a fixed length $L$ be in the upper half-plane $(y > 0)$ with the end points $(0, 0)$ and $(b, 0)$. Let $J[y]$ be the area enclosed by this curve and the $x$-axis. We want to find $\hat{y}(x)$ which maximizes $J[y]$.

    (a) Formulate the variational problem. (Obtain the Lagrangian and specify the end conditions).

    (b) In the general case, $b$ is not specified and the problem is not of the form for which the Euler–Lagrange equation was established in [II.2]. Show that by using the arclength of the curve $y(x)$ measured from the point $(0, 0)$ as the independent variable, the problem can be reformulated as a standard basic problem to which the Euler–Lagrange equation is applicable.

11. Find all admissible $C^2$ extremals for the basic problem with

$$F(x, y_1, y_2, y_1', y_2') = 4y_1^2 + y_2^2 + y_1'y_2'$$

and

$$y_1(0) = 1, \; y_1(\pi/4) = 0, \; y_2(0) = 0, \; y_2(\pi/4) = 1.$$

12. (a) What are the Euler–Lagrange equations of

$$J[\varphi(t), \theta(t)] = \int_0^T \left\{ \frac{1}{2} m\ell^2 (\dot{\theta}^2 + \sin^2\theta \, \dot{\varphi}^2) + mg\ell \cos\theta \right\} dt,$$

where $(\dot{\;}) = d(\;\;)/dt$?

(b) Obtain (Hamilton's equations of motion for the above pendulum in space as) the Euler DE.

13. (a) The Newtonian theory of (the two-body) planetary motion for the Earth's orbit around the sun is most conveniently described by polar coordinates $(r, \theta)$ in the plane of motion $(z = 0)$. Deduce the two equations of motion for the Earth's position as the Euler DE of

$$J[r(t), \theta(t)] = \int_0^T \left\{ \frac{m}{2} (\dot{r}^2 + r^2\dot{\theta}^2) + \frac{k}{r} \right\},$$

with $k = GMm$, where $m$ and $M$ are the mass of the Earth and the sun, respectively, and $G$ is a universal constant.

(b) Deduce a conservation law from one of the above Euler DE. What is being conserved?

(c) Upon elimination of $\theta$, the other Euler DE is a second order ODE for $r$ alone. Let $u(\theta) = 1/r(t)$ and transform this ODE into

$$\frac{d^2u}{d\theta^2} + u = A_0$$

where $A_0$ is a known constant.

14. Find the extremals of the basic problems with

(a) $F = (y_1')^2 + (y_2')^2 + y_1'y_2'$;

(b) $F = 2y_1y_2 - 2y_1^2 + (y_1')^2 - (y_2')^2$

15. (a) Formulate the problem of minimal surface of revolution in parametric form.

(b) Obtain the vertical segments of the PWS extremal.

(c) Comment on the parametrization of the horizontal segment of the PWS extremal.

# 3

# Modifications of the Basic Problem

## 1. The Variational Notation

Applications of the calculus of variations in science and engineering are conventionally done in the notation of "variations". This notation has the distinct advantage of being analogous to that used in differential calculus. In this section, we introduce this popular variational notation and make use of it in subsequent sections whenever there is an advantage to do so.

Suppose again that $\widehat{y}(x)$ furnishes a weak local minimum for the basic problem. As before, we compare $J[\widehat{y}]$ with $J[y]$ for admissible comparison functions $y(x) = \widehat{y}(x) + \varepsilon h(x)$. The change $\varepsilon h(x)$ is called the *variation* of $\widehat{y}(x)$ and is conventionally denoted by $\delta y$:

$$\delta y \equiv \varepsilon h(x). \tag{3.1}$$

It is important to note the difference between $\delta y$ and the more familiar quantity $dy$. The variation $\delta y$ is a small change in the shape of the function $\widehat{y}$ for all $x$ in $[a, b]$, whereas the differential $dy$ is a small change in the value of $\widehat{y}$ at a point $x$ due to a small change in $x$.

Corresponding to the change $\delta y$ in $\widehat{y}(x)$, the quantity $F(x, y, y')$ at a particular value of $x$ changes by an amount

$$\begin{aligned} \Delta F &= F(x, \widehat{y}(x) + \varepsilon h(x), \widehat{y}'(x) + \varepsilon h'(x)) - F(x, \widehat{y}(x), \widehat{y}'(x)) \\ &= \widehat{F}_{,y}(x)\varepsilon h(x) + \widehat{F}_{,y'}(x)\varepsilon h'(x) + 0(\varepsilon^2) \end{aligned} \tag{3.2}$$

where $O(\varepsilon^2)$ indicates a collection of terms, each proportional to $\varepsilon^2$ or higher powers of $\varepsilon$. Analogous to the definition of a differential, the first two terms in the right-hand member of (3.2) are defined to be the (first) *variation of F*:

$$\delta F = \widehat{F}_{,y}(x)(\varepsilon h) + \widehat{F}_{,y'}(x)(\varepsilon h') = \widehat{F}_{,y}(x)\delta y + \widehat{F}_{,y'}(x)(\delta y)'.$$

In the special case when $F(x, y, y') = y'$, we have

$$\delta(y') = (\delta y)' \equiv \delta y' \tag{3.3}$$

so that differentiation with respect to the independent variable and the $\delta$ operator are commutative [given that $\delta(y') = \varepsilon h' = (\varepsilon h)' = (\delta y)'$] and there is no ambiguity in writing $\delta y'$. The definition of $\delta F$ can then be written as

$$\delta F = \widehat{F}_{,y}(x)\,\delta y + \widehat{F}_{,y'}(x)\,\delta y' \tag{3.4}$$

Note that a more complete analogy with the definition of a differential would be $\delta F = \widehat{F}_{,x}(x)\,\delta x + \widehat{F}_{,y}(x)\,\delta y + \widehat{F}_{,y'}(x)\,\delta y'$. But the first term on the right drops out because $x$ is not varied so that $\delta x = 0$. The *second variation* is defined to be the group of terms proportional to $\varepsilon^2$ in (3.2). The term "variation" is often used for the first variation.

From definition (3.4), it is a straightforward calculation to verify that the laws of variation of sums, products, quotient, powers, etc., are completely analogous to the corresponding laws of differentiation. We have, in particular,

$$\begin{aligned} &\delta(c_1F_1 \pm c_2F_2) = c_1\,\delta F_1 \pm c_2\,\delta F_2 \\ &\delta(F_1F_2) = F_1\,\delta F_2 + F_2\,\delta F_1, \qquad \delta\left(\frac{F_1}{F_2}\right) = \frac{F_2\,\delta F_1 - F_1\,\delta F_2}{F_2^2} \end{aligned} \tag{3.5}$$

If $x$ and $y$ are both functions of an independent variable $t$, then we have by the chain rule

$$\frac{dy}{dx} = \frac{dy/dt}{dx/dt} = \frac{y^\bullet}{x^\bullet}, \qquad (\ )^\bullet \equiv \frac{d(\ )}{dt}.$$

It follows that

$$\delta\left(\frac{dy}{dx}\right) = \delta\left(\frac{y^\bullet}{x^\bullet}\right) = \frac{x^\bullet\,\delta y^\bullet - y^\bullet\,\delta x^\bullet}{(x^\bullet)^2} \tag{3.6a}$$

But the operators $\delta(\ )$ and $d(\ )/dt$ commute; we can write the above relation as

$$\delta\left(\frac{dy}{dx}\right) = \frac{(\delta y)^\bullet}{x^\bullet} - \frac{dy}{dx}\frac{(\delta x)^\bullet}{x^\bullet} = \frac{d}{dx}(\delta y) - \frac{dy}{dx}\frac{d}{dx}(\delta x) \tag{3.6b}$$

We recover (3.3) only if $x$ does not vary.

When $a$ and $b$ are fixed constants, we take the (first) variation of the definite integral $J[y]$ in (1.1) to be

$$\begin{aligned} \delta J &= \delta\int_a^b F(x, y, y')\,dx = \int_a^b \delta F\,dx \\ &= \int_a^b [\widehat{F}_{,y}\,\delta y + \widehat{F}_{,y'}\,\delta y']\,dx. \end{aligned} \tag{3.7}$$

It follows from [I.2] or [II.2] that if $\hat{y}$ extremizes (minimizes or maximizes) $J[y]$ subject to the end conditions $y(a) = A$ and $y(b) = B$, then $\delta J = 0$.

If $\hat{F}_{,y} - (\hat{F}_{,y'})'$ is continuous, then integrating the second term of (3.7) by parts gives

$$\delta J = [\hat{F}_{,y'}\delta y]_a^b + \int_a^b [\hat{F}_{,y} - (\hat{F}_{,y'})'] \, \delta y \, dx. \tag{3.8a}$$

Now, we have $\delta y = 0$ at the two boundary points for the basic problem because $y$ cannot vary from its prescribed values at these points. By [I.2], we conclude that

*At a smooth extremal* $\hat{y}(x)$ *of* $J[y]$, *the first variation* $\delta J$ *vanishes for any weak variaton* $\delta y$.

This alternative statement of [I.2] is the variational analogue of the vanishing of the differential $df$ at an extreme point $\hat{x}$ of an ordinary function $f(x)$.

In a theory for PWS functions, $F_{,y} - (F_{,y'})'$ may not be continuous, and we should work with

$$\begin{aligned} \delta J &= [\varphi(x)\delta y]_a^b + \int_a^b [\hat{F}_{,y'} - \varphi] \, \delta y' \, dx \\ &= \int_a^b [\hat{F}_{,y'} - \varphi - c] \, \delta y' \, dx \end{aligned} \tag{3.8b}$$

instead, where

$$\varphi(x) = \int_a^x \hat{F}_{,y}(t) \, dt, \qquad c = \frac{1}{b-a}\int_a^b [\hat{F}_{,y'} - \varphi] \, dx. \tag{3.8c}$$

Hence, [II.2] is equivalent to the vanishing of $\delta J$ at an extremal $\hat{y}(x)$ of $J$ for any weak variation $\delta y$.

Given the linearity of $\delta J$ in $\delta y$ and $\delta y'$, we see immediately that the necessary condition $\delta J = 0$ actually holds whether or not we have the end conditions. With

$$J[\hat{y} + \delta y] - J[\hat{y}] = \delta J + \delta^2 J + \delta^3 J + \ldots \tag{3.9a}$$

where

$$\delta^2 J = \frac{1}{2!}\int_a^b [\hat{F}_{,yy}(x)(\delta y)^2 + 2\hat{F}_{,yy'}(x)\delta y \delta y' + \hat{F}_{,y'y'}(x)(\delta y')^2] \, dx, \tag{3.9b}$$

etc. The second and higher-order variations, $\delta^k J$ $(k \geq 2)$, involve higher powers of $\delta y$ and $\delta y'$ and are smaller in magnitude by at least one power of $\varepsilon$ (keeping in mind $\delta y = \varepsilon h$ and $\delta y' = \varepsilon h'$). For sufficiently small $|\delta y|$ and $|\delta y'|$, the sign of the right-hand side of (3.9a) is determined by $\delta J = \delta J[\delta y]$. Suppose we have $\delta J[\delta y] > 0$, for some choice of $\delta y$. Then for any $\alpha > 0$, however small, we have

$$\delta J[-\alpha\delta y] = \int_a^b [\widehat{F}_{,y}(x)(-\alpha\,\delta y) + \widehat{F}_{,y'}(x)(-\alpha\,\delta y')]\,dx$$
$$= -\alpha\int_a^b [\widehat{F}_{,y}(x)\,\delta y + \widehat{F}_{,y'}(x)\,\delta y']\,dx < 0\,. \tag{3.10}$$

Hence, $J[y] - J[\widehat{y}]$ can be made to have either sign for sufficiently small $\delta y$ unless $\delta J = 0$. The same argument applies to the case of a vector unknown $\widehat{\mathbf{y}}(x)$ and allows us to conclude:

**[III.1]** ***If*** $\widehat{\mathbf{y}}(\mathbf{x})$ ***extremizes*** $\mathbf{J}[\mathbf{y}]$ ***(whether or not we have prescribed end conditions), the first variation***

$$\delta \mathbf{J} = \int_a^b [\widehat{\mathbf{F}}_{,\mathbf{y}}\cdot\delta\mathbf{y} + \widehat{\mathbf{F}}_{,\mathbf{y}'}\cdot\delta\mathbf{y}']\,d\mathbf{x} \tag{3.7'}$$

***must vanish.***

This result will be useful in the next few sections.

## 2. Euler Boundary Conditions

Suppose the end condition at $x = b$ of our variational problem is unspecified. This would certainly affect its solution. For example, the surface area of the surface of revolution discussed in section of Chapter 2 depends on the value $B$ and the shape of $\widehat{y}$ varies with $B$. In general, we expect the minimum value of $J[y]$ to decrease (or at least not increase) when an end constraint is removed. Although the extremals which meet the previously prescribed end constraint remain available, the removal of the constraint makes available many new admissible comparison functions which may furnish a smaller $J[y]$. The question is how to pick the best solution among the even larger collection of candidates available to us now.

For $\widehat{y}$ to furnish a (weak) local minimum for our new problem, we again consider $J[y] = J[\widehat{y} + \varepsilon h] = J[\widehat{y} + \delta y]$ where now $\varepsilon h(b) = \delta y(b)$ does not have to vanish because $\widehat{y}(b)$ is now unrestricted. In the variational notation, the condition for $\widehat{y}$ to render $J[y]$ an extremum requires $\delta J = 0$ to hold for all PWS functions $\delta y$ for which $\delta y(a) = 0$. We now reexamine the application of the DuBois–Reymond lemma to take into account the fact that $\delta y(b)$ does not necessarily vanish. With [III.1], (3.8b) and $\delta y(a) = 0$, we have

$$\varphi(b)\,\delta y(b) + \int_a^b \{\widehat{F}_{,y'} - \varphi\}\,\delta y'\,dx = 0 \tag{3.10a}$$

or

$$[\varphi(b) + c]\,\delta y(b) + \int_a^b \{\widehat{F}_{,y'} - \varphi - c\}\,\delta y'\,dx = 0 \tag{3.10b}$$

with $\varphi(x)$ as defined in (3.8c). Because this condition must hold for the PWS variations $\delta y$ which vanish at $x = b$, we deduce as before the existence of $c$ for which the Euler–Lagrange equation (2.8) continues to hold for our new problem. But the condition (3.10b) now applies to $\delta y$ with $\delta y(b) \neq 0$ as well; in that case, we must have the end condition $\varphi(b) + c = 0$ as an additional requirement. The Euler–Lagrange equation may then be used to express this condition more conveniently as

$$\widehat{F}_{,y'}(b) \equiv F_{,y'}(b, \widehat{y}(b), \widehat{y}'(b)) = 0\,. \tag{3.11}$$

Thus, we have

**[III.2]** ***If $\widehat{y}(x)$ satisfies the single end constraint $\widehat{y}(a) = A$ and furnishes a (weak) local minimum for $J[y]$, then $\widehat{y}(x)$ must satisfy the Euler (or natural) boundary condition (3.11) as well as the Euler–Lagrange equation (2.8) in $(a, b)$.***

**EXAMPLE 1:** Find the shortest distance between the point $y(a) = A$ and the line $x = b$.

This problem is a modified version of the geodesic problem between two points in the $x,y$-plane described in Chapter 1. The integral to be minimized is (1.4) for $n = 0$ and $k = 1$ with only one end condition prescribed. The Euler–Lagrange equation for this problem is $F_{,y'} = y'/\sqrt{1 + (y')^2} = c_0$ and the Euler boundary condition (BC) requires that the left-hand side be zero at $x = b$. This implies $c_0 = 0$, which, in turn, requires $\widehat{y}'(x) = 0$. It follows that $\widehat{y}(x) = c_1$ and the end condition $\widehat{y}(a) = A$ requires $c_1 = A$, leaving us with $\widehat{y}(x) = A$. The horizontal straight line from $(a,A)$ to $(b,A)$ is, therefore, an admissible extremal for our problem. It is the only one because Erdmann's corner conditions eliminate the possibility of a broken extremal. We know a minimum exists for our problem; $J[\widehat{y}]$ must, therefore, be the global minimum.

**EXAMPLE 2:** Find the minimal surface of revolution with only one fixed end point $\widehat{y}(a) = A$.

The missing end condition at $x = b$ is replaced by the Euler BC

$$\widehat{F}_{,y'}(b) = \frac{\widehat{y}(b)\widehat{y}'(b)}{\sqrt{1+(\widehat{y}'(b))^2}} = 0 \quad \text{or} \quad \widehat{y}(b)\widehat{y}'(b) = 0 .$$

This condition can be satisfied by either $\widehat{y}(b) = 0$ or $\widehat{y}'(b) = 0$. The second option, applied to the extremal $\widehat{y} = c_0 \cosh(c_1 + x/c_0)$, requires $c_1 = -b/c_0$ so that

$$\widehat{y}(x) = c_0 \cosh\left(\frac{x-b}{c_0}\right)$$

with $c_0$ determined by $\widehat{y}(a) = A$ or $\cosh(t) = \sigma t$ as before where $t = (b-a)/c_0$ and $\sigma = A/(b-a)$. The first option on the other hand requires that $\widehat{y}(x) = 0$ for the segment of the $x$-axis $(a <)\bar{x} \le x \le b$. It is left to the reader to construct an admissible extremal which also satisfies the end condition at $x = a$.

Evidently, the end condition at $x = a$ or both end conditions may be unspecified (instead of just the one at $x = b$). In all cases, a missing end constraint in the basic problem (1.1) and (1.2) is replaced by an Euler BC at the same end point. There should always be a sufficient number of boundary conditions to determine the unknown constants of integration in the solution of the Euler–Lagrange equation (or the Euler DE).

We will now extend the above discussion to allow for $J[y]$ to depend explicitly on the end value(s) of the unknown. For example, there may be a salvage value at delivery time for the extra goods produced, or there may be a special reward (such as payment for patent rights) for completing a task. For these kinds of situations, the quantity to be minimized is the sum of an integral and a term which is a function of the end value of $y$. With no loss in generality, we focus our attention on the case where $y(a) = A$ is prescribed and we want to minimize

$$J[y] = G(y(b)) + \int_a^b F(x, y(x), y'(x))\,dx, \tag{3.12}$$

where $G$ is at least continuously differentiable, i.e., $G$ is at least $C^1$. The term $G(y(b))$ has different interpretations depending on the phenomenon being investigated. It is often called the *terminal payoff* or *salvage value*. With $y = \widehat{y} + \delta y$ and

$$J[y] = G(\widehat{y}(b) + \delta y(b)) + \int_a^b F(x, \widehat{y} + \delta y, \widehat{y}' + \delta y')\,dx, \tag{3.13}$$

the condition $\delta J = 0$ is again necessary for $J[\widehat{y}]$ to be a minimum (or maximum) as shown in an exercise. This, in turn, requires that the Euler–Lagrange equation be satisfied. But now instead of the Euler BC (3.11), we have the new Euler BC

$$\widehat{F}_{,y'}(b) + \widehat{G}_{,y}(b) \equiv F_{,y'}(b, \widehat{y}(b), \widehat{y}'(b)) + G_{,y}(\widehat{y}(b)) = 0. \qquad (3.11')$$

So instead of [III.2], we have now the more general result:

**[III.2′]** *If* $\widehat{y}(x)$ *satisfies the only end constraint* $y(a) = A$ *and extremizes the performance index (3.12), then* $\widehat{y}$ *satisfies the Euler–Lagrange equation (2.8) in* $(a, b)$ *and the Euler BC (3.11′) at* $x = b$.

**EXAMPLE 3:** For the minimum production cost problem in Section 2 of Chapter 1, suppose there is no holding cost so that $c_h = 0$. Suppose also that there is a penalty on a delivery at time $T$ less than the actual order $B$ and a bonus for a delivery in excess of $B$. Let both penalty and bonus be given by $\alpha[y(b) - B]^m/B^m$, where $m$ is an odd integer and $\alpha$ is positive constant. Obtain the admissible extremals.

The relevant $J[y]$ for this problem is

$$J[y] = \int_0^T k(y')^2 dt - \alpha\left[\frac{y(b)}{B} - 1\right]^m$$

which gives the total production cost plus the penalty (or minus the bonus). For this particular Lagrangian, the extremals are straight lines (as Erdmann's conditions show that there cannot be any corner): $\widehat{y}(t) = c_1 t + c_0$. The end condition $\widehat{y}(0) = 0$ requires that $c_0 = 0$. By [III.2′], the Euler BC at $t = T$ is

$$2k\widehat{y}'(b) - \frac{\alpha m}{B^m}[\widehat{y}(b) - B]^{m-1} = 0$$

which determines the remaining unknown constant $c_1$.

For $m = 1$, the Euler BC simplifies to

$$2kc_1 - \frac{\alpha}{B} = 0 \quad \text{or} \quad c_1 = \frac{\alpha}{2kB}$$

so that

$$\widehat{y}(t) = \frac{\alpha t}{2kB}.$$

This is to be compared with the unique admissible extremal $\widehat{y}(t) = Bt/T$ for the corresponding basic problem [where we have the end constraint $y(T) = B$ for a prescribed value $B$]. For the special case $\alpha = 0$, the admissible extremal of the modified problem is not expected to reduce to the admissible extremal for the basic problem. When there is no penalty for not delivering, and no bonus for delivering more than $B$ (as in the case of $\alpha = 0$), the minimum production cost is achieved by not producing (which gives rise to no production cost at all). When delivery is not stipulated and an alternative use of the production machinery is not possible, a meaningful mathematical model for the production problem must also consider the actual profit, not just the cost.

For $m = 3$, the Euler BC becomes

$$\frac{2kB^2}{3\alpha T}\left(c_1\frac{T}{B}\right) - \left(c_1\frac{T}{B} - 1\right)^2 = 0.$$

The two solutions of this quadratic equation in $c_1$ are

$$\frac{T}{B}c_1 = \left(1 + \frac{kB^2}{3\alpha T}\right) \pm \sqrt{\left(1 + \frac{kB^2}{3\alpha T}\right)^2 - 1}\,.$$

Both values of $c_1$ are real and non-negative. For small $\alpha$, we have

$$\frac{T}{B}c_1 = \frac{kB^2}{3\alpha T}\left\{\left[1 + \frac{3\alpha T}{kB^2}\right] \pm \left[1 + \frac{3\alpha T}{kB^2} - \frac{1}{2}\left(\frac{3\alpha T}{kB^2}\right)^2 + \dots\right]\right\}.$$

For $c_1$ to tend to zero with $\alpha$, we should take the negative sign to get

$$\frac{T}{B}c_1 = \left(1 + \frac{kB^2}{3\alpha T}\right) - \sqrt{\left(1 + \frac{kB^2}{3\alpha T}\right)^2 - 1}$$

with

$$c_1 \approx \frac{3\alpha}{2kB}$$

for $3\alpha T/kB^2 \ll 1$. On the other hand, we have for $kB^2/3\alpha T \ll 1$,

$$\frac{T}{B}c_1 \approx 1 - \sqrt{\frac{2kB^2}{3\alpha T}}$$

so that $\widehat{y}(t)$ tends to $Bt/T$ as $\alpha T$ tends to infinity or $kB^2$ tends to zero. In this case, the penalty is too high for delinquency and the bonus too small compared to the unit production cost to stimulate overproduction.

## 3. Free Boundary Problems

In some variational problems, the end value $y(b) = B$ is specified at the right end $b$ of the solution domain, but the boundary point $b$ itself is not specified and must be determined as a part of the solution. An example of this so-called *free boundary problem* is the following R & D (Research and Development) problem:

**EXAMPLE 4:** For many R & D projects, faster spending leads only to diminishing returns (because haste makes waste, etc.). Let $z(t)$ be the rate of spending at time $t$ and $y(t)$ be the cumulative effective effort devoted to a particular project. Suppose $y' = \sqrt{z}/\mu$ for some positive constant $\mu$, and the total effective effort required to complete this project is $B$. In that case, we have $y(0) = 0$ and $y(T) = B$, where $T$ is the completion time for the project. The optimal $z(t)$, $y(t)$, and $T$ are to be determined by the solution process. Suppose there is a known reward $R$ (such as patent value or sale profit) generated at the completion of the project. The present value (at $t = 0$) of the profit from completing the project is

$$J[y] = e^{-rt}R - \int_0^T e^{-rt} z(t)\, \mathrm{dt} = e^{-rt}R - \int_0^T e^{-rt}[\mu y'(t)]^2\, dt,$$

where $r$ is the discount rate for future income or expenditure. Determine the optimal spending rate $\hat{z}(t)$ and optimal completion time $\hat{T}$ so that $J[\hat{y}]$ is a maximum (or so that $-J[\hat{y}]$ is a minimum).

In this example, we know $B$ but not $b = T$. This type of problem is in contrast to the problem with an unspecified end value $B$ discussed in the last section. There, the end point $b$ was prescribed, but $B$ was not.

For the new problem with $b$ unspecified, we need an auxiliary condition to determine the optimal value of $b$ which, along with $\hat{y}(x)$, renders $J[y]$ a minimum. For a little more generality, we will include a terminal payoff in $J[y]$:

$$J[y] = G(b) + \int_a^b F(x, y, y')\, dx \tag{3.14}$$

with

$$y(a) = A, \tag{3.15a}$$

$$y(b) = B, \tag{3.15b}$$

where $a, A$, and $B$ are known constants. The needed necessary conditions to determine admissible extremals are most easily deduced by working with a parametric form of the problem. Let

$$\begin{cases} x = x(t) \\ \\ y = y(t) \end{cases} \qquad (\alpha \le t \le \beta) \tag{3.16}$$

be the parametric representation of the unknown function $y(x)$ with

$$x(\alpha) = a, \qquad y(\alpha) = A,$$
$$x(\beta) = b, \qquad y(\beta) = B, \tag{3.17}$$

where $b$ is an unknown parameter. We can then write (3.14) as

$$J = G(x(\beta)) + \int_\alpha^\beta f(x, y, x^\bullet, y^\bullet)\, dt \equiv \bar{J}\,[x, y] \tag{3.18a}$$

where

$$f(x, y, x^\bullet, y^\bullet) = F\left(x, y, \frac{y^\bullet}{x^\bullet}\right) x^\bullet \tag{3.18b}$$

with $(\ )^\bullet \equiv (\ )/dt$.

In its parametric form, the present problem is one with an unspecified end value $x(\beta)$. We have from Exercise 6 of this chapter that

$$\begin{aligned}\delta J &= G_{,x}(\widehat{x}(\beta))\,\delta x(\beta) + \int_\alpha^\beta [\widehat{f}_{,x^\bullet}\delta x^\bullet + \widehat{f}_{,y^\bullet}\delta y^\bullet + \widehat{f}_{,x}\delta x + \widehat{f}_{,y}\delta y]\, dt \\ &= [G_{,x}(\widehat{x}(\beta)) + \widehat{f}_{,x^\bullet}(\beta)]\,\delta x(\beta) = 0,\end{aligned} \tag{3.19}$$

where we have made use of [II.3] and $\delta x(\alpha) = \delta y(\beta) = 0$ to simplify the expression for $\delta J$. The Euler boundary condition for the parametric solution of the problem is

$$\widehat{f}_{,x^\bullet}(\beta) + \widehat{G}_{,x}(\beta) = 0. \tag{3.20}$$

By (2.22b), we have

$$f_{,x^\bullet} = F - y' F_{,y'}\,.$$

Hence, the Euler BC (3.20) implies the following result for the original problem:

**[III.3]** ***For $\widehat{y}(x)$ and $\widehat{b}$ to extremize $J[y]$ as given in (3.14) subject to the end conditions (3.15), it is necessary that the*** **transversality** ***condition***

$$\widehat{F}(\widehat{b}) - \widehat{y}'(\widehat{b})\widehat{F}_{,y'}(\widehat{b}) + G_{,b}(\widehat{b}) = 0 \tag{3.21}$$

***be satisfied in addition to the Euler–Lagrange equation, where $G(b)$ is the terminal payoff in $J[y]$.***

**EXAMPLE 4: (continued):**

To apply [III.3] to the R & D problem in Example 4, we first obtain the extremal for that problem by integrating the associated *Euler–Lagrange* equation to get

$$e^{-rt}\hat{y}'(t) = c_1 \qquad \text{or} \qquad \hat{y} = \frac{c_1}{r}e^{rt} + c_2 .$$

The two constants of integration $c_1$ and $c_2$ are determined by the end conditions, $\hat{y}(0) = 0$ and $\hat{y}(\hat{T}) = B$, to be $c_1 = -rc_2$ and $c_2 = \beta/(1 - e^{r\hat{T}})$. To determine the remaining unknown $\hat{T}$, we specialize (3.21) to obtain

$$\hat{y}'(\hat{T}) = \sqrt{rR/\mu^2} \tag{3.22}$$

for the present problem. Condition (3.22) requires

$$\hat{T} = -\frac{1}{r}\ln\left[1 - \sqrt{rB^2\mu^2/R}\right] \tag{3.23}$$

For this solution to be meaningful, we need $R > rB^2\mu^2$.

For an interpretation of expression (3.23) for $\hat{T}$, we note that $\mu^2B^2/T^2$ is the average spending rate over the period $[0, T]$ so that $\mu^2B^2/T$ is the total spending at that rate. The quantity $r\mu^2B^2$ is then the total (uncompounded) interest earned by the amount $\mu^2B^2/T$ over the period $[0, T]$. Therefore, the expression (3.23) tells us that

> *For the R & D project to be worthwile (feasible), the terminal payoff should exceed the (uncompounded) interest normally earned by the total spending.*

It suggests that we should not undertake the (investiment on the) project otherwise.

## 4. Free and Constrained End Points

If both $b$ and $B$ are not specified, we need two auxiliary conditions to determine $\hat{b}$ and $\hat{B}$. It is not difficult to extend the analysis of the last section to a parametric solution for the performance index (1.1) with both $b$ and $B$ unspecified. With $B$ also unspecified, we have from of Exercise 6 that

$$\delta J = \hat{f}_{,x'}(\beta)\,\delta x(\beta) + \hat{f}_{,y'}(\beta)\,\delta y(\beta) = 0 . \tag{3.24}$$

The relations of (2.22a) and (2.22b) can be used to write (3.24) as

$$\hat{F}_{,y'}(\hat{b})\,dB + [\hat{F}(\hat{b}) - \hat{y}'(\hat{b})F_{,y'}(\hat{b})]\,db = 0 , \tag{3.25}$$

where we have written $db$ and $dB$ for $\delta x(\beta)$ and $\delta y(\beta)$, respectively. If the boundary point $b$ is fixed so that $db = 0$ but the end value $B$ is not specified, we have from (3.25), $\hat{F}_{,y'}(\hat{b}) = 0$ in agreement with the necessary condition (3.11) of [III.2] in Section 2. If the end value is prescribed so that $dB = 0$ but the end point $b$ is not specified, then we recover from (3.25) the transversality condition (3.21) of [III.3] without the terminal payoff term $G(b)$.

When both $b$ and $B$ are not specified and allowed to be chosen freely, then (3.25) requires that the coefficients of $dB$ and $db$ both vanish and, therefore, both (3.11) and (3.21) hold. In addition, condition (3.11) allows some simplification of (3.21), leaving us with the two conditions,

$$\widehat{F}_{,y'}(\widehat{b}) = 0, \tag{3.26a}$$

$$\widehat{F}(\widehat{b}) = 0, \tag{3.26b}$$

for $G(b) = 0$. We may, if we wish, think of (3.26a) as a replacement for the end condition $y(b) = B$ (which now serves merely as a definition of the symbol $B$) and call it an Euler boundary condition. The remaining condition (3.26b) will be used to determine $\widehat{b}$ and will be called a *transversality condition.*

If a terminal payoff term $G(b, y(b)) \equiv G(b, B)$ is included in $J[y]$, then we have

$$J[y] = G(b, B) + \int_a^b F(x, y, y')\, dx \tag{3.27a}$$

with $b$ and $B \equiv y(b)$ both unspecified, the results in (3.26) will be modified. We have now

$$\begin{aligned}\delta J = {} & [\widehat{F}_{,y'}(\widehat{b}) + G_{,B}(\widehat{b}, \widehat{B})]\, dB \\ & + [\widehat{F}(\widehat{b}) - \widehat{y}'(\widehat{b})\widehat{F}_{,y'}(\widehat{b}) + G_{,b}(\widehat{b}, \widehat{B})]\, db = 0,\end{aligned} \tag{3.27b}$$

where $\widehat{B} = \widehat{y}(\widehat{b})$. It follows that

**[III.4]** ***For*** $\widehat{y}(x)$, $\widehat{b}$, ***and*** $\widehat{B}$ ***to minimize*** $J[y]$ ***(as given by (3.27a)) subject to*** $y(a) = A$, ***it is necessary for*** $\widehat{y}$ ***to satisfy the Euler–Lagrange equation and, together with*** $\widehat{b}$ ***and*** $\widehat{B}$, ***to satisfy the auxiliary conditions***

$$\widehat{F}_{,y'}(\widehat{b}) + G_{,B}(\widehat{b}, \widehat{B}) = 0 \tag{3.28a}$$

$$\widehat{F}(\widehat{b}) + \widehat{y}'(\widehat{b})G_{,B}(\widehat{b}, \widehat{B}) + G_{,b}(\widehat{b}, \widehat{B}) = 0 \tag{3.28b}$$

For many problems in science and engineering, the parameters $b$ and $B$, though not specified, are not allowed to vary freely. They may be restricted to lie on a particular curve, $B = g(b)$ [or $\bar{g}(b, B) = 0$, in the $b, B$-plane. It follows from (3.27b) and $dB = g'(\widehat{b})\, db$ that

**[III.5]** ***For the perfomance index*** $J$ ***of (3.27a) to be stationary so that*** $\delta J = 0$, ***it is necessary for*** $\widehat{y}(x)$, $\widehat{b}$, ***and*** $\widehat{B}$ ***to satisfy***

$$\widehat{F}(\widehat{b}) + [g'(\widehat{b}) - \widehat{y}'(\widehat{b})]\widehat{F}_{,y'}(\widehat{b}) + [g'(\widehat{b})G_{,B}(\widehat{b}, \widehat{B})] = 0, \tag{3.29a}$$

***where*** $\widehat{B} = g(\widehat{b})$. ***In the case of no terminal payoff so that*** $G(b, B) = 0$, ***(3.29a) simplifies to***

$$\widehat{F}(\widehat{b}) + [g'(\widehat{b}) - \widehat{y}'(\widehat{b})]\widehat{F}_{,y'}(\widehat{b}) = 0. \tag{3.29b}$$

**EXAMPLE 5:** $F = \sqrt{1+(y')^2}$, $g(x) = mx + n$, and $G = 0$.

The extremals are straight lines with $(a, A)$ as one end point: $\hat{y}(x) = c(x-a)+A$ [see Fig. (3.1)]. The transversality condition (3.29b) requires that the constant of integration $c$ satisfy

$$\sqrt{1+c^2} + \frac{c[m-c]}{\sqrt{1+c^2}} = 0$$

or $c = -1/m$. The result corresponds to the fact that the shortest distance from the point $(a, A)$ to the nearest point on the line $y = mx + n$ is in the direction orthogonal to that line. The two relations $\hat{B} = \hat{y}(\hat{b})$ and $\hat{B} = g(\hat{b})$ determine $\hat{b}$ and $\hat{B}$.

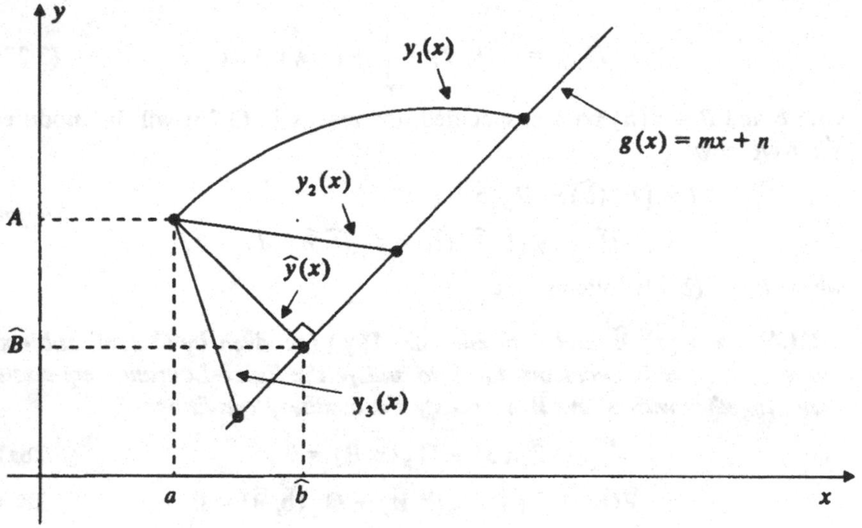


**Figure 3.1**

**EXAMPLE 6:** Consider the integral $J[y]$ given by (1.4) with $k = 1$, $n = -1$, and $y(0) = 0$. Determine the extremals for this problem if the point $(b, B)$ is restricted to lie on the circle $B^2 + (b-9)^2 = 9$ in the upper half-plane.

It can be verified that the *Erdmann* conditions require the extremals to be smooth in $0 < x < b$ and $\hat{F}_{,y'y'}$ not to vanish there. Because $F$ does not depend explicitly on $x$; we have $\hat{F} - \hat{y}'\hat{F}_{,y'} = c_0$ or, with $u = c_0\hat{y}$,

$$\frac{1}{c_0}u' = \sqrt{\frac{1-u^2}{u^2}}.$$

We solve this separable first-order ODE to get (for $d = 1/c_0$)

$$c_0^2\hat{y}^2 + (c_0x + 1)^2 = 1 \quad \text{or} \quad \hat{y}^2 + (x - d)^2 = d^2 \tag{3.30}$$

where we have specified one of the two constants of integration so that $\hat{y}(0) = 0$ is satisfied. The remaining constant $d$ will be determined simultaneously with $\hat{B}$ and $\hat{b}$. These three costants must be chosen to satisfy the following three conditions: (i) the relation (3.30) evaluated at $x = \hat{b}$ and $\hat{y} = \hat{B}$, (ii) the end point constraint $\hat{B}^2 + (\hat{b} - 9)^2 = 9$, and (iii) the transversality condition (3.29b).

With

$$\hat{F}_{,y'} = \frac{y'}{y\sqrt{1 + (y')^2}}, \qquad F - y'F_{,y'} = \frac{1}{y\sqrt{1 + (y')^2}},$$

the condition (3.29b) may be taken in the form

$$\hat{y}'(\hat{b})\hat{g}'(\hat{b}) + 1 = 0.$$

In terms of $\hat{B}, \hat{b}$, and $d$, the three conditions become

$$\hat{B}^2 = 2d\hat{b} - \hat{b}^2, \qquad \hat{B}^2 = 9 - (\hat{b} - 9), \qquad \hat{B}^2 = (\hat{b} - 9)(d - \hat{b}),$$

where we have made use of the fact $g'(\hat{b}) = -(\hat{b} - 9)/\hat{B}$ (obtained by differentiating the end point constraint) and $\hat{y}'(\hat{b}) = -(\hat{b} - d)/\hat{B}$ [obtained by differentiating (3.30)]. The solution of these three simultaneous equations is $d = 4, \hat{b} = 36/5$, and $\hat{B} = 12/5$ [for $(\hat{b}, \hat{B})$ to be in the upper half-plane]. Geometrically,

> *The end point* $(\hat{b}, \hat{B})$ *is the intersection of the circle of radius 4 centered at* $(4,0)$ *and the circle of radius 3 centered at* $(9,0)$ *in the first quadrant of the* $x, y$*-plane (see Fig. 3.2).*

## 5. Higher Derivatives

The treatment of the basic problem in Chapters 1 and 2 can also be extended to handle Lagrangians which depend on higher derivatives of the unknown $y(x)$. Such problems arise naturally in a variety of applications. One example is the bending of an elastic beam on an elastic foundation, the classical mathematical model for rails (Fig. 3.3) In dimensionless form, this problem seeks $\hat{y}(x)$ to minimize the following performance index:

$$J[y] = \frac{1}{2}\int_a^b [(y'')^2 + ky^2 - 2py]\, dx,$$

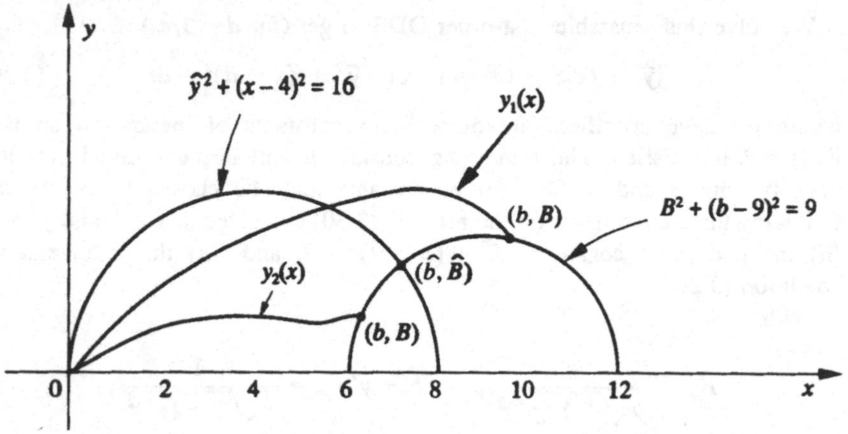


Figure 3.2

where $k$ is the dimensionless spring constant of the elastic foundation and $p$ is the known dimensionless vertical pressure load distribution. In effect, the energy approach to the mechanics of deformable bodies postulates that the behavior of the rail under the prescribed load is the consequence of minimizing the total energy in the rail.

The energy integral $J[y]$ in (3.31) is of the form

$$J[y] = \int_a^b F(x, y, y', y'')\, dx. \tag{3.32}$$

For a beam of finite lenght with both ends constrained from any displacement, we have

$$y(a) = 0, \qquad y(b) = 0 \tag{3.33}$$

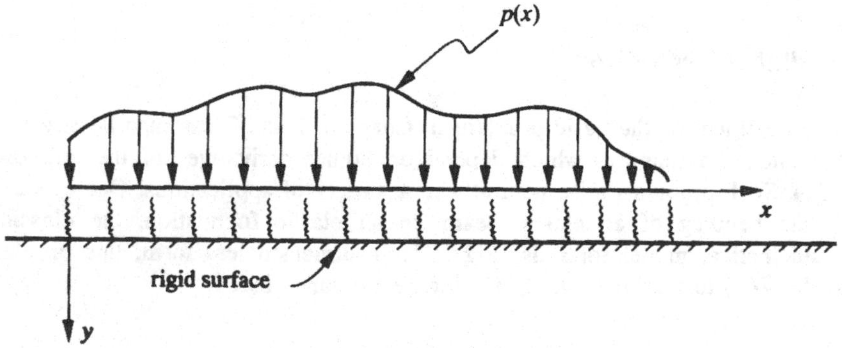


Figure 3.3

as the end conditions. Physically, we may also constrain the beam from a rotation in the $x,y$-plane at the ends (so that the slope of the beam axis remains horizontal). In that case, we have also

$$y'(a) = 0, \qquad y'(b) = 0. \tag{3.34}$$

The end conditions (3.33) and (3.34) for the beam problem suggest that the appropriate end conditions for the general basic problem associated with (3.32) should be

$$y(a) = A, \qquad y(b) = B, \qquad y'(a) = A_1, \qquad y'(b) = B_1. \tag{3.35}$$

For simplicity, we will assume the minimizing function $\widehat{y}(x)$ of the basic problem defined by (3.32) and (3.35) to be $C^4$. The first variation of the perfomance index (3.32) is then

$$\begin{aligned} \delta J &= \int_a^b \{\widehat{F}_{,y}\delta y + \widehat{F}_{,y'}\delta y' + \widehat{F}_{,y''}\delta y''\}\, dx \\ &= [\widehat{F}_{,y''}\delta y' + \{\widehat{F}_{,y'} - (\widehat{F}_{,y''})\}'\delta y]_a^b + \int_a^b \{\widehat{F}_{,y} - [\widehat{F}_{,y'}]' + [\widehat{F}_{,y''}]''\}\, \delta y\, dx. \end{aligned} \tag{3.36}$$

By the end conditions (3.35), we must have $\delta y(a) = \delta y(b) = \delta y'(a) = \delta y'(b) = 0$. Thus, terms outside the integral sign in (3.36) vanish for the basic problem. A more general version of the basic lemma [I.1] (see Exercise 25 at the end of this chapter) can then be used to deduce the following result from expression (3.36) for $\delta J$:

**[III.6]** ***Any $C^4$ function which satisfies the end conditions (3.35) and minimizes (3.32) must satisfy the Euler DE***

$$[\widehat{F}_{,y''}]'' - [\widehat{F}_{,y'}]' + \widehat{F}_{,y} = 0. \tag{3.37}$$

As usual, we have assumed the Lagrangian $F$ to be as continuously differentiable in all its arguments as we need it to be. A solution of the Euler DE (3.37) is called a $C^4$ *extremal* of the basic problem. An extremal which also satisfies the end constraints is again called an *admissible extremal*.

**EXAMPLE 7:** The beam bending problem of (3.31).

The corresponding Euler DE is

$$[\widehat{y}'']'' + k\widehat{y} = p. \tag{3.37'}$$

This is a fourth-order ODE and requires four auxiliary conditions for the complete determination of $\widehat{y}(x)$. The four end conditions (3.35) provide just the right number of boundary conditions for this equation. They are also physically meaningful, corresponding to the ends of the beam being constrained from vertical displacements and rotation about an axis normal to the $x, y$-plane. (As a model for rails, the beam is so long that the effect of trains away from the ends is localized and does not reach the two ends of the rail). The solution process for the boundary-value problem is straightforward and will not be discussed here (see the Exercises).

If $F$ does not depend on $x$ explcitly, a rearrangement of (3.37) similar to that which yielded [I.5] gives us the following first integral of the Euler DE:

**[III.7]** *For* $F = F(y, y', y'')$, *the associated Euler DE can be integrated once to give*

$$\hat{y}'[\hat{F}_{,y''}] - \hat{y}''\hat{F}_{,y''} - \hat{y}'\hat{F}_{,y'} + \hat{F} = C, \tag{3.38}$$

*where C is an arbitrary constant of integration.*

It is worth noting that the first integral (3.38) does not offer as much of an advantage as [I.5] does for the case $F = F(y, y')$. The latter reduces the second-order Euler DE to a first-order ODE for which there are many special methods of solution. There are considerably fewer special techniques for a third-order ODE such as (3.38). Moreover, the application of (3.38) to the beam problem gives

$$\hat{y}'y''' - \frac{1}{2}(\hat{y}'')^2 + \frac{1}{2}k\hat{y}^2 - py = C. \tag{3.38'}$$

This is a third-order *nonlinear* ODE and is, thus, considerably more difficult to solve than the linear fourth-order Euler DE (3.37'). In general, the first integral (3.38) should not be used if the Euler DE is linear. The same rule of thumb also applies to the case $F = F(y, y')$; but not observing this rule in this case generally leads to much less serious consequences given that the first integral (1.20) is a first-order ODE and is usually separable. Still, the difference can be seen by solving the Euler DE for $F = [(y')^2 - y^2]/2$ by the two different solution processes.

It may have occurred to the reader that the basic problem involving higher derivatives can be transformed into a problem of several unknowns. For (3.32), we can set $z = y'$ so that $y'' = z'$ and $F = F(x, y, z, z')$. Unfortunately, the results of Section 4 of Chapter 2 are not directly applicable as they do not take into account the fact that $z$ and $y$ are related by $z = y'$. The relation between $z$ and $y$, which must be satisfied by the actual solution, serves as a constraint on the comparison functions. The subject of constrained optimization will be discussed later in a number of chapters, including Chapters 10–14.

For the general Lagrangian $F = F(x, y, y', y'', \ldots, y^{(n)})$, it is not difficult to see that the appropriate $2n$ end conditions, for the associated basic problem are

$$y^{(k)}(a) = A_k, \qquad y^{(k)}(b) = B_k, \qquad (k = 0, 1, 2, \ldots, n-1). \tag{3.39}$$

For this more general problem, it is straightforward to deduce the following:

**[III.8]** ***If a $C^{2n}$ function $\hat{y}(x)$ satisfies the 2n end constraints (3.39) and minimizes the integral $J[y]$ in which F is a function of the derivatives of y up to order n, $\hat{y}(x)$ must satisfy the Euler DE***

$$\begin{aligned} &(-1)^n \frac{d^n}{dx^n}[\hat{F}_{,y^{(n)}}] + (-1)^{n-1}\frac{d^{n-1}}{dx^{n-1}}[\hat{F}_{,y^{(n-1)}}] + \ldots \\ &+ \frac{d^2}{dx^2}[\hat{F}_{,y''}] - \frac{d}{dx}[\hat{F}_{,y'}] + \hat{F}_{,y} = 0. \end{aligned} \tag{3.40}$$

Note that (3.40) is an ODE of order ($2n$). Its solution is again called a $C^{2n}$ *extremal* of the basic problem. The ($2n$) end conditions (3.39) provide just the right number of auxiliary conditions for the Euler DE (3.40) to be a well-defined BVP. A solution of this BVP is again called an *admissible extremal* of the basic problem.

It is also possible to deduce a first integral of (3.40) if $F$ does not depend on $x$ explicitly. However, such a first integral offers little benefit for the case $n > 1$ and will not be given here.

If some or all of the end conditions for a basic problem involving higher derivatives are not prescribed, the missing end conditions are replaced by a corresponding number of Euler boundary conditions. These Euler BC are obtained by a process similar to that described for the simpler basic problem analyzed in the last two chapters. To illustrate, suppose some or all of the end conditions at $x = b$ are not prescribed for a Lagrangian which depends on $x, y, y'$, and $y''$. Instead of (3.36), the expression for the first variation of $J$ is now given by

$$\begin{aligned} \delta J &= \hat{F}_{,y''}\delta y'(b) + \{\hat{F}_{,y'}(b) - \hat{F}'_{,y''}(b)\}\delta y(b) \\ &+ \int_a^b [\hat{F}_{,y} - \hat{F}'_{,y'} + \hat{F}''_{,y''}]\,\delta y\, dx \end{aligned} \tag{3.41}$$

because $\delta y(a) = \delta y'(a) = 0$. For $\delta J = 0$ to hold for those comparison functions with $\delta y(b) = \delta y'(b) = 0$, the Euler DE (3.37) must again be a necessary condition for $\hat{y}(x)$ to minimize $J[y]$. This reduces (3.41) to

$$\delta J = \hat{F}_{,y''}(b)\delta y'(b) + \{\hat{F}_{,y'}(b) - \hat{F}'_{,y''}(b)\}\delta y(b) \tag{3.42}$$

which must also vanish for other types of admissible comparison functions [for which $\delta y(b)$ and/or $\delta y'(b)$ may not be zero]. If $y'(b)$ is not prescribed, we must have the Euler BC

$$\widehat{F}_{,y''}(b) = 0 \tag{3.43}$$

[whether or not $y(b)$ is prescribed]. Similarly, we need the Euler BC

$$\widehat{F}_{,y'}(b) - \widehat{F}'_{,y''}(b) = 0 \tag{3.44}$$

as a necessary condition for $J[\hat{y}]$ to be a minimum if $y(b)$ is not prescribed. We summarize these results for $F(x, y, y', y'')$ in the following theorem:

**[III.9]** *Any $C^4$ function which satisfies the end conditions $y(a) = A_0$ and $y'(a) = A_1$ and minimizes (3.32) must satisfy the Euler BC (3.44) if $y(b)$ is not prescribed and must satisfy the Euler BC (3.43) if $y'(b)$ is not prescribed.*

A perfomance index $J[y]$ which includes salvage value $G(y(b), y'(b))$ can also be handled similarly. The presence of $G$ can either complicate or simplify the corresponding Euler BC. It is not difficult to see that our treatment of Euler BC for $F(x, y, y', y'')$ can be extended to Lagrangians involving still higher-order derivatives.

## 6. Other End Conditions

By definition, the comparison functions of the basic problem of the calculus of variations (1.1) and (1.2) must take on the prescribed value at each end of the solution domain $[a, b]$. The two conditions are just what we need for the Euler DE to define a two-point BVP in ODE. Anyone who is familiar with ODEs knows that there are other types of end conditions which are also appropriate for the formulation of a BVP or an IVP. Yet, we have thus far not considered any combination of end conditions other than $y(a) = A$ and $y(b) = B$ for the problem of minimizing (1.1). We have omitted one or the other (or both) of these two conditions, but we have not replaced them.

To see whether we can have a variational problem with $y$ and $y'$ prescribed at the same end $x = a$, say

$$y(a) = A, \tag{3.45a}$$

$$y'(a) = U \tag{3.45b}$$

for a given pair of $A$ and $U$, we note that, in the case of a smooth extremal, the first variation of (1.1) is

$$\delta J = \int_a^b \{\widehat{F}_{,y}(x)\delta y + \widehat{F}_{,y'}(x)\delta y'\}\, dx$$
$$= [\widehat{F}_{,y'}(x)\delta y]_a^b + \int_a^b \{\widehat{F}_{,y}(x) - [\widehat{F}_{,y'}(x)]'\}\,\delta y\, dx. \tag{3.46}$$

The same argument used in Section 2 allows us to deduce the Euler DE as a necessary condition for $J[\widehat{y}]$ to be a (local) minimum. With $y(a) = A$ prescribed so that $\delta y(a) = 0$, the condition $\delta J = 0$ reduces to

$$\widehat{F}_{,y'}(b)\delta y(b) = 0. \tag{3.47}$$

It follows that the Euler BC $\widehat{F}_{,y'}(b) = 0$ is a necessary condition for a minimum $J[y]$; this is the same as the Euler BC (3.11) of [III.2] for problems with a single end condition $y(a) = A$. But now, the extremal $\widehat{y}$ must satisfy the two prescribed initial conditions at $x = a$ as well as the Euler BC at $x = b$. Given that the extremal is the solution of a second-order Euler DE, these three auxiliary conditions overdetermine the unknown $\widehat{y}(x)$. Hence, the variational problem of minimizing $J[y]$ of (1.1) subject to the two initial conditions (3.45) is generally not a properly posed problem.

**EXAMPLE 8:** $F = (y')^2 - y^2 \quad (0 \le x \le 1), \quad y(0) = 1, \quad y'(0) = 0.$

The extremals are all of the form $\widehat{y} = c_1 \cos x + c_2 \sin x$. The two initial conditions determine $c_1 = 1$ and $c_2 = 0$. The Euler BC for this problem requires $y'(b) = 0$. But $\widehat{y}'(1) = -\sin(1) \neq 0$; hence, the problem has no solution.

Note that if we have $b = \pi$ instead, then $\widehat{y}'(b) = \widehat{y}'(\pi) = -\sin(\pi) = 0$. The solution is no longer overdetermined. However, we will see from the results of Chapter 4 that the problem with the new solution domain does not meet the strengthened Jacobi condition for $J[\widehat{y}]$ to be a minimum.

Evidently, the same conclusion applies to variational problems with the two prescribed conditions $y(b) = B$ and $y'(b) = V$ and to the case in which $y$ is prescribed at one end and $y'$ at the other. As long as $y(x)$ is not presribed at an end, an Euler BC will be required at that end. Consequently, prescribing $y'$ at an end generally leads to too many boundary conditions for the Euler DE and, therefore, overdetermines $\widehat{y}$. There are, however, some exceptional situations to be described below.

**EXAMPLE 9:** $F = [(y') - y^2]/2$, with $y'(0) = 0$, and $y(1) = 1$.

The Euler BC at $x = 0$ is $y'(0) = 0$ which is identical to the prescribed initial condition. Thus, the prescription of $y'(x)$ at $x = 0$ does not lead to any difficulty.

If $y'(0) = 0$ is replaced by $y'(0) = V(\neq 0)$ for this example, the Euler BC would be inconsistent with the prescribed condition and we would have an ill-posed problem. The inconsistency would not occur if the expression for $J[y]$ is replaced by

$$J[y] = \int_0^1 \frac{1}{2}[(y')^2 - y^2]\, dx + Vy(0). \tag{3.48}$$

For a given variational problem for a Langrangian $F(x, y, y')$, we see that prescribing $y'$ at an end of the interval $[a, b]$ is generally inappropriate except for special cases. On the other hand, for many problems in applications, we are given one or more differential equations and end conditions and want to construct a corresponding variational formulation for which the given DE's are the Euler DE's. In that case, we may choose a performance index suitable for the given end conditions to avoid having too many or incosistent BC's. For Example 9 with $y'(0) = V$ [instead of $y'(0) = 0$], we can use (3.48) [which contains a terminal payoff $Vy(0)$] to avoid any inconsistency.

**EXAMPLE 10:** Construct a variational formulation of (or principle for) the BVP

$$y'' + y = 0, \qquad y'(0) = V, \qquad y(1) = 1.$$

We know the given DE is the Euler DE of the Langrangian $F = [(y')^2 - y^2]/2$. We can take (3.48) and the perfomance index for the variational formulation of the BVP to obtain an acceptable set of prescribed and Euler BCs.

**EXAMPLE 11:** Construct a variational principle for the IVP

$$y'' + y = 0, \qquad y(0) = A, \qquad y'(0) = V.$$

One method is to take the perfomance index for this problem to be

$$J[y] = \frac{1}{2}\int_0^1 [y'' + y]^2 dx.$$

With $v = y'' + y$, the vanishing of the first variation of $J$ gives

$$\begin{aligned} \delta J &= \int_0^1 v(\delta y'' + \delta y)\, dx \\ &= [v\delta y' - v'\delta y]_0^1 + \int_0^1 (v'' + v)\,\delta y\, dx \\ &= [v(1)\delta y'(1) - v'(1)\delta y(1)] + \int_0^1 (v'' + v)\,\delta y\, dx, \end{aligned}$$

where we have made use of the fact $\delta y(0) = \delta y'(0) = 0$ to simplify $\delta J$. Because $y(1)$ and $y'(1)$ are not prescribed, we have the Euler BC's

$$v(1) = v'(1) = 0$$

in addition to the Euler DE

$$v'' + v = 0.$$

This "initial" (or more correctly, final or terminal) value problem for $v(x)$, $0 \le x \le 1$, has the trivial solution $v(x) = 0$ as its unique solution so that the Euler DE is reduced to the given DE

$$y'' + y = 0.$$

We have already used up the Euler BC on $v$ and $v'$ at $x = 1$, leaving us with the prescribed initial conditions which are appropriate for the reduced Euler DE $y'' + y = 0$.

More generally, for a given *linear* ODE

$$L[y] = 0$$

with an appropriate set of initial conditions to define a well-posed IVP, the method of Example 11 suggests that the quantity

$$J[y] = \int_a^b (L[y])^2 \, dx$$

is generally an appropriate perfomance index for a variational principle for the IVP. The same approach may also be used for nonlinear ODE's but its appropriateness must be verified in each case because the IVP for the corresponding $v$ may now have a nontrivial solution.

## 7. Exercises

1. (a) For the integrand $F = 1 + x + y + y^2$, calculate $dF$ at $x = 0$ for $y = \sin(x)$ and $dx = \varepsilon$ and then $\delta F$ at $x = 0$ for $y = \sin(x)$ and $\delta y = \varepsilon(x + 1)$.

   (b) For the integrand $F = x^2 - y^2 + (y')^2$, calculate a few terms of $\Delta F$ for $y = x$ and $\delta y = \varepsilon x^2$ and then $\delta F$ for the same $y$ and $\delta y$.

2. Let $y = 1 + x^2$ and both $x$ and $y$ are functions of an independent variable $t$.

   (a) Calculate $\delta(dy/dx)$ and $d(\delta y)/dx$ when $x = 1/t$ and $\delta x = \varepsilon t^2$.

   (b) Verify $\delta(dy/dx) = d(\delta y)/dx - (dy/dx)\, d(\delta x)/dx$ for our problem.

3. Obtain a $C^2$ admissible extremal for

$$J[y] = \int_0^b [y']^2 dx, \quad y(0) = 2, \quad B = \sin(b).$$

What can be said about the optimal value $\hat{b}$ if $B$ is specified but not $b$?

4. Obtain a $C^2$ admissible extremal for

$$J[y] = \int_0^b y^2 (y')^2 dx, \quad y(0) = 0, \quad B^2 - b^2 = r^2,$$

where $r$ is a known number.

5. Find a $C^2$ extremal for $F = \sqrt{1 + (y')^2}\,/y$ with end points lying on the two nonintersecting circles in the upper half-plane $(x-5)^2 + (y-5)^2 = 1$ and $(x-2)^2 + (y-3)^2 = 1$.

6. Find the shortest distance between $y = x$ and $y^2 = x - 1$.

7. Obtain a $C^2$ extremal for

$$J[y] = \int_0^b [(y')^2 + 4(4y - b)]\,dx, \quad y(0) = 2, \quad y(b) \equiv B = b^2.$$

8. (a) Show that if $\hat{\mathbf{y}}(x) = (\hat{y}_1(x), \ldots, \hat{y}_n(x))^T$ extremizes

$$J[\mathbf{y}] = G(\mathbf{y}(b)) + \int_a^b F(x, \mathbf{y}, \mathbf{y}')\,dx,$$

then

$$\delta J = \hat{G}_{,\mathbf{y}}(b)\delta\mathbf{y} + \int_a^b [\hat{F}_{,\mathbf{y}}\delta\mathbf{y} + \hat{F}_{,\mathbf{y}'}\delta\mathbf{y}']\,dx = 0.$$

where $G_{,\mathbf{y}}$ is the gradient row vector of $G$ in the $\{y_i\}$ space:

$$G_{,\mathbf{y}} \equiv (G_{,y_1}, \cdots, G_{,y_n}).$$

(b) Deduce from $\delta J = 0$ the vector Euler BC

$$\hat{F}_{,\mathbf{y}'}(b) + G_{,\mathbf{y}}(b) = 0$$

which is equivalent to the following set of scalar Euler BC's

$$\hat{F}_{,y_i'}(b) + \hat{G}_{,y_i}(b) = 0 \quad (i = 1, 2, \ldots, n).$$

9. (a) Find the admissible extremals for

$$J[y] = \int_0^{\pi/4} [(y')^2 - y^2]\,dx, \quad y(0) = 0$$

with $y$ unspecified at the other end point $x = \pi/4$.

(b) Find the admissible extremals for

$$J[y] = \int_0^1 [(y')^2 - 2yy' + 2y' + 2y]\,dx$$

with both $y(0)$ and $y(1)$ unspecified.

10. For the monopolist problem of Chapter 1, Exercise 3, suppose there is no discounting ($r = 0$) and the current unit price for the goods is $p(0) = A$.
    (a) Solve for an admissible extremal if $p(T)$ is fixed to be $B$.
    (b) Solve for an admissible extremal if $p(T)$ is not specified.

11. For the price taker producer problem of Chapter 1 of Exercise 14, we know the initial capital stock $k(0) = k_0$.
    (a) Suppose the price taker would like to have a stock of capital $k_T$ at the end of the planning period. Formulate a BVP for determining $k(t)$.
    (b) If we leave $k(T)$ unspecified, deduce the relevant Euler BC.
    (c) Assuming $f(0) = 0$ (you need capital to produce goods) and $c(0) = 0$ (there is no cost when there is no infusion of new capital), what does the Euler BC of part (b) tell us about $k(T)$?

12. For a minimum cost production problem in Section 2 of Chapter 1, suppose the client does not insist on the delivery of a fixed quantity of goods at time $T$. Instead a certain minimum quantity $B$ is desired; the producer will be penalized for any delivery shortfall and given a bonus for any surplus production in excess of $B$. Both bonus and penalty are to be calculated from the formula $D_0[1 - \exp(\{B - y(T)\}/Y]$ where $D_0$ and $Y$ are know constants.
    (a) Obtain the Euler BC for this problem.
    (b) Obtain all extremals satisfying $y(0) = 0$.
    (c) Set up the condition for the determination of the remaining constant of integration.

13. Find all $C^2$ admissible extremals for the basic problem with the Lagrangian $F = (y')^2 - 2\alpha yy' - \beta y'$ ($\alpha$ and $\beta$ are know constants) and with

(a) $y(0) = 0$ and $y(1) = 1$;

(b) $y(0) = 0$ and $y(1)$ unspecified;

(c) $y(1) = 1$ and $y(0)$ unspecified;

(d) no end conditions specified.

14. Find $C^2$ extremal for

$$J[y] = \int_0^b [y']^2 dx + [y(1)]^2, \qquad y(0) = 1.$$

15. Find all $C^4$ admissible extremals of the basic problem with
$$F = yy' + (y'')^2, \quad y(0) = 0, \quad y'(0) = 1, \quad y(1) = 2, \quad y'(1) = 4.$$

16. Find all $C^4$ admissible extremals of the basic problem with
$$F = y^2 + (y')^2 + (y'' + y')^2, \qquad y(0) = 1, \qquad y'(0) = 2, \qquad y(\infty) = 0,$$
and $y'(\infty) = 0$.

17. (a) Find the Euler DE and the Euler BC at $x = b$ for
$$F = [y + y' + xy'']^2, \qquad y(a) = A_0, \qquad y'(a) = A_1.$$
(b) Comment on the prescribed and Euler BC's of the problem with
$$F = y^2 + 2(y')^2 + (y'')^2, \qquad y''(a) = A_2, \qquad y'''(a) = A_3.$$

18. Derive the first integral (3.38) for the Lagrangian $F(y, y', y'')$.

19. Obtain the Euler DE and the Euler BC for the basic problem with
$$F(x, y, y', y'', z, z', z'', z''') = y''z' + xyz'' + z'''y^2 \quad (a < x < b),$$
where $(\ )' \equiv d(\ )/dx$.

20. The equilibrium configuration of a slender elastic column of lenght $L$ under equal and opposite axial forces of magnitude $P$ is characterized by a lateral displacement function $y(x)$ where the $x$-axis coincides with the central axis of the column. The deformed shape function $y(x)$ minimizes the total energy of the column given by.

$$J[y] = \frac{1}{2}\int_0^L [EI(y'')^2 - P(y')^2]\, dx,$$

where $E$ is Young's modulus and $I$ is the moment of inertia of the column's cross section.

(a) Obtain the $C^4$ extremals which satisfy $y(0) = y(L) = y'(0) = y'(L) = 0$.

(b) Find all the natural boundary conditions (Euler BCs) if only $y(0) = 0$ and $y(L) = 0$ are prescribed.

(c) Find all the Euler BCs if only $y(0) = y'(0) = 0$ are prescribed.

21. (a) If the vendor in the minimum production cost problem of Section 2, Chapter 1 can negotiate the delivery time $T$, show that the optimal time $T$ which minimizes the production cost is $\widehat{T} = 2\sqrt{Bk/c_k}$.

(b) If $T$ is fixed but production does not have to begin immediately, find $\hat{t}_0 > 0$ which minimizes total production cost.

22. A monopolist of a mine is trying to maximize the current (present) value of the total profit from the sale of the mineral extracted [given a constant instantaneous discount rate $r$ as defined in Exercise 8 of Chapter 1]. Let $y(t)$ be the cumulative amount of a mineral sold by time $t$ and $p(y^\bullet)$ be the *net profit* (price minus cost of extraction) with $p' = dp/dy^\bullet < 0$, e.g.,

$$p(y^\bullet) = \frac{p_0}{y^\bullet}(1 - e^{-ky^\bullet}) \qquad (k > 0).$$

Suppose $B$ is the total mineral deposit and $T$ is time to exhaustion

(a) Formulate the variational problem which determines the optimal extraction policy [characterized by $\hat{y}(t)$] and the optimal exhaustion time $\widehat{T}$.

(b) Obtain the admissible extremal(s) for your variational problem with the specific profit function given above.

(b) Show that the optimal time of exhaustion for the given $p(\cdot)$ is $\widehat{T} = \sqrt{2kB/r}$.

(d) For a general $p(\cdot)$, show that the average profit at $\widehat{T}$ per unit of extraction equals the marginal profit $p'(y^\bullet(\widehat{T}))$ at that time.

23. Suppose the net profit function of Exercise 18 depends on $y$ as well so that $p = p(y, y^\bullet)$ (given that mining may get more difficult and therefore more expensive as the digging goes deeper). An example of $p(y, y^\bullet)$ for $B - y \ll B$ is

$$p = k_0 - k_1 y - k_2 y^\bullet.$$

(a) Obtain the admissible extremal(s) for the simple $p(\cdot)$ above.

(b) Set up the condition(s) for determining $\widehat{T}$ (but do not solve for $\widehat{T}$).

24. Suppose the aging monopolist is due to retire at a prescribed time $T$ and wants to exhaust the deposit by $T$. Suppose there is no commitment to begin extraction at $t = 0$. Determine the optimal starting time $\widehat{t}_0$ which maximizes the present value of the monopolist's total profit for the specific profit function given in Exercise 18.

25. State and prove a version of the fundamental lemma [I.1] for the basic problems with $F = F(x, y, y, y'')$.

26. Solve the BVP defined by (3.37′) and (3.35).

# 4

# A Weak Minimum

## 1. The Legendre Condition

In the first three chapters, we established the most useful necessary conditions for a weak minimum of the basic problem in the calculus of variations and its variants. Among these conditions, the ***Euler–Lagrange*** equation enables us to narrow our search for a local minimum point to a small group of PWS functions called extremals. However, the fact that $\hat{y}(x)$ satisfies the Euler–Lagrange equation does not necessarily make it the solution of our problem, just as the condition $f'(x_0) = 0$ does not necessarily imply that $f(x_0)$ is a local minimum for $f(x)$. For the problem of extremizing an ordinary function $f(x)$, we must examine $f''(x_0)$ to decide whether $x_0$ is a minimum point, a maximum point, or a point of inflection of $f(x)$. Analogously, we need to examine the second variation to learn whether an extremal $\hat{y}(x)$ is a minimum point of our variational problem.*

To gain some insight into sufficient conditions for a minimum in the calculus of variations, we consider first the special case $F = F(y')$. For this case, the extremals are of the form of $\hat{y} = c_1 x + c_2$ [with $c_1 = (B - A)/(b - a)$ and $c_2 = (bA - aB)/(b - a)$ for a smooth extremal] and

$$\left.\frac{d^2 J}{d\varepsilon^2}\right|_{\varepsilon=0} = \int_a^b F_{,y'y'}(\hat{y}')(h')^2 dx = F_{,y'y'}(c_1)\int_a^b (h')^2 dx.$$

The integrand $F(x,y,y')$ of $J[y]$ is said to satisfy the ***Legendre condition*** at $x$ if $\hat{F}_{,y'y'}(x) \geq 0$. It satisfies the ***strengthened Legendre condition*** at $x$ if $\hat{F}_{,y'y'}(x) > 0$. It follows from the expression for $J^{\prime\prime}(0)$ above that $J[y]$ is

* If only a stationary value of $J[y]$ is sought, this chapter and the next may be bypased, as stationarity is assured by the Euler–Lagrange equation and, when appropriate, Euler BCs and/or transversality conditions.

minimized by $\hat{y}$ if $F(\hat{y}')$ satisfies the *strengthened Legendre* condition in $[a, b]$. More precisely, we have

**[IV.1]** ***If for a particular extremal*** $\hat{y}(x)$, $F(y')$ ***satisfies the strengthened Legendre condition in*** $[a, b]$, ***then*** $J[\hat{y}]$ ***is a weak local minimum.***

The result applies to broken extremals, i.e., extremals with corners, as well as the unique smooth (straight-line) extremal.

Without additional information about the Lagrangian and the extremal, it is generally not possible to conclude whether a weak local minimum is the true minimum for our problem. For the special case of $F(x,y,y') \equiv F(y')$ however, it is possible to decide whether a local minimum is also a global minimum of the basic problem on the basis of a Legendre-type condition for $F$. By Taylor's theorem, we have

$$J[\hat{y} + th] - J[\hat{y}] = J(t) - J(0)$$

$$= t\frac{dJ}{dt}\bigg|_{t=0} + \frac{1}{2}t^2\frac{d^2J}{dt^2}\bigg|_{t=\bar{t}} = \frac{1}{2}t^2\int_a^b F_{,y'y'}(\hat{y}' + \bar{t}h')(h')^2\,dx$$

$$= \frac{1}{2}t^2\int_a^b F_{,y'y'}(c_1 \pm \bar{t}h')(h')^2 dx.$$

for some $\bar{t}$ in $[0, t]$. To emphasize that there is no restriction on $th$ being small in magnitude, we have used the parameter $t$ instead of $\varepsilon$. The following result is an immediate consequence of the Taylor expansion above:

**[IV.1']** ***If*** $\hat{y}$ ***is an admissible extremal for the basic problem with*** $F = F(y')$, ***then*** $J[\hat{y}]$ ***is a global minimum if*** $F^{\bullet\bullet}(v) > 0$ ***for all*** $v$.

With $F^{\bullet}(v) = dF/dv$, the requirement $F^{\bullet\bullet}(v) > 0$ for $|v| < \infty$ is not any kind of Legendre's condition because $F^{\bullet\bullet}$ is not evaluated at an extremal.

The *Legendre condition* also plays an important role in the basic problem involving a general Lagrangian $F = F(x,y,y')$. For the general case, it is straightforward to show

$$\frac{d^2J}{d\varepsilon^2}\bigg|_{\varepsilon=0} = \int_a^b [h^2\hat{F}_{,yy} + 2hh'\hat{F}_{,yy'} + (h')^2\hat{F}_{,y'y'}]\,dx.$$

The integral on the right-hand side can be made negative if $\hat{F}_{,y'y'} < 0$ for any noncorner point $x$ in $[a, b]$. Suppose $\hat{F}_{,y'y'}(x_0) = -K^2 < 0$. Then by continuity, we have $\widehat{F}_{,y'y'} < -\frac{1}{2}K^2$ for a small interval around $x_0$, say $|x - x_0| < \delta$. Now, choose

$$h(x) = \begin{cases} 0 & (a \le x \le x_0 - \delta) \\ 1 - \frac{1}{\delta}(x_0 - x) & (x_0 - \delta \le x < x_0) \\ 1 - \frac{1}{\delta}(x - x_0) & (x_0 \le x \le x_0 + \delta) \\ 0 & (x_0 + \delta \le x < b). \end{cases}$$

In other words, $h(x)$ vanishes outside $x_0 - \delta \le x \le x_0 + \delta$ and has the shape of an isosceles triangle of unit height inside the interval (Fig. 4.1). In particular, we have $|h| \le 1$, $|h'| = 1/\delta$, and $(h')^2\widehat{F}_{,y'y'} < (1/\delta)^2(-K^2/2)$ in $[x_0 - \delta, x_0 + \delta]$. Now, $\widehat{F}_{,yy}(x)$ and $\widehat{F}_{,yy'}$ are bounded for all $x$ in $[a, b]$. With $|\widehat{F}_{,yy}| \le k_1$ and $|\widehat{F}_{,yy'}| \le k_2$, we have

$$\left.\frac{d^2J}{d\varepsilon^2}\right|_{\varepsilon=0} < \int_{x_0-\delta}^{x_0+\delta}\left[|h|^2|\widehat{F}_{,yy}| + 2|h||h'||\widehat{F}_{,yy'}| - \frac{1}{2}K^2\delta^{-2}\right]dx$$

$$< \int_{x_0-\delta}^{x_0+\delta}\left[k_1 + 2k_2\delta^{-1} - \frac{1}{2}K^2\delta^{-2}\right]dx = 2k_1\delta + 4k_2 - K^2\delta^{-1}.$$

The right-hand side can be made negative by choosing $\delta$ sufficiently small.

A similar argument leading to the same conclusion applies to a corner point $x_0$ after writing $J[y]$ as the sum of two integrals, one from $a$ to $x_0$ and the other from $x_0$ to $b$. We summarize these results in the following theorem:

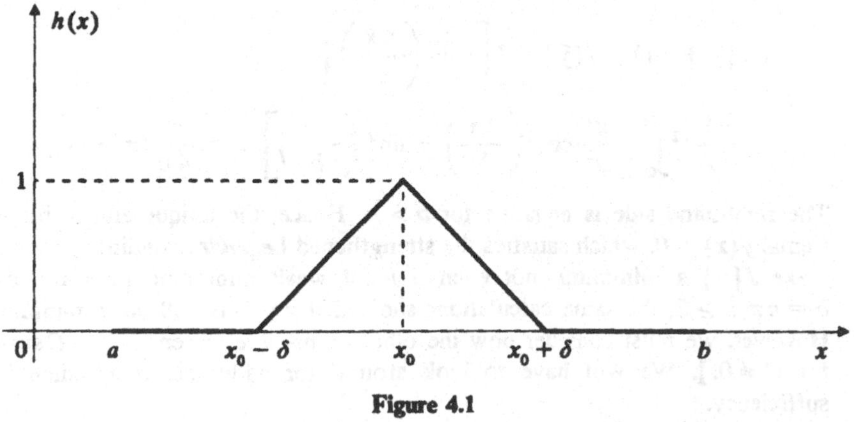


Figure 4.1

**[IV.2]** ***For a general Lagrangian*** $F(x,y,y')$, ***the Legendre condition*** $\widehat{F}_{,y'y'}(x) \geq 0$ ***for all*** $x$ ***in*** $[a,b]$ ***is a necessary condition for*** $\widehat{y}$ ***to minimize*** $J[y]$.

The following examples illustrate the applications of the theoretical results [IV.1], [IV.1′], and [IV.2].

**EXAMPLE 1:** $F(y') = (y')^2 + (y')^3$ with $y(0) = 0$ and $y(1) = k$.

We learned from Example (3) in Chapter 2 that $\widehat{y}(x) = kx$ is the only admissible extremal for this problem. Here, we have $\widehat{F}_{,y'y'}(x) = 2 + 6k$ so that the unique admissible extremal satisfies the strengthened Legendre condition for $k > -1/3$. If a weak minimum exists for the problem in this range of $k$ values, then $J[\widehat{y}]$ is that weak minimum (because $\widehat{y}$ is the only admissible extremal). On the other hand, *Legendre's* condition $\widehat{F}_{,y'y'} \geq 0$ is not satisfied for any $x$ in $[a,b]$ if $k < -1/3$. By [IV.2], $\widehat{y} = kx$ does *not* minimize $J[y]$ for $k < -1/3$.

Unlike the situation in minimizing an ordinary function, the strengthened *Legendre* condition by itself generally does not assure us that $J[\widehat{y}]$ is any kind of a minimum for a general $F(x,y,y')$. This is demonstrated by the following counter-example.

**EXAMPLE 2:** $F(x,y,y') = [(y')^2 - y^2]/2$ with $y(0) = 0$ and $y(b) = 0$.

With $F_{,y'y'} = 1$, the strengthened *Legendre* condition is satisfied for all $x$ in $[0,b]$. The extremals for this problem are $\widehat{y} = c_1\cos(x) + c_2\sin(x)$, with $c_1 = 0$ from $y(0) = 0$. The other end condition requires $c_2\sin(b) = 0$ so that $\widehat{y} = 0$ is an admissible extremal. For $b \neq n\pi$, $n = 1,2,3,\ldots$, it is the only one. In that case, we have $J[\widehat{y}] = 0$ for the unique (PWS) extremal. By taking $h(x) = \sin(\pi x/b)$, we have for this case

$$J[\widehat{y} + \varepsilon h] - J[\widehat{y}] = J\left[\varepsilon \sin\left(\frac{\pi x}{b}\right)\right]$$

$$= \frac{1}{2}\varepsilon^2 \int_0^b \left[\frac{\pi^2}{b^2}\cos^2\left(\frac{\pi x}{b}\right) - \sin^2\left(\frac{\pi x}{b}\right)\right] dx = \frac{\varepsilon^2}{4b}(\pi^2 - b^2).$$

The right-hand side is negative for $b > \pi$. Hence, the unique admissible extremal $\widehat{y}(x) = 0$, which satisfies the strengthened *Legendre* condition, does not make $J[y]$ a minimum, not even a local weak minimum. [For the case $b = n\pi, n \geq 2$, the same calculations show that $\widehat{y} = 0$ is still not a minimum. However, we must consider now the other admissible extremals $\widehat{y} = C\sin(x)$ for $C \neq 0$.]. We will have to look around for additional requirements of sufficiency.

## 2. Jacobi's Test

One reason why *Legendre's* condition, strengthened or not, does not guarantee any kind of a minimum is that it is local in nature. It only specifies the behavior of the Lagrangian at each point of the solution domain; it does not relate the behavior of $\widehat{y}(x)$ at different points throughout $[a, b]$. On the other hand, $J[y]$ depends on $y$ throughout the solution domain; its value is not affected by a change of $y$ at a point if it is appropriately compensated at other points in $[a, b]$. Some kind of global condition involving $\widehat{y}$ at more than a single point is apparently needed to pick out the minimum point from the collections of admissible comparison functions which satisfy the same point condition requirements (such as the Euler DE, the Euler BC, and the Legendre conditions).

To see what such a global condition may entail, consider the problem of the shortest path between two points on a sphere. The extremals for this problem are portions of a great circle. For any two prescribed points $P_i$ and $P$, there are two admissible extremals corresponding to the two segments of the great circle which passes through the two points (see Fig. 4.2). The shorter arc is the solution of our problem, but we have as yet no way to pick out this correct solution with the theory we have developed so far. To find a way to do this, we note that great circles (extremals) emanating from a point $P_i$ on the spherical surface all meet again at the same point $P_f$ diametrically opposite to $P_i$. It is clear that any arc which contains $P_f$ is not the shortest arc between $P_i$ and another point $P$ which is not $P_f$ itself. We discuss in this section a necessary condition of *Jacobi* which is analogous to requiring the shortest arc between $P_i$ and $P$ not to contain the *conjugate point* $P_f$.

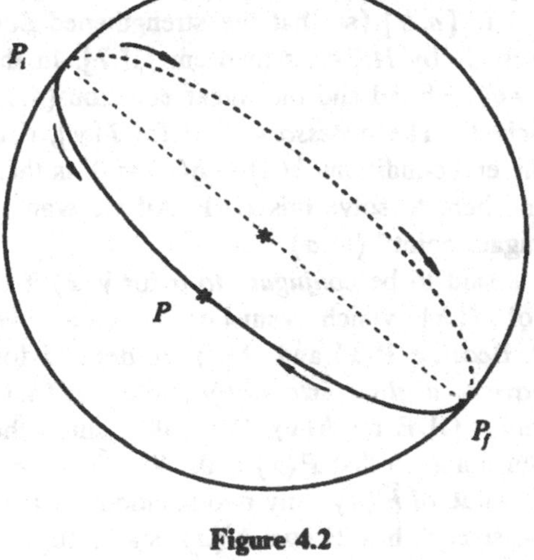


Figure 4.2

For *Jacobi's* condition, we assume that the extremal $\hat{y}$ for the problem is smooth. Recall from a previous calculation

$$J^{\bullet\bullet}(0) \equiv \left.\frac{d^2J}{d\varepsilon^2}\right|_{\varepsilon=0}$$

$$= \int_a^b \left\{ \hat{F}_{,yy}(x)h^2(x) + 2\hat{F}_{,yy'}(x)h(x)h'(x) + \hat{F}_{,y'y'}(x)[h'(x)]^2 \right\} dx \equiv I[h].$$

For a given admissible extremal $\hat{y}(x)$, the value of the integral $J^{\bullet\bullet}(0)$ depends on our choice of $h(x)$. This dependence is indicated by writing $I[h]$ for $J^{\bullet\bullet}(0)$. It immediately suggests a new variational problem in which we chose an admissible $h(x)$ to minimize $I[h]$. This new basic problem [with $h(a) = h(b) = 0$] is called the *accessory problem* for $J[y]$; its solution will provide useful information about the minimum value of $J[y]$. For example, if the minimum of $I[h]$ for all nontrivial $h(x)$ is positive, then $J[\hat{y}]$ is at least a weak local minimum. We will settle, however, for the less ambitious result [IV.3] [than obtaining the sign of $I[\hat{h}]$] to be derived below.

The basic problem of minimizing $I[h]$ with $h(x)$ satisfying $h(a) = h(b) = 0$ can be treated by the techniques we have developed so far. In particular, the Euler DE for $I[h]$ is

$$[P(x)h'(x)]' - Q(x)h(x) = 0 \tag{4.1}$$

where

$$P(x) = \hat{F}_{,y'y'}(x), \qquad Q(x) = \hat{F}_{,yy}(x) - [\hat{F}_{,yy'}(x)]'. \tag{4.2}$$

The differential equation (4.1) is called *Jabobi's (accessory) equation*. We know from [IV.2] that we must have $P(x) \geq 0$ for $J[\hat{y}]$ to be a minimum value. If $P(x) > 0$ in $[a, b]$ (so that the strengthened *Legendre* condition is satisfied), then $\hat{y}$ is $C^2$ by *Hilbert's* theorem [II.7]. In that case, $[\hat{F}_{,yy'}(x)]'$ and $[P(x)]'$ are well defined and the linear equation (4.1) is, therefore, with continuous coefficients. The *accessory BVP* for $J[y]$, defined by the Jacobi ODE (4.1) and the end conditions $h(a) = h(b) = 0$, is then properly posed. It is not our intention here to solve this BVP. All we want to know is whether $\hat{y}(x)$ has a "conjugate point" (to $a$).

A point $c > a$ is said to be *conjugate to a* for $\hat{y}(x)$ if there is a *nontrivial* extremal $\hat{h}(x)$ of $I[h]$ which vanishes at $c$ as well as at $a$, i.e., $\hat{h}(a) = \hat{h}(c) = 0$. Because $P(x)$ and $Q(x)$ are defined for a particular $\hat{y}(x)$, conjugacy is *relative to a given extremal* $\hat{y}$. Note that (4.1) is a *linear* homogeneous second-order ODE for $h(x)$. We will assume that the strengthened Legendre condition holds so that $P(x) > 0$. With $\hat{h}(a) = 0$, $\hat{h}(x)$ is uniquely determined by the value of $\hat{h}'(a)$. Any two solutions of the ODE are multiples of each other because each has its own $\hat{h}'(a)$ as a multiplicative factor. To test for points conjugate to $a$, it suffices to consider any nontrivial $\hat{h}(x)$ which vanishes at $a$. The subsequent zeros (if any) of this $\hat{h}(x)$ are the conjugate points of $a$ for $\hat{y}(x)$.

We can now establish the following necessary condition of Jacobi for $J[\hat{y}]$ to be a weak local minimum:

**[IV.3]** *Let* $J[y]$ *attain a weak local minimum with the admissible extremal* $\hat{y}(x)$ *and the strengthened Legendre condition holds in* $[a, b]$ *for* $\hat{y}(x)$*. Then there are no points in* $(a, b)$ *conjugate to* $a$ *relative to* $\hat{y}$*.*

Suppose that this were not so and there were a nontrivial solution $h_0(x)$ of (4.1) with

$$h_0(a) = h_0(c) = 0$$

for some $c$ in $(a, b)$. We would have to have $h_0'(c-) \neq 0$; otherwise the "initial" value problem defined by (4.1) and $h_0(c) = h_0'(c) = 0$ would require $h_0(x) \equiv 0$ for $a \leq x \leq c$. Consider the new function

$$\tilde{h}(x) = \begin{cases} h_0(x) & (a \leq x \leq c) \\ 0 & (c \leq x \leq b) \end{cases}$$

which is a PWS admissible comparison function for $I[h]$ (see Fig. 4.3). It vanished at both $a$ and $b$ and has a corner at $x = c$ because we have $\tilde{h}'(c-) = h_0'(c-) \neq 0$ and $\tilde{h}'(c+) = h_0'(c+) = 0$. We will show presently that $\tilde{h}$ minimizes $I[h]$ and must, therefore, be an admissible extremal.

We have from a previous expression for $J^{\bullet\bullet}(0)$ and $\tilde{h} = 0$ for $x \geq c$ that

$$I[\tilde{h}] = \int_a^c \left\{ P(x)[h_0']^2 + \widehat{F}_{,yy}(x)[h_0]^2 \right\} dx + \int_a^c \widehat{F}_{,yy'}(x)\frac{d}{dx}[h_0^2(x)]\,dx$$

$$= [Ph_0'h_0 + \widehat{F}_{,yy'}h_0^2]_a^c + \int_a^c \left\{ Qh_0 - [Ph_0']' \right\} h_0\,dx = 0,$$

where the conditions $h_0(a) = h_0(c) = 0$ and the ODE (4.1) have been used. By our hypothesis, $\hat{y}$ is a (weak local) minimum point; this requires $J^{\bullet\bullet}(0) = I[h] \geq 0$ for any admissible $h(x)$. It follows that $\tilde{h}(x)$ minimizes $I[h]$ because $I[\tilde{h}] = 0$ and $I[h]$ cannot be smaller. The PWS function $\tilde{h}(x)$ is, therefore, an extremal of $I[h]$. By Erdmann's first corner condition for the accessory problem, $2\widehat{F}_{,yy'}h\forall + 2\widehat{F}_{,y'y'}\tilde{h}'$ should be continuous; this, in turn, requires that $\tilde{h}'$ be continuous because all other terms in the expression are continuous. But this contradicts the definition of $\tilde{h}$ which has $c$ as a corner point! Hence, $\tilde{h}(c)$ cannot vanish for any $c$ in $(a, b)$.

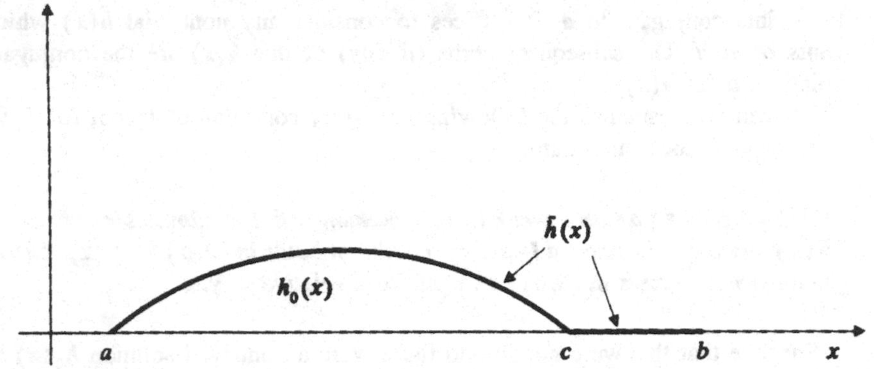


**Figure 4.3**

**EXAMPLE 3:** $F(x,y,y') = (y')^3 + (y')^2$ with $y(0) = 0$ and $y(1) = k$.

For this problem, we know already that $\hat{y} \equiv kx$ is the only admissible extremal. For this extremal, we have $P = 2 + 6k$, $Q \equiv 0$, and the strengthened Legendre condition is satisfied for $k > -1/3$. In that case, the solution of the Jacobi accessory equation is $h(x) = c_1x + c_0$. Now $h(0) = 0$ requires $c_0 = 0$; hence, any nontrivial $h(x)$ has no conjugate points in $(a, b)$ (in fact for all $x > 0$). The Jacobi (necessary) condition [IV.3] is, therefore, met by $kx$ for $k > -1/3$ (and also for $k < -1/3$).

**EXAMPLE 4:** $F(x,y,y') = [(y')^2 - y^2]/2$ with $y(0) = 0$ and $y(b) = 0$.

For this problem, we have $P = 1$ and $Q = -1$ so that Jacobi's accessory equation takes the form $h'' + h = 0$. For $h(0) = 0$, the solution is $h(x) = c_1 \sin(x)$. Jacobi's necessary condition is not met if $b > \pi$ [because $\sin(\pi) = 0$]. Note that the strengthened Legendre condition is satisfied by any extremal $\hat{y}$ of this problem given $\hat{F}_{,y'y'} = 1 > 0$. Hence, the strengthened Legendre condition alone does not guarantee a (local weak) minimum.

## 3. Conjugate Points

For any given point $P_i$ on a sphere, all great circles emanating from that point intersect at the same point $P_f$ on the sphere diametrically opposite the initial point. Given two points $P_i$ and $P$ on the sphere, an arc of the great circle connecting these two points is not the shortest path between them if the arc also passes through $P_f$ as an intermediate point; going around the other segment of the same great circle is shorter (Fig. 4.2). Thus, the point $P_f$ of the geodesic problem on the sphere is analogous to a conjugate point for the general basic problem, given that the great circles emanating from $P_i$ correspond to the extremals of the problem with $P_i$ as the initial point. The above observation suggests the following:

**[IV.4]** ***Suppose all extremals of*** $J[y]$ ***with the same initial point meet again at a point*** $c > a$***, then*** $c$ ***must be a conjugate point.***

Although [IV.4] does not provide a practical method for locating a conjugate point, it does provide a rather visual geometrical description of an important concept.

To verify [IV.4], suppose a family of extremals with the same initial point $(a, A)$ is given by $y = g(x, \alpha)$ so that $g(a, \alpha) = A$ independent of the parameter $\alpha$. A particular admissible extremal $\widehat{y}(x)$ for our basic problem corresponds to $g(x, \alpha)$ for a particular value $\alpha = \widehat{\alpha}$ so that $\widehat{y}(x) = g(x, \widehat{\alpha})$. For example, the smooth extremals of $F(y')$ are linear functions $c_1 x + c_2$. Those which which satisfy $y(a) = A$ are $y = \alpha(x - a) + A$, where $\alpha$ is an unknown parameter (with $c_1 = \alpha$ and $c_2 = A - c_1 a$). A family of extremals is generated by varying the parameter $\alpha$. Similarly, all extremals of $F = (y')^2 - y^2$ are of the form $c_1 \sin x + c_2 \cos x$. Those which satisfy $y(a) = A$ require $c_2 = (A - c_1 \sin a)/\cos a$ so that $y = \{c_1 \sin(x - a) + A \cos x\}/\cos a$. We may take $\alpha = c_1$ (or $\alpha = c_1/\cos a$) in this case.

We are interested in conjugate points of $a$ relative to $\widehat{y} = g(x, \widehat{\alpha})$ and need a solution for the Jacobi accessory equation for this purpose. It is rather remarkable that

*The function* $g_{,\alpha}(x, \alpha)$ *solves the Jacobi ODE for any* $\alpha$.

Assuming sufficient differentiability, we differentiate the *Euler* DE for the family of extremals $y = g(x, \alpha)$ with respect to the parameter $\alpha$ to get

$$\frac{d}{dx}\left\{F_{,y'y'}\frac{d}{dx}[g_{,\alpha}(x, \alpha)]\right\} + \frac{d}{dx}[F_{,y'y}]g_{,\alpha} + F_{,y'y}g_{,\alpha x} = F_{,yy}g_{,\alpha} + F_{,yy'}g_{,x\alpha}.$$

Upon setting $\alpha = \widehat{\alpha}$, we obtain

$$\frac{d}{dx}\left\{P\frac{d}{dx}[g_{,\alpha}(x, \widehat{\alpha})]\right\} - Q[g_{,\alpha}(x, \widehat{\alpha})] = 0, \tag{4.3}$$

where $P$ and $Q$ are as defined by (4.2). Therefore, $g_{,\alpha}(x, \widehat{\alpha})$ solves *Jacobi's* equation.

If $g_{,\alpha}(x, \widehat{\alpha})$ does not vanish identically, we differentiate the condition $g(a, \widehat{\alpha}) = A$ with respect to $\alpha$ to get $g_{,\alpha}(a, \alpha) = 0$. In particular, we have

$$g_{,\alpha}(a, \widehat{\alpha}) = 0. \tag{4.4a}$$

Also, we know from the hypothesis of [IV.4] that all extremals of the family $g(x, \alpha)$ meet again at the same point $c > a$ (Fig. 4.4), i.e., $g(c, \alpha) = C$ independent of $\alpha$. It follows that $g_{,\alpha}(c, \alpha) = 0$ for all $\alpha$ and, in particular,

$$g_{,\alpha}(c, \hat{\alpha}) = 0. \tag{4.4b}$$

Evidently, the point $c$ is a conjugate point of $a$ relative to $\hat{y}(x) = g(x, \hat{\alpha})$ (with $h(x) \equiv g_{,\alpha}(x, \hat{\alpha})$ being the solution of the Jacobi accessory equation which vanishes at $a$ and again at $c$) as indicated in [IV.4].

Thus, a point where the entire pencil of extremals from $(a, A)$ meets again is also a conjugate point relative to any extremal which satisfies the strengthened Legendre condition. The example of the shortest distance between two points on the spherical surface suggests that any extremal which contains a conjugate point $c < b$ does not minimize $J[y]$. Theorem [IV.3] confirms this conclusion.

**EXAMPLE 5:** $F = (y')^2 - y^2$, $y(0) = 0$, and $y(1) \doteq 1$.

The family of extremals which satisfies the end condition at $x = 0$ is $y(x) = \alpha \sin x \equiv g(x, \alpha)$. The extremal which also satisfies the other end condition is for $\alpha = 1/\sin(1) \equiv \hat{\alpha}$. The Jacobi accessory equation for this problem is $h'' + h = 0$. It is satisfied by $g_{,\alpha}(x, \alpha) = \sin x$ (which is independent of $\alpha$ for our problem). Now, all members of $g(x, \alpha)$ meet again at $x = \pi\,(2\pi, 3\pi$, etc.) with $g(\pi, \alpha) = 0$ for all $\alpha$. The pencil tip $x = \pi$ is also a conjugate point of $a = 0$ as $h(\pi) = g_{,\alpha}(\pi, \hat{\alpha}) = \sin \pi = 0$.

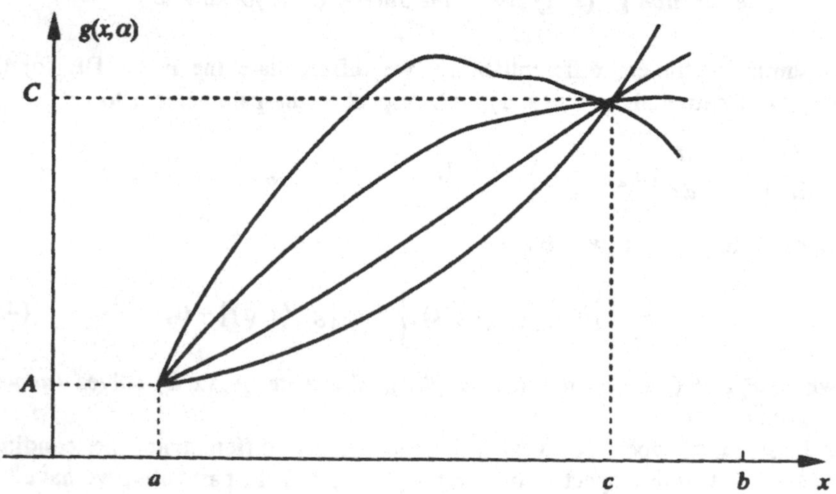


Figure 4.4

Other families of extremals [emanating from a given point $(a, A)$] do not meet again at a point. For them, there may be another geometrical description of conjugate points. An *envelope* of a family of plane curves is a curve $e(x)$ which is not a member of the given family but tangent to each member exactly once. For example, the family of straight lines at unit distance from the origin has the unit circle as its envelope (Fig. 4.5a). In many cases, a family of extremals with the same initial point $(a, A)$ has an envelope. This is the case for the family of catenaries which are the extremals for the minimal surface of revolution (soap bubble) problem (Fig. 4.5b). The following result relates the points on an envelope of extremals to conjugate points:

**[IV.5]** ***The point of (tangential) contact between an extremal and the envelope*** $e(x)$ ***for the associated family of extremals is a conjugate point of the extremal.***

For a particular $\alpha$, let the contact point between the extremal $g(x, \alpha)$ and the envelope $e(x)$ of the family be $x_\alpha = \bar{x}(\alpha)$, i.e., $e(\bar{x}(\alpha)) = g(\bar{x}(\alpha), \alpha)$. Upon differentiating both sides with respect to $\alpha$, we obtain

$$e'(\bar{x}(\alpha))\bar{x}^{\bullet}(\alpha) = g_{,x}(\bar{x}(\alpha), \alpha)\bar{x}^{\bullet}(\alpha) + g_{,\alpha}(\bar{x}(\alpha), \alpha)$$

where $(\ )' = d(\ )/dx$ and $(\ )^{\bullet} = d(\ )/d\alpha$. Because tangency means $e'(x_\alpha) = g_{,x}(x_\alpha, \alpha)$, the contact point $x_\alpha = \bar{x}(\alpha)$ is, therefore, characterized by $g_{,\alpha}(\bar{x}(\alpha), \alpha) = 0$ for the extremal corresponding to $\alpha$. On the other hand, $g_{,\alpha}(x, \hat{\alpha})$ is known to satisfy Jacobi's accessory equation and $g_{,\alpha}(a, \hat{\alpha}) = 0$ [see (4.3) and (4.4a)]. Hence, $x_{\hat{\alpha}} \equiv \bar{x}(\hat{\alpha})$ is a conjugate point for the extremal $g(x, \hat{\alpha})$ and the conclusion [IV.5] follows.

Some typical extremals of the soap bubble problem and their envelope are shown in Figure (4.5b). Exactly two catenaries of the family of extremals pass through each point above the envelope. Both are solutions of the *Euler–Lagrange* equation, but one touches the envelope at an intermediate point (whereas the other does not). This point of contact is now known to be a conjugate point by [IV.5]. According to [IV.3], the catenary having this tangential contact with the envelope between end points should be rejected. We are then left with just one catenary to furnish a minimum $J[y]$. Note that the Goldschmidt curve of Section 6, Chapter 2, remains a candidate for a minimum point; it is not a smooth extremal.

If the final point $(b, B)$ lies on the envelope, then there is only one admissible smooth extremal (and the Goldschmidt curve). If $(b, B)$ lies below the envelope, there is no admissible extremal except the Goldschmidt curve.

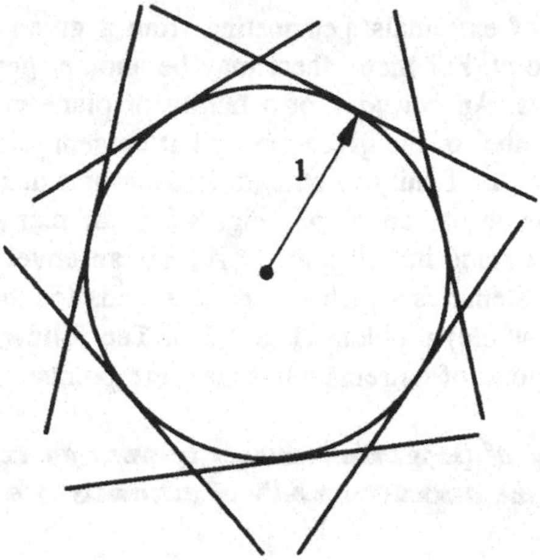


Figure 4.5a

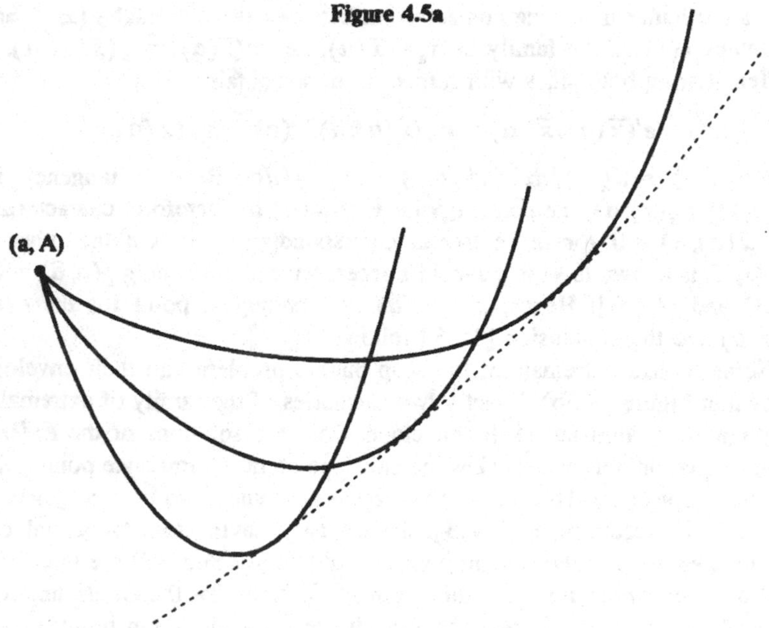


Figure 4.5b

## 4. Sufficiency

By now, we have a number of necessary conditions for $\widehat{y}(x)$ to be a solution of the basic problem. They are conditions $\widehat{y}(x)$ must satisfy if it is to render $J[y]$ a weak local minimum subject to $y(a) = A$ and $y(b) = B$. Among these necessary conditions, three are distinctly different from each other. We loosely summarize them here in a form suitable for later reference:

(I) $\widehat{y}(x)$ is an (admissible) extremal. (Euler–Lagrange equation)
(II) $\widehat{F}_{,y'y'}(x) \geq 0$ on $[a, b]$. (Legendre's condition)
(III) $\widehat{y}(x)$ has no conjugate points to $a$ in $(a, b)$. (Jacobi's test)

It turns out that a *strengthened* form of these three conditions, denoted by (I′), (II′) and (III′) will be useful in developing sufficiency conditions for minimizing $J[y]$:

(I′) $\widehat{y}(x)$ is a *smooth* (admissible) extremal. (Euler DE)
(II′) $\widehat{F}_{,y'y'}(x) > 0$ on $[a, b]$. (Strengthened Legendre condition)
(III′) $\widehat{y}(x)$ has no conjugate points to $a$ in $(a, b]$. (Strengthened Jacobi condition)

To develop a sufficiency condition based on (I′)-(III′), consider the quantity $J(\varepsilon) \equiv J[\widehat{y} + \varepsilon h]$ with

$$J(\varepsilon) - J(0) = J[\widehat{y} + \varepsilon h] - J[\widehat{y}] = \varepsilon J^{\bullet}(0) + \frac{\varepsilon^2}{2} J^{\bullet\bullet}(0) + \dots. \quad (4.5)$$

Now we have $J^{\bullet}(0) = 0$ because $\widehat{y}$ is an admissible extremal. For small variations $\varepsilon h$, the main contribution to the difference on the left comes from the second variation $\varepsilon^2 J^{\bullet\bullet}(0)/2$ with

$$\varepsilon^2 J^{\bullet\bullet}(0) = \int_a^b [\widehat{F}_{,y'y'}(x)(\delta y')^2 + 2\widehat{F}_{,yy'}(x)\delta y\, \delta y' + \widehat{F}_{,yy}(x)(\delta y)^2]\, dx$$

$$= \int_a^b [P(\delta y')^2 + Q(\delta y)^2]\, dx.$$

The second variation would be non-negative if the integrand is a perfect square. We can make it a perfect square by adding a term such as

$$\int_a^b [w(\delta y)^2]'\, dx = [w(\delta y)^2]_a^b = 0$$

without altering its value [because $h(x)$ vanishes at both end points]. With the added term, we have

$$J^{\bullet\bullet}(0) = \int_a^b [P(h')^2 + Q(h)^2 + 2whh' + w'(h)^2]$$

$$= \int_a^b P[(h')^2 + \left(\frac{2w}{P}\right)hh' + \left(\frac{[Q + w']}{P}\right)(h')^2]\, dx.$$

We now choose $w$ so that the expression inside the brackets is a perfect square for all $x$ in $[a, b]$. This requires $w$ to be a solution of the *Riccati* equation

$$w' = \frac{1}{P} w^2 - Q \tag{4.6}$$

The general method for solving the above first-order nonlinear ODE is to transform it into the linear second-order ODE

$$-[Pr'] + Qr = 0 \tag{4.7a}$$

for another function $r(x)$ by setting $w = -r'(x)P/r(x)$.

Rather remarkably, (4.7a) is just the Jacobi accessory equation with $P(x)$ being continuous and positive [by (I′) and (II′)]. The theory of linear second-order ODEs ensures that this equation has a nontrivial solution. But for $w = -r'P/r$ to be well defined, we want $r(x)$ to be nonvanishing in $[a, b]$. Fortunately, a well-defined $w(x)$ is ensured by the following result from the theory of linear ODEs.

> *For any sufficiently small number* $\sigma$ *and a continuous and positive function* $P(x)$, *there is always a* $C^2$ *solution* $\bar{r}(x)$ *of the equation*
>
> $$-[(P - \sigma^2)r']' + Qr = 0 \tag{4.7b}$$
>
> *Furthermore,* $\bar{r}(x)$ *does not vanish in* $[a, b]$ *if* (III′) *holds.*

We choose a $\sigma$ small enough so that not only do we have $P(x) > 0$ but also $P - \sigma^2 > 0$ for all $x$ in $[a, b]$. For this $\sigma$, we have

$$J^{\bullet\bullet}(0) = \int_a^b [(P - \sigma^2)(h')^2 + \sigma^2(h')^2 + Q(h)^2 + 2wh' + w'(h)^2]\,dx$$

$$= \int_a^b (P - \sigma^2)\left\{(h')^2 + \left[\frac{2w}{P - \sigma^2}\right]h'h + \left[\frac{Q + w'}{P - \sigma^2}\right](h)^2\right\}dx + \sigma^2 \int_a^b (h')^2 dx.$$

We can make the integrand of the first integral into a perfect square by choosing $w = -\bar{r}'(P - \sigma^2)/\bar{r}$ (instead of $-r'P/r$ as we did before) with $\bar{r}(x)$ being a solution of (4.7b) [instead of (4.7a)]. Let $\hat{h}(x)$ be the extremal of $I[h]$ with $\hat{h}(a) = 0$ and $\hat{h}'(a) = 1$. By (III′), $h(x) \neq 0$ for $a < x \leq b$. To have $\bar{r} \neq 0$ for all $x$ in $[a, b]$, take $\bar{r}$ to be the solution of (4.7b) with $\bar{r}(a) = \varepsilon, \bar{r}'(a) = 1$. For $|\varepsilon|$ sufficiently small, $\bar{r}(x)$ can be made as close to $\hat{h}(x)$ as we wish because the solution of (4.7b) depends continuously on its initial data) and, hence, $\bar{r}(x) \neq 0$ for all $x$ in $[a, b]$. For this choice of $\bar{r}(x)$, $w(x)$ is well defined and $J^{\bullet\bullet}(0)$ becomes

$$J^{\bullet\bullet}(0) = \int_a^b (P - \sigma^2)\left[h' + \frac{w}{P - \sigma^2}h\right]^2 dx + \sigma^2 \int_a^b (h')^2 dx.$$

Hence, $J^{\bullet\bullet}(0)$ is non-negative as both integrals are non-negative. By taking $|h|$ and $|h'|$ small enough, it can be shown by some simple estimates of the remainder of the Taylor's expansion (4.5) for $J[\hat{y} + \varepsilon h]$ that higher-order variations of $J$ beyond $\delta^2 J$ contribute at worst a negative value no larger in magnitude than the contribution of half of the second integral of $J^{\bullet\bullet}(0)$ above. We have then

$$J[\hat{y} + \varepsilon h] - J[\hat{y}] \geq \frac{\varepsilon^2}{2} J^{\bullet\bullet}(0) - \frac{\sigma^2 \varepsilon^2}{4} \int_a^b (h')^2 dx$$

$$= \frac{\varepsilon^2}{2} \int_a^b (P - \sigma^2)[h' + wh(P - \sigma^2)]^2 dx + \frac{\sigma^2 \varepsilon^2}{4} \int_a^b (h')^2 dx > 0$$

for $\sigma^2 > 0$ and $(h')^2$ not identically zero in $[a, b]$. We omit the technical details leading to this last inequality and conclude from it that

**[IV.6]** ***An admissible extremal of the basic problem which satisfies (I′), (II′) and (III′) renders*** **J[y]** ***(at least) a weak local minimum.***

EXAMPLE 6: $F(x, y, y') = (y')^2 + (y')^3$ with $y(0) = 0$ and $y(1) = k(> -1/3)$.

From Examples 1 and 2 of this chapter, we see that $\hat{y} = kx$ satisfies (I′), (II′), and (III′) for $k > -1/3$ and is, therefore, a weak local minimum by [IV.6]. Of course, an even stronger result can be obtained from [IV.1] for this case.

## 5. Several Unknowns

The sufficiency conditions of [IV.6] can be extended to basic problems with several unknowns:

$$J[\mathbf{y}] = \int_a^b F(x, \mathbf{y}, \mathbf{y}')\,dx, \qquad \mathbf{y}(a) = \mathbf{A}, \qquad \mathbf{y}(b) = \mathbf{B} \tag{4.8}$$

where $\mathbf{y} = (y_1, y_2, \ldots, y_n)^T$. We know from [II.3] that the solution of the problem $\hat{\mathbf{y}}(x)$ satisfies the Euler–Lagrange equation, i.e.,

(V) $\hat{\mathbf{y}}$ is an admissible extremal.

For an analogue of [IV.6] for several unknowns, we introduce

(V′) $\hat{\mathbf{y}}$ is a *smooth* (admissible) extremal.

Experience with quadratic forms suggests that the analogue of the Legendre condition for several unknowns should be the non-negatve definiteness of the Hessian matrix $[F_{,y_i'y_k'}]$ of the Lagrangian $F(x,\mathbf{y},\mathbf{y}')$. On the other hand, it is not immediately obvious what the analogue of a conjugate point in the Jacobi test should be for the more general case.

From the Taylor expansion of $J[\widehat{\mathbf{y}}+\varepsilon\mathbf{h}]\equiv J(\varepsilon)$ about $\varepsilon=0$, where $\mathbf{h}=(h_1,\ldots,h_n)^T$ and $\mathbf{h}(a)=\mathbf{h}(b)=\mathbf{0}$, we have

$$J^{\bullet\bullet}(0)=\int_a^b\sum_{i=1}^n\sum_{k=1}^n\left\{\widehat{F}_{,y_iy_k}h_ih_k+2\widehat{F}_{,y_iy_k'}h_ih_k'+\widehat{F}_{,y_i'y_k'}h_i'h_k'\right\}dx. \quad (4.9)$$

When $n=1$, we integrate $\widehat{F}_{,y_1y_1'}h_1h_1'$ by parts to get

$$\int_a^b 2\widehat{F}_{,y_1y_1'}h_1h_1'dx=\int_a^b\widehat{F}_{,y_1y_1'}(h_1^2)'dx=-\int_a^b(\widehat{F}_{,y_1y_1'})'h_1^2dx.$$

This result can be used in (4.9) to obtain (for $n=1$)

$$J^{\bullet\bullet}(0)=\int_a^b[P(h_1')^2+Qh_1^2]dx,$$

where $P(x)=\widehat{F}_{,y_1'y_1'}(x)$ and $Q(x)=\widehat{F}_{,y_1y_1}(x)-[\widehat{F}_{,y_1y_1'}(x)]'$. For the $n$ unknown case, we also wish to eliminate the term

$$\sum_{i=1}^n\sum_{k=1}^n 2\widehat{F}_{,y_iy_k'}h_ih_k'=\sum_{i=1}^n\sum_{k=1}^n\left[\widehat{F}_{,y_iy_k'}h_ih_k'+\widehat{F}_{,y_ky_i'}h_kh_i'\right].$$

We limit our discussion here to the important class of problems in applications for which $\widehat{F}_{,y_iy_k'}=\widehat{F}_{,y_ky_i'}$. For these problems, (4.9) may be transformed by integration by parts into

$$\begin{aligned}J^{\bullet\bullet}(0)&=\int_a^b\sum_{i=1}^n\sum_{k=1}^n\left\{\widehat{F}_{,y_i'y_k'}h_i'h_k'+[\widehat{F}_{,y_iy_k}-(\widehat{F}_{,y_iy_k'})']h_ih_k\right\}dx\\&=\int_a^b[P\mathbf{h}'\cdot\mathbf{h}'+Q\mathbf{h}\cdot\mathbf{h}]dx\equiv I[\mathbf{h}],\end{aligned} \quad (4.10)$$

where $P(x)$ is the Hessian matrix $[\widehat{F}_{,y_i'y_k'}(x)]$ evaluated at $\mathbf{y}=\widehat{\mathbf{y}}(x)$ and $Q(x)$ is the matrix $[\widehat{F}_{,y_iy_k}-\widehat{F}'_{,y_iy_k'}]$. Similar to the scalar case, we want the quadratic form $P\mathbf{h}'\cdot\mathbf{h}'$ to be (at least) non-negative:

$$P\mathbf{h}'\cdot\mathbf{h}'=\sum_{i=1}^n\sum_{k=1}^n\widehat{F}_{,y_i'y_k'}h_i'h_k'\geq 0.$$

Hence, the $n$ unknown version of the Legendre condition is seen to be

(VI) The Hessian matrix $P(x)\equiv[\widehat{F}_{,y_i'y_k'}(x)]$ is non-negative definite.

For an analogue of [IV.6], we need the strengthened Legendre condition instead:

(VI′) The Hessian matrix $P(x) \equiv [\hat{F}_{y_i'y_k'}(x)]$ is positive definite.

To formulate the analogue of Jacobi's condition, we observe that a term of the form

$$0 = \int_a^b [W\mathbf{h}\cdot\mathbf{h}]'\,dx = \int_a^b [W'\mathbf{h}\cdot\mathbf{h} + 2W\mathbf{h}\cdot\mathbf{h}']\,dx,$$

where $W$ is an $n \times n$ matrix function of $x$, can always be added to (4.9) without altering its value. When this is done, the expression for $J^{\bullet\bullet}(0)$ becomes

$$J^{\bullet\bullet}(0) = \int_a^b [P\mathbf{h}'\cdot\mathbf{h}' + 2W\mathbf{h}\cdot\mathbf{h}' + Q\mathbf{h}\cdot\mathbf{h} + W'\mathbf{h}\cdot\mathbf{h}]\,dx.$$

Because $P$ is (symmetric) positive definite, a "root matrix" $R$ exists with $R^2 = P$. We can then write $J^{\bullet\bullet}(0)$ as

$$J^{\bullet\bullet}(0) = \int_a^b [(R\mathbf{h}')\cdot(R\mathbf{h}') + 2(R^{-1}W\mathbf{h})\cdot(R\mathbf{h}') + (Q + W')\mathbf{h}\cdot\mathbf{h}]\,dx$$

$$= \int_a^b (R\mathbf{h}' + R^{-1}W\mathbf{h})\cdot(R\mathbf{h}' + R^{-1}W\mathbf{h}) + (Q + W' - WP^{-1}W)\mathbf{h}\cdot\mathbf{h}]\,dx.$$

The first term in the last expression is a perfect square. We now choose the matrix $W$ so that it satisfies the matrix Riccati equation

$$W' + Q = WP^{-1}W. \tag{4.11}$$

Hence,

$$J^{\bullet\bullet}(0) = \int_a^b (R\mathbf{h}' + R^{-1}W\mathbf{h})\cdot(R\mathbf{h}' + R^{-1}W\mathbf{h})\,dx. \tag{4.10′}$$

is non-negative.

For a solution of the Riccati equation (4.11), we use a matrix version of the transformation $w = -Pu'/u$ for the scalar case and set

$$W = -PU'U^{-1}. \tag{4.12}$$

Upon substituting (4.12) into (4.11), we obtain, after some cancellation,

$$-(PU')' + QU = 0. \tag{4.13}$$

Any invertible smooth solution of (4.13) reduces (4.10) to (4.10′); hence, $J^{\bullet\bullet}(0)$ is non-negative if such a solution for $U$ can be found.

Similar to the scalar case, (4.13) is the matrix analogue of the vector Euler DE (4.7a) for $I[\mathbf{h}] \equiv J^{\bullet\bullet}(0)$ in (4.10):

$$-(P\mathbf{h}')' + Q\mathbf{h} = 0. \tag{4.13'}$$

Let $\mathbf{h}_i \equiv (h_{1i}, h_{2i}, \dots, h_{ni})^T$ be the solution of the second-order linear ODE (4.13′) which satisfies the initial conditions

$$\mathbf{h}_i(a) = 0, \qquad \mathbf{h}_i'(a) = (\delta_{i1}, \dots, \delta_{ii}, \dots, \delta_{in})^T, \tag{4.14}$$

where $\delta_{ij} = 0$ for $i \neq j$ and $\delta_{ii} = 1$. Because $P$ is continuous and positive definite for all $x$ in $[a, b]$, such a solution exists and is unique for $i = 1, 2, \dots, n$. We now let

$$U_0 = [\mathbf{h}_1, \mathbf{h}_2, \dots, \mathbf{h}_n] = [h_{ki}]. \tag{4.15}$$

It is not difficult to see that $U_0$ satisfies (4.13) because each column of $U$ satisfies (4.13′). In addition, $U_0$ satisfies the initial condition

$$U_0(a) = 0, \qquad U_0'(a) = I. \tag{4.16}$$

Now, for $U_0(x)$ to be invertible for $x > a$, we only need the determinant of $U_0$ to be nonvanishing. Hence, we adopt the following definition of a conjugate point for several unknowns: *A* point $c\,(\neq a)$ is said to be ***conjugate*** to the end point $a$ relative to the (smooth) extremal $\widehat{\mathbf{y}}(x)$ if the determinant $|h_{ij}(x)|$ vanishes at $x = c$. The Jacobi condition for several unknowns then takes the form:

(VII) The smooth extremal $\widehat{\mathbf{y}}(x)$ has no conjugate point to the point $a$ in $(a, b)$.

For the $n$ unknown analogue of [IV.6], we need the strengthened Jacobi condition:

(VII′) The smooth extremal $\widehat{\mathbf{y}}(x)$ has no conjugate point to the point $a$ in $(a, b]$.

The above development toward an analogue of [IV.6] is still incomplete in two ways. First, the strengthened Jacobi condition (VII′) only assures us that $W(x)$ is well defined for $a < x \leq b$ (by taking $U = U_0$) but not at $x = a$ where $|U_0| = 0$ [because $U_0(a) = 0$]. However, given (V′), (VI′), and (VII′), the theory of ODE guarantees that for a sufficiently small value of $\varepsilon$, there is a unique solution of the IVP defined by (4.13′) and

$$\begin{aligned} \bar{\mathbf{h}}_i(a) &= \varepsilon(\delta_{1i}, \dots, \delta_{ni})^T, \\ \bar{\mathbf{h}}_i'(a) &= (\delta_{1i}, \dots, \delta_{ni})^T \end{aligned} \tag{4.14'}$$

which is as close to $\mathbf{h}_i(x)$ as we wish in $(a, b]$. We now take

$$U(x) = [\bar{h}_{ij}(x)]$$

which is nonsingular in $[a, b]$ [because $U(a) = \varepsilon I$ and $|U(x)| \cong |U_0(x)| \neq 0$ for $(a, b]$.

Second, for an actual (local weak) minimum, we need $J^{\cdot\cdot}(0) > 0$, not just non-negative. To get this, we set

$$P\mathbf{h}' \cdot \mathbf{h}' = (P - \sigma^2 I)\mathbf{h}' \cdot \mathbf{h}' + \sigma^2 \mathbf{h}' \cdot \mathbf{h}'$$

in (4.10) and repeat the development after (4.10) with $(P - \sigma^2 I)\mathbf{h}' \cdot \mathbf{h}'$ instead of $P\mathbf{h}' \cdot \mathbf{h}'$ to transform $J^{\cdot\cdot}(0)$ into

$$J^{\cdot\cdot}(0) = \int_a^b [(R_\sigma \mathbf{h}' + R_\sigma^{-1} W\mathbf{h}) \cdot (R_\sigma \mathbf{h}' + R_\sigma^{-1} W\mathbf{h}) + \sigma^2 \mathbf{h}' \cdot \mathbf{h}']\, dx, \quad (4.17)$$

where now $R_\sigma^2 = P - \sigma^2 I$. Following the rest of the argument leading to [IV.6], we obtain the following analogue of that result for several unknowns:

**[IV.6']** ***Let*** $\hat{\mathbf{y}}(\mathbf{x})$ ***be an admissible extremal of the basic problem (4.8) with*** $F_{,y_i y_k'} = F_{,y_k y_i'}$. ***Suppose (V'), (VI'), and (VII') are satisfied by this particular extremal. Then*** $J[\hat{\mathbf{y}}]$ ***is (at least) a weak local minimum.***

The symmetry condition $F_{,y_i y_k'} = F_{,y_k y_i'}$ which does not appear in the scalar case is somewhat restrictive. However, many problems in the physics of matters are known to satisfy this condition. The following compound pendulum problem (see Fig. 4.6) serves to illustrate that the condition is sometimes (but not always) satisfied. For this problem, we have from Hamilton's principle (see Section 2, Chapter 6, and also Exercise 4 of that chapter) $F(\varphi_1, \varphi_2, \varphi_1', \varphi_2') = K - V$, where

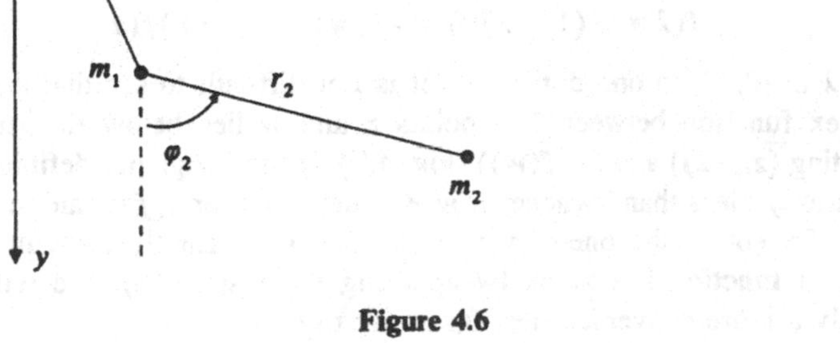

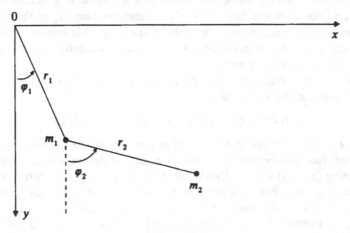

Figure 4.6

$$V = -(m_1 + m_2) g \ell_1 \cos\varphi_1 - m_2 g \ell_2 \cos\varphi_2,$$

$$K = \frac{1}{2}[(m_1 + m_2)\ell_1^2(\varphi_1')^2 + m_2\ell_2^2(\varphi_2')^2 + 2m_2\ell_1\ell_2\varphi_1'\varphi_2' \cos(\varphi_1 - \varphi_2)]$$

are the potential and kinetic energies of the pendulum system, respectively. It follows from these expressions that

$$F_{,\varphi_1\varphi_2'} = -m_2\ell_1\ell_2\varphi_1' \sin(\varphi_1 - \varphi_2), \qquad F_{,\varphi_2\varphi_1'} = m_2\ell_1\ell_2\varphi_2' \sin(\varphi_1 - \varphi_2).$$

The two quantities are not equal; hence, the symmetry condition $F_{,y_iy_k'} = F_{,y_ky_i'}$ is not satisfied. On the other hand, if the motion of the pendulum masses is of very small amplitude, we may adequately approximate $F$ by retaining only quadratic terms of the unknowns so that

$$F \cong \frac{1}{2}[(m_1 + m_2)\ell_1^2(\varphi_1')^2 + m_2\ell_2^2(\varphi_2')^2 + 2m_2\ell_1\ell_2\varphi_1'\varphi_2']$$
$$- \frac{1}{2}[(m_1 + m_2)g\ell_1\varphi_1^2 + m_2 g\ell_2\varphi_2^2].$$

In this case, we have

$$F_{,\varphi_1\varphi_2'} = 0 = F_{,\varphi_2\varphi_1'}$$

so that the symmetry condition is satisfied for the same problem.

## 6. Convex Integrand

There is one situation where any admissible extremal is guaranteed to be a minimum. For a smooth function $g(x)$, a stationary point $x_0$ of the function is guaranteed to be a minimum point if the graph of $g(x)$ is known to be convex. The analogous situation in the calculus of variations requires that the integrand $F(x, y, v)$ be *convex* in $y$ and $v$.

A scalar-value function $f(\mathbf{y}) = f(y_1, \ldots, y_n)$ is said to be *convex* if for any pair of points $\mathbf{z}$ and $\mathbf{w}$, we have

$$f(\lambda \mathbf{w} + (1-\lambda)\mathbf{z}) \leq \lambda f(\mathbf{w}) + (1-\lambda) f(\mathbf{z}) \tag{4.18}$$

for all $\lambda$ in $[0, 1]$. In one dimension, it is not difficult to see that the graph of a convex function between two points $z$ and $w$ lies below the straight line connecting $(z, f(z))$ and $(w, f(w))$ (Fig. 4.7). If the inequality defining convexity is strictly "less than" whenever $w \neq z$ and $\lambda \neq 0$ or 1, $f$ is said to be *strictly convex*. Except in the one-dimensional case, it is usually difficult to decide whether a function is convex by applying the test (4.18). We will develop presently a more convenient test for convexity.

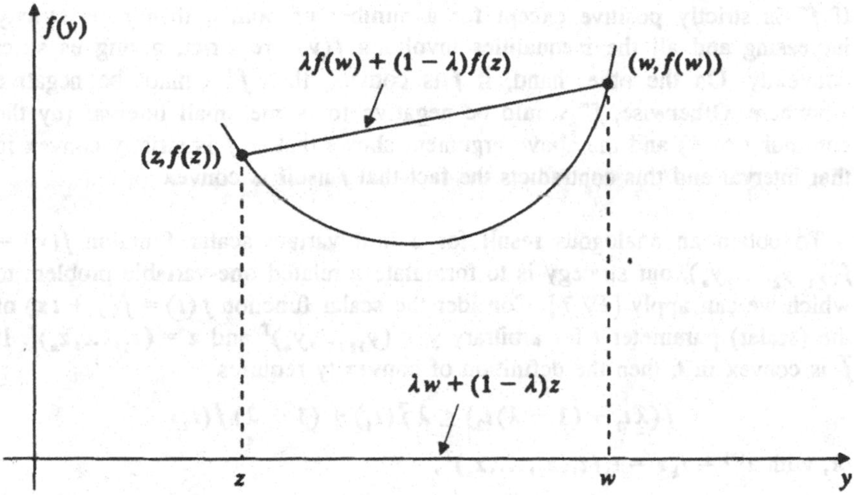


**Figure 4.7**

Because a positive second derivative ensures a stationary value to be a minimum for a scalar function of one variable, it should bear some relation to convexity in one dimension. To see this relation, we note that, for a scalar function of one variable $f$, its derivative $f'$ is increasing if $f''$ is non-negative (and not identically zero). In that case, we have

$$f(y) - f(w) = \int_w^y f'(t)\,dt \le \int_w^y f'(y)\,dt = f'(y)(y - w),$$

$$f(z) - f(y) = \int_y^z f'(t)\,dt \ge \int_y^z f'(y)\,dt = f'(y)(z - y).$$

With $y = \lambda w + (1 - \lambda)z$, these inequalities become

$$f(y) \le f(w) + (1 - \lambda)\,f'(y)(z - w),$$

$$f(y) \le f(z) - \lambda\,f'(y)(z - w).$$

From the combination $\lambda f(y) + (1 - \lambda)f(y) = f(y)$, we get (4.18) and thereby conclude that

**[IV.7]** ***A scalar value function of one variable $f(y)$ is convex if and only if $f''$ is everywhere non-negative. Furthermore, $f$ is strictly convex if $f''$ is positive except possibly for a number of points.***

If $f''$ is strictly positive except for a number of points, then $f'$ is strictly increasing and all the inequalities involving $f(y)$ are strict, giving us strict convexity. On the other hand, if $f$ is convex, then $f''$ cannot be negative anywhere. Otherwise, $f''$ would be negative for some small interval (by the continuity of $f$) and the above argument shows that $-f$ is strictly convex in that interval and this contradicts the fact that $f$ itself is convex.

To obtain an analogous result for a multivariate scalar function $f(\mathbf{y}) = f(y_1, y_2, \ldots, y_n)$, our strategy is to formulate a related one-variable problem to which we can apply [IV.7]. Consider the scalar function $\bar{f}(t) = f(\mathbf{y} + t\mathbf{z})$ of the (scalar) parameter $t$ for arbitrary $\mathbf{y} = (y_1, \ldots, y_n)^T$ and $\mathbf{z} = (z_1, \ldots, z_n)^T$. If $\bar{f}$ is convex in $t$, then the definition of convexity requires

$$\bar{f}(\lambda t_1 + (1-\lambda)t_2) \le \lambda \bar{f}(t_1) + (1-\lambda)\bar{f}(t_2)$$

or, with $\mathbf{z}^{(k)} = t_k \mathbf{z} = t_k(z_1, z_2, \ldots, z_n)^T$,

$$f(\mathbf{y} + \lambda \mathbf{z}^{(1)} + (1-\lambda)\mathbf{z}^{(2)}) \le \lambda f(\mathbf{y} + \mathbf{z}^{(1)}) + (1-\lambda) f(\mathbf{y} + \mathbf{z}^{(2)}).$$

Given $\mathbf{y} + \lambda\mathbf{z}^{(1)} + (1-\lambda)\mathbf{z}^{(2)} = \lambda(\mathbf{y} + \mathbf{z}^{(1)}) + (1-\lambda)(\mathbf{y} + \mathbf{z}^{(2)})$, $f$ is evidently convex if $\bar{f}(t)$ is convex for arbitrary $\mathbf{y}$ and $\mathbf{z}$.

Now, we have

$$\bar{f}'(t) = z_1 f_{,1} + \ldots + z_n f_{,n} = \nabla f(\mathbf{y} + t\mathbf{z})\mathbf{z}, \tag{4.19}$$

where $f_{,j} = \partial f / \partial y_j$, the row vector $\nabla f = (f_{,1}, f_{,2}, \ldots, f_{,n})$ is the gradient (row vector) of $f$, and

$$\begin{aligned} \bar{f}''(t) &= z_1^2 f_{,11} + 2 z_1 z_2 f_{,12} + z_2^2 f_{,22} + \ldots + 2 z_1 z_n f_{,1n} + \ldots + z_n^2 f_{,nn} \\ &= \sum_{i=1}^{n} \sum_{j=1}^{n} z_i z_j f_{,ij}(\mathbf{y} + t\mathbf{z}) = \mathbf{z}^T H(\mathbf{y} + t\mathbf{z})\mathbf{z}, \end{aligned} \tag{4.20}$$

where $H(\mathbf{y})$ is the *Hessian matrix* of $f(\mathbf{y})$,

$$H(\mathbf{y}) = [f_{,ij}(\mathbf{y})]. \tag{4.21}$$

From [IV.7] and (4.20), we have

**[IV.8]** $\mathrm{f}(\mathbf{y})$ *is convex if and only if the Hessian matrix (4.11) is non-negative definite so that* $\mathbf{u}^T \mathrm{H}(\mathbf{y})\mathbf{u} \ge 0$ *for any* $\mathbf{u}$. *Furthemore,* f *is strictly convex if* H *is positive definite except possibly for a number of points.*

In the case of two variables, we have for any $C^2$ function $f(y_1, y_2)$

$$H = \begin{bmatrix} f_{,11} & f_{,12} \\ f_{,21} & f_{,22} \end{bmatrix}.$$

For $H$ to be non-negative definite, we need

$$f_{,11} \geq 0 \quad \text{and} \quad \begin{vmatrix} f_{,11} & f_{,12} \\ f_{,21} & f_{,22} \end{vmatrix} = f_{,11}f_{,22} - f_{,12}^2 \geq 0$$

which are the familiar requirements for a (local) minimum value of a scalar function of two variables. In the more general case of $n$ variables, alternative tests of a non-negative definite matrix can be found in most books on matrix algebra, e.g., G. Strang (1980).

## 7. Global Minimum

We are now in a position to describe the importance of convexity for an extremum of a scalar function and an extreme value of an integral. For any convex function $f$ differentiable at a point $\mathbf{y}$, we set $\bar{f}(t) \equiv f(\mathbf{y} + t(\mathbf{z} - \mathbf{y}))$ and observe the relation

$$\begin{aligned} f(\mathbf{z}) - f(\mathbf{y}) &= \bar{f}(1) - \bar{f}(0) = \int_0^1 \bar{f}'(t)\,dt \geq \int_0^1 \bar{f}'(0)\,dt \\ &= \bar{f}(0) = \nabla f(\mathbf{y})(\mathbf{z} - \mathbf{y}) \end{aligned} \tag{4.22}$$

which holds for all $\mathbf{z}$ because $\bar{f}'$ is increasing. It follows immediately from (4.22) that

**[IV.9]** ***A convex function*** $\mathbf{f(y)}$ ***which is*** $\mathbf{C^2}$ ***attains a global minimum at a point*** $\mathbf{y}$ ***if and only if*** $\nabla \mathbf{f} = \mathbf{0}$ ***at*** $\mathbf{y}$.

The importance of convexity is now seen to be realized in two ways. First, any candidate for an extremum of a convex function, i.e., any stationary point, is automatically a minimum. Second, any local (relative) minimum of a convex function is automatically a global (absolute) minimum. These two observations are rather evident in one- and two-dimensional cases where we can visualize the graph of a convex function. [IV.9] gives us the same results for higher dimensions. Furthermore, the development leading up to these results now enables us to derive a sufficient condition for an extremal $\hat{y}(x)$ to render $J[y]$ a minimum value.

From the *Euler–Lagrange* equation, we have

$$p(x) \equiv F_{,y'}(x, \hat{y}, \hat{y}') = c + \int_a^x F_{,y}(t, \hat{y}(t), \hat{y}'(t))\,dt,$$

where we have introduced an abbreviation $p(x)$ for $F_{,y'}(x, \hat{y}(x), \hat{y}'(x))$ which will also be useful in our discussion of Hamiltonian systems in Chapter 6. By the Erdmann first corner condition, $p(x)$ is at least PWS. At noncorner points, $p(x)$ satisfies

$$(p'(x), p(x)) = (\hat{F}_{,y}, \hat{F}_{,y'}) = \nabla F(x, \hat{y}, \hat{y}'),$$

where the gradient operator $\nabla$ is on the arguments $y$ and $y'$. The second part $p = \hat{F}_{,y'}$ is just the definition of $p$, whereas the first part $p' = \hat{F}_{,y}$ follows from the Euler DE.

For any admissible comparison function $y(x)$, we have by (4.22) for any differentiable $F(x, y, v)$ convex in $y$ and $v$,

$$\begin{aligned} J[y] - J[\hat{y}] &= \int_a^b [F(x, y, y') - F(x, \hat{y}, \hat{y}')]\, dx \\ &\geq \int_a^b [(y - \hat{y})\hat{F}_{,y} + (y' - \hat{y}')\hat{F}_{,y'}]\, dx \\ &= \int_a^b [(y - \hat{y})p' + (y' - \hat{y}')p]\, dx \\ &= \int_a^b \frac{d}{dx}[p(y - \hat{y})]\, dx = 0. \end{aligned}$$

In the presence of corners, the integral over $[a, b]$ is understood to be a sum of integrals over subintervals between corner points or between a corner point and an end point. It follows that

**[IV.10]** ***For differentiable functions*** $F(x, y, y')$ ***convex in*** $y$ ***and*** $y'$***, any admissible extremal of the basic problem renders*** $J[y]$ ***a global minimum and conversely.***

The conclusion applies to PWS extremals, not just the smooth ones as in [IV.6]. For $F(x, y, v)$ convex only in a region of the $(y, v)$ space, [IV.10] holds relative to all admissible comparison functions which lie in that region.

Unfortunately, many integrands of basic problems in calculus of variations are not convex in $y$ and $y'$. We must, therefore, continue our effort to establish sufficient conditions for a minimum (or maximum) $J[y]$.

**EXAMPLE 8:** $F = \dfrac{1}{2}[(y')^2 - y^2]$.

This Lagrangian is not convex because $F_{,yy} = -1 < 0$. It is also not concave because $F_{,y'y'}F_{,yy} - F_{,yy'}^2 = -1 < 0$. (A function $f$ is concave if $-f$ is convex).

## 8. Exercises

1. (a) For the plane geodesic problem with $F = \sqrt{1 + (y')^2}$, prove that the unique admissible extremal for the basic problem renders $J[y]$ a weak local minimum by showing directly $J^{\bullet\bullet}(0) > 0$.

   (b) Obtain the same result by invoking [II.1].

2. (a) Obtain the unique admissible extremal for the basic problem with

$$F = (y')^3, \quad y(0) = 2, \quad y(1) = 3.$$

   (b) Does it render $J[y]$ a weak local minimum? A global minimum? Justify your answers.

   (c) Suppose $y(1) = 3$ is replaced by $y(1) = 1$. Repeat part (b).

   (d) $\hat{y}(x) = 0$ is the unique admissible extremal for the same problem with $y(0) = 0$ and $y(1) = 0$ as end conditions. Is $J[\hat{y}]$ any kind of a minimum in this case?

3. (a) For the basic problem with $F = y'(1 + x^2 y')$, $y(-1) = 1$, and $y(2) = 1$, show that the strengthened Legendre condition (II′) is not satisfied.

   (b) Show that the unique admissible extremal $\hat{y}(x)$ nevertheless renders $J[y]$ a global minimum. [Hint: Consider $J[\hat{y} + \varepsilon h] - J[\hat{y}]$.]

   (c) Show that the admissible extremal for the more general end conditions $y(a) = A$ and $y(b) = B$ does not render $J[y]$ a global minimum if $B \neq A$, i.e., $J[\hat{y} + \varepsilon h] < J[\hat{y}]$ for some $h(x)$.

   (d) If $0 < a < b$ (or $a < b < 0$), show that $J[\hat{y}]$ of part (c) is a weak local minimum even if $B \neq A$.

4. (a) Show that the extremals of any basic problem where $F$ does *not* depend on $y$ admits no conjugate points.

   (b) For $F = (y')^2 - y^2 + 4y\cos x$, show that Jacobi equation is the same for any PWS $y(x)$ (not necessarily an extremal).

Determine whether each of the next three basic problems has a weak local minimum or maximum:

5. $F = 4y^2 - (y')^2 + 8y, \quad y(0) = -1, \quad y(\pi/4) = 0.$

6. $F = x^2(y')^2 + 12y^2, \quad y(1) = 1, \quad y(2) = 8.$

7. $F = (y')^2 + y^2 + 2ye^{2x}$, $y(0) = 1/3$, $y(1) = e^2/3$.

8. Let $\hat{y}(x)$ be the unique admissible extremal for the Lagrangian $F = (y')^2 + 2yy' + 2y' + 2y$ with unspecified end conditions at the ends of a fixed interval $[a, b]$. Is $J[\hat{y}]$ a weak local minimum? Justify your answer.

9. For the aircraft flight-path problem of Chapter 1, the unique admissible extremal for $D < \pi H$ [with $y(0) = y(D) = 0$] is

$$\hat{y}(x) = H \ln\left[\frac{\cos([D - 2x]/2H)}{\cos(D/2H)}\right].$$

   (a) Show that there is no conjugate point for the point $a = 0$ (relative to $\hat{y}$).

   (b) Is $J[\hat{y}]$ a weak local minimum? Justify your answer.

10. Show that $\hat{y} = 0$ is an admissible extremal of the basic problem with $F = (y')^2 - 4y(y')^3 + 2x(y')^4$, $y(0) = 0$, and $y(1) = 0$, and that it renders $J[y]$ a weak local minimum.

11. Suppose $\hat{y}(x) = c_0$ (a constant) is the extremal for the basic problem with $F = F(y, y')$ and $\hat{F}_{,y'y'} > 0$:

   (a) If $\hat{F}_{,yy} \geq 0$, show that there are no conjugate points relative to $\hat{y}$.

   (b) If $\hat{F}_{,yy} < 0$, show that there are conjugate points and determine the distance between two consecutive conjugate points.

12. For the basic problem with $F = y/(y')^2$, $y(0) = 1$, and $y(b) = B$ where $0 < b < 1$ and $0 < B < 1$:

   (a) Show that there are two smooth admissible extremals.

   (b) Show that there is an admissible extremal $\hat{y}_c(x; \bar{x})$ with one corner in any location $\bar{x}$ in $(0, b)$.

   (c) Show that the choice of $\bar{x}$, which minimizes $J[\hat{y}_c(x; \bar{x})]$, reduces $\hat{y}_c(x; \bar{x})$ to one of the two smooth admissible extremals.

   (d) Which extremals render $J[y]$ a weak local minimum?

13. Determine whether the following functions are convex:

$$\text{(a)}\quad f(s, v) = e^s e^v; \qquad \text{(b)}\quad f(s, v) = e^{sv}.$$

14. Show that a finite sum of convex functions is also convex.

15. Show that $\widehat{y}(x) = x$ is the global minimum of the basic problem with $F(x, y, y') = e^y e^{y'}$, $y(0) = 0$, and $y(1) = 1$.

16. Suppose a mine contains an amount $D$ of a resource deposit (such as coal, copper, or fossil fuel). Let $P = P_0 \ln[y'(t)]$ be the rate of profit that can be earned from selling the resource at the rate $y'(t)$, where $y(t)$ is the cumulative amount of resource sold by time $t$ with $y(0) = 0$. The resource will have no net value beyond time $T$ (when a new technology will be available); hence, the entire deposit should be sold by $T$ so that $y(T) = D$. The present value of the total profit from the sale of the resource is given by

$$J[y] = \int_0^T e^{-rt} P \, dt = P_0 \int_0^T e^{-rt} \ln[y'(t)] \, dt,$$

where $r$ is the constant discount rate and $P_0$ is a known constant. Determine the optimal extraction-sales program $\widehat{y}(t)$ which maximizes $J[y]$. (Explain why $J[\widehat{y}]$ is a global maximum).

17. $J[y] = \int_a^b [y^2 + 2\alpha y y' + \beta (y')^2] \, dx, \qquad y(a) = A, \qquad y(b) = B.$

(a) If $\beta = 0$, show that $\widehat{y}(x) \equiv 0$ is the only extremal and that it minimizes $J[y]$ provided $A = B = 0$.

(b) For what range of $\alpha$ and $\beta (\neq 0)$ can we conclude by convexity that any admissible extremal makes $J$ a global minimum? Justify your answer. [You do not have to obtain $\widehat{y}(x)$ explicitly].

(c) Show that any admissible extremal $\widehat{y}$ for the problem is at least a weak local minimum whether or not the restriction on $\alpha$ and $\beta$ of part (b) is satisfied.

(d) Without obtaining $\widehat{y}$ explicitly, show that $\widehat{y}$ does not depend on $\alpha$.

18. (a) Show that an admissible extremal $\widehat{y}(x)$ of any basic problem with $F(x, y, y') = f(x)\sqrt{1 + (y')^2}$ always satisfies Jacobi's necessary condition if $f(x) > 0$.

(b) Show that $J[\widehat{y}]$ is a global minimum if $f(x) > 0$.

19. Without obtaining $\widehat{y}(x)$, show that any admissible extremal of the basic problem for the minimum production cost Lagrangian (see Section 2 of Chapter 1) $F = k(y')^2 + c_h y$ renders $J[y]$ a weak local minimum.

20. $J[y] = \int_{3/2}^{2} y^2(1 - y')^2\,dx, \qquad y(3/2) = 0, \qquad y(2) = 1.$

A smooth extremal for this problem was found in Chapter 1 to be

$$\hat{y}^2 + \frac{9}{16} = \left(x - \frac{3}{4}\right)^2 \qquad \left(\frac{2}{3} \leq x \leq 2\right).$$

We wish to decide whether $J[\hat{y}]$ is at least a weak (local) minimum.

(a) Is [IV.6] applicable for this purpose?

(b) Is $F(x, y, y') = y^2(1 - y')$ a convex function in $y$ and $y'$?

# 5

# A Strong Minimum

## 1. A Weak Minimum May Not Be A True Minimum

Consider the basic problem

$$J[y] = \int_0^1 [(y')^2 + (y')^3]\, dx, \qquad y(0) = 0, \quad y(1) = 0.$$

The extremals for this problem are special cases of $\hat{y} = c_1 x + c_2$ (and must be smooth by the result of Example 3 of Chapter 2). The end conditions are met by $c_1 = c_2 = 0$ so that $\hat{y}(x) = 0$ is the unique admissible extremal with $J[\hat{y}] = 0$. The strengthened Legendre condition is satisfied as $\hat{F}_{,y'y'}(x) = 2 + 3\hat{y}' = 2 > 0$. With $P = 2$ and $Q = 0$, the Jacobi accessory equation has as its only nontrivial solution $h(x) = c_3 x$ [where we have made use of $h(0) = 0$ to eliminate one constant of integration]. Hence, the conditions (I′), (II′), and (III′) are all satisfied by $\hat{y}$ and we conclude from [IV.6] that $J[\hat{y}] = 0$ is a weak local minimum. Moreover, because $\hat{y}$ is the only admissible extremal, $J[\hat{y}]$ appears to be the only (weak) minimum point possible.

Consider, however, the PWS admissible comparison function $y_\theta$ given by the polygonal line PQR shown in Figure 5.1, where the angle at $Q$ is a right angle. Evidently, the lengths of the segments $QR$ and $SR$ are given by

$$QR = 1 \cdot \sin\theta, \qquad SR = 1 - x_0 = QR \sin\theta = \sin^2\theta$$

so that $x_0 = 1 - \sin^2\theta = \cos^2\theta$. It follows that

$$J[y_\theta] = \int_0^{x_0} (\tan^2\theta + \tan^3\theta)\, dx + \int_{x_0}^1 (\cot^2\theta - \cot^3\theta)\, dx$$

$$= \sin^2\theta(1 + \tan\theta) + \cos^2\theta(1 - \cot\theta) = 1 - 2\cot(2\theta).$$

Note that $y_\theta$ tends to $\hat{y} \equiv 0$ as $\theta$ tends to zero (with the angle at $Q$ kept at 90°) as $J[y_\theta]$ tends to $-\infty$. Hence, $J[\hat{y}] = 0$ is not even a local minimum because

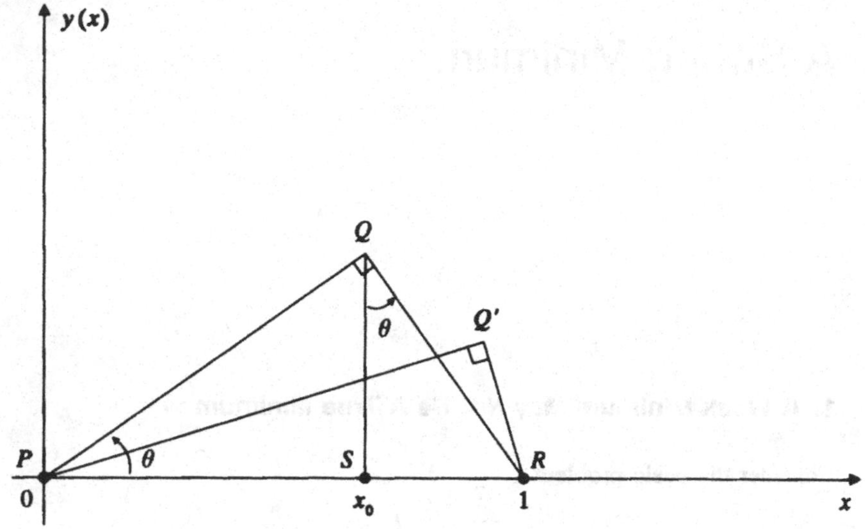


**Figure 5.1**

many nearby PWS functions furnish a negative $J$! Moreover, there does not exist a (finite) minimum because by choosing $\theta$ small enough, we can make $J$ as negatively large as we wish.

It appears that we have two contradictory results: a mathematical theorem [IV.6] (along with Erdmann's corner conditions [II.4] and [II.5]) assuring us that $\hat{y} = 0$ furnishes the only (weak) minimum point on the one hand, and specific counter-examples (one for a different $\theta$ value) giving different values for $J[y]$ less than $J[\hat{y}] = 0$ on the other hand! A clue to the resolution of this apparent contradiction lies in the observation that the slope of $QR$, i.e., the part of $y_\theta(x)$ to the right of $Q$ becomes more and more negative (with $|y'_+(t)| \to \infty$ as $\theta \to 0$). Such a comparison function was not among the admissible comparison functions in our previous search for a minimum. Up to now, we compared $\hat{y}$ only with those functions which may be written as $y = \hat{y} + \varepsilon h$ (sometimes using $\delta y$ to denote $\varepsilon h$) where $\varepsilon$ is a small parameter and $\hat{y}$ and $h$ are PWS functions with $|\varepsilon h| \ll 1$. In fact, we compare $\hat{y}$ only with "nearby" PWS functions which satisfy the condition

$$\left|y(x) - \hat{y}(x)\right| < \varepsilon \tag{5.1}$$

as well as

$$\left|y'(x) - \hat{y}'(x)\right| < \varepsilon \tag{5.2}$$

in $[a, b]$ for $0 < \varepsilon \ll 1$. Condition (5.2) was mentioned briefly in Section 1 of Chapter 2 and has been tacitly assumed up to now. In particular, the relation $|y' - \hat{y}'| = |\varepsilon h'|$ was tacitly assumed to be of the same order of magnitude as $|\varepsilon h|$ in the proof of [IV.6]. This tacit assumption is now explicitly given as (5.2) which allows us to give a precise definition of a *weak minimum*: An admissible comparison function $\hat{y}$ for the basic problem renders $J[y]$ a *weak local minimum* if $J[y] \geq J[\hat{y}]$ for all admissible $y$ satisfying both (5.1) and (5.2) in $[a,b]$. At a corner point, a derivative $f'$ in (5.2) is replaced by $f'_+$ or $f'_-$. For a given $\varepsilon$, the set of all (PWS) comparison functions $y$ satisfying (5.1) and (5.2) is called a *weak neighborhood* of $\hat{y}$ (of radius $\varepsilon$) and the corresponding $\delta y$ *weak variations.*

Because $h$ is PWS and $h'$ PWC, they are both bounded in $[a,b]$ so that for sufficiently small $t$ the PWS function $y = \hat{y} + th$ satisfies (5.1) and (5.2) and, therefore, lies inside a weak neighborhood of $\hat{y}$. On the other hand, the function $y_\theta$ shown in Figure 5.1 does not lie in any given weak neighborhood of $\hat{y} = 0$ if $\theta$ is made small enough. Whereas $y_\theta$ tends to $\hat{y}$, $y'_\theta$ tends to $-\infty$ near the point $R$ (and, in fact, along the whole segment QR) as $\theta$ tends to zero so that (5.2) cannot be met for any fixed $\varepsilon$. An admissible comparison function $\hat{y}$ is said to render $J[y]$ a *strong local minimum* if $J[y] \geq J[\hat{y}]$ for any admissible $y$ satisfying (5.1) in $[a,b]$ (and no condition is imposed on $y'$). The set of all comparison functions satisfying only (5.1) is called a *strong* neighborhood of $\hat{y}$ (of radius $\varepsilon$) and the corresponding $\delta y$ *strong variations.*

## 2. The Weierstrass Excess Function

In speaking of a global minimum, a strong local minimum, or a weak local minimum, the admissible extremal $\hat{y}(x)$ is compared to members of succeedingly smaller sets of PWS comparison functions. It follows that a global minimum must also be a strong local minimum, and the latter must, in turn, be a weak local minimum. Consequently, a condition such as the *Euler–Lagrange* equation that is necessary for a weak local minimum is also necessary for a strong local minimum. In turn, a condition that is necessary for a strong local minimum must also be satisfied by a global minimum.

On the other hand, an admissible extremal may make $J[y]$ a weak local minimum but not a strong local minimum. The illustrative problem of the last section is one such example. Naturally, we would like to know what requirements should be imposed on an admissible extremal for it to furnish a strong local minimum of the basic problem. In one way, we already have an answer in the sufficient condition [IV.10], which guarantees a global minimum. But not all Lagrangians are convex in $y$ and $y'$. For those which are not, we may ask what additional requirements are needed beyond (I′), (II′), and (III′) for $J[\hat{y}]$ to be a strong local minimum.

For $J[\hat{y}]$ to be a strong minimum, the difference $J[y] - J[\hat{y}]$ must be non-negative for any neighboring admissible comparison function $y$ for which $|y' - \hat{y}'|$ may not be small. The difference $J[y] - J[\hat{y}]$ can be written as

$$J[y] - J[\hat{y}] = \int_a^b \{F(x,y,y') - F(x,\hat{y},\hat{y}')\}\,dx$$

$$= \int_a^b \{F(x,y,y') - F(x,\hat{y},y') - (y' - \hat{y}')\,F_{,y'}(x,\hat{y},\hat{y}')$$

$$+ (y' - \hat{y}')\,F_{,y'}(x,\hat{y},\hat{y}') + F(x,\hat{y},y') - F(x,\hat{y},\hat{y}')\}\,dx$$

$$= \int_a^b \{F_{,y}(x,\bar{y},y')(y - \hat{y}) + (y' - \hat{y}')\,F_{,y'}(x,\hat{y},\hat{y}') + E(x,\hat{y},\hat{y}',y')\}\,dx$$

where $\bar{y}(x)$ is a point between $y(x)$ and $\hat{y}(x)$ and

$$E(x,y,v,w) = F(x,y,w) - F(x,y,v) - (w - v)F_{,y'}(x,y,v). \tag{5.3}$$

The quantity $E(x,y,v,w)$ is called the *Weierstrass excess function*. We may integrate the first term in the expression for $J[y] - J[\hat{y}]$ by parts to get

$$J[y] - J[\hat{y}] = [\bar{\varphi}(x)(y - \hat{y})]_a^b$$

$$+ \int_a^b \{[F_{,y'}(x,\hat{y},\hat{y}') - \bar{\varphi}(x)](y' - \hat{y}') + E(x,\hat{y},\hat{y}',y')\}\,dx,$$

where

$$\bar{\varphi}(x) = \int_a^x F_{,y}(x,\bar{y},y')\,dx + c.$$

The first term in brackets vanishes because of the end constraints. If $\bar{y}(x)$ were $\hat{y}(x)$, then the term of the integrand in brackets would vanish by the *Euler-Lagrange* equation and $J[\hat{y}]$ would be a local minimum if $E(x,\hat{y},\hat{y}',y')$ should be non-negative. However, $\bar{y}$ is not $\hat{y}$; the first term in the integrand, $[\hat{F}_{,y'} - \bar{\varphi}](y' - \hat{y}')$, may be negative. Hence, we have even more reason to want $E(x,\hat{y},\hat{y}',w)$ to be non-negative for all $w$. It is shown rigorously in the appendix of this chapter that we have the following result:

**[V.1]** *If, for the admissible extremal* $\hat{y}$, $J[\hat{y}]$ *is a strong local minimum for* $J[y]$, *it is necessary that*

$$\text{(IV)} \quad E(x,\hat{y}(x),\hat{y}'(x),w) \geq 0$$

*for all* $x$ *in* $[a,b]$ *and for all* $|w| < \infty$. *At a corner point, this equality is to hold for both* $\hat{y}'_+$ *and* $\hat{y}'_-$.

We designate the necessary condition $E[x,\hat{y}(x),\hat{y}'(x),w] \geq 0$ for all $x$ in $[a,b]$ and for all $|w| < \infty$ as condition (IV) and add it to the list of necessary conditions given previously in Section 4 of Chapter 4.

EXAMPLE 1: $J[y] = \int_0^b [(y')^2 - y^2]\,dx,\quad y(0) = 0$, and $y(b) = 0$.

We have $F_{,y'}(x,\hat{y},\hat{y}') = 2\hat{y}'$ so that

$$E(x,\hat{y},\hat{y}',w) = [w^2 - \hat{y}^2] - [(\hat{y}')^2 - \hat{y}^2] - 2\hat{y}'(w - \hat{y}') = (w - \hat{y}')^2 \geq 0$$

for all $|w| < \infty$. However, we learned in Example 2 of Chapter 4 that $J[\hat{y} = 0] = 0$ is not even a weak local minimum for $b > \pi$. Hence, (IV) by itself does not guarantee $J[\hat{y}]$ to be a minimum.

EXAMPLE 2: $J[y] = \int_0^1 [(y')^2 + (y')^3]\,dx,\quad y(0) = 0$, and $y(1) = k$.

For this problem, we have $\hat{y}' = kx$ and

$$\begin{aligned} E[x,\hat{y},\hat{y}',w] &= (w^2 + w^3) - (k^2 + k^3) - (w - k)k(2 + 3k) \\ &= (w - k)^2(1 + w + 2k) \end{aligned}$$

with $E < 0$ for all $w < -(2k + 1)$. The Weierstrass condition (IV) is not satisfied and $\hat{y}$ cannot be a strong minimum. [Note that the strengthened Legendre condition (II′) is satisfied if $k > -1/3$ and the strengthened Jacobi test (III′) is satisfied for all $k$.]

## 3. The Figurative

It is possible to give a geometrical interpretation of the necessary condition [V.1] on $E(x,y,v,w)$. For a given $\hat{y}$ and a fixed value of $x$, we show in Figure 5.2 a graph of $F(x,\hat{y}(x),v)$ as a function of $v$; the graph is called the *figurative* of $F$.

Suppose a given $x$ is not a corner point and we draw the tangent line to the graph of the figurative at $v = \hat{y}'(x)$. The slope of the tangent is $F_{,y'}(x,\hat{y}(x),\hat{y}'(x))$ and the equation for the tangent itself is given by

$$S(v) = \hat{F}(x) + (v - \hat{y}'(x))\hat{F}_{,y'}(x).$$

The intersection of this tangent line and the vertical line at $v = w$ is, therefore, equal to $\hat{F}(x) + (w - \hat{y}'(x))\hat{F}_{,y'}(x)$. The point $F(x,\hat{y}(x),w)$ on the figurative is above or below this intersection by a distance $E(x,\hat{y}(x),\hat{y}'(x),w)$. Whether it is above or below depends on the sign of the excess function at $w$. We have therefore the following geometrical consequence of [V.1]:

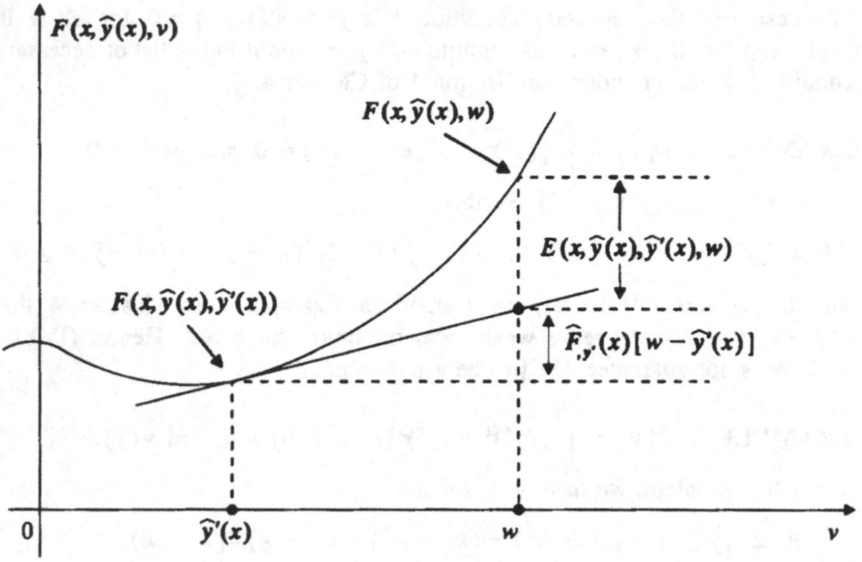


**Figure 5.2**

(i) *The Weierstrass necessary condition for a strong (local) minimum requires that* $\mathrm{F}(\mathrm{x}, \widehat{\mathrm{y}}(\mathrm{x}), \mathrm{w})$ *be at or above the tangent line emanating from the point on the figurative at* $\widehat{\mathrm{y}}'(\mathrm{x})$.

Suppose $x$ is now a corner point. By Erdmann's first corner condition, we have

$$F_{,y'}(x, \widehat{y}, \widehat{y}'_+) = F_{,y'}(x, \widehat{y}, \widehat{y}'_-) \equiv m \tag{5.4}$$

where it is understood that $\widehat{y}$ and $\widehat{y}'$ are evaluated at $x_+$ or $x_-$ and $y'_\pm = \widehat{y}'(x_\pm)$. Therefore:

(ii) *The tangent lines to the figurative at* $\mathrm{v} = \widehat{\mathrm{y}}'_-$ *and* $\mathrm{v} = \widehat{\mathrm{y}}'_+$ *are parallel, with the same slope* m.

In fact, the tangent line at $\widehat{y}'_-$ is given by

$$S_-(v) \equiv F(x, \widehat{y}, \widehat{y}'_-) + m(v - \widehat{y}'_-) = mv + [F(x, \widehat{y}, \widehat{y}'_-) - m\widehat{y}'_-], \tag{5.5}$$

whereas the one at $y'_+$ is given by

$$S_+(v) \equiv F(x, \widehat{y}, \widehat{y}'_+) + m(v - \widehat{y}'_+) = mv + [F(x, \widehat{y}, \widehat{y}'_+) - m\widehat{y}'_+]. \tag{5.6}$$

By the second Erdmann corner condition, the quantities in brackets on the right-hand side of (5.5) and (5.6) are equal. Hence,

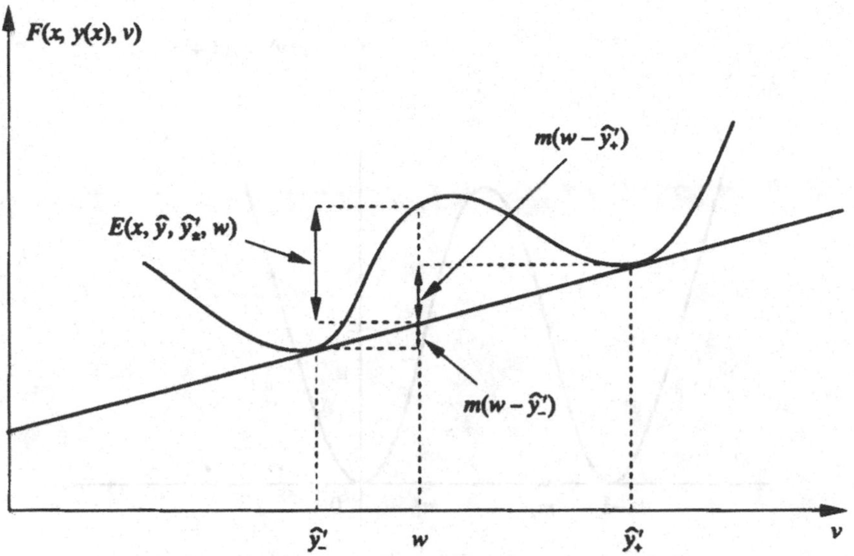


**Figure 5.3**

(iii) *The two tangent lines at $y'_-$ and $y'_+$ actually coincide (Fig. 5.3); the Weierstrass necessary condition requires the figurative to be above this double tangent line.*

We summarize observations (i), (ii), and (iii) as

**[V.2]** ***Suppose $J[\hat{y}]$ is a strong local minimum. Then, at non-corner points of $\hat{y}$, the figurative of the basic problem must lie above its tangent line through the point $(\hat{y}'(x), \hat{F}(x))$. At corner points, the figurative must lie above its double tangent line through $(\hat{y}'_-, F(x, \hat{y}(x), \hat{y}'_-))$ and $(\hat{y}'_+, F(x, \hat{y}(x), \hat{y}'_+))$.***

That the two points $(\hat{y}'(x-), \hat{F}(x-))$ and $(\hat{y}'(x+), \hat{F}(x+))$ of the figurative share the same tangent line is an intrinsic property of extremals. This geometrical feature is a consequence of the Erdmann corner conditions; it has nothing to do with the Weierstrass condition on the excess function.

**EXAMPLE 3:** $F = (y')^2(1 + y')^2$.

The figurative is independent of $x$ and $y$ in this case. Figure 5.4 displays a graph of $F(v) = v^2(1 + v)^2$. The figurative is shaped like a $W$ with two minimum values on the $v$ axis at the points $v = -1$ and $v = 0$. It follows immediately from [V.2] that corners of $\hat{y}(x)$ can only occur with slopes 0 on one side of the corner and $-1$ on the other side. It is much more difficult to

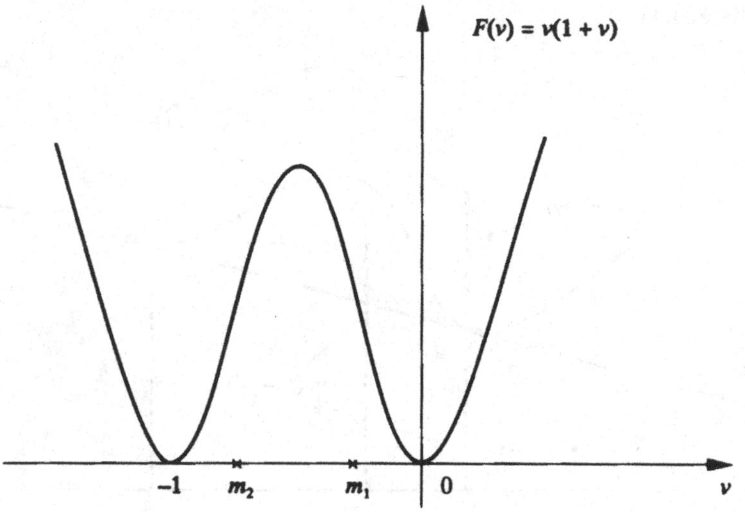


Figure 5.4

obtain the same result from the two Erdmann conditions which may be arranged to read

$$p(1+p)(1+2p) = q(1+q)(1+2q),$$

$$p^2(1+p)(1+3p) = q^2(1+q)(1+3q).$$

It is easy to see that $p = -1$ and $q = 0$ (or the reverse) satisfy these conditions; but it is less obvious that there are no other possible combinations.

## 4. Fields of Extremals

The *Weierstrass* excess function plays an important role in sufficient conditions for a strong minimum. We will establish one such condition in the next section. This condition involves the notion of a *field of extremals* associated with a given basic problem. For example, the extremals of $F = (y')^2$ are linear functions of $x$, namely, $y = mx + c$. For a given $m$, the extremals corresponding to different values of $c$ are parallel lines; they cover *any* region of the $x,y$-plane (Fig. 5.5a). Another set of extremals of the same F is generated by fixing $y(a) = A$ and varying $m$. The extremals are now (nonvertical) straight lines in the $x,y$-plane intersecting at the point $(a, A)$ (Fig. 5.5b). The vertical line at $(a, A)$ is excluded as $y$ is not well defined along that line (unless we use a parametric representation). This set of extremals covers any region of the $x, y$-plane which does not include a part of the vertical line through $(a, A)$.

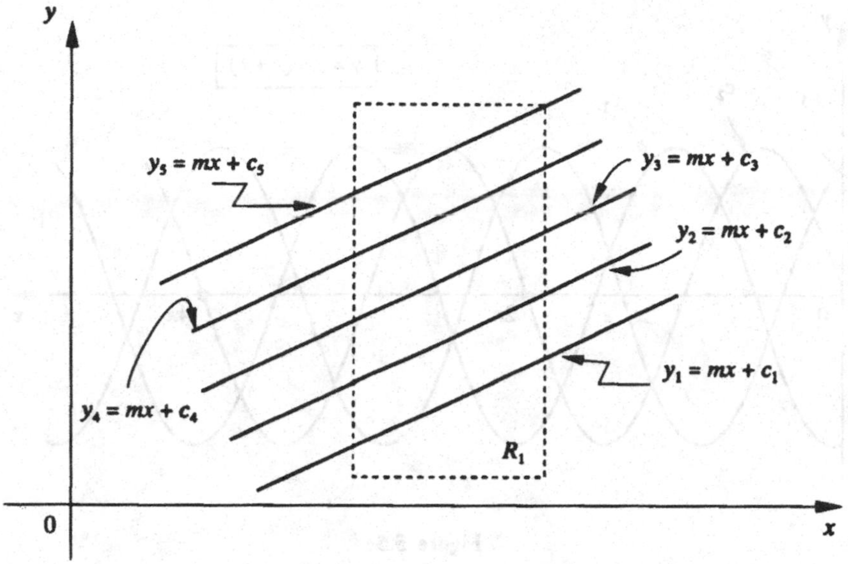


Figure 5.5a

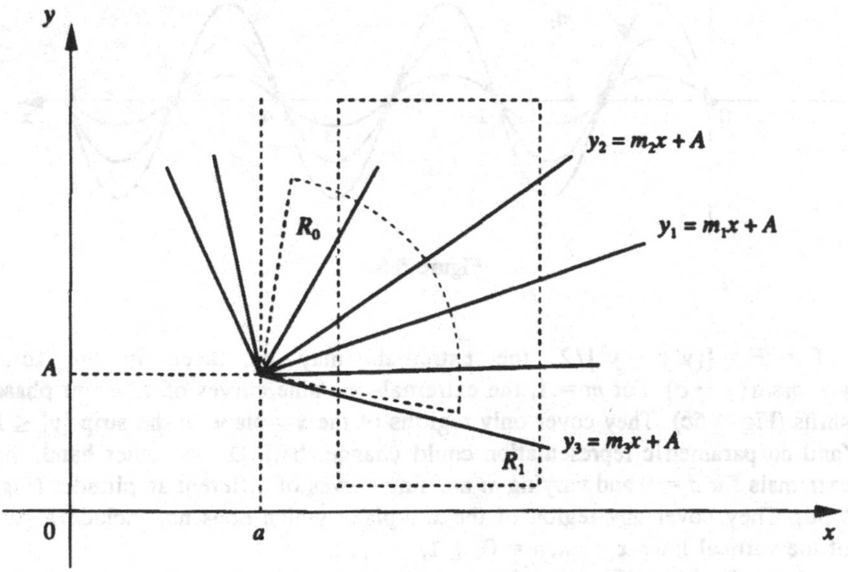


Figure 5.5b

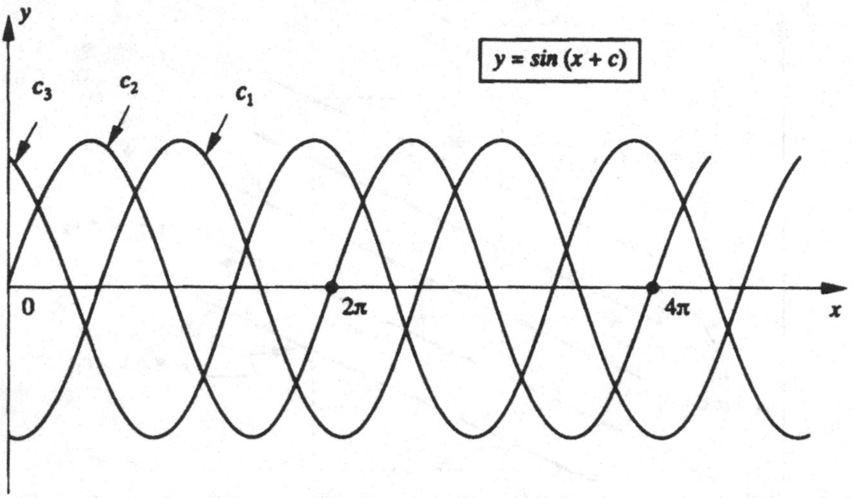


Figure 5.5c

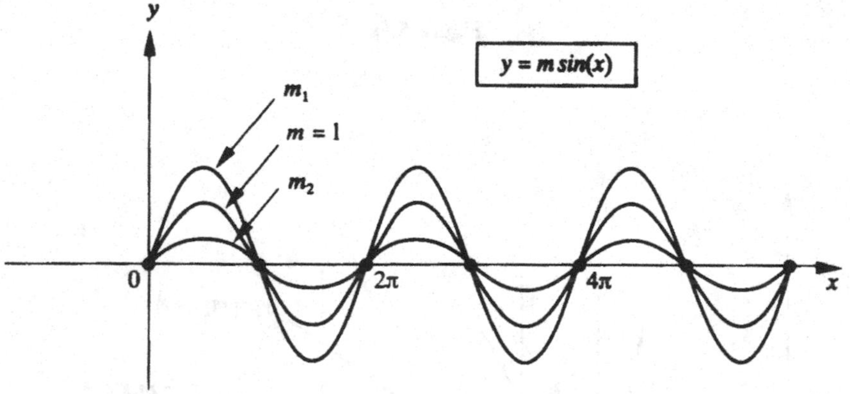


Figure 5.5d

For $F = [(y')^2 - y^2]/2$, the extremals may be taken in the form $y = m\sin(x + c)$. For $m = 1$, the extremals are sine curves of different phase shifts (Fig. 5.5c). They cover only regions of the $x$-$y$-plane in the strip $|y| \leq 1$ (and no parametric representation could change that). On the other hand, the extremals for $c = 0$ and varying $m$ are sine curves of different amplitudes (Fig. 5.5d). They cover any region of the $x, y$-plane which does not include a part of the vertical lines $x = n\pi, n = 0, \pm 1, \pm 2, \ldots$.

Of the four specific sets of extremals generated above, the set given by $y = \sin(x + c)$ differs from the other three in one important aspect: There are two extremals passing through any point in a region covered by this set. In contrast, only a single extremal of the set $y = mx + c$ with a fixed $m$ passes through any point in the $x,y$-plane. The same is true for the set $y = m(x - a) + A$ except at the point $(a,A)$. Finally, only the points $x = n\pi$ are the intersections of more than one extremal of the set $y = m\sin(x)$. For a family of extremals which covers a point in a region only once (similar to the family $y = mx + c$ with a fixed $m$), a unique *slope function* $s(x,y)$ can be defined at each interior point $(x,y)$ of the region covered by the extremals. More specifically, $s(x,y)$ is defined to be *the slope of the unique extremal passing through the point* $(x,y)$. This would not be possible for the set $y(x) = \sin(x + c)$. It would be possible for $\hat{y} = m\sin x$ except $x = n\pi$ $(n = 0, \pm 1, \pm 2, \ldots)$ and for $\hat{y} = mx + (A - ma)$ except for the point $(a,A)$.

In general, a simply-connected open region $R$ is said to be covered by a *field of extremals* if there is a (one-parameter) family of *smooth* extremals such that exactly one extremal of the family passes through any point $(x,y)$ of $R$. For example, the region $R_1$ in Figure 5.5a is covered by a field of extremals $\hat{y}_c = x + c$ of the Lagrangian $F = (y')^2$. Similarly, the extremals $\hat{y}_m = m(x - a) + A$ for different values of $m$ constitute a field over the region $R_1$ in Figure 5.5b. We denote by $s(x,y)$ the derivative of the uniquely determined extremal at the point $(x,y)$. Unless specifically indicated otherwise, we limit our discussion to fields for which $s(x,y)$ is continuously differentiable in both arguments for all points in $R$.

One way to establish $J[\hat{y}]$ as a strong local minimum involves embedding $\hat{y}$ in a field of extremals. More specifically, we want to have $\hat{y}$ as a member of a field of extremals covering an open region $R$ which contains the rectangle $R_Y = \{a < x < b, |y - \hat{y}(x)| < Y\}$ for some $Y > 0$. Such a smooth extremal $\hat{y}$ is said to be *embeddable* in a (proper) field over $R$. If ($F$ is sufficiently differentiable and) an admissible extremal $\hat{y}$ of the basic problem exists, we know from the theory of differential equations that $\hat{y}$ depends continuously on the initial condition $y(a) = A$ [or $y(b) = B$]. In principle, a field of extremals can then be constructed by varying $A$ (or $B$) continuously. [Embedding theorems with less stringent requirements on $F(x,y,y')$ have also been established in Hestenes (1966)].

The actual construction of the embedding field of extremals often depends on the nature of the given extremal. We saw earlier that the line joining two fixed points $(a,A)$ and $(b,B)$ [for $F = (y')^2$] can be embedded in the one-parameter family of straight lines parallel to the fixed line (by taking $c$ as the parameter in $\hat{y}_c = mx + c$ with $m = (B - A)/(b - a)$) as shown in Figure 5.5a. As an alternative, we can regard the fixed line as one of a pencil of lines [corresponding to $\hat{y}_m = m(x - a) + A$ for different values of the parameter $m$] with the center, i.e., the pencil tip, $(a,A)$ being outside the region of interest.

To allow for this last alternative, the notion of a *central field* over a region

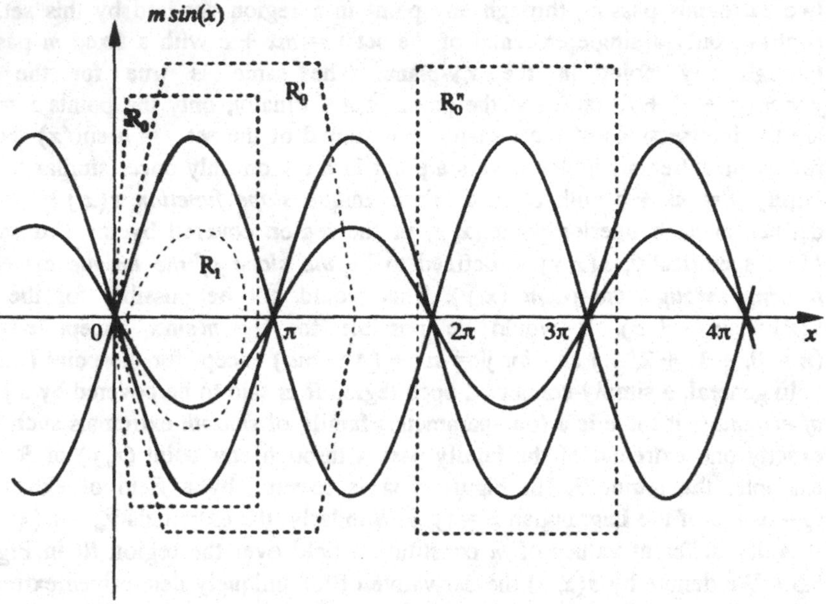


Figure 5.5e

$R_0$ is sometimes introduced. A family of smooth extremals covering an open simply-connected region $\overline{R}$ is said to be a *central field* of extremals over $\overline{R}$ if exactly one extremal of the family passes through any point of the region except for one boundary point of $\overline{R}$ (which is not in $\overline{R}$ because it is open) where the slope function $s(x, y)$ is not unique. This boundary point is called the *pencil point* of the central field. For example, the extremals $\widehat{y}_m = m(x - a) + A$ for different values of $m$ constitute a central field over $R_0$ in Figure 5.5b with the point $(a, A)$ being the pencil tip. On the other hand, the family $y = m \sin x$ constitutes neither a (proper) field nor a central field over $R_0'$ or $R_0''$ in Figure 5.5e; it is however a central field over $R_0$ in the same figure. As we shall see, the embedding of an extremal may be either over a proper field or a central field because the region $R_Y$ is open.

## 5. Sufficiency

Suppose $\widehat{y}$ is an admissible extremal for our basic problem and $\tilde{y}$ is any admissible comparison function, not necessarily an extremal. For $\widehat{y}$ to render $J[y]$ a strong minimum, we must have

$$J[\bar{y}] - J[\widehat{y}] = \int_a^b [F(x,\bar{y},\bar{y}') - F(x,\widehat{y},\widehat{y}')]\,dx \geq 0. \tag{5.7}$$

We assume that $\widehat{y}$ can be embedded in an appropriate field of extremals in the region of the $x,y$-plane of interest. In that case, the integrand of the above integral can be written in terms of the *Weierstrass* excess function:

$$\begin{aligned} F(x,\bar{y},\bar{y}') - F(x,\widehat{y},\widehat{y}') &= \{F(x,\bar{y},\bar{y}') - F(x,\bar{y},\bar{s}) - (\bar{y}' - \bar{s})F_{,y'}(x,\bar{y},\bar{s})\} \\ &\quad + \{F(x,\bar{y},\bar{s}) - \bar{s}F_{,y'}(x,\bar{y},\bar{s})\} - \{F(x,\widehat{y},\widehat{s}) \\ &\quad - \widehat{s}F_{,y'}(x,\widehat{y},\widehat{s})\} + \{\bar{y}'F_{,y'}(x,\bar{y},\bar{s}) - \widehat{y}'F_{,y'}(x,\widehat{y},\widehat{s})\} \\ &= E(x,\bar{y},\bar{s},\bar{y}') + \{F(x,\bar{y},\bar{s}) - \bar{s}F_{,y'}(x,\bar{y},\bar{s})\} \\ &\quad - \{F(x,\widehat{y},\widehat{s}) - \widehat{s}F_{,y'}(x,\widehat{y},\widehat{s})\} \\ &\quad + \{\bar{y}'F_{,y'}(x,\bar{y},\bar{s}) - \widehat{y}'F_{,y'}(x,\widehat{y},\widehat{s})\} \end{aligned} \tag{5.8}$$

where we have set

$$\widehat{s} = s(x,\widehat{y}(x)) \equiv \widehat{y}'(x) \qquad \text{and} \qquad \bar{s} = s(x,\bar{y}(x))$$

with $\bar{s}$ being the slope of the unique extremal passing through the point $(x,\bar{y}(x))$. The integral in (5.7) can then be written as

$$\begin{aligned} J[\bar{y}] - J[\widehat{y}] = \\ &\int_{\bar{C}} \{[F(x,\bar{y},s(x,\bar{y})) - s(x,\bar{y})F_{,y'}(x,\bar{y},s(x,\bar{y}))]\,dx + F_{,y'}(x,\bar{y},s(x,\bar{y}))\,d\bar{y}\} \\ &- \int_{\widehat{C}} \{[F(x,\widehat{y},s(x,\widehat{y})) - \widehat{s}(x,\widehat{y})F_{,y'}(x,\widehat{y},s(x,\widehat{y}))]\,dx + F_{,y'}(x,\widehat{y},s(x,\widehat{y}))\,d\widehat{y}\} \\ &+ \int_a^b E(x,\bar{y}(x),s(x,\bar{y}(x)),\bar{y}'(x))\,dx. \end{aligned} \tag{5.9}$$

The two line integrals in $J[\bar{y}] - J[\widehat{y}]$ can be combined into a loop integral:

$$\begin{aligned} \int_{\bar{C}} - \int_{\widehat{C}} = \oint_C \{[F(x,y,s(x,y)) - s(x,y)F_{,y'}(x,y,s(x,y))]\,dx \\ + F_{,y'}(x,y,s(x,y))\,dy\} + \int_a^b E(x,\bar{y}(x),s(x,\bar{y}(x)),\bar{y}'(x))\,dx. \end{aligned} \tag{5.10}$$

where $\bar{C}$ is the path from $(a,A)$ to $(b,B)$ along the graph of $\bar{y}(x)$, $\widehat{C}$ is the path from $(a,A)$ to $(b,B)$ along the graph of $\widehat{y}(x)$, and $C$ is the path from $(a,A)$ to $(b,B)$ along $\bar{C}$ and from $(b,B)$ to $(a,A)$ along $\widehat{C}$ (Fig. 5.6).

Let $M(x,y)$ and $N(x,y)$ be defined as the coefficient of $dx$ and $dy$ in (5.10), respectively:

$$M = F(x,y,s(x,y)) - s(x,y)F_{,y'}(x,y,s(x,y)),$$

$$N = F_{,y'}(x, y, s(x, y)) \tag{5.11}$$

where $y$ is taken to be the value of the (unique) extremal $y(x)$ passing through $(x, y)$ and $s(x, y)$ is its slope. It is not difficult to verify that $M_{,y} = N_{,x}$ for all $(x, y)$. Given $N_{,x} = F_{,y'x} + s_{,x}F_{,y'y'}$ and $M_{,y} = F_{,y} - sF_{,y'y} - ss_{,y}F_{,y'y'}$, we use the Euler differential equation in ultra-differentiated form

$$F_{,y} = \frac{d}{dx}[F_{,y'}] = F_{,y'x} + F_{,y'y}s(x, y) + F_{,y'y'}\frac{ds}{dx}$$

$$= F_{,y'x} + sF_{,y'y} + F_{,y'y'}[s_{,x} + ss_{,y}] \tag{5.12}$$

to eliminate $F_{,y}$ and simplify $M_{,y}$ to $F_{,y'x} + s_{,x}F_{,y'y'}$. In the simplification of $M_{,y}$, we have made use of $y' = s(x, y)$ for the extremal $y(x)$ passing through $(x, y)$ (so that $y'$ is itself $C^1$) and correspondingly

$$[s(x, y(x))]' = s_{,x} + s_{,y}y' = s_{,x} + ss_{,y}. \tag{5.13}$$

Here is the important result from these calculations: The simplified expression for $M_{,y}$ is just $N_{,x}$.

With $M_{,y} = N_{,x}$, the loop integral in (5.10) vanishes by Green's theorem. We are then left with

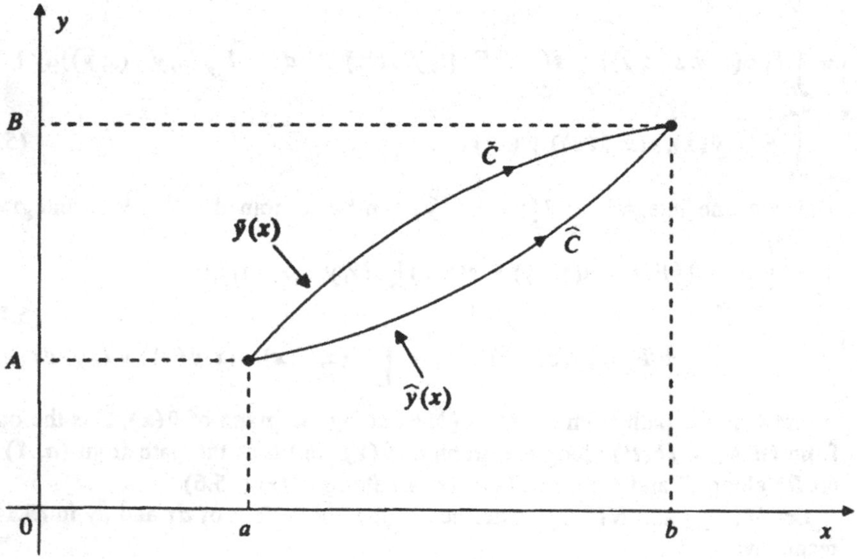


Figure 5.6

$$J[\bar{y}] - J[\hat{y}] = \int_a^b E(x, \bar{y}(x), s(x, \bar{y}(x)), \bar{y}'(x))\,dx. \qquad (5.14)$$

(If $\widehat{C}$ and $\bar{C}$ cross each other one or more times, we get the same result by rearranging the two line integrals in (5.9) as several loop integrals, and applying Green's theorem to each). This allows us to conclude:

**[V.3]** ***If the admissible extremal $\hat{y}$ is embeddable in a field of extremals over an open simply-connected (and convex) region R (as specified in Section 4) and if $F_{,y'y'}(x, y, v) \geq 0$ for all $(x, y)$ in R and $|v| < \infty$, then $J[\hat{y}]$ is a strong local minimum.***

This theorem follows from (5.14) by applying Taylor's theorem to $E[x, \bar{y}, s(x, \bar{y}), \bar{y}']$ so that

$$E[x, \bar{y}, s(x, \bar{y}), \bar{y}'] = \frac{1}{2!}[\bar{y}' - s(x, \bar{y})]^2 F_{,y'y'}(x, \bar{y}, v) \qquad (5.15)$$

where $|v - s(x, \bar{y})| \leq |\bar{y}' - s(x, \bar{y})|$. Given the hypothesis on $F_{,y'y'}(x, y, v)$, condition (5.7) holds and $J[\hat{y}]$ is, therefore, a strong local minimum. It is important to distinguish the condition on $F_{,y'y'}(x, y, v)$ in [V.3] from the Legendre condition. The latter, $F_{,y'y'}(x, \hat{y}(x), \hat{y}'(x)) \geq 0$, refers to an extremal $\hat{y}(x)$.

We discussed rather loosely in Section 3 how a $C^n$ function $F$ for a sufficiently large $n$ would ensure the existence of a field of extremals adjacent to $\hat{y}$. The region covered by this field of extremals generally varies with the structure of $F$. For the derivation of (5.14), we need the region $R$ to be a strip containing (at least) the graph of $\hat{y}(x)$, $a < x < b$. For [V.3] to be useful, we need some way to ensure the existence of a field of extremals over such a region. Rather remarkably, the desired embedding field of extremals is assured by the same conditions (I′), (II′), and (III′) which renders $J[\hat{y}]$ a local weak minimum for any admissible extremal $\hat{y}$.

Given the smooth extremal $\hat{y}(x), a \leq x \leq b$, let $R_Y$ be the region in the $x, y$-plane defined by $a < x < b$ and $|y - \hat{y}(x)| < Y$. We can now give the following formal definition of embeddability of $\hat{y}(x)$: The smooth extremal $\hat{y}(x)$ on $[a, b]$ is said to be *embeddable in a field (of extremals) over a region R*, if for some positive number $Y$, the region $R_Y$ is contained in $R$ and if $R$ is covered by a field (or a central field) of extremals including $\hat{y}$. With some help from the theory of differential equations, it is then possible to prove that

**[V.4]** ***$\hat{y}$ is embeddable in a field over R if F is $C^4$ and $\hat{y}$ satisfies (I′), (II′), and (III′).***

This result is a consequence of the continuous dependence of the solutions of ODE systems on the initial data and system parameters (see F. Clarke, 1980).

We designate as condition (IV′) the condition

$$\text{(IV}'\text{)}\quad E(x,u,v,w)=F(x,u,w)-F(x,u,v)-(w-v)F_{,y'}(x,u,v)\geq 0$$

for all $(u,v)$ near $(\hat{y},\hat{y}')$ and all $|w|<\infty$. Obviously, condition (IV′) is a strengthened form of condition (IV) of Section 2 and is called the *strengthened Weierstrass condition.*

**EXAMPLE 4:** $F(x,y,y')=(y')^2-y^2$.

We have for this example

$$E[x,u,v,w]=(w^2-u^2)-(v^2-u^2)-(w-v)2v=(w-v)^2\geq 0$$

for all $v$ and $w$. Hence, the strengthened Weierstrass condition (IV′) is satisfied. However, (IV′) by itself does not imply a strong minimum as we saw from Example 2 of Chapter 4. For the end conditions $y(0)=y(b)=0$, $J[\hat{y}]=0$ is not even a weak local minimum for this problem if $b>\pi$.

It is now possible to deduce the following sufficient condition for a strong local minimum as a consequence of [V.4]:

**[V.5]** *If F is $C^4$ and the admissible extremal $\hat{y}$ satisfies (I′), (II′), (III′), and (IV′), then $J[y]$ is a strong local minimum.*

The strengthened Legendre condition means that any extremal is at least $C^2$ or, equivalently, $s(x,y)$ is $C^1$. If $\bar{y}$ lies in a strong neighborhood of $\hat{y}$ of sufficiently small radius, then we have $\bar{s}(x)-\hat{s}(x)=[\bar{y}(x)-\hat{y}(x)]s_{,y}(x,\tilde{y})$ with $|\tilde{y}(x)-\hat{y}(x)|\leq|\bar{y}(x)-\hat{y}(x)|$. In other words, $s(x,\bar{y}(x))$ will be uniformly near $s(x,\hat{y}(x))$. It follows from this and the condition (IV′) that

$$E[x,\bar{y}(x),s(x,\bar{y}(x)),\bar{y}'(x)]\geq 0. \tag{5.16}$$

[V.5] is an immediate consequence of the inequality above and (5.14).

## 6. An Illustrative Example

Consider the basic problem

$$J[y]=\int_0^1 (y')^2(1+y')^2\,dx,\qquad y(0)=0,\quad y(1)=m.$$

With $F=F(y')$, the extremals of this problem are straight lines and possibly broken (polygonal) lines. The only smooth admissible extremal is easily seen to be $\widehat{y} = mx$. We will show that it is also the only admissible extremal if $m < -1$ or $m > 0$.

Figure 5.4, which gives a graph of $F(x, y, v) = v^2(1+v)^2$ as a function of $v$, shows only one double tangent with two points of tangency at $v = -1$ and $v = 0$ for the figurative of $F$. It follows from [V.2] that for a broken extremal to render $J[y]$ a strong minimum for this problem, the slopes at a corner must be 0 and $-1$, precisely the values of $\widehat{y}'$ which make the integrand vanish. No polygonal path with a combination of 0 and $-1$ slopes can get from $(0,0)$ to $(1,m)$ if $m \geq 0$ or $m \leq -1$. Hence, the only admissible extremal for these ranges of values of $m$ is the *smooth extremal* $\widehat{y} = mx$.

The straight extremal $\widehat{y} = mx$, in fact, renders $J[y]$ a strong minimum for $m \geq 0$ and $m \leq -1$ (and, hence, a global minimum as it is the only admissible extremal in these ranges of $m$). It is smooth and, therefore, satisfies (I′). For Legendre's condition, we note that

$$\widehat{F}_{,y'y'} = 2(6m^2 + 6m + 1) = 2(m - m_1)(m - m_2),$$

where

$$\begin{pmatrix} m_1 \\ m_2 \end{pmatrix} = -\frac{1}{2} \pm \frac{\sqrt{3}}{6}.$$

It is not difficult to see that both $m_1$ and $m_2$ are negative with $-1 < m_2 < m_1 < 0$ and $\widehat{F}_{,y'y'} > 0$ for $m < m_2$ and for $m > m_1$. Therefore, (II′) is satisfied for $m < -1$ and $m > 0$. The Jacobi accessory equation for our problem is

$$(12m^2 + 12m + 2)\, h'' = 0.$$

As long as (II′) holds, so does the strengthened Jacobi condition (III′) Finally, the Weierstrass excess function $E(x, u, v, w)$ for the problem can be rearranged to read

$$\begin{aligned} E(x, u, v, w) &= w^2(1+w)^2 - v^2(1+v)^2 - (w - v)(4v^3 + 6v^2 + 2v) \\ &= [w(1+w) - v(1+v)]^2 + 2v(v+1)(w-v)^2. \end{aligned}$$

The quantity $v(v+1)$ is positive if $v < -1$ or $v > 0$ for all $|w| < \infty$. Hence, the strengthened Weierstrass condition (IV′) is also satisfied for $m > 0$ and $m < -1$ (because we can always have $m > v > 0$ and $m < v < -1$, respectively). It follows from [V.5] that $\widehat{y} = mx$ is a strong local minimum (and, hence, a global minimum given that it is the unique admissible extremal) for these ranges of $m$.

Conditions (I′), (II′), and (III′) are also satisfied in $-1 < m < m_2 (< 0)$ and $(-1 <) m_1 < m < 0$; but (IV′) is not. It follows from [IV.6] that $\widehat{y} = mx$ is a weak local minimum. A similar argument applied to $-J[y]$ shows that

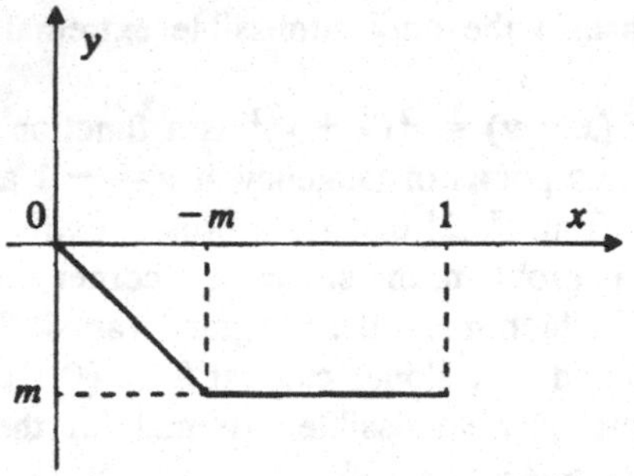


Figure 5.7a

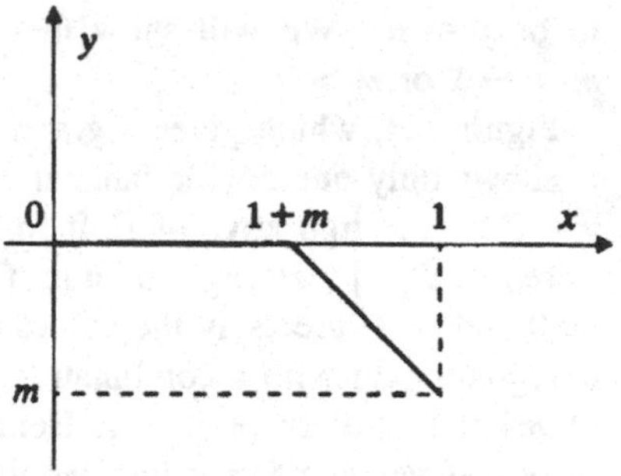


Figure 5.7b

$J[mx]$ is a weak local maximum for $m_2 < m < m_1$. However, neither of these results is of much interest because there are infinitely many admissible broken extremals $\hat{y}$ in the range $-1 \le m \le 0$ for which $J[\hat{y}] = 0$. Two such admissible broken extremals (Fig. 5.7) are

$$\text{a)} \quad \hat{y} = \begin{cases} -x & (0 \le x \le -m) \\ m & (-m \le x \le 1), \end{cases}$$

$$\text{b)} \quad \hat{y} = \begin{cases} 0 & (0 \le x \le 1 + m) \\ 1 + m - x & (1 + m \le x \le 1). \end{cases}$$

Given $J[y] \ge 0$, each of these two broken admissible extremals is a global minimum of the problem.

## 7. Hilbert's Integral

There is another way to motivate and derive the sufficient condition [V.5] for a strong minimum. This alternate approach makes use of the path-independent nature of the loop integral in (5.10). An integral

$$I[y] = \int_a^b G(x, y, y')\,dx \tag{5.17}$$

is said to be *path independent* if, for two arbitrary functions $y_1(x)$ and $y_2(x)$ with $y_1(a) = y_2(a)$ and $y_1(b) = y_2(b)$, we have $I[y_1] = I[y_2]$. The value of the integral $I$, therefore, depends only on the end points of the curve $y(x)$ and not on the path of integration determined by the choice of $y(x)$ and not on the path of integration determined by the choice of $y(x)$.

A trivial example of a path-indenpendent integral is

$$I[y]\int_a^b y'(x)\,dx = y(b) - y(a)$$

with $G(x, y, y') = y'$. Another example is

$$G(x, y, y') = m(x, y) + n(x, y)y'$$

with $m_{,y} = n_{,x}$. For this example, the integral (5.17) becomes

$$I[y] \equiv \int_a^b (m + y'n)\,dx = \int_C [m(x, y)\,dx + n(x, y)\,dy],$$

where the path of integration, $C$, of the line integrals is the graph of $y(x)$ connecting the two end points $(a, y(a))$ and $(b, y(b))$. Because $m_{,y} = n_{,x}$ and all functions involved are assumed to be well defined in a simply-connected region $R$ of the $x, y$-plane which contains $C$, we know from Green's theorem that $I[y]$ depends on the end points of $C$ only and not on $C$ [*or* $y(x)$] itself.

Suppose the admissible extremal $\widehat{y}(x)$ minimizes $J[y]$ and there is a path-independent integral $I[y]$ for which $J[\widehat{y}] = I[\widehat{y}]$. With $I[\widehat{y}] = I[y]$ for any admissible $y(x)$ of $I[y]$, we have

$$\begin{aligned} J[y] - J[\widehat{y}] &= J[y] - I[\widehat{y}] = J[y] - I[y] \\ &= \int_a^b [F(x, y, y') - G(x, y, y')]\,dx \end{aligned} \tag{5.18}$$

Because $\widehat{y}$ minimizes $J[y]$, we must have

$$J[y] - I[y] \geq 0. \tag{5.19}$$

Thus, we have replaced a comparison of $J[y]$ along two different paths of integration between the same two end points by one for two integrals $J$ and $I$ along the same path. It turns out that the latter is more tractable for our purpose.

To work with (5.19), we need a suitable path-independent integral $I[y]$. The development of Section 5, particularly (5.9), suggests that we consider the Hilbert integral

$$I_H[y] = \int_C [M(x, y)\,dx + N(x, y)\,dy] \tag{5.20a}$$

where

$$M(x, y) = F(x, y, s(x, y)) - s(x, y)F_{,y'}(x, y, s(x, y)) \tag{5.20b}$$

$$N(x, y) = F_{,y'}(x, y, s(x, y)) \tag{5.20c}$$

Assuming the existence of an appropriate field of extremals in which $\hat{y}(x)$ can be embedded, we take $s(x,y)$ to be the slope of the unique extremal which passes through the point $(x,y)$. It has already been shown in Section 5 that $M_{,y} = N_{,x}$ for this special case. By Green's theorem, $I_H[y]$ is path independent in the region of the field of extremals.

For any PWS function $y$, we may write $I_H[y]$ as

$$I_H[y] = \int_a^b [M(x,y(x)) + N(x,y(x))y'(x)]\,dx$$

$$\equiv \int_a^b [G_H(x,y(x),y'(x))\,dx. \tag{5.21}$$

With

$$\begin{aligned} G_H(x,\hat{y},\hat{y}') &= M(x,\hat{y}(x)) + N(x,\hat{y}(x))\hat{y}'(x) \\ &= F(x,\hat{y}(x), s(x,\hat{y}(x)) - s(x,\hat{y}(x))F_{,y'}(x,\hat{y}'(x), s(x,\hat{y}(x)) \\ &\quad + F_{,y'}(x,\hat{y}(x),s(x,\hat{y}(x)))\hat{y}'(x) \end{aligned}$$

and $s(x,\hat{y}(x)) = \hat{y}'(x)$, we have

$$G_H(x,\hat{y}(x),\hat{y}'(x)) = F(x,\hat{y}(x),\hat{y}'(x)) \tag{5.22}$$

so that

$$I_H[\hat{y}] = J[\hat{y}] \tag{5.23}$$

It follows from (5.23) and $I_H$ being path independent that

$$\begin{aligned} J[y] - J[\hat{y}] &= J[y] - I_H[y] \\ &= \int_a^b [F(x,y,y') - G_H(x,y,y')]\,dx \\ &= \int_a^b [F(x,y(x),y'(x)) - F(x,y(x),s(x,y(x))) \\ &\qquad - \{y'(x) - s(x,y(x))\}F_{,y'}(x,y(x),s(x,y(x)))]\,dx \\ &= \int_a^b E(x,y(x),s(x,y(x)),y'(x))\,dx \end{aligned} \tag{5.24}$$

which is just (5.14). Hence, the development in Section 5 following that equation may be invoked to give the sufficiency theorem [V.5] for a strong minimum.

## 8. Several Unknowns

For a Lagrangian $F(x, \mathbf{y}, \mathbf{y}')$ in several unknowns $\mathbf{y} = (y_1, \ldots, y_n)^T$, the Weierstrass excess function is still as defined in (5.3) except now $F_{,v}(x, y, v)$ is replaced by the gradient

$$F_{,\mathbf{v}}(x, \mathbf{y}, \mathbf{v}) = (F_{,v_1}, F_{,v_2}, \ldots, F_{,v_n}) \tag{5.25}$$

where $\mathbf{v} = (v_1, \ldots, v_n)^T$, and $(w - v)F_{,y'}(x, y, v)$ is replaced by the scalar product

$$F_{,\mathbf{v}}(x, \mathbf{y}, \mathbf{v})(\mathbf{w} - \mathbf{v}) = \sum_{i=1}^{n} (w_i - v_i) F_{,v_i}(x, \mathbf{y}, \mathbf{v}). \tag{5.26}$$

In this way, we have as the definition of the excess function

$$E(x, \mathbf{y}, \mathbf{v}, \mathbf{w}) = F(x, \mathbf{y}, \mathbf{w}) - F(x, \mathbf{y}, \mathbf{v}) - F_{,\mathbf{y}'}(x, \mathbf{y}, \mathbf{v})(\mathbf{w} - \mathbf{v}). \tag{5.27}$$

The analogue of [V.1] is then

**[V.1′]** *If $\hat{\mathbf{y}}(\mathbf{x})$ is an admissible extremal of the basic problem for several unknowns and $\mathbf{J}[\hat{\mathbf{y}}]$ is a strong local minimum, then we must have*

(VIII) $$\mathbf{E}(\mathbf{x}, \hat{\mathbf{y}}(\mathbf{x}), \hat{\mathbf{y}}'(\mathbf{x}), \mathbf{w}) \geq 0$$

*for all $\mathbf{x}$ in $[\mathbf{a}, \mathbf{b}]$ and all finite vector $\mathbf{w}$ [with the usual convention at a corner point of $\hat{\mathbf{y}}(x)$].*

EXAMPLE 5: $F(x, \mathbf{y}, \mathbf{y}') = [(y_1')^2 + (y_2')^2] - [k_{11}y_1^2 + k_{12}y_1y_2 + k_{22}y_2^2]$.

For this Lagrangian, we have

$$\begin{aligned} E(x, \hat{\mathbf{y}}, \hat{\mathbf{y}}', \mathbf{w}) &= (w_1^2 + w_2^2) - [(\hat{y}_1')^2 + (\hat{y}_1')^2] - (\mathbf{w} - \hat{\mathbf{y}}')\cdot(2\hat{\mathbf{y}}') \\ &= (w_1^2 + w_2^2) + [(y_1')^2 + (y_2')^2] - (2w_1y_1' + 2w_2y_2') \\ &= (w_1 - y_1')^2 + (w_2 - y_2')^2 \end{aligned}$$

which is non-negative for all finite vectors $\mathbf{w} = (w_1, w_2)^T$. Hence, the Weierstrass condition (VIII) is satisfied by any extremal $\hat{\mathbf{y}}$ of the given Lagrangian.

It is also true in the case of $n$ unknowns that for a smooth extremal which can be imbedded in a (central or proper) field over a simply-connected region $R$, we can write $J[\mathbf{y}] - J[\hat{\mathbf{y}}]$ in terms of the excess function (see Gelfand and Fomin, 1963):

$$J[\mathbf{y}] - J[\hat{\mathbf{y}}] = \int_a^b E(x, \mathbf{y}(x), \mathbf{s}(x, \mathbf{y}(x)), \mathbf{y}'(x))\, dx, \tag{5.28}$$

where $s_i(x, \mathbf{y}(x))$ is the slope of the unique $\hat{y}_i(x)$ passing through the point $(x, \mathbf{y}(x))$. We have then the following analogue of [V.3]:

**[V.3′]** ***If the smooth extremal*** $\hat{\mathbf{y}}$ ***is embeddable in a (central or proper) field over a simply-connected*** R ***and if*** $F_{,\mathbf{y}'\mathbf{y}'}(\mathbf{x}, \mathbf{y}, \mathbf{w})$ ***is non-negative definite for all*** $x$, **y** ***in*** R ***and all finite vector*** **w**, ***then*** $J[\hat{\mathbf{y}}]$ ***is a strong local minimum.***

This result follows from an application of Taylor's theorem to $F(x, \mathbf{y}, \mathbf{y}')$ in (5.28) to get

$$E(x, \mathbf{y}(x), \mathbf{s}(x, \mathbf{y}(x)), \mathbf{y}'(x))$$

$$= \frac{1}{2} F_{,\mathbf{y}'\mathbf{y}'}(x, \mathbf{y}(x), \mathbf{w})[\mathbf{y}' - \mathbf{s}(x, \mathbf{y}(x))] \cdot [\mathbf{y}' - \mathbf{s}(x, \mathbf{y}(x))]$$

$$= \frac{1}{2} \sum_{i=1}^{n} \sum_{k=1}^{n} F_{,y_i'y_k'}(x, \mathbf{y}(x), \mathbf{w})\, [y_i'(x) - s_i(x, \mathbf{y}(x))]\, [y_k'(x) - s_k(x, \mathbf{y}(x))]$$

for some vector **w** which satisfie $|w_i - s_i(x, \mathbf{y}(x))| \leq |y_i'(x) - s_i(x, \mathbf{y}(x))|$, $i = 1, \ldots, n$. The non-negative definiteness of the Hessian matrix $F_{,\mathbf{y}'\mathbf{y}'}(x, \mathbf{y}(x), \mathbf{w})$ implies

$$E(x, \mathbf{y}(x), s(x, \mathbf{y}(x)), \mathbf{y}'(x)) \geq 0$$

so that $J[\mathbf{y}] \geq J[\hat{\mathbf{y}}]$ as required.

As in the single unknown case, embeddability is assured by the following analogue of [V.4]:

**[V.4′]** $\hat{\mathbf{y}}(x)$ ***is embeddable in a (central or proper) field of extremals over a simply-connected region*** R ***if*** F ***is*** $C^4$ ***and*** $\hat{\mathbf{y}}$ ***satisfies (V′), (VI′), and (VII′).***

The strengthened Weierstrass condition for several unknowns is defined by

(VIII′) For all $x$ in $[a, b]$, all **u** close to $\hat{\mathbf{y}}(x)$, all **v** close to $\hat{\mathbf{y}}'(x)$ and all finite vector **w**, we have

$$E(x, \mathbf{u}, \mathbf{v}, \mathbf{w}) \geq 0.$$

The following is an analogue of [V.5]:

**[V.5′]** ***If*** F ***is*** $C^4$ ***and the admissible extremal*** $\hat{\mathbf{y}}(\mathbf{x})$ ***satisfies (V′), (VI′), (VII′), and (VIII′), then*** $J[\hat{\mathbf{y}}]$ ***is a strong local minimum.***

The proof of [V.4′] and [V.5′] is similar to the corresponding proof for the scalar case.

**EXAMPLE 5 (continued):**

For this Lagrangian, we have

$$E(x, \mathbf{u}, \mathbf{v}, \mathbf{w}) = (w_1 - v_1)^2 + (w_2 - v_2)^2 \geq 0$$

for all $|w_k| < \infty$ and all $|v_l| < \infty$. Hence, the strengthened Weierstrass condition is satisfied.

## 9. Exercises

1. (a) Verify the Weierstrass and Legendre conditions for the Brachistochrone problem.

   (b) Show that the extremals of the basic problem with

   $$F = f(x,y)\sqrt{1+(y')^2} \quad \text{and} \quad f(x,y) > 0$$

   for all $(x, y)$ satisfy both the Legendre and Weierstrass conditions.

2. (a) Does the basic problem with $F = (y')^3$ admit a *strong* local minimum or maximum with an admissible extremal which has a corner?

   (b) Show that the unique extremal for $F = 4y^2 - (y')^2 + 8y$, $y(0) = -1$, and $y(\pi/4) = 0$ makes $J[y]$ a strong minimum.

3. (a) Show that $J[\hat{y}]$ with $\hat{y} = 0$ is not a strong local minimum for the basic problem with $F = (y')^2 - 4y(y')^3 + 2x(y')^4$, $y(0) = 0$, and $y(1) = 0$.

   (b) Show that $\hat{y} = 0$ satisfies (I), (II), (III), and (IV) for the basic problem above.

4. Show that $J[\hat{y}]$ with $\hat{y} = Bx/b$ is a weak minimum but not a strong minimum for $F = (\hat{y}')^3$, $y(0) = 0$, and $y(b) = B > 0$.

5. Consider the basic problem with $F = 6(y')^2 - (y')^4 + yy'$, $y(0) = 0$, and $y(b) = B > 0$.

   (a) Show that there is exactly one admissible extremal for a strong local minimum or maximum for this problem if $B/b \geq \sqrt{3}$.

   (b) Analyze the situation for $0 < B/b < \sqrt{3}$.

6. Discuss the possibility of a strong minimum (or maximum) for the basic problem in Exercises 5, 6, and 7 of Chapter 4.

7. Suppose $F = y^2(y')^2$ in the basic problem with $y(0) = 0$ and $y(1) = 1$. Exhibit an appropriate field for the embedding of the smooth admissible extremal of the problem and that the extremal renders $J[y]$ a strong local minimum.

8. Show that the basic problem with $F = (y')^2 + y^2 + 2ye^{2x}$ has a strong minimum with $\hat{y}(x) = e^{2x}/3$ joining the end points $y(0) = 1/3$ and $y(1) = e^2/3$.

9. Show that the unique admissible extremal of the minimum cost flight path of Chapter 1, Exercise 5 (with $D < \pi H$) makes $J[y]$ a strong local minimum.

10. Suppose $F = x^2(y')^2 + 12y^2$ for a basic problem with $y(-1) = -1$ and $y(1) = 1$. Show that $J[y]$ has a strong minimum at $\hat{y} = x^3$. How do the results change if the end conditions are replaced by $y(1) = 1$ and $y(2) = 8$?

11. (a) Find the curve joining the points $(a,A) = (1,3)$ and $(b,B) = (2,5)$ which minimizes $J[y]$ with $F = y'(1 + x^2y')$. What is the nature of the minimum?
    (b) Repeat part (a) with $(a,A) = (-1,1)$ and $(b,B) = (2,1)$.

12. A particle moves in the plane in such a way that its speed is proportional to the square of its distance from the origin. What is the minimum time path $\hat{y}(x)$ between two points distinct from the origin and not colinear with it? Prove that $J[\hat{y}]$ is a strong local minimum. [Hint: One approach uses polar coordinates.]

13. For the basic problem with $F = (y')^2 - y^2$, $y(0) = 0$, and $y(n\pi) = 0$ where $n$ is a positive integer, $\hat{y}_0(x) \equiv 0$ is an admissible extremal.
    (a) Show that $\hat{y}_0$ satisfies the necessary conditions of Legendre and Weierstrass.
    (b) Show that $\hat{y}_0$ satisfies the strengthened conditions of Legendre and Weierstrass.
    (c) Show that $J[\hat{y}_0]$ is not even a weak local minimum for $n > 1$.

14. $F(x, y, y') = (y')^2 - y^2 + 4y\cos x, \qquad y(0) = 0, \qquad y(b) = B.$
    (a) Obtain a family of smooth extremals $y(x, \alpha)$.
    (b) Verify that $g_{,\alpha}(x, \hat{\alpha})$ solves the Jacobi accessory equation [relative to $\hat{y}(x)$].
    (c) Show that the points conjugated to $a$ are the zeros of the function $g_{,\alpha}(x, \hat{\alpha})$.

## Appendix: Proof of the Weierstrass Necessary Condition

To prove [V.1], we consider a subinterval $[c,d]$ of $[a,b]$ (with $a<c<d<b$) in which $\widehat{y}$ has no corner. Let $z$ be a point in $[c,d]$ and

$$y(x)=\begin{cases} Y(x) & (c\le x\le z) \\ \psi(x;z) & (z\le x\le d) \\ \widehat{y}(x) & (\text{elsewhere in } [a,b], \end{cases}$$

where

$$Y(x)=\widehat{y}(c)+w(x-c),$$

$$\psi(x;z)=\widehat{y}(x)+[Y(z)-\widehat{y}(z)]\frac{d-x}{d-z}$$

for any real number $w$. In that case, we have

$$\Phi(z)\equiv J[y]-J[\widehat{y}]=\int_c^d [F(x,y,y')-F(x,\widehat{y},\widehat{y}')]\,dx$$

$$=\int_c^z [F(x,Y,Y')-F(x,\widehat{y},\widehat{y}')]\,dx+\int_z^d [F(x,\psi,\psi_{,x})-F(x,\widehat{y},\widehat{y}')]\,dx.$$

Note that as a function of $z$, we have

$$\Phi(c)=\int_c^d [F(x,\psi(x;c),\psi_{,x}(x;c))-F(x,\widehat{y}(x),\widehat{y}'(x))]\,dx=0$$

which follows from

$$Y(c)=\widehat{y}(c)\quad\text{so that}\quad \psi(x;c)=\widehat{y}(x)\quad\text{and}\quad \psi_{,x}(x;c)=\widehat{y}'(x).$$

By hypotheses, $\widehat{y}$ minimizes $J[y]$ so that $\Phi(z)\ge 0$. With $\Phi(c)=0$, $\Phi(z)$ must be non decreasing at $z=c$. In other words, we must have $\Phi^{\bullet}(c)\ge 0$. It remains to show that $\Phi^{\bullet}(c)=E(c,\widehat{y}(c),\widehat{y}'(c),w)$. Now, we have

$$\Phi^{\bullet}(z)=\frac{d}{dz}\int_c^z [F(x,Y(x),Y'(x))-F(x,\widehat{y}(x),\widehat{y}'(x))]\,dx$$

$$+\frac{d}{dz}\int_z^d [F(x,\psi(x;z),\psi_{,x}(x;z))-F(x,\widehat{y}(x),\widehat{y}'(x))]\,dx$$

$$=[F(z,Y(z),Y'(z))-\widehat{F}(z)]-[F(z,\psi(z;z),\psi_{,x}(z;z))-\widehat{F}(z)]$$

$$+\int_z^d [F_{,y}(x,\psi,\psi_{,x})\psi_{,z}+[F_{,y'}(x,\psi,\psi_{,x})\psi_{,xz}]\,dx$$

$$= F(z, Y(z), w) - F(z, Y(z), \psi_{,x}(z;z)) + [F_{,y'}(x, \psi, \psi_{,x})\psi_{,z}(x;z)]_z^d$$

$$+ \int_z^d \{F_{,y}(x, \psi, \psi_{,x}) - [F_{,y'}(x, \psi, \psi_{,x})]'\}\psi_{,z}(x;z)\,dx.$$

At $z = c$, we have

$$Y(c) = \widehat{y}(c), \qquad \psi_{,x}(c;c) = \left[\widehat{y}'(z) - \frac{Y(z) - \widehat{y}(z)}{\mathrm{d} - \mathrm{z}}\right]_{z=c} = \widehat{y}'(c),$$

$$\psi_{,z}(d;c) = 0, \qquad \psi_{,z}(c,c) = w - \widehat{y}'(c)$$

and therewith

$$\begin{aligned}\Phi'(c) &= F(c, \widehat{y}(c), w) - F(c, \widehat{y}(c), \widehat{y}'(c)) \\ &\quad - [F_{,y'}(z, \psi(z;z), \psi_{,x}(z;z), \psi_{,z}(z;z)]_{z=c} \\ &\quad + \int_c^d \{F_{,y}(x, \psi(x;c), \psi_{,x}(x;c)) \\ &\qquad - [F_{,y'}(x, \psi(x;c), \psi_{,x}(x;c))]'\}\psi_{,z}(x;c)\,dx \\ &= E(c, \widehat{y}(c), \widehat{y}'(c), w) + \int_c^d \{\widehat{F}_{,y}(x) - [\widehat{F}_{,y'}(x)]'\}\psi_{,z}(x;c)\,dx.\end{aligned}$$

The integrand vanishes as $\widehat{y}(x)$ satisfies Euler DE. [V.1] for a noncorner point $c$ then follows from the fact that $\Phi(z)$ is nondecreasing at $z = c$.

The Weierstrass excess function $E(x, \widehat{y}(x), \widehat{y}'(x), w)$ is a continuous function of $x$ between corner points or between a corner point and an end point. By taking the appropriate one-sided limits, we get [V.1] also for a corner point or an end point.

# 6

# The Hamiltonian

## 1. The Legendre Transformation and Hamiltonian Systems

The partial derivative of the Lagrangian $F(x, y, v)$ with respect to $v$, $F_{,v}(x, y, v)$ or $F_{,y'}(x, y, y')$, has played an important role in the development of the theory of the calculus of variations up to this point. It enters into the Euler–Lagrange equation and both Erdmann corner conditions as well as Euler boundary conditions and transversality conditions. Traditionally, mathematicians, scientists and engineers have found it convenient to give it a separate label by setting.

$$p = F_{,y'}(x, y, y'). \tag{6.1}$$

It is often desirable to work with $(x, y, p)$ instead of $(x, y, y')$. In that case, we solve (6.1) for $y'$ in terms of $p, x$, and $y$ and write

$$y' = v(x, y, p). \tag{6.2}$$

**EXAMPLE 1:** $F(x, y, y') = f(x, y)\sqrt{1 + (y')^2}$.

We have for this case

$$F_{,y'} = \frac{y'f}{\sqrt{1 + (y')^2}} = p \qquad \text{or} \qquad (y')^2(f^2 - p^2) = p^2,$$

so that

$$y' = \pm\frac{p}{\sqrt{f^2 - p^2}}.$$

For $f(x, y) > 0$, the expression for $F_{,y'}$ requires $y'$ and $p$ to have the same sign; we, therefore, set

$$y' = \frac{p}{\sqrt{f^2 - p^2}} \equiv v(x, y, p). \tag{6.3}$$

We now introduce the quantity $H(x, y, p)$ by the defining relation

$$H(x, y, p) \equiv py' - F(x, y, y') = pv(x, y, p) - F(x, y, v(x, y, p)), \tag{6.4}$$

where (6.2) has been used to express $y'$ in terms of $x$, $y$, and $p$. $H(x, y, p)$ is called the *Hamiltonian* of the problem and relation (6.4) is called the ***Legendre transformation***. Note that the expression on the right hand side of (6.4) appears in the alternative form (2.23) of the Euler DE (see also (1.20)), the second Erdmann condition and various transversality conditions (see (3.21) and (3.29)).

**EXAMPLE 1 (continued):**

We have from (6.3) and (6.4)

$$H(x, y, p) = pv - F(x, y, v) = \frac{p^2}{\sqrt{f^2 - p^2}} - f(x, y)\sqrt{1 + \frac{p^2}{f^2 - p^2}}$$

$$= -\sqrt{[f(x, y)]^2 - p^2}.$$

The aim of this chapter is to exploit the advantages of working with the quantity $p$ instead of $y'$ and to rephrase developments in the calculus of variations in terms of the Hamiltonian (instead of the Lagrangian) with the help of the Legendre transformation (6.4). We begin by writing the Euler differential equation for a *smooth extremal* $\hat{y}(x)$ in terms of the Hamiltonian:

**[IV.1]** *In terms of the Hamiltonian* $\mathrm{H}(\mathrm{x}, \mathrm{y}, \mathrm{p})$, *the smooth extremal* $\hat{\mathrm{y}}(\mathrm{x})$ *of the Lagrangian* $\mathrm{F}(\mathrm{x}, \mathrm{y}, \mathrm{y}')$ *and the corresponding* $\hat{\mathrm{p}}(\mathrm{x}) = \mathrm{F}_{,y'}(\mathrm{x}, \hat{\mathrm{y}}(\mathrm{x}), \hat{\mathrm{y}}'(\mathrm{x})) \equiv \hat{\mathrm{F}}_{,y'}(\mathrm{x})$ *satisfy the Hamiltonian system*

$$\hat{\mathrm{y}}' = \hat{\mathrm{H}}_{,p}, \qquad \hat{\mathrm{p}}' = -\hat{\mathrm{H}}_{,y}.$$

*where* $\hat{\mathrm{f}}(\mathrm{x}) \equiv \mathrm{f}(\mathrm{x}, \hat{\mathrm{y}}(\mathrm{x}), \hat{\mathrm{p}}(\mathrm{x}))$.

To verify [VI.1], we simply compute $H_{,p}$ and $\hat{H}_{,y}$ to get

$$\hat{H}_{,p} = \hat{v} + \hat{p}\hat{v}_{,p} - \hat{F}_{,y'}\hat{v}_{,p} = \hat{v} = \hat{y}',$$

$$\hat{H}_{,y} = \hat{p}\hat{v}_{,y} - \hat{F}_{,y} - \hat{F}_{,y'}\hat{v}_{,y} = -\hat{F}_{,y} = -\frac{d}{dx}[\hat{F}_{,y'}] = -\hat{p}',$$

where (6.2) was used in the first relation and the Euler DE was used in the second.

The Hamiltonian system is customarily called the canonical form of the Euler differential equation and $\{y, p\}$ the *canonical variables*.

**EXAMPLE 2:** $F = \frac{1}{2}[m(y')^2 - ky^2]$.

This Lagrangian corresponds to a spring-mass system with the Euler differential equation $m\widehat{y}'' + k\widehat{y} = 0$ being the equation of motion of the (point) mass. (The independent variable $x$ here is time). From the expression for $F$, we obtain $F_{,y'} = my' = p$ or $y' = p/m \equiv v(x, y, p)$ and, correspondingly,

$$H = pv - \frac{1}{2}[m(v)^2 - ky^2] = \frac{p^2}{m} - \left[\frac{1}{2}\frac{p^2}{m} - ky^2\right] = \frac{1}{2}\left[\frac{p^2}{m} + ky^2\right].$$

It follows from [VI.1] that

$$\widehat{y}' = \widehat{H}_{,p} = \frac{\widehat{p}}{m}, \qquad -\widehat{p}' = \widehat{H}_{,y} = k\widehat{y}.$$

Note that upon differentiating the first relation and using the result to eliminate $\widehat{p}'$ from the second, we get $m\widehat{y}'' + k\widehat{y} = 0$, which is identical to the Euler DE for the given Lagrangian.

In Example 2, $p$ is the momentum and $H$ is the total energy of the mass in motion. This is also the case for any Lagrangian associated with a dynamical system of mass particles. With this observation.

*The Erdmann first and second corner condition can be interpreted as conservation of momentum and energy, respectively, across a discontinuity of the relevant velocity.*

More generally, whenever the evolution of a dynamic phenomenon is characterized by two variables $y(x)$ and $p(x)$ (which may be vector quantities) and governed by the system of first-order ODE

$$y' = H_{,p}, \qquad -p' = H_{,y} \tag{6.5}$$

for some $H = H(x, y, p)$, the dynamical system is called a *Hamiltonian system*. Hamiltonian systems arise naturally in mathematical modeling of dynamic phenomena which are not necessarily related to Newtonian mechanics or variational considerations. Hence, a theory for Hamiltonian systems will also find application in areas other than the calculus of variations. Because of their conventional role in particle dynamics, $p$ and $H$ are sometimes referred to as the (generalized) momentum and energy of the system whatever the actual modeled phenomenon may be.

Among all dynamical systems, those characterized by autonomous ODE have some special properties. (When the independent variable $x$ does not appear explicitly, a system of ordinary differential equations is called an *autonomous* system). For autonomous Hamiltonian systems, we have the following conservation law:

**[VI.2]** ***If the Hamiltonian does not depend on*** $\mathbf{x}$ ***explicitly so that*** $\mathrm{H} = \mathrm{H}(\mathrm{y}, \mathrm{p})$, ***and if the pair*** $\{\bar{\mathrm{y}}(\mathrm{x}), \bar{\mathrm{p}}(\mathrm{x})\}$ ***is a solution of the Hamiltonian system (6.5), then*** $\bar{\mathrm{H}}(\mathrm{x}) = \mathrm{H}(\bar{\mathrm{y}}(\mathrm{x}), \bar{\mathrm{p}}(\mathrm{x}))$ ***is a constant, i.e.,*** $\bar{\mathrm{H}}(\mathrm{x}) = \mathrm{H}_0$ ***for some constant*** $\mathrm{H}_0$.

This follows immediately from

$$\frac{d\overline{H}}{dx} = \overline{H}_{,y}\bar{y}' + \overline{H}_{,p}\bar{p}' = \overline{H}_{,y}\overline{H}_{,p} + \overline{H}_{,p}(-\overline{H}_{,y}) = 0.$$

Conclusion [VI.2] should not be surprising given that

$$H = pv - F = -[F - y'F_{,y'}]$$

and the right-hand side is known from [I.4] to be a constant when the Lagrangian $F$ does not depend on $x$ explicitly. Given that $H$ is the (generalized) energy of the dynamical system, [VI.2] simply asserts that the energy of the system is conserved. A dynamical system with a Hamiltonian not explicitly dependent on $x$ is called a *conservative* system.

## 2. Hamilton's Principle

One of the most important application areas of the calculus of variations is classical mechanics, the study of the evolution (and stability) of dynamic phenomena involving objects which may be idealized as particles or rigid bodies. (In later chapters of this book, we will discuss the applications of the calculus of variations to continuum mechanics, the study of the mechanical behavior of deformable bodies). The state of an evolving phenomenon is characterized by a number of dynamic variables such as position and velocity. The evolution of the phenomenon is determined by a system of ODE (and suitable initial conditions) governing these dynamic variables. The appropriate ODE may be deduced from the physical laws (such as Newton's laws of motion) postulated for the phenomenon. We are interested here in an alternative method of obtaining the governing equations using some minimum (maximum or stationary) principle expected of the phenomenon. One of the most useful variational principles in classical (analytical) mechanics, which can be generalized to include many areas of physics, is Hamilton's principle:

> *The evolution of a dynamical system from time* $\mathrm{t} = \mathrm{t}_1$ *to a later time* $\mathrm{t} = \mathrm{t}_2$ *is always such that the integral of the difference between kinetic and potential energy of the system over the elapsed time interval* $[\mathrm{t}_1, \mathrm{t}_2]$ *is stationary.*

The stationary value is actually a minimum if the inteval $[t_1, t_2]$ does not contain a point conjugate to $t_1$.

To illustrate the application of this principle (wich is just a postulate), consider the two-dimensional motion of a single particle whose position in the plane of motion is characterized by the coordinate functions $y^{(1)}(t)$ and $y^{(2)}(t)$ with $\mathbf{y} = (y^{(1)}, y^{(2)})^T$. The kinetic energy of this particle at time $t$ is $KE = m\{[\dot{y}^{(1)}]^2 + [\dot{y}^{(2)}]^2\}/2 = m\dot{\mathbf{y}}\cdot\dot{\mathbf{y}}/2 \equiv m\|\dot{\mathbf{y}}\|^2/2$ where $(\dot{\ }) \equiv d(\ \ )/dt$. (The *norm* of a vector $\mathbf{z}$, denoted by $\|\mathbf{z}\|$, is the positive square root of the scalar product $\mathbf{z}\cdot\mathbf{z}$). The potential energy is generally some scalar function of $\mathbf{y}$, $\dot{\mathbf{y}}$ and $t$, so that $PE = V(t, y^{(1)}, y^{(2)}, \dot{y}^{(1)}, \dot{y}^{(2)}) = V(t, \mathbf{y}, \dot{\mathbf{y}})$. In this section, we limit our discussion to potentials which depend only on $\mathbf{y} = (y^{(1)}, y^{(2)})^T$ for 2-dimensional motion of a single particle so that $V = V(\mathbf{y})$. Dynamical systems with such a potential will be seen to be *conservative* systems. Hamilton's principle postulates for the two-dimensional motion of this single-mass particle:

> ***The motion of a system evolving from an initial state at time $t_1$ to another state at time $t_2$ renders the quantity***
>
> $$J[\mathbf{y}] = \int_{t_1}^{t_2} \left[\frac{1}{2} m\|\dot{\mathbf{y}}\|^2 - V(\mathbf{y})\right] dt, \qquad (6.6)$$
>
> ***a stationary value, i.e., $\delta J = 0$.***

In other words, the actual motion of a dynamical system is given by an admissible extremal of $J[\mathbf{y}]$ and the two Euler DE determining the extremal are the equations of motion for the system. The independent variable $t$ does not appear explicitly in these equations. The version of Hamilton's principle as stated above applies also to the motion of a single mass in one or three dimensions if we interpret $\mathbf{y}$ to have one or three components.

To construct the Hamiltonian corresponding to the Lagrangian of (6.6), we observe that

$$F_{,\dot{y}^{(k)}} = m\dot{y}^{(k)} \equiv p^{(k)} \qquad \text{or} \qquad \dot{y}^{(k)} = \frac{p^{(k)}}{m} \equiv v^{(k)}(t, \mathbf{y}, \mathbf{p}).$$

In that case, we have $\mathbf{p} \equiv (p^{(1)}, p^{(2)})^T$, $\mathbf{v} \equiv (v^{(1)}, v^{(2)})^T = (p^{(1)}, p^{(2)})^T/m$, and $\mathbf{y} = (y^{(1)}, y^{(2)})^T$ so that

$$H = \mathbf{p}\cdot\mathbf{v} - F = \frac{1}{m}\sum_{k=1}^{2}[p^{(k)}]^2 - \frac{1}{2m}\sum_{k=1}^{2}[p^{(k)}]^2 + V(y^{(1)}, y^{(2)})$$

$$= \frac{1}{2m}\{[p^{(1)}]^2 + [p^{(2)}]^2\} + V(\mathbf{y}) \equiv \frac{1}{2m}\|\mathbf{p}\|^2 + V(\mathbf{y}).$$

For the more general three-dimensional motion of $n$ particles of mass $m_k$, $k=1,\dots,n$, we let $\mathbf{y}_k=(y_k^{(1)},y_k^{(2)},y_k^{(3)})^T$ and $\dot{\mathbf{y}}_k=\mathbf{v}_k=\mathbf{p}_k/m_k=(p_k^{(1)},p_k^{(2)},p_k^{(3)})^T/m_k$ be the position and velocity vector of the $k$-th particle. Then the kinetic and potential energy of the dynamical system (in a conservative field) are

$$KE=\frac{1}{2}\sum_{i=1}^{n}m_i\|\mathbf{v}_i\|^2=\frac{1}{2}\sum_{i=1}^{n}m_i\,[(v_i^{(1)})^2+(v_i^{(2)})^2+(v_i^{(3)})^2], \tag{6.7a}$$

$$PE=V(\mathbf{y})\equiv V(\mathbf{y}_1,\mathbf{y}_2,\dots,\mathbf{y}_n)\equiv V(y_1^{(1)},y_1^{(2)},y_1^{(3)},y_2^{(1)},\dots,y_n^{(3)}), \tag{6.7b}$$

respectively. Let

$$J[\mathbf{y}]=\int_{t_1}^{t_2}[KE-PE]\,dt. \tag{6.7c}$$

The integrand of $J[\mathbf{y}]$ is the excess of kinetic energy over potential energy and is called the ***Lagrangian function*** or simply the ***Lagrangian*** in honor of Lagrange who contributed so much to classical mechanics. The term has since been used for the integrand of $J[\mathbf{y}]$ for any calculus of variations problem. Hamilton's principle asserts that the actual motion $\widehat{\mathbf{y}}(t)$ of the particles evolving from $\mathbf{y}(t_1)=\mathbf{A}$ to $\mathbf{y}(t_2)=\mathbf{B}$ makes $J$ a stationary value, that is, $\widehat{\mathbf{y}}^T=(\widehat{\mathbf{y}}_1^T,\dots,\widehat{\mathbf{y}}_n^T)$ is an extremal of the basic problem for $J[\mathbf{y}]$.

For the case where the force field is gravity in the $-y^{(3)}$ direction of the physical space, the potential energy can be taken as

$$V(\mathbf{y})=-\sum_{i=1}^{n}g\,m_i\,y_i^{(3)}, \tag{6.8}$$

where $g$ is the gravitational acceleration. The corresponding Hamiltonian system in vector form is

$$-\dot{\mathbf{p}}_i=(0,0,m_i g)^T,\qquad \dot{\mathbf{y}}_i=\frac{\mathbf{p}_i}{m_i}\quad (i=1,2,\dots,n). \tag{6.9}$$

The second set of equations show that $\mathbf{p}_i$ is the momentum vector of the $i$th particle. Upon using these equations to eliminate $\mathbf{p}_i$, we get $\ddot{\mathbf{y}}_i=(0,0,-g)^T$ for $i=1,2,\dots,n$. The second-order vector ODE for $\mathbf{y}_i$ is just Newton's second law of motion in a gravity field.

For the $n$-particle system above, we have

$$F_{,\dot{y}_k^{(i)}}=m_k\dot{y}_k^{(i)}=p_k^{(i)}\quad\text{or}\quad \dot{y}_k^{(i)}=p_k^{(i)}/m_k \tag{6.10}$$

and therewith

$$H=\sum_{i=1}^{n}\frac{1}{2}m_i\|\mathbf{v}_i\|^2+V(\mathbf{y})=\sum_{i=1}^{n}\frac{\|\mathbf{p}_i\|^2}{2m_i}+V(\mathbf{y})\equiv KE+PE. \tag{6.11}$$

Thus, the Hamiltonian $H$ is, in fact, the total energy of the $n$-particle system. Because $H$ as given by (6.11) does not depend on $t$ explicitly, we have, from [VI.2], $\dot{H} = 0$ for this conservative system. This allows us to write

$$\begin{aligned} J[y] &= \int_{t_1}^{t_2} (KE - PE)\,dt = \int_{t_1}^{t_2} 2KE\,dt - \int_{t_1}^{t_2} (KE + PE)\,dt \\ &= 2\int_{t_1}^{t_2} KE\,dt - \int_{t_1}^{t_2} H\,dt = 2\int_{t_1}^{t_2} KE\,dt + (t_2 - t_1)H_0. \end{aligned} \tag{6.12}$$

Because the therm $(t_2 - t_1)H_0$ is a fixed constant, Hamilton's principle for a conservative system now extremizes the kinetic energy $KE$ while the total energy $PE + KE$ remains fixed. In this form, it coincides with Maupertuis' *principle of least action.*

With the machinery developed in the last two chapters, it is possible to show (see Exercises) that Hamilton's principle is, in fact, a minimum principle for a sufficiently short time interval (before the occurrence of a conjugate point).

## 3. Canonical Transformations

By Example 2, the equation of motion for the simple harmonic oscillator can be taken in the form of a Hamiltonian system:

$$-p' = ky = H_{,y}, \tag{6.13a}$$

$$y' = \frac{p}{m} = H_{,p}. \tag{6.13b}$$

with

$$H = \frac{1}{2}\left[\frac{p^2}{m} + ky^2\right]. \tag{6.13c}$$

Suppose we want to work with a new set of variables $Y$ and $P$ instead of $y$ and $p$ where

$$Y = f(x, y, p), \qquad P = g(x, y, p). \tag{6.14}$$

Upon writing the Hamiltonian system (6.13) in terms of $Y$ and $P$, we get a new system of ODE. For example, the change of variables

$$Y = p, \qquad P = -y, \tag{6.15}$$

transforms system (6.13) into

$$Y' = kP, \qquad -P' = \frac{Y}{m} \tag{6.13'}$$

System (6.13′) is a Hamiltonian system with a new Hamiltonian

$$H^* = \frac{1}{2}\left(\frac{Y^2}{m} + kP^2\right). \tag{6.16}$$

For the present problem, $H^*(x, Y, P)$ can be obtained from $H(x, y, p)$ by a direct substitution of (6.15) into expression (6.13c) for $H$. In general, it is constructed from the right-hand side of the transformed system [such as (6.13′)].

A change of variables of the form (6.14) does not always preserve the Hamiltonian structure of an ODE system. As an example, the change of variables

$$Y = e^y, \qquad P = p \tag{6.17}$$

transforms system (6.13) into

$$Y' = \frac{1}{m}PY, \qquad -P' = k\ln Y. \tag{6.13″}$$

System (6.13″) is not a Hamiltonian system. It certainly cannot be derived from

$$H(x, y, p) = \frac{1}{2}\left[\frac{p^2}{m} + ky^2\right] = \frac{1}{2}\left[\frac{P^2}{m} + k(\ln Y)^2\right].$$

If there were a Hamiltonian $H^*(Y, P)$ for which $H^*_{,P} = PY/m$ and $H^*_{,Y} = k\ln Y$, the second relation would require

$$H^* = kY(\ln Y - 1) + h(P)$$

for some function $h$ of the variable $P$ only. Now $H^*_{,P} = dh/dP$ must be a function of $P$ only. The first relation of (6.13″) then requires $H^*_{,P} = PY/m = dh/dP$ which is not possible for any function $h(P)$. It follows that (6.13″) is not a Hamiltonian system.

We call transformation (6.14) a *canonical transformation* if it transforms a Hamiltonian system (in the $y, p$ variables) into another Hamiltonian system (in the new variables $Y$ and $P$). In other words, *a canonical transformation preserves the Hamiltonian structure of the ODE system.* The transformation (6.15) is canonical, whereas the transformation (6.17) is not.

Suppose instead of (6.15) and (6.17), we consider a different change of variables

$$Y = \frac{1}{2}\left[ky^2 + \frac{p^2}{m}\right], \tag{6.18a}$$

$$P = Q(y, p)\left[\sqrt{mk}\,\sin^{-1}\left(\frac{y}{Q}\right) - kx\right] \tag{6.18b}$$

for the simple harmonic oscillator (6.13) where

$$Q(y,p) = \sqrt{y^2 + \frac{p^2}{mk}}. \tag{6.19}$$

To rewrite the Hamiltonian system (6.13) in terms of the new variables $Y$ and $P$, we differentiate both sides of (6.18a) and (6.18b) with respect to $x$ to get

$$Y' = kyy' + \frac{1}{m}pp' = ky\left(\frac{p}{m}\right) + \frac{p}{m}(-ky) = 0, \tag{6.20a}$$

$$P' = Q'\left[\sqrt{mk}\,\sin^{-1}\left(\frac{y}{Q}\right) - kx\right] + Q\left[-k + \frac{\sqrt{mk}}{Q^2}\,\frac{Qy' - yQ'}{\sqrt{1 - \frac{y^2}{Q^2}}}\right].$$

With $Q' = (yy' + pp'/mk)/Q = 0$ [where use has been made of (6.13)], the expression for $P'$ further simplifies to

$$P' = -kQ + \frac{\sqrt{mk}\,Qy'}{\sqrt{Q^2 - y^2}} = kQ\left[m\frac{y'}{p} - 1\right] = 0, \tag{6.20b}$$

where (6.13) has been used for the last equality. Thus, not only is the transformation (6.18) canonical, but the transformed system is so simple that its solution is trivial, namely,

$$Y = Y_0, \qquad P = P_0. \tag{6.21}$$

The solution of the original system now follows immediateluy from (6.18) and (6.21):

$$y^2 + \frac{p^2}{mk} = \frac{2Y_0}{k}, \qquad P_0 = \sqrt{\frac{2Y_0}{k}}\left[\sqrt{mk}\,\sin^{-1}\left(\frac{y}{\sqrt{2Y_0/k}}\right) - kx\right]$$

or

$$y = \sqrt{\frac{2Y_0}{k}}\,\sin\left(\sqrt{\frac{k}{m}}\,x + \mu\right), \qquad p = \sqrt{2Y_0 m}\,\cos\left(\sqrt{\frac{k}{m}}\,x + \mu\right), \tag{6.22}$$

where $\mu = P_0/\sqrt{2mY_0}$.

The above example suggests a different (and somewhat indirect) method for solving the Euler DE of a variational problem or Hamiltonian systems in general:

> ***Find a canonical transformation which transforms the given Hamiltonian system into another Hamiltonian system of a particularly***

*simple form. If the new Hamiltonian does not depend on the new (and old) canonical variables so that* $H^*(x\,Y,P) \equiv H^*(x)$, *then we have* $Y = Y_0$ *and* $P = P_0$.

Unless stated otherwise, we take transformation (6.14) to be invertible so that we may also write.

$$y = \bar{f}(x, Y, P), \qquad p = \bar{g}(x, Y, P). \tag{6.23}$$

Upon differentiating both sides with respect to $x$, we get

$$\begin{aligned} y' &= \bar{f}_{,Y}Y' + \bar{f}_{,P}P' + \bar{f}_{,x} = G_1(x, y, p) = \overline{G}_1(x, Y, P), \\ p' &= \bar{g}_{,Y}Y' + \bar{g}_{,P}P' + \bar{g}_{,x} = G_2(x, y, p) = \overline{G}_2(x, Y, P) \end{aligned} \tag{6.24}$$

with $G_1 = H_{,p}$, $G_2 = -H_{,y}$, and $\overline{G}_1$ and $\overline{G}_2$ obtained from $G_1$ and $G_2$, respectively, by using (6.23) to eliminate $y$ and $p$. We take (6.24) as a linear system for $Y'$ and $P'$ which may be solved to give

$$Y' = \overline{G}_3(x, Y, P), \qquad P' = \overline{G}_4(x, Y, P).$$

For transformation (6.14) or (6.23) to be canonical, we need $\overline{G}_3$ and $\overline{G}_4$ to be derivable from a Hamiltonian $H^*(x, Y, P)$ with $\overline{G}_3 = H^*_{,P}$ and $\overline{G}_4 = -H^*_{,Y}$. For the solution of the original system, we will seek in the next section a transformation for which H* is independent of $Y$ and $P$.

## 4. The Hamilton–Jacobi Equation

To find a special canonical transformation which gives the new Hamiltonian system $Y' = P' = 0$, we will work with a variational version of the problem. The relevant variational problem here is for the two unknowns $y$ and $p$

$$J[y, p] \equiv \int_a^b [py' - H(x, y, p)]\,dx = \int_a^b \overline{F}(x, y, p, y', p')\,dx \tag{6.25}$$

(although $\overline{F}$ actually does not depend on $p'$). The perfomance index $J[y, p]$ may be obtained from the perfomance index $J[y]$ of (1.1) by expressing $F(x, y, y')$ in terms of $H(x, y, p)$ by the Legendre transformation (6.4). Omitting the usual $\frown$ designation for an extremal, we have as the Euler DE for $J[y, p]$

$$\frac{d}{dx}[\overline{F}_{,y'}] = \overline{F}_{,y}, \qquad \frac{d}{dx}[\overline{F}_{,p'}] = \overline{F}_{,p}$$

or

$$p' = -H_{,y}, \qquad 0 = y' - H_{,p}. \tag{6.5}$$

The development above proved the following theorem:

**[VI.3]** ***The solutions of the Hamiltonian system (6.5) are the extremals of*** $J[y,p]$ ***in (6.25).***

Similarly, the transformed Hamiltonian system

$$-P' = H^*_Y, \qquad Y' = H^*_P \tag{6.26}$$

are the Euler DE of

$$\bar{J}[Y,P] = \int_a^b [PY' - H^*(x,Y,P)]\,dx. \tag{6.27}$$

We may use (6.14) to express the integrand of (6.27) in terms of $y$ and $p$ to get

$$\bar{J}[Y,P] = \int_a^b G(x,y,p,y',p')\,dx. \tag{6.28}$$

It is not necessary for $G(x,y,p,y',p')$ and $\bar{F}(x,y,p,y',p') = py' - H(x,y,p)$ to be identical for the system (6.5) to be transformed into (6.26) by way of the canonical transformation (6.14). It suffices that (6.25) and (6.27) have the same extremals.

The integrands $F(x,\mathbf{z},\mathbf{z}')$ and $G(x,\mathbf{z},\mathbf{z}')$ of

$$J_F[\mathbf{z}] = \int_a^b F(x,\mathbf{z},\mathbf{z}')\,dx, \qquad J_G[\mathbf{z}] = \int_a^b G(x,\mathbf{z},\mathbf{z}')\,dx \tag{6.29}$$

are said to be ***variationally equivalent*** if, for some $C^2$ function $\varphi(x,\mathbf{z})$, we have

$$F(x,\mathbf{z},\mathbf{v}) = G(x,\mathbf{z},\mathbf{v}) + \varphi_{,x}(x,\mathbf{z}) + \varphi_{,\mathbf{z}}(x,\mathbf{z})\,\mathbf{v}. \tag{6.30}$$

With $\mathbf{z} = (z_1, z_2, \ldots, z_n)^T$ and $\mathbf{v} = (v_1 \ldots, v_n)^T$, the (row) vector $\varphi_{,\mathbf{z}} = (\varphi_{,z_1} \cdots, \varphi_{,z_n})$ is the gradient in the $\mathbf{z}$ space and $\varphi_{,\mathbf{z}}\mathbf{v}$ is the scalar product of the two vector quantities. The following result furnishes the reason for the definition of variational equivalence of $F$ and $G$:

**[VI.4]** ***If*** **F** ***and*** **G** ***are variationally equivalent, they generate the same extremals.***

This conclusion follows from

$$J_F[\mathbf{z}] = \int_a^b F(x,\mathbf{z},\mathbf{z}')\,dx = \int_a^b [G(x,\mathbf{z},\mathbf{z}') + \varphi_{,x}(x,\mathbf{z}) + \varphi_{,\mathbf{z}}(x,\mathbf{z})\mathbf{z}']\,dx$$
$$= [\varphi(x,\mathbf{z}(x))]_a^b + J_G[\mathbf{z}] = [\varphi(b,\mathbf{B}) - \varphi(a,\mathbf{A})] + J_G[\mathbf{z}].$$

The term $\varphi(b,\mathbf{B}) - \varphi(a,\mathbf{A})$ does not affect the form of the Euler DE of $J_F[\mathbf{z}]$. Hence, $J_F$ and $J_G$ have the same extremals.

For the Lagrangian $\overline{F}(x, y, p, y', p') = py' - H(x, y, p)$ associated with the Hamiltonian system (6.5), we have $z = (z_1, z_2)^T = (y, p)^T$. Suppose the function $G(x, y, p, y', p')$ is the Lagrangian of the transformed Hamiltonian system (6.26) expressed in terms of $(x, y, p)$ with the help of the canonical transformation (6.14). For $G$ and $\overline{F}$ to have the same Hamiltonian system (6.5) as Euler DE, it suffices for $G$ and $\overline{F}$ to be variationally equivalent, i.e.,

$$\overline{F}(x, y, p, y', p') = G(x, y, p, y', p') + \varphi_{,x}(x, y, p) + y'\varphi_{,y}(x, y, p) + p'\varphi_{,p}(x, y, p)$$

for some $C^2$ function $\varphi = \varphi(x, y, p)$. From the expression $\overline{F} = py' - H(x, y, p)$ and $G(x, y, p, y,' p') = PY' - H^*$, the relation between $\overline{F}$ and $G$ above can be written as

$$py' - H(x, y, p) = PY' - H^*(x, Y, P) + \varphi_{,x} + \varphi_{,y}y' + \varphi_{,p}p'$$

or

$$p\,dy - H\,dx = P\,dY - H^*dx + d\varphi. \tag{6.31}$$

The form of (6.31) suggests that we take $y$ and $Y$ as the primary variables (instead of $y$ and $p$) with (6.14) [or (6.23)] rearranged to read $p = \tilde{g}_1(x, y, Y)$ and $P = \tilde{g}_2(x, y, Y)$. In that case, we take $d\varphi = \varphi_{,x}dx + \varphi_{,y}dy + \varphi_{,Y}dY$ so that (6.31) becomes

$$(p - \varphi_{,y})dy - (P + \varphi_{,Y})dY - (H - H^* + \varphi_{,x})dx = 0. \tag{6.32}$$

Equation (6.32) is satisfied if

$$p = \varphi_{,y}, \tag{6.33a}$$

$$-P = \varphi_{,Y}, \tag{6.33b}$$

$$H^* = H + \varphi_{,x}. \tag{6.33c}$$

Given $\varphi(x, y, Y)$, (6.33a) can generally be solved to give $Y$ in terms of $x$, $y$, and $p$. The result can be used to eliminate $Y$ from the right-hand side of (6.33b). Together, we then have $Y$ and $P$ in terms of $x$, $y$, and $p$; they determine the form of $f$ and $g$ in (6.14). For this reason, $\varphi$ is called the ***generating function*** of the transformation (6.14). We assume that (6.14) can be inverted to give (6.23). The inverted relations enable us to eliminate $y$ and $p$ from the right-hand side of (6.33c) to give $H^*$ as a function of $x$, $Y$, and $P$. Transformation (6.14) so induced by the generating function $\varphi(x, y, Y)$ is canonical because, by construction, the perfomance index $J_G$ associated with $H^*$ has the same extremal(s) as the perfomance index $J_F$ associated with the given Hamiltonian $H$.

We summarize the preceding discussion in the following statement:

**[IV.5]** ***If, with*** $\mathbf{p} = \varphi_{,y}$ ***and*** $\mathbf{P} = -\varphi_{,Y}$***, the*** $C^2$ ***function*** $\varphi(x, y, Y)$ ***induces a transformation (6.14) which can be inverted to give*** $y$ ***and*** $p$ ***in terms of*** $x, Y,$ ***and*** $P$***, then the transformation is canonical. It transforms the Hamiltonian system (6.5) into another Hamiltonian system with its Hamiltonian*** $H^*$ ***given by*** $H + \varphi_{,x}$ ***(expressed in terms of*** $x$, $Y$, ***and*** $P$***).***

The requirement that $\varphi$ be $C^2$ is related to the invertibility of (6.33a) and (6.33b).

If we wish, we can take $y$ and $P$ as the primary variables (instead of $y$ and $Y$) and write (6.31) as

$$pdy - Hdx = d(PY) - TdP - H^*dx + d\varphi = d\psi - YdP - H^*dx,$$

where $\psi = PY + \varphi$, or

$$(p - \psi_{,y})dy + (Y - \psi_{,P})dP - (H - H^* + \psi_{,x})dx = 0. \tag{6.34}$$

Equation (6.34) is satisfied by

$$p = \psi_{,y}, \tag{6.35a}$$

$$Y = \psi_{,P}, \tag{6.35b}$$

$$H^* = H + \psi_{,x}. \tag{6.35c}$$

The first two relations determine $P$ and $Y$ in terms of $y$ and $p$ and thereby fix the form of $f$ and $g$ in (6.14). $H^*$ is then given by (6.35c) and the inverted relations (6.23).

For an easy solution of the transformed system, we would like the new Hamiltonian $H^*$ to be independent of $Y$ and $P$. In fact, it would be nice if $H^* = 0$. With $y$ and $Y$ as the primary unknowns, (6.33c) now gives

$$H(x, y, p) + \varphi_{,x}(x, y, Y) = 0$$

or, with (6.33a),

$$H(x, y, \varphi_{,y}(x, y, Y)) + \varphi_{,x}(x, y, Y) = 0. \tag{6.36}$$

Equation (6.36) is a first order nonlinear partial differential equation (PDE) for the unknown function $\varphi(x, y, Y)$. Note that $Y$ does not appear explicitly in (6.36) and the solution $\varphi$ is only determined as a function of $x$ and $y$ by the PDE (6.36) with $Y$ carried along as a parameter. For a given *Hamiltonian* $H(x, y, p)$, the PDE

$$H(x, y, \varphi_{,y}) + \varphi_{,x} = 0 \tag{6.37a}$$

is called the *Hamilton–Jacobi equation* for the corresponding *Hamiltonian* system.

It is important to realize that we only need a particular solution of (6.37a) for our purpose. Any nontrivial solution would allow us to complete our solution process for the original Hamiltonian system (or the extremals of the corresponding Lagrangian). This key result of the development of the last two sections is summarized below as [ VI.6 ] for easy reference:

**[VI.6]** *The Hamilton–Jacobi equation (6.37a) determines the generation function* $\varphi(x, y, Y)$ *and therewith the special canonical transformation (6.33a) and (6.33b) for the solution of the Hamiltonian system (6.5). The actual solution of (6.5) is obtained by solving (6.33b), i.e.,* $P = \varphi_{,Y}(x, y, Y)$, *for* $y$ *to get* $y = \hat{y}(x, Y, P)$ *and then using this result in (6.33a) to get* $p = \hat{p}(x, Y, P)$, *where* $Y$ *and* $P$ *are two arbitrary constants.*

It is not difficult to see that the result above also holds for the case where $y$ and $p$ (and correspondingly, $Y$ and $P$ also) are vectors [ see also Section 7 of this Chapter ].

We will not be concerned here with theoretical issues such as the existence of the solution of the Hamilton–Jacobi equation; such issues properly belong to the subject of partial differential equations. Instead, we will turn to a discussion of the solution of this first-order PDE and of the corresponding Hamiltonian system in the next two sections.

## 5. Solutions of the Hamilton–Jacobi Equation

For transformation (6.14) induced by $\varphi(x, y, Y)$ to be truly one and one (invertible), the solution of the Hamilton–Jacobi equation must depend on the parameter $Y = (Y_1, \ldots, Y_n)^T$. Note that $\varphi(x, y, Y)$ is determined up to an arbitrary function of $Y$. If $\varphi$ is a solution of the Hamilton–Jacobi equation, so is $u = \varphi + c_0(Y)$, where $c_0$ is an arbitrary $C^1$ function.

If $H$ is independent of $x$, we set $\varphi = \Phi(y, Y) - Ex$, where E is a constant which generally depends on the parameters $Y_1, Y_2, \ldots, Y_n$ so that $E = E(Y)$. In terms of $\Phi(y, Y)$, the Hamilton–Jacobi equation becomes

$$H(y, \Phi_{,y}) = E. \qquad (6.37b).$$

For a scalar $y$, the transformed Hamilton–Jacobi equation (6.37b) is an ODE with $E$ as a parameter. It is much easier to solve an ODE than a PDE.

EXAMPLE 2 (continued):

For the simple harmonic oscillator, the Hamiltonian was found earlier to be $H = (ky^2 + p^2/m)/2$. Because $H$ does not depend on $x$, we may work with the reduced form (6.37b) and obtain

$$mky^2 + \Phi_{,y}^2 = 2mY, \tag{6.38}$$

where we have set $E = Y (= Y_1)$ because $y$ is a scalar (hence, so is $Y$). The separable first-order ODE (6.38) for $\Phi$ can be written as

$$\Phi_{,y} = \sqrt{2mY - mky^2} \qquad \text{or} \qquad \Phi = \int^y \sqrt{2mY - mky^2}\, dy$$

so that

$$\varphi(x, y, Y) = \int^y \sqrt{2mY - mky^2}\, dy \; - \; Yx.$$

It follows from (6.33) that

$$p = \varphi_{,y} = \sqrt{2mY - mky^2}, \tag{6.39a}$$

$$P = -\varphi_{,Y} = -m \int^y \frac{dy}{\sqrt{2mY - mky^2}} + x$$

$$= x - \sqrt{\frac{m}{k}} \sin^{-1}\left(\sqrt{\frac{k}{2Y}}\, y\right), \tag{6.39b}$$

$$-H^* = Y - H(y, \Phi_{,y}) = Y - \frac{1}{2}\left(ky^2 + \frac{p^2}{m}\right) = 0. \tag{6.39c}$$

Relation (6.39c) is simply a first integral of the equation of motion (the Euler DE) for the problem:

$$Y = \frac{1}{2}\left(ky^2 + \frac{p^2}{m}\right). \tag{6.40}$$

To verify that we, in fact, have the desired results, we compute $P'$ and $Y'$ to get

$$Y' = kyy' + \frac{p}{m} p' = ky\left(\frac{p}{m}\right) + \frac{p}{m}\left(-ky\right) = 0,$$

$$P' = 1 - \sqrt{\frac{m}{k}} \frac{\sqrt{k/2Y}\, y'}{\sqrt{1 - ky^2/2Y}} = 1 - \sqrt{m} \frac{p/m}{\sqrt{2Y - ky^2}}$$

$$= 1 - \frac{p/\sqrt{m}}{\sqrt{ky^2 + (p^2/m) - ky^2}} = 0$$

where we have used the Hamiltonian system (6.13) to eliminate $y'$ and $p'$. The solution of the original problem can, therefore, be obtained from (6.39b) and the fact that $P$ is actually an arbitrary constant, so that

$$y(x) = \sqrt{\frac{2Y}{k}} \sin\left(\sqrt{\frac{k}{m}}\,[x - P]\right), \tag{6.41}$$

where $Y$ is a second arbitrary constant.

EXAMPLE 3: $F = f(y)\sqrt{1 + (y')^2}$ .

We have in this case

$$F_{,y'} = \frac{f(y)y'}{\sqrt{1 + (y')^2}} = p \qquad \text{or} \qquad y' = \frac{p}{\sqrt{f^2(y) - p^2}} .$$

The corresponding Hamiltonian is

$$H = pv - F(x, y, v(x, y, p))$$

$$= \frac{p^2}{\sqrt{f^2 - p^2}} - f\sqrt{1 + \frac{p^2}{f^2 - p^2}} = -\sqrt{f^2 - p^2} .$$

This expression for $H$ allows us to write down the following Hamilton–Jacobi equation for the problem:

$$H(x, y, \varphi_{,y}) + \varphi_{,x} = -\sqrt{[f(y)]^2 - \varphi_{,y}^2} + \varphi_{,x} = 0,$$

or

$$\varphi_{,x}^2 + \varphi_{,y}^2 = [f(y)]^2. \tag{6.42}$$

Because $H$ does not depend on $x$ explicitly, we set $\varphi = \Phi(y, Y) - Yx$ so that (6.48) becomes an ODE:

$$\Phi_{,y} = \pm\sqrt{f^2 - Y^2}$$

or

$$\Phi = \pm \int^y \sqrt{[f(t)]^2 - Y^2}\, dt$$

where the constant of integration plays no role in the final results and has been omitted. We then use (6.33) to obtain

$$p = \varphi_{,y} = \Phi_{,y} = \pm\sqrt{f^2 - Y^2} \tag{6.43a}$$

and

$$P = -\varphi_{,Y} = x - \Phi_{,Y} = x \mp \int^y \frac{Y\,dt}{\sqrt{[f(t)]^2 - Y^2}}. \tag{6.43b}$$

Given $f(y)$, the right hand side of (6.43b) is a known quantity. The fact that $P$ and $Y$ are constants allows us to use (6.43b) to give $y$ as a function of $x$ with

$P$ and $Y$ as parameters (to be determined by appropriate auxiliary conditions). If we wish, we can then use (6.43a) to determine $p$ as a function of $x$ with $P$ and $Y$ as parameters.

## 6. The Method of Additive Separation

It should be observed that the Hamilton–Jacobi approach does not necessarily simplify the process of obtaining the extremal(s) for a given Lagrangian. Instead of solving a set of Euler DE which are ordinary differential equations, we now have to solve the Hamilton–Jacobi equation which is a partial differential equation. In general, PDEs are much more difficult to solve than ODEs. In fact, several useful methods of solution for PDEs (including the method of characteristics) effectively reduce a PDE problem to a problem of solving a system of ODEs. It would seem then that the Hamilton–Jacobi procedure offers very little pratical advantage for the solution of the Euler DE (though it does lead to new theoretical developments which enrich our uderstanding of Hamiltonian systems). Fortunately, there is a nontrivial and practically important class of problems for which a solution of the Hamilton–Jacobi equation is relatively straightforward. This class of problems has the feature that $\varphi(x, \mathbf{y}, \mathbf{Y})$ is a sum of functions each involving only one component of $\mathbf{y}$.

For simplicity, we limit our discussion to conservative systems so that $H$ does not depend explicitly on $x$. In that case, we may set $\varphi(x, \mathbf{y}, \mathbf{Y}) = \Phi(\mathbf{y}, \mathbf{Y}) - E(\mathbf{Y})x$ and work with the reduced Hamilton–Jacobi equation (6.37b):

$$H(\mathbf{y}, \Phi_{,\mathbf{y}}) \equiv H(y_1, y_2, \ldots, y_n, \Phi_{,1}, \ldots, \Phi_{,n}) = E,$$

where $\Phi_{,k} \equiv \partial \Phi / \partial y_k$. For many Hamiltonians, the solution of the above equation is of the form

$$\Phi(\mathbf{y}; \mathbf{Y}) = \Phi_1(y_1; \mathbf{Y}) + \Phi_2(y_2; \mathbf{Y}) + \ldots + \Phi_n(y_n; \mathbf{Y}), \qquad (6.44)$$

where $\mathbf{Y} = (Y_1, \ldots, Y_n)^{\mathrm{T}}$ assumes the role of a parameter (vector). In other cases, only some of the variables $\{y_k\}$ are additively separated in $\Phi(\mathbf{y}; \mathbf{Y})$.

To see when an additively separated solution of the reduced Hamilton–Jacobi equation (6.37b) is possible, consider the case where $H$ depends on $y_1$ and $\Phi_{,1}$ only through a combination $h_1(y_1, \Phi_{,1})$ :

$$H(\mathbf{y}, \Phi_{,\mathbf{y}}) = H(h_1(y_1, \Phi_{,1}), y_2, \ldots, y_n, \Phi_{,2}, \ldots, \Phi_{,n}), \qquad (6.45)$$

For this case, we seek a solution for $\Phi$ in the form

$$\Phi(\mathbf{y}; \mathbf{Y}) = \Phi_1(y_1; \mathbf{Y}) + \bar{\Phi}(y_2, \ldots, y_n; \mathbf{Y}) \qquad (6.44')$$

so that (6.37b) becomes

$$H(h_1(y_1, \Phi_{1,1}), y_2, \ldots, y_n, \tilde{\Phi}_{,2}, \ldots, \tilde{\Phi}_{,n}) = E. \tag{6.37c}$$

The equation above must be valid for all $y_1$ while all other $y_k$ are kept fixed. When $y_1$ changes, only $h_1(y_1, \Phi_{1,1})$ is effected in $H$. It follows that $h_1(y_1, \Phi_{1,1}(y_1))$ must be independent of $y_1$, i.e.,

$$h_1(y_1, \Phi_1^\bullet) = a_1(Y),$$

where $(\ )^\bullet$ indicates differentiation with respect to the argument of $(\ )$ (which is $y_1$ in this case) and $a_1$ is a constant parameter. Equation (6.37c) now becomes

$$H(a_1, y_2, \ldots, y_n, \tilde{\Phi}_{,2}, \ldots, \Phi_{,n}) = \mathrm{E}.$$

If $y_2$ and $\tilde{\Phi}_{,2}$ also appear in $H$ only in the combination $h_2(y_2, \tilde{\Phi}_{,2})$, the same argument allows us to conclude that variable $y_2$ may also be (additively) separable in $\Phi(\mathbf{y}; \mathbf{Y})$ so that

$$\Phi(\mathbf{y}; \mathbf{Y}) = \Phi_1(y_1; \mathbf{Y}) + \Phi_2(y_2; \mathbf{Y}) + \overline{\Phi}(y_3, \ldots, y_n; \mathbf{Y})$$

with $\Phi_1$ and $\Phi_2$ determined by the ODE

$$h_k(y_k, \Phi_k^\bullet) = a_k(\mathbf{Y}) \quad (k = 1, 2), \tag{6.46}$$

whereas (6.37b) becomes a PDE for $\overline{\Phi}(y_3, \ldots, y_n; \mathbf{Y})$:

$$H(a_1, a_2, y_3, \ldots, y_n, \overline{\Phi}_{,3}, \ldots, \overline{\Phi}_{,n}) = E. \tag{6.37d}$$

It is now evident that a solution of (6.37b) in the form (6.44) is possible if $H$ depends on all $y_k$ and $\Phi_{,k}$ only through the $n$ combinations $h_k(y_k, \Phi_{,k})$, $k = 1, 2, \ldots, n$.

**EXAMPLE 4:** The motion of a particle in space under the action of a uniform field in the $y_3$ direction. (An example of such a system is the motion of a mass particle under gravity.)

By Hamilton's principle, the Lagrangian for this system is

$$F = \frac{1}{2} m(\dot{y}_1^2 + \dot{y}_2^2 + \dot{y}_3^2) - mgy_3.$$

The corresponding Hamiltonian is

$$H = \frac{1}{2m}(p_1^2 + p_2^2 + p_3^2) + mgy_3,$$

where $p_k = \dot{y}_k / m$. Because $H$ does not depend on $t$ explicitlly, the reduced Hamilton–Jacobi equation (6.37b) is appropriate for this problem. The equation itself becomes

$$\left[mgy_3 + \frac{1}{2m}\Phi_{,3}^2\right] + \left[\frac{1}{2m}\Phi_{,2}^2\right] + \left[\frac{1}{2m}\Phi_{,1}^2\right] = E. \tag{6.47}$$

The above equation is of the form $h_1(y_1, \Phi_{,1}) + h_2(y_2, \Phi_{,2}) + h_3(y_3, \Phi_{,3}) = E$. It follows that $\Phi$ may be taken in the form (6.44) with

$$\Phi_{k,k} = a_k \qquad (k = 1, 2)$$

and

$$\Phi_{3,3}^2 + 2m^2 gy_3 + a_1^2 + a_2^2 = 2mE.$$

These relations may be integrated to give

$$\begin{aligned} \Phi_k &= a_k(y_k - y_{k0}) \quad (k = 1, 2), \\ \Phi_3 &= \int_{y_{30}}^{y_3} \sqrt{2mE - a_1^2 - a_2^2 - 2m^2gz}\, dz. \end{aligned} \tag{6.48}$$

The constants of integration have been chosen for the initial state $y_k(0) = y_{k0}$, $k = 1, 2, 3$. The parameters $a_1$, $a_2$ and $E$ may be relabeled as $Y_1$, $Y_2$, and $Y_3$, respectively, to be consistent with the development of the previous two sections.

With (6.48), we obtain from (6.33a, b)

$$p_1 = Y_1, \quad p_2 = Y_2, \quad p_3 = -mgy_3 + R(y_3; \mathrm{Y}),$$

$$-P_k = y_k - Y_k \int_{y_{30}}^{y_3} \frac{dz}{R(z; \mathrm{Y})} \quad (k = 1, 2),$$

$$-P_3 = -t + m \int_{y_{30}}^{y_3} \frac{dz}{R(z; \mathrm{Y})},$$

where

$$R(y_3; \mathrm{Y}) = \sqrt{2mY_3 - Y_1^2 - Y_2^2 - 2m^2 gy_3}.$$

The initial conditions $y_k(0) = y_{k0}$ = require $P_3 = 0$ and $P_k = -y_{k0}$ (k = 1, 2) so that the relations above give

$$m\int_{y_{30}}^{y_3} \frac{dz}{R(z; \mathrm{Y})} = \left[-\frac{1}{mg} R(z; \mathrm{Y})\right]_{y_{30}}^{y_3} = t \tag{6.49a}$$

and

$$y_k = y_{k0} + \frac{Y_k}{m} t \qquad (k = 1, 2). \tag{6.49b}$$

The expression (6.49a) may be further simplifed to give

$$y_3(t) = y_{30} + \frac{1}{m} R(y_{30}; \mathbf{Y})t - \frac{1}{2} gt^2. \tag{6.49c}$$

The three remaining unknown constants $\{Y_1, Y_2, Y_3\}$ (which are the components of $\mathbf{Y}$) are determined by the remaining initial conditions $\dot{y}_k(0) = v_{k0}$ $(k = 1, 2, 3)$.

For the simple problem above, the Hamilton–Jacobi approach to the motion of the particle seems unnecessarily complicated compared to a straightforward solution of the Euler DE for the same problems. That this is not always the case can be seen from problems involving planar motion of a particle in a central force field. The best-known example of such a problem is the motion of a single planet around the sun (ignoring the effect of other planets).

**EXAMPLE 5:** Planar motion of a particle under the action of a central force field characterized by $V(\mathbf{y}) \equiv V(r)$, where $r$ is the radial distance from the origin of the reference frame.

In polar coordinates $(r, \theta)$ in the plane of motion, the Hamiltonian for this problem is

$$H = \frac{1}{2m}\left(p_r^2 + \frac{1}{r^2} p_\theta^2\right) + V(r)$$

which follows from Hamilton's principle and the Legendre transformation. $H$ is independent of $t$ so that the reduced Hamilton–Jacobi equation (6.37b) again applies:

$$H(r, \theta, \Phi_{,r}, \Phi_{,\theta}) = \frac{1}{2m}\left[\Phi_{,r}^2 + 2m V(r)\right] + \frac{1}{2mr^2} \Phi_{,\theta}^2 = E$$

with the generating function given by $\varphi(t, r, \Phi, Y_1, Y_2) = \Phi(r, \theta, Y_1, Y_2) - Et$. We note that $\theta$ and $\Phi_{,\theta}$ appear in $H$ only in the combination $(\Phi_{,\theta})^2$ and may be set equal to a constant $a_\theta^2$.

A more useful approach is to observe that $H$ does not depend on $\theta$ explicitly. In that case, we call the coordinate $\theta$ a *cyclic* (or *ignorable*) coordinate. For a cyclic coordinate, we may write

$$\Phi(r, \theta; Y) = \overline{\Phi}(r; Y) + a_\theta \theta$$

and transform (6.37b) into

$$r^2\left[\overline{\Phi}_{,r}^2 + 2m V(r)\right] + a_\theta^2 = 2mEr^2.$$

This equation may be rearranged to read

$$\overline{\Phi}_{,r} = \sqrt{2mY_1 - 2mV(r) - \frac{Y_2^2}{r^2}} \equiv R(r; \mathbf{Y})$$

where we identify $Y_2 = a_\theta$ and $Y_1 = E$, so that

$$\overline{\Phi}(r;\mathbf{Y}) = \int_{r_0}^{r} R(\xi;\mathbf{Y})\,d\xi + \beta_r,$$

where $r_0 = r(t = 0)$ and $\beta_r$ is a constant of integration which has no role in the motion of the particle and will be omitted. In that case, we have

$$\varphi(t, r, \theta; \mathbf{Y}) = Y_2\theta - Y_1 t + \int_{r_0}^{r} R(\xi;\mathbf{Y})\,d\xi.$$

By (6.39), we obtain

$$p_r = R(r;\mathbf{Y}), \qquad p_\theta = a_\theta = Y_2,$$

$$-P_r = -t + \int_{r_0}^{r} \frac{m}{R(\xi;\mathbf{Y})}\,d\xi, \tag{6.50}$$

$$-P_\theta = \theta - \int_{r_0}^{r} \frac{Y_2}{R(\xi;\mathbf{Y})\xi^2}\,d\xi.$$

The initial state of the particle $r(0) = r_0$ and $\theta(0) = \theta_0$ requires $P_r = 0$ and $P_\theta = -\theta_0$. The expression for $P_r$ then becomes an equation for $t$ as a function of $r$:

$$t = \int_{r_0}^{r} \frac{m}{R(\xi;\mathbf{Y})}\,d\xi$$

which can be inverted to give $r = r(t)$ under suitable conditions. The expression for $P_\theta$ gives $\theta$ as a function of $r$

$$\theta - \theta_0 = \int_{r_0}^{r} \frac{p_\theta}{\xi^2 R(\xi;\mathbf{Y})}\,d\xi$$

where we have written $p_\theta$ for $Y_2$ [given that $p_\theta = Y_2$ by (6.50)]. In other words, the new canonical coordinate $Y_2$ is now identified to be the angular momentum of the particle and the result $Y_2' = 0$ (guaranteed by our choice of $\varphi$) implies that

*The angular momentum of a particle in motion under the action of a central force is conserved.*

Kepler's second law of planetary motion is a particular realization of this theorem.

The expressions for $t$ and $\theta$ are known once $V(r)$ is prescribed. For planetary motion, the inverse square force corresponds to $V(r) = c_0/r$ so that

$$\theta - \theta_0 = \int_{r_0}^{r} \frac{p_\theta \, d\xi}{\xi^2 \sqrt{2mY_1 + \dfrac{2mc_0}{\xi} - \dfrac{p_\theta^2}{\xi^2}}}$$

$$= \int_{r^{-1}}^{r_0^{-1}} \frac{du}{\sqrt{\dfrac{2m}{p_\theta^2}(Y_1 + c_0 u) - u^2}}$$

where we have set $u = 1/\xi$ with $du = -d\xi/\xi^2$. The integral can now be evaluated by writing $2mp_\theta^{-2}(Y_1 + c_0 u) - u^2 = c^2 - v^2$ with $v = u - mc_0 p_\theta^{-2}$ and $c^2 = 2mp_\theta^{-2}Y_1 + (mc_0 p_\theta^{-2})^2$. The result is an equation which describes an ellipse; it corresponds to *Kepler's first law* of planetary motion (which was originally inferred from observational data).

The example above gives some indication of the power and elegance of the Hamilton–Jacobi method. In a few short steps, we obtained $r$ as a function of $t$ and the equation for the shape of the orbit, $r = r(\theta)$. To obtain the same result by working with the Euler DE of the corresponding Hamilton's principle (or directly with the equations of motion in Newtonian mechanics) is known to require considerably more effort and ingenuity.

## 7. Hamilton's Principal Function

Our attempt to obtain the solution of Hamiltonian systems by canonical transformations stimulated a study of the Hamilton–Jacobi equation in Section 4. The development leading to the main result given in [VI.6] is direct but tedious. In this section, we discuss a simpler alternative derivation of the same partial differential equation but now for a Lagrangian (or Hamiltonian) with several unknowns. At first, the development does not appear to be directly related to the solution of any Hamiltonian system, but it will turn out to be so related in the end.

Suppose we have a basic problem for $n$ unknowns with the general Lagrangian $F(x, \mathbf{y}, \mathbf{v})$ for $\mathbf{y} = (y_1, \ldots, y_n)^T$ and $\mathbf{v} = (v_1, \ldots, v_n)^T$ and with end points $\mathbf{A} = (x_a, \mathbf{y}(x_a)) \equiv (x_a, \mathbf{y}^a)$ and $\mathbf{B} = (x_b, \mathbf{y}(x_b)) \equiv (x_b, \mathbf{y}^b)$. Consider the related *Hamilton's principal function* of the end point $\mathbf{B}$:

$$\varphi(x_b, \mathbf{y}^b) \equiv \int_{x_a}^{x_b} F(x, \mathbf{y}, \mathbf{y}')\, dx, \qquad \mathbf{y}(x_a) = \mathbf{y}^a, \qquad \mathbf{y}(x_b) = \mathbf{y}^b. \tag{6.51}$$

Along the extremal of $F$, the Hamilton principal function $\varphi(x, \hat{\mathbf{y}}(x))$ will be shown to satisfy the Hamilton–Jacobi equation. Let $\hat{\mathbf{y}}$ be an extremal of this problem for a particular set of values of $x_a, x_b, \mathbf{y}^a$ and $\mathbf{y}^b$. We wish to compare

$\Phi(x_b, \mathbf{y}^b)$ with $\Phi(x_b + \Delta x, \mathbf{y}^b + \Delta \mathbf{y})$. Because the end value $x_b$ is also allowed to change, the situation is similar to a variable end point problem. The development of Chapter 2 suggests that such problems are most effectively treated by reparametrizing $(x, \mathbf{y})$ so that $x = \xi(t)$ and $\mathbf{y}(x) = \boldsymbol{\eta}(t)$, $\alpha \le t \le \beta$, for fixed $\alpha$ and $\beta$. The integral $\varphi(x_b, \mathbf{y}^b)$ then becomes

$$\varphi(x_b, \mathbf{y}^b) = \int_\alpha^\beta G(t, \xi, \boldsymbol{\eta}, \xi^\bullet, \boldsymbol{\eta}^\bullet)\, dt, \qquad (\ )^\bullet \equiv \frac{d(\ )}{dt} \tag{6.52a}$$

with

$$G(t, \xi, \boldsymbol{\eta}, \xi^\bullet, \boldsymbol{\eta}^\bullet) = \xi^\bullet F(\xi, \boldsymbol{\eta}, \boldsymbol{\eta}^\bullet / \xi^\bullet), \qquad \xi(\beta) = x_b, \quad \boldsymbol{\eta}(\beta) = \mathbf{y}^b \tag{6.52b}$$

(and of course $\xi(\alpha) = a$ and $\boldsymbol{\eta}(\alpha) = \mathbf{y}^a$ at the fixed end). Note that $G$ actually does not depend on $t$ explicitly.

The first variation of $\varphi$ is given by

$$\delta\varphi = \int_\alpha^\beta \left\{ \sum_{k=1}^n \left[ -(\widehat{F}_{,y'_k})' + \widehat{F}_{,y_k} \right] \delta y_k + \left[ -\left( \widehat{F} - \sum_{k=1}^n \widehat{y}'_k \widehat{F}_{,y'_k} \right)' + \widehat{F}_{,x} \right] dx \right\} \widehat{x}^\bullet\, dt$$
$$+ \left[ \left( \widehat{F} - \sum_{k=1}^n \widehat{y}'_k \widehat{F}_{,y'_k} \right) \delta x + \sum_{k=1}^n \widehat{F}_{,y'_k} \delta y_k \right]_{t=\beta}$$

because $x$ and $\mathbf{y}$ are fixed at $t = \alpha$. The coefficients of $\delta x$ and $\delta y_k$ in the integrand of the integral from $\alpha$ to $\beta$ vanishes because of the Euler DE for the extremal $(\widehat{x}(t), \widehat{\mathbf{y}}(t))$. We are then left with.

$$\delta\varphi = \left[ \widehat{F}(x_b) - \sum_{k=1}^n \widehat{y}'_k(x_b) \widehat{F}_{,y'_k}(x_b) \right] \delta x(\beta) + \sum_{k=1}^n \widehat{F}_{,y'_k}(x_b) \delta y_k(\beta), \tag{6.53}$$

where $\widehat{F}(x_b) = F(x_b, \mathbf{y}^b, \widehat{\mathbf{y}}'(x_b))$ and $\widehat{F}_{,y'_k}(x_b) = F_{,y'_k}(x_b, \mathbf{y}^b, \widehat{\mathbf{y}}'(x_b)) = p_k(x_b, \mathbf{y}^b)$ are both functions of the prescribed value of $x_b$ and $\mathbf{y}^b$ ($\widehat{\mathbf{y}}'(x_b)$ itself depends on $x_b$ and $\mathbf{y}^b$). The coefficient of $\delta x(\beta)$ in (6.53) is the negative of the Hamiltonian associated with $F(x, \mathbf{y}, \mathbf{y}')$:

$$\widehat{F}(x_b) - \sum_{k=1}^n \widehat{y}'_k(x_b) \widehat{F}_{,y'_k}(x_b) = -\left[ \sum_{k=1}^n \widehat{y}'_k(x_b) p_k(x_b, \mathbf{y}^b) - F(x_b, \mathbf{y}^b, \widehat{\mathbf{y}}'(x_b)) \right]$$
$$= -H(x_b, \mathbf{y}^b, \mathbf{p}(x_b, \mathbf{y}^b)).$$

If the problem were one with a variable end point **B**, then the coefficients of $\delta x(\beta)$ and $\delta x_k(\beta)$ must vanish separately so that $\delta\Phi = 0$. But that is not the case here. Instead, $\widehat{\mathbf{y}}(x)$ is actually the extremal for the fixed end point problem as specified in (6.51); we are merely comparing $\varphi(x_b + \Delta x, \mathbf{y}^b + \Delta \mathbf{y})$ and $\varphi(x_b, \mathbf{y}^b)$ with

$$\Delta\varphi \equiv \varphi(x_b + \Delta x, \mathbf{y}^b + \Delta \mathbf{y}) - \varphi(x_b, \mathbf{y}^b)$$
$$\cong \varphi_{,x} \delta x(\beta) + \varphi_{,y_k} \delta y_k(\beta) = \delta\varphi \tag{6.54}$$

to leading order. From (6.53) and (6.54), we obtain the following set of relations for $\varphi(x, \mathbf{y})$ after replacing $(x_b, \mathbf{y}^b)$ by $(x, \mathbf{y})$ to simply our notation:

$$\frac{\partial\varphi}{\partial x} = -H(x, \mathbf{y}, \mathbf{p}(x, \mathbf{y})), \qquad \frac{\partial\varphi}{\partial y_k} = p_k(x, \mathbf{y}) \tag{6.55}$$

where we have omitted the $\frown$ symbol associated with the extremal. If we now use the second equation to eliminate $p_k$ from the first, we obtain

$$\frac{\partial\varphi}{\partial x} + H\left(x, \mathbf{y}, \frac{\partial\varphi}{\partial y_1}, \ldots, \frac{\partial\varphi}{\partial y_n}\right) = 0 \tag{6.56}$$

which is just the Hamilton–Jacobi equation of Section 4. Note that the derivation of (6.56) makes no use of the Hamiltonian system associated with $F$.

A first integral of the Hamiltonian system associated with $F$ can now be obtained immediately from $\varphi$ (see (6.51)):

**[VI.7]** *Let* $\varphi = \Phi(x, \mathbf{y}; \alpha_1, \ldots, \alpha_m)$ $(m \leq n)$ *be a solution of* (6.56) *with* m *parameters* $(\alpha_1, \ldots, \alpha_m)$. *Then* $\partial\varphi/\partial\alpha_j$ *is a constant along an extremal of* $F(x, \mathbf{y}, \mathbf{v})$.

To prove [VI.7], we need only to show that $d(\partial\varphi/\partial\alpha_j)/dx = 0$ along an extremal. Upon differentiating (6.56) partially with respect to $\alpha_j$, we we obtain

$$\frac{\partial^2\varphi}{\partial\alpha_j\, \partial x} = -\sum_{k=1}^{n} H_{,p_k} \frac{\partial^2\varphi}{\partial\alpha_j\, \partial y_k},$$

This relation is then used to eliminate $\varphi_{,\alpha_j x}$ from

$$\frac{d}{dx}\left(\frac{\partial\varphi}{\partial\alpha_j}\right) = \frac{\partial^2\varphi}{\partial x\, \partial\alpha_j} + \sum_{k=1}^{n} \frac{\partial^2\varphi}{\partial y_k\, \partial\alpha_j} \frac{dy_k}{dx}$$

to get

$$\frac{d}{dx}\left(\frac{\partial\varphi}{\partial\alpha_j}\right) = \sum_{k=1}^{n} \frac{\partial^2\varphi}{\partial y_k\, \partial\alpha_j}\left[\frac{dy_k}{dx} - H_{,p_k}\right] = 0 \tag{6.57}$$

because $y_k' = H_{,p_k}$ along an extremal.

The fact that $\partial\varphi/\partial\alpha_j$ must be a constant can be used to solve the Hamiltonian system for the given $F(x, \mathbf{y}, \mathbf{v})$ provided that $\varphi$ contains enough parameters $\{\alpha_1, \ldots, \alpha_n\} = \{\boldsymbol{\alpha}\}$. Suppose we have a solution of (6.56) with $n$ parameters, $\varphi = \varphi(x, \mathbf{y}; \boldsymbol{\alpha})$. We know from [VI.7] that $\partial\varphi/\partial\alpha_j$ is a constant, i.e.,

$$\psi_j(x, \mathbf{y}, \boldsymbol{\alpha}) = \frac{\partial}{\partial\alpha_j}[\varphi(x, \mathbf{y}, \boldsymbol{\alpha}] = \beta_j \quad (j = 1, 2, \ldots, n).$$

We can normally solve these to find $\mathbf{y}$ in terms of $x$, $\boldsymbol{\alpha}$, and $\boldsymbol{\beta}$, so that $\mathbf{y} = \hat{\mathbf{y}}(x;\boldsymbol{\alpha},\boldsymbol{\beta})$. From the second half of (6.55), we have

$$\frac{\partial}{\partial y_k}[\varphi(x,\mathbf{y};\boldsymbol{\alpha})] = p_k(x,\mathbf{y};\boldsymbol{\alpha})$$

which, upon setting $\mathbf{y} = \hat{\mathbf{y}}(x;\boldsymbol{\alpha},\boldsymbol{\beta})$, becomes $p_k = \hat{p}_k(x;\boldsymbol{\alpha},\boldsymbol{\beta})$. The functions $\{\hat{y}_k(x;\boldsymbol{\alpha},\boldsymbol{\beta})\}$ and $\{\hat{p}_i(x;\boldsymbol{\alpha},\boldsymbol{\beta})\}$ will be shown to satisfy the Hamiltonian system for the given Lagrangian $F(x,\mathbf{y},\mathbf{v})$.

**[VI.8]** *Let $\varphi = \varphi(x,y;\alpha_1,\ldots,\alpha_n)$ be a solution of the Hamilton-Jacobi equation (6.56) with n parameters. Suppose the determinant of the n × n matrix*

$$\mathbf{M} = [\mathbf{M}_{ij}] = \left[\frac{\partial^2\varphi}{\partial y_k\,\partial\alpha_j}\right]$$

*is nonzero. Then $\{\hat{y}_k(x;\alpha,\beta)\}$ and $\{\hat{p}_i(x;\alpha,\beta)\}$ as defined above satisfy the Hamiltonian system for the given Lagrangian.*

The nonvanishing of the determinant of the matrix $M$ ensures the solvability of $\psi_j(x,\mathbf{y};\boldsymbol{\alpha}) = \beta_j$ $(j = 1,2,\ldots,n)$ for $\mathbf{y} = \hat{\mathbf{y}}(x;\boldsymbol{\alpha},\boldsymbol{\beta})$. It also allows us to conclude from (6.57) for $j = 1,2,\ldots,n$, that we have

$$\frac{d\hat{y}_k}{dx} = H_{,p_k}(x,\hat{\mathbf{y}}(x),\hat{\mathbf{p}}(x)). \tag{6.58}$$

This result is immediately used in

$$\frac{d\hat{p}_i}{dx} = \frac{d}{dx}\left(\frac{\partial\varphi}{\partial y_i}\right) = \frac{\partial^2\varphi}{\partial x\,\partial y_i} + \sum_{k=1}^{n}\frac{\partial^2\varphi}{\partial y_k\,\partial y_i}\frac{d\hat{y}_k}{dx}$$

$$= \frac{\partial^2\varphi}{\partial x\,\partial y_i} + \sum_{k=1}^{n}\frac{\partial^2\varphi}{\partial y_k\,\partial y_i}H_{,p_k}(x,\hat{\mathbf{y}}(x),\hat{\mathbf{p}}(x)).$$

At same time, differentiating (6.56) partially with respect to $y_i$ gives

$$\frac{\partial^2\varphi}{\partial y_i\,\partial x} = -\frac{\partial H}{\partial y_i} - \sum_{k=1}^{n}\frac{\partial^2\varphi}{\partial y_i\,\partial y_k}H_{,p_k}(x,\hat{\mathbf{y}}(x),\hat{\mathbf{p}}(x)).$$

We can use this relation to eliminate $\partial^2\varphi/\partial y_i\partial x$ from the espression for $\hat{p}_i'$ to obtain

$$\frac{d\hat{p}_i}{dx} = -H_{,y_i}(x,\hat{\mathbf{y}}(x),\hat{\mathbf{p}}(x)) \qquad (i = 1,2,\ldots,n). \tag{6.59}$$

The relations (6.58) and (6.59) shows that $\{\hat{p}_i\}$ and $\{\hat{y}_k\}$ derived from the solution of the Hamilton–Jacobi equation, in fact, solve the associated Hamiltonian systems.

By considering an incremental change of Hamilton's principal function, we have obtained by a simple and short derivation a Hamilton–Jacobi equation for the principal function itself. It is worth emphasizing, however, that the Hamiltonian system plays no role in this particular derivation. That the solution of (6.56) may be used to solve the corresponding Hamiltonian system as described in [VI.8] is rather unexpected. For this reason, it would be pedagogically unsound to introduce the subject of the Hamilton–Jacobi equation for the first time by way of the Hamilton principal function.

## 8. Exercises

1. $F = (y' + ky)^2 \qquad (k \neq 0)$.
   (a) Find the Hamiltonian of the above Lagrangian.
   (b) Find the extremals of the corresponding basic problem by solving the Hamiltonian system.
   (c) Find the extremals by solving the Hamilton-Jacobi equation.

2. Repeat Exercise 1 for $F = e^{-y}\sqrt{1 + (y')^2}$.

3. Suppose $F(x, \mathbf{y}, \mathbf{v})$ is strictly convex in $\mathbf{v} = (v_1, v_2, \ldots, v_n)^T$ for each pair of $(x, \mathbf{y}) = (x, y_1, y_2, \ldots, y_n)$. Prove that the associated Hamiltonian is given by

$$H(x, \mathbf{y}, \mathbf{p}) = \sup_{\mathbf{v}} \{\mathbf{p} \cdot \mathbf{v} - F(x, \mathbf{y}, \mathbf{v})\}$$

   where the sup ($\equiv$ supremum) is taken over all $\mathbf{v}$ in $R^n$.

4. Obtain the Hamiltonian and the Hamiltonian system for the following Lagrangian:

   (a) $F(x, y, y') = \frac{1}{2}(y')^2 - yy'$.

   (b) $F(x, y, y') = \frac{1}{2}(y')^2 - xy$.

   (c) $F(t, r, \theta, \dot{r}, \dot{\theta}) = \frac{1}{2}m(\dot{r}^2 + r^2\dot{\theta}^2) + \frac{m\lambda}{r}$.

5. (a) The transformation $\{Y = p, P = -y\}$ is canonical. Find an appropriate generating function $\varphi$ for this transformation.

   (b) Use the Hamilton-Jacobi equation to find the extremals of the basic problem with $F(x, y, y') = \sqrt{x^2 + y^2}\sqrt{1 + (y')^2}$. [Hint: Try $\alpha x^2 + 2\beta xy + \gamma y^2$.]

6. (a) Obtain the Hamilton–Jacobi equation for $F = (y')^2$.

   (b) Obtain a complete solution of the above Hamilton–Jacobi equation.

   (c) Use the solution in part (b) to generate a canonical transformation for $F$ with $H^* \equiv 0$.

   (d) Deduce the extremals of the basic problem for $F$ with $A \neq B$.

7. (a) Obtain the Hamiltonian for the Lagrangian

$$F = \frac{m}{2}(\dot{r}^2 + r^2\dot{\theta}^2 + r^2\sin^2\theta\dot{\psi}^2) - V(r, \theta, \psi).$$

   (b) Solve the Hamilton–Jacobi equation for $\varphi(t, r, \theta, \psi, Y_r, Y_\theta, Y_\psi)$ with $V = -k/r$.

   (c) Find the extremals for $F$ from the results in $(b)$.

8. Solve the Hamilton–Jacobi equation of Exercise 7 for

$$V = -kr^{-2}\cos\theta.$$

9. Solve the Hamilton–Jacobi equation for

$$H(\mathbf{y}, \mathbf{p}) = \frac{f_1(y_1, p_1) + f_2(y_2, p_2) + \ldots + f_n(y_n, p_n)}{g_1(y_1, p_1) + \ldots + g_n(y_n, p_n)}$$

10. (a) Solve the Hamilton–Jacobi equation for the Lagrangian

$$F = \frac{1}{2}(\dot{y}_1^2 + \dot{y}_2^2)(y_1^2 + y_2^2) - (y_1^2 + y_2^2)^{-1}.$$

   (b) Obtain the motion of the dynamical system [characterized by $\{y_1(t), y_2(t)\}$] in terms of quadratures (definite integrals).

11. Obtain by way of the Hamilton–Jacobi equation the solution of the Hamiltonian system with $H = p + \alpha y^2$.

12. Use the Hamilton–Jacobi theory to show that the path of motion for a Hamiltonian system with

$$H = \frac{1}{2}(p_1^2 + p_2^2)\,(y_1^2 + y_2^2)^{-1} + (y_1^2 + y_2^2)^{-1}$$

is a conic in the $y_1, y_2$ -plane.

13. A mass particle moves in a force field characterized by the potential

$$V(r, \theta, \varphi) = f(r) + \frac{1}{r^2}\,g(\theta).$$

Obtain by the Hamilton–Jacobi method a general solution for the path of motion.

# 7

# Lagrangian Mechanics

## 1. Generalized Coordinates

We indicated in the last chapter that the calculus of variations provides an alternative characterization of the motion of mass particles in the form of Hamilton's principle. Often the variational characterization is actually more effective for determining the motion of a dynamical system. For many problems, use of particle trajectories in Cartesian coordinates is not efficient for the solution process. For the particular example to be analyzed below, it is more advantageous to use cylindrical polar coordinates. For other problems, the primary dependent variables may not even be length or geometrical quantities. In the next few sections, we will show how the equations of motion in non-Cartesian (or generalized) coordinates can be obtained very efficiently with the help of the calculus of variations.

Consider a bead of mass $m$ sliding frictionless along a smooth wire in the shape of general helix. In Cartesian coordinates $(x, y, z)$ (of an inertial frame), the general helical curve is given by

$$\frac{x^2}{a^2} + \frac{y^2}{b^2} = 1, \qquad z = c\cos^{-1}\left(\frac{x}{a}\right) \tag{7.1}$$

for three positive constants $a, b$, and $c$. The only forces acting on the bead are (1) gravity in the negative $z$-direction, and (2) the normal reaction of the wire $\mathbf{N}$. The unknown force $\mathbf{N}$ is orthogonal to the tangent vector of the curve and, therefore, normal to the velocity vector of the bead:

$$\dot{x}N_x + \dot{y}N_y + \dot{z}N_z = 0. \tag{7.2}$$

Newton's second law of motion requires

$$m\ddot{x} = N_x, \qquad m\ddot{y} = N_y, \qquad m\ddot{z} = -mg + N_z, \tag{7.3}$$

where $(\dot{\ })$ indicates differentiation with respect to time $t$, and $N_s$ is the component of $\mathbf{N}$ in the $s$ direction. System (7.3) consists of three second order ODEs for six unknowns $\{x, y, z, N_x, N_y, N_z\}$. Three of the unknowns, say $y, z$, and $N_x$ in (7.3), may be expressed in terms of $N_y, N_z$, and $x$ by the three relations in (7.1) and (7.2). The resulting three equations together with appropriate initial conditions of the bead determine its subsequent motion. In practice, it is difficult to carry out this solution process. Fortunately, a much simpler alternative approach is possible.

In cylindrical coordinates $(r, \theta, z)$, the general helix (7.1) is more naturally described by

$$x = a\cos\theta, \qquad y = b\sin\theta, \qquad z = c\theta \tag{7.4}$$

Note that the two relations in (7.1) are identically satisfied by (7.4). In terms of $\theta$, the first and second derivatives of $\{x, y, z\}$ are given by

$$\dot{x} = -a\dot{\theta}\sin\theta, \qquad \dot{y} = -b\dot{\theta}\cos\theta, \qquad \dot{z} = -c\dot{\theta}, \tag{7.5}$$

and

$$\begin{aligned} \ddot{x} &= -a\dot{\theta}^2\cos\theta - a\ddot{\theta}\sin\theta, \\ \ddot{y} &= -b\dot{\theta}^2\sin\theta - b\ddot{\theta}\cos\theta, \\ \ddot{z} &= -c\ddot{\theta}. \end{aligned} \tag{7.6}$$

To get a single equation for the unknown $\theta(t)$, we work with the vector equation of motion

$$(m\dot{\mathbf{r}})^{\cdot} = \mathbf{N} - mg\mathbf{i}_z \tag{7.7}$$

with

$$\mathbf{r} = x(t)\mathbf{i}_x + y(t)\mathbf{i}_y + z(t)\mathbf{i}_z, \qquad \mathbf{N} = N_n\mathbf{n} + N_b\mathbf{b}, \tag{7.8}$$

where $\mathbf{i}_s$ is the unit directional vector in the $s$ direction and $\mathbf{n}$ and $\mathbf{b}$ are the unit normal and binormal vector of the general helix. The component representation of $\mathbf{N}$ is consistent with the fact that the normal reaction does not have a component tangent to the general helical wire. The unit tangent vector to the general helix is given by

$$\begin{aligned} \mathbf{t} &= \frac{\dot{\mathbf{r}}}{|\dot{\mathbf{r}}|} = \frac{1}{|\dot{\mathbf{r}}|}[\dot{x}\mathbf{i}_x + \dot{y}\mathbf{i}_y + \dot{z}\mathbf{i}_z] \\ &= \frac{1}{|\dot{\mathbf{r}}|}[-a\dot{\theta}\sin\theta\mathbf{i}_x + b\dot{\theta}\cos\theta\mathbf{i}_y + c\dot{\theta}\mathbf{i}_z] \end{aligned} \tag{7.9}$$

The component of the vector equation (7.7) in the direction of $\mathbf{t}$ is obtained by taking the dot product (of both sides) of (7.7) with $\mathbf{t}$. In this way, obtain

$$(a^2\sin^2\theta + b^2\cos^2\theta + c^2)\ddot{\theta} + (a^2 - b^2)\dot{\theta}^2\cos\theta\sin\theta = -cg. \tag{7.10}$$

This is a single second-order autonomous equation which may be reduced to a first-order equation for $\omega \equiv \dot{\theta}$, using $\theta$ as the independent variable:

$$(a^2\sin^2\theta + b^2\cos^2\theta + c^2)\omega\frac{d\omega}{d\theta} + (a^2 - b^2)\omega^2\cos\theta\sin\theta + cg = 0 \tag{7.11}$$

By inspection, (7.11) is a first-order linear ODE for $\omega^2$. The exact solution of this equation gives a separable first-order equation for $\theta(t)$. Along with the prescribed initial conditions, this ODE completely determines $\theta(t)$ and, hence, the motion of the bead [by way of (7.4)].

Even without actually obtaining the exact solution of the original problem, it should be evident that using the cylindrical coordinate system has significantly simplified the determination of the motion of the mass particle. Instead of limiting ourselves to the Cartesian coordinates $(x, y, z)$ [or $(y^{(1)}, y^{(2)}, y^{(3)})$ in the form used in Chapter 6], we should allow for other possible choices of variables $(q_1, q_2, q_3)$ in describing the motion of a mass particle. For the dynamics of $N$ particles, we should, allow for other combinations of $3N$ coordinates $\{q_1, q_2, q_3, \ldots, \ldots q_{3N}\}$ and not just the Cartesian coordinates $\{y_1^{(1)}, y_1^{(2)}, y_1^{(3)}, y_2^{(1)}, y_2^{(2)}, \ldots, y_N^{(1)}, y_N^{(2)}, y_N^{(3)}\}$. For some problems, the optimal choice of some of the $\{q_n\}$ may not even be geometrical quantities. The $3N$ variables $\{q_1, \ldots, q_{3N}\}$ are called ***generalized coordinates*** if they uniquely specify the Cartesian coordinates of the $N$ particles. For brevity, the adjective "generalized" will sometimes be omitted so that the $\{q_n\}$ will just be called the coordinates of the problem.

As an $N$-particle system evolves with time, each particle traces out a curve in space. The motion of the system can, therefore, be visualized by a set of $N$ curves in the physical space. However, as long as we are working with the $3N$ coordinates $\{q_1, \ldots, q_{3N}\}$ without specifically identifying certain coordinates with a particular particle, it is actually more convenient and desirable to treat the $N$ curves as a single orbit (or trajectory) in a space of $3N$-dimensions. We call this $3N$-dimensional space the ***configuration space***. Each point of the configuration space corresponds to a particular configuration of the system at a particular time. As time evolves, successive configurations trace out a "curve" in the $3N$-dimensional configuration space.

When the coordinates $\{q_i\}$ are actually the Cartesian coordinates with $q_{3k+1} = y_{k+1}^{(1)}$, $q_{3k+2} = y_{k+1}^{(2)}$ and $q_{3k+3} = y_{k+1}^{(3)}$, $k = 0, 1, 2, \ldots, N$, the equations of motion may be written as

$$\mu_i\ddot{q}_i = F_i \quad (i = 1, 2, \ldots, 3N), \tag{7.12}$$

where $\mu_{3k+1} = \mu_{3k+2} = \mu_{3k+3} = m_{k+1}$ and $F_{3k+i} = F_{k+1}^{(i)}$ for $i = 1, 2, 3$ and $k = 0, 1, 2, \ldots, (N-1)$. The solution of the system (7.12) depends on the $6N$

initial values of $\{q_i\}$ and $\{\dot{q}_m\}$. Through each point of the configuration space, there is a different orbit for each choice of the velocity $\dot{\mathbf{q}}$.

Just as in the case of the one-dimensional motion of a single particle, the motion of the $N$ particles can be captured and visualized more simply in the *phase space*. Let

$$\dot{q}_i = w_i \quad (i = 1, 2, \ldots, 3N) \tag{7.13a}$$

and write (7.12) as

$$\mu_i \dot{w}_i = F_i \quad (i = 1, 2, \ldots, 3N). \tag{7.13b}$$

Instead of (7.12), we may work with the system (7.13a) and (7.13b) of $6N$ first-order ODEs for the $6N$ unknowns $\{q_1, \ldots, q_{3N}\}$ and $\{w_1, \ldots, w_{3N}\}$. The $N$ physical trajectories traced out by the $N$ particles and the corresponding $N$ velocity curves are more usefully treated as a single orbit in a $6N$-dimensional space, called the *phase space* of the system. Each point in the phase space corresponds to a particular state of motion of the dynamical system, giving the position and velocity of each of the $N$ particles at a particular time. By the uniqueness of the solution of initial-value problems in ODE, the orbits in the phase space do not cross each other.

## 2. Coordinate Transformations

For more complex problems, the calculations leading from (7.7) to the simplified equation of motion (7.10) can be very tedious or untractable. The expressions for the second time derivatives corresponding to (7.6) are usually the source of difficulty. The calculus of variations can, in fact, help us to circumvent the need for that type of calculations and provide a simpler road to the desired equations of motion in generalized coordinates. We will introduce the essential ideas of the variational approach using the bead-wire example.

In terms of the angular variable $\theta$, the kinetic energy KE and the potential energy PE of bead-wire system are given by

$$\begin{aligned} \text{KE} &= \frac{1}{2} m\dot{\mathbf{y}} \cdot \dot{\mathbf{y}} = \frac{1}{2} m\dot{\theta}^2 \left[ a^2 \sin^2\theta + b^2 \cos^2\theta + c^2 \right] \equiv T(\theta, \dot{\theta}), \\ \text{PE} &= mgy^{(3)} = mgc\theta \equiv V(\theta). \end{aligned} \tag{7.14}$$

With only $\theta(t)$ varying independently, the Euler DE of

$$J[\mathbf{y}] = \int_{t_1}^{t_2} [T(\theta, \dot{\theta}) - V(\theta)]\, dt \equiv I[\theta] \tag{7.15}$$

is easly found to be the second-order ODE for $\theta(t)$

$$\frac{d}{dt}\left[\frac{\partial T}{\partial \dot{\theta}}\right] - \frac{\partial}{\partial \theta}[T - V] = 0. \tag{7.16}$$

By Hamilton's principle, it is the equation of motion for the bead-wire system. It is not difficult to verify that (7.16) is identical to (7.10).

The advantage of the variational approach above for obtaining the equation of motion of the bead is the absence of second time-derivative terms in the Lagrangian $T - V$, leaving further differentiation of terms involving velocity to the end, namely, in the Euler DE. This suggests that there should be considerable advantage in the use of Hamilton's principle with an appropriate set of generalized coordinates $\mathbf{Q} = (Q_1, Q_2, ..., Q_m)^T$, $m \le n$, for the analysis of complex dynamical systems. In general, the Lagrangian, $L \equiv KE - PE$, depends on $\mathbf{q} = (q_1, q_2, ..., q_n)^T$, $\dot{\mathbf{q}} = (\dot{q}_1, \dot{q}_2, ..., \dot{q}_n)^T$ and $t$, i.e., $L = L(t, \mathbf{q}, \mathbf{w})$ with $\mathbf{w} = \dot{\mathbf{q}}$. Given the coordinate transformation from $\mathbf{q}$ to $\mathbf{Q}$, we need the expression $\dot{q}_i(t)$ in terms of $\mathbf{Q}, \dot{\mathbf{Q}}$, and $t$ for any orbit $\mathbf{q}(t)$ in the configuration space.

In contrast to Chapter 6, we are interested here in coordinate transformations of the type

$$q_i = G_i(\mathbf{Q}, t) = G_i(Q_1, Q_2, ..., Q_m, t) \quad (i = 1, 2, ..., n). \tag{7.17}$$

Normally, $m$ is equal to $n$ and the transformation is *invertible*. In other cases such as the bead-wire system, we take advantage of the constraints in the system to reduce the number of unknown quantities so that $m < n$. For an orbit $\mathbf{Q}(t)$ in the configuration space, we have by the chain rule

$$\dot{q}_i = \frac{\partial G_i}{\partial t} + \sum_{k=1}^{m} \frac{\partial G_i}{\partial Q_k} \dot{Q}_k \quad (i = 1, 2, ..., n). \tag{7.18}$$

We can then use (7.17) and (7.18) to eliminate $\mathbf{q}$ and $\dot{\mathbf{q}}$ from $L$ to get

$$J[\mathbf{q}] = \int_{t_1}^{t_2} L(t, \mathbf{q}, \dot{\mathbf{q}})\, dt = \int_{t_1}^{t_2} L(t, \mathbf{Q}, \dot{\mathbf{Q}})\, dt \equiv I[\mathbf{Q}]. \tag{7.19}$$

The equations of motion for the new coordinates $\{Q_1, ..., Q_m\}$ are then obtained as the Euler DE of $I[\mathbf{Q}]$ with the $m$ components of $\delta\mathbf{Q}$ varying independently.

**EXAMPLE 1:** A mass particle (of mass $m$) moves under the influence of a conservative force $\mathbf{F}$. We want to work in spherical polar coordinates $(r, \varphi, \theta)$ instead of the usual Cartesian coordinates $(x, y, z)$ with

$$x = r \sin\varphi \cos\theta, \qquad y = r \sin\varphi \sin\theta, \qquad z = r \cos\varphi. \tag{7.20}$$

Let $q_1 = x$, $q_2 = y$, $q_3 = z$, $Q_1 = r$, $Q_2 = \varphi$ (the azimuthal angle) and $Q_3 = \theta$ (the polar angle). We may use (7.20) to write KE and PE as

$$\text{KE} = \frac{1}{2} m(\dot{r}^2 + r^2\dot{\varphi}^2 + r^2\dot{\theta}^2 \sin^2\varphi)$$

$$= \frac{1}{2} m(W_1^2 + Q_1^2 W_2^2 + Q_1^2 W_3^2 \sin^2 Q_2) \equiv T_Q(\mathbf{Q}, \mathbf{W}), \tag{7.21a}$$

$$\text{PE} = V(\mathbf{y}) = V_Q(\mathbf{Q}) \tag{7.21b}$$

with $\mathbf{W} = (\dot{r}, \dot{\varphi}, \dot{\theta}) = \dot{\mathbf{Q}}$. The equations of motion in the new coordinate system are obtained as the Euler DE of $J[\mathbf{y}] \equiv J_Q[\mathbf{Q}]$. With the Lagrangian $L = T_Q - V_Q$, Euler differential equations are

$$\frac{d}{dt}\left(\frac{\partial T_Q}{\partial \dot{Q}_k}\right) - \frac{\partial T_Q}{\partial Q_k} = -\frac{\partial V_Q}{\partial Q_k} \quad (k = 1, 2, 3) \tag{7.22}$$

or

$$m[\ddot{r} - r(\dot{\varphi}^2 + \dot{\theta}^2 \sin^2\varphi)] = -\frac{\partial V_Q}{\partial r}, \tag{7.23a}$$

$$m[r^2\ddot{\varphi} + 2r\dot{r}\dot{\varphi} - r^2\dot{\theta}^2 \sin\varphi \cos\varphi] = -\frac{\partial V_Q}{\partial \varphi}, \tag{7.23b}$$

$$m[r^2\dot{\theta}\sin^2\varphi]^{\cdot} = -\frac{\partial V_Q}{\partial \theta} \tag{7.23c}$$

For the general case, we need to show that the extremals of $I[\mathbf{Q}]$ of (7.19) are also the extremals of $J[\mathbf{q}]$ and conversely, at least for the case $m = n$ and an invertible transformation. For this purpose, consider a curve $\mathbf{Q}(t) = (Q_1(t), Q_2(t), \ldots, Q_m(t))^T$ in the $m$-dimensional configuration space with end points $\mathbf{C}$ and $\mathbf{D}$, i.e., with $\mathbf{Q}(a) = \mathbf{C}$ and $\mathbf{Q}(b) = \mathbf{D}$. Let $\mathbf{q}(t)$ be the corresponding curve in a different configuration space with the set of coordinates $(q_1, q_2, \ldots, q_n)^T = \mathbf{q}, m \leq n$, given in terms of $\mathbf{Q}$ by $q_i = G_i(t, \mathbf{Q})$, $i = 1, 2, \ldots, n$. At the end points, we write $\mathbf{q}(a) = \mathbf{A} = \mathbf{G}(a, \mathbf{C})$ and $\mathbf{q}(b) = \mathbf{B} = \mathbf{G}(b, \mathbf{D})$ with $\mathbf{G} = (G_1, \ldots, G_n)^T$.

We denote the generalized velocity $\dot{\mathbf{q}}(t)$ by $\mathbf{w}(t) = (w_1(t), \ldots, w_n(t))^T$ and observe that

$$w_i(t) = \dot{q}_i(t) = \frac{\partial G_i}{\partial t} + \sum_{j=1}^{m} \frac{\partial G_i}{\partial Q_j} \dot{Q}_j(t) \equiv \frac{\partial G_i}{\partial t} + \sum_{j=1}^{m} \frac{\partial G_i}{\partial Q_j} W_j(t), \tag{7.24}$$

where $W_j(t) \equiv \dot{Q}_j(t)$. A smooth extremal $\mathbf{q}(t)$ of the Lagrangian $F(t, \mathbf{q}, \mathbf{w})$ satisfies the Euler DE

$$\frac{d}{dt}(\widehat{F}_{,\dot{q}_i}) - \widehat{F}_{,q_i} = 0 \quad (i = 1, 2, \ldots, n). \tag{7.25}$$

Similarly, the same Lagrangian in the $\{Q_j\}$ coordinates, $\mathcal{F}(t, \mathbf{Q}, \mathbf{W}) = F(t, \mathbf{q}, \mathbf{w})$, has as its Euler DEs

$$\frac{d}{dt}(\widehat{\mathcal{F}}_{,\dot{Q}_i}) - \widehat{\mathcal{F}}_{,Q_i} = 0 \quad (i = 1, 2, \ldots, m) \tag{7.26}$$

for an extremal $\widehat{\mathbf{Q}}(t)$. The following result is important for changing coordinates:

**[VII.1]** *The two sets of Euler DE's (7.25) and (7.26) are related by*

$$\frac{d}{dt}\left(\frac{\partial \mathcal{F}}{\partial W_i}\right) - \frac{\partial \mathcal{F}}{\partial Q_i} = \sum_{k=1}^{n} \frac{\partial q_k}{\partial Q_i}\left[\frac{d}{dt}\left(\frac{\partial F}{\partial w_k}\right) - \frac{\partial F}{\partial q_k}\right], \tag{7.27}$$

$i = 1, 2, 3, \ldots, m$. *Both sides are evaluated at the extremal.*

The proof is strictly a matter of book-kepping. For a smooth extremal $\mathbf{Q} = \mathbf{Q}(t)$ and $\mathbf{W} = \dot{\mathbf{Q}}(t)$, we have from (7.17) that $G_k$ does not depend on $W_j$ so that

$$\begin{aligned}\frac{\partial \mathcal{F}}{\partial W_i} &= \sum_{k=1}^{n}\left[\frac{\partial F}{\partial q_k}\frac{\partial q_k}{\partial W_i} + \frac{\partial F}{\partial w_k}\frac{\partial w_k}{\partial W_i}\right] = \sum_{k=1}^{n}\frac{\partial F}{\partial w_k}\frac{\partial \dot{q}_k}{\partial W_i} \\ &= \sum_{k=1}^{n}\left[\frac{\partial F}{\partial w_k}\frac{\partial}{\partial W_i}\left(\frac{\partial G_k}{\partial t} + \sum_{j=1}^{m}\frac{\partial G_k}{\partial Q_j}\dot{Q}_j\right)\right]\end{aligned} \tag{7.28}$$

or (again by (7.17))

$$\begin{aligned}\frac{\partial \mathcal{F}}{\partial W_i} &= \sum_{k=1}^{n}\frac{\partial F}{\partial w_k}\sum_{j=1}^{m}\frac{\partial G_k}{\partial Q_j}\frac{\partial W_j}{\partial W_i} \\ &= \sum_{k=1}^{n}\sum_{j=1}^{m}\frac{\partial F}{\partial w_k}\frac{\partial G_k}{\partial Q_j}\delta_{ij} = \sum_{k=1}^{n}\frac{\partial F}{\partial w_k}\frac{\partial G_k}{\partial Q_i}\end{aligned} \tag{7.29}$$

and therewith

$$\begin{aligned}\frac{d}{dt}(\mathcal{F}_{,W_i}) &= \sum_{k=1}^{n}\frac{d}{dt}\left(\frac{\partial F}{\partial w_k}\frac{\partial G_k}{\partial Q_i}\right) \\ &= \sum_{k=1}^{n}\left\{\frac{d}{dt}\left(\frac{\partial F}{\partial w_k}\right)\frac{\partial G_k}{\partial Q_i} + \frac{\partial F}{\partial w_k}\left[\frac{\partial^2 G_k}{\partial t\,\partial Q_i} + \sum_{j=1}^{n}\frac{\partial^2 G_k}{\partial Q_i \partial Q_j}W_j\right]\right\}.\end{aligned} \tag{7.30}$$

With the help of (7.24), we have also

$$\frac{\partial \mathcal{F}}{\partial Q_i} = \sum_{k=1}^{n} \left[ \frac{\partial F}{\partial q_k} \frac{\partial q_k}{\partial Q_i} + \frac{\partial F}{\partial w_k} \frac{\partial w_k}{\partial Q_i} \right]$$

$$= \sum_{k=1}^{n} \left[ \frac{\partial F}{\partial q_k} \frac{\partial q_k}{\partial Q_i} + \frac{\partial F}{\partial w_k} \left[ \frac{\partial^2 G_k}{\partial t \partial Q_i} + \sum_{j=1}^{m} \frac{\partial^2 G_k}{\partial Q_i \partial Q_j} W_j \right] \right\}. \qquad (7.31)$$

Relation (7.27) follows on substracting (7.31) from (7.30).

The following fact pertaining to the extremals of $F(t, \mathbf{q}, \mathbf{w})$ and $\mathcal{F}(t, \mathbf{Q}, \mathbf{W})$ is an immediate consequence of [VII.1]:

**[VII.2]** *If* $\mathbf{q}(t) \equiv \mathbf{G}(t, \mathbf{Q}(t))$ *is an extremal of* $\mathbf{F}(t, \mathbf{q}, \mathbf{w})$, *then* $\mathbf{Q}(t)$ *is an extremal of*

$$\mathcal{F}(t, \mathbf{Q}, \mathbf{W}) \equiv \mathbf{F}\left(t, \mathbf{G}, \mathbf{G}_{,t} + \sum_{j=1}^{m} \frac{\partial \mathbf{G}}{\partial Q_j} W_j\right).$$

*Furthermore, if* $m = n$ *(and* $\mathbf{G}$ *is invertible), then* $\mathbf{q}(t)$ *is also an extremal of* $\mathbf{F}(t, \mathbf{q}, \mathbf{w})$ *if* $\mathbf{Q}(t)$ is an extremal of $\mathcal{F}(t, \mathbf{Q}, \mathbf{W})$.

**EXAMPLE 2:** The bead-wire system of Section 1.

The Lagrangian for this problem is given by

$$L(t, \mathbf{y}, \dot{\mathbf{y}}) = \frac{1}{2} m (\dot{x}^2 + \dot{y}^2 + \dot{z}^2) - mgz.$$

With the coordinate transformation (7.4), we have

$$L(t, \mathbf{y}, \dot{\mathbf{y}}) = \frac{1}{2} m\dot{\theta}^2 (a^2 \sin^2\theta + b^2 \cos^2\theta + c^2) - mgc\theta \equiv \mathcal{L}(t, \theta, \dot{\theta}).$$

The Euler DE for $\mathcal{L}(t, \theta, \dot{\theta})$ is identical to the equation of motion for the problem as given in (7.11).

The same Euler equation can also be obtained by a direct application of [VII.1] instead because we already have the equations of motion in the $\mathbf{q} \equiv (x, y, z)^T$ coordinate system. We will show how this is accomplished for the general case.

For a dynamical system in a Cartesian inertial reference frame, the $q_k$-component of the momentum vector is given by $\partial T / \partial w_k \equiv \partial T / \partial \dot{q}_k$ where $T(t, \dot{\mathbf{q}})$ is the kinetic energy. Because $T$ does not depend on $\mathbf{q}$ in Cartesian coordinates, the equations of motion can be written as

$$\frac{d}{dt}\left(\frac{\partial T}{\partial w_k}\right) - \frac{\partial T}{\partial q_k} = f_k \quad (k = 1, 2, \ldots, n), \qquad (7.32)$$

where $\mathbf{f} = (f_1, \ldots, f_n)^T$ is the external force vector with $f_k$ as its components in the Cartesian inertial frame, whether or not the external force is conservative. For a new set of coordinates $Q_i = g_i(t, \mathbf{q})$, $= 1, 2, \ldots, n$, where the transformation $\mathbf{g}$ is invertible, (7.27) may be taken in the form

$$\frac{d}{dt}\left(\frac{\partial T}{\partial w_k}\right) - \frac{\partial T}{\partial q_k} = \sum_{i=1}^{n} \frac{\partial g_i}{\partial q_k}\left[\frac{d}{dt}\left(\frac{\partial \tau}{\partial W_i}\right) - \frac{\partial \tau}{\partial Q_i}\right],$$

where $\tau(t, \mathbf{Q}, \mathbf{W}) = T(t, \dot{\mathbf{q}})$ by (7.24) with $\mathbf{W} = \dot{\mathbf{Q}} = (\dot{Q}_1, \ldots, \dot{Q}_n)^T$. We can then write the left-hand side of the $n$ equations of motion (7.32) in terms of $\tau$ and form

$$\sum_{k=1}^{n}\sum_{i=1}^{n} \frac{\partial g_i}{\partial q_k}\left[\frac{d}{dt}\left(\frac{\partial \tau}{\partial W_i}\right) - \frac{\partial \tau}{\partial Q_i}\right]\frac{\partial q_k}{\partial Q_j} = \sum_{k=1}^{n} f_k \frac{\partial q_k}{\partial Q_j} \equiv \bar{f}_j. \tag{7.33}$$

With

$$\frac{\partial g_i}{\partial q_k}\frac{\partial q_k}{\partial Q_j} = \frac{\partial Q_i}{\partial q_k}\frac{\partial q_k}{\partial Q_j} = \delta_{ij},$$

we obtain from (7.32) and (7.33) the following result for the equations of motion in the new coordinate system:

**[VII.3]** *Let* $\mathrm{T}(\mathrm{t}, \dot{\mathbf{q}})$ *be the KE of a dynamical system relative to a Cartesian inertial reference frame. Suppose* $\mathrm{Q}_\mathrm{j} = \mathrm{g}_\mathrm{i}(\mathrm{t}, \mathbf{q})$, $\mathrm{i} = 1, 2, \ldots, \mathrm{n}$, *with* $\mathrm{q}_\mathrm{k} = \mathrm{G}_\mathrm{k}(\mathrm{t}, \mathbf{Q})$, $\mathrm{k} = 1, 2, \ldots, \mathrm{n}$, *as the inverse transformation and* $\mathrm{T}(\mathrm{t}, \dot{\mathbf{q}}) = \tau(\mathrm{t}, \mathbf{Q}, \dot{\mathbf{Q}})$. *Then the equations of motion in the generalized coordinates* $\mathbf{Q} = (\mathrm{Q}_1, \mathrm{Q}_2, \ldots, \mathrm{Q}_\mathrm{n})^T$ are

$$\frac{\mathrm{d}}{\mathrm{dt}}\left(\frac{\partial \tau}{\partial \mathrm{W}_\mathrm{j}}\right) - \frac{\partial \tau}{\partial \mathrm{Q}_\mathrm{j}} = \bar{\mathrm{f}}_\mathrm{j}, \quad (\mathrm{j} = 1, 2, \ldots, \mathrm{n}) \tag{7.34}$$

*where the generalized forces* $\{\bar{\mathrm{f}}_\mathrm{j}\}$ *are defined in (7.33) and* $\{\mathrm{W}_\mathrm{j} = \dot{\mathrm{Q}}_\mathrm{j}\}$ *are the generalized velocities.*

**EXAMPLE 3:** Same as Example 1 except that the external force $\mathbf{f}(t)$ is not necessarily conservative.

Suppose in the Cartesian coordinate system $\mathbf{q} = (q_1, q_2, q_3)^T = (x, y, z)^T$, and $\mathbf{f}$ has components $(f_1, f_2, f_3)^T$. In terms of the new coordinates $(r, \varphi, \theta)$, we have

$$\bar{f}_1 \equiv \bar{f}_r = \sum_{k=1}^{3} \frac{\partial q_k}{\partial r} f_k = f_1 \sin\varphi\cos\theta + f_2 \sin\varphi\sin\theta + f_3\cos\varphi,$$

$$\bar{f}_2 \equiv \bar{f}_\varphi = \sum_{k=1}^{3} \frac{\partial q_k}{\partial \varphi} f_k = r(f_1\cos\varphi\cos\theta + f_2\cos\varphi\sin\theta - f_3\sin\varphi),$$

$$\bar{f}_3 \equiv \bar{f}_\theta = \sum_{k=1}^{3} \frac{\partial q_k}{\partial \theta} f_k = r(-f_1 \sin\varphi \sin\theta + f_2 \sin\varphi \cos\theta).$$

The equations of motion in spherical coordinates are then given by (7.23) with the right-hand side replaced by the expression for $\{\bar{f}_j\}$ above.

For a conservative system so that $f_k = -\partial U/\partial q_k$ with $U = U(t, \mathbf{q})$, we have

$$\bar{f}_i = \frac{\partial U}{\partial q_k}\frac{\partial q_k}{\partial Q_i} = -\frac{\partial \mu}{\partial Q_i}, \qquad \mu(t, \mathbf{Q}) = U(t, \mathbf{G}(t, \mathbf{Q})).$$

It follows that the equations of motion in the new coordinate system become

$$\frac{d}{dt}\left(\frac{\partial \tau}{\partial W_j}\right) - \frac{\partial \tau}{\partial Q_j} + \frac{\partial \mu}{\partial Q_j} = 0$$

or, in terms of the new Lagrangian $\mathcal{L}(t, \mathbf{Q}, \dot{\mathbf{Q}}) \equiv \tau - \mu = T - U = L(t, \mathbf{q}, \dot{\mathbf{q}})$,

$$\frac{d}{dt}\left(\frac{\partial \mathcal{L}}{\partial W_j}\right) - \frac{\partial \mathcal{L}}{\partial Q_j} = 0 \quad (j = 1, 2, \ldots, n) \tag{7.35}$$

with $W_i = \dot{Q}_i$. The $n$ equations of (7.35) are the Euler DE of the Lagrangian $L = \mathrm{T} - \mathrm{U}$ expressed in the new coordinates $\{Q_i\}$.

## 3. Holonomic Constraints

In many mechanical problems, the dynamic system of interest is not completely free to evolve. Rather, its motion is subject to certain costraints. These constraints are sometimes geometrical in nature. The bead-wire system of Section 1 is one such problem with (7.1) being the (two) geometrical constraints. The bead is not free to move in space; it must be confined to move along the general helical curve. Another example is the motion of a mass particle on a smooth spherical surface of radius $a$ under the influence of gravity. In Cartesian coordinates, the single constraint $x^2 + y^2 + z^2 = a^2$ is again geometrical. Analysis of these dynamical systems by way of Hamilton's principle gives rise to problem in the calculus of variations with auxiliary contraints for each instance of time. The are known as *point constraints*.

In the two examples given above, the constraints are of the form $f_i(\mathbf{q}) = 0, i = 1, 2, \ldots, n - m$, for some $m$ (The new functions $\{f_i\}$ are not to be confused with the forces acting on the dynamical system in the previous section). For the bead-wire system, we have $n - m = 2$ (and therewith $m = 1$ because $n = 3$). For the "particle on the sphere" problem, there is only one costraint so that $m = n - 1 = 2$. In general, we call constraints of the form

$$f_i(t, \mathbf{q}) = 0 \quad (i = 1, 2, \ldots, n - m) \tag{7.36}$$

*holonomic constraints*; they do not inovolve the (generalized) velocities $\{\dot{q}_i\}$. Those which do are said to be *nonholonomic*. General variational problems with auxiliary constraints will be analyzed by the method of Lagrange multipliers later in Chapters 10-14. For some holonomic constraints, however, a more direct method based on coordinate transformations is often effective. The discussion of coordinate transformations in Section 2 was motivated by the application of this method for the bead-wire problem.

Suppose a basic problem in the calculus of variations with a general Lagrangian $F(t, \mathbf{q}, \dot{\mathbf{q}})$ is subject to $(n - m)$ holonomic constraints (7.36). In general, these constraints should reduce the number of degress of freedom of the problem from $n$ to $m$. This conclusion is based on the observation that we can solve the $n - m$ relations of (7.36) for $n - m$ of the $n$ coordinates $\{q_i\}$ in terms of the remaining $m$ coordinates if (7.36) are $n - m$ "independent" relations. More precisely, a set of $n - m$ constraints (7.36) is said to be *independent* if the $(n - m) \times n$ matrix $[\partial f_i / \partial q_k]$ has maximal rank for each point in the configuration space. The following theorem forms the basis of our method for variational problems with holonomic constraints:

**[VII.4]** ***If the holonomic constraints (7.36) are independent and the quantities $\{f_i\}$ are $C^1$ functions, then there is always a system of generalized coordinates $Q = (Q_1, \ldots, Q_n)^T$ in which the constraints are $Q_{m+i} = 0$, $i = 1, \ldots, n - m$.***

Because the matrix $[\partial f_j / \partial q_k]$ is of maximal rank, we may assume with no loss of generality that the Jacobian matrix

$$\frac{\partial(f_1, \ldots, f_{n-m})}{\partial(q_{m+1}, \ldots, q_n)} \equiv \begin{bmatrix} \dfrac{\partial f_1}{\partial q_{m+1}} & \dfrac{\partial f_1}{\partial q_{m+2}} & \cdots & \cdots & \dfrac{\partial f_1}{\partial q_n} \\ \dfrac{\partial f_2}{\partial q_{m+1}} & \cdots & \cdots, & \cdots & \dfrac{\partial f_2}{\partial q_n} \\ \vdots & & & & \\ \vdots & & & & \\ \dfrac{\partial f_{n-m}}{\partial q_{m+1}} & \cdots & \cdots & \cdots & \dfrac{\partial f_{n-m}}{\partial q_n} \end{bmatrix} \tag{7.37}$$

is nonsingular in the region of the configuration space of interest. By the implicit function theorem, we can solve for $(q_{m+1}, \ldots, q_n)$ in terms of $(q_1, \ldots, q_m)$ and $t$ so that $q_{m+i} = h_i(t, q_1, \ldots, q_m)$, $i = 1, \ldots, n - m$. We may then take as our new generalized coordinates $Q_i = g_i(t, q_1, ..., q_m)$, $i = 1, ..., m$, for some suitable $\{g_i\}$ and $Q_j = f_j(t, \mathbf{q}) = \bar{f}_j(t, q_1, ..., q_m)$, $j = m+1, ..., n$.

The integer $m$ is called the *residual degree of freedom* for the system and the new coordinates $\{Q_1, ..., Q_n\}$ are said to be *adapted to the constraints*. In pratice, we often choose $q_i = G_i(t, \mathbf{Q}), i = 1, 2, ..., m$, instead of $Q_i = g_i(t, q_1, ..., q_m)$.

**EXAMPLE 2 (continued):**

For the bead-wire system of Section 1, we take the constraints (7.1) in the form

$$f_1 \equiv \frac{q_2^2}{a^2} + \frac{q_3^2}{b^2} - 1, \quad f_2 \equiv \frac{q_2}{a} - \cos\left(\frac{q_1}{c}\right).$$

The Jacobian matrix

$$\frac{\partial(f_1, f_2)}{\partial(q_2, q_3)} = \begin{bmatrix} \dfrac{2q_2}{a^2} & \dfrac{2q_3}{b^2} \\ \dfrac{1}{a} & 0 \end{bmatrix}$$

is nonsingular for $q_3 \neq 0$. We take $Q_1$ to be the polar angle $\theta$ and $q_2 = a\cos\theta = a\cos Q_1$. To satisfy the constraints, we now set $q_3 = b\sin Q_1$ (so that $f_1 = 0$) and $q_1 = cQ_1$ (so that $f_2 = 0$).

**EXAMPLE 4:** A particle of mass $m$ moves on (the surface of) a smooth sphere of radius $a$ with gravity as the only external force.

The single constraint in this case is $f_1 \equiv q_1^2 + q_2^2 + q_3^2 - a^2 = 0$. The Jacobian $\partial f_1 / \partial q_i \neq 0$ for any $i$ (because the origin is excluded). Instead of solving $f_1 = 0$ for $q_i$ in terms of the other two coordinates, it is more natural to introduce spherical coordinates and set

$$q_1 = (Q_3 + a)\sin Q_1 \cos Q_2, \quad q_2 = (Q_3 + a)\sin Q_1 \sin Q_2, \quad q_3 = (Q_3 + a)\cos Q_1.$$

Note that the constraint $f_1 = 0$ is satisfied by $Q_3 = 0$. In terms of the new coordinates, we have

$$T(t, \dot{\mathbf{q}}) \equiv \tau(t, \dot{\mathbf{Q}}) = \frac{1}{2} ma^2(\dot{\varphi}^2 + \dot{\theta}^2 \sin^2\varphi),$$

$$U(t, \mathbf{q}) \equiv \mu(t, \mathbf{Q}) = mga\cos\varphi.$$

With $\mathcal{L} = \tau - \mu$, the equations of motion are easily obtained upon applications of (7.35) for $j = 1$ and 2 (as there are only two residual degrees of freedom) to be

$$\frac{d}{dt}(ma^2\dot{\varphi}) - ma^2\dot{\theta}^2\cos\varphi\sin\varphi - mga\sin\varphi = 0$$

and

$$\frac{d}{dt}(ma^2\dot{\theta}\sin^2\varphi) = 0,$$

**EXAMPLE 5:** Consider the motion of a bead sliding frictionlessly along a smooth circular wire of radius $a$. The wire is rotating with constant angular velocity $\omega_0$ about an axis through the center of the circle and inclined at an angle $[(\pi/2 - \alpha)]$ to the plane of the wire for an acute angle $\alpha$, $0 < \alpha < \pi/2$.

Choose a rotating Cartesian reference frame with its origin at the center of the circle and with $\mathbf{j}_3$ normal to the plane of the circle. Let $\mathbf{j}_1$ be along the intersection of the plane of the circle and the plane spanned by $\mathbf{j}_3$ and the axis of the rotation $\boldsymbol{\omega}$. Finally, take $\mathbf{j}_2 = \mathbf{j}_3 \times \mathbf{j}_1$. In that case, we have $\boldsymbol{\omega} = \omega_0(\sin\alpha\,\mathbf{j}_1 + \cos\alpha\,\mathbf{j}_3)$. With $q_k$ being the coordinate in the $\mathbf{j}_k$ direction, the constraints on the motion of the bead are

$$f_1 \equiv q_1^2 + q_2^2 - a^2 = 0, \qquad f_2 \equiv q_3 = 0$$

The matrix

$$\begin{bmatrix} \dfrac{\partial f_1}{\partial q_1} & \dfrac{\partial f_1}{\partial q_2} & \dfrac{\partial f_1}{\partial q_3} \\[2ex] \dfrac{\partial f_2}{\partial q_1} & \dfrac{\partial f_2}{\partial q_2} & \dfrac{\partial f_2}{\partial q_3} \end{bmatrix} = \begin{bmatrix} 2q_1 & 2q_3 & 0 \\ 0 & 0 & 1 \end{bmatrix}$$

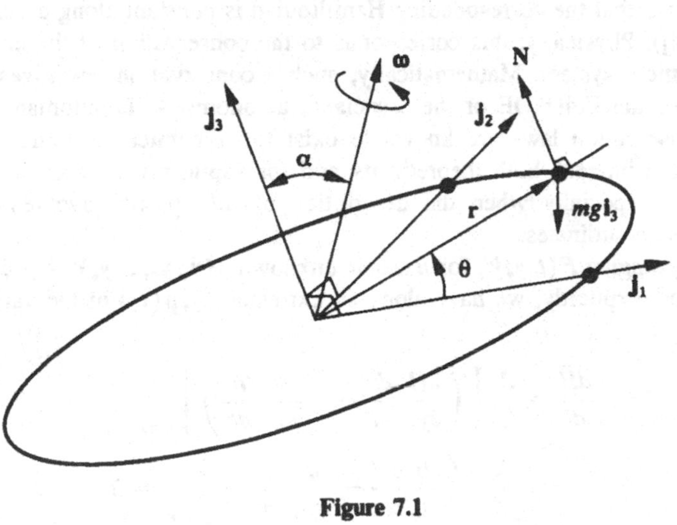


Figure 7.1

is of maximal rank (because $\mathbf{q} = 0$ is excluded). The residual degree of freedom is, therefore, one. We take this to be the polar angle $\theta$ in the plane of the wire so that the position vector of the bead is $\mathbf{r} = a(\cos\theta\,\mathbf{j}_1 + \sin\theta\,\mathbf{j}_2)$.

The velocity of the bead relative to the inertial frame is

$$\dot{\mathbf{r}} + \boldsymbol{\omega} \times \mathbf{r} = -a\sin\theta(\dot{\theta} + \omega_0\cos\alpha)\mathbf{j}_1 + a\cos\theta(\dot{\theta} + \omega_0\cos\alpha)\mathbf{j}_2 + \omega_0 a\sin\theta\,\sin\alpha\,\mathbf{j}_3 .$$

The kinetic energy is given by $m|\dot{\mathbf{r}} + \boldsymbol{\omega} \times \mathbf{r}|^2/2$ so that

$$T(t,\dot{\mathbf{q}}) = \tau(t,\dot{\theta}) = \frac{1}{2}ma^2[\dot{\theta}^2 + 2\omega_0\dot{\theta}\cos\alpha + \omega_0^2(\sin^2\alpha\,\sin^2\theta + \cos^2\alpha)] .$$

The potential energy due only to gravity is

$$U(t,\mathbf{q}) = \mu(t,\theta) = mga\sin\alpha\cos\theta .$$

The single equation of motion in the new coordinates is obtained from (7.35) with $\mathcal{L} = \tau(t,\dot{\theta}) - \mu(t,\theta)$ to be

$$\ddot{\theta} - \omega_0^2\sin^2\alpha\,\sin\theta\,\cos\theta - \frac{g}{a}\sin\alpha\,\sin\theta = 0 .$$

## 4. Poisson Brackets

When the Lagrangian $F(x,y,y')$ does not depend on $x$ explicitly, we learned in Chapter 6 that the corresponding Hamiltonian is constant along an extremal (see [VI.2]). Physically, this corresponds to the conservation of the energy of the dynamical system. Mathematically, such a conservation law gives a first integral for the Euler DE or the associated autonomous Hamiltonian system. Other conservation laws are known to exist for dynamical systems. It is of considerable interest, both theoretically and for applications, to know how to find them, especially when the description of the system involves several generalized coordinates.

If a Lagrangian $F(t,\mathbf{y},\dot{\mathbf{y}})$ for $n$ scalar unknowns $(y_1,y_2,...y_n)^T = \mathbf{y}$ does not depend on $t$ explicitly, we have along an extremal $(\hat{\mathbf{y}},\hat{\mathbf{p}}(t))$ of the variational problem,

$$\frac{d\hat{H}}{dt} = \sum_{k=1}^{n}\left[\left(\frac{\partial H}{\partial y_k}\frac{dy_k}{dt} + \frac{\partial H}{\partial p_k}\frac{dp_k}{dt}\right)\right]_{(\hat{\mathbf{y}},\hat{\mathbf{p}})}$$
$$= \sum_{k=1}^{n}\left[\left(\frac{\partial H}{\partial y_k}\frac{\partial H}{\partial p_k} - \frac{\partial H}{\partial p_k}\frac{\partial H}{\partial y_k}\right)\right]_{(\hat{\mathbf{y}},\hat{\mathbf{p}})} = 0,$$

because the corresponding Hamiltonian does not depend on $t$ explicitly. Hence, *H is a constant along an extremal of a conservative dynamical system* and $\widehat{H}(t) = H_0$ is said to be a *first integral* of the autonomous Hamiltonian system. (Note that it is possible to have $H$ not dependent on $t$ explicitly, but still to have $\partial F/\partial t \neq 0$.) We have then a generalization of [VI.2] to $n$ unknowns.

Suppose $K(\mathbf{y}, \mathbf{p})$ is another first integral of the same autonomous Hamiltonian system. We must have $d\widehat{K}/dt = 0$ along an extremal of the system. But, a similar calculation gives

$$\frac{dK}{dt} = \sum_{k=1}^{n} \left[ \frac{\partial K}{\partial y_k} \frac{\partial H}{\partial p_k} - \frac{\partial K}{\partial p_k} \frac{\partial H}{\partial y_k} \right] \equiv [K, H]$$

which should vanish when evaluated at an extremal. The espression $[K, H]$ is called the *Poisson bracket* of the functions $K(\mathbf{y}, \mathbf{p})$ and $H(\mathbf{y}, \mathbf{p})$. We have then the following elegant characterization of a first integral of an autonomous Hamiltonian system:

**[VII.5]** ***If*** $\mathrm{K}(\mathbf{y}, \mathbf{p})$ ***is a first integral of an autonomous Hamiltonian system with Hamiltonian*** $\mathrm{H}(\mathbf{y}, \mathbf{p})$***, then the Poisson bracket*** $[\mathrm{K}, \mathrm{H}]$ ***vanishes identically for any solution of the Hamiltonian system.***

For the more general case when $H$ depends on $t$ as well, $K(t, \mathbf{y}, \mathbf{p})$ is a first integral of the Hamiltonian system if

$$\frac{dK}{dt} = \frac{\partial K}{\partial t} + [K, H] = 0.$$

If the Poisson bracket vanishes identically in this case, $K$ must not depend on $t$ explicitly (even if $H$ does).

It can be shown (see Exercises) that a Poisson bracket is skew-symmetric:

$$[K, H] = -[H, K] \tag{7.38}$$

and satisfies the following Jacobi identity for any three $C^1$ functions $H(\mathbf{y}, \mathbf{p})$, $K(\mathbf{y}, \mathbf{p})$, and $M(\mathbf{y}, \mathbf{p})$:

$$[H, [K, M]] + [K, [M, H]] + [M, [H, K]] = 0. \tag{7.39}$$

Also, Hamiltonian systems themselves can be written in terms of Poisson brackets:

$$\dot{y}_k = [H, y_k], \qquad \dot{p}_k = [H, p_k]. \tag{7.40}$$

The properties (7.38)–(7.40) of Poisson brackets are valid whether or not the functions involved depends on $t$ explicity. Given (7.40) and [VII.5], it is not surprising that the theory of canonical transformations and the Hamilton–Jacobi equation leading to the solution of Hamiltonian systems can be developed by way of Poisson brackets. Such an approach would seek transformations of coordinates which preserve Poisson brackets. We will not pursue a discussion along this line but will continue to develop a method for finding first integrals of general Hamiltonian systems leading to Noether's theorem in Section 6.

## 5. Variationally Invariant Lagrangians

To develop a systematic method for finding first integrals of Euler DEs, we consider the effect of a different kind of transformations on the Lagrangian of a variational problem and on the variational problem itself. Suppose we have the following one-parameter family of transformations

$$X = \varphi(x, y; \varepsilon), \qquad Y = \psi(x, y; \varepsilon) \tag{7.41}$$

with

$$\varphi(x, y; 0) = x, \qquad \psi(x, y; 0) = y. \tag{7.42}$$

The functions $\varphi$ and $\psi$ are at least $C^2$. An example of such a family is

$$X = x + \varepsilon, \qquad Y = y \tag{7.43}$$

Another is

$$X = x\cos\varepsilon + y\sin\varepsilon, \qquad Y = x\sin\varepsilon + y\cos\varepsilon. \tag{7.44}$$

We denote the Jacobian matrix of $\varphi$ and $\psi$ by $j(\varepsilon)$ with

$$j(\varepsilon) = \begin{vmatrix} \varphi_{,x} & \varphi_{,y} \\ \psi_{,x} & \psi_{,y} \end{vmatrix}, \qquad j(0) = 1. \tag{7.45}$$

It follows that we may invert the transformation (7.41) to obtain

$$x = \Phi(X, Y; \varepsilon), \qquad y = \Psi(X, Y; \varepsilon) \tag{7.46}$$

for $|\varepsilon| \ll 1$. The inverted relations for the first example are

$$x = X - \varepsilon, \qquad y = Y, \tag{7.43'}$$

whereas those for the second example are

$$x = X\cos\varepsilon - Y\sin\varepsilon, \qquad y = X\sin\varepsilon + Y\cos\varepsilon. \tag{7.44'}$$

Given a PWS function $\bar{y}(x)$, we have for the first set of transformations (7.43)

$$Y = \bar{y}(x) = \bar{y}(X - \varepsilon) \equiv \widehat{Y}_\varepsilon(X). \tag{7.47}$$

For the specific example $\overline{y}(x) = e^x$, relationship (7.47) becomes

$$Y = \overline{y}(x) = e^x = e^{x-\varepsilon} \equiv \overline{Y}_\varepsilon(X) \tag{7.48}$$

For $\overline{y}(x)$ defined in an interval $t \le x \le \tau$, we have $t_\varepsilon \equiv t + \varepsilon < X < \tau + \varepsilon \equiv \tau_\varepsilon$ and $\overline{Y}_\varepsilon(X)$ is, therefore, defined in $[t_\varepsilon, \tau_\varepsilon]$.

More generally, we have for a PWS $\overline{y}(x)$ defined in $[t, \tau]$,

$$\varphi(t, \overline{y}(t); \varepsilon) = t_\varepsilon, \qquad \varphi(\tau, \overline{y}(\tau); \varepsilon) = \tau_\varepsilon \tag{7.49}$$

with $t_0 = t$ and $\tau_0 = \tau$. If $\varepsilon$ is sufficiently small, we have $t_\varepsilon < \tau_\varepsilon$. For every $X$ in $[t_\varepsilon, \tau_\varepsilon]$, we set

$$\overline{Y}_\varepsilon(X) = \psi(x, \overline{y}(x); \varepsilon), \tag{7.50}$$

where $x$ is the solution of $X = \varphi(x, \overline{y}(x); \varepsilon) \equiv \overline{\varphi}(x; \varepsilon)$. The solution $x = \overline{\Phi}(X; \varepsilon)$ is well defined given that $d\overline{\varphi}/dx = \varphi_{,x} + \varphi_{,y}\overline{y}'$ is unity at $\varepsilon = 0$ and near unity for $0 < |\varepsilon| \ll 1$.

The Lagrangian $F(x, y, y')$ of $J[y]$ is said to be *variationally invariant over* $[a, b]$ *under the one-parameter family of* $C^2$ *transformations (7.41) if for*

(i) all pairs $[t, \tau]$ in $[a, b]$ with $t < \tau$,

(ii) all PWS functions $\overline{y}(x)$ on $[t, \tau]$,

(iii) all $\varepsilon$ near 0,

we have

$$\int_t^\tau F(x, \overline{y}(x), \overline{y}'(x))\, dx = \int_{t_\varepsilon}^{\tau_\varepsilon} F(X, \overline{Y}_\varepsilon(X), \overline{Y}_\varepsilon^{\bullet}(X))\, dX, \tag{7.51}$$

where $t_\varepsilon = \varphi(t, \overline{y}(t); \varepsilon), \tau_\varepsilon = \varphi(\tau, \overline{y}(\tau); \varepsilon)$, and $(\ )^{\bullet} = d(\ )/dX$. In other words, we have $J_F[\overline{y}(x)] = J_F[\overline{Y}_\varepsilon(X)]$ if $F$ is variationally invariant in $[a, b]$ under (7.41). The integral $J[y]$ is said to be *invariant under the transformation* (7.41) if (7.51) holds.

EXAMPLE 6: $F = \sqrt{1 + (y')^2}$, $X = \Phi(x, y; \varepsilon) = x\sqrt{1 - \varepsilon^2} - y\varepsilon$, and $Y = \psi(x, y; \varepsilon) = x\varepsilon + y\sqrt{1 - \varepsilon^2}$.

For any PWS function $y = \overline{y}(x)$, we have $X = x\sqrt{1 - \varepsilon^2} - \overline{y}(x)\varepsilon \equiv X_\varepsilon(x)$ and $Y = x\varepsilon + \overline{y}(x)\sqrt{1 - \varepsilon^2} \equiv Y_\varepsilon(x)$. We can solve the first relation $X = X_\varepsilon(x)$ to get $x = \overline{\Phi}(X; \varepsilon)$ and use it to eliminate $x$ from the second expression $Y = Y_\varepsilon(x)$ to get $Y = \overline{Y}_\varepsilon(X)$. With

$$\frac{dX_\varepsilon}{dx} = \sqrt{1 - \varepsilon^2} - \overline{y}'(x)\varepsilon$$

and

$$\frac{d\overline{Y}_\varepsilon}{dX} = \frac{d\psi/dx}{dX/dx} = \frac{\psi_{,x} + \psi_{,y}\overline{y}'}{\sqrt{1-\varepsilon^2} - \varepsilon\overline{y}'} = \frac{\varepsilon + \sqrt{1-\varepsilon^2}\,\overline{y}'}{\sqrt{1-\varepsilon^2} - \varepsilon\overline{y}'},$$

the expression for $J[\overline{Y}_\varepsilon]$ may be written in terms of $x$ and $\overline{y}(x)$ to get

$$J[\overline{Y}_\varepsilon] = \int_{a_\varepsilon}^{b_\varepsilon} \sqrt{1 + (d\overline{Y}_\varepsilon/dX)^2}\, dX = \int_a^b \sqrt{1 + \frac{(\varepsilon + \sqrt{1-\varepsilon^2}\,\overline{y}')^2}{(\sqrt{1-\varepsilon^2} - \varepsilon\overline{y}')^2}}\, \frac{dX_\varepsilon}{dx}\, dx$$

$$= \int_a^b \sqrt{1 + (\overline{y}')^2}\, \frac{dX_\varepsilon/dx}{\sqrt{1-\varepsilon^2} - \varepsilon\overline{y}'}\, dx = \int_a^b \sqrt{1 + (\overline{y}')^2}\, dx = J[\overline{y}].$$

Hence, $J[y]$ is variationally invariant under the given one-parameter family of transformations.

**EXAMPLE 7a:** $F = F(y, y')$ with $X = x + \varepsilon$ and $Y = y$.

Evidently, all Lagrangians which do not depend on $x$ explicitly are variationally invariant under a simple translation in $x$, (7.43), over $[a, b]$ because $x = X - \varepsilon, Y = y(X - \varepsilon) \equiv Y_\varepsilon(X)$, and therewith

$$\int_{t_\varepsilon}^{\tau_\varepsilon} F(\overline{Y}_\varepsilon(X), \overline{Y}_\varepsilon'(X))\, dX = \int_{t+\varepsilon}^{\tau+\varepsilon} F\left(\overline{y}(X-\varepsilon), \overline{y}'(X-\varepsilon)\frac{dx}{dX}\right) dX$$

$$= \int_t^\tau F(\overline{y}(x), \overline{y}'(x))\, dx.$$

In particular, $F = (y')^2$ is variationally invariant under (7.43) as it is of the form $F(y')$.

**EXAMPLE 7b:** $F = x(y')^2$ is not variationally invariant under (7.43).

This is seen from

$$J_F[\overline{Y}_\varepsilon(X)] = \int_{t_\varepsilon}^{\tau_\varepsilon} X\left(\frac{d\overline{Y}_\varepsilon}{dX}\right)^2 dX = \int_{t+\varepsilon}^{\tau+\varepsilon} X\left(\frac{d[\overline{y}(X-\varepsilon)]}{dX}\right)^2 dX$$

$$= \int_t^\tau (x+\varepsilon)\left(\frac{d[\overline{y}(x)]}{dx}\right)^2 dx = J_F[\overline{y}] + \varepsilon\int_t^\tau [\overline{y}'(x)]^2\, dx.$$

It is not difficult to see from Example 7b that $F$ is generally not variationally invariant under (7.43) if $F$ depends on $x$ explicitly.

## 6. Noether's Theorem

When the Lagrangian $F$ does not depend on $x$ explicitly, we know from [I.4] that the Euler DE has a first integral or, what is the same, $H$ is a constant (see also [VI.1]). When $F$ does depend on $x$ explicitly, we know that $H$ is generally not a constant (energy is not conserved). Together with Examples 7a and 7b, they suggest that there is a correlation between $F$ being variationally invariant under (7.43) on the one hand $H$ being a constant (and the Euler DE having a first integral) on the other. Suppose $F(x, y, y')$ is variationally invariant under some transformation other than (7.43). Is there some other first integral of the Euler DE if it is not $H =$ constant? Is there anyting being conserved? In other words, is there always a conservation law associated with the variationally invariant property of the Lagrangian (for some family of transformations)? The answer is provided by the following result of *Noether*:

**[VII.6]** *If* $F(x, y, y')$ *is variationally invariant under a family of* $C^2$ *trasformations of the form (7.41), then the Euler DE for an extremal of the corresponding* $J[y]$ *has as its first integral*

$$\widehat{F}_{,y'}(x)\,\eta(x, \widehat{y}(x)) + [\widehat{F}(x) - \widehat{y}'(x)\,\widehat{F}_{,y'}(x)]\,\xi(x, \widehat{y}(x)) = C_0, \quad (7.52)$$

*where* $\eta(x, y) = \psi_{,\varepsilon}(x, y; 0)$ *and* $\xi(x, y) = \varphi_{,\varepsilon}(x, y; 0)$ *are known as the generators of the transformations (7.41).*

The linear approximation pair

$$\begin{aligned} X &\simeq \varphi(x, y; 0) + \varepsilon\varphi_{,\varepsilon}(x, y; 0) = x + \varepsilon\xi(x, y), \\ Y &\simeq \psi(x, y; 0) + \varepsilon\psi_{,\varepsilon}(x, y; 0) = y + \varepsilon\eta(x, y) \end{aligned} \quad (7.53)$$

is known as the *infinitesimal transformation* approximation of (7.41).

To derive the first integral (7.52), we think of the perfomance index $J[y]$ [with the Lagrangian $F(X, \widehat{Y}_\varepsilon, \widehat{Y}_\varepsilon^{\bullet})$, $(\ )^{\bullet} = d(\ )/dX$] as a function of the parameter $\varepsilon$, namely, $J(\varepsilon)$. Given that $F(x, y, y')$ is variationally invariant for the given family of transformations, the value of $J(\varepsilon)$ must be independet of the value of $\varepsilon$:

$$J(\varepsilon) = \int_{t_\varepsilon}^{\tau_\varepsilon} F(X, \widehat{Y}_\varepsilon(X), \widehat{Y}_\varepsilon^{\bullet}(X))\, dX = \int_t^{\tau} F(x, \widehat{y}(x), \widehat{y}'(x))\, dx.$$

In other words, we must have

$$\begin{aligned} \frac{dJ}{d\varepsilon} &= \int_{t_\varepsilon}^{\tau_\varepsilon} \left[ F_{,y}(X, \widehat{Y}_\varepsilon, \widehat{Y}_\varepsilon^{\bullet}) \frac{\partial \widehat{Y}_\varepsilon}{\partial \varepsilon} + F_{,y'}(X, \widehat{Y}_\varepsilon, \widehat{Y}_\varepsilon^{\bullet}) \frac{\partial \widehat{Y}_\varepsilon^{\bullet}}{\partial \varepsilon} \right] dX \\ &\quad + F(\tau_\varepsilon, \widehat{Y}_\varepsilon(\tau_\varepsilon), \widehat{Y}_\varepsilon^{\bullet}(\tau_\varepsilon)) \frac{\partial \tau_\varepsilon}{\partial \varepsilon} - F(t_\varepsilon, \widehat{Y}_\varepsilon(t_\varepsilon), \widehat{Y}_\varepsilon^{\bullet}(t_\varepsilon)) \frac{\partial t_\varepsilon}{\partial \varepsilon} \\ &= \int_{t_\varepsilon}^{\tau_\varepsilon} \left[ \widehat{F}_{,y}(X) - \frac{d}{dX} \widehat{F}_{,y'}(X) \right] \frac{\partial \widehat{Y}_\varepsilon}{\partial \varepsilon}\, dX + \left[ \widehat{F}_{,y'}(X) \frac{\partial \widehat{Y}_\varepsilon}{\partial \varepsilon} + \widehat{F}(X) \frac{\partial X}{\partial \varepsilon} \right]_{t_\varepsilon}^{\tau_\varepsilon} = 0 \end{aligned} \quad (7.54)$$

for the given Lagrangian to be variationally invariant over $[a, b]$ under (7.41). This is true in particular at $\varepsilon = 0$. From (7.42), we have

$$\left[F_{,y}\left(X, \widehat{Y}_\varepsilon(X), \frac{d\widehat{Y}_\varepsilon}{dX}\right)\right]_{\varepsilon=0} = \left[F_{,y}\left(X, \widehat{Y}_\varepsilon(X), \frac{d\widehat{Y}_\varepsilon}{dx}\frac{dx}{dX}\right)\right]_{\varepsilon=0} = F_{,y}(x, \widehat{y}(x), \widehat{y}'(x));$$

and

$$\frac{d}{dX}\left[F_{,y'}\left(X, \widehat{Y}_\varepsilon(X), \frac{d\widehat{Y}_\varepsilon}{dX}\right)\right]_{\varepsilon=0} = \frac{d}{dx}\left[F_{,y'}(x, \widehat{y}(x), \widehat{y}'(x))\right].$$

Hence, the integrand in (7.54) vanishes for $\varepsilon = 0$ by the Euler DE of $J[y]$. It remains to evaluate $\partial \widehat{Y}_\varepsilon/\partial\varepsilon$, $\partial\tau_\varepsilon/\partial\varepsilon$, and $\partial t_\varepsilon/\partial\varepsilon$ at $\varepsilon = 0$.

With $Y_\varepsilon$ being a function of $X$ and $\varepsilon$, we have from (7.41), (7.42) and (7.53)

$$\begin{aligned}\left[\frac{\partial \widehat{Y}_\varepsilon}{\partial\varepsilon}\right]_{\varepsilon=0} &= \left[\widehat{\psi}_{,x}\frac{\partial x}{\partial\varepsilon} + \widehat{\psi}_{,y}\widehat{y}'\frac{\partial x}{\partial\varepsilon} + \widehat{\psi}_{,\varepsilon}\right]_{\varepsilon=0} \\ &= \left[\widehat{y}'\frac{\partial x}{\partial\varepsilon}\right]_{\varepsilon=0} + \eta(x, \widehat{y}(x))\end{aligned} \tag{7.55}$$

with $x$ treated as a function of the independent variable $X$ and the parameter $\varepsilon$. Upon differentiating $X = \varphi(x, \widehat{y}(x); \varepsilon)$ with respect $\varepsilon$, we get

$$0 = \widehat{\varphi}_{,x}\frac{\partial x}{\partial\varepsilon} + \widehat{\varphi}_{,y}\widehat{y}'\frac{\partial x}{\partial\varepsilon} + \widehat{\varphi}_{,\varepsilon}.$$

At $\varepsilon = 0$, $\widehat{\varphi}_{,y} = [x]_{,y} = 0$, and $\widehat{\varphi}_{,x} = 1$ leaving us with

$$\left.\frac{\partial x}{\partial\varepsilon}\right|_{\varepsilon=0} = -\varphi_{,\varepsilon}(x, \widehat{y}(x); 0) = -\xi(x, \widehat{y}(x)). \tag{7.56}$$

It follows that (7.55) becomes

$$\begin{aligned}\left[\frac{\partial \widehat{Y}_\varepsilon}{\partial\varepsilon}\right]_{\varepsilon=0} &= \psi_{,\varepsilon}(x, \widehat{y}(x); 0) - \widehat{y}'(x)\,\varphi_{,\varepsilon}(x, \widehat{y}(x); 0) \\ &= \eta(x, \widehat{y}(x)) - \widehat{y}'(x)\,\xi(x, \widehat{y}(x)).\end{aligned} \tag{7.57}$$

For $\partial t_\varepsilon/\partial\varepsilon$ and $\partial\tau_\varepsilon/\partial\varepsilon$ at $\varepsilon = 0$, it is straightforward to find

$$\left[\frac{\partial t_\varepsilon}{\partial\varepsilon}\right]_{\varepsilon=0} = \left[\frac{\partial}{\partial\varepsilon}\varphi(t, \widehat{y}(t); \varepsilon)\right]_{\varepsilon=0} = \varphi_{,\varepsilon}(t, \widehat{y}(t); 0) = \xi(t, \widehat{y}(t)), \tag{7.58}$$

$$\left[\frac{\partial \tau_\varepsilon}{\partial\varepsilon}\right]_{\varepsilon=0} = \left[\frac{\partial}{\partial\varepsilon}\varphi(\tau, \widehat{y}(\tau)\ \ \varepsilon)\right]_{\varepsilon=0} = \varphi_{,\varepsilon}(\tau, \widehat{y}(\tau); 0) = \xi(\tau, \widehat{y}(\tau)), \tag{7.59}$$

With (7.57)-(7.59), condition (7.54) evaluated at $\varepsilon = 0$ becomes

$$[F(x)\xi(x,\hat{y}(x)) + \hat{F}_{,y'}(x)\{\eta(x,\hat{y}(x)) - \hat{y}'(x)\xi(x,\hat{y}(x))\}]_t^\tau$$
$$= [\hat{F}_{,y'}(x)\eta(x,\hat{y}(x)) + \{\hat{F}(x) - \hat{y}'(x)\hat{F}_{,y'}(x)\}\xi(x,\hat{y}(x))]_t^\tau = 0.$$

Because $t$ and $\tau$ are arbitrary, the expression inside the brackets must be a constant function so that (7.52) holds.

**EXAMPLE 8:** $F = F(y, y')$.

This Lagrangian is variationally invariant under the "time"-translation $X = x + \varepsilon$, $Y = y$. For this transformation, we have $\psi_{,\varepsilon} = 0$ and $\varphi_{,\varepsilon} = 1$. The first integral (7.52) becomes $\hat{F} - \hat{y}'\hat{F}_{,y'} = c_0$ or, in terms of the Hamiltonian, $H(\hat{y},\hat{p}) = c$ with $H(y,p)$ obtained from $F(y,y')$ in the usual way.

**EXAMPLE 9:** $F(x,y,y') = x^2[(y')^2 - \frac{1}{3}y^6]/2$.

It is straightforward to verify that $F$ is variationally invariant under the transformations $\varphi = (1+\varepsilon)x$ and $\psi = y/(1+\varepsilon)^{1/2}$. The Euler DE of the corresponding basic problem is the equation

$$\hat{y}'' + \frac{2}{x}\hat{y}' + \hat{y}^5 = 0.$$

The generators are $\varphi_{,\varepsilon}(x,y;0) \equiv \xi(x,y) = x$ and $\psi_{,\varepsilon}(x,y;0) \equiv \eta(x,y) = -y/2$. A first integral for this problem is, therefore,

$$x^2\hat{y}'\left(-\frac{\hat{y}}{2}\right) + \frac{x^2}{2}\left[(\hat{y}')^2 - \frac{1}{3}\hat{y}^6 - 2(\hat{y}')^2\right]x = c_0$$

or

$$\frac{x^3}{6}\hat{y}^6 + \left[\frac{x^3}{2}(\hat{y}')^2\right] + \frac{x^2}{2}\hat{y}\hat{y}' = C.$$

Noether's theorem may be expressed in terms of the Hamiltonian:

**[VII.7]** *Let* $H(x,y,p)$ *be the Hamiltonian associated with the Lagrangian* F *which is variationally invariant under the transformations (7.41). Then the conservation law*

$$\hat{p}(x)\eta(x,\hat{y}(x)) - \hat{H}(x)\xi(x,\hat{y}(x)) = C_0. \qquad (7.60)$$

*must hold for an extremal* $\hat{y}(x)$ *of* F.

Expression (7.60) follows from (7.52) and the definition of $p$ and $H$.

It is not difficult to show that Noether's theorem can be extended to several unknowns. In terms of p and $H(x, y, p)$, we have following extension of [VII.7]:

**[VII.7′]** ***If*** $\mathbf{F}(\mathbf{x},\mathbf{y},\mathbf{y}')$ ***is variationally invariant under the transformation***

$$\mathbf{X} = \varphi(\mathbf{x},\mathbf{y};\varepsilon), \qquad \mathbf{Y} = \psi(\mathbf{x},\mathbf{y};\varepsilon) = (\psi_1,\ldots,\psi_n)^T, \tag{7.41'}$$

***then its extremal satisfies***

$$\hat{\mathbf{p}}(\mathbf{x})\cdot\eta(\mathbf{x},\hat{\mathbf{y}}(\mathbf{x})) - \hat{\mathbf{H}}(\mathbf{x})\xi(\mathbf{x},\hat{\mathbf{y}}(\mathbf{x})) = \mathbf{C}_0 \tag{7.60'}$$

***where*** $\eta = \psi_{,\varepsilon}(\mathbf{x},\mathbf{y};0)$, $\hat{\mathbf{p}} = \mathbf{F}_{,\mathbf{y}'}(\mathbf{x},\hat{\mathbf{y}},\hat{\mathbf{y}}')$, ***and*** $\mathbf{H}$ ***is the Hamiltonian associated with*** $\mathbf{F}$.

**EXAMPLE 10:** $F(x,\mathbf{y};\mathbf{y}') = T(\mathbf{y}') - V(x,\mathbf{y}) \equiv L(x,\mathbf{y},\mathbf{y}')$. **Obtain the first integral, if any, associated with the transformation**

$$X = x, \qquad Y_1 = y_1 + \varepsilon, \qquad Y_2 = y_2, \qquad Y_3 = y_3,$$

**where $x$ is time and $\mathbf{y} = (y_1, y_2, y_3)^T$.**

**It is straightforward to verify that the given Lagrangian $F$ is variationally invariant under the given transformation. With $\xi = 0$, $\eta_1 = 1$, and $\eta_2 = \eta_3 = 0$, we have from (7.60′)**

$$\hat{p}_1 = c_1$$

**as a first integral of the Euler DE. It corresponds to the conservation of momentum in the $y_1$ direction. (Similarly, it can be shown that $F$ is variationally invariant under a parallel-displacement transformation in $y_2$ or $y_3$ direction. It follows that the momentum is also conserved in these directions.)**

## 7. Generators for Variationally Invariant Lagrangians

The usefulness of Noether's theorem hinges on knowing the (Lie) group of transformations under which $F$ is variationally invariant. We now develop a systematic way to determine such transformations for a given Lagrangian. We begin with the known relation

$$\frac{dJ}{d\varepsilon} = \frac{d}{d\varepsilon}\int_{t_0}^{T_\varepsilon} F(X,\overline{Y}_\varepsilon,\overline{Y}_\varepsilon^{\bullet})\,dX = 0, \qquad (\ )^{\bullet} \equiv \frac{d(\ )}{dX}.$$

In contrast to the way we established Noether's theorem, we now do the integration with respect to the original independent variable $x$ instead (of $X$) so that

$$\frac{dJ}{d\varepsilon} = \frac{d}{d\varepsilon}\int_{t}^{T} F(X_\varepsilon,\overline{Y}_\varepsilon,\overline{Y}_\varepsilon^{\bullet})\,\frac{dX_\varepsilon}{dx}\,dx$$

$$= \int_{t}^{T} \frac{d}{d\varepsilon}\left[F(X_\varepsilon,\overline{Y}_\varepsilon,\overline{Y}_\varepsilon^{\bullet})\,\frac{dX_\varepsilon}{dx}\right] dx = 0.$$

The integral vanishes for any interval $[t, \tau]$ in $[a, b]$; it follows that the integrand itself must vanish (by the continuity of the integrand):

$$\frac{d}{d\varepsilon}\left[F(X_\varepsilon, \overline{F}_\varepsilon, \overline{Y}_\varepsilon^\bullet)\frac{dX_\varepsilon}{dx}\right] = 0$$

or

$$\frac{dX_\varepsilon}{dx}\left[F_{,x}\frac{dX_\varepsilon}{d\varepsilon} + F_{,y}\frac{d\overline{Y}_\varepsilon}{d\varepsilon} + F_{,y'}\frac{d\overline{Y}_\varepsilon^\bullet}{d\varepsilon}\right] + F\frac{d}{d\varepsilon}\left(\frac{dX_\varepsilon}{dx}\right) = 0. \quad (7.61)$$

For $\varepsilon = 0$, we have $dX_\varepsilon/d\varepsilon = \varphi_{,\varepsilon}(x, \overline{y}; 0) = \xi(x, \overline{y})$, $d\overline{Y}_\varepsilon/d\varepsilon = \psi_{,\varepsilon}(x, \overline{y}; 0) = \eta(x, \overline{y})$, $d(dX_\varepsilon/dx)/d\varepsilon = d(dX_\varepsilon/d\varepsilon)/dx = d[\varphi_{,\varepsilon}(x, \overline{y}; 0)]/dx = \xi' = \xi_{,x} + \xi_{,y}\overline{y}'$, and

$$\left.\frac{d\overline{Y}_\varepsilon^\bullet}{d\varepsilon}\right|_{\varepsilon=0} = \left.\frac{d}{d\varepsilon}\left[\frac{d\overline{Y}_\varepsilon/dx}{dX_\varepsilon/dx}\right]\right|_{\varepsilon=0} = \left[\frac{d(d\overline{Y}_\varepsilon/dx)/d\varepsilon}{dX_\varepsilon/dx} - \frac{(d\overline{Y}_\varepsilon/dx)\{d(dX_\varepsilon/dx)/d\varepsilon\}}{(dX_\varepsilon/dx)^2}\right]_{\varepsilon=0}$$

$$= \frac{d(\psi_{,\varepsilon}(x, \overline{y}; 0))/dx}{1} - \frac{\overline{y}'d(\varphi_{,\varepsilon}(x, \overline{y}; 0))/dx}{(1)^2}$$

$$= \eta' - \xi'\overline{y}'.$$

where we have made use of the fact that $x_\varepsilon = x$ *at* $\varepsilon = 0$. Hence, the identity (7.61) at $\varepsilon = 0$ becomes

$$\overline{F}_{,x}\xi + \overline{F}_{,y}\overline{\eta} + \overline{F}_{,y'}(\overline{\eta}' - \overline{y}'\overline{\xi}') + \overline{F}\overline{\xi}' = 0,$$

where $\overline{g}(x) \equiv g(x, \overline{y}(x))$. After tightening up some technical details, we get

**[VII.8]** ***The Lagrangian*** $F(x, y, y')$ ***is variationally invariant under the (Lie) transformations (7.41) with generators*** $\xi$ ***and*** $\eta$ ***if and only if***

$$F_{,x}\xi + F_{,y}\eta + F_{,y'}(\eta' - y'\xi') + F\xi' = 0, \quad (7.62)$$

***where*** $\eta' = \eta_{,x} + \eta_{,y}y'$ ***and*** $\xi' = \xi_{,x} + \xi_{,y}y'$ ***for any PWS function*** $y(x)$ ***in*** $[a, b]$.

Among the many applications of [VII.8] is the determination of the generators $\xi(x, y)$ and $\eta(x, y)$ of the transformations under which $F$ is variationally invariant. This is best illustrated by examples:

**EXAMPLE 11:** $F = (y'^2 - y')/2$.

For this Lagrangian, we have $F_{,x} = 0$, $F_{,y} = -y$, and $F_{,y'} = y'$. Hence, (7.62) becomes

$$-y\eta + y'[(\eta_{,x} + \eta_{,y}y') - y'(\xi_{,x} + \xi_{,y}y')] + (\xi_{,x} + \xi_{,y}y')F = 0$$

or

$$(y')^3\left[-\frac{1}{2}\xi_{,y}\right]+(y')^2\left[\eta_{,y}-\frac{1}{2}\xi_{,x}\right]+y'\left[\eta_{,x}-\frac{1}{2}y^2\xi_{,y}\right]-\left[y\eta+\frac{1}{2}y^2\xi_{,x}\right]=0.$$

This condition is to hold for any PWS $y(x)$ on $[a,b]$. The requirement can be met by choosing $\xi$ and $\eta$ so that the coefficients of different powers of $y'$ vanish independently. The resulting four conditions on $\xi$ and $\eta$ may be simplified to

$$\xi=\xi(x), \qquad \eta_{,y}=\frac{1}{2}\xi'(x),$$

$$\eta_{,x}=0, \qquad \eta=-\frac{1}{2}y\xi'(x).$$

The last two relations require $\xi''=0$ or

$$\xi=c_0+c_1x, \qquad \eta=-\frac{1}{2}c_1y.$$

But the second, $\eta_{,y}=\xi'(x)/2$, requires $-c_1/2=c_1/2$ or $c_1=0$. Altogether, we have

$$\xi=c_0, \qquad \eta=0.$$

The corresponding first integral from (7.60) is $-c_0H(x,y,p)=C_0$ or

$$H(x,y,p)=C_1$$

as we would expect by [VI.2].

**EXAMPLE 12:** $F(x,y,y')=\frac{1}{2}x^2[(y')^2-\frac{1}{3}y^6]$.

For this Lagrangian, we have $F_{,x}=x[(y')^2-y^6/3]$, $F_{,y}=-x^2y^5$, and $F_{,y'}=x^2y'$. Hence, (7.62) becomes

$$(y')^3\left[-\frac{1}{2}x^2\xi_{,y}\right]+(y')^2\left[x\xi-\frac{1}{2}x^2\xi_{,x}+x^2\eta_{,y}\right]+y'\left[\eta_{,x}x^2-\frac{1}{6}x^2y^6\xi_{,y}\right]$$
$$-\left[\frac{1}{3}x\xi y^6+\frac{1}{6}x^2y^6\xi_{,x}+x^2y^5\eta\right]=0.$$

This condition can be met by requiring the coefficients of different powers of $y'$ to vanish independently. The resulting four conditions determine $\eta$ and $\xi$ up to a constant $c_0$:

$$\xi=c_0x, \qquad \eta=-\frac{1}{2}c_0y.$$

By setting $c_0 = 1$, we recover the generators in Example 9 which were obtained from the given transformations for which $F$ is variationally invariant. With [VII.8], we can now deduce the generators and the relevant conservation laws without any knowledge of the appropriate transformations a priori.

## 8. Relativistic Mechanics

Suppose an observer located at the origin of a Cartesian inertial reference frame at $t = 0$ moves at constant speed $v$ along the $x_1$-axis. At $t = 0$, a light source instantaneously emits rays in all directions with speed $c$ relative to the inertial frame. After time $t$, the equation of the spherical wave front formed by the light rays will be

$$x_1^2 + x_2^2 + x_3^2 - c^2t^2 = 0 \tag{7.63}$$

and the coordinates of the moving observer will be $(vt, 0, 0)$. Relative to a moving Cartesian frame centered at the position of the moving observer, however, the wave front is still a sphere of radius $ct$ in a conventional Galilean world but its center has coordinates $(-vt, 0, 0)$. In terms of the moving coordinates $(\bar{x}_1, \bar{x}_2, \bar{x}_3)$, the wave front is located at

$$(\bar{x}_1 + vt)^2 + (\bar{x}_2)^2 + (\bar{x}_3)^2 - c^2t^2 = 0 \tag{7.64}$$

in Galilean kinematics.

Evidence from the Michelson and Morley experiments and others do not agree with (7.64). The discrepancy has led to the development of the *theory of relativity*. This theory asserts that

> *The speed of light must be the same for all observers independently of their individual velocities; the distance of the wave front from the moving observer must be* ct *in all directions.*

With this postulate, the equations of the wave front in the space-time reference frame of the moving observer must be of the same form as (7.63) and not (7.64). This is possible if we allow (as the theory of relativity does) the measurements of time to depend also on the velocity of the observer, just as the measurement of quantities such as velocity. Such a concept of time is radically different from the conventional Galilean view that time is independent of the velocity of any observer so that same $t$ enters into (7.63) and (7.64).

Because the observer is moving only in the $x_1$ direction with constant speed, we expect the measurement in the $\bar{x}_2$ and $\bar{x}_3$ directions to be the same as those in the inertial frame. The postulate of time transformation then is met by setting

$$x_1 = a(\bar{x}_1 + v\bar{t}), \qquad x_2 = \bar{x}_2, \qquad x_3 = \bar{x}_3,$$
$$t = \beta_1\bar{x}_1 + \beta_2\bar{x}_2 + \beta_3\bar{x}_3 + \gamma\bar{t}. \tag{7.65}$$

For the wave front in the moving frame to be spherical and centered at the moving observer, we need

$$a^2(\bar{x}_1 + v\bar{t})^2 + \bar{x}_2^2 + \bar{x}_3^2 - c^2(\beta_1\bar{x}_1 + \beta_2\bar{x}_2 + \beta_3\bar{x}_3 + \gamma\bar{t})^2$$
$$= \bar{x}_1^2 + \bar{x}_2^2 + \bar{x}_3^2 - c^2\bar{t}^2 \tag{7.66}$$

to hold for all $\bar{x}_k$ and $\bar{t}$. It is easy to see immediately that we must have $\beta_2 = \beta_3 = 0$. For the remaining quadratic relation

$$a^2(\bar{x}_1 + v\bar{t})^2 - c^2(\beta_1\bar{x}_1 + \gamma\bar{t})^2 = \bar{x}_1^2 - c^2\bar{t}^2$$

to hold for all $\bar{x}_1$ and $\bar{t}$, we must have

$$a^2 - c^2\beta_1^2 = 1, \qquad a^2v^2 - \gamma^2c^2 = -c^2, \qquad a^2v = \gamma\beta_1c^2.$$

The positive solution of this system is

$$\gamma = a = \left(1 - \frac{v^2}{c^2}\right)^{-1/2}, \qquad \beta_1 = \frac{v}{c^2}\left(1 - \frac{v^2}{c^2}\right)^{-1/2} \tag{7.67}$$

so that (7.65) becomes

$$x_1 = a(\bar{x}_1 + v\bar{t}), \qquad x_2 = \bar{x}_2, \qquad x_3 = \bar{x}_3, \qquad t = a\left(\bar{t} + \frac{v}{c^2}\bar{x}_1\right). \tag{7.68}$$

The inverse transformation is

$$\bar{x}_1 = a(x_1 - vt), \qquad \bar{x}_2 = x_2, \qquad \bar{x}_3 = x_3, \qquad \bar{t} = a\left(t - \frac{v}{c^2}x_1\right). \tag{7.69}$$

For an observer moving with a nonconstant velocity $v(t)$ in the $x_1$ direction of the inertial Cartesian reference frame, the results of (7.68) and (7.69) still hold locally. When the moving observer reaches the point $(x_1, x_2, x_3)$ at time $t$, consider a time interval $(t, t + dt)$ so short that $v$ is pratically constant. At this later time, the observer has moved to the point $(x_1 + dx_1, x_2 + dx_2, x_3 + dx_3)$. In that case, a relativistic theory asserts that the relativity transformations from the inertial frame to the accelerating frame must change from $dx_1^2 + dx_2^2 + dx_3^2 - c^2\,dt^2$ to $d\bar{x}_1^2 + d\bar{x}_2^2 + d\bar{x}_3^2 - c^2\,d\bar{t}^2$. This, in turn, requires

$$dx_1 = a(d\bar{x}_1 + v\,d\bar{t}), \qquad dx_2 = d\bar{x}_2, \qquad dx_3 = d\bar{x}_3, \qquad dt = a\left(d\bar{t} + \frac{v}{c^2}\,d\bar{x}_1\right) \tag{7.70}$$

and

$$d\bar{x}_1 = a(dx_1 + v\,dt), \quad d\bar{x}_2 = dx_2, \quad d\bar{x}_3 = dx_3, \quad d\bar{t} = a\left(dt - \frac{v}{c^2}\,dx_1\right), \quad (7.71)$$

the infinitesimal version of (7.68) and (7.69), respectively.

Corresponding to this relativistic theory of kinematics is a relativistic theory of dynamics of mass particles (and rigid bodies). Although there are many interesting features of this theory (such as a different measurement of mass and energy), we will consider only one aspect of it which is germane to the calculus of variations, namely, the form of Hamilton's principle in the new theory. The stationarity condition

$$\delta \int_{t_A}^{t_B} L\,dt = 0$$

with fixed end points $(t_A, \mathbf{A})$ and $(t_B, \mathbf{B})$ is equivalent to

$$\int_{t_A}^{t_B} (\delta L)\,dt = 0$$

in the classical Newtonian theory for which the time variable is not allowed to vary. For a relativistic theory, there should be no distinction in the roles of spatial and temporal variables; time should be allowed to vary as well. Does this change the form of the equations of motion for the dynamical system? Does Hamilton's principle have to be modified to give the correct set of Euler equations?

For an answer, we again observe that a suitable parametrization of both space and time variables is appropriate for our problem. Let

$$t = \tau(x), \qquad q_i(t) = Q_i(x) \quad (a \le x \le b, i = 1, 2, \ldots, n) \qquad (7.72)$$

for fixed $a$ and $b$ with $\mathbf{Q}(a) = \mathbf{A}$, $\mathbf{Q}(b) = \mathbf{B}$, $t(a) = t_A$, and $t(b) = t_B$. We express the action integral in terms of $\mathbf{Q}$, $\tau$, and $x$ to obtain

$$J[\mathbf{q}] = \int_{t_A}^{t_B} L(t, \mathbf{q}, \dot{\mathbf{q}})\,dt = \int_a^b \mathcal{L}(x, \mathbf{Q}, \tau, \mathbf{Q}', \tau')\,dx \equiv I[\mathbf{Q}, \tau] \qquad (7.73a)$$

with $(\ )' \equiv d(\ )/dx$ and

$$\mathcal{L} = \tau' L\left(\tau, \mathbf{Q}, \frac{\mathbf{Q}'}{\tau'}\right). \qquad (7.73b)$$

The Euler DE (with $\delta Q_i$ and $\delta\tau$ varying independently) may now be obtained from [II.3′] to be

$$-\left(\frac{\partial \mathcal{L}}{\partial Q_i'}\right)' + \frac{\partial \mathcal{L}}{\partial Q_i} = 0, \qquad -\left(\frac{\partial \mathcal{L}}{\partial \tau'}\right)' + \frac{\partial \mathcal{L}}{\partial \tau} = 0 \qquad (7.74)$$

for $i = 1, 2, \ldots, n$. In terms of $L$, we have from the first $n$ equations of (7.74)

$$-\frac{d}{dt}\left(\frac{\partial L}{\partial \dot{q}_i}\right)+\frac{\partial L}{\partial q_i}=0 \quad (i=1,2,\ldots,n) \tag{7.75}$$

which are identical to the Euler DEs of the nonrelativistic Hamilton's principle, i.e., the equations of motion in Newtonian mechanics. It would appear then that there is an additional requirement in the form of a new Euler DE given by the last equation of (7.74) in the relativistic world. However, the development in Section 4 of Chapter 2 suggests the opposite conclusion:

**[VII.9]** ***In a relativistic theory of mechanics, the usual form of Hamilton's principle, i.e., the stationarity of the action integral,***

$$\mathrm{J}[\mathbf{q}]=\int_{t_1}^{t_2}[\mathrm{T}-\mathrm{V}]\,dt \equiv \int_{t_1}^{t_2} \mathrm{L}(t,\mathbf{q},\dot{\mathbf{q}})\,dt \tag{7.76}$$

***along an extremal with fixed end points, implies the usual form of the equations of motion and no other conditions on the extremals.***

Though the result is implicit in the discussion leading up to [II.4], we will prove it explicitly. We have from (7.73b)

$$\begin{aligned}\frac{d}{dx}\left(\frac{\partial \mathcal{L}}{\partial \tau'}\right)&=\tau'\frac{d}{dt}\left[L+\tau'\sum_{i=1}^{n}\frac{\partial L}{\partial \dot{q}_i}\left\{-\frac{Q'_i}{(\tau')^2}\right\}\right]\\&=\tau'\frac{d}{dt}\left[L-\sum_{i=1}^{n}\frac{\partial L}{\partial \dot{q}_i}\dot{q}_i\right]\\&=\tau'\left[\frac{\partial L}{\partial t}+\sum_{i=1}^{n}\left\{\frac{\partial L}{\partial q_i}\dot{q}_i+\frac{\partial L}{\partial \dot{q}_i}\ddot{q}_i-\frac{d}{dt}\left(\frac{\partial L}{\partial \dot{q}_i}\dot{q}_i\right)\right\}\right].\end{aligned}$$

We now use (7.75) to express $\partial L/\partial q_i$ in terms of $d(\partial L/\partial \dot{q}_i)/dt$ so that

$$\frac{d}{dx}\left(\frac{\partial \mathcal{L}}{\partial \tau'}\right)-\frac{\partial \mathcal{L}}{\partial \tau}=\tau'\sum_{i=1}^{n}\left[\frac{d}{dt}\left(\frac{\partial L}{\partial \dot{q}_i}\right)\dot{q}_i+\frac{\partial L}{\partial \dot{q}_i}\ddot{q}_i-\frac{d}{dt}\left(\frac{\partial L}{\partial \dot{q}_i}\dot{q}_i\right)\right] \tag{7.77}$$

The right-hand side of (7.77) is identically zero. Hence the additional Euler DE in (7.75) imposes no new restriction on the extremals of $L(t,\mathbf{q},\dot{\mathbf{q}})$.

## 9. Exercises

1. A particle of mass $m$ is subject to an external force $\mathbf{F}(t)$. Obtain the equations of motion in cylindrical coordinates.

2. Two particles of the same mass $m$ are moving under their mutual gravitational attraction (so that the potential $V$ is equal to $-\gamma m/2r$, where $2r$, is the distance between them and $\gamma$ is a constant). Find the equation of motion in terms of the Cartesian coordinates $(X, Y, Z)$ of the center of mass of the two-particle system and the spherical coordinates $(r, \varphi, \theta)$ of one particle relative to the center of mass.

3. Given a Lagrangian $L(t, \mathbf{q}, \dot{\mathbf{q}})$ and any scalar function $f(t, \mathbf{q})$, show that the new Lagrangian

$$\mathcal{L} = L(t, \mathbf{q}, \dot{\mathbf{q}}) + \frac{\partial f}{\partial t} + \sum_{k=1}^{n} \frac{\partial f}{\partial q_k} \dot{q}_k$$

generates the same dynamics as $L$.

4. For the Lagrangian

$$L(t, \mathbf{q}, \dot{\mathbf{q}}) = \frac{1}{2} m \sum_{i,j=1}^{n} T_{ij}(\mathbf{q})\, \dot{q}_i\, \dot{q}_j\,.$$

(a) Show that $dL/dt = 0$.

(b) Show that the new Lagrangian $\mathcal{L} = f(L)$ for any real-valued function $f(\cdot)$ generates the same dynamics as $L$.

5. A particle of mass $m$ is constrained to move under gravity on the surface of a smooth right circular cone of semivertex angle $\pi/4$. The axis of the cone is vertical with the vertex downward. Find the equations of motion in terms of the height above the vertex $z$ and the angular coordinates around the circular cross section $\theta$.

(a) Show that $\dot{z}^2 + gz + h^2/2z^2 = E$, where $E$ and $h$ are constants.

(b) Sketch and interpret the orbits in the $z$, $\dot{z}$-plane for a fixed value of $h$.

6. Two bobs of mass $m_1$ and $m_2$ form a double pendulum as shown in Figure 7.2. If $(x_1, y_1)$ and $(x_2, y_2)$ denote the positions of the bobs, then the kinetic energy of the system is $[m_1(\dot{x}_1^2 + \dot{y}_1^2) + m_2(\dot{x}_2^2 + \dot{y}_2^2)]/2$.

(a) Show that, in the generalized coordinates $\varphi_1$ and $\varphi_2$, the kinetic energy is

$$\{m_1 r_1^2 \dot{\varphi}_1^2 + m_2 [r_1^2 \dot{\varphi}_1^2 + r_2^2 \dot{\varphi}_2^2 + 2r_1 r_2 \dot{\varphi}_1 \dot{\varphi}_2 \cos(\varphi_2 - \varphi_1)]\}/2.$$

(b) Find the potential energy of the system (under gravity).

(c) Apply Hamilton's principle to obtain the equations of motion in terms of $\varphi_1$ and $\varphi_2$.

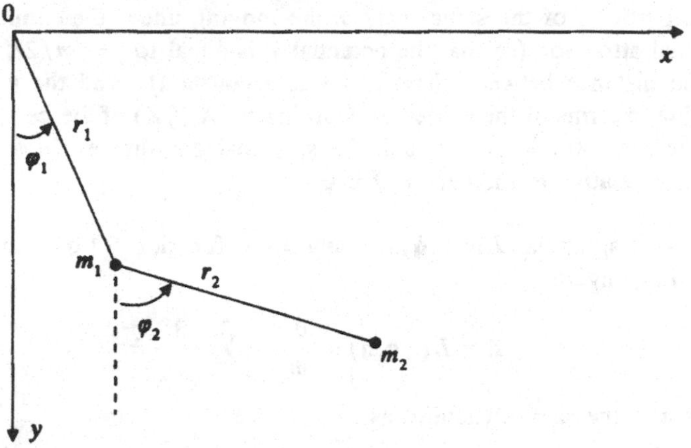


**Figure 7.2**

7. (a) Show that $F = x(y')^2$ is not variationally invariant under the time translation $X = x + \varepsilon, Y = y$.

   (b) Show that the same $F$ is variationally invariant under the spatial translation $X = x$, $Y = y + \varepsilon$, and find a first integral of the Euler DE for $F$ by Noether's theorem.

   (c) Is the same $F$ variationally invariant under $X = (1 + \varepsilon)x$, $Y = y$? Justify your answer and obtain the corresponding first integral if it is.

8. $$J[r, \theta] = \int_a^b \left\{ \frac{1}{2} m(\dot{r}^2 + r^2\dot{\theta}^2) + \frac{k}{r} \right\} dt, \qquad (\ )^{\cdot} = \frac{d(\ )}{dt}.$$

   The Euler DEs of $J[r, \theta]$ are the equations for planetary motion in the $x, y$-plane and $(r, \theta)$ are the corresponding polar coordiantes, i.e., equations of motion for a particle of mass $m$ in a central force field.

   (a) Show that the Lagrangian of $J[r, \theta]$ is variationally invariant under $T = t + \varepsilon$, $R = r$, $H = \theta$. Obtain a first integral of the Euler DE of $F$.

   (b) Show that $F$ is also variationally invariant under $T = t$, $R = r$, $H = \theta + \varepsilon$. Obtain the corresponding first integral of the Euler DE.

9. (a) Verify that the basic problem with $F = (y' + ky)^2$ is variationally invariant under the transformation $X = x + \varepsilon$, $Y = y + \varepsilon a e^{-kx}$.

   (b) Use Noether's theorem to produce a constant of motion, i.e., a first integral of the Euler DE.

(c) Verify that the extremal(s) found in Exercise 1 of Chapter 6 satisfies the conclusion of part (b).

10. (a) Obtain the generators of $F = x(y')^2$.

(b) Use the results of part (a) and Noether's theorem to obtain the most general first integral possible for the relevant Euler DE.

(c) Specialize the result of part (b) to recover the two first integrals in Exercise 7. Are there others implied by the result of part (b)?

11. (a) Show that all motions of mass particles in a conservative force field are smooth.

(b) Suppose there is no point in $(t_1, t_2)$ conjugate to $t_1$. Show that $J[\hat{y}]$ is a weak local minimum for any admissible extremal $\hat{y}$ in a conservative force field.

(c) Under the same hypotheses as part (b), show that $J[\hat{y}]$ is actually a strong *local minimum*. [Hint: For simplicity, work with (6.6) and not (6.7) of Chapter 6.]

12. Suppose the Lagrangian $L = T(\dot{y}) - V(y)$ is variationally invariant under rotations about the $z \equiv y^3$ axis. For a single mass particle, the type of transformation involved is given by

$$Y^1 = y^1 \cos\varepsilon + y^2 \sin\varepsilon, \qquad Y^2 = -y^1 \sin\varepsilon + y^2 \cos\varepsilon, \qquad Y^3 = y^3.$$

(For more particles, only subscripts need to be added.) Deduce an appropriate conservation law for the dynamical system. What is conserved physically?

13. Repeat Exercise 12 for a dynamical system of $N$ mass particles.

14. A mass particle is attracted to a fixed point in space by some force which does not vary with time. Show that energy and angular momentum of the particle are conserved but not its (linear) momentum.

15. A mass particle is attracted to a homogeneous linear mass distribution lying on the $z$-axis. Determine what physical quantities of the particle are conserved.

# 8

# Direct Methods

## 1. The Rayleigh–Ritz Method

Exact solutions for problems in the calculus of variations are available only for relatively simple problems. Fortunately, accurate approximate solutions often suffice for many applications. One approach to approximate solutions is to approximately solve the boundary-value problem (BVP) which determines the admissible extremals; the minimum value of $J$, if it exists, must be attained at such an extremal. Some basic numerical methods for BVPs of ODEs will be discussed in the Appendix of this book. Perturbation and asymptotic methods for these problems will be illustrated in Chapter 14 [see also Kevorkian and Cole (1981) and references therein]. We can also obtain an approximate solution by solving a discrete analogue of the basic problem similar to that formulated at the end of Section 3, Chapter 1. For a fine subdivision of the interval $[a, b]$, the result is expected to provide a good approximation for the exact solution of the variational problem.

In this chapter, we explore still another approach to implement the optimization process approximately. This approach seeks the minimum value of $J[y]$ relative to a smaller class of comparison functions than the entire class of PWS admissible functions. A *direct* search for the actual variational problem was deemed impractical previously. In fact, the Euler–Lagrange necessary condition was developed to avoid a direct search and to reduce the problem to solving (integro-) differential equations instead. However, as long as we are willing to settle for an approximate solution, a limited direct search for a local minimum will be seen to offer a feasible alternative to numerical methods involving discretization of the basic problem or the corresponding Euler DE. The key to a viable limited direct-search method is to search within a sufficiently small class of comparison functions so that we can always find the local minimum relative to the comparison functions allowed. At the same time, it should be possible to enlarge the group of comparison functions in a natural way so that it would eventually include the exact solution of the problem, at least in the limit.

To illustrate the feasibility of the general idea described above, consider the basic problem

$$J[y] = \int_0^1 \frac{1}{2}[(y')^2 - y^2 - 2f(x)y]\,dx, \qquad y(0) = y(1) = 0. \tag{8.1}$$

Suppose we take $y(x)$ to be proportional to $\varphi_1(x) = x(1-x)$, i.e.,

$$y = a_1\varphi_1(x) \equiv y_1(x), \qquad \varphi_1(x) = x(1-x), \tag{8.2}$$

where $a_1$ is an unknown constant. Note that we have $\varphi_1(0) = \varphi_1(1) = 0$ so that $y(x)$ satisfies the end conditions (and, hence, is *admissible*). The problem now becomes one of finding the particular choice of $y_1$, denoted by $\widehat{y}_1$, which gives a minimum $J[y]$ for comparison functions of the type (8.2). This amounts to finding an optimal constant $a_1$ and can be done efficiently by observing

$$\begin{aligned} J[y_1] &= \int_0^1 \frac{1}{2}[(a_1\varphi_1')^2 - (a_1\varphi_l)^2 - 2f(x)a_1\varphi_1]\,dx \\ &= \frac{1}{2}a_1^2\,\alpha_{11} - a_1\alpha_{10} \equiv J(a_1); \end{aligned} \tag{8.3}$$

where

$$\alpha_{11} = \int_0^1 [(\varphi_1')^2 - (\varphi_l)^2]\,dx = \frac{3}{10}, \qquad \alpha_{10} = \int_0^1 f(x)\varphi_1\,dx. \tag{8.4}$$

For $f(x) = \sin(\pi x)$ for example, we have $\alpha_{10} = 4/\pi^3$. Thus, by restricting the admissible comparison functions to those of the form (8.2), the minimization problem of the calculus of variations is reduced to the *classical minimization* of an ordinary function $J(a_1)$. The minimum value of $J(a_1)$ as a function of $a_1$ is found to be attained at the stationary point $\widehat{a}_1$ by a straightforward calculation, with

$$J^{\bullet}(\widehat{a}_1) = \widehat{a}_1\alpha_{11} - \alpha_{10} = 0 \qquad \text{or} \qquad \widehat{a}_1 = \alpha_{10}/\alpha_{11}. \tag{8.5}$$

The stationary value $J(\widehat{a}_1)$ is, in fact, a minimum because

$$J^{\bullet\bullet}(\widehat{a}_1) = \frac{3}{10} > 0. \tag{8.6}$$

As there is no other stationary point, the minimum is global [relative to the class of admissible comparison functions (8.2)] so that for any $a_1$ we have

$$J[y_1] \geq J[\widehat{y}_1] \equiv \mu_1, \qquad \widehat{y}_1 = \widehat{a}_1\varphi_1(x) = \frac{\alpha_{11}}{\alpha_{10}}\varphi_1(x). \tag{8.7}$$

Evidently, $\mu_1$ may not be the true minimum of $J[y]$ given that it is obtained relative to only a small subset of the original set of admissible comparison functions. For example, $\mu_1$ cannot be less than, and is most likely greater than, the minimum relative to the larger class of comparison functions of the form

$$y_2 = b_1\varphi_1 + b_2\varphi_2, \qquad \varphi_2 = x^2(1-x) \tag{8.8}$$

and $\varphi_1$ is as given in (8.2). We calculate $J[y_2]$ to otbain

$$J[y_2] = \frac{1}{2}\int_0^1 [(b_1\varphi_1' + b_2\varphi_2')^2 - (b_1\varphi_1 + b_2\varphi_2)^2 - 2f(x)(b_1\varphi_1 + b_2\varphi_2)]\,dx$$

$$= \frac{1}{2}[b_1^2\alpha_{11} + 2b_1b_2\alpha_{12} + b_2^2\alpha_{22}] - b_1\alpha_{10} - b_{20}\alpha_{20} \equiv J(b_1, b_2), \tag{8.9a}$$

where $\alpha_{11}$ is as given in (8.4) and

$$\alpha_{12} = \int_0^1 (\varphi_1'\varphi_2' - \varphi_1\varphi_2)\,dx = \frac{3}{20},$$

$$\alpha_{22} = \int_0^1 [(\varphi_1')^2 - \varphi_2^2]\,dx = \frac{13}{105}, \tag{8.9b}$$

$$\alpha_{k0} = \int_0^1 f(x)\varphi_k(x)\,dx.$$

Among different pairs of $(b_1, b_2)$, the one which minimizes $J(b_1, b_2)$, denoted by $(\hat{b}_1, \hat{b}_2)$, renders it stationary, i.e.,

$$J_{,b_1}(\hat{b}_1, \hat{b}_2) = \alpha_{11}\hat{b}_1 + \alpha_{12}\hat{b}_2 - \alpha_{10} = 0,$$

$$J_{,b_2}(\hat{b}_1, \hat{b}_2) = \alpha_{12}\hat{b}_1 + \alpha_{22}\hat{b}_2 - \alpha_{20} = 0.$$

This system is nonsingular and can be solved for $\hat{b}_1$ and $\hat{b}_2$. The Hessian matrix for $J(b_1, b_2)$

$$H = \begin{bmatrix} \alpha_{11} & \alpha_{12} \\ \alpha_{21} & \alpha_{22} \end{bmatrix}$$

is non-negative definite because $\alpha_{11} = 3/10 > 0$ and $\alpha_{11}\alpha_{22} - \alpha_{12}^2 = 41/2800 > 0$. Hence, $J[\hat{y}_2 = \hat{b}_1\varphi_1 + \hat{b}_2\varphi_2]$ is a local minimum relative to the class of comparison functions (8.8). Moreover, we have $J[\hat{y}_1] \geq J[\hat{y}_2]$ because $\hat{y}_1$ is a special case of $\hat{y}_2$ (with $b_2 = 0$ and $b_1 = \hat{a}_1$).

We can continue this process and take

$$y_n = \sum_{k=1}^{n} c_k\varphi_k(x), \tag{8.10}$$

where $\{\varphi_k(x)\}$ is a set of $n$ different smooth (or PWS) functions which satisfy the homogeneous end conditions. Some examples of $\varphi_k(x)$ are $x^k(1-x)$, $x(1-x)^k$, $x^k(1-x)^k$, and $\sin(k\pi x)$. By taking $y = y_n$ in (8.1) and carrying out the integration, we obtain $J[y_n] = J(c_1, c_2, ..., c_n) \equiv J(\mathbf{c})$. We then choose $\{c_k\}$ to minimize $J(\mathbf{c})$. If a local minimum is attained with $c_k = \hat{c}_k$, $k = 1, 2, \ldots, n$, then we have

$$J[y_n] \geq J[\hat{y}_n] \equiv \mu_n, \qquad \hat{y}_n = \sum_{k=1}^{n} \hat{c}_k \varphi_k(x)$$

where the coefficients $\{\hat{c}_k\}$ are determined by the stationary conditions

$$J_{,i}(\hat{\mathbf{c}}) \equiv \left.\frac{\partial J}{\partial c_i}\right|_{\mathbf{c}=\hat{\mathbf{c}}} = \alpha_{i1}\hat{c}_1 + \ldots + \alpha_{in}\hat{c}_n - \alpha_{i0} = 0, \quad i = 1, 2, \ldots, n. \qquad (8.11)$$

Note again that $J[\hat{y}_{n-1}] \geq J[\hat{y}_n]$ as $\hat{y}_{n-1}$ is a special choice of $y_n$. It would appear that by an appropriate choice of *coordinate functions* $\{\varphi_k\}$, the value $\mu_n = J[\hat{y}_n]$ would provide an accurate approximation of the exact minimum $J[\hat{y}] \equiv \mu$ if $n$ is sufficiently large. We will show in the next section that, in a well-defined way, this is, in fact, the case.

## 2. Completeness and Minimizing Sequence

Suppose a solution of the variational problem is assured by an existence theorem [see Ewing, (1985) and references therein] or the physics of the problem, i.e., a minimizing function $\hat{y}(x)$ exists with $J[\hat{y}] = \mu$ for some finite value of $\mu$. It is then important that we have a way of obtaining an approximate solution for $\hat{y}$ and $\mu$ because an exact solution is not possible for most problems. As our illustrative examples in the last section suggest, this may be done by constructing a minimizing sequence $\{\hat{y}_n(x)\}$ and a corresponding sequence of minimum values $\{\mu_n \equiv J[\hat{y}_n]\}$ in such a way that $\mu_n \to \mu$ and $\hat{y}_n(x) \to \hat{y}(x)$ as $n \to \infty$.

To construct a minimizing sequence $\{\hat{y}_n(x)\}$, we first consider basic problems with prescribed *homogeneous boundary conditions* so that $y(a) = y(b) = 0$. Let $\varphi_k(x)$, $k = 1, 2, \ldots$, be PWS coordinate functions which vanish at the end points $x = a$ and $x = b$ and let $y_n(x) = c_1\varphi_1(x) + \ldots + c_n\varphi_n(x)$ as in (8.10). We choose $c_k, k = 1, 2, \ldots, n$, to be the set of values $\hat{c}_k$ which minimizes $J[y_n] = J(c_1, c_2, \ldots, c_n)$ with $\hat{y}_n = \hat{c}_1\varphi_1 + \ldots + \hat{c}_n\varphi_n$ and $J[\hat{y}_n] = \mu_n$. From the discussion of the last section, we know that the sequence $\{\mu_n\}$ is non increasing as $n$ increases. Even if the existence of $\hat{y}$ is assured, the Rayleigh–Ritz method described there is only as good as our choice coordinate functions $\{\varphi_k(x)\}$ permits. Unless $\hat{y}(x)$ can be approximated closely by (8.10) for a sufficiently large $n$, accuracy cannot be expected.

The sequence of coordinate functions $\{\varphi_k(x)\}$ with $\varphi_k(a) = \varphi_k(b) = 0$ is said to be *complete* (in the space of PWS functions on $[a, b]$ vanishing at the end points) if for any admissible comparison function $y(x)$ and any $\varepsilon > 0$, there is an $n$ (which depends on $\varepsilon$) such that

$$\|y_n - y\| \equiv \int_a^b |y_n'(x) - y'(x)|dx < \varepsilon \tag{8.12}$$

for a set of coefficients $\{c_1, \dots, c_n\}$. Note that we have from (8.12)

$$\varepsilon > \int_a^b |y_n'(t) - y'(t)|dt \geq \int_a^{\bar{x}} |y_n'(t) - y'(t)|dt$$
$$\geq \left| \int_a^{\bar{x}} [y_n'(t) - y'(t)]\, dy \right| = |y_n(\bar{x}) - y(\bar{x})|$$

for any $\bar{x}$ in $[a, b]$. Therefore, (8.12) implies that $y_n(x)$ is close to $y(x)$ for all $x$ in $[a, b]$:

$$\max_{a \leq x \leq b} |y_n(x) - y(x)| < \varepsilon. \tag{8.13}$$

We can now construct an accurate approximate solution with the help of any complete sequence of coordinate functions $\{\varphi_k(x)\}$:

**[VIII.1]** ***Suppose the infimum $\mu$ of $J[y]$ over all admissible comparison functions is a finite number. If $F(x, y, v)$ is at least $C^1$ in $y$ and $v$ and if $\{\varphi_k\}$ is a complete sequence, then for any $\varepsilon > 0$, there is an $N(\varepsilon)$ for which $0 < \mu_n - \mu < \varepsilon$ for all $n > N$ where $\mu_n = J[\hat{y}_n]$.***

In other words, we can choose $n$ sufficiently large that an approximate solution of the form (8.10) can be used to get $J[\hat{y}_n]$ as close to $\mu$ as we wish. We will prove this below by squeezing $\mu_n$ (for sufficiently large $n$) between $\mu$ and $\mu + \varepsilon$.

For any $\varepsilon > 0$, there is always an admissible comparison function $\bar{y}(x)$ for which $J[\bar{y}] < \mu + \varepsilon/2$; this is so by the definition of $\mu$. Because $\{\varphi_k(x)\}$ is complete, we can always find an $N(\delta)$ so that $\|\hat{y}_n - \bar{y}\| < \delta$ for all $n \geq N$ and any $\delta > 0$. So, for any $\delta > 0$, we have

$$|J[\hat{y}_n] - J[\bar{y}]| \leq \int_a^b |F(x, \hat{y}_n, \hat{y}_n') - F(x, \bar{y}, \bar{y}')|dx$$
$$= \int_a^b |F_{,y}(x, \xi, \eta)(\hat{y}_n - \bar{y}) + F_{,y'}(x, \xi, \eta)(\hat{y}_n' - \bar{y}')|dx$$
$$\leq M \int_a^b [\,|\hat{y}_n - \bar{y}| + |\hat{y}_n' - \bar{y}'|\,]dx \leq M\{\delta(b - a) + \delta\},$$

where $\xi(x) = \bar{y}(x) + \lambda_1[\hat{y}_n(x) - \bar{y}(x)]$ and $\eta(x) = \bar{y}' + \lambda_2(\hat{y}_n' - \bar{y}')$ for some $0 < \lambda_k < 1$ (by Taylor's theorem) and $M = \max[|F_{,y}(x,\xi,\eta)|, |F_{,y'}(x,\xi,\eta)|]$ over all $x$ in $[a,b]$. Note that we have made use of (8.13) to get the last inequality. Take

$$\delta = \frac{\varepsilon}{2M[1 + (b-a)]}$$

and we obtain

$$|J[\hat{y}_n] - J[\bar{y}]| < \varepsilon/2 \qquad \text{or} \qquad J[\bar{y}] - \varepsilon/2 < J[\hat{y}_n] < J[\bar{y}] + \varepsilon/2.$$

Given $\mu \le J[\hat{y}_n]$ and $J[\bar{y}] < \mu + \varepsilon/2$, it follows that we have

$$\mu \le J[\hat{y}_n] < \mu + \varepsilon$$

for all $n > N(\delta)$ as we set out to prove.

As long as $J[y]$ has a finite minimum value $\mu$ attained at $\hat{y}(x)$, [VIII.1] assures us that a ***minimizing sequence*** $\{\hat{y}_n(x)\}$ can theoretically be found so that $J[\hat{y}_n] = \mu_n$ tends to $\mu$ from above. However. it is not always the case that the sequence $\{\hat{y}_n(x)\}$ itself tends to the minimizing admissible extremal $\hat{y}(x)$. In fact, $\hat{y}_n(x)$ may not even tend to a limit function as $n \to \infty$. To illustrate, consider the example

$$J[y] = \int_{-1}^{1} (xy')^2 dx, \qquad y(-1) = -1, \qquad y(1) = 1. \tag{8.14}$$

The integrand is non-negative; the infimum (greatest lower bound) of $J[y]$ is, therefore, zero. Let

$$\hat{y}_n(x) = \frac{\tan^{-1}(nx)}{\tan^{-1}(n)} \tag{8.15}$$

with $\hat{y}_n' = n/[\tan^{-1}(n)(1 + n^2x^2)]$. Evidently, we have $\hat{y}_n(\pm 1) = \pm 1$; hence, each $\hat{y}_n(x)$ is an admissible comparison function. Furthermore, they form a minimizing sequence as

$$\mu_n = J[\hat{y}_n] = \frac{1}{[\tan^{-1}n]^2} \int_{-1}^{1} \frac{(nx)^2}{[1 + (nx)^2]^2} dx < \frac{1}{[\tan^{-1}n]^2} \int_{-1}^{1} \frac{dx}{1 + (nx)^2}$$

$$= \frac{1}{(\tan^{-1}n)^2} \left[\frac{\tan^{-1}(nx)}{n}\right]_{-1}^{1} = \frac{2}{n[\tan^{-1}(n)]} \equiv \bar{\mu}_n \ge 0.$$

With $\tan^{-1}z \to \pi/2$ as $z \to \infty$, we have $\bar{\mu}_n \to 0$ as $n \to \infty$ so that $\mu_n (\ge 0)$ also tends to zero, which is the infimum of $J[y]$. On the other hand, $\hat{y}_n(x)$ does not tend to a smooth or PWS function (see Fig. 8.1); instead, we have

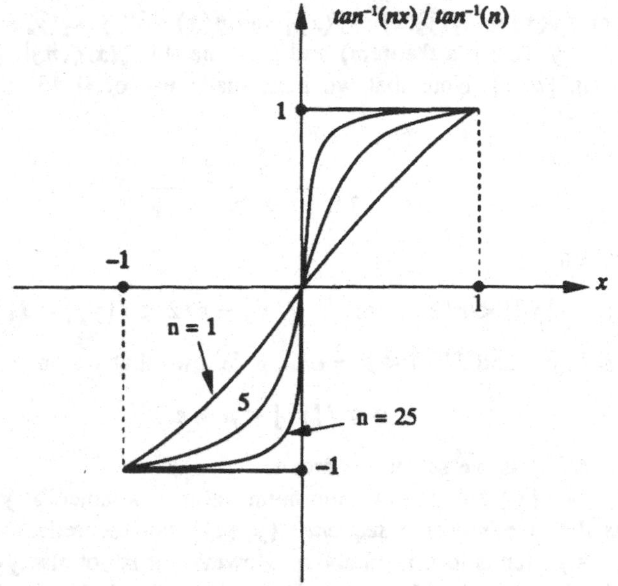


**Figure 8.1**

$$\lim_{n\to\infty} \hat{y}_n(x) = Y(x) \equiv \begin{cases} -1 & (x<0) \\ 0 & (x=0) \\ 1 & (x>0). \end{cases}$$

We may say that $\hat{y}_n(x)$ tends to $Y(x) = 2H(x) - 1$, where $H(x)$ is the Heaviside unit step function with $H(0)$ taken to be 1/2.

Because [VIII.1] deals only with the convergence of $\{\mu_n\}$ to $\mu$, the following result on the convergence of $\{\hat{y}_n(x)\}$ to $\hat{y}(x)$ should be of interest in view of the illustrative example above:

**[VIII.2]** *If $\{\varphi_k\}$ is complete and the variational problem has a solution $\hat{y}$ among the PWS functions, then for any positive $\varepsilon$ and $\varepsilon'$, there is an* $N(\varepsilon, \varepsilon')$ *such that* $\|\hat{y}_n - \hat{y}\| < \varepsilon'$ and $0 < \mu_n - \mu < \varepsilon$ *for all* $n > N$.

Choose $N_0(\varepsilon)$ so that $\|\hat{y}_n - \hat{y}\| < \delta = \varepsilon/\{M[(b-a)+1]\}$ to get $\mu_n$ within an $\varepsilon$ neighborhood of $\mu$ for $n > N_0$ (by the same argument used in [VIII.1]). If $\delta < \varepsilon'$, we are done (with $N = N_0$) because (8.13) follows from (8.12). If not, find a sufficiently large $N(\varepsilon, \varepsilon') > N_0$, so that $\|\hat{y}_n - \hat{y}\| < \varepsilon'$ for $n > N$ (by the completeness of $\{\varphi_k\}$). Because $N > N_0$, we continue to have $0 < \mu_n - \mu < \varepsilon$ for $n > N$.

It is evident that the infimum $\mu = 0$ of the basic problem for (8.14) is *not* attained in the class of PWS functions, although we can get as close to it as we wish. Similar to the maximum rocket height problem of Chapter 2, the example demonstrates once more that a given variational problem may not have a solution in the space of PWS comparison functions and that, strictly speaking, the theory developed in Chapter 2 does not apply. For other problems with no true finite minimum such as the example in Section 1 of Chapter 5, it would be a waste of time to seek an approximate solution. In some cases, we may even run the risk of being misled by a seemingly reasonable result if we should persist. Together, these examples illustrate the importance of existence theorems for approximate solutions of variational problems.

## 3. A Weighted Least-Squares Approximation

The determination of the coefficients $\{\widehat{c}_k\}$ in the trial function $y_N$ of (8.10) by the method of Rayleigh–Ritz generally involves the solution of a set of nonlinear equations. For the following normalized version of the brachistochrone problem,

$$J[y] = \int_0^1 \sqrt{\frac{1+(y')^2}{y}}\, dx, \qquad y(0)=0, \;\; y(1)=1,$$

even a one-term trial solution

$$y_1 = x + c_1\varphi_1(x) = x + c_1 x(1-x)$$

requires a numerical solution for $\widehat{c}_1$. The integral $J[y]$ for this case becomes

$$J[y_1] = \int_0^1 \sqrt{\frac{2+2c_1(1-2x)+c_1^2(1-2x)^2}{x[1+c_1(1-x)]}}\, dx \equiv J(c_1).$$

For $J(\widehat{c}_1)$ to be a stationary value of $J(c_1)$, we require that

$$\left.\frac{dJ}{dc_1}\right|_{c_1=\widehat{c}_1} \equiv J^{\bullet}(\widehat{c}_1) = 0$$

or $\widehat{c}_1$ to be a root of the nonlinear equation

$$g(c_1) \equiv \int_0^1 \frac{c_1^2(1-x)(1-2x)^2 + 2c_1(1-2x)^2 - 2x}{\sqrt{x[1+c_1(1-x)]^3}\,\sqrt{2+2c_1(1-2x)+c_1^2(1-2x)^2}}\, dx = 0.$$

There is no method for an explicit solution of the equation $g(c_1) = 0$. To obtain a root $\widehat{c}_1$, we will have to use some kind of iterative numerical scheme such as Newton's method. For each iteration of that method, the integral involved will have to be evaluated numerically anew. The same process extends to $\widehat{y}_N$ for $N > 1$, but now we must deal with a system of nonlinear equations to obtain the $N$ unknown parameters $\{\widehat{c}_1, \ldots, \widehat{c}_N\}$. The numerical computation required can be simplified somewhat by working with localized coordinate functions to be discussed in the next few sections. Still, the Rayleigh–Ritz method is more attractive for Lagrangians which involves only quadratic and linear terms in the unknowns and its derivatives (giving a linear Euler DE). For the scalar case, the most general basic problem of this type (involving only $x$, $y$ and $y'$) can be cast in the form

$$J[y] = \int_0^1 \frac{1}{2}\,[p(x)(y')^2 + q(x)y^2 - 2f(x)y]\,dx \tag{8.16a}$$

$$y(0) = y(1) = 0. \tag{8.16b}$$

For many problems of the form (8.16) in application areas, we have $p(x) > 0$ and $q(x) \geq 0$ for all $x$ in $[0,1]$. For these problems, the Rayleigh–Ritz solution can be interpreted as a weighted least-squares type approximation of the exact solution of the problem. To see this, we let

$$e_N(x) = \widehat{y}(x) - y_N(x) \tag{8.17a}$$

and

$$\|e_N\|_{pq} = \int_0^1 \frac{1}{2}\,[p(x)(e_N')^2 + q(x)e_N^2\,]\,dx. \tag{8.17b}$$

Given $p(x) > 0$ and $q(x) \geq 0$, $\|e_N\|_{pq}$ is non-negative and only vanishes if $e_N \equiv 0$. Upon writing

$$\int_0^1 p(x)(e_N')^2 dx = [p(x)e_N' e_N]_0^1 - \int_0^1 [p(x)e_N'\,]' e_N\,dx$$

and observing the end conditions so that $e_N(0) = e_N(1) = 0$, we transform (8.17b) into

$$\|e_N\|_{pq} = \frac{1}{2}\int_0^1 \{-[p(x)(e_N')]' + q(x)e_N\}e_N\,dx$$

$$= \frac{1}{2}\int_0^1 \{[-(py_N')' + qy_N]y_N - [-(py_N')' + qy_N]\widehat{y}$$

$$- [-(p\widehat{y}')' + q\widehat{y}]y_N + [-(p\widehat{y}')' + q\widehat{y}]\widehat{y}\}\,dx$$

with the integral of the last bracketed terms being $\|\widehat{y}\|_{pq}$.

By repeated integration by parts, we have

$$\int_0^1 [-(py_N')' + qy_N]\hat{y}\,dx$$
$$= [-(p\hat{y}_N')\hat{y} + (p\hat{y}')\hat{y}_N]_0^1 + \int_0^1 [-(p\hat{y}')' + q\hat{y}]y_N\,dx$$
$$= \int_0^1 [-(p\hat{y}')' + q\hat{y}]y_N\,dx$$

The relation above can be used to simplify the expression for $\|e_N\|_{pq}$ further to

$$\|e_N\|_{pq} = \frac{1}{2}\int_0^1 \{[-(py_N')' + qy_N]y_N - 2[-(p\hat{y}')' + q\hat{y}]y_N\}\,dx + \|\hat{y}\|_{pq}$$
$$= \frac{1}{2}\int_0^1 [-(py_N')' + qy_N - 2f(x)]y_N\,dx + \|\hat{y}\|_{pq}$$
$$= \frac{1}{2}\int_0^1 [p(y_N')^2 + qy_N^2 - 2fy_N]\,dx + \|\hat{y}\|_{pq}$$
$$= J[y_N] + \|\hat{y}\|_{pq}.$$

Given that $\|\hat{y}\|_{pq}$ is a fixed number, $\|e_N\|_{pq}$ is minimized if $y_N$ is chosen to minimize $J[y_N]$. Hence,

**[VIII.3]** ***A Rayleigh–Ritz solution $\hat{y}_N$ for the basic problem (8.16) with $p(x) > 0$ and $q(x) \geq 0$ minimizes the sum of the weighted mean-square error of $\hat{y}_N$ and $\hat{y}_N'$ as defined by (8.17b).***

Because $F(x, y, y') = p(y')^2 + qy^2 - 2fy$ is convex in $y$ and $y'$, the stationary point $(\hat{c}_1, \ldots, \hat{c}_N)$ minimizes $J[y_N]$.

## 4. Inhomogeneous End Conditions

For basic problems with at least one inhomogeneous end condition so that $A \neq 0$ or $B \neq 0$, the linear combination (8.10) would not satisfy the given inhomogeneous end conditions(s). On the other hand, by choosing the coordinate functions to satisfy $y(a) = A$ individually, we would have $y_n(a) = (c_1 + c_2 + \ldots + c_n)A$ which is generally *not* equal to $A$. One possible way of handling inhomogeneous end conditions is to subtract an *admissible* comparison function $y_0(x)$ from $y$ and denote the difference by $\hat{z}(x) = y - y_0$. For example, we may take $y_0(x) \equiv A + (B - A)(x - a)/(b - a)$. With $y_0(a) = A$ and $y_0(b) = B$, $z(x)$ evidently satisfies the homogeneous end conditions $z(a) = z(b) = 0$. We may, therefore, set $z_n = c_1\varphi_1 + \ldots + c_n\varphi_n$ for any complete sequence of $\{\varphi_k\}$ with $\varphi_k(a) = \varphi_k(b) = 0, k = 1, 2, \ldots$, as in the previous section.

To illustrate, consider the basic problem (I) of minimizing

$$\text{(I)} \qquad J[y] = \int_0^1 (y')^2 dx, \qquad y(0) = 1, \qquad y(1) = 0.$$

The function $y_0(x) = \cos(\pi x/2)$ is an *admissible* comparison function and $y(x) - y_0(x) \equiv z(x)$ satisfies the homogeneous end conditions $z(0) = z(1) = 0$. In terms of $z(x)$, the basic problem (I) for $y(x)$ becomes

$$J[y] = \int_0^1 (z' + y_0')^2 dx, \qquad z(0) = 0, \qquad z(1) = 0.$$

We now take $\{\varphi_k(x)\}$ to be $\{\sin(k\pi x)\}$ and

$$z_n(x) = y_n(x) - y_0(x) = \sum_{k=1}^{n} c_k \varphi_k(x)$$

with $J[y] = J(c_1, c_2, \ldots, c_n)$.

For $n = 0$, we have $z_0 = 0$ and

$$J_0 \equiv J[y_0] = \int_0^1 (2/\pi^2)^2 \sin^2(2/\pi x) dx = 8/\pi^2 \cong 1.234.$$

For $n = 1$, we have

$$J[y_1] = J(c_1) = 8/\pi^2 + c_1(3/2\pi) + c_1^2(2/\pi^2).$$

The minimum of $J(c_1)$ is attained at $c_1 = -2/3\pi \equiv \widehat{c}_1$ with

$$J_1 \equiv J(\widehat{c}_1) \cong 1.012, \qquad \widehat{y}_1 = \cos\left(\frac{\pi x}{2}\right) - \frac{2}{3\pi}\sin(\pi x).$$

We may continue the process to get $\widehat{y}_2 = \widehat{y}_1 - \sin(2\pi x)/15\pi$ and $J_2 \cong 1.003$, etc. These should be compared with the exact solution $\widehat{y}(x) = 1 - x$ and $J[\widehat{y}] = 1$. As Table 8.1 [from F. Clarke (1980)] shows, $y_2(x)$ is already an excellent approximation for $\widehat{y}(x) = 1 - x$ (see Fig. 8.2).

Convergence of $\{\widehat{y}_n(x)\}$ to the exact solution $\widehat{y}(x)$ is not always as rapid as indicated by the example above. Consider the basic problem (II):

$$\text{(II)} \qquad J[y] = \frac{1}{2}\int_0^1 [\varepsilon^2 (y')^2 + y^2] dx, \qquad y(0) = 1, \qquad y(1) = 0.$$

The global minimum of this problem is attained at the unique extremal

$$\widehat{y}(x) = \cosh\left(\frac{x}{\varepsilon}\right) - \coth\left(\frac{1}{\varepsilon}\right)\sinh\left(\frac{x}{\varepsilon}\right)$$

**Table 8.1. Convergence of $\{\mu_n\}$ and $\{\widehat{y}_n(x)\}$ for Basic Problem (I)**

| $\widehat{y}_n$ | $x$ | | | | | | $\mu_n = J[\widehat{y}_n]$ |
|---|---|---|---|---|---|---|---|
| | 0 | 0.2 | 0.4 | 0.6 | 0.8 | 1.0 | |
| $\widehat{y}_0$ | 1. | 0.95 | 0.81 | 0.59 | 0.31 | 0. | 1.234 |
| $\widehat{y}_1$ | 1. | 0.83 | 0.61 | 0.39 | 0.18 | 0. | 1.012 |
| $\widehat{y}_2$ | 1. | 0.81 | 0.59 | 0.40 | 0.20 | 0. | 1.003 |
| $\widehat{y}$ | 1. | 0.80 | 0.60 | 0.40 | 0.20 | 0. | 1.000 |

with

$$J[\widehat{y}] = \frac{\varepsilon}{2}\,\frac{e^{2/\varepsilon} + 1}{e^{2/\varepsilon} - 1} = \mu.$$

Suppose we again take $y_0 = \cos(\pi x/2)$ and $\varphi_n(x) = \sin(n\pi x)$. It is a straightforward (though tedious) calculation to obtain the following results:

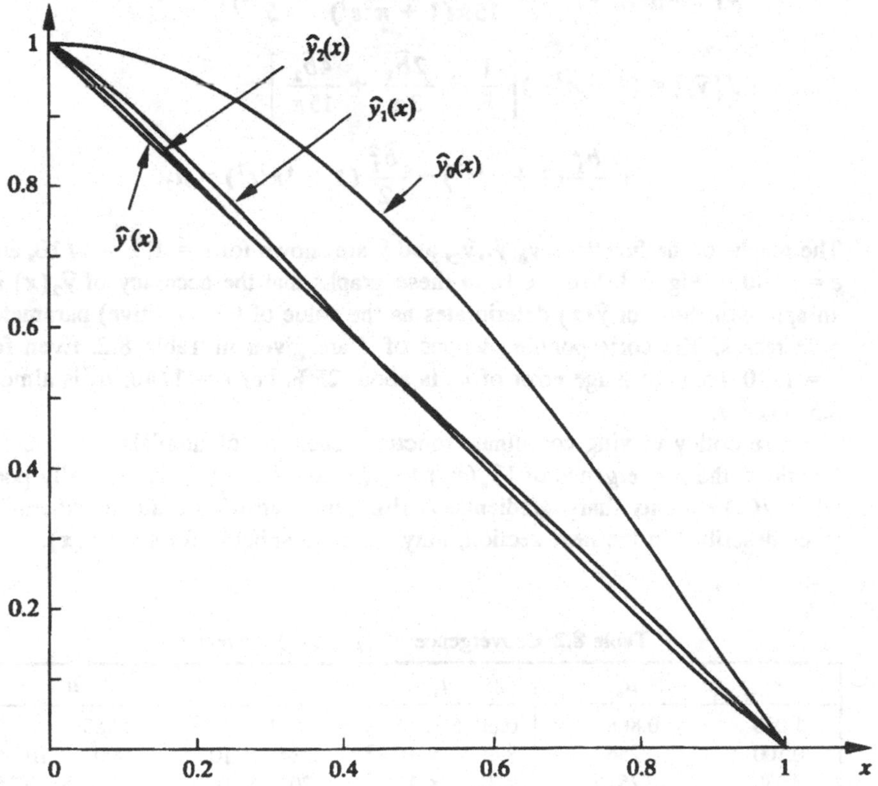


Figure 8.2

$$n=0:\quad y_0(x)=\cos\left(\frac{\pi x}{2}\right),$$

$$J[y_0]=\frac{1}{4}\left[1+\frac{\pi^2\varepsilon^2}{4}\right]=\mu_0;$$

$$n=1:\quad \hat{y}_1(x)=y_0(x)+\hat{a}_1\sin(\pi x),\qquad \hat{a}_1=-\frac{2(4+\pi^2\varepsilon^2)}{3\pi(1+\pi^2\varepsilon^2)},$$

$$J[\hat{y}_1]=\frac{1}{4}\left\{\left[1+\frac{\pi^2\varepsilon^2}{4}\right]\left[1+\frac{16\hat{a}_1}{3\pi}\right]+\hat{a}_1^2[1+\pi^2\varepsilon^2]\right\}=\mu_1;$$

$$n=2:\quad \hat{y}_2(x)=y_0(x)+\hat{b}_1\sin(\pi x)+\hat{b}_2\sin(2\pi x)$$

$$\hat{b}_1=\hat{a}_1,\qquad \hat{b}_2=-\frac{4(4+\pi^2\varepsilon^2)}{15\pi(1+\pi^2\varepsilon^2)}=\frac{2}{5}\hat{a}_1;$$

$$J[\hat{y}_2]=(4+\pi^2\varepsilon^2)\left[\frac{1}{8}+\frac{2\hat{b}_1}{3\pi}+\frac{4\hat{b}_2}{15\pi}\right]$$

$$+\frac{\hat{b}_1^2}{2}(1+\pi^2\varepsilon^2)+\frac{\hat{b}_2^2}{2}(1+4\pi^2\varepsilon^2)=\mu_2.$$

The graphs of the functions $y_0, \hat{y}_1, \hat{y}_2$, and $\hat{y}$ are shown for $\varepsilon=1$, $\varepsilon=1/10$, and $\varepsilon=1/40$ in Fig. 8.3. We see from these graphs that the accuracy of $\hat{y}_2(x)$ as an approximation for $\hat{y}(x)$ deteriorates as the value of the (positive) parameter $\varepsilon$ decreases. The corresponding values of $J$ are given in Table 8.2. Even for $\varepsilon=1/10$, the percentage error of $\mu_2$ is about 25%. For $\varepsilon=1/40$, $\mu_2$ is almost 3.5 times $\mu$.

For smoothly varying coordinate functions such as polynomials and circular functions, the convergence of $\{\hat{y}_n(x)\}$ to $\hat{y}(x)$ and $\{\mu_n\}$ *to* $\mu$ *is generally poor* when $\hat{y}(x)$ exhibits sharp gradients. A different kind of coordinate functions, to be described in the next section, may be more suitable for such $\hat{y}(x)$.

**Table 8.2** Convergence of $\{\mu_n\}$ for Different $\varepsilon$

| $\varepsilon$ | $\mu_0$ | $\mu_1$ | $\mu_2$ | $\mu$ |
|---|---|---|---|---|
| 1.000 | 0.8669 | 0.6676 | 0.6591 | 0.6565 |
| 0.100 | 0.2562 | $8.403 \times 10^{-2}$ | $6.234 \times 10^{-2}$ | $5.000 \times 10^{-2}$ |
| 0.050 | 0.2515 | $7.358 \times 10^{-2}$ | $4.702 \times 10^{-2}$ | $2.500 \times 10^{-2}$ |
| 0.025 | 0.2504 | $7.081 \times 10^{-2}$ | $4.260 \times 10^{-2}$ | $1.2500 \times 10^{-2}$ |

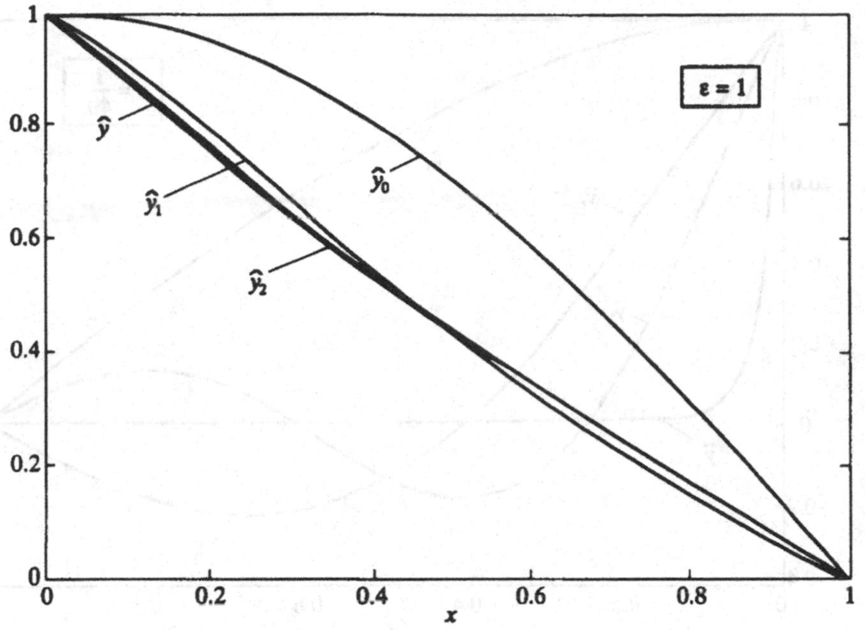


**Figure 8.3a**

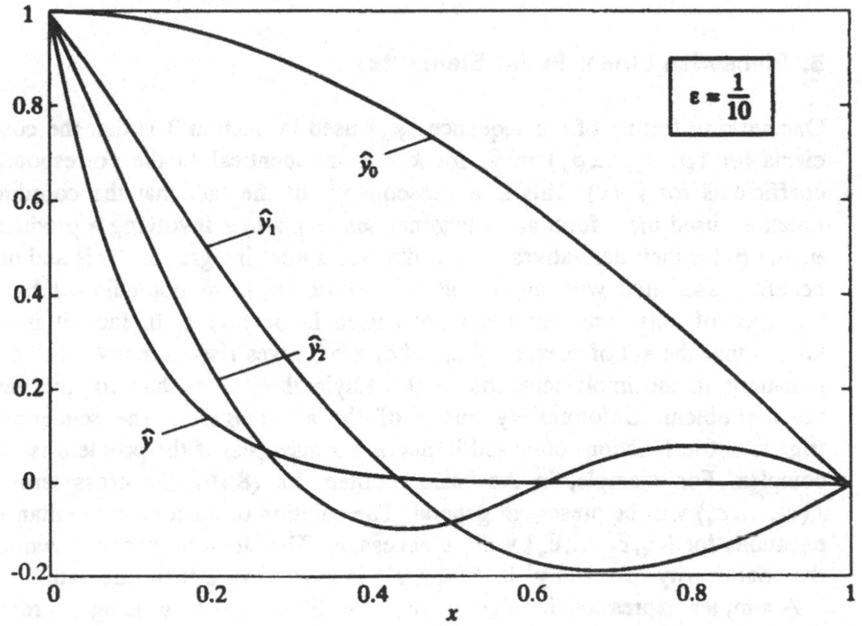


**Figure 8.3b**

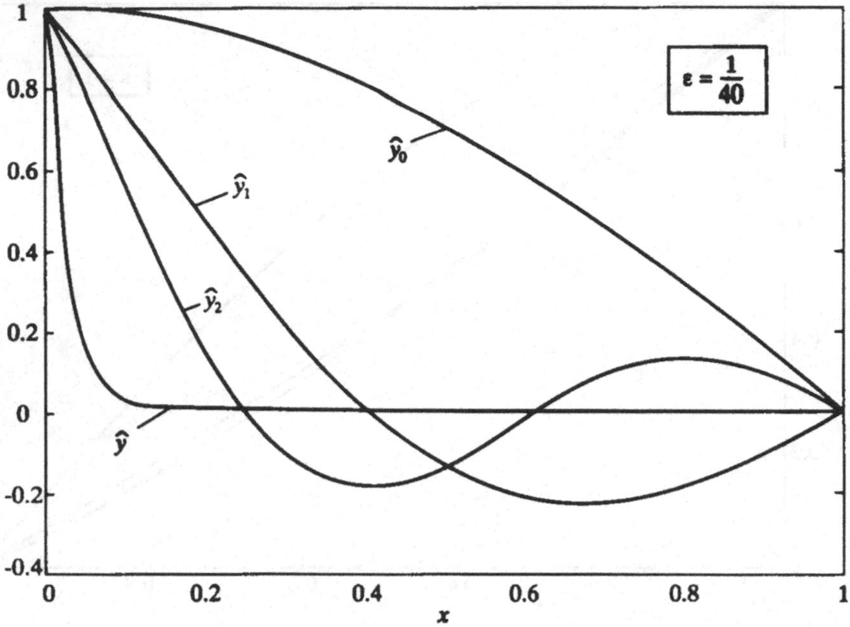


Figure 8.3c

## 5. Piecewise Linear Finite Elements

One striking feature of the sequence $\{\varphi_k\}$ used in Section 3 is that the coefficients for $\{\varphi_1, \varphi_2, \ldots, \varphi_n\}$ in $\widehat{y}_k$ for $k > n$ are identical to the corresponding coefficients for $\widehat{y}_n(x)$. This is a consequence of the fact that the coordinate functions used there form an orthogonal set; terms in $J$ involving a product of $\varphi_k$ and $\varphi_j$ (or their derivatives), $k \neq j$, disappear after integration. This and other benefits associated with an orthogonal set of $\{\varphi_k\}$ are not enjoyed by the sequence of polynomial trial functions used in Section 1. In fact, it is well known that the set of powers of $x, \varphi_k(x) = x^k$, gives rise to many other complications in the implementation of the Rayleigh–Ritz method for the variational problem. Unfortunately, many of the advantages of the sequence of trigonometric functions observed in Section 3 disappear if the problem is more complex. For example, in the basic problem for (8.16), the cross terms in $J(c_1, \ldots, c_n)$ will be present in general. The solution of a set of $n$ simultaneous equations for $\{\widehat{c}_1, \widehat{c}_2, \ldots, \widehat{c}_n\}$ will ne necessary. The situation becomes worse if the nonlinearity of $y$ and $y'$ in $F(x, y, y')$ is more severe than quadratic.

A simpler expression for $J(c_1, \ldots, c_n)$ can be obtained by using coordinate functions which vanish outside a small sub-interval of $[a, b]$. As in the finite difference solution, we introduce a set of mesh points $x_0 = a, x_1, \ldots, x_N, x_{N+1} = b$.

Although it is not necessary to do so, we take them to be evenly spaced so that $x_k = a + kh$ with $h = (b-a)/(N+1)$. We now introduce a set of *finite element coordinate functions* $\{\varphi_k^h\}$ with each $\varphi_k^h$ being a polynomial function of $x$ in the interval $(x_{k-1}, x_{k+1})$ taking on the value 1 at $x_k$ and zero outside the interval. For $\varphi_k^h$ to be PWS, it has to vanish at $x_{k-1}$ and $x_{k+1}$ and, therefore, cannot be a constant function inside $(x_{k-1}, x_{k+1})$. The simplest PWS polynomial function possible is the piecewise linear (or roof) function (see Fig. 8.4):

$$\varphi_k^h(x) = \begin{cases} 1 - \dfrac{1}{h}|x - x_k| & (x_{k-1} \le x \le x_{k+1}) \\ 0 & (x < x_{k-1}, x > x_{k+1}) \end{cases} \qquad (k = 1, \ldots, N). \tag{8.18}$$

The *trial functions,* given by

$$y_N(x) = c_1\varphi_1^h + c_2\varphi_2^h + \ldots + c_N\varphi_N^h \qquad (N = 1, 2, 3, \ldots), \tag{8.19}$$

satisfy the homogeneous end conditions $y(a) = y(b) = 0$ automatically, and the constant $c_j$ is just the value of $y_N$ at the mesh point $x_j$. Furthermore, we have the "orthogonality" relations

$$\int_a^b p(x)(\varphi_i^h)'(\varphi_j^h)'\,dx = 0, \qquad \int_a^b q(x)\varphi_i^h\varphi_j^h dx = 0, \tag{8.20}$$

for any PWS functions $p(x)$ and $q(x)$ if $|i - j| \ge 2$.

To see the benefit of this particular choice of trial functions, we consider problem (8.1) again. By approximating $y(x)$ by $y_N(x)$ from (8.19), we have

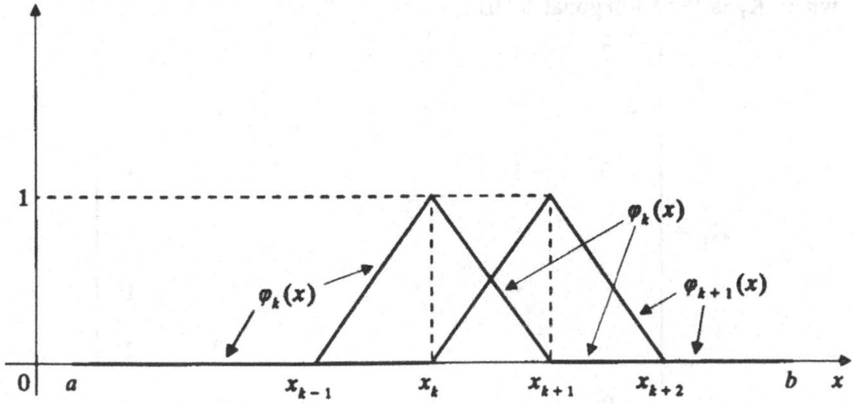


Figure 8.4

$$\int_0^1 (y_N')^2 dx = \int_0^1 [c_1(\varphi_1^h)' + c_2(\varphi_2^h)' + \ldots + c_N(\varphi_N^h)']^2 dx$$
$$= [c_1^2 V_{11} + c_2^2 V_{22} + \ldots + c_N^2 V_{NN}] + 2[c_1 c_2 V_{12} + c_2 c_3 V_{23} + \ldots$$
$$+ c_k c_{k+1} V_{k(k+1)} + \ldots + c_{N-1} c_N V_{(N-1)N}],$$

where

$$V_{ij} = \int_0^1 (\varphi_i^h)'(\varphi_j^h)' dx.$$

Note that the first relation of (8.20) implies $V_{ij} = 0$ for $|i-j| \geq 2$. It is also not difficult to verify

$$V_{ii} = \int_{x_{i-1}}^{x_{i+1}} [(\varphi_i^h)']^2 dx = \frac{2}{h},$$

$$V_{i,i+1} = \int_{x_i}^{x_{i+1}} [(\varphi_i^h)' (\varphi_{i+1}^h)'] dx = -\frac{1}{h}$$

so that

$$\int_0^1 (y_N')^2 dx = \frac{2}{h}\left[\sum_{i=1}^{N} c_i^2 - \sum_{i=1}^{N-1} c_i c_{i+1}\right].$$

The right-hand side can be expressed as a quadratic form for $\mathbf{c} = (c_1, c_2, \ldots, c_N)^T$:

$$\int_0^1 (y_N')^2 dx = \frac{1}{h} \mathbf{c}^T K_1 \mathbf{c},$$

where $K_1$ is the tridiagonal matrix

$$K_1 = \begin{bmatrix} 2 & -1 & 0 & \cdot & \cdot & \cdot & 0 \\ -1 & 2 & -1 & \cdot & \cdot & \cdot & \cdot \\ 0 & -1 & & & & \cdot & \cdot \\ \cdot & \cdot & & & & & \cdot \\ \cdot & \cdot & & & & & 0 \\ \cdot & & & \cdot & \cdot & 2 & -1 \\ 0 & \cdot & \cdot & \cdot & 0 & -1 & 2 \end{bmatrix}$$

Similarly, we have

$$\int_0^1 y_N^2 dx = (c_1^2 U_{11} + \ldots + c_N^2 U_{NN}) + 2(c_1 c_2 U_{12} + c_2 c_3 U_{23} + \ldots + c_{N-1} c_N U_{(N-1)N}),$$

where

$$U_{ij} = \int_0^1 \varphi_i^h \varphi_j^h dx$$

with $U_{ij} = 0$ for $|i - j| \geq 2$ by the second relation of (8.20). It is not difficult to verify

$$U_{ii} = \int_{x_{i-1}}^{x_{i+1}} (\varphi_i^h)^2 dx = \frac{2h}{3},$$

$$U_{i(i+1)} = \int_{x_i}^{x_{i+1}} \varphi_i^h \varphi_{i+1}^h dx = \frac{h}{6}$$

so that

$$\int_0^1 y_N^2 dx = \frac{h}{6}\left[4\sum_{i=1}^{N} c_i^2 + 2\sum_{i=1}^{N-1} c_i c_{i+1}\right] = \frac{h}{6}\mathbf{c}^T \mathbf{K}_0 \mathbf{c},$$

where $K_0$ is also a tridiagonal matrix:

$$K_0 = \begin{bmatrix} 4 & 1 & 0 & \cdot & \cdot & \cdot & \cdot & 0 \\ 1 & 4 & 1 & 0 & \cdot & & & \cdot \\ 0 & 1 & & & & & & \cdot \\ \cdot & \cdot & & & & & \cdot & \cdot \\ \cdot & & & & & & 1 & 0 \\ \cdot & & & & \cdot & 1 & 4 & 1 \\ 0 & \cdot & \cdot & \cdot & \cdot & 0 & 1 & 4 \end{bmatrix}$$

For the last term in $F(x, y, y')$, we have

$$\int_0^1 f(x) y_N(x) dx = \frac{1}{h}[c_1 F_1 + \ldots + c_N F_N] = \frac{1}{h}\mathbf{F} \cdot \mathbf{c},$$

where

$$F_k = h\int_{x_{k-1}}^{x_{k+1}} f(x)\varphi_k^h(x)\,dx \tag{8.21}$$

In general, the integrals for $F_k$, $k = 1, 2, \ldots, N$, will have to be evaluated numerically. Upon combining these results, we get the following expression for $J[y_N]$:

$$J[y_N] = \frac{1}{2h}\mathbf{c}^T K\mathbf{c} - \frac{1}{h}\mathbf{F}\cdot\mathbf{c},$$

where

$$K = K_1 - \frac{1}{6}h^2 K_0.$$

Given that $K$ is a symmetric positive definite matrix, the minimum of $J[y_N]$ is attained at the stationary point $\hat{\mathbf{c}}$ which is the solution of the linear system

$$K\hat{\mathbf{c}} = \mathbf{F}. \tag{8.22}$$

An important advantage of this finite element formulation is that the "stiffness" matrix $K$ is tridiagonal and the computational effort required to solve the linear system involves only $O(N)$ multiplications. This is two orders of magnitude less than the $O(N^3)$ multiplications needed for the generally dense $K$ matrix of the Rayleigh–Ritz method with a set of nonlocal basis functions [see G. Strang (1980)].

Another advantage of the finite element coordinate functions is that they are localized. This property allows for an efficient approximation of functions with sharp gradients. The localized roof functions will be applied to Problem (II) of Section 4 and problem (14) of Exercises.

## 6. The Finite Element Method

We consider now the following variational problem associated with a general second-order linear Euler DE:

$$J[y] = \frac{1}{2}\int_a^b \{p(x)(y')^2 + q(x)y^2 - 2f(x)y\}\,dx, \qquad y(a) = y(b) = 0. \tag{8.23}$$

We again use the local coordinate functions (8.18) for this problem. In that case, the term

$$\begin{aligned}\int_a^b p(x)(y_N')^2\,dx &= \int_a^b p(x)[c_1(\varphi_1^h)' + \ldots c_N(\varphi_N^h)']^2\,dx \\ &= (c_1^2 V_{11} + \ldots + c_N^2 V_{NN}) + 2(c_1c_2V_{12} + c_2c_3V_{23} + \ldots + c_{N-1}c_N V_{(N-1)N})\end{aligned} \tag{8.24}$$

involves

$$V_{ij} = \begin{cases} \displaystyle\int_{x_i}^{x_{i+1}} p(x)\,(\varphi_i^h)'(\varphi_{i+1}^h)'\,dx = -\frac{2}{h^2}\int_{x_i}^{x_{i+1}} p(x)\,dx & (j = i+1) \\ \displaystyle\int_{x_{i-1}}^{x_{i+1}} p(x)\,[(\varphi_i^h)']^2\,dx = -\frac{2}{h^2}\int_{x_{i-1}}^{x_{i+1}} p(x)\,dx & (j = i). \end{cases} \tag{8.25}$$

Correspondingly, the term

$$\int_a^b q(x)(y_N)^2\,dx = (c_1^2 U_{11} + \dots + c_N^2 U_{NN}) + 2(c_1 c_2 U_{12} + \dots + c_{N-1} c_N U_{(N-1)N}) \tag{8.26}$$

involves

$$U_{ij} = \begin{cases} \displaystyle\int_{x_i}^{x_{i+1}} q(x)\,\varphi_i^h\,\varphi_{i+1}^h\,dx & (j = i+1) \\ \displaystyle\int_{x_{i-1}}^{x_{i+1}} q(x)\,[\varphi_i^h]^2\,dx & (j = i). \end{cases} \tag{8.27}$$

Similar to the expressions for $F_k$ in (8.21), integrals in (8.25) and (8.27) usually cannot be evaluated exactly; some kind of numerical integration scheme (usually by Gaussian quadratures) will have to be used to obtain $V_{ij}$, $U_{ij}$, and $F_k$ approximately. However, this additional complication does not detract from the merit of the present (*finite element*) method because the same complication occurs for other choices of the coordinate functions $\{\varphi_k\}$ as well and the integration involved in nonlocal coordinate functions are often over the entire interval $[a,b]$, not just over small subintervals. Meanwhile, other advantages of the piecewise linear local coordinate functions known as *linear elements* seen in the last section carry over to the more general problem in (8.23). Among them, (i) the relevant stiffness matrix $K$ in (8.22) for the solution of $\{\widehat{c}_k\}$ continues to be tridiagonal and symmetric positive definite, (ii) the trial functions $\{y_N\}$ automatically allow for the possibility of a genuinely PWS extremal, and (iii) sharp gradient features can be efficiently captured by trial functions constructed from localized coordinate functions.

For a fixed $h$, the finite element formulation also lends itself to an analysis of the error in the approximate solution: $e_N(x) = y(x) - y_N(x)$. Let the *2-norm* be defined as

$$\|v(x)\|_2 = \int_a^b [v(x)]^2\,dx. \tag{8.28}$$

Evidently, the mean-square error $\|e_N\|_2 \equiv \|\hat{y}(x) - \hat{y}_N(x)\|_2$ measures how close $\hat{y}_N(x)$ is to the true solution $\hat{y}(x)$ over the whole interval $[a, b]$. For problem (8.23), it is possible to show [see G. Strang and G. Fix (1973)] that the rms (root-mean-square) error of the Rayleigh–Ritz solution $\hat{y}_N(x)$ for the linear element coordinate functions is proportional to $h^2$. More precisely, we have the following error bound for $\hat{y}_N(x)$:

**[VIII.4]** *There is a positive constant* $c_0$ *for which*

$$\|e_N\|_2 = \|\hat{y} - \hat{y}_N\|_2 \le c_0 h^2 \|f\|_2, \tag{8.29}$$

*where* $f(x)$ *is the known function in (8.23).*

Hence, $e_N$ goes to zero in the mean-square sense as $h \to 0$ and the convergence is quadratic. The result (8.29) holds with no continuity requirement on the forcing term $f(x)$, only that it be square integrable. The finite element scheme itself requires only the existence of the integral (8.21). To obtain a similar bound on $e_N$ for the *maximum norm* (i.e., for the pointwise errors), we would have to impose some continuity condition on the data $f(x)$.

The *finite element method* also allows for the use of more general localized coordinate functions consisting of higher-order polynomials over subintervals of $[a, b]$ which need not be of equal length. The added smoothness may be important when the derivatives of the unknown(s) are physically significant. Also, elements of higher-degree polynomials are necessary when the Lagrangian $F$ depends on higher derivatives of $y$ than the first. It is reasonable to expect that for a given $F$, there will be an improvement in the accuracy of the corresponding approximate solution with a higher-degree polynomial coordinate functions. In fact, it has been established that, under a suitable smoothness assumption on the exact solution, $\|e_N\|_2$ is proportional to $h^m$ for elements with polynomials of degree $(m - 1)$. Further development of the method of finite elements can be found in *Strang and Fix* (1973) and elsewhere. The discussion of this and the last section is intended as an introduction to this attractive method in conjunction with the direct method of calculus of variations. We will discuss the finite element method again in Chapter 15 on problems in higher spatial dimensions where the advantage of the method over others is even more substantial.

## 7. Duality

It is not always practical to obtain an accurate approximate solution for a given variational problem. The various available approximate methods of solution which should work in principle may turn out to be too tedious or may take an unacceptably large amount of computation to arrive at an acceptable approximation of the true solution. For such an occasion, it is important to know that there is a different but equivalent problem which may be more tractable whenever the Lagrangian of the given problem is convex. In this section, we formulate this ***dual problem*** for the ***primal*** variational problem

$$J[y] = \gamma y(b) + \int_a^b F(x, y, y')\,dx, \qquad y(a) = A. \tag{8.30}$$

where $F$ is ***strictly convex*** in $y$ and $y'$ and $y(b)$ is not specified. The ***primal problem*** is to minimize $J[y]$ over all ***smooth*** comparison functions on $[a, b]$ which satisfy $y(a) = A$ for a given pair of constants $\gamma$ and $A$. Although we limit our discussion to the case of a scalar unknown $y(x)$, the development to follow can be extended to the case of a vector unknown. For the primal problem (8.30), we know from Chapter 3 that there is an Euler boundary condition

$$\hat{F}_{,y'}(b) \equiv F_{,y'}(b, \hat{y}(b), \hat{y}'(b)) = -\gamma \tag{8.31}$$

as a second necessary condition (in addition to the Euler DE or the Euler–Lagrange equation) to be satisfied by any minimizing $\hat{y}(x)$.

For the dual problem of (8.30), we define a new function $G(x, p, q)$ by

$$G(x, p, q) \equiv \min_{y,v} \{F(x, y, v) - pv - qy\} \tag{8.32}$$

where minimization is for all real $y$ and $v$. We will assume that the minimum exists (and is finite) for the given $F$. The minimum is attained at the unique stationary point $(\bar{y}, \bar{v})$ of $F - pv - qy$:

$$F_{,y}(x, \bar{y}, \bar{v}) - q = 0, \qquad F_{,v}(x, \bar{y}, \bar{v}) - p = 0. \tag{8.33}$$

Now, $F$ is strictly convex (so that its Hessian is positive definite); we can, therefore, solve for $\bar{y}$ and $\bar{v}$ as functions of $x$, $p$, and $q$ to obtain

$$\bar{y} = \varphi(x, p, q), \qquad \bar{v} = \psi(x, p, q), \tag{8.34}$$

and the resulting expression

$$G(x, p, q) = F(x, \varphi(x, p, q), \psi(x, p, q)) - p\psi(x, p, q) - q\varphi(x, p, q)$$

is $C^1$. Evidently, $G$ is a kind of (generalized) Legendre transformation in both $y$ and $y'$.

With

$$G_{,p} = (F_{,v} - p)\psi_{,p} + (F_{,y} - q)\varphi_{,p} - \psi = -\bar{v}, \tag{8.35a}$$

$$G_{,q} = (F_{,v} - p)\psi_{,q} + (F_{,y} - q)\varphi_{,q} - \varphi = -\bar{y} \tag{8.35b}$$

[where use has been made of (8.33)], we can show

**[VIII.5]** *$G(x,p,q)$ is concave for each x if F is strictly convex.*

It suffices to show that the Hessian of $G$ in $p$ and $q$ is nonpositive definite. We need expressions for $G_{,pp} = -\bar{v}_{,p}, G_{,pq} = -\bar{v}_{,q}, G_{,qp} = -\bar{y}_{,p}$ and $G_{,qq} = -\bar{y}_{,q}$ [see (8.35)]. We get them by differentiating (8.33) with respect to $p$ and $q$. For example, we have

$$\bar{F}_{,yy}\bar{y}_{,p} - \bar{F}_{,yv}\bar{v}_{,p} = 0, \quad \bar{F}_{,vy}\bar{y}_{,p} - \bar{F}_{,vv}\bar{v}_{,p} = 1,$$

where $\bar{f} \equiv f(x, \bar{y}, \bar{v})$; these can be solved to give

$$\bar{y}_{,p} = -\bar{F}_{,yv}/H_F, \quad \bar{v}_{,p} = \bar{F}_{,yy}/H_F,$$

where $H_F = \bar{F}_{,yy}\bar{F}_{,vv} - \bar{F}_{,yv}\bar{F}_{,vy}$ is the Hessian of $F$ for a fixed $x$. Similarly, we get

$$\bar{y}_{,q} = -\bar{F}_{,vv}/H_F, \quad \bar{v}_{,q} = \bar{F}_{,vy}/H_F.$$

*It follows from these results that* $G_{,pp} = -\bar{F}_{,yy}/H_F < 0, G_{,qq} = -\bar{F}_{,vv}/H_F < 0$, and the Hessian of $G$ as a function of $p$ and $q$ for a fixed $x$ is

$$H_G = G_{,pp}G_{,qq} - G_{,pq}G_{,qp} = (\bar{F}_{,yy}\bar{F}_{,vv} - \bar{F}_{,yv}\bar{F}_{,vy})/H_F^2 > 0.$$

The *dual problem* of (8.30) may now be stated. It is the problem of *maximizing*

$$I[p] \equiv -Ap(a) + \int_a^b G(x, p, p')\,dx \tag{8.36a}$$

over all comparison functions $p(x)$ on $[a, b]$ which satisfy

$$p(b) = -\gamma. \tag{8.36b}$$

If the admissible comparison function $\hat{p}(x)$ solves the dual problem, we know from the results of Chapter 2 that $\hat{p}(x)$ must be an extremal for $G(x, p, p')$ and

$$\hat{G}_{,p'}(a) \equiv G_{,p'}(a, \hat{p}(a), \hat{p}'(a)) = -A. \tag{8.37}$$

It is rather remarkable that

**[VIII.6]** *The maximum* $I[\hat{p}]$ *for the dual problem is equal to the minimum* $J[\hat{y}]$ *for the primal.*

To see this, we note that the solution $\hat{y}(x)$ of the primal problem (which is assumed to be smooth and is, therefore, $C^2$ because we have $\hat{F}_{,y'y'} > 0$ in $[a, b]$) satisfies the Euler DE

$$\frac{d}{dx}[\widehat{F}_{,y'}(x)] = F_{,y}(x, \widehat{y}(x), \widehat{y}'(x)) \tag{8.38a}$$

and the Euler BC (8.31)

$$\widehat{F}_{,y'}(b) = -\gamma. \tag{8.38b}$$

Let

$$\widehat{p}_0(x) \equiv F_{,y'}(x, \widehat{y}(x), \widehat{y}'(x)), \qquad \widehat{q}_0(x) \equiv F_{,y}(x, \widehat{y}(x), \widehat{y}'(x)). \tag{8.39}$$

We will show that $\widehat{p}_0(x)$ is, in fact, an admissible extremal of the dual problem.

The Euler differential equation (8.38a) implies $\widehat{q}_0 = \widehat{p}_0'$. With $p = \widehat{p}_0$ and $q = \widehat{p}_0'$ in (8.33), $(\widehat{y}, \widehat{y}')$ is the minimum point of the right-hand side of (8.32) and, hence, defines $G(x, \cdot, \cdot)$, i.e., $\overline{y} = \widehat{y}$ and $\overline{v} = \widehat{y}'$. It follows from (8.35a) and (8.35b) that

$$G_{,p}(x, \widehat{p}_0, \widehat{p}_0') = -\widehat{y}', \qquad G_{,q}(x, \widehat{p}_0, \widehat{p}_0') = G_{,p'}(x, \widehat{p}_0, \widehat{p}_0') = -\widehat{y} \tag{8.40}$$

which, in turn, imply

$$\frac{d}{dx}[G_{,p'}(x, \widehat{p}_0, \widehat{p}_0')] = G_{,p}(x, \widehat{p}_0, \widehat{p}_0'). \tag{8.41}$$

In other words, $\widehat{p}_0(x)$ as defined by (8.39) is an extremal for the Lagrangian $G(x, p, p')$ [normally denoted by $\widehat{p}(x)$]. It is an admissible extremal because of (8.38b) and it is smooth because $\widehat{q}_0(x) = F_{,y}(x)$ is continuous [for a smooth $\widehat{y}(x)$]. Also, the condition $\widehat{y}(a) = A$ can now be written by the second relation in (8.40) as

$$G_{,p'}(a, \widehat{p}_0(a), \widehat{p}_0'(a)) = -A. \tag{8.42}$$

Thus, the admissible extremal $\widehat{p}_0(x)$ satisfies the Euler boundary condition for the dual problem as well. It follows from the concavity of $G(x, p, q)$ in $p$ and $q$ that $\widehat{p}_0(x)$ solves the dual problem. Moreover, we have

$$\begin{aligned}
\min_y \{J[y]\} &= \gamma\widehat{y}(b) + \int_a^b F(x, \widehat{y}, \widehat{y}')\,dx \\
&= -\widehat{p}(b)\widehat{y}(b) + \int_a^b F(x, \widehat{y}, \widehat{y}')\,dx \\
&= -\widehat{p}(a)\widehat{y}(a) + \int_a^b \left\{F(x, \widehat{y}, \widehat{y}') - \frac{d}{dx}(\widehat{p}\widehat{y})\right\}dx \\
&= -A\widehat{p}(a) + \int_a^b \left\{F(x, \widehat{y}, \widehat{y}') - \widehat{p}'\widehat{y} - \widehat{p}\widehat{y}'\right\}dx
\end{aligned}$$

$$= -A\hat{p}(a) + \int_a^b G(x, \hat{p}, \hat{p}')dx = \max_p \{I[p]\}.$$

EXAMPLE 1: $J[y] = \gamma y(1) + \int_0^1 \frac{1}{2}[(y')^2 + y^2]dx, \quad y(0) = A.$

Consider $G = F(x, y, v) - pv - qy = \frac{1}{2}(v^2 + y^2) - pv - qy$. For a minimum point, we look for the stationary point of $G$ From

$$(G_{,y}, G_{,v}) = (y - q, v - p) = 0,$$

we get $y = q$ and $v = p$. Upon using these to eliminate $y$ and $v$, we obtain $G(x, p, q) = -\frac{1}{2}(p^2 + q^2)$. This is the minimum value for $G$ because in the $(y, v)$ space

$$H[G] = H[F] = \begin{bmatrix} 1 & 0 \\ 0 & 1 \end{bmatrix}$$

is positive definite. The dual problem is then given by

$$\max_p \{I[p] \,|\, p(b) = -\gamma\},$$

where

$$I[p] = -Ap(0) + \int_0^1 \left\{-\frac{1}{2}[p^2 + (p')^2]\right\}dx.$$

The importance of [VIII.6] is well beyond the need for a feasible alternative to solving the primal problem. In actual applications, we need to know the of accuracy of any approximate solution for a given problem when the exact solution is not known. Having the dual problem, we are in a position to compute both an upper and lower bound for the minimum value of $J[y]$. The *upper bound* is provided by $J[\tilde{y}]$, where $\tilde{y}$ is an approximate admissible extremal of the primal obtained by any method. The *lower bound* is provided by $I[\bar{p}]$, where $\bar{p}$ is any approximate admissible extremal of the dual problem. Together, these bounds give us some idea on how close $J[\tilde{y}]$ and $I[\bar{p}]$ are from the true minimum of the primal (or the true maximum of the dual problem). If the gap $J[\tilde{y}] - I[\bar{p}]$ as a fraction of $J[\tilde{y}]$ or $I[\bar{p}]$ is not small compared to one, we should improve upon either $\tilde{y}$ or $\bar{p}$ or both. Therefore, it is significant and fortunate that the kind of duality described in this section in fact permeates throughout the calculus of variations (and the more general subject of optimization).

## 8. The Inverse Problem

Introduction of the finite element type trial functions has made direct methods of solution such as the Rayleigh–Ritz method attractive for finding approximate solutions of variational problems. It also offers a useful approach for solving general boundary-value problems which arise naturally and directly in areas unrelated to the calculus of variations. Some major sources of boundary-value problems for differential equations include mechanics, electromagnetics, heat conduction, optics, economic growth theory, population dynamics, etc. To take advantage of direct methods for these problems, we need to find an appropriate Lagrangian for which the given differential equation is the Euler DE. This is known as the *inverse problem* of the calculus of variations. A desire to use the direct method of the calculus of variations (with finite element or other trial functions) for an approximate solution of a given BVP is only one of the many motivations for the inverse problem. For problems involving PDE, the application of semidirect methods (to be described in later chapters) is an even stronger motivation.

For simple differential equations, we can try to construct the Lagrangian by reversing the process which leads to the Euler DE Take the ODE $-y'' - y + f(x) = 0$. To get $F$, we work with variational notations and form the integral

$$-\int_a^b [y'' + y - f(x)]\,\delta y\,dx = 0$$

or, after integration by parts,

$$[-y'\delta y]_a^b + \int_a^b [y'\delta y' - y\delta y + f(x)\delta y]\,dx = 0.$$

The first term vanishes because $\delta y = 0$ at the two end points. The remaining integral can be written as the first variation of an integral $J[y]$:

$$\delta J = \delta\int_a^b \frac{1}{2}[(y')^2 - y^2 + 2f(x)y]\,dx.$$

Thus, we have found an integral $J[y]$ [with an appropriate Lagrangian $F(x, y, y')$] whose Euler DE is the given ODE.

The process above for finding a Lagrangian for a given ODE does not always work. Consider the equation of motion for a damped oscillator $-my'' - ay' - ky + f(x) = 0$ which gives the following integral relation:

$$0 = -\int_a^b [my'' + ay' + ky - f(x)]\,\delta y\,dx$$

$$= -[(my' + ay)\delta y]_a^b + \int_a^b [my'\delta y' + ay\,\delta y' - ky\,\delta y + f(x)\,\delta y]\,dx$$

$$= \delta \int_a^b \frac{1}{2}[m(y')^2 - ky^2 + 2f(x)y]\,dx + \int_a^b ay\,\delta y'\,dx.$$

As the second integral is not the first variation of anything, we have failed to find by this process an appropriate Lagrangian $F(x, y, y')$ whose Euler DE is the given equation of motion.

On the other hand, the same equation for the damped oscillator, multiplied through by $e^{ax/m}$, may be written as

$$-e^{ax/m}[my'' + ay' + ky - f(x)] = -(me^{ax/m}y')' - e^{ax/m}ky + f(x)e^{ax/m} = 0.$$

An appropriate Lagrangian can now be found for the transformed equations by the same process used for the undamped case:

$$-\int_a^b [(me^{ax/m}y')' + e^{ax/m}ky - e^{ax/m}f(x)]\,\delta y\,dx$$

$$= -[me^{ax/m}y'\,\delta y]_a^b + \int_a^b [my'\,\delta y' - ky\,\delta y + f(x)\,\delta y]\,e^{ax/m}\,dx$$

$$= \delta \int_a^b \frac{1}{2}[m(y')^2 - ky^2 + 2f(x)y]\,e^{ax/m}\,dx = 0.$$

This example suggests the following result for the general second-order linear ODE

$$-[y'' + \bar{p}(x)y' + \bar{q}(x)y] + \bar{f}(x) = 0. \qquad (8.44)$$

**[VIII.7]** *The ODE (8.44) may be written as*

$$-[p(x)y']' - q(x)y + f(x) = 0 \qquad (8.45)$$

*where*

$$\{p(x), q(x), f(x)\} = \{1, \bar{q}(x), \bar{f}(x)\}\,exp\left\{\int^x \bar{p}(t)\,dt\right\} \qquad (8.46)$$

*Moreover, (8.45) is the Euler DE associated with the Lagrangian* $F(x, y, y') = [p(x)(y')^2 - qy^2 + 2f(x)y]/2.$

The validity of [VIII.6] can be confirmed by direct verification.

For problems involving the more general second-order ODE

$$y'' = f(x, y, y'),$$

there is also a systematic method which finds one or more of the needed Lagrangians, at least in principle. Suppose that the desired Lagrangian is $F(x, y, y')$ with the Euler DE $(F_{,y'})' = F_{,y}$. To match the Euler DE with the given ODE, we use the ultradifferentiated form

$$F_{,y'y'}y'' + F_{,y'y}y' + F_{,y'x} = F_{,y},$$

or, upon using the given ODE to eliminate $y''$,

$$F_{,y} - F_{,y'x} - y'F_{,y'y} - f(x, y, y')F_{,y'y'} = 0. \tag{8.47}$$

This is now a partial differential equation for $F(x, y, y')$ and a solution can, in principle, be attempted using known analytical technique for such equations. Unfortunately, it is a second-order nonlinear PDE for which there are very few useful methods for its exact solution. A first-order PDE would be more tractable. It is, in fact, possible to transform (8.47) into a first order PDE by differentiating (8.47) partially with respect to $y'$ and setting

$$u(x, y, y') \equiv F_{,y'y'}. \tag{8.48a}$$

This process results in the following first-order *linear* PDE for $u(x, y, v)$:

$$f(x, y, v)u_{,v} + f_{,v}u + vu_{,y} + u_{,x} = 0. \tag{8.48b}$$

We summarize these observations in the following statement:

**[VIII.8]** ***The ODE $y'' = f(x, y, y')$ is the Euler DE for the Lagrangian $F(x, y, y')$ obtained from (8.48a) by two integrations with u determined by (8.48b) and (8.47).***

**EXAMPLE 2:** $y'' = -y + \bar{f}(x)$.

The PDE (8.48b) simplifies to

$$(-y + \bar{f})u_{,y'} + y'u_{,y} + u_{,x} = 0.$$

By inspection, an exact solution of this equation is $u = u_0$ (a constant) so that we get from $F_{,y'y'}(x, y, y') = u_0$

$$F = \frac{1}{2}u_0(y')^2 + c_1(x, y)y' + c_0(x, y),$$

where $c_1$ and $c_0$ are two arbitrary functions of $x$ and $y$. To determine $c_1$ and $c_0$, we substitute the expression for $F$ into (8.47) to obtain

$$c_{0,y} - c_{1,x} - [-y + \bar{f}]u_0 = 0.$$

It is easy to see that this equation is satisfied by

$$c_1 = 0, \qquad 2c_0 = [-y^2 + 2\bar{f}(x)y]u_0,$$

so that

$$F = \frac{u_0}{2}[(y')^2 - y^2 + 2\bar{f}(x)y].$$

The scale factor $u_0$ may be set equal to one without any effect on the corresponding Euler DE. As such, we obtain the same Lagrangian as before by the new method. However, the new method offers other possible Lagrangians with the same Euler DE. An obvious alternative is $c_0 \equiv 0$ and $c_1 = [xy - g(x)]u_0$, where $g'(x) = \bar{f}(x)$ so that $F = u_0[y'\{xy - g(x)\} + (y')^2/2]$. There are many others.

A systematic application of the *method of characteristics* (Kevorkian, 1990) to (8.48b) leads to only one nontrivial relation

$$\frac{du}{dx} = -f_{,y'}u \tag{8.49}$$

(and, of course, the original ODE itself). The complete solution of (8.48) by this method generally requires the explicit solution of the original ODE [see Leitmann, (1981)]. Hence, the method of characteristics is generally not useful. An exception is when $f_{,y'}$ is only a function of $x$, including the special case $f_{,y'} = 0$ as in the case of the simple harmonic oscillator treated above as well as the damped oscillator treated earlier. Below, we show how the method works for the damped oscillator.

EXAMPLE 3: $y'' = -\frac{a}{m}y' - \frac{k}{m}y + \frac{1}{m}\bar{f}$.

For this case, (8.49) becomes $du/dx = (a/m)u$ or $u = u_0 exp(ax/m)$. From $u = F_{,y'y'}$, we then get

$$F(x, y, y') = \frac{1}{2}(y')^2 u_0 e^{ax/m} + u_1(x, y)y' + u_2(x, y).$$

To determine the two unknowns $u_1(x, y)$ and $u_2(x, y)$, we substitute the above expression for $F$ into (8.47). After some cancellation, we get

$$u_{2,y} - u_{1,x} = \frac{u_0}{m}e^{ax/m}[\bar{f}(x) - ky]$$

which can again be solved by the method of characteristics. However, because any particular solution suffices, we get one solution by taking $u_1 \equiv 0$ which then requires

$$u_2 = \frac{u_0}{m}e^{ax/m}[y\bar{f}(x) - \frac{1}{2}ky^2]$$

and therewith

$$F(x, y, y') = \frac{u_0}{2m} e^{ax/m} \{m(y')^2 - ky^2 + 2y\bar{f}(x)\}.$$

The multiplicative constant $u_0$ may be set equal to $m$ without any effect on the solution. The resulting Lagrangian is identical to what we found previously by an ad hoc method which led up to [VIII.7].

If we are looking for any Lagrangian for the ODE [and, therefore, any particular solution of (8.48b)], it is sometimes possible (and more effective) to obtain it by comparing the Euler DE $F_{,y'y'}y'' + F_{,y'y}y' + F_{,y'x} - F_{,y} = 0$ with a multiple of the given ODE $g(x)[y'' - f(x, y, y')] = 0$ and matching terms. We illustrate the process with several examples:

**EXAMPLE 4:** $f = -y + \bar{f}(x)$.

For this $f$, we have $g(x)[y'' + y - \bar{f}(x)] = 0$ for which

$$F_{,y'y'} = g(x), \qquad F_{,y'y} = 0, \qquad F_{,y'x} - F_{,y} = g(x)(y - \bar{f}).$$

It follows from the first condition that

$$F(x, y, y') = \frac{1}{2} g(x)(y')^2 + c_1(x, y)y' + c_0(x, y).$$

The second condition requires

$$c_{1,y} = 0 \quad \text{or} \quad c_1 = c_1(x).$$

The third condition then requires

$$g'(x)y' + c_1' - c_{0,y} = g(x)[y - \bar{f}(x)]$$

which, in turn, implies $g'(x) = 0$ or $g = g_0$. We may now take $c_1' = 0$ and $c_{0,y} = g_0(\bar{f} - y)$ or $c_0(x, y) = g_0[\bar{f}(x)y - y^2/2]$. Altogether, we have as a possible Lagrangian for the given ODE

$$F(x, y, y') = \frac{1}{2} g_0 \{(y')^2 - y^2 + 2\bar{f}(x)y\}$$

as we expect. The constant $g_0$ may be set equal to 1.

**EXAMPLE 5:** $y'' = -\frac{1}{2}(y')^3(x^2 + y^2)$.

For this problem, the method of matching does not work for any $g(x)$ as $F_{,y'y'} = g(x)$ implies $F = [g(x)(y')^2/2] + c_1(x, y)y' + c_0(x, y)$. Such a Lagrangian does not allow us to match the term proportional to $(y')^3$ on the right-hand side of the ODE. A more general multiplicative factor $g(x, y, y')$ which depends on $y$ and $y'$ as well is required. To simplify the presentation, we will use $g = 2/(y')^3$ as a rather natural first attempt; it removes the necessity of matching $y'F_{,y'y}$ with terms involving higher powers of $y'$. For this $g$, we have

$$F_{,y'y'} = \frac{2}{(y')^3} \quad \text{or} \quad F = \frac{1}{y'} + c_1(x,y)y' + c_0(x,y).$$

The rest of the matching requires

$$F_{,y'y} = c_{1,y} = 0,$$
$$F_{,y'x} - F_{,y} = c_{1,x} - (c_{1,y}y' + c_{0,y}) = x^2 + y^2.$$

The first condition implies $c_1 = c_1(x)$ and the second may be satisfied by taking $c_0 = -(x^2y + y^3/3)$ and $c_1 = 0$ so that

$$F = \frac{1}{y'} - \left(x^2y + \frac{1}{3}y^3\right).$$

It is not difficult to verify that the condition $\delta J = 0$ with the above expression for $F$ as the Lagrangian does lead to the given ODE as the Euler DE.

## 9. Weak Solutions

For more general nonlinear ODE; it is not always possible to identify a Lagrangian for a given equation or system of equations. For these problems, a class of approximate methods has been developed to exploit the conditions that some weighted average of the given ODE is satisfied, analogous to the requirement of $\delta J = 0$ imposed on an Euler DE. To illustrate these so-called Galerkin methods, we consider the general second-order ODE $y'' = f(x, y, y')$. For this equation, the integrated condition

$$\int_a^b [y'' - f(x,y,y')]\, \delta y \, dx = 0 \tag{8.50}$$

holds for all admissible variations $\delta y$ on $[a, b]$. Condition (8.50) is analogous (or identical) to $\delta J = 0$ in the case where a Lagrangian exists for which $y'' = f(x, y, y')$ is the Euler DE. For an approximate solution based on (8.50) and without the benefit of a Lagrangian, we again take a trial function $y \cong y_n = c_1\varphi_1 + \ldots + c_n\varphi_n$, where $\{\varphi_k\}$ is a set of coordinate (or basis) functions previously defined in Section 8.2 and $\{c_k\}$ is a set of $n$ unknown constants to be determined. As $y_n$ is generally not the exact solution of the ODE, $r_n(x) \equiv y_n'' - f(x, y_n, y_n')$ will not be zero. We let $\delta y$ be a member of another set of functions called *test functions* $\{\psi_k(x)\}$ with $\psi_k(a) = \psi_k(b) = 0$, $k = 1, 2, \ldots, n$ and require that (8.50) be satisfied for all $n$ choices of $\psi_k$. For each $\psi_k$, the integral gives one equation for the $n$ unknown coefficients $\{c_j\}$.

The $n$ different test functions give just enough equations for the $n$ coefficients. By using a solution of these equations in the trial function (8.10) gives an approximate solution for $y(x)$. Equation (8.50) is called the *weak form* of the BVP for $y$, and the solution obtained through (8.50) is called the *weak solution*.

With (8.50) written in the form

$$\int_a^b [y'' - f(x, y, y')]\,\psi_k\,dx = 0 \quad (k = 1, 2, \ldots, n), \tag{8.51}$$

the Galerkin method effectively requires the residual $r_n(x)$ to be "orthogonal" to the $n$ test functions. The residual will be more and more restricted as the number of the functions increase.

If both $\{\varphi_k\}$ and $\{\psi_k\}$ are complete, we expect $\widehat{y}_n$ to tend to an exact solution for $\widehat{y}$ in the limit as $n \to \infty$. If the sequence $\{\psi_k\}$ is the same as $\{\varphi_k\}$, then we have the Rayleigh–Ritz solution discussed in the earlier sections of this chapter whenever there exists a Lagrangian for the ODE. The orthogonality condition corresponding to (8.51) provides another interpretation of the Rayleigh–Ritz method (see Exercises). Other choices of $\{\psi_k\}$ lead to different types of approximate solutions. For example, the choice $\{\varphi_k = \delta(x - x_k)\}$, where $\delta(x - x_k)$ is the Dirac delta function centered at $x = x_k$, corresponds to *collocations* at $\{x_1, \ldots, x_n\}$. *

If $\psi_k$ is $C^1$, then we have from (8.50)

$$\int_a^b \{y'' - f(x, y, y')\}\,\psi_k dx = [y'\psi_k]_a^b - \int_a^b [y'\psi_k' + f(x, y, y')\,\psi_k]\,dx$$

$$= -\int_a^b [y'\psi_k' + f(x, y, y')\,\psi_k]\,dx = 0, \tag{8.52}$$

where we have used the end conditions $\psi_k(a) = \psi_k(b) = 0$ for the test functions. This transformed version of (8.50) no longer requires that $y$ be $C^2$ and is, therefore, more suitable for the finite element method. For higher-order equations, repeated integration by parts reduces the order of the highest derivative in the integrand (as much as half of the original order of the ODE).

* Operationally, the Dirac delta function may be taken to be the limit as $\varepsilon \to 0$ of the function (Fig. 8.5)

$$\delta_\varepsilon(x) = \begin{cases} \dfrac{1}{2\varepsilon} & (-\varepsilon < x < \varepsilon) \\ 0 & (x < -\varepsilon,\ x > \varepsilon). \end{cases}$$

We often make use of the following two consequences of this operational definition:

$$\int_{-\infty}^{\infty} \delta(t)\,dt = \lim_{\varepsilon \to 0} \frac{1}{2\varepsilon} \int_{-\varepsilon}^{\varepsilon} dt = 1, \qquad \int_{-\infty}^{\infty} \delta(t - x) f(t)\,dt = f(x).$$

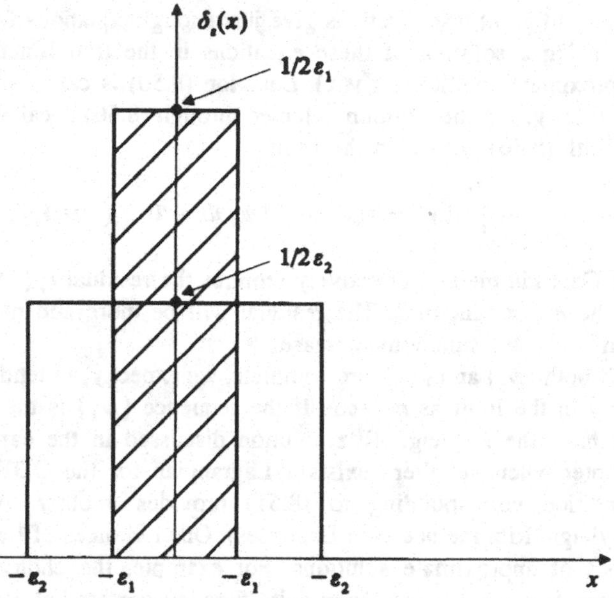


Figure 8.5

## 10. Exercises

1. Use the coordinate functions $\{x(x-1)^2, x^2(x-1)^2, \ldots\}$ to find an approximate solution to the problem of minimizing the functional

$$J[y] = \int_0^1 (x^3 y''^2 + 100xy^2 - 20xy)\,dx,$$

with the end conditions $y(0) = y'(1) = 0$.

2. Use the coordinate functions $\{(x-1)(x-2), (x-1)^2(x-2)^2, \ldots\}$ to find an approximate solution for the problem of minimizing

$$J[y] = \int_1^2 x(y'^2 - y^2)\,dx,$$

with inhomogeneous end conditions $y(1) = 1$, $y(2) = 2$, by calculating $\widehat{y}_0$ and $\widehat{y}_1$ as well as the corresponding $J[\widehat{y}_0]$ and $J[\widehat{y}_1]$. Also, set up the conditions for determining the two constants in $\widehat{y}_2$.

3. Find an approximate solution of the BVP $y'' + x^2 y = x$, $y(0) = y(1) = 0$ by identifying the ODE as the Euler–Lagrange equation of a calculus of variations problem and using the trial function method. Determine $\widehat{y}_1$ and

$\widehat{y}_2$, and compare their values at $x = 0.25, 0.5, 0.75$. To test how close these are to solving the differential equation, compute $r(x) \equiv y'' + x^2 y - x$ for the approximate solutions at the same points. Use

(a) $\varphi_k(x) = x^k(x-1)$,

(b) $\varphi_k(x) = \sin(k\pi x)$.

4. (a) Find the dual variational problem (involving the integral $I[p]$) of

$$\min_{y \text{ in } C^1} \{J[y] \equiv y(1) + \int_0^1 [y^2 - 6y + 2(y')^2]\, dx \,|\, y(0) = 3\}.$$

(b) Evaluate $J[y]$ and $I[p]$ for a pair of reasonable admissible $y_0(x)$ and $p_0(x)$ and calculate the difference $J[y_0] - I[p_0]$.

5. $$\min_{y \text{ in } C^1} \left\{ J[y] \equiv \int_0^1 \left[ \frac{1}{2}(y')^2 + y^6 + y \right] dx \,|\, y(0) = 0 \right\}.$$

(a) Find the dual problem.

(b) Find the minimum value of $J[y]$ to within $\pm 0.03$. (No lengthy calculations; just prove your answer is in this range.)

6. Use [VIII.6] to find a Lagrangian for a damped oscillator. You may take the equation of motion of the damped oscillator to be $y'' + py' + qy = f(x)$ where $p$ and $q$ are constants.

7. Match the ODE for the damped oscillator with the general Euler DE (8.47) $F_{,y'y'}y'' + F_{,yy'}y' + F_{,y'x} - F_{,y} = 0$ to get a Lagrangian.

8. Find, by the method of matching, a Lagrangian for $y'' + py' + qy = 0$ where $p$ and $q$ are given functions of $x$.

9. Find a Lagrangian $F(x, y, y')$ which has the equation $y'' + (2/x)y' + y^5 = 0$ as its Euler DE.

10. Find a Lagrangian for $y'' = w(x, y)(y')^2$, where $w(x, y)$ is a prescribed function.

11. Find a Lagrangian for $y'' + p(x)y' + q(x)g(y) = 0$ where $p(\cdot)$, $q(\cdot)$ and $g(\cdot)$ are known functions.

12. Use a finite element method with the linear elements of (8.18) and four subintervals to solve the BVP

$$y'' + xy' + x^2 y = \sin x, \qquad y(0) = y(1) = 0.$$

(a) Find the Lagrangian $F(x, y, y')$ for the ODE.

(b) Apply the Rayleigh–Ritz method to the correspondingly basic problem.

13. Let

$$J(c) = \int_a^b F(x, y_N, y_N')\,dx, \qquad y_N(a) = y_N(b) = 0,$$

where $y_N$ is the trial function (8.10).

(a) Obtain the relations $\partial J / \partial c_k = 0$, $k = 1, 2, \ldots, N$.

(b) Obtain from the results in part (*a*)

$$\int_a^b [(-F_{,y'}^{(N)})' + F_{,y}^{(N)}]\varphi_k\,dx = 0 \quad (k = 1, 2, \ldots, n),$$

where $f^{(N)}(x) = f(x, y_N(x), y_N'(x))$.

(c) Interpret the result of (b).

14. $$J[y] = \frac{1}{2}\int_0^1 [\varepsilon^2 (y')^2 + y^2]\,dx, \qquad y(0) = 1, \;\; y(1) = 0.$$

Let $h = 1/(N+1), x_k = kh, k = 0, 1, \ldots, (N+1)$ (with $x_0 = 0$ and $x_{N+1} = 1$), $\{\varphi_k\}$ be the roof functions of (8.18), and

$$\varphi_0^h(x) = \begin{cases} \dfrac{1}{h}(x_1 - x) & (x_0 \le x \le x_1) \\ 0 & (x > x_1). \end{cases}$$

(a) For $N = 0$, take $y_0 = \varphi_0^h(x)$ and calculate $J[y_0]$.

(b) For $N = 1$, take $y_1 = \varphi_0^h + A_1\varphi_1^h$. Determine $\hat{y}_1$ and evaluate $J[\hat{y}_1]$.

(c) For $N = 2$, take $y_2 = \varphi_0^h + b_1\varphi_1^h + b_2\varphi_2^h$. Determine $\hat{y}_2$ and evaluate $J[\hat{y}_2]$.

(d) Compare $\mu_0$, $\mu_1$, and $\mu_2$ with $\mu$ (see Table (8.2)) for $\varepsilon = 1$, 1/10, 1/20, and 1/40. Graph $\hat{y}_0, \hat{y}_1, \hat{y}_2$, and $\hat{y}$ for $\varepsilon = 1$ and $\varepsilon = 1/40$.

# 9

# Dynamic Programming

## 1. The Shortest Route Problem

In this chapter, we describe a completely different approach to approximate solutions for the basic problem of the calculus of variations. This new approach makes use of the technique of ***dynamic programming*** in discrete optimization. For some variational problems, it is known to be superior to the approaches discussed in Chapter 8 and the Appendix of this book. In the first few sections of this chapter, we introduce the essential ideas and algorithms in dynamic programming for problems in discrete optimization. The variational problem will then be recast in a form suitable for the application of dynamic programming in the last section.

To introduce the basic idea of dynamic programming, consider the following typical "***shortest route***" problem:

> An inspection team has to make three inspection stops between city *A* and city *J* (see Fig. 9.1). Each stop is for a region which covers plants in several cities. The team may inspect plants in city *B, C,* or *D* at the first stop, *E, F,* or *G* at the second stop, and *H* or *I* at the third stop. The distance between any two consecutive stops is given in Figure 9.1 (in units of 100 miles). Find the shortest inspection route.

The reason for wanting the shortest route may be a desire to minimize the travel time or a need to minimize travel costs. One solution technique for this problem is obvious: We simply calculate the total distance traveled (or the total cost incurred) for each of the 18 possible routes and pick out the shortest (cheapest) one. This brute force approach would require a total of $4 \times 18 = 72$ additions and 18 comparisons. These calculations can be done by a computer in no time at all or, with a little patience, even by hand.

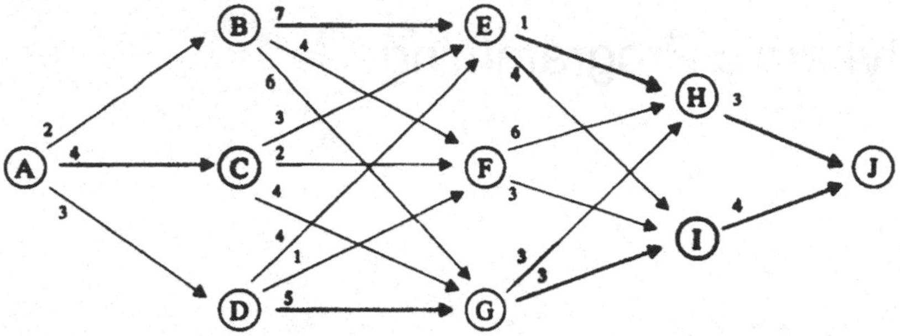


Figure 9.1

For substantially larger versions of our simple shortest route problem, the same approach would keep even the fastest computer available busy for hours or days. Medium-size versions of this problem are solved routinely by business and government thousands of times each week and for different sets of system parameter values, e.g., changing the distance between stops each time. The cost for computing the solutions of such problems by the brute force method can easily exceed the resources available for obtaining them. Therefore, it is worthwhile to consider different (and hopefully more efficient) methods of solution. A preliminary analysis of the problem will, in fact, lead to an alternative method of solution which substantially reduces the computing required and, therefore, makes the solution process more cost-effective. For the given problem, the shortest route can, in fact, be found with only 18 additions and 11 comparisons.

To work out this more efficient method of solution, suppose we have arrived at the third (i.e., the last in-between) stop at either site $H$ or $I$. In either case, there is no alternative; the next stop has to be $J$ (Fig. 9.2). From site $H$, our best (of the only) route requires another three units (of 100 miles); from $I$, it would take 4. These conclusions require no addition or comparison.

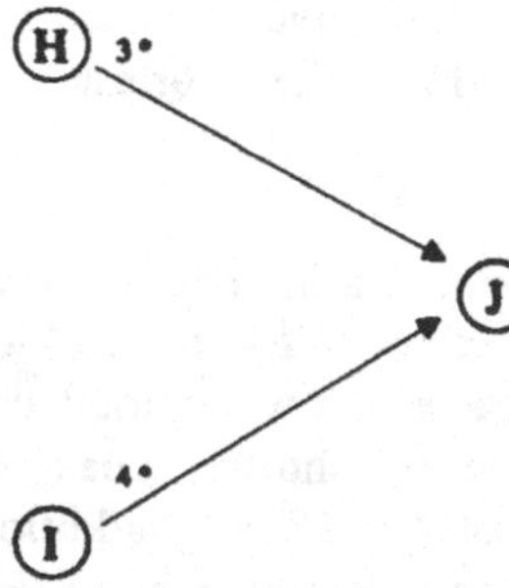


Figure 9.2

Now, suppose we have arrived only at the second stop: site $E$, $F$, or $G$. If we should be at $E$ and choose to go to $H$, then the shortest route to $J$ would take $1 + 3^* = 4$ units. An asterisk in this section indicates the shortest distance from the next stop to $J$. On the other hand, if we should go (from $E$) to $I$ instead, then the shortest route to $J$ would be $4 + 4^* = 8$. A simple comparison indicates

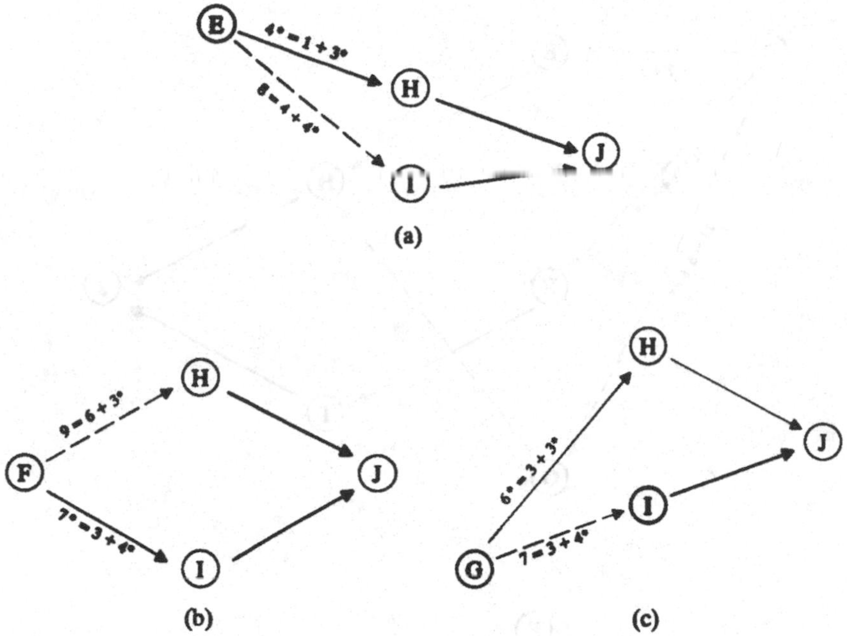


**Figure 9.3**

that we should not go to $I$; the best strategy from $E$ would be to go to $H$ (see Fig. 9.3a). Similar considerations indicate that we should go to $I$ from $F$ and to $H$ from $G$ with a total distance to $J$ equal to $3 + 4^* = 7$ and $3 + 3^* = 6$, respectively (see Fig. 9.3b and 9.3c). Note that we arrived at the best strategy for all three sites of this second stop with a total of six additions and three comparisons.

For the first stop at site $B$, $C$, or $D$, we repeat the process to obtain the best strategy for each site. The best route from $B$ to $J$ would be to go to either $E$ or $F$ for a total distance of 11 units for the entire route to $J$ (Fig. 9.4a). The best route from $C$ would be to go to $E$, for a total of seven units to $J$ (Fig. 9.4b). From $D$, the best route would be to go to either $E$ or $F$ for a total of eight units (Fig. 9.4c). The operations needed for these solutions consist of nine additions and six comparisons, a third of each for each current site.

We are now ready for the solution of the original problem. From the starting point $A$, we conclude after 3 additions and 2 comparisons that the best route to $J$ would be to go to $C$ or $D$ for a total distance of 11 units for the entire route to $J$ (Fig. 9.5). The total operations for all the stops add up to 18 $(= 6 + 9 + 3)$ additions and 11 $(= 3 + 6 + 2)$ comparisons. These are no-

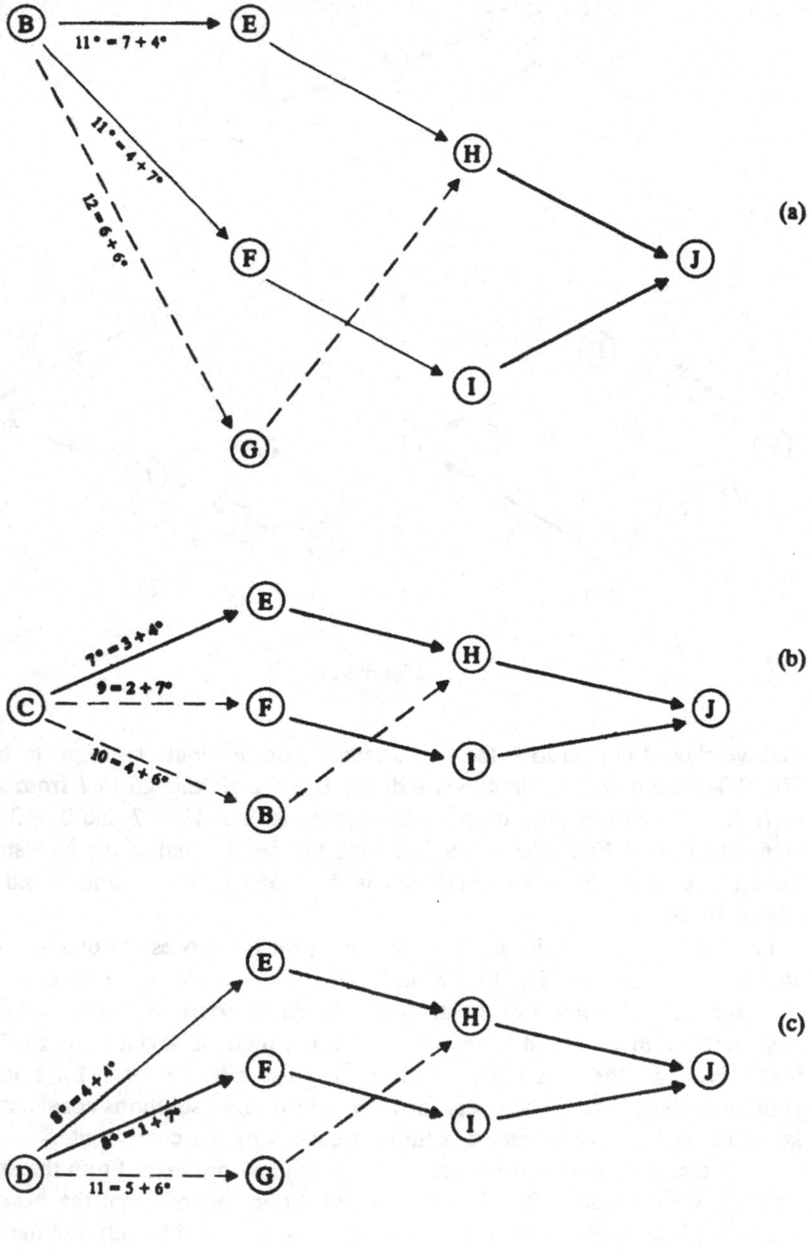
B
E
F
G
H
I
J
11* = 7 + 4*
11* = 4 + 7*
12 = 6 + 6*
(a)
C
E
F
B
H
I
J
7* = 3 + 4*
9 = 2 + 7*
10 = 4 + 6*
(b)
D
E
F
G
H
I
J
8* = 4 + 4*
8* = 1 + 7*
11 = 5 + 6*
(c)

**Figure 9.4**

tably less than the 72 additions and 18 comparisons needed for the brute force method. The gain in efficiency increases geometrically for more complex problems. Note that the shortest route generally does not head for the nearest next stop; such a route has a total distance of 13 units. Sometimes, it pays to sacrifice a little at the start for closer stops later. In an exercise at the end of the chapter, we will see that the backward nearest-neighbor approach (working backward from the end site) is also not optimal in general.

With high-speed computing facilities readily available to our scientists and engineers, there is a tendency for an individual to do "*computer simulations*" for all but the simplest problems. The above shortest route example shows that it often pays handsomely to give some thought to the simulation procedure. The efficiency gained from a preliminary analysis usually helps to reduce the computing cost for the problem; this cost reduction is often critical to the feasibility of the solution process (given limited computing funds). There may also be a significant gain on a different front. In our efficient solution process to obtain the shortest route, the following striking feature forces itself to our attention: ***The optimal strategy from a given stop does not depend on how we arrive at that stop.*** It is this so-called *Markov* property which allows us to work backward from the end stop, eliminating the non optimal route at each stop and thereby the associated calculations thereafter. This property, the analogue of which plays a dominant role in probability theory, is common to a very large class of problems cutting across many disciplines. In this broader context, the property is often stated in the form of the following version of the ***Principle of Optimality*** for systems evolving through a discrete number of ***stages*** (stops) and being in one of a number of possible ***states*** (sites) at each stage:

**Principle of Optimality I:** ***An optimal strategy must have the property that regardless of how we enter a particular state (site) of the***

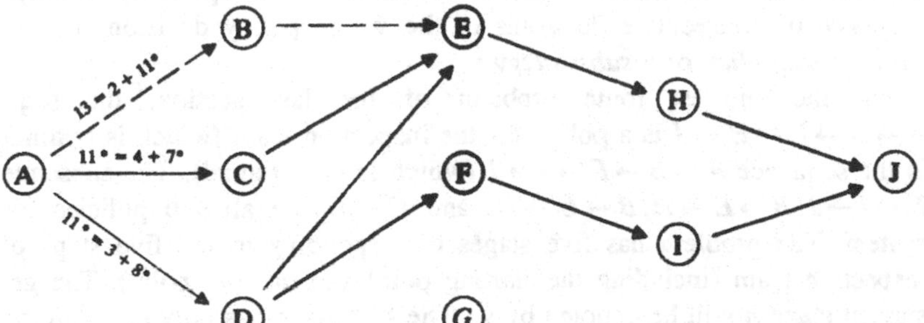


**Figure 9.5**

***system, the remaining decisions must constitute an optimal (sub-) policy for leaving that state.***

A more complete statement of this theorem and a simple proof will be given later in Section 4.

The solution process for the optimal strategy which makes use of the above principle of optimality is known as *dynamic programming* [see *Hillier* and *Lieberman* (1974) for a more general discussion]. Dynamic programming has been found useful in many other classes of problems such as integer programming problems and knapsack problems which will be discussed in later sections of this chapter. The applicability of the solution technique for the shortest route problem to these seemingly unrelated problems makes the analysis of the shortest route problem valuable well beyond the gain in computing efficiency for this one problem. It illustrates the benefit of analysis prior to numerical simulation. Ideally, machine computation should be done only to confirm quantitatively what we already know qualitatively.

## 2. Backward Recursion

To extract an algorithm from the solution process of the shortest route problem for other applications, we observe that the entire plant inspection process may be considered as a system consisting of a number of *stages* corresponding to the different stops in the shortest route problem. At the $n$th stage, the system will be in one of several different possible *states* $\{s_n^{(1)}, s_n^{(2)}, \dots, s_n^{(i_n)}\}$ (corresponding to the different sites at the $n$th stage of the shortest route problem). A policy *decision* (or *move*) $x_n$ will have to be made among the several alternatives $\{x_n^{(1)}, x_n^{(2)}, \dots, x_n^{(k_n)}\}$ to get from $s_n$ to one of the possible states $\{s_{n+1}^{(1)}, s_{n+1}^{(2)}, \dots, s_{n+1}^{(i_{n+1})}\}$ of the next stage. We have a *policy* or *strategy* for the system after all the necessary decisions are made (optimally or not). Any sequence of consecutive decisions of the whole policy decision sequence is called a *subpolicy* or a *substrategy*.

For the shortest route problem of the last section, the sequence $A \to D \to E \to H \to J$ is a policy for the inspection team (which is optimal); as is the sequence $A \to B \to E \to I \to J$ (which is not optimal). The subsequences $F \to I \to J, B \to E \to H, B \to G \to H$, and $C \to G$ are all sub policies for the system. The problem has five stages corresponding to the five stops of the inspection team (including the starting point and the end point). The generic state at stage $n$ will be denoted by $s_n$. The first stage has only one state so that $\{s_1^{(1)}, \dots, s_1^{(i_1)}\} = \{A\} \equiv \{s_1^{(1)}\}$ and $s_1$ is necessarily $s_1^{(1)}$. The second stage has three possible states, namely, $B, C$, or $D$, so that $s_2$ can be one of $\{B, C, D\} \equiv \{s_2^{(1)}, s_2^{(2)}, s_2^{(3)}\}$, and so on. The final state, $s_5$, must be $J = s_5^{(1)}$.

When the system is in one of the possible states $\{s_n^{(1)},\ldots,s_n^{(i_n)}\}$ at stage $n$, i.e., $s_n = s_n^{(i)}$, and a *decision* $x_n$ is made to move the system to one of the possible states at the next stage, i.e., $s_{n+1} = s_{n+1}^{(k)}$, $k = 1,\ldots,$ or $i_{n+1}$, there is a *penalty* (or *payoff*) associated with that decision denoted by $p_n$. For the shortest route problem, the penalty is the distance traveled and it obviously depends on the actual state $s_n^{(i)}$ of the system at state $n$ and the particular decision $x_n$ made on the course of action at that stage, $x_n$ being one of $\{x_n^{(1)},\ldots,x_n^{(k_n)}\}$, so that $p_n = p_n(s_n, x_n)$. The choice of $x_n$, in turn, fixes the state of the system at the $(n + 1)$st stage. In the backward recursion solution process for the shortest route problem, the *best* (optimal) *cumulative penalty*, $f_n^*(s_n)$, for the system in a particular state $s_n$ $(= s_n^{(i)})$ is the smallest total penalty incurred in getting from $s_n$ to the last state $s_N$:

$$f_n^*(s_n) = \min_{x_n}\{f_n(s_n, x_n)\} \equiv \min_{1\le k\le k_n}\{f_n(s_n, x_n^{(k)})\} \tag{9.1}$$

where $f_n(s_n^{(i)}, x_n^{(k)})$ is the *optimal cumulative penalty* for the choice $x_n^{(k)}$, given the system is in state $s_n^{(i)}$. Minimum is replaced by maximum in problems with payoffs (instead of penalties).

For the shortest route problem, the optimal cumulative penalty at stage $n$ was seen to be the total distance to site $J$ from $s_n$. At stage 4, for instance, $x_4$ must be the site $J$; there is no other choice. On the other hand, $s_4$ may be site $H$ or site $I$ with $f_4(H,J) = 3$ and $f_4(I,J) = 4$. For this stage, we have trivially $f_4^*(H) = f_4(H,J) = 3$ and $f_4^*(I) = f_4(I,J) = 4$ as the optimal cumulative penalty to $J$.

A more typical situation is stage 3. Here we have to look at $f_3(E,x_3)$, $f_3(F,x_3)$ and $f_3(G,x_3)$ separately with $f_3(s_3,x_3) = p_3(s_3,x_3) + f_4^*(s_4)$ and in each case the choice for $x_3$ is $H$ or $I$. For example, we have

$$f_3(E,H) = p_3(E,H) + f_4^*(H) = 1 + 3 = 4,$$

$$f_3(E,I) = p_3(E,I) + f_4^*(I) = 4 + 4 = 8,$$

where the term $p_3(s_3,x_3)$ in each case can be read off Fig. 9.1 whereas $f_4^*(s_4)$ has just been obtained above. The minimum penalty for being at site $E$ in stage 3 is

$$f_3^*(\mathrm{E}) = \min_{x_3}\{f_3(E,x_3)\} = \min\{4,8\} = 4 = f_3(E,H)$$

with the last equality showing that $f_3^*(E)$ is attained at $x_3 = H = x_3^*$. Similarly, we have $f_3(F,H) = 9$, $f_3(F,I) = 7$ so that $f_3^*(F) = 7$ (with $x_3^* = I$) and $\{f_3(G,H) = 6, f_3(G,I) = 7\}$ so that $f_3^*(G) = 6$ (with $x_3^* = H$). We can continue the process to earlier stages. The key step at each stage is to obtain

$f_n^*(s_n)$. This is done by first obtaining for each possible $s_n$ (i.e., for $s_n^{(i)}$ for each $i$) the least cummulative penalty

$$f_n(s_n, x_n) = p_n(s_n, x_n) + f_{n+1}^*(s_{n+1}(x_n)), \tag{9.2}$$

where $s_{n+1}(x_n) = x_n$ for the shortest route problem, and then optimizing $f_n(s_n, x_n)$ over $x_n$ as indicated in (9.1).

We are now ready to formalize the algorithm for the exact and efficient solution of the shortest route problem:

**[IX.1] Backward Recursion:** *Start at the last stage,* n = N, *and work backward to the first stage. At each stage* (n), *use (9.2) and then (9.1) to calculate the optimal policy associated with each of the* $i_n$ *possible states* $s_n^{(i)}$, *each over* $k_n$ *possible decisions* $x_n^{(k)}$. *The optimal policy for the single state at the first stage dictates the (sub-) policy of all subsequent stages and thereby the solution(s) of the problem.*

That the solution should be optimal is a consequence of theorem [IX.3] to be proved later.

To illustrate the workings of the backward recursion scheme, we solve the shortest route problem again using the new terminology and a tableau format (instead of the graphical presentation of the last section) to keep track of the intermediate results. We begin at stage 4 because there is nothing more to do at stage 5. The result is summarized in Table 9.1a below. For stage 5, we have $f_5^*(s_5(x_4)) = f_5^*(J) = 0$; there is no more distance to travel beyond site $J$. The penalty $p_4$ is found from Figure 9.1. For each possible $s_4$, there is only one possible decision, namely, $x_4 = J$; so the optimization over $x_n$ is trivial for this stage. For the present problem, we have $s_5(x_4^*) = x_4^*$ so that the last column is really not necessary; but, in general, this will not be the case.

**Table 9.1a**

| $s_4$ \ $x_4$ | $J$ | | | |
|---|---|---|---|---|
| $H$ | $3+0=3$ | 3 | $J$ | $J$ |
| $I$ | $4+0=4$ | 4 | $J$ | $J$ |
| | $p_4 + f_5^*(s_5(x_4)) = f_4(s_4, x_4)$ | $f_4^*(s_4)$ | $x_4^*$ | $s_5(x_4^*)$ |

Stage 3 is more interesting and the result of our backward recursion algorithm for this stage is summarized in the tableau in Table 9.1b. Unlike stage 4, we have two choices of action for each possible $s_n$ and one choice gives a better result than the other. The entry in the column $f_3^*(s_3)$ for each row is the

Table 9.1b

| $s_3$ \ $x_3$ | H | I | | | |
|---|---|---|---|---|---|
| E | 1 + 3 = 4 | 4 + 4 = 8 | 4 | H | H |
| F | 6 + 3 = 9 | 3 + 4 = 7 | 7 | I | I |
| G | 3 + 3 = 6 | 3 + 4 = 7 | 6 | H | H |
| | $p_3 + f_4^*(s_4(x_3)) = f_3(s_3, x_3)$ | | $f_3^*(s_3)$ | $x_3^*$ | $s_4(x_3^*)$ |

Table 9.1c

| $s_2$ \ $x_2$ | E | F | G | | | |
|---|---|---|---|---|---|---|
| B | 7 + 4 = 11 | 4 + 7 = 11 | 6 + 6 = 12 | 11 | E,F | EF |
| C | 3 + 4 = 7 | 2 + 7 = 9 | 4 + 6 = 10 | 7 | E | E |
| D | 4 + 4 = 8 | 1 + 7 = 8 | 5 + 6 = 11 | 8 | E,F | EF |
| | $p_2 + f_3^*(s_3(x_2)) = f_2(s_2, x_2)$ | | | $f_2^*(s_2)$ | $x_2^*$ | $s_3$ |

Table 9.1d

| $s_1$ \ $x_1$ | B | C | D | | | |
|---|---|---|---|---|---|---|
| A | 2 + 11 = 13 | 4 + 7 = 11 | 3 + 8 = 11 | 11 | C,D | C,D |
| | $p_1 + f_2^*(s_2(x_1)) = f_1(s_1, x_1)$ | | | $f_1^*(s_1)$ | $x_1^*$ | $s_2$ |

minimum among those entries in the columns for $f_3(s_3, x_3)$ over the possible decisions $x_3$ for that state. For each possible state $s_3$, the corresponding $p_3$ is again found from Figure (9.1) whereas $f_4^*(s_4(x_3))$ is from Table 9.1a. The results for the remaining two stages are given in Tables 9.1c and 9.1d.

Starting with the single state *A* in stage 1, the optimal subpolicy in Table 9.1d requires that the inspection team go to *C* or *D* ($= s_2$). From *C*, the team should go to *E* (by Table 9.1c), then to *H* (by Table 9.1b), and finally to *J* with no other choice. This sequence gives us one optimal policy for the shortest route, $A \to C \to E \to H \to J$, for a total distance of 11 units. By taking the alternative optimal subpolicy (to *D*) from *A*, we end up with two more (equally good) optimal policies: $A \to D \to E \to H \to J$ and $A \to D \to F \to I \to J$. Altogether, we have three solutions for our problem. The total distance is 11 units in each case and is strictly less than that of any other policy (route). Therefore, all three are optimal under the minimum total distance criterion of optimality for our problem.

The basic dynamic programming algorithm [IX.1] is seen to reproduce the solutions found by our more or less ad hoc and intuitive argument of Section

1. In the next section, we will see how it can be applied to another class of problems seemingly unrelated to the shortest route problem. Further modifications and extensions of the basic algorithm [IX.1] will then be discussed in preparation for the application of dynamic programming to approximate solutions for problems in the calculus of variations in the last section of the chapter.

It should be noted that dynamic programming may not be the most efficient method for solving the shortest route problem. We have shown that it is more efficient than the brute force enumeration method for our example. Efficiency becomes even more important when the number of cities (vertices) $V$ and routes (edges) $E$ connecting them are large (see Exercises). However, an increase in the number of states significantly increases the number of evaluations of the different alternatives at each state. This number may become unacceptably large for problems with thousands of state variables.

It is not known what characterizes problems which can be effectively solved by dynamic programming and which cannot be. However, there is a general agreement that dynamic programming is not applicable to certain classes of problems. These include all linear programming problems for which the coefficients of the linear inequality constraints are not of one sign. There are also many problems for which dynamic programming is less (or no more) efficient than other available methods. The shortest route problem with a large number of vertices and edges is a case in point [see Sedgewick (1988)]. We do not wish to pursue a discussion of the computational efficiency of dynamic programming relative to other techniques for discrete optimization. Our goal is merely to offer an alternative method for approximate solutions for problems in the calculus of variations.

## 3. The Knapsack Problem

Suppose we have three kinds of packaged cargos. Their package size $v_i$ (in cubic feet) and package worth $q_i$ (in millions of dollars) are given in Table 9.2. We wish to load a knapsack with a combination of these cargos so as to achieve a maximum total knapsack worth, without exceeding a prescribed total volume

Table 9.2

| Cargo $n$ | $v_n$ (volume per pkg.) | $q_n$ (worth per pkg.) |
|---|---|---|
| 1 | 10 | 17 |
| 2 | 41 | 72 |
| 3 | 20 | 35 |

capacity, say $V = 50$ ft$^3$. Similar but much larger versions of this problem occur regularly in applications.

Let $x_i$ be the number of packages of cargo $i$ in the combination chosen. Then the total worth of the knapsack is $P = 17x_1 + 72x_2 + 35x_3$. The size limitation requires $10x_1 + 41x_2 + 20x_3 \leq 50$. It is not possible to choose less than zero units of any cargo; hence, $x_i$ must be a non-negative integer.

To achieve a maximum knapsack value, it would be natural to pick as many units of cargo 2 as possible, namely, 50/41. Unfortunately, fractional packages are not allowed. Rounding 50/41 down to the nearest integer (namely, 1) for a total knapsack worth of 72 is not optimal as the combination of 3 units of cargo 1, and 1 unit of cargo 3 gives a total knapsack worth of 86. The requirement that the $x_i$'s be non-negative integers poses an *integer programming* problem. It is known to be more difficult than the same problem without the requirement that $x_i$ be an integer. In this section, we use the knapsack problem to demonstrate how the method of dynamic programming can be used to solve integer programming problems. At present, dynamic programming is the only known sure fire method for the exact solution of this type of optimization problems. Unfortunately, the method is practical only if the problem is not too elaborate (as we shall see in Section 5).

We begin by defining the stages and identifying the states of the knapsack problem so that we may apply the backward recursion algorithm. Let stage $n$ be the time when we choose cargo $n$, and let state $n$ be the knapsack capacity still available for cargos not yet selected. At stage 1, nothing has been chosen and all 50 ft$^3$ of the knapsack are still available; hence, there is only a single possible state 1, namely, $s_1 = \{50\}$. The number of cargo 1 packages we can choose may be $x_1 = \{0, 1, 2, 3, 4, \text{ or } 5\}$, leaving us with $s_2 = \{50, 40, 30, 20, 10, \text{ or } 0\}$. Note that the maximum $x_n$ is the largest integer $\leq V/v_n$ and the subsequent states are determined by $s_{n+1} = s_n - x_n v_n$. For each of the six states of stage 2, we have to identify all the possible decisions $x_2$ which give us a set of $s_3$, etc. It is easy to see that there are four stages and we will have arrived at the solution at stage 4 as there will be no other cargos to choose. We summarize all possible states and decisions in the tree in Table 9.3.

At stage 3, we see from the tree that the possible states for $s_3$ are $0, 9, 10, 20, 30, 40$, and $50$, whereas the possible choices of $x_3$ are 0, 1, and 2. The tableau of the backward recursion for this stage is given in Table 9.4a. The payoff $p_n$ for picking $x_n$ packages of cargo $n$ is $q_n x_n$. At the last stage, we have $f_4^*(s_4) = 0$, as no more cargos can be added. With $s_{n+1}(x_n) = s_n - v_n x_n$, the entries of the last column are obtained from $s_4(x_3) = s_3 - 20x_3$; they are the unused knapsack capacities associated with the optimal subpolicy for the different 3-states.

At stage 2, we see from the tree of Table 9.3 that the possible states at stage 2 are $\{0, 10, 20, 30, 40, 50\}$ and the possible decisions are $\{0, 1\}$. The back-

Table 9.3

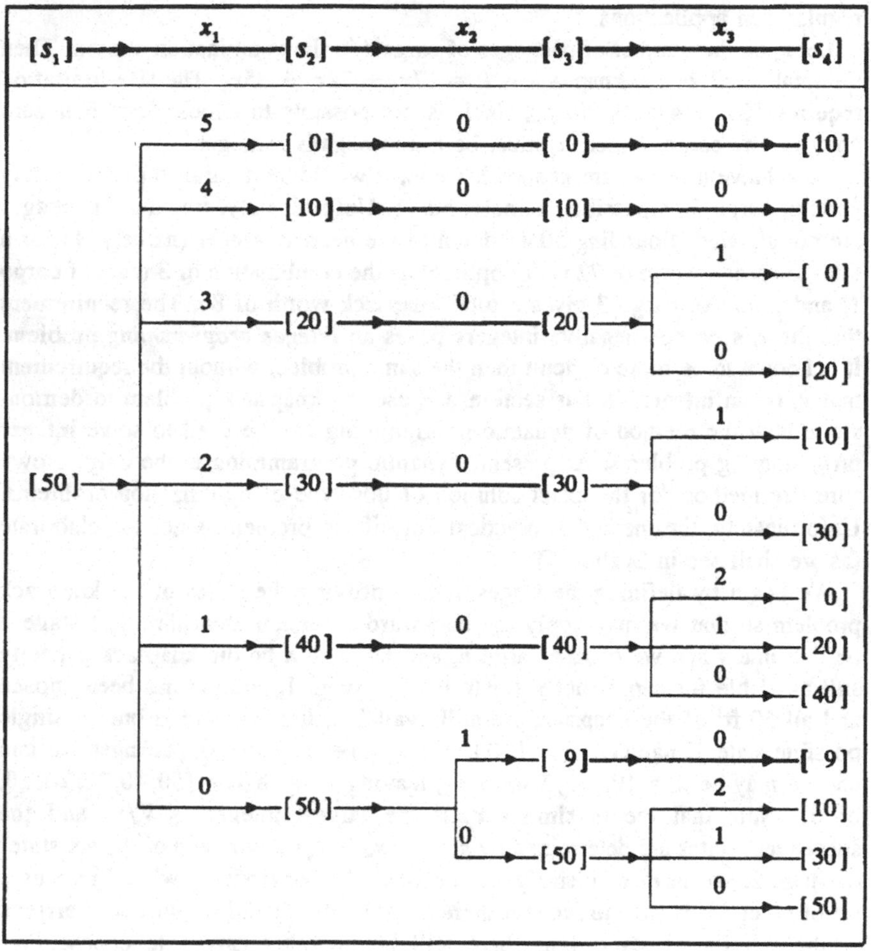


ward recursion tableau for this stage is given in Table 9.4b with the last column obtained from the formula $s_3(x_2) = s_2 - 41x_2$.

For stage 1, we have from Table 9.3 $s_1 = \{50\}$ and $x_1$ may be anyone of $\{0,1,2,3,4,5\}$. The backward recursion tableau for this stage is given in Table 9.4c with the entries of the last column determined by the formula $s_2(x_1) = s_1 - 10x_1$. From the three tableaus, we obtain the package combination of 1,0, and 2 as the optimal solution which gives a maximum knapsack value of 87:

$$(x_1^*, x_2^*, x_3^*) = (1,0,2), \qquad P^* = q_1x_1^* + q_2x_2^* + q_3x_3^* = 87.$$

**Table 9.4a**

| $s_3$ \ $x_3$ | 0 | 1 | 2 | | | |
|---|---|---|---|---|---|---|
| 0 | 0 | — | — | 0 | 0 | 0 |
| 9 | 0 | — | — | 0 | 0 | 9 |
| 10 | 0 | — | — | 0 | 0 | 10 |
| 20 | 0 | 35 | — | 35 | 1 | 0 |
| 30 | 0 | 35 | — | 35 | 1 | 10 |
| 40 | 0 | 35 | 70 | 70 | 2 | 0 |
| 50 | 0 | 35 | 70 | 70 | 2 | 10 |
| | $p_3 + f_4^*(s_4) = q_3x_3 = f_3(s_3, x_3)$ | | | $f_3^*(s_3)$ | $x_3^*$ | $s_4(x_3^*)$ |

**Table 9.4b**

| $s_2$ \ $x_2$ | 0 | 1 | | | |
|---|---|---|---|---|---|
| 0 | 0 + 0 = 0 | — | 0 | 0 | 0 |
| 10 | 0 + 0 = 0 | — | 0 | 0 | 10 |
| 20 | 0 + 35 = 35 | — | 35 | 0 | 20 |
| 30 | 0 + 35 = 35 | — | 35 | 0 | 30 |
| 40 | 0 + 70 = 70 | — | 70 | 0 | 40 |
| 50 | 0 + 70 = 70 | 72 + 0 = 72 | 72 | 1 | 9 |
| | $p_2 + f_3^*(s_3) = q_2x_2 + f_3^*(s_3) = f_2(s_2, x_2)$ | | $f_2^*(s_2)$ | $x_2^*$ | $s_3(x_2^*)$ |

**Table 9.4c**

| $s_1$ \ $x_1$ | 0 | 1 | 2 | 3 | 4 | 5 | | | |
|---|---|---|---|---|---|---|---|---|---|
| 50 | 72 | 87 | 69 | 86 | 68 | 85 | 87 | 1 | 40 |
| | $p_1 + f_2^*(s_2) = q_1x_1 + f_2^*(s_2) = f_1(s_1, x_1)$ | | | | | | $f_1^*(s_1)$ | $x_1^*$ | $s_2(x_1^*)$ |

It should be noted that the final result does not depend on the order in which we choose the cargo.

## 4. Forward Recursion

For a particular loading order, the solution algorithm does not have to be a backward recursion. For the knapsack problem, we can set up a forward recursion algorithm by setting the state $s_n$ to be total knapsack capacity already allocated through stage $n$ with the leftover assigned to $s_0$ so that $s_3 = V = 50$.

The policy decision $x_n$ at stage $n$ will again be taken as the number of packages of cargo $n$ loaded at that stage. At stage $n$, the knapsack capacity still available can be as high as $V$; hence, the largest value of $x_n$ possible is the largest integer less than or equal to $V/v_n$. The possible states at each stage are determined by $s_n = s_{n+1} - x_{n+1}v_{n+1}$ with $s_3 = V = 50$ and $s_0$ being the leftover weight capacity. We may then use:

**[IX.2] Forward Recursion:** *Start with the first stage,* n = 1, *and work forward to the last stage,* n = N. *At each stage, calculate the optimal policy associated with each possible* $s_n$ *over all possible decision* $x_n$. *The optimal subpolicy of the single state at* n = N *dictates the optimal subpolicy of all previous stages and thereby the solution(s) of the problem.*

For the knapsack problem treated in the last section, the maximum value of $x_n$ possible is 5, 1, and 2 for stages 1, 2, and 3, respectively. Thus, the possible choices at each stage are $x_1 = \{0,1,2,3,4,5\}$, $x_2 = \{0,1\}$, and $x_3 = \{0,1,2\}$. The possible states at each stage are determined by $s_n = s_{n+1} - x_{n+1}v_{n+1}$, $n = 0,1,2,3$, with $s_3 = V = 50$ by our definition of $s_n$. It follows that $s_2 = \{50 - 20x_2\}$ (which may be 50, 30, or 10) and $s_1 = \{s_2 - 41x_2\}$ (which may be $50 - 41x_2, 30 - 41x_2$, or $10 - 41x_2$) so that the possible values for $s_1$ are 50, 30, 10, or 9 because $s_n$ cannot be negative. We are now ready for the forward recursion.

The tableau in Table 9.5a gives the results of the solution process at stage 1. As indicated earlier, the formula $s_1 - v_1x_1^* = s_1 - 10x_1^*$ was used to calculate $s_0(x_1^*)$ for the last column of the table. Also, we have $f_0^*(s_0) = 0$ because, by definition, the leftover cannot be used to increase the value of the knapsack.

**Table 9.5a**

| $s_1$ \ $x_1$ | 0 | 1 | 2 | 3 | 4 | 5 | $f_1^*(s_1)$ | $x_1^*$ | $s_0(x_1^*)$ |
|---|---|---|---|---|---|---|---|---|---|
| 9 | 0 | — | — | — | — | — | 0 | 0 | 9 |
| 10 | 0 | 17 | — | — | — | — | 17 | 1 | 0 |
| 30 | 0 | 17 | 34 | 51 | — | — | 51 | 3 | 0 |
| 50 | 0 | 17 | 34 | 51 | 68 | 85 | 85 | 5 | 0 |
| | $17x_1 + f_0^*(s_1 - 10x_1) = f_1(s_1, x_1)$ | | | | | | | | |

Similarly, we have the two tableaus in Table 9.5b and 9.5c for stages 2 and 3 of the forward recursion. Starting with stage 3 (Table 9.5c), we trace the optimal subpolicy at each stage and the corresponding previous state to get the following optimal overall policy:

Table 9.5b

| $s_2$ \ $x_2$ | 0 | 1 | | | |
|---|---|---|---|---|---|
| 10 | 17 | — | 17 | 0 | 10 |
| 30 | 51 | — | 51 | 0 | 30 |
| 50 | 85 | 72 | 85 | 0 | 50 |
| | $72x_2 + f_1^*(s_2 - 41x_2) = f_2(s_2, x_2)$ | | $f_2^*(s_2)$ | $x_2^*$ | $s_1(x_2^*)$ |

Table 9.5c

| $s_3$ \ $x_3$ | 0 | 1 | 2 | | | |
|---|---|---|---|---|---|---|
| 50 | 85 | 86 | 87 | 87 | 2 | 10 |
| | $35x_3 + f_2^*(s_3 - 20x_3) = f_3(s_3, x_3)$ | | | $f_3^*(s_3)$ | $x_3^*$ | $s_2(x_3^*)$ |

$$[s_3 = 50] \xrightarrow{x_3^* = 2} [s_2(x_3^*) = 10] \xrightarrow{x_2^* = 0} [s_1(x_2^*) = 10] \xrightarrow{x_1^* = 1} [s_0(x_1^*) = 0].$$

With $(x_1^*, x_2^*, x_3^*) = (1,0,2)$, we get, as the optimal total knapsack value, 87 as before.

The forward recursion algorithm is based on another version of the principle of optimality symmetric to the one observed in Section 9.1:

> **Principle of Optimality II:** *Regardless of how we leave a particular state of the system, the decisions already made must constitute an optimal (sub)policy for entering that state.*

Together, the two versions of the principle of optimality observed suggest the following theorem:

**[IX.3] The Principle of Optimality:** *An optimal policy must contain only optimal subpolicies.*

This theorem is easily proved by contradiction. Suppose a subpolicy between $s_i$ and $s_j$ of an optimal policy were not an optimal subpolicy. Then there would exist a better one and the original (optimal) policy could be improved by replacing its subpolicy between $s_i$ and $s_j$ by the new and better one. This contradicts the assertion that the given overall policy is optimal.

Adroit usage of this theorem has led to the solution of many difficult optimization problems. In particular, it follows from [IX.3] that any overall policy generated by either of the two algorithms [IX.1] and [IX.2] must be optimal.

## 5. Intermediate Knapsack Capacities

One important difference between the knapsack problem and the shortest route problem is in the states of the intermediate stages of the problem. At any stage of the shortest route problem, all admissible states are known explicitly. At stage 2 of the problem shown in Figure 9.1 for example, the only admissible states are cities *B, C*, and *D* and no others. They are explicitly prescribed as a part of the problem statement. In contrast, the states at any stage after the first of the backward recursion algorithm for the knapsack problem are not known explicitly in advance; they have to be worked out as we did in Section 3 to arrive at Table 9.3. Similarly, the states for all but the last stage of the forward recursion also have to be worked out from the given information. For problems with a large number of cargos, the process of obtaining the unknown states for all the stages may be complicated or time-consuming (or both). A different kind of algorithm which does not require a separate determination of the admissible states at all stages is sometimes preferrable for these problems.

One such algorithm often used in practice obtains the optimal combination of cargos for all intermediate knapsack capacities up to and including the prescribed capacity, and for successively more cargo types (starting with one) until all types are included. For the knapsack problem considered in the last two sections, the results by this algorithm are given in Table 9.6. The first row of the table lists all possible intermediate knapsack capacities *J* up to and including the actual capacity (50). The next two rows give two important quantities when only cargo 1 is available (I = 1). For each intermediate capacity *J*, the third row gives the optimal *type* of cargo for the last unit of cargo added [denoted by OPT(J)], whereas the second row gives the corresponding cumulative value of the knapsack after the added unit [denoted by CUM(J)]. The fourth and fifth rows give the same two quantities when only the first two types of cargos are available (I = 2). Finally, the last two rows give CUM(J) and OPT(J) when all three types of cargos are available for selection (I = 3).

Note that only one column is assigned to knapsack capacities 1 through 9 because the same results apply to each. The situation is the same for intermedi-

**Table 9.6a. Optimal Solution for Intermediate Knacksack Capacities**

| | J = | 1-9 | 10-19 | 20-29 | 30-39 | 40 | 41-49 | 50 |
|---|---|---|---|---|---|---|---|---|
| I = 1: | CUM(J) = | 0 | 17 | 34 | 51 | 68 | 68 | 85 |
| | OPT(J) = | — | 1 | 1 | 1 | 1 | 1 | 1 |
| I = 2: | CUM(J) = | 0 | 17 | 34 | 51 | 68 | 72 | 85 |
| | OPT(J) = | — | 1 | 1 | 1 | 1 | 2 | 1 |
| I = 3: | CUM(J) = | 0 | 17 | 35* | 52 | 70* | 72 | 87* |
| | OPT(J) = | — | 1 | 3* | 1 | 3* | 2 | 1* |

**Table 9.6b. A FORTRAN Program for the Intermediate Capacity Algorithm**

```
      DO 1 J = 1,M
      CUM(J) = 0
1     OPT(J) = 0
      ASSIGN 2 TO NG
      DO 4 I = 1,N
      K = 0
      DO 3 J = 2,M
      J1 = MOD(J,W(I))
      IF J1 = K, GO TO NG
      K = J1
      IF CUM(J - 1) + P(I) < CUM(J), GO TO 3
      CUM(J) = CUM(J - 1) + P(I)
      OPT(J) = I
      GO TO 3
2     CUM(J) = CUM(J - 1)
      OPT(J) = OPT(J -1)
3     CONTINUE
4     ASSIGN 3 TO NG
```

ate capacities 10-19, 20-29, 30-39, and 41-49. In the actual implementation of the algorithm, we may omit the listing and calculations for the knapsack capacities 1-9 and begin with J = 10; it is clear that no cargo of any kind can be chosen unless the knapsack can accommodate at least 10 of cargo volume. The backward and forward recursion of the last two sections take advantage of the known admissible states (from a separate calculation) to avoid doing the unnecessary computation for the 44 different duplicated results. They are, therefore, more efficient, assuming the efficiency gained is not dissipated by the calculation needed for the determination of the admissible states.

When only one cargo is available, the optimal strategy is clearly to pack as many units of that cargo as possible (and, hence, the first two rows of results in Table 9.6a). With two cargos available, we must compare alternatives. There are no alternatives to more cargo 1 until we reach a capacity of 41 or larger. For 41-49, we may have 4 units of cargo 1 for a total knapsack value of 68 or 1 unit of cargo 2 for a value of 72. The 5th row of Table 9.6a lists the optimal choice for that capacity range. At a capacity of 50, 5 units of cargo 1 is better than 1 unit of cargo 2.

With all three cargos available, there are alternatives to picking only cargo 1 even for the capacity range 20-40. The optimal combinations for different capacities in that range are no longer the same as those for I = 2. For the capacity range 41-50, there are also more possible cargo combinations. The final results recorded in Table 9.6a were obtained after comparing all these combinations.

By construction, the column for a 50-ft$^3$ knapsack shows that the last optimal selection is one unit of cargo 1. We are then left with a knapsack capacity of 40 ft$^3$ (= 50 − (1 × 10)) for more selections. The principle of optimality allows us to find from the column for a 40 cu.ft. knapsack the optimal choice of one unit of cargo 3, leaving 20 cu.ft. ( = 40 − (1 × 20)) for still more selections. The column for a 20-ft$^3$ knapsack shows that we should select another unit of cargo 3 which uses up all of the remaining knapsack capacity. The total value of the knapsack is therefore 35 + 35 + 17 = 87 as we know from the result of the last two sections.

The *intermediate capacity algorithm* for dynamic programming problems described above applies to a general knapsack problem with a knapsack capacity $M$ (volume) units and $N$ different types of cargos of different sizes [$W$(I)] and worths [$Q$(I)]. The simple (pseudo-FORTRAN) program shown in Table 9.6b is easily coded to implement the general algorithm.

The following result on the efficiency of the intermediate capacity algorithm is a simple consequence of the program above.

**[IX.4]** *For a knapsack problem with* M *units of knapsack capacity and* N *different types of cargos (where* M *and* N *are both positive integers), the amount of computing time required by the intermediate capacity algorithm is proportional to* NM.

It is evident from [IX.4] that the intermediate capacity algorithm is practical when $M$ is not extremely large. $N$ is considerably smaller than $M$ in most applications.

The general idea of calculating the optimal solution for all intermediate states in preparation for the actual solution applies to other dynamic programming problems. Some examples of such applications can be found in the exercises at the end of the chapter. On the other hand, the algorithm is applicable only when $N$ and $M$ are integers.

## 6. Vector- and Continuous-State Variables

By a few simple examples, we will show in this section how the dynamic programming technique can be extended to allow for vector- and continuous-state variables. The extensions are needed for obtaining approximate solutions of many problems in the calculus of variations.

### *Vector variables*

Consider a knapsack problem with only two cargos but with a weight limitation of 8 units as well as a volume limitation of 12 units. The unit package weights

$\{w_1, w_2\}$, volumes $\{v_1, v_2\}$, and payoffs $\{p_1, p_2\}$, are given in Table 9.7a. The integer programming problem now is to choose $x_1$ and $x_2$ to maximize $P \equiv 7x_1 + 4x_2$ subject to the weight restriction $5x_1 + 4x_2 \leq 8$, the volume constraint $5x_1 + 6x_2 \leq 12$, and the requirement that $x_1$ and $x_2$ be non-negative integers. Now, the maximum value of $x_n$ is the largest integer less than or equal to the minimum of $W/w_n$ and $V/v_n$. It follows that, for the present problem, the collection of possible $x_1$ is $\{0, 1\}$ and the collection for $x_2$ is $\{0, 1, 2\}$.

For backward recursion, we let the state $s_n$ be the combination of weight capacity $W_n$ and volume capacity $V_n$ still available for cargos $n, n+1, \ldots, N$ (with $N = 2$ for our problem). Evidently, $s_1$ is the combination (8, 12) which can be treated as a vector-state variable. As in the last section, we have $s_2 = s_1 - x_1(w_1, v_1) = (8,12) - x_1(5,5) = (8,12) - \{0,1\}(5,5) = \{(8,12), (3,7)\}$. The backward recursion then gives the tableau in Table 9.7b for the last stage with the last column obtained from $s_3 = s_2 - x_2^*(w_2, v_2) = s_2 - x_2^*(4,6)$. Note that $f_3^*(s_3(x_2)) = 0$, as there are no more cargos to be added to the knapsack after stage 2.

Similarly, we have the tableau in Table 9.7c for stage 1 with the last column obtained from $s_2 = s_1 - x_1^*(5,5)$. The optimal combination of cargos is given by $(x_1^*, x_2^*) = (0,2)$ with a total knapsack value of $7x_1^* + 4x_2^* = 8$.

**Table 9.7a**

| Cargo $i$ | $w_i$ | $v_i$ | $p_i$ |
|---|---|---|---|
| 1 | 5 | 5 | 7 |
| 2 | 4 | 6 | 4 |

**Table 9.7b**

| $s_2$ \ $x_2$ | 0 | 1 | 2 | | | |
|---|---|---|---|---|---|---|
| (8,12) | 0 | 4 | 8 | 8 | 2 | (0,0) |
| (3,7) | 0 | — | — | 0 | 0 | (3,7) |
| | $4x_2 + f_3^*(s_3(x_2)) = f_2(s_2, x_2)$ | | | $f_2^*(s_2)$ | $x_2^*$ | $s_3(x_2^*)$ |

**Table 9.7c**

| $s_1$ \ $x_1$ | 0 | 1 | | | |
|---|---|---|---|---|---|
| (8,12) | 8 | 7 | 8 | 0 | (8,12) |
| | $7x_1 + f_2^*(s_1 - x_1(5,5)) = f_1(s_1, x_1)$ | | $f_1^*(s_1)$ | $x_1^*$ | $s_2(x_1^*)$ |

### *Continuous variables*

(a) *Maximum Product of N Parts*

Problems in dynamic programming sometimes involve a continuous-state variable. To illustrate, suppose we want a subdivision of 1 into $N$ parts $\{z_1, z_2, \ldots, z_N\}$ (so that $z_1 + z_2 + \ldots + z_N = 1$) which maximizes the product of the $N$ parts, i.e.,

$$\max_{\{z_j\}} \left\{ Z \equiv \prod_{k=1}^{N} z_k = z_1 \ldots z_N \,\middle|\, \sum_{k=1}^{N} z_k = 1, 0 < z_k < 1 \right\}. \tag{9.3}$$

This problem can be solved by the backward recursion technique of dynamic programming.

Let $s_n$ be the fraction of 1 still available for the remaining parts $z_n, z_{n+1}, \ldots, z_N$ and $x_n$ be the value chosen for $z_n$. Clearly, we must have $0 < x_n < s_n$, $1 \le n \le N$; otherwise, we would get the undesirable result $Z = z_1 z_2 \ldots z_N = 0$. At stage 1, the only possible state is $s_1 = 1$ because the whole unit is still available for subdivision. At the later stages, we have generally

$$s_{n+1} = s_n - x_n = s_{n-1} - (x_{n-1} + x_n) = \ldots$$

$$= s_1 - (x_1 + x_2 + \ldots + x_n) = 1 - \sum_{k=1}^{n} x_k.$$

For $n < N, x_n$ may be any positive value less than $s_n$. (Otherwise, we have $x_n = s_n$ which requires $s_{n+1} = 0$ and $x_{n+1} = 0$ and therewith $Z = 0$.) Hence, $s_{n+1}$ can be any positive number inside the interval $(0, s_n)$. It follows that $s_n < s_{n-1} < s_{n-2} < \ldots < s_2 < s_1 = 1$ and the range of each $s_n$ $(n \ge 2)$ is $0 < s_n < 1$.

At the last stage, the situation is a little different. We may use up what is left so that $0 < x_N \le s_N$; there are no remaining parts to be assigned. The Table 9.8a summarizes the backward recursion algorithm at stage $N$, with the payoff being the partial product consisting only of the last factor of the entire product, namely $x_N$.

Stage $N-1$ is more typical. In particular, the positive value $x_{N-1}$ must be strictly less than $s_{N-1}$. The tableau of the backward recursion algorithm for this step is given in Table 9.8b. We note in particular that the payoff at stage $N-1$ is another partial product consisting of the multiplicative factor $x_{N-1}$ multiplying the optimal partial product (subpolicy) $f_N^*(s_{N-1}(x_{N-1}))$ of the remaining stage. This gives $x_{N-1} f_N^*(s_{N-1}(x_{N-1})) = x_{N-1} x_N^* = x_{N-1}(s_{N-1} - x_{N-1})$ for the last two stages. For the best of all such partial products, we need the maximum value of $x_{N-1}(s_{N-1} - x_{N-1})$. This is just a calculus problem of maximizing

**Table 9.8a**

| $x_N$ \ $s_N$ | $0 < x_N \leq s_N$ | | | |
|---|---|---|---|---|
| $0 < s_N < 1$ | $x_N$ | $s_N$ | $s_N$ | $0$ |
| | $x_N = f_N(s_N, x_N)$ | $f_N^*(s_N)$ | $x_N^*$ | $s_{N+1}(x_N^*)$ |

**Table 9.8b**

| $x_N - 1$ \ $s_N - 1$ | $0 < x_{N-1} < s_{N-1}$ | | | |
|---|---|---|---|---|
| $0 < s_{N-1} < 1$ | $x_{N-1}x_N^* = x_{N-1}s_N = x_{N-1}(s_{N-1} - x_{N-1})$ | $\frac{1}{4}s_{N-1}^2$ | $\frac{1}{2}s_{N-1}$ | $\frac{1}{2}s_{N-1}$ |
| | $x_{N-1}f_N^*(s_N(x_{N-1})) = f_{N-1}(s_{N-1}, x_{N-1})$ | $f_{N-1}^*(s_{N-1})$ | $x_{N-1}^*$ | $s_N(x_{N-1}^*)$ |

$x_{N-1}(s_{N-1} - x_{N-1}) \equiv g_{N-1}(x_{N-1})$ with $s_{N-1}$ as a parameter. The stationary point $x_{N-1}^*$ for this problem is given by

$$\frac{dg_{N-1}}{dx_{N-1}} = s_{N-1} - 2x_{N-1}^* = 0 \quad \text{or} \quad x_{N-1}^* = \frac{1}{2}s_{N-1}. \tag{9.4}$$

The stationary value $g_{N-1}(x_{N-1}^*) = s_{N-1}^2/4$ is, in fact, a maximum because

$$\frac{d^2g_{N-1}}{dx_{N-1}^2} = -2 < 0.$$

This maximum value is entered as the value of $f_{N-1}^*(s_{N-1})$ in Table 9.8b and $x_{N-1}^* = s_{N-1}/2$ itself is the entry for the next column. Finally, we have $s_N(x_{N-1}^*) = s_{N-1} - x_{N-1}^* = s_{N-1} - s_{N-1}/2 = s_{N-1}/2$ which gives the entry to the last column.

We may continue to obtain similar tableaus for stage, $N-2, N-3$, etc.; the optimization problem for determining $f_n^*(s_n)$ in each case is a simple classical parameter optimization problem for a different $g_n(x_n)$. For a typical stage $n$, we have again $0 < s_n < 1$ and $0 < x_n < s_n$. Now, the optimal subpolicy from stage $n$ to the end is

$$f_n(s_n, x_n) = x_n f_{n+1}^*(s_{n+1}(x_n)) = x_n\left(\frac{s_{n+1}}{N-n}\right)^{N-n} = x_n\left(\frac{s_n - x_n}{N-n}\right)^{N-n} \tag{9.5}$$

This formula can be established by induction. It is true for $n = N - 1$ (which is more typical and easier to verify than the $n = N$ case). Suppose it is true for $n + 1$; then we have

$$f_n(s_n, x_n) = x_n f^*_{n+1}(s_{n+1}(x_n)) = x_n\left\{\max_{x_{n+1}} f_{n+1}(s_{n+1}, x_{n+1})\right\}$$

$$= x_n\left\{\max_{x_{n+1}}\left[x_{n+1}\left(\frac{s_{n+1} - x_{n+1}}{N - n - 1}\right)^{N-n-1}\right]\right\}.$$

A routine calculation gives as the unique solution of the classical maximization problem for the function

$$g_{n+1}(x_{n+1}) \equiv x_{n+1}[(s_{n+1} - x_{n+1})/(N - n - 1)]^{N-n-1}$$

the following expressions for $x^*_{n+1}$ and $g_{n+1}(x^*_{n+1})$

$$x^*_{n+1} = \frac{s_{n+1}}{N - n}, \qquad g_{n+1}(x^*_{n+1}) = \left(\frac{s_{n+1}}{N - n}\right)^{N-n} = f^*_{n+1}(s_{n+1}) \tag{9.6}$$

with $s_{n+1}(x^*_n) = s_n - x^*_n = (N - n)s_n/(N - n + 1)$. It follows from these results that formula (9.5) for $f_n(s_n, x_n)$ is true for stage $n$ as well. This completes the induction proof.

The backward recursion tableau for stage $n$ may now be put together as shown in Table 9.8c. The entries to the columns $f^*_n(s_n)$ and $x^*_n$ are obtained by the same calculation which leads to $g_{n+1}(x^*_{n+1})$ and $x^*_{n+1}$ in (9.6) [or simply replacing $n$ by $n - 1$ in (9.6)]. For $n = 1$, we know $s_1 = 1$. Upon specializing Table 9.8c to $n = 1$, we get Table 9.8d for the first stage.

We can now proceed forward to get from the tableau for stage $n$: $x^*_2 = s_2/(N - 2 + 1) = [(N - 1)/N]/(N - 1) = 1/N, x^*_3 = \ldots = 1/N$, etc. In fact, we have for $1 \le n \le N$

**Table 9.8c**

| $s_n$ \ $x_n$ | $0 < x_n < s_n$ | | | |
|---|---|---|---|---|
| $0 < s_N < 1$ | $x_n\left(\frac{s_n - x_n}{N - n}\right)^{N-n}$ | $\left(\frac{s_n}{N - n + 1}\right)^{N-n+1}$ | $\frac{s_n}{N - n + 1}$ | $\frac{N - n}{N - n + 1}s_n$ |
| | $x_n f^*_{n+1}(s_{n+1}(x_n)) = f_n(s_n, x_n)$ | $f^*_n(s_n)$ | $x^*_n$ | $s_{n+1}(x^*_n)$ |

**Table 9.8d**

| $s_1$ \ $x_1$ | $0 < x_1 < 1$ | | | |
|---|---|---|---|---|
| 1 | $x_1\left(\frac{1 - x_1}{N - 1}\right)^{N-1}$ | $\left(\frac{1}{N}\right)^N$ | $\left(\frac{1}{N}\right)$ | $1 - \frac{1}{N}$ |
| | $x_1 f^*_2(s_2(x_1)) = f_1(s_1, x_1)$ | $f^*_1(s_1)$ | $x^*_1$ | $s_2(x^*_1)$ |

$$x_n^* = \frac{s_n}{N-n+1} = \frac{1}{N-n+1}\frac{N-n+1}{N-n+2}s_{n-1}$$

$$= \frac{1}{N-n+2}\frac{N-n+2}{N-n+3}s_{n-2}$$

$$= \dots$$

$$= \frac{1}{N-2}\frac{N-2}{N-1}s_2 = \frac{1}{N-1}\frac{N-1}{N}s_1 = \frac{1}{N}$$

so that the maximum $Z(x_1^*, x_2^*, \dots, x_N^*) \equiv Z^*$ is given by

$$Z^* = \left(\frac{1}{N}\right)\dots\left(\frac{1}{N}\right) = \left(\frac{1}{N}\right)^N$$

which is the solution of our optimization problem.

The problem of maximizing the product of $N$ parts of unity can be solved in a more straightforward manner as a simple calculus problem. Write the product $Z$ as $z_1 z_2 \dots z_{N-1}(1 - z_1 - z_2 \dots - z_{N-1})$ by using the constraint $z_1 + z_2 + \dots + z_N = 1$ to eliminate $Z_N$ from the product. We then seek the stationary point of $Z$ by solving the system of $N-1$ equations for $z_1, \dots, z_{N-1}$:

$$\frac{\partial Z}{\partial z_1} = z_2 z_3 \dots z_{N-1}(z_N - z_1) = 0,$$

$$\frac{\partial Z}{\partial z_2} = z_1 z_3 \dots z_{N-1}(z_N - z_2) = 0,$$

- - - - - - - - - - - - - - - - - - - - - - - -

$$\frac{\partial Z}{\partial z_n} = z_1 z_2 \dots z_{n-1} z_{n+1} \dots z_{N-1}(z_N - z_n) = 0,$$

- - - - - - - - - - - - - - - - - - - - - - - -

$$\frac{\partial Z}{\partial z_{N-1}} = z_1 z_2 \dots z_{N-2}(z_N - z_{N-1}) = 0.$$

Because we can always do better than $Z = 0$, it follows from the system above that the unique stationary point of $Z$ has as its $N$ parts $z_1^* = z_2^* = \dots = z_{N-1}^* = z_N^* = 1/N$ which fortuitously satisfies the constraints $0 < z_k < 1$. (To show that this stationary value is, in fact, a maximum requires a little more work).

The fact that the maximum product can be found by calculus method in no way diminishes the pedagogical value of a solution by dynamic programming. The same dynamic programming technique applies also to many (more complex) optimization problems with continuous-state variables which cannot be solved by classical methods. The following nonlinear programming problem is one example.

(b) *Nonlinear Programming*

Choose *non-negative* values $z_1$ and $z_2$ to minimize the quantity

$$Z = (2z_1 - 5)^2 + (2z_2 - 1)^2$$

subject to the inequality constraint $z_1 + 2z_2 \leq 2$. The function $Z$ has only one stationary point $(\bar{z}_1, \bar{z}_2) = (5/2, 1/2)$ for which $Z$ vanishes and, therefore, attains its minimum. However, $\bar{z}_1 + 2\bar{z}_2 = (5/2) + 1 = 7/2$ violates the prescribed inequality constraint. If we should allow $z_1 + 2z_z$ to be as close to $7/2$ as permitted by taking $z_1 + 2z_2 = 2$ and use the equality constraint to eliminate $z_2$ from $Z$ to get $Z(z_1) = (2z_1 - 5)^2 + (1 - z_1)^2$, the minimum of $Z$ would be attained at $\hat{z}_1 = 11/5$ and $\hat{z}_2 = -1/10$. However, this is not the solution of our problem as it does not satisfy the non-negativity constraint. Evidently, classical techniques is not appropriate for this problem.

On the other hand, the backward recursion technique in dynamic programming can be applied to this problem in a straightforward way. To use this technique, we let $s_n$ be what is left of 2 (the maximum "capacity" available) at stage $n$ for the remaining items $z_n, z_{n+1}, \ldots\ z_N$ (with $N = 2$ for our problem). Let $x_n$ be the value chosen for $z_n$. By inspection, we have $s_1 = 2$, $0 \leq x_1 \leq s_1/1 = 2$, and $0 \leq x_2 \leq s_2/2$. The backward recursion tableau for stage 2 can now be constructed and is shown in Table 9.9a. There is no more penalty to be added to $Z$ beyond stage 2; $f_3^*(s_3)$ is, therefore, zero. The minimum of $f_2(s_2, x_2) = (2x_2 - 1)^2$ is zero (with $x_2^* = 1/2$) unless $s_2 < 1$, in which case we do the best we can by setting $x_2^* = s_2/2$ and $f_2^*(s_2) = (s_2 - 1)^2$. In general, we should take

$$x_2^* = \min(1/2, s_2/2)$$

with $f_2^*(s_2) = (2x_2^* - 1)$ which may be $(s_2 - 1)$ if $s_2 < 1$ or zero if $s_2 \geq 1$.

For stage 1, the backward recursion applies similarly. The corresponding tableau is given in Table 9.9b. To determine the entry to the columns for $x_1^*$ and $f_1^*(s_1 = 2)$, we note that $x_2^* = \min(s_2/2, 1/2 = \min((s_1 - x_1)/2, 1/2) = \min((2 - x_1)/2, 1/2)$ with

$$f_2^*(s_2) = f_2^*(s_1 - x_1) = \begin{cases} 0 & (2 - x_1 \geq 1) \\ (1 - x_1)^2 & (2 - x_1 < 1) \end{cases}$$

so that

**Table 9.9a**

| $s_2$ \ $x_2$ | $0 \le x_2 \le s_2/2$ | | | |
|---|---|---|---|---|
| $0 \le s_2 \le 2$ | $(2x_2-1)^2$ | $(2x_2^*-1)^2$ | $min\left(\dfrac{1}{2},\dfrac{s_2}{2}\right)$ | $s_2-2x_2^*$ |
| | $(2x_2-1)^2+f_3^*(s_3)=f_2(s_2,x_2)$ | $f_2^*(s_2)$ | $x_2^*$ | $s_3(x_2^*)$ |

**Table 9.9b**

| $s_1$ \ $x_1$ | $0 \le x_1 \le s_1 = 2$ | | | |
|---|---|---|---|---|
| 2 | $(2x_1-5)^2+(2x_2^*-1)^2$ | 2 | 2 | 0 |
| | $(2x_1-5)^2+f_2^*(s_2=s_1-x_1)=f_1(s_1,x_1)$ | $f_1^*(s_1)$ | $x_1^*$ | $s_2(x_1^*)=s_1-x_1^*$ |

$$f_1(s_1,x_1)=\begin{cases}(2x_1-5)^2 & (2-x_1\ge 1)\\ (2x_1-5)^2+(1-x_1)^2 & (2-x_1<1)\end{cases}$$

The minimum of $(2x_1-5)^2$ is attained at $x_1^*=5/2$ which is not in the range $2-x_1>1$. On the other hand, the minimum of $(2x_1-5)^2+(1-x_1)^2$ is attained at $x_1^*=2.2$ which satisfies $(2-x_1)<1$. Given $x_1 \le s_1 \le 2$, we do the best we can by taking $x_1^*=2$. This choice of $x_1$ corresponds to $s_2=2-x_1^*=0$ which, in turn, gives $f_1^*(s_1)=f_1(s_1,x_1^*)=2$. Thus, the minimum $Z$ is 2 and the combination of $(z_1^*,z_2^*)$ which gives this value is $(2,0)$. It can be verified that the optimal choice of $(z_1,z_2)$ does not depend on the order in which we choose $z_1$ and $z_2$.

## 7. The Variational Problem

Our primary interest in this volume is in the calculus of variations. Although dynamic programming provides a natural method of solution for many problems in optimization unrelated to this calculus, it is introduced in this chapter mainly as another approach to obtaining approximate solutions for variational problems. We illustrate this approach with the following familiar basic problem: Choose $y(t)$ to minimize

$$J[y]=\int_0^{\pi/2}\frac{1}{2}\,[(y^\bullet)^2-y^2]\,dt,\qquad (\;)^\bullet\equiv\frac{d(\;)}{dt},$$

subject to the end constraints $y(0) = 0$ and $y(\pi/2) = 1$. (Note that we are using $t$ as the independent variable here to avoid possible confusion, given the notation already adopted for the decision variable in dynamic programming).

For an approximate solution by dynamic programming, we introduce a set of mesh points in $[0, \pi/2]$. To simplify our presentation, we consider only the case of three mesh points $\{t_1, t_2, t_3\} = \{0, \pi/4, \pi/2,\}$. At each *interior* mesh point $t_n$ (and there is only one for our chosen mesh), the unknown value $y(t_n)$ is to be determined by the solution process. Although this quantity, or its approximation $y_n$, may take on any real value, we limit it to a small number of positive values. In view of the end conditions $y(0) = 0$ and $y(\pi/2) = 1$, we will, as a first attempt, consider only $y_n$ in the range $[0, 1]$. Moreover, we avoid the more difficult case of continuous-state and decision variables by allowing only the values $\{0, 1/4, 1/2, 3/4, \text{ or } 1\}$ for $y_n$ at any interior mesh point $t_n$. If necessary, we can always repeat the solution process with a more refined set of possible $y_n$ values (including a continuous range) as well as more mesh points in the interval $[0, \pi/2]$ to get a more accurate approximate solution.

For backward recursion, we designate stage $n$ to be the instance when the integration of $J$ has reached the mesh point $t_n$ and we are ready to choose the value for $y_{n+1}$, the approximation of $y$ at the next mesh point $t_{n+1}$. The decision variable $x_n$ at stage $n$ is, therefore, $y_{n+1}$ and the state of the system at stage $n$ is the value assumed by $y_n$. The total "penalty" integral $J[y]$ may be written as a sum of two integrals:

$$J[y] = \int_{t_1}^{t_2} F(t, y, y^{\bullet})\, dt + \int_{t_2}^{t_3} F(t, y, y^{\bullet})\, dt.$$

Within the interval $[t_n, t_{n+1}]$, we approximate $y(t)$ by the average of $y_{n+1}$ and $y_n$ and $y^{\bullet}(t)$ by the difference quotient:

$$y(t) = \frac{1}{2}(y_{n+1} + y_n), \qquad y^{\bullet}(t) = \frac{1}{\Delta t}(y_{n+1} - y_n) \quad (t_n < t < t_{n+1}),$$

where $\Delta t = \pi/4$ for our choice of mesh points. [Other methods of approximation may also be used. In an exercise at the end of this chapter, the approximation for $y(t)$ will be replaced by $y(t) = y_n + m_n(t - t_n)$, where $m_n = (y_{n+1} - y_n)/\Delta t$ and $y^{\bullet}(t) = m_n$ as before.] The penalty for the segment $(t_n, t_{n+1})$ is given by

$$\begin{aligned} p_n(s_n, x_n) &= \int_{t_n}^{t_{n+1}} \frac{1}{2}\{m_n^2 - [(y_{n+1} + y_n)/2]^2\}\, dt \\ &= \frac{1}{2\Delta t}\left\{(y_{n+1} - y_n)^2 - \left(\frac{\Delta t}{2}\right)^2 (y_{n+1} + y_n)^2\right\}, \end{aligned}$$

keeping in mind $s_n = y_n$ and $x_n = y_{n+1}$.

For stage 2, $s_2 = y_2$ may be any one of the five possible values $\{0, 1/4, 1/2, 3/4, 1\}$ (we allowed), whereas there is only one choice of $x_n = y_{n+1} = 1$ [because $y(t_3) = y(\pi/2) = 1$ by one prescribed end condition]. The penalty for the interval $(t_2, t_3) = (\pi/4, \pi/2)$ is, therefore,

$$p_2(s_2, x_2) = \frac{2}{\pi}\left\{(1 - y_2)^2 - \left(\frac{\pi}{8}\right)^2 (1 + y_2)^2\right\}$$

which takes on a different value depending on the value of $y_2$. The backward recursion tableau for this stage is given in Table 9.10a. Note that there is no further penalty beyond the mesh point $t_3 = \pi/2$; hence, we have $f_3^*(s_3) = 0$ and $f_2^*(s_2) = f_2(s_2, x_2) = p_2(s_2, x_2)$.

**Table 9.10a**

| $x_2 = y_3$ / $s_2 = y_2$ | 1 | | | |
|---|---|---|---|---|
| 0 | $\frac{2}{\pi}\left[1 - \left(\frac{\pi}{8}\right)^2\right]$ | $f_2(s_2, 1)$ | 1 | 1 |
| 1/4 | $\frac{2}{\pi}\left[\left(\frac{3}{4}\right)^2 - \left(\frac{\pi}{8}\right)^2\left(\frac{5}{4}\right)^2\right]$ | $f_2(s_2, 1)$ | 1 | 1 |
| 1/2 | $\frac{2}{\pi}\left[\left(\frac{1}{2}\right)^2 - \left(\frac{\pi}{8}\right)^2\left(\frac{3}{2}\right)^2\right]$ | $f_2(s_2, 1)$ | 1 | 1 |
| 3/4 | $\frac{2}{\pi}\left[\left(\frac{1}{4}\right)^2 - \left(\frac{\pi}{8}\right)^2\left(\frac{7}{4}\right)^2\right]$ | $f_2(s_2, 1)$ | 1 | 1 |
| 1 | $-\frac{\pi}{8}$ | $f_2(s_2, 1)$ | 1 | 1 |
| | $p_2(s_2, x_2) = f_2(s_2, x_2)$ | $f_2^*(s_2)$ | $x_2^*$ | $s_3(x_2^*) = x_2^*$ |

For stage, 1, $s_1 = y_1$ must be zero by the end condition $y(t_1) = y(0) = 0$. Now, $x_1 = y_2$ may be any one of the five possible values (we allowed): $\{0, 1/4, 1/2, 3/4, 1\}$. The penalty for the interval $(t_1, t_2)$ is

$$p_1(s_1, x_1) = \frac{2}{\pi}\left\{(y_2 - 0)^2 - \left(\frac{\pi}{8}\right)^2 (y_2 - 0)^2\right\} = \frac{2}{\pi}\left[1 - \left(\frac{\pi}{8}\right)^2\right] y_2^2. \quad (9.7)$$

Given $f_2^*(s_2) = p_2(s_2, x_2)$, the total penalty from stage 1 to the end is

$$f_1(s_1, x_1) = p_1(s_1, x_1) + f_2^*(s_2(x_1))$$

$$= \frac{2}{\pi}\left[1 - \left(\frac{\pi}{8}\right)^2\right] y_2^2 + \frac{2}{\pi}\left[(1 - y_2)^2 - \left(\frac{\pi}{8}\right)^2 (1 + y_2)^2\right] \quad (9.8)$$

as we have only a two-stage process (for our choice of mesh points). The first term of this expression for $f_1(s_1, x_1)$ is positive because $1-(\pi/8)^2>0$. To minimize the contribution of this term to $f_1(s_1,x_1)$, we should make $y_2$ as small as possible (which is zero in our set up). However, making $y_2$ small would increase the contribution of the positive part $(2/\pi)(1-y_2)^2$ in the second term and decrease the contribution of the negative part $-(\pi/32)(1+y_2)^2$. In fact, we do better with $y_2=1$ (which gives $f_1=(2/\pi)[1-5(\pi/8)^2]$) than with $y_2=0$ (which gives $f_1=(2/\pi)[1-4(\pi/8)^2]$). If $y_2$ should be allowed to range over $[0,1]$, a simple calculation would show that $f_1$ as given by (9.8) attains its minimum at $y_2=[1+(\pi/8)^2]/2[1-(\pi/8)^2]<1$. But $y_2$ can only be one of the five allowed values; the actual minimum will have to be obtained from the backward recursion tableau for stage 1 as shown in Table 9.10b with the numerical values of $f_1$ for different values of $x_1$ calculated from (9.8). The optimal choice $\hat{y}_2=x_1^*=3/4$, compares favorably with the exact solution $\hat{y}(\pi/4)=\sin(\pi/4)=0.7071\ldots$ .

**Table 9.10b**

| $x_1=y_2$ / $s_1=y_1$ | 0 | 1/4 | 1/2 | 3/4 | 1 | | | |
|---|---|---|---|---|---|---|---|---|
| 0 | 0.2439 | 0.2384 | 0.0729 | 0.0420 | 0.1458 | 0.0420 | 3/4 | 3/4 |
| | $p_1(s_1,x_1)+f_2^*(s_2(x_1))=f_1^*(s_1,x_1)$ | | | | | $f_1^*(s_1)$ | $x_1^*$ | $s_2(x_1^*)$ |

The approximate minimum obtained by the backward recursion algorithm is seen from the tableau for stage 1 to be $J[y]\simeq 0.042$ with the corresponding approximate $y(x)$ given only at three points $\hat{y}(0)=0, \hat{y}(\pi/4)\simeq 3/4$, and $\hat{y}(\pi/2)=1$. The exact solution of the problem is $J[\hat{y}]=0$. Considering the crude approximations we have made to arrive at the approximate solution, the accuracy achieved by our dynamic programming solution is rather remarkable. As previously indicated, more accurate approximate solutions can be obtained using more mesh points, a larger (and possibly different) set of allowable values for each $y_n$, and a better approximation scheme for $y(x)$ and $y'(x)$ within each subinterval $(t_n, t_{n+1})$. If necessary, the calculations involved can be done on a programmable computer.

From the results of the simple example above, we see how the backward recursion solution technique may be applied to obtain an approximate solution of other basic problems of the calculus of variations. To distill the main ideas of this solution process, we let

$$f_n^*(s_n)=\min_y\left\{\int_{t_n}^{b}F(t,y,y^{\prime})\,dt\,\middle|\,y(t_n)=s_n, y(b)=B\right\}.$$

The following recursive relation is an immediate consequence of this definition:

$$f_n^*(s_n) = \min_{y, s_{n+1}} \left\{ f_{n+1}^*(s_{n+1}) + \int_{t_n}^{t_{n+1}} F(t, y, y') \, dt \right\}.$$

If we discretize the integral as before [with $\Delta t = (b - a)/N, t_n = a + n\Delta t$, and $s_j$ taking on only a discrete set of allowable values], then we have

$$f_n^*(s_n) = \min_{s_{n+1}} \left\{ f_{n+1}^*(s_{n+1}) + \Delta t F\left(t_n + \frac{1}{2}\Delta t, \frac{1}{2}[s_{n+1} + s_n], \frac{1}{\Delta t}[s_{n+1} - s_n]\right) \right\}. \tag{9.9}$$

Equation (9.9) is a discrete version of the Hamilton–Jacobi equation introduced earlier in Chapter 6. With $f_N^*(s_N) = f_N^*(B) = 0$, we calculate $f_{N-1}^*(s_{N-1})$ to be

$$f_{N-1}^*(s_{N-1}) = \Delta t F\left(t_{N-1} + \frac{1}{2}\Delta t, \frac{1}{2}[B + s_{N-1}], \frac{1}{\Delta t}[B - s_{N-1}]\right). \tag{9.10}$$

By recursion, we can then obtain $f_{N-2}^*(s_{N-2}), f_{N-3}^*(s_{N-3}), \dots, f_1^*(s_1) = f_1^*(A)$ as we did for the example $F = [(y')^2 - y^2]/2$. At each stage, the minimization process indicated in (9.9) must be carried out.

From (9.9), we have

$$f_{n+1}^*(s_{n+1}) - f_n^*(s_n) + \Delta t F\left(t_n + \frac{1}{2}\Delta t, \frac{1}{2}[s_{n+1} + s_n], \frac{1}{\Delta t}[s_{n+1} - s_n]\right) \geq 0 \tag{9.11}$$

which is a discrete version of the Hamilton–Jacobi inequality to be discussed in Chapter 12.

## 8. Exercises

1. By the graphical method, find the shortest route from $S$ to $T$ in Figure 9.6.

2. Solve Exercise 1 using backward recursion.

3. Use the graphical method to find the shortest route between $A$ and $B$ of the problem in Figure 9.7. Be sure to show your work for each of the stages. [Hint: It may help if fictitious cities with a zero cost (distance) are introduced at suitable places.]

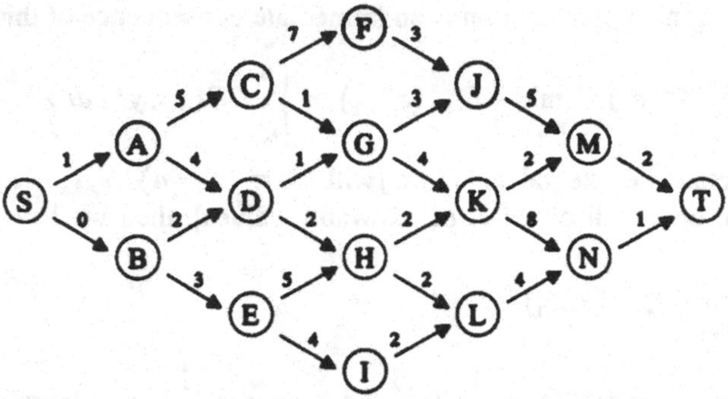


**Figure 9.6**

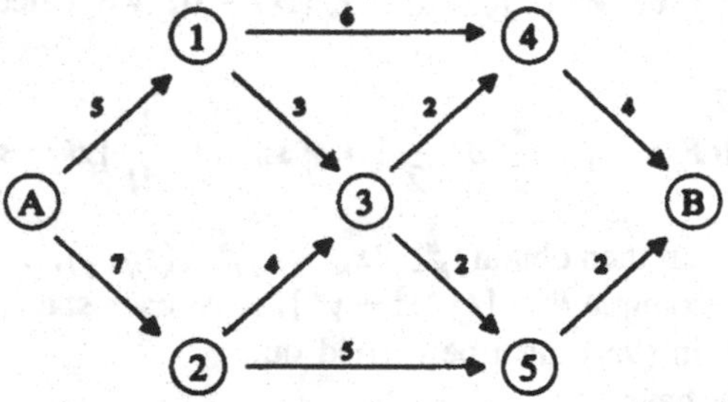


**Figure 9.7**

4. Consider a road network configuration of the type shown in Figure 9.6 but now of $N + 1$ stages where $N$ is an even integer. (Fig. 9.6 is of seven stages so that $N = 6$.)

   (a) Show that the brute force enumeration method requires $(N - 1)\binom{N}{N/2}$ additions and $\binom{N}{N/2} - 1$ comparisons.

   (b) Show that a total of $N + N^2/2$ additions and $N^2/4$ comparisons are needed for the dynamic programming solution.

   (c) Compare the additions and comparisons required for $N = 6$ and $N = 20$.

5. Solve by the backward recursion the following knapsack problem for $V = 17$:

| Cargo $i$ | $v_i$ | $q_i$ |
|---|---|---|
| 1 | 3 | 4 |
| 2 | 4 | 5 |
| 3 | 7 | 10 |
| 4 | 8 | 11 |
| 5 | 9 | 13 |

6. Solve the same knapsack problem (Exercise 5) by the intermediate capacity algorithm.

7. Suppose $A, B, C, D, E,$ and $F$ are $(4 \times 2)$, $(2 \times 3)$, $(3 \times 1)$, $(1 \times 2)$, $(2 \times 2)$, and $(2 \times 3)$ matrices.

   (a) Show that multiplying them from left to right to get a $(4 \times 3)$ matrix $G = (((((A \times B) \times C) \times D) \times E) \times F)$ requires 84 scalar multiplications.

   (b) Show that $H = (A \times (B \times (C \times (D \times (E \times F))))) = G$ requires only 69 scalar multiplication.

   (c) Find the optimal order of multiplication which requires only 36 scalar multiplications.

8. A company has three factories, all being considered for possible expansion. The total capital (in millions of dollars) available for expansion is 5. Each factory has a set of alternative plans for expansion as given below. Each plan involves a capital allocation $C$ for the factory and an estimated revenue $R$ from its expansion. Use a backward recursion to find an allocation of the available capital to the three factories which gives the highest total estimated revenue for this capital budgeting problem.
   [Note: There are three plans for factory 1, four plans for factory 2, and two plans for factory 3.]

| | Factory 1 | | Factory 2 | | Factory 3 | |
|---|---|---|---|---|---|---|
| Plan | $C_1$ | $R_1$ | $C_2$ | $R_2$ | $C_3$ | $R_3$ |
| 1 | 0 | 0 | 0 | 0 | 0 | 0 |
| 2 | 1 | 5 | 2 | 8 | 1 | 3 |
| 3 | 2 | 6 | 3 | 9 | — | — |
| 4 | — | — | 4 | 12 | — | — |

9. Use a forward recursion algorithm to solve the following nonlinear programming problem:

$$\max_{x_i} [x_0 \equiv 7x_1^2 + 6x_1 + 5x_2^2 | x_1 + 2x_2 \leq 10, x_1 - 3x_2 \leq 9, x_i \geq 0].$$

10. A chemical plant *must purify* all 20,000 pounds of a certain chemical compound in a period of 4 weeks. Two methods are available for the job. Method I completely purifies $z$ pounds of the compound in a week at a cost of $\$0.6z^2$. On the other hand, method II purifies only 70% of $w$ pounds of the compound processed in a week at the lower cost of $\$0.12w^2$ (and leaves 30% of $w$ to be processed again during a subsequent week just like the yet unprocessed compound.) Use a backward recursion algorithm to determine a policy (specifying the amount of chemical processed by each of the two methods in each of the four weeks) which minimizes the total cost ot the operation. [Note: The problem involves a continuous-state variable and, unfortunately, you will have to slug it out with the help of your pocket calculator. Keep in mind also that a policy may not be optimal unless $z + w = 20{,}000$.]

11. Solve the following integer programming problem by dynamic programming (with a vector-state variable):

$$\max_{x_i} [x_0 \equiv 8x_1 + 7x_2 | 2x_1 + x_2 \leq 8, 5x_1 + 2x_2 \leq 15,$$
$$x_i = \text{non-negative integer}].$$

12. Solve the following integer-value nonlinear programming problem by dynamic programming:

$$\max_{x_i} [x_0 \equiv 3x_1(2 - x_1) + 2x_2(2 - x_2) | x_1 + x_2 \leq 3,$$
$$x_k = \text{non-negative integer}].$$

13. Solve Exercise 8 with the non-negative integer-value constraints replaced by $x_k \geq 0, k = 1,2$.

14. Use the method of dynamic programming to find a combination of non-negative integers $m_1, ..., m_4$ which gives the maximum value of $M \equiv (m_1 + 2)^2 + (m_2 m_3) + (m_4 - 5)^2$ with $m_1 + m_2 + m_3 + m_4 \leq 5$.

15. Solve the nonlinear programming problem of Section 6 by forward recursion.

16. Solve the basic variational problem with $F(t,y,y^{\bullet}) = [(y^{\bullet})^2 - y^2]/2$, $y(0) = 0$, and $y(3\pi/4) = 1/2$ by using a mesh of $\{0, \pi/4, \pi/2, 3\pi/4\}$ and the admissible set of values $\{0, 1/4, 1/2, 3/4, 1\}$ for $y(t)$ at the mesh points.

17. Find a dynamic programming solution for the brachistochrone problem with $y(0) = 0$ and $y(4) = -4$ using four stages and the set of admissible values $\{0, -1, -2, -3, -4\}$ for $y(t)$.

18. Consider the optimal flight-path problem of Chapter 1 (see Exercise 5 of Chapter 1) with $H = 2, y(0) = 0$, and $y(4) = -1$ and additional constraints $|y'| \leq 2$ and $y(x) \geq 0$ for $0 < x < 2$, and $y(x) \geq -1$ for $2 < x < 4$.

    (a) Use the dynamic programming formalism with four stages ($N = 4$) and integer-valued $y$ at the mesh points to approximate the optimal path and cost. [Hint: The constraints limit the possible values of $y$ between mesh points.]

    (b) What approximation would result from the additional condition $y(1) = 0$?

    (c) What are the difficulties in applying the trial function method to this problem?

# 10

# Isoperimetric Constraints

## 1. The Shape of the Hanging Chain

To analyze the shape of a chain hung from two pegs, we model the chain as a line of length $\gamma$ and constant mass per unit length $\rho$. The two ends of the line hang from two fixed points $(a, A)$ and $(b, B)$ in the $x, y$-plane with gravity in the negative $y$ direction [and *with* $\sqrt{(b-a)^2 + (B-A)^2} < \gamma$ and $b - a < \gamma$]. Let the hanging shape of the chain be described by the plane curve $y(x)$ with the coordinate axes adjusted so that $y(x) > 0$ for all $x$ in $[a, b]$ (see Fig. 10.1). From mechanics of deformable bodies, we expect the chain to hang in such a way that its potential energy is a minimum. A differential chain element of lenght $ds$ at $(x, y)$ has mass $\rho\, ds$ and potential energy (relative to $y = 0$) $\rho g y\, ds$. The total potential energy of the chain hanging in the shape $y(x)$ is given by

$$J[y] = \int_a^b \rho g y \sqrt{1 + (y')^2}\, dx. \tag{10.1}$$

The actual chain shape $\hat{y}(x)$ is expected to minimize $J[y]$ subject to the fact that the length of the chain is fixed so that

$$C[y] = \int_a^b \sqrt{1 + (y')^2}\, dx = \gamma \tag{10.2}$$

and to the fixed end conditions.

Many variational problems with integral constraints of the form (10.2) arise in conjunction with constant length restrictions. Queen Dido's problem described in Section 2, Chapter 1 is reputed to be the first problem of this type. For this reason, it is customary to call variational problems with integral *isoperimetric problems* and to call integral constraints of the form (10.2) *isoperimetric constraints*. Typically, we want to minimize

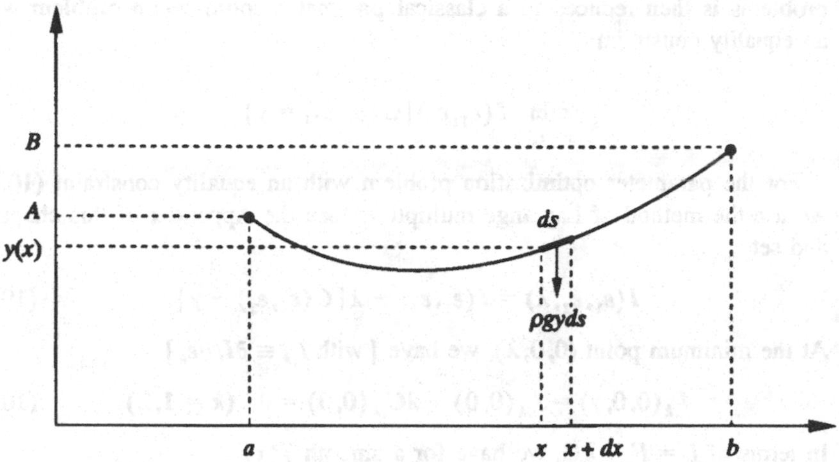


Figure 10.1

$$J[y] = \int_a^b F(x, y, y')\, dx \tag{10.3}$$

subject to the end constraints

$$y(a) = A, \quad y(b) = B \tag{10.4}$$

and the isoperimetric (or integral) constraints

$$C_i[y] = \int_a^b G(x, y, y')\, dx = \gamma_i \quad (i = 1, 2, \ldots, m). \tag{10.5}$$

For simplicity, we discuss mainly the case $m = 1$.

To obtain necessary conditions which help us to efficiently narrow down possible candidates for $\hat{y}$, it would be natural to proceed as before for the (unconstrained) basic problem by writing an admissible extremal as $y = \hat{y} + \varepsilon h$ to reduce $J[y]$ to $J(\varepsilon)$ and $C(\varepsilon)$, respectively. Unfortunately, this approach is inappropriate for the present problem. The comparison function $\hat{y} + \varepsilon h$ generally does not satisfy the isoperimetric constraint even if $\hat{y}$ does. To ensure that (10.5) is satisfied (for $m = 1$), we introduce a second parameter which may be chosen to meet the constraint without restricting the first parameter. More specifically, suppose $\hat{y}(x)$ is the actual solution of our problem. Let

$$y(x) = \hat{y}(x) + \varepsilon_1 h_1(x) + \varepsilon_2 h_2(x), \tag{10.6}$$

where $h_i(x)$ is any comparison function with $h_i(a) = h_i(b) = 0, i = 1, 2$. In that case, we have $J[y] \equiv J(\varepsilon_1, \varepsilon_2)$ and $C[y] \equiv C(\varepsilon_1, \varepsilon_2)$. The variational problems is then reduced to a classical parameter optimization problem with an equality constraint:

$$\min_{(\varepsilon_1,\varepsilon_2)} \{J(\varepsilon_1,\varepsilon_2) \mid C(\varepsilon_1,\varepsilon_2) = \gamma\}. \tag{10.7}$$

For the parameter optimization problem with an equality constraint (10.7), we use the method of Lagrange multipliers (see the Appendix of this chapter) and set

$$I(\varepsilon_1,\varepsilon_2,\lambda) = J(\varepsilon_1,\varepsilon_2) - \lambda[C(\varepsilon_1,\varepsilon_2) - \gamma]. \tag{10.8}$$

At the minimum point $(0,0,\lambda)$, we have [with $I_{,k} \equiv \partial I/\partial\varepsilon_k$]

$$I_{,k}(0,0,\gamma) = J_{,k}(0,0) - \lambda C_{,k}(0,0) = 0 \quad (k = 1,2). \tag{10.9}$$

In terms of $L = F - \lambda G$, we have for a smooth $\widehat{y}(x)$

$$\begin{aligned} I_{,k}(0,0,\lambda) &= \int_a^b [L_{,y}(x,\widehat{y},\widehat{y}')h_k + L_{,y'}(x,\widehat{y},\widehat{y}')h_k']\,dx \\ &= \int_a^b \left[\widehat{L}_{,y} - \frac{d}{dx}(\widehat{L}_{,y'})\right] h_k(x)\,dx = 0, \end{aligned} \tag{10.10}$$

where $\widehat{f}(x) \equiv f(x,\widehat{y}(x),\widehat{y}'(x))$. We obtain from (10.10) the same Euler DE

$$\widehat{L}_{,y} = (\widehat{L}_{,y'})' \tag{10.11}$$

for both $k = 1$ and $k = 2$.

The general solution of the second-order ODE (10.11) contains two constants of integration and the unknown constant (Lagrange multiplier) $\lambda$. These three constants are to be chosen to satisfy the two end conditions (10.4) and the isoperimetric constraint (10.5). Note that for the purpose of obtaining (10.11), we may omit the term $-\gamma$ in (10.8) for $I$. However, by including $-\gamma$ in $I$, we may formally recover the prescribed constraint from the stationarity of $I$ with respect to $\lambda$, $I_{,\lambda} = 0$.

For the hanging chain problem, we have

$$L = \rho g y\sqrt{1 + (y')^2} - \lambda\sqrt{1 + (y')^2}\,.$$

As $L$ does not depend on $x$ explicitly, the Euler DE (10.11) has a first integral $\widehat{L} - \widehat{y}'\widehat{L}_{,y'} = c_1$ or, for the given $L$,

$$\frac{\rho g\widehat{y} - \lambda}{\sqrt{1 + (\widehat{y}')^2}} = c_1. \tag{10.12}$$

This first-order ODE can be rearranged to read

$$\frac{d\widehat{y}}{\sqrt{(\rho g\widehat{y}-\lambda)^2-c_1^2}}=\frac{1}{c_1}\,dx$$

which can be integrated to give

$$\widehat{y}=\frac{\lambda}{\rho g}+\frac{c_1}{\rho g}\cosh\left(\frac{\rho g x}{c_1}+c_2\right) \tag{10.13}$$

where $c_1$ and $c_2$ are two constants of integration.

To simplify our calculations for $c_1$ and $c_2$, we limit ourselves to the case $a=0$, $b=1$, and $B=A$. For this case, we have

$$\begin{aligned}\widehat{y}(0)&=\frac{\lambda}{\rho g}+\frac{c_1}{\rho g}\cosh(c_2)=A,\\ \widehat{y}(1)&=\frac{\lambda}{\rho g}+\frac{c_1}{\rho g}\cosh\left(\frac{\rho g}{c_1}+c_2\right)=A.\end{aligned} \tag{10.14}$$

The two conditions in (10.14) imply

$$\cosh(c_2)=\cosh\left(\frac{\rho g}{c_1}+c_2\right)$$

or

$$c_2=-\frac{\rho g}{2c_1} \tag{10.15}$$

and therewith

$$\widehat{y}=\frac{\lambda}{\rho g}+\frac{c_1}{\rho g}\cosh\left[\frac{\rho g(2x-1)}{2c_1}\right]. \tag{10.16}$$

We have not exhausted the content of (10.14) and will return to these two conditions later after we have satisfied the isoperimetric constraint (10.2). With $\sqrt{1+(\widehat{y}')^2}=[\rho g\widehat{y}-\lambda]/c_1$, condition (10.2) becomes

$$\begin{aligned}\gamma&=\int_a^b\sqrt{1+(\widehat{y}')^2}\,dx=\int_0^1\cosh\left(\frac{\rho g\{2x-1\}}{2c_1}\right)dx\\ &=\left[\frac{c_1}{\rho g}\sinh\left(\frac{\rho g\{2x-1\}}{2c_1}\right)\right]_0^1=\frac{2c_1}{\rho g}\sinh\left(\frac{\rho g}{2c_1}\right)\end{aligned}$$

which may be written as

$$\gamma t=\sinh(t),\quad t=\frac{\rho g}{2c_1}. \tag{10.17b}$$

Because the length of the chain $\gamma$ is a positive number ($> b - a = 1$), there is exactly one $t(>0)$ which satisfies the above equation. We take it to be $\rho g/2c_1 = \hat{t}$ (with $c_2 = -\hat{t}$); then (10.16) becomes

$$\hat{y}(x) = \frac{\lambda}{\rho g} + \frac{1}{2\hat{t}}\cosh[\hat{t}(2x-1)].$$

Finally, we can determine $\lambda$ from the first equation of (10.14)

$$\hat{y}(0) = \frac{\lambda}{\rho g} + \frac{1}{2\hat{t}}\cosh(-\hat{t}) = A.$$

Hence, the unique admissible extremal for our problem is

$$\hat{y}(x) = A + \frac{1}{2\hat{t}}[\cosh(\hat{t}\{2x-1\}) - \cosh(\hat{t})] \tag{10.18}$$

with $J[\hat{y}]$ obtained by substituting the above expression into (10.1) and evaluating the integral.

For $\gamma \leq 1$ (and $b - a = 1$), whe have $\hat{t} = 0$, and $\hat{y} = \lambda/\rho g$ so that the chain is a straight horizontal line.

## 2. Normal Isoperimetric Problems and a Duality

Consider the problem of minimizing

$$J[y] = \int_0^1 (y')^2 dx \tag{10.19}$$

with

$$y(0) = 0, \qquad y(1) = 1, \tag{10.20}$$

and

$$C[y] \equiv \int_0^1 \sqrt{1 + (y')^2}\, dx = \gamma. \tag{10.21}$$

The Lagrangian of $I[y] \equiv J[y] - \lambda(C[y] - \gamma)$ is independent of $x$ and $y$; hence, the smooth extremals of $I[y]$ are linear functions $\hat{y} = c_1 x + c_0$. The end conditions determine the two constants to be $c_1 = 1$ and $c_0 = 0$. Note that the admissible extremal $\hat{y}(x) = x$ does not depend on the multiplier $\lambda$. For this admissible extremal, we have

$$C[\hat{y}] = \int_0^1 \sqrt{1 + (\hat{y}')^2}\, dx = \sqrt{2}.$$

It follows that the isoperimetric constraint can only be met if $\gamma = \sqrt{2}$. We conclude by [IV.2] that $J[y]$ has at least a weak local minimum at $\widehat{y} = x$ if $\gamma = \sqrt{2}$ with the Lagrange multiplier $\lambda$ not determined by the solution process. If $\gamma \neq \sqrt{2}$, the smooth extremal $\widehat{y}(x)$ does not satisfy the constraint and is not a solution of the problem. With $\widehat{y} = x$ being an extremal of $C[y]$ [and therefore $(0,0)$ being a stationary point of $C(\varepsilon_1, \varepsilon_2)$], the problem appears to be analogous to the abnormal case in parameter optimization (see Appendix of this chapter). We should, therefore, attempt to identify and characterize the abnormal isoperimetric problems in a way similar to what has been done for parameter optimization.

With (10.6) transforming the variational problem into a problem of classical optimization given by (10.7), the theory for such a problem (see Appendix at the end of this chapter) indicates that we should work with

$$I_0[y, \lambda, \lambda_0] \equiv \lambda_0 J[y] - \lambda(C[y] - \gamma) \equiv I_0(\varepsilon_1, \varepsilon_2; \lambda, \lambda_0). \qquad (10.22)$$

Proposition [X.B] in the Appendix of this chapter then tells us that if $(0,0)$ is not a stationary point of $C(\varepsilon_1, \varepsilon_2)$, we may set $\lambda_0 = 1$ and proceed with the usual solution process of the multiplier method. In particular, there will be an appropriate value $\lambda$ which renders $I_0$ ( = I) stationary at the expected minimum point of our isoperimetric problem, $(\widehat{\varepsilon}_1, \widehat{\varepsilon}_2) = (0,0)$. We rephrase this result in the language of the calculus of variations as follows:

**[X.1]** ***Let $\widehat{y}$ be a smooth admissible comparison function which minimizes $J[y]$ subject to the isoperimetric constraint $C[y] = \gamma$. If $\widehat{y}$ is not an extremal of $C[y]$, then the problem is normal and we may take $\lambda_0 = 1$ in (10.22). Moreover there exists a value $\lambda$ for which $\{\widehat{y}(x,\lambda), \lambda\}$ satisfies the Euler DE***

$$\frac{d}{dx}(\widehat{F}_{,y'} - \lambda \widehat{G}_{,y'}) = (\widehat{F}_{,y} - \lambda \widehat{G}_{,y}), \qquad (10.23)$$

***and the isoperimetric constraint $C[\widehat{y}] = \gamma$.***

For the hanging chain problem of the last section, the extremal (10.16) [with $c_1$ chosen so that $\widehat{y}(0) = A$] is not an extremal for $C[y]$. The constraint functional $C[y]$ has only extremals in the form of linear functions. Hence, we have a normal problem and, by [X.1], may set $\lambda_0 = 1$ to work with (10.8) as we did.

On the other hand, the admissible extremal $\widehat{y}(x) = x$ of $I_0[y; \lambda, \lambda_0]$ for the example (10.19)-(10.21) is also an extremal of $C[y]$ because $G(x, y, y') = \sqrt{1 + (y')^2}$ is independent of both $x$ and $y$. In the context of the corresponding parameter optimization setting, $(\widehat{\varepsilon}_1, \widehat{\varepsilon}_2) = (0,0)$ is a stationary point of $C(\hat{\varepsilon}_1, \hat{\varepsilon}_2) = C[\hat{y} + \varepsilon_1 h_1 + \varepsilon_2 h_2]$. It is, therefore, consistent to call the original calculus of variations problem an *abnormal* problem. Analogous to the parameter optimization situation, we need to set $\lambda_0 = 0$ to have a solution for an abnormal problem in the calculus of variations. For the variational problem (10.19)-(10.21), $\hat{y}(x)$ is also an extremal of $J[y]$ (or $(0,0)$ is a stationary point of $J(\varepsilon_1, \varepsilon_2) \equiv J[\hat{y} + \varepsilon_1 h_1 + \varepsilon_2 h_2]$). In that ease neither $\lambda_0$ nor $\lambda$ is determined by the problem which is consistent with [X.B] of the Appendix of this chapter.

The same argument applies to the general case (10.3)-(10.5) to obtain the following result:

**[X.2]** ***If a smooth extremal $\hat{y}(x)$ minimizes $J[y]$ locally subject to the isoperimetric constraint $C[y] = \gamma$, then $\hat{y}(x)$ is a smooth extremal of $I_0$ of (10.22) for two real multipliers $\lambda_0$ and $\lambda$, not both zero. For the normal case, [X.1] applies. For the abnormal case, we take $\lambda_0 = 0$ unless $\hat{y}$ is also an extremal of $J[y]$, in which case neither $\lambda_0$ nor $\lambda$ is specified.***

For the problem (10.19)-(10.21), $\hat{y}(x) = x$ satisfies the constraint $C[y] = \gamma$ only if $\gamma = \sqrt{2}$. It is not difficult to see that the problem has no solution if $\gamma < \sqrt{2}$ because the shortest path between the points $(0,0)$ and $(1,1)$ is $\sqrt{2}$. For $\gamma > \sqrt{2}$, the solution may exist in the form of a polygonal path; PWS extremals between the end points with total path length being $\gamma$ should be considered.

For a normal isoperimetric problem with only one constraint, we may write

$$I[\hat{y}, \lambda] = \int_a^b \left\{F(x, \hat{y}, \hat{y}') + \lambda\left[\frac{\gamma}{b-a} - G(x, \hat{y}, \hat{y}')\right]\right\} dx \equiv \hat{I}(\gamma), \qquad (10.24)$$

where $\hat{y} = \hat{y}(x;\gamma)$ and $\lambda = \lambda(\gamma)$ depend on the parameter $\gamma$ (as well as $a, b, A$ and $B$) with $\hat{y}' = \partial\hat{y}/\partial x$ whenever we treat $\hat{y}$ as a function of two variables $x$ and $\gamma$. We now differentiate $I$ with respect to $\gamma$ and integrate by parts to get

$$\frac{d\hat{I}}{d\gamma} = \int_a^b \left\{\left[(F_{,y} - \lambda\hat{G}_{,y}) - (F_{,y'} - \lambda\hat{G}_{,y'})'\right]\frac{\partial\hat{y}}{\partial\gamma}\right.$$

$$\left. + \frac{d\lambda}{d\gamma}\left[\frac{\gamma}{b-a} - G(x, \hat{y}, \hat{y}')\right]\right\} dx + \lambda = \lambda, \qquad (10.25)$$

because the first group of bracketed terms vanishes by the Euler DE and the remaining integral vanishes because of the isoperimetric constraint. We have from (10.25) the following interpretation of $\lambda$ as in classical optimization problems with constraints.

**[X.3]** ***The Lagrange multiplier λ is the rate of change (or marginal value) of the minimum J[y] with respect to γ.***

Let $j$ be the minimum value of $J[y]$ subject to $C[y] = \gamma$. The relation

$$j = J[\hat{y}] - \lambda(C[\hat{y}] - \gamma) \qquad (10.26)$$

may be written as

$$\gamma = C[\hat{y}] - \bar{\lambda}(J[\hat{y}] - j), \qquad \bar{\lambda} = \frac{1}{\lambda}. \qquad (10.27)$$

The integral $C[y] - \bar{\lambda}J[y]$ is the negative of $J[y] - \lambda C[y]$ except for a multiplicative factor which does not affect its extremals. The minimum associated with the Lagrangian $F - \lambda G$ (for $J[y] - \lambda C[y]$), therefore, corresponds to the maximum associated with the Lagrangian $G - \bar{\lambda}F$ (for $C[y] - \bar{\lambda}J[y]$). The following dual (or reciprocal) relation is an immediate consequence of this observation:

**[X.4]** ***If $\hat{y}$ minimizes (maximizes) $J[y]$ subject to $C[y] = \gamma$ with $J[\hat{y}] = j$, then $\hat{y}$ also maximizes (minimizes) $C[y]$ subject to $J[y] = j$.***

As an example of this duality, the solution $\hat{y}(x)$ of the problem of a hanging chain of fixed length also maximizes the length of a hanging chain of fixed potential energy.

The duality described in [X.4] also applies to classical optimization problems with an equality constraint. For example, the minimum of $J(z_1, z_2) \equiv z_1^2 + z_2^2$ subject to $C(z_1, z_2) \equiv z_1 z_2 = \frac{1}{2}\,(=\gamma)$ is attained at $\hat{z}_1 = \hat{z}_2 = 1/\sqrt{2}$ with $j \equiv J(\hat{z}_1, \hat{z}_2) = 1$. It is not difficult to verify that $(\hat{z}_1, \hat{z}_2) = (1/\sqrt{2}, 1/\sqrt{2})$ also maximizes $z_1 z_2$ subject to $z_1^2 + z_2^2 = 1\ (=j)$ with the maximum being $\hat{z}_1\hat{z}_2 = \frac{1}{2}\,(=\gamma)$.

## 3. Eigenvalue Problems and Mechanical Vibration

Boundary-value problems (BVP) in ODE are known to be more difficult to analyze than initial-value problems. The more complex nature of BVP can be seen in part from the following simple linear homogeneous problem:

$$y'' + \lambda y = 0, \qquad y(0) = 0, \qquad y(1) = 0. \qquad (10.28)$$

The general solution of the differential equation is

$$y = c_1\cos(\sqrt{\lambda}x) + c_2\sin(\sqrt{\lambda}x). \tag{10.29}$$

The *end condition* $y(0) = 0$ requires $c_1 = 0$. If, instead of $y(1) = 0$ as in (10.28), $y'(0) = 0$ were prescribed (so that we have an initial-value problem (IVP), then we would have $c_2 = 0$ as well. In that case, the only solution of the IVP would be the trivial solution $y(x) \equiv 0$. For problem (10.28), the end condition $y(1) = 0$ requires

$$c_2\sin(\sqrt{\lambda}) = 0 \tag{10.30}$$

which can be satisfied by $c_2 = 0$ so that we again have the trivial solution. However, the homogeneous boundary condition at $x = 1$ is automatically satisfied if $\lambda = n^2\pi^2$ for any integer $n$ (which may be taken to be positive with no loss in generality). In that case, there is no restriction on $c_2$ so that there is a one-parameter family of solutions denoted by $y_n(x) = c_n\varphi_n(x)$, where $c_n$ is an arbitrary constant and $\varphi_n(x) \equiv \sqrt{2}\sin(n\pi x)$. The values of $\lambda$ for which a nontrivial solution of the linear homogeneous boundary-value problem is possible are called *eigenvalues* of the problem. The function $\varphi_n(x)$ is called the *eigenfunction* associated with the eigenvalue $\lambda_n = n^2\pi^2$; the multiplicative factor $\sqrt{2}$ is chosen (for convenience) to normalize $\varphi_n(x)$ so that

$$\int_0^1 \varphi_n^2(x)\,dx = 1. \tag{10.31}$$

The determination of the (infinitely many positive) values of $\lambda$ for which the given homogeneous BVP has a nontrivial solution is called an *eigenvalue problem*.

The eigenfunctions $\{\varphi_n(x)\}$ of (10.28) have a number of interesting properties which are important in applications. For $m \neq n$, we have by direct verification

$$\int_0^1 \sin(m\pi x)\sin(n\pi x)\,dx = 0 \qquad (m \neq n). \tag{10.32}$$

We call this the *orthogonality* property of the eigenfunctions.

For each eigenvalue $\lambda_n$ of problem (10.28), there is only one eigenfunction (up to a multiplicative constant factor). We contrast this to the situation where the same ODE is supplemented by the two auxiliary conditions $y(0) = y(1)$ and $y'(0) = y'(1)$ instead. The general solution of the new BVP is the same as in (10.29). For $\lambda \neq 0$, the two periodic boundary conditions require

$$\begin{aligned} c_1 &= c_1\cos(\sqrt{\lambda}) + c_2\sin(\sqrt{\lambda}), \\ c_2 &= -c_1\sin(\sqrt{\lambda}) + c_2\cos(\sqrt{\lambda}). \end{aligned} \tag{10.33}$$

From these conditions, the eigenvalues are seen to be $\lambda_n = (2n\pi)^2$, $n = 1,2,3,\ldots$. (Note that $\lambda_n = n^2\pi^2$ will not do if $n$ is an odd integer.) For each $n$, the solution of the BVP is

$$y_n(x) = c_{1n}\cos(2n\pi x) + c_{2n}\sin(2n\pi x) \tag{10.34}$$

for arbitrary $c_{1n}$ and $c_{2n}$. Two different eigenfunctions, $\varphi_n^{(1)} \equiv \sqrt{2}\cos(2n\pi x)$ and $\varphi_n^{(2)} \equiv \sqrt{2}\sin(2n\pi x)$, are obtained by setting $c_{2n}$ and $c_{1n}$ equal to zero, respectively.

Other important properties of eigenfunctions are concerned with their zeros, i.e., where they vanish inside the solution interval [which is $(0,1)$ for the present problem]. It is well known that $\sin(n\pi x)$ has $(n-1)$ zeros in $(0,1)$ and that these zeros are simple (so that the derivative does not vanish there also). It is not difficult to see from the properties of $\sin(n\pi x)$ that the zeros of $\varphi_n(x)$ and $\varphi_{n-1}(x)$ interlace. For example, the zeros of $\varphi_3(x)$ are at $x = 1/3$ and $2/3$, whereas the zeros of $\varphi_4(x)$ are at $x = 1/4$, $1/2$, and $3/4$ with $1/4 < 1/3 < 1/2 < 2/3 < 3/4$.

Finally, it is known from the theory of Fourier series that, with minimal restrictions (to be described below), a function $f(x)$ defined on the solution interval of our problem with $f(0) = f(1) = 0$ can be expressed as a linear combination of the eigenfunctions of our homogeneous boundary-value problem:

$$f(x) = \sum_{n=1}^{\infty} a_n\varphi_n(x). \tag{10.35}$$

The appropriate Fourier coefficients $\{a_n\}$ in the representation (10.35) are determined very efficiently by the orthogonality property (10.32) to be

$$a_n = \int_0^1 f(x)\varphi_n(x)\,dx \tag{10.36}$$

where we have used the fact that the eigenfunctions are normalized [see (10.31)].

The equality in the *Fourier expansion* (10.35) is known to hold pointwise for all PWS functions $f(x)$ with $f(0) = f(1) = 0$. If $f(x)$ is PWS except for a finite jump discontinuity at $\bar{x}$, the Fourier series converges to $[f(\bar{x}+) + f(\bar{x}-)]/2$ at $\bar{x}$. For a function $f(x)$ which is only square-integrable, its Fourier series representation (10.35) and (10.36) holds only in the sense of *convergence in the mean*, i.e.,

$$\lim_{m\to\infty} \int_0^1 [f(x) - S_m(x)]^2\,dx = 0, \tag{10.37}$$

where the *partial sum* $S_m(x)$ is given by

$$S_m(x) = \sum_{n=1}^{m} a_n \varphi_n(x). \tag{10.38}$$

The convergence in the mean (10.37) is often abbreviated as the *limit in the mean* (l.i.m.):

$$\lim_{m \to \infty} S_m(x) = f(x). \tag{10.39}$$

Among their many applications, eigenvalue problems arise naturally in mechanical vibration including the free and forced vibration of strings and membranes [see Chapter 9 of Wan, (1989)]. For example, the problem of the small-amplitude free vibration of a uniform string of unit length (which lies at rest along the interval $0 \le x \le 1$ prior to $t = 0$) is governed by the linear partial differential equation (PDE)

$$w_{,tt} = c^2 w_{,xx} \qquad (0 < x < 1, t < 0), \tag{10.40}$$

where $(\ )_{,z}$ denotes $\partial(\ )/\partial z$, $w(x,t)$ is the transverse displacement of the point $x$ along the string at time $t$, and $c$ is the speed of signal transmission associated with the particular string material. For a string held fixed at both ends, we have

$$w(0,t) = w(1,t) = 0 \qquad (t > 0). \tag{10.41}$$

We may seek a solution of the problem in the form

$$w(x,t) = [a\cos(\omega t) + b\sin(\omega t)]u(x)$$

for some frequency $\omega$ and amplitude factors $a$ and $b$ (in expectation of a simple harmonic motion at any given point along the length of the string). In that case, the PDE (10.40) becomes an ODE for $u(x)$:

$$u'' + \lambda u = 0, \qquad \lambda = \frac{\omega^2}{c^2}, \tag{10.42}$$

and the end conditions (10.41) become

$$u(0) = u(1) = 0. \tag{10.43}$$

The homogeneous BVP defined by (10.42) and (10.43) has only a trivial solution if $\lambda$ is not one of its eigenvalues. As we do expect the string to vibrate, we want to know all the eigenvalues and eigenfunctions of the problem. This information allows us to see (among other things) at what frequencies ($\omega$) a point on the string can oscillate. From the results for problem (10.28), we are assured of the existence of an infinite number of possible nontrivial solutions; each corresponds to a different frequency. For a particular allowed frequency $\omega_n = n\pi c$ (corresponding to $\lambda_n = n^2\pi^2$), the eigenfunction

$$\varphi_n(x) = \sqrt{2}\sin(n\pi x)$$

gives the relative magnitude of the displacements experienced by the different points of the string and, hence, the deformed shape of the string at any instant in time. For this reason, different eigenfunctions describing different modes of vibration are known as the *normal modes* of the vibrating string. The locations of the zeros of a specific mode do not move during the vibratory motion and are known as *nodes* of the vibration. Different modes have a different number of nodes and no two nodes of consecutive modes coincide. Finally, the *expansion theorem* (10.35) (to be stated more precisely later) tells us that any vibratory motion of a string is a linear combination (i.e., superposition) of the modes available. Hence, the analysis for an understanding of the small-amplitude vibration of strings is reduced to a complete characterization of the eigenvalues and eigenfunctions. In actual application, the string is set in motion by some initial displacement and initial velocity so that

$$w(x,0) = w_0(x) \quad \text{and} \quad w_{,t}(x,0) = v_0(x). \tag{10.44}$$

For arbitrary $w_0(x)$ and $v_0(x)$, these two conditions cannot be satisfied by any particular normal mode of vibration

$$w_n(x,t) = [a_n\cos(\omega_n t) + b_n\sin(\omega_n t)]\varphi_n(x). \tag{10.45}$$

However, the linearity of the problem allows for a superposition of any number of modes. The linear combination

$$w(x,t) = \sum_{n=1}^{\infty}\{a_n\cos(\omega_n t) + b_n\sin(\omega_n t)\}\varphi_n(x) \tag{10.46}$$

clearly satisfies the boundary conditions (10.41) and, assuming termwise differentiability, also the PDE (10.40). The two initial conditions now require

$$\sum_{n=1}^{\infty} a_n\sin(n\pi x) = w_0(x), \qquad \sum_{n=1}^{\infty}\omega_n b_n\sin(n\pi x) = v_0(x). \tag{10.47}$$

The expansion theorem of (10.35) assures us that (10.47) can be satisfied and (10.37) tells us how to compute $\{a_n\}$ and $\{b_n\}$.

## 4. Variational Formulation of Sturm–Liouville Problems

The homogeneous BVP (10.28) is a special case of the general *Sturm–Liouville problem* defined by the linear ODE

$$-[p(x)y']' + q(x)y = \lambda r(x)y \quad (a < x < b) \tag{10.48}$$

and the boundary conditions

$$\alpha_a y(a) + \beta_a y'(a) = 0, \qquad \alpha_b y(b) + \beta_b y'(b) = 0. \tag{10.49}$$

In (10.48), $p(x)$ is a positive $C^1$ function and both $q(x)$ and $r(x)$ are at least continuous with $r(x) > 0$ in $[a, b]$. To simplify the presentation, our development of the corresponding variational problem will, at first, be limited to the simpler boundary conditions

$$y(a) = 0, \qquad y(b) = 0. \tag{10.50}$$

The simple example (10.28) corresponds to the case with $r = p = 1$ and $q = 0$. What we found for that problem suggests that the homogeneous BVP (10.48) and (10.49) may have nontrivial solutions. For many important applications in science and engineering, we want to determine the values of $\lambda$ for which this is the case as well as the corresponding nontrivial solutions. We are, therefore, interested in the eigenvalue problem associated with (10.48) and (10.49). We will not devote time here to develop the general theory for Sturm–Liouville problems for the purpose of obtaining specific properties such as those observed for (10.28). We will merely summarize in [X.5] the most important properties of (10.48) and (10.49) which we have already identified for the special case of (10.28):

**[X.5]** *For the eigenvalue problem defined by (10.48) and (10.49), the following properties hold:*

*(i) There are a (countably) infinite sequence of real eigenvalues $\lambda_1, \lambda_2, \lambda_3, \ldots$, ordered in increasing magnitude with $\lambda_1$ denoting the smallest. The sequence increases without bound.*

*(ii) The eigenvalues are simple in that there is exactly one eigenfunction (determined up to a multiplicative factor) for each eigenvalue. The eigenfunctions may be normalized by*

$$\int_a^b r(x)\varphi_n^2(x)\,dx = 1. \tag{10.51}$$

*(iii) The eigenfunctions are mutually orthogonal relative to $r(x)$ so that*

$$\int_a^b \varphi_n(x)\varphi_m(x)r(x)\,dx = 0 \quad (n \neq m). \tag{10.52}$$

*(iv) $\varphi_n(x)$ has $(n-1)$ simple zeros.*

*(v) The zeros of $\varphi_n(x)$ and $\varphi_{n+1}(x)$ interlace.*

*(vi) The (normalized) eigenfunctions $\{\varphi_n(x)\}$ form a complete set so that any square integrable function $f(x)$ may be represented by a linear*

*combination of the eigenfunctions in the sense of convergence in the mean with respect to the weight function* $r(x)$

$$\lim_{n\to\infty}\int_a^b r(x)\left[f(x)-\sum_{n=1}^{m} a_n\varphi_n(x)\right]^2 dx = 0 \tag{10.53}$$

*with coefficients* $\{a_n\}$ *determined by (10.51) and (10.52) to be*

$$a_n = \int_a^b f(x)\varphi_n(x)r(x)\,dx. \tag{10.54}$$

A detailed development of this theory can be found in Churchill and Brown (1978) and will not be pursued here. Instead, we will discuss a variational formulation of the Sturm–Liouville problem. It is generally not possible to obtain an explicit solution of a Sturm–Liouville problem in terms of simple functions. A variational formulation of the same problem will enable us to obtain approximate solutions of the problem by the direct methods of Chapter 8 and, more importantly, to deduce qualitative information about the solution.

In many applications involving vibration and stability of deformable bodies, the most important piece of information required is the lowest (smallest) eigenvalue. For most problems, this information allows the designer to (re-) design a system with a sufficiently high $\lambda_1$ so that it is not reached under normal operating conditions. Therefore it is important that we have the following variational characterization for the lowest eigenvalue:

**[X.6]** *The lowest eigenvalue* $\lambda_1$ *of (10.48) and (10.50) is the minimum of*

$$J[y] \equiv \int_a^b [p(y')^2 + qy^2]\,dx \tag{10.55}$$

*subject to the end constraints*

$$y(a) = y(b) = 0 \tag{10.56}$$

*and the isoperimetric constraint*

$$C[y] \equiv \int_a^b r(x)y^2 dx = 1. \tag{10.57}$$

*The corresponding extremal* $\hat{y}$ *is the eigenfunction* $\varphi_1(x)$*, normalized with respect to* $r(x)$ *by (10.57).*

To prove [X.6], suppose $\hat{y}$ minimizes $J[y]$ subject to the given constraints. By [X.1], $\hat{y}$ satisfies (10.23) with

$$L(x,y,y';\lambda) = F(x,y,y') - \lambda G(x,y,y')$$
$$= p(y')^2 + qy^2 - \lambda r y^2, \tag{10.58}$$

for some multiplier $\lambda$. Upon substituting (10.58) into (10.23) and keeping in mind the end conditions $\hat{y}(a) = \hat{y}(b) = 0$, we conclude that $\hat{y}(x)$ is a nontrivial solution, i.e., an eigenfunction, of the eigenvalue problem (10.48) and (10.50) with the multiplier $\lambda$ being the corresponding eigenvalue. This eigenvalue will be shown presently to be equal to $\hat{J} = J[\hat{y}]$.

Multiply the ODE (10.48) for $\hat{y}$ by $\hat{y}$ and integrate the result over $[a,b]$ by parts to get

$$-[p\hat{y}'\hat{y}]_a^b + \int_a^b [p(\hat{y}')^2 + q\hat{y}^2]\,dx = \lambda \int_a^b r\hat{y}^2\,dx. \tag{10.59}$$

With the end conditions $\hat{y}(a) = \hat{y}(b) = 0$ and the isoperimetric constraint (10.57), the relations (10.59) simplifies to

$$\lambda = \int_a^b [p(\hat{y}')^2 + q\hat{y}^2]\,dx = \hat{J}. \tag{10.60}$$

As $\hat{J} = J[\hat{y}]$ is the minimum value attained by $J[y]$ subject to (10.56) and (10.57), $\lambda = \hat{J}$ is necessarily the *smallest* eigenvalue possible and $\hat{y}$ the corresponding normalized eigenfunction so that $(\hat{y},\lambda) = (\varphi_1,\lambda_1)$. More formally, let $(\varphi,\lambda)$ be any other eigenpair, with $\varphi$ normalized as in (10.51). Then

$$J[\varphi] = \int_a^b [p(\varphi')^2 + q\varphi^2]\,dx$$
$$= [(p\varphi')\varphi]_a^b + \int_a^b \{-(p\varphi')' + q\varphi\}\varphi\,dx$$
$$= \int_a^b \lambda r\varphi^2\,dx = \lambda. \tag{10.61}$$

Because $J[\hat{y}]$ is the minimum of all $J[y]$ with the same given constraints, we must have $\hat{J} \le \lambda$. Hence, we have $\hat{J} = \lambda_1$ and $\hat{y} = \varphi_1(x)$ as asserted by [X.6].

As a by-product of the development above, we now know that the eigenpairs $(\varphi_n,\lambda_n)$, $n = 1,2,\ldots$, of (10.48) and (10.50) render the performance index $J[y]$ of (10.55) stationary subject to the end conditions (10.56) and the isoperimetric constraint (10.57).

## 5. The Rayleigh Quotient

For additional theoretical development as well as the efficient application of direct methods, it is desirable to cast the isoperimetric problem of [X.6] for the lowest eigenpair of our Sturm–Liouville problem in a different form. If we do not normalize the minimizing extremal $\widehat{y}(x)$, the relation (10.59) may be written as

$$\lambda_1 = \frac{J[\widehat{y}]}{C[\widehat{y}]}. \tag{10.62}$$

The normalization condition (10.57) is not essential for the determination of the lowest eigenvalue. This is evident from the solution process for the simple Sturm–Liouville problem (10.28). It can also be seen from the fact that $\widehat{y}$ may be adjusted by an amplitude factor without altering the value of $\lambda_1$ because both $J[y]$ and $C[y]$ are quadratic in $y$. The ratio $J[y]/C[y] \equiv R[y]$ is called the *Rayleigh quotient* for the Sturm–Liouville problem. The following result due to Lord Rayleigh is an expected consequence of these observations:

**[X.7]** ***The lowest eigenvalue $\lambda_1$ of the Sturm–Liouville problem, (10.48) and (10.50), is the minimum of the Rayleigh quotient*** $R[y] \equiv J[y]/C[y]$ ***over all nontrivial admissible comparison functions.***

To prove [X.7], suppose $\widehat{y}$ minimizes $R[y]$ with $\rho = R[\widehat{y}]$ and $y = \widehat{y} + \varepsilon h$ so that $R[y] = R(\varepsilon)$. It is a straightforward calculation to show

$$\left.\frac{dR}{d\varepsilon}\right|_{\varepsilon=0} = \frac{2}{C(0)}\int_a^b [-(p\widehat{y}')' + q\widehat{y} - \rho r\widehat{y}]\,h\,dx = 0,$$

where $C(0) = C[\widehat{y}] > 0$. The usual argument requires that $\widehat{y}$ be the solution of (10.48) with $\lambda = \rho$ [and, of course, we must have $\widehat{y}(a) = \widehat{y}(b) = 0$ for $\widehat{y}(x)$ to be admissible]. In other words, the minimum point $\widehat{y}(x)$ of $R[y]$ is an eigenfunction of the Sturm–Liouville problem and the minimum value $\rho$ of the Rayleigh quotient is the corresponding eigenvalue.

It remains to show that $\rho$ is the lowest eigenvalue. For any other eigenpair $(\varphi, \lambda)$, we have from (10.48)

$$\int_a^b \varphi\{-(p\varphi')' + q\varphi\}\,dx = \int_a^b \lambda r(x)\varphi^2\,dx. \tag{10.63}$$

The left-hand side of (10.63) may be integrated by parts to give

$$[-\varphi p\varphi']_a^b + \int_a^b [p(\varphi')^2 + q\varphi^2]\,dx = \int_a^b [p(\varphi')^2 + q\varphi^2]\,dx$$

so that, after observing the end conditions $\varphi(a) = \varphi(b) = 0$, relation (10.63) becomes

$$\int_a^b [p(\varphi')^2 + q\varphi^2]\,dx = \lambda\int_a^b r\varphi^2\,dx$$

or

$$\lambda = \frac{J[\varphi]}{C[\varphi]} = R[\varphi] \geq \rho$$

because $\rho = R[\hat{y}] = \min R[y]$. In other words, any eigenvalue cannot be smaller than $\rho$ and, therefore, $\rho$ is the lowest eigenvalue $\lambda_1$ of the Sturm–Liouville problem.

One immediate application of the Rayleigh quotient is to obtain upper and lower bounds for the lowest eigenvalue of the Sturm–Liouville problem. Any admissible comparison function $\bar{y}$ will give an *upper bound* for $\lambda_1$ because

$$R[\bar{y}] \geq R[\hat{y}] = \rho = \lambda_1 . \tag{10.64}$$

To obtain a lower bound, we construct a *comparison Sturm–Liouville problem* defined by

$$-[\tilde{p}(x)y']' + \tilde{q}(x)y = \tilde{\lambda}\tilde{r}(x)y \tag{10.65a}$$

and (10.56) with

$$p(x) \geq \tilde{p}(x) > 0, q(x) \geq \tilde{q}(x), \text{ and } 0 < r(x) \leq \tilde{r}(x) \tag{10.65b, c, d}$$

for all $x$ in $[a,b]$. It follows immediately from (10.65b)-(10.65d) that

$$\int_a^b [\tilde{p}(y')^2 + \tilde{q}y^2]\, dx \leq \int_a^b [p(y')^2 + qy^2]\, dx \tag{10.66}$$

and

$$\int_a^b [\tilde{r}(x)y^2]\, dx \geq \int_a^b [r(x)y^2]\, dx . \tag{10.67}$$

Let $\tilde{R}[y]$ be the Rayleigh quotient associated with the comparison problem and $\tilde{\lambda}_1$ its minimum value attained at $\tilde{y}(x)$. Inequalities (10.66) and (10.67) imply

$$\tilde{R}[y] = \frac{\int_a^b [\tilde{p}(y')^2 + \tilde{q}y^2]\, dx}{\int_a^b \tilde{r}y^2\, dx} \leq \frac{\int_a^b [p(y')^2 + qy^2]\, dx}{\int_a^b ry^2\, dx} = R[y] \tag{10.68}$$

for any admissible comparison function $y(x)$. In particular, we have

$$\tilde{R}[\hat{y}] \leq R[\hat{y}] = \lambda_1 . \tag{10.69}$$

Inequality (10.69) is still not useful because we normally do not know $\hat{y}$ (which is the reason why we need bounds). An additional observation which makes it useful is the fact that $\tilde{R}[\hat{y}] \geq \tilde{\lambda}_1$ the minimum value of $\tilde{R}[y]$. Altogether, we have the following result which gives a lower bound for $\lambda_1$:

**[X.8]** ***The minimum Rayleigh quotient*** $\bar{R}[y]$ ***of the comparison problem, as defined by (10.65) and (10.56), is not less than*** $\lambda_1$ **so that**

$$\bar{\lambda}_1 \le \lambda_1. \tag{10.70}$$

For the bound (10.70) to be useful, we must find a comparison problem which admits a simple solution. A simple solution is likely if the ODE (10.65) is of constant coefficients. Hence, we should take

$$\bar{p} = \min_x p(x), \qquad \bar{q} = \min_x q(x), \qquad \bar{r} = \max_x r(x), \tag{10.71}$$

so that (10.65a) becomes

$$y'' + (\bar{\lambda} r_0 - q_0)y = 0, \tag{10.72}$$

where $q_0 = \bar{q}/\bar{p}$ and $r_0 = \bar{r}/\bar{p}$. Its exact solution which satisfies $\bar{y}(a) = 0$ is

$$\bar{y}(x) = C\sin(\mu[x-a]), \qquad \mu^2 = \bar{\lambda} r_0 - q_0.$$

The boundary condition $\bar{y}(b) = 0$ then takes the form

$$\bar{y}(b) = C\sin(\mu[b-a]) = 0,$$

which requires $\mu(b-a) = n\pi$ for a nontrivial solution. In terms of $\bar{\lambda}$, this condition becomes

$$\mu_n^2 = \bar{\lambda}_n r_0 - q_0 = \frac{n^2\pi^2}{(b-a)^2} \quad \text{or} \quad \bar{\lambda}_n = \frac{1}{r_0}\left[\frac{n^2\pi^2}{(b-a)^2} + q_0\right].$$

The lowest eigenvalue of the comparison problem is then

$$\bar{\lambda}_1 = \frac{1}{\bar{r}}\left[\frac{\bar{p}\pi^2}{(b-a)^2} + \bar{q}\right]. \tag{10.73}$$

By [X.8], we know the lowest eigenvalue of the actual Sturm–Liouville problem of interest is larger than $\bar{\lambda}_1$; the latter then serves as a lower bound for $\lambda_1$.

## 6. Higher Eigenvalues

Having obtained a variational characterization for the lowest eigenvalue of the Sturm–Liouville problem, (10.48) and (10.50), we would like to do something similar for the higher eigenvalues of the same problem. We know already from the proof of [X.6] that any eigenpair $(\varphi_n, \lambda_n)$ renders the $J[y]$ of (10.55) stationary, subject to (10.56) and (10.57). However, it is more desirable to have a minimum principle for the Rayleigh quotient similar to [X.7]. In principle, this task is reasonably straightforward. By [X.5], the eigenfunctions are mutually orthogonal relative to $r(x)$. This suggests that to get $\lambda_2$, we only have to find the minimum of the Rayleigh quotient $R[y] = J[y]/C[y]$ over all admissible comparison functions [with $y(a) = y(b) = 0$] which are orthogonal to $\varphi_1(x)$ relative to $r(x)$. More generally, we have

**[X.9]** *$\lambda_n$ is the minimum of $R[y]$ take over all comparison functions with $y(a) = y(b) = 0$ and*

$$\int_a^b ry\varphi_k \, dx = 0 \quad (k < n). \tag{10.74}$$

The proof of this assertion will be left as an exercise.

The result in [X.9] is impractical as we often do not have $\varphi_k$, $k = 1, 2, \ldots, (n-1)$ exactly (which is one of the main reasons why we are interested in a variational formulation of the Sturm–Liouville problem). If we only know some of these eigenfunctions approximately, then the orthogonality constraints on $y$ would not be of much use to us. Fortunately, the need to know the eigenfunctions for the lower eigenvalues exactly may be circumvented with the help of the following result:

**[X.10]** *Let $\bar{\lambda}_2$ be the minimum Rayleigh quotient $R[y]$ subject to*

$$\int_a^b ryz \, dx = 0 \tag{10.75}$$

*for some admissible comparison function $z(x)$. Then we have*

$$\bar{\lambda}_2 \leq \lambda_2. \tag{10.76}$$

To see this, we let $Y(x) = \alpha\varphi_1 + \beta\varphi_2$ and choose $\beta$ so that (10.75) is satisfied for $y = Y$. For this particular $Y(x)$, we have

$$R[Y] = \frac{[pY'Y]_a^b + \int_a^b [-(pY')' + qY] Y \, dx}{\int_a^b rY^2 \, dx}$$

$$= \frac{\int_a^b [\alpha\lambda_1 r\varphi_1 + \beta\lambda_2 r\varphi_2][\alpha\varphi_1 + \beta\varphi_2] \, dx}{\int_a^b r(\alpha^2\varphi_1^2 + 2\alpha\beta\varphi_1\varphi_2 + \beta^2\varphi_2^2) \, dx} = \frac{\lambda_1\alpha^2 + \lambda_2\beta^2}{\alpha^2 + \beta^2} \leq \lambda_2.$$

Because $\bar{\lambda}_2$ is the minimum of $R[y]$ over all admissible $y$ which satisfies (10.75), we have $\bar{\lambda}_2 \leq R[Y]$ and hence (10.76).

The inequality (10.76) says that $\lambda_2$ is an upper bound for $\bar{\lambda}_2$. In that case, $\lambda_2$ should be the maximum of $\bar{\lambda}_2$ over all test functions $z$. This establishes

**[X.11] (The max-min principle for $\lambda_2$)**

$$\lambda_2 = \max_{z}\left[\min_{y}\{R[y] \mid y(a)=y(b)=0, \int_a^b r(x)yz\,dx = 0\}\right]. \qquad (10.77)$$

The argument leading to [X.11] can be extended to establish

**[X.12] (The max-min principle for Higher Eigenvalues)**

$$\lambda_{n+1} = \max_{(z_1,\ldots,z_n)} [\tilde{\lambda}_{n+1}], \qquad (10.78)$$

*where, for a given set of* $(z_1,\ldots,z_n)$,

$$\tilde{\lambda}_{n+1} = \min_{y}\left\{R[y] \mid y(a)=y(b)=0, \int_a^b ryz_j\,dx = 0, j=1,\ldots,n\right\}.$$

The proof is left as an exercise.

## 7. Mixed End Conditions

So far, we have limited our discussion of Sturm–Liouville problems to those with the simple end conditions (10.50). Analogous results can be obtained for the general (regular) Sturm–Liouville problems with the mixed end conditions (10.49).

For the analogue of [X.6], we see from (10.29) that the integral (10.55) is not the appropriate performance index for the general problem (10.48) and (10.49) (when $\alpha_a \neq 0$ or $\alpha_b \neq 0$); the first term in (10.59) cannot be eliminated as $y(x)$ no longer vanishes at the end points. To get the appropriate $J[y]$ (for $\beta_a \neq 0$ and $\beta_b \neq 0$), we use the end conditions (10.49) to eliminate $y'(a)$ and $y'(b)$ from the boundary terms of (10.59) to get

$$\left[\frac{\alpha}{\beta}py^2\right]_a^b + \int_a^b [p(y')^2 + qy^2]\,dx = \lambda\int_a^b ry^2\,dx, \qquad (10.79)$$

where $\alpha/\beta = \alpha_a/\beta_a$ at $x=a$ and $\alpha/\beta = \alpha_b/\beta_b$ at $x=b$. We take the left-hand side of (10.79) as the performance index of the new problem and set the integral on the right equal to unity as the isoperimetric constraint:

$$J[y] \equiv \int_a^b [p(y')^2 + qy^2]\,dx + \left[\frac{\alpha}{\beta} py^2\right]_a^b, \tag{10.80}$$

$$C[y] \equiv \int_a^b ry^2\,dx = 1 \tag{10.81}$$

with

$$\frac{\alpha}{\beta} = \begin{cases} \dfrac{\alpha_a}{\beta_a} & (x = a) \\[2ex] \dfrac{\alpha_b}{\beta_b} & (x = b) \end{cases} \tag{10.82}$$

It is not difficult to verify that the stationary condition

$$\delta I = \delta J - \lambda\,\delta C = 0$$

requires

$$[p(\hat{y}' + \frac{\alpha}{\beta}\hat{y})\delta y]_a^b + \int_a^b [-(p\hat{y}')' + q\hat{y} - \lambda r\hat{y}]\,\delta y\,dx = 0. \tag{10.83}$$

Note that $\delta y$ is not required to vanish at the end points as $y$ is not prescribed there. From (10.83) follow (10.48) as the Euler DE and (10.49) as the Euler BCs. Hence, the extremal which minimizes (10.80) subject to (10.81) is a solution of the Sturm–Liouville problem (10.48) and (10.49), normalized relative to $r(x)$.

It is now possible to prove the following analogue of [X.6]:

**[X.6′]** ***The lowest eigenvalue $\lambda_1$ of (10.48) and (10.49) is the minimum of (10.80) subject to the isoperimetric constraint (10.57). The corresponding extremal $\hat{y}$ is the associated eigenfunction $\varphi_1(x)$ normalized with respect to $r(x)$.***

The proof of this result is similar to that for [X.6] and will be left as an exercise. Results analogous to [X.7] and [X.12] can also be established similarly, with (10.80) replacing (10.55) in all cases. The elimination of $\hat{y}'$ in the boundary term of (10.59) is a critical step for all these results.

## 8. Optimal Harvesting of a Uniform Forest

In this section, we discuss an isoperimetric problem in forest management which has some unusual features. A forest of trees of uniform age and with a total tree biomass of $W_0$ units is to be harvested at the rate of $h(t)$ units of biomass per unit time. The lumber harvested at time $t$ can be sold at the price of $p(t)$ dollars per unit of biomass. At a very young age, a tree is not worth much. Its unit price increases with age but at a decreasing rate until the onset of biological decay. Beyond that point, the unit price may actually decline.

A harvesting cost $c(t, h)$ is incurred for each unit of tree biomass harvested. The graph of $c$ as a function of $h$ is generally U shaped. At low harvest rate, $c$ is high due to the fixed setup cost. At high harvest rate, $c$ is also high because of overtime pay and excessive wear on the machinery.

Suppose logging begins at $t = t_1$ with the entire forest completely harvested at $t = t_2$. In that case we have

$$\int_{t_1}^{t_2} h(t)\,dt = W_0. \tag{10.84}$$

If there is no replanting or alternative use of the forest land after harvest, the present value of the total profit from harvesting the entire forest is given by

$$P[h] = \int_{t_1}^{t_2} e^{-rt}[p(t) - c(t, h)]h\,dt, \tag{10.85}$$

where $r$ is the constant discount rate. We wish to choose the harvest rate $h(t)$, the starting time $t_1$, and the terminal time $t_2$ to maximize $P[h]$. The maximization is subject to the isoperimetric constraint (10.84). Note that the value of $h(t)$ is not specified at the two ends $t_1$ and $t_2$ other than the general requirement of $h(t) \geq 0$.

Neither the Lagrangian $F(t, h, h') = e^{-rt}[p(t) - c(t_1, h)]h$ nor the integrand $G(t_1, h, h') = h$ of the integral constraint (10.84) depends on $h'$. The augmented Lagrangian $F - \lambda G$ is, therefore, of type $d$ in the classification of Section 4 in Chapter 1. We saw there that the corresponding Euler DE is

$$F_{,h} - \lambda G_{,h} = e^{-rt}[p(t) - c(t, h) - hc_{,h}(t, h)] - \lambda = 0 \tag{10.86}$$

which can, in principle, be solved for $h$ to get $h = h(t; \lambda)$. Because $F - \lambda G$ does not depend on $h'$ and $P[h]$ contains no terminal payoff, there is also no Euler BC at the two end times $t_1$ and $t_2$ even when $h$ is not prescribed there.

The two end points being free, [III.3] can be invoked to require that the transversality condition (3.21) on the augmented Lagrangian $F - \lambda G$ be satisfied at both $t_1$ and $t_2$. For the present problem, these two conditions are simply

$$F(t_1, h(t_1)) - \lambda G(h(t_1)) = 0, \tag{10.87a}$$

$$F(t_2, h(t_2)) - \lambda G(h(t_2)) = 0. \tag{10.87b}$$

Equations (10.87a) and (10.87b) and the isoperimetric constraint (10.84) provide the three conditions for the determination of the three unknown constants $t_1$ and $t_2$ and $\lambda$ of the problem.

The example above illustrates again that variational problems with a Lagrangian of the form $F(x,y)$ do have meaningful applications. This type of problems even occurs with an isoperimetric constraint and free end points. A special case of the forest management problem discussed below offers additional challenge and requires some further analysis.

For the case in which $c$ does not vary with $h$, the total discounted profit (10.85) becomes

$$P[h] = \int_{t_1}^{t_2} e^{-rt} v(t) h(t)\, dt, \tag{10.88}$$

where $v(t) \equiv p(t) - c(t)$ is a known concave function. Formally, the Euler DE now takes the form $e^{-rt}v(t) - \lambda = 0$ for $t_1 < t < t_2$. This requirement is generally not met as $\lambda$ is a constant [whereas $e^{-rt}v(t)$ generally is not] and $h(t)$ is not involved in the "Euler DE."

When the capacity to harvest is abundant so that there is no restriction on the harvesting rate $h(t)$, the solution of our problem is evident from the simple dependence of the integrand of (10.88) on $h$. The maximum $P$ is attained by waiting until the instance $t_m$ when the discounted profit for a unit harvest $e^{-rt}v(t)$ reaches its maximum value and harvesting the entire forest at once at that instance. Note that $e^{-rt}v(t)$ attains its maximum when

$$[e^{-rt}v(t)]'_{t=t_m} = 0 \quad \text{or} \quad \frac{v'(t_m)}{v(t_m)} = r. \tag{10.89}$$

To harvest the entire forest at $t_m$, we formally set

$$h(t) = W_0 \delta(t - t_m), \tag{10.90}$$

where $\delta(\tau)$ is the Dirac delta function (see the last section of Chapter 8) so that (10.84) is satisfied. The optimal harvest time $t_m$ is the well-known *Fisher age* for trees of a once-and-for-all forest to be harvested for maximum discounted profit when instantaneous clear-cutting is feasible.

It is important to note that the solution obtained above is not PWS or continuous; in fact, $h(t) = W_0\delta(t - t_m)$ is not an ordinary function. We have then another example of meaningful and well-posed problems for which the calculus of variations for PWS functions developed in this volume is, strictly speaking, not adequate.

In practice, the harvesting capacity of a firm is limited by its logging crew size and available machinery. This sets an inequality constraint on the harvesting rate

$$0 \le h(t) \le h_m, \tag{10.91}$$

and the optimal instantaneous harvest strategy (10.90) is not feasible. What then is the optimal harvesting strategy? As in the case with $h_m = \infty$, the answer can again be obtained by an ad hoc argument with the correct solution involving a PWC harvesting rate. We will postpone the discussion of the solution to this case until Chapter 12.

## 9. Exercises

1. (a) Minimize $J(x,y) \equiv x^2 + y^2$ subject to $C(x,y) \equiv 4x^2 + 3y^2 = 12$.

   (b) What point(s) on the surface $z^2 + xy = 1$ is closest to the origin?

   (c) Find the maximum of $J(x,y,z) = x + 2y - 3z^2$ subject to

   $$C_1(x,y,z) \equiv x - y = 2 \text{ and } C_2(x,y,z) \equiv x + 2y = 4.$$

   (d) Minimize $J(x,y) = x^3 - \cos y$ subject to $C(x,y) \equiv x^2 + \sin^2 y = 0$.

2. (a) If $\nabla C(\widehat{x},\widehat{y}) \neq 0$, show that both $\nabla C$ and $\nabla J$ at a stationary point $\widehat{\mathbf{z}} = (\widehat{x},\widehat{y})$ are orthogonal to the same incremental change $d\mathbf{z}$ from $\widehat{\mathbf{z}}$.

   (b) Deduce from the result of (a) the multiplier rule in the form $\nabla\widehat{J} + \lambda\nabla\widehat{C} = 0$.

3. $$\min_y \left\{ J[y] = \int_0^1 [y']^2 dx \right\}$$

   subject to $y(0) = 0$, $y(1) = 2$, and $C[y] \equiv \int_0^1 y\,dx = L$.

4. [Queen Dido's problem]

   $$\max_y \{ J[\overline{y}] \equiv \int_0^b y\,dx \mid y(0) = 0,\ y(b) = 0, \text{ and}$$

   $$C[y] \equiv \int_0^b \sqrt{1 + (y')^2}\,dx = L \}.$$

5. (a) $\min_y \{ J[y] \equiv \int_0^b \sqrt{1 + y^2}\,dx \mid C[y] \equiv \int_0^b y\,dx = L\ (L > 0, b > 0) \}$.

   (b) $\max_y \{ J[\overline{y}] \equiv \int_0^1 (2y - y^2)\,dx \mid C[y] \equiv \int_0^1 xy\,dx = 1 \}$.

6. Consider a curve $y(x)$ joining two points $(a,A)$ and $(b,B)$ in the upper half-plane so that the solid of revolution generated by rotating the curve about the $x$-axis has a fixed surface area. Find $\widehat{y}(x)$ which maximizes the solid volume.

(a) Formulate the isoperimetric problem [but do *not* attempt to determine $\hat{y}(x)$ as it is expressed in terms of elliptic functions].

(b) Show that any extremal must be smooth.

(c) If $\rho$ is the radius of curvature at the point $(x, y)$ of an extremal, then prove that $\rho y =$ constant at each point of the extremal. [Hint: Set $\cos\psi = [1+(y')^2]^{-1/2}$ so that $d\psi/ds = 1/\rho$. Also use first integral of the Euler DE]

7. $\min_y \{J[y] \equiv \int_0^b e^{-rx} y\,dx \mid \int_0^b y^{1/2}\,dx = A\}.$

(a) Solve the isoperimetric problem by way of a Lagrange multiplier.

(b) Reformulate the problem as a basic problem without the isoperimetric constraint. [Hint: Set $z(x) \equiv \int_0^x \sqrt{y}\,dt$.]

8. $\min_y \{J[y] \equiv \int_0^1 (y')^2 dx \mid y(0) = y(1) = 0,\ C[y] \equiv \int_0^1 \sqrt{1+(y')^2}\,dx = L\}.$

(a) Show that the extremals of $I[y] \equiv J[y] - \lambda C[y]$ may have corner if $\lambda \geq 2$.

(b) Sketch (at least) three different possible nonsmooth extremals (for $\lambda > 2$). Show that $I[y] = -1 - \lambda^2/4$ in all cases.

(c) For $0 \leq \lambda \leq 2$, show that the only extremal possible is $\hat{y}(x) = 0$. For what range of values of $L$ is this solution applicable?

(d) For $0 \leq \lambda \leq 2$, show that $I[y] \geq I[0]$. [Hint: Set $y' = \tan\psi$ where $\psi = \psi(x)$.]

(e) For $\lambda > 2$, again set $y' = \tan\psi$ and show that $I[y]$ has a minimum value when $\sec\psi = \lambda/2$.

9. Given two points $(a,A)$ and $(b,B)$ in the $x, y$-plane, let $\Gamma$ be a fixed curve $\bar{y}(x)$ joining them. Among all (other) curves of length $L$ joining the same two points, find the curve which together with $\Gamma$ encloses the greatest area.

10. Find the admissible extremals of the following isoperimetric problem:

$$\min_y \{J[y] \equiv \int_0^1 [(y')^2 + x^2]\,dx \mid y(0) = 0,\ y(1) = 0,$$

$$\text{and } C[y] \equiv \int_0^1 y^2\,dx = 2\}.$$

11. The small-amplitude free vibration with frequency $\omega$ of a taut string fixed at both ends is governed by the eigenvalue problem

$$(Ty')' + \rho\omega^2 y = 0, \qquad y(0) = y(1) = 0,$$

where $\omega^2$ is the eigenvalue parameter, $\rho$ is the mass density per unit string length, and $T$ is the tension in the string.

(a) Obtain the Rayleigh quotient for this problem.

(b) Determine the natural frequencies $\{\omega_n\}$ (or the eigenvalues $\{\lambda_n = \omega_n^2\}$) and the normal modes (eigenfunctions) $\{\varphi_n(x)\}$ for the case of a uniform string [with constant $T$ and $\rho$].

(c) With the help of $\varphi_1(x)$ from part (b), obtain an approximate expression for the lowest frequency $\omega_1$ (or $\lambda_1 = \omega_1^2$) for a nonuniform string [*with* $T = T(x)$ and $\rho = \rho(x)$].

(d) For $T = T_0(1 + \alpha x)$ and $\rho_0(1 + \beta x)$, show that

$$\omega_1 \cong \frac{\pi}{\ell}\sqrt{\frac{\overline{T}}{\overline{\rho}}}$$

(with $\omega^2$ being the eigenvalue parameter) where $\overline{T}$ and $\overline{\rho}$ are the mean values of $T$ and $\rho$, respectively.

12. (a) Find the Euler DE which determines the extremals of

$$J[y] \equiv \int_a^b \{s(x)(y'')^2 - p(x)(y')^2 + q(x)y^2\}\,dx$$

subject to the isoperimetric constraint

$$\int_a^b r(x)y^2\,dx = 1,$$

and $y(a) = A$, $y(b) = B$, $y'(a) = U$, and $y'(b) = V$.

(b) Suppose there are no end conditions at $x = b$; what are the corresponding Euler boundary conditions?

(c) Obtain the relevant Rayleigh quotient for this problem.

13. The small amplitude $y(x)$ of an elastic beam in free harmonic vibration with frequency $\omega$ satisfies the differential equation

$$(EIy'')'' - \rho\omega^2 y = 0,$$

where $E$ is the Young's modulus of elasticity, I (not to be confused with $I[y]$) is the moment inertia of the cross section of the beam, and $\rho$ is the mass density per unit beam length.

(a) Determine the natural frequencies $\{\omega_n\}$ (or the eigenvalues $\omega_n^2 = \lambda_n$) and the normal modes (eigenfunctions) for the case of a uniform hinged beam so that $E$, $I$, and $\rho$ are constants and

$$y(0) = y''(0) = y(1) = y''(1) = 0.$$

(b) If $\rho$ and $I$ are both functions of $x$ (so that the beam is not uniform), obtain an approximate expression for $\omega_1^2$ using the first eigenfunction from part (a).

(c) For $I = I_0 x^m$ and $\rho = \rho_0 x^n$ (but $E$ remains constant), obtain from the results of part (*b*) an approximate value for $\omega_1^2$ for $m = 1$ and $n = 2$. What happens when $m = n = 1$?

14. Consider the Sturm–Liouville (eigenvalue) problems

(a) $y'' - xy = -\lambda_a y, \quad y(0) = y(1) = 0;$

(b) $y'' = -\lambda_b y, \quad y(0) = y(1) = 0.$

Show that the two minimal eigenvalues $\lambda_a$ and $\lambda_b$ satisfy

$$\lambda_b \le \lambda_a \le \lambda_b + 1.$$

Deduce that $9.87 \le \lambda_a \le 10.87$. Can you improve the upper bound on $\lambda_a$ (see Exercice 15)?

15. For the Sturm–Liouville problem $(py') - qy + \lambda ry = 0$, $y(a) = y(b) = 0$, we may apply the Rayleigh–Ritz trial function method to

$$I[y] \equiv J[y] - \bar{\lambda}C[y] = \int_a^b [p(y')^2 + qy^2]\,dx - \bar{\lambda}\int_a^b ry^2\,dx$$

for all $C^2$ functions satisfying the end constraints. For $p = 1$, $q = x$, $r = 1$, $a = 0$, and $b = 1$, let $\bar{y} = c_1\varphi_1(x) \equiv c_1 x(1 - x)$ so that $I[\bar{y}] = c_1^2(a_{11} - \bar{\lambda}b_{11})$. Calculate $a_{11}$ and $b_{11}$ and determine $\bar{\lambda}$ for a nontrivial solution for $\bar{y}$. (The exact solution for $\lambda_1$ is about 10.4; does $\bar{y}$ lead to a better upper bound for $\lambda_1$ than 10.87 obtained previously in Exercise 14?)

16. Show that the lowest eigenvalue of the Sturm–Liouville problem

$$\frac{d}{dx}\{x^2 y'\} = -\lambda x^2 y, \qquad y(1/2) = 0, \qquad y(1) = 0$$

satisfies the inequality $\lambda_1 \le 40$ using the trial function method with $\varphi_1(x) = (2x - 1)(1 - x)/x$. (The exact value of $\lambda_1$ is $4\pi^2 \simeq 39.5$.)

17. If, in the standard form of the Sturm–Liouville problem, we have $q = 0$, $a = 0$, $b = \ell$, and constant $p$ and $r$, let $\bar{y} = c_1\varphi_1(x) + c_2\varphi_2(x)$, where

$\varphi_1 = x(\ell - x)$ and $\varphi_2 = x(\ell^2 - x^2)$ so that $I[\bar{y}] = c^T(A - \bar{\lambda}B)c$ where $I[y]$ is as in Exercise 15 and $c = (c_1, c_2)^T$.

(a) Obtain the two 2 × 2 matrices $A$ and $B$.

(b) For $I(c_1, c_2)$ to be stationary, $c$ must be a nontrivial solution of (the generalized matrix eigenvalue problem) $(A - \bar{\lambda}B)c = 0$. Show that this is possible only if $\bar{\lambda} = \bar{\lambda}_1 = 10p/r\ell^2$ or $\bar{\lambda} = \bar{\lambda}_2 = 42p/r\ell^2$.

18. Reformulate the hanging chain problem of Section 3 in parametric form by taking $y = y(t)$ and $x = x(t)$. Solve the problem again in this form.

19. The formulation of Queen Dido's problem with $y$ as a function of $x$ may rule out curves which are multivalued functions in that formulation. Solve the problem again using a parametric formulation.

20. Obtain the admissible extremal for the abnormal case of

$$\min_y \{J[y] = \int_0^1 \frac{1}{2}[(y')^2 - y^2]\,dx \,|\, y(0) = 0, y(1) = 1 \text{ and}$$

$$C[y] \equiv \int_0^1 \sqrt{1 + (y')^2}\,dx = \sqrt{2}\}.$$

21. Prove the max-min principle for higher eigenvalues [X.12].

22. State and prove the analogue of [X.7] for the general Sturm–Liouville problem (10.48) and (10.49) for $\alpha_a \alpha_b \beta_a \beta_b \neq 0$.

23. State and prove the analogue of [X.11] for the same general Sturm–Liouville problem as in Exercise (22).

## Appendix. Lagrange Multipliers in Parameter Optimization

### a. The Normal Case

The analysis of the basic problem of the calculus of variations was seen in the earlier chapters to be based on a study of the corresponding classical parameter optimization problem. Similarly, the analysis of variational problems with isoperimetric or other equality point constraints has also been reduced to a study of parameter optimization with equality constraints. In this Appendix, we will review the method of Lagrange multipliers for equality constraints in parameter optimization.

Consider the problem of minimizing a function of $n$ variables $J[\mathbf{z}] \equiv J(z_1, z_2, \ldots, z_n)$, subject to $m$ ($< n$) equality constraints $C_i(z_1, z_2, \ldots, z_n) = \gamma_i$, $i = 1, 2, \ldots, m$ where $\{\gamma_1, \ldots, \gamma_m\}$ are known constants. These constants will be set to zero (as they may be included in the functions $\{C_i(\mathbf{z})\}$) except when we seek an interpretation of the Lagrange multipliers later in this Appendix. Hence, we will write

$$\min_{\mathbf{z}} \{J(\mathbf{z}) \mid C_i(\mathbf{z}) = 0, \ i = 1, 2, \ldots, m\} \tag{A.1}$$

with no restrictions on the range of $\mathbf{z}$. In principle, we can use the $m$ constraints to eliminate $m$ of the variables, say $z_{n-m+1}, \ldots, z_n$, to reduce $J(\mathbf{z})$ to a function of $n - m$ variables:

$$J(\mathbf{z}) \equiv \tilde{J}(z_1, \ldots, z_{n-m}).$$

We then have an ordinary minimization problem in the $n - m$ dimensional (parameter) space without constraints.

**EXAMPLE A:** Find the rectangular parallelopiped of fixed volume (and finite edge lengths) with a minimum total surface area.

For this problem, we have $J[\mathbf{z}] = 2(z_1 z_2 + z_2 z_3 + z_3 z_1)$ and $C_1(\mathbf{z}) \equiv z_1 z_2 z_3 - \gamma_1 = 0$, where $z_1$, $z_2$ and $z_3$ are the edges of the rectangular parallelopiped in three different directions and $\gamma_1$ is its volume. We can use the constraint to express $z_3$ as a function of $z_1$ and $z_2$, i.e., $z_3 = \gamma_1 / z_2 z_1$, and then eliminate $z_3$ from $J[\mathbf{z}]$ to get

$$J[\mathbf{z}] = 2\left[z_1 z_2 + \frac{\gamma_1}{z_1} + \frac{\gamma_1}{z_2}\right] \equiv \tilde{J}(z_1, z_2).$$

The usual calculations lead to a unique stationary point $\hat{z}_1 = \hat{z}_2 = \hat{z}_3 = \gamma_1^{1/3}$. The second-order conditions confirm that $J(\hat{\mathbf{z}}) = 6\gamma_1^{2/3}$ is a minimum.

Unfortunately, it is not always possible to solve the $m$ equality constraints explicitly for $m$ variables in terms of the remaining ones. For a more useful solution process, we note that we must have $dJ = 0$ at a minimum point $\hat{\mathbf{z}}$ of $J(\mathbf{z})$ where

$$dJ = J_{,1}(\hat{\mathbf{z}})\, dz_1 + \ldots + J_{,n}(\hat{\mathbf{z}})\, dz_n \tag{A.2}$$

with $(\ )_{,k} \equiv \partial(\ )/\partial z_k$. If there were no constraints, a necessary condition for $\hat{\mathbf{z}}$ to be a minimum point of $J(\mathbf{z})$ would be the vanishing of the coefficients of the different increments $\{dz_i\}$, i.e., $J_{,k}(\hat{\mathbf{z}}) = 0$ for all $k = 1, 2, \ldots, n$. [If $J_{,1}(\hat{\mathbf{z}}) \neq 0$, we can make $dJ < 0$ by choosing $dz_k = 0$, $k > 1$, and $dz_1$ to be such that $J_{,1}(\hat{\mathbf{z}})dz_1 < 0$.] With equality constraints, not all $J_{,k}(\hat{\mathbf{z}})$ have to vanish as the increments $\{dz_i\}$ can no longer be assigned independently. They are related by

$$\begin{aligned} dC_1 &= C_{1,1}(\hat{\mathbf{z}})dz_1 + \ldots + C_{1,n}(\hat{\mathbf{z}})dz_n = 0, \\ &\text{-------------------------} \\ &\text{-------------------------} \\ dC_m &= C_{m,1}(\hat{\mathbf{z}})dz_1 + \ldots + C_{m,n}(\hat{\mathbf{z}})dz_n = 0. \end{aligned} \tag{A.3}$$

Normally, we would solve the linear system (A.3) for $m$ of the $\{dz_i\}$ in terms of the others. For example, if the $m \times m$ matrix $[C_{i,j}]$ for $i = 1,\ldots,m$, $j = n - m + 1,\ldots,n$, is of rank $m$, we can solve for $dz_{n-m+1},\ldots,dz_n$ in terms of $dz_1,\ldots,dz_{n-m}$. We then use the results to eliminate $\{dz_{n-m+1},\ldots,dz_n\}$ from the expression for $dJ$ in (A.2) to obtain $dJ$ as a linear combination of $\{dz_1,\ldots,dz_{n-m}\}$:

$$dJ = \alpha_1(\hat{\mathbf{z}})dz_1 + \ldots + \alpha_{n-m}(\hat{\mathbf{z}})dz_{n-m}; \tag{A.4}$$

Because $dz_1,\ldots,dz_{n-m}$ are now arbitrary, we must have

$$\alpha_k(\hat{\mathbf{z}}) = 0, \quad k = 1,\ldots,(n-m), \tag{A.5}$$

at a minimum point $\hat{\mathbf{z}}$. The $(n - m)$ equations (A.5) and the $m$ constraints $C_i(\hat{\mathbf{z}}) \equiv \hat{C}_i = 0$ can now be solved simultaneously to determine the point(s) $\hat{\mathbf{z}}$. If necessary, this can be done numerically, by Newton's method for example.

**EXAMPLE A (continued):**

For this problem, we have $n = 3$, $m = 1$, and

$$dJ = 2(\hat{z}_2 + \hat{z}_3)dz_1 + 2(\hat{z}_1 + \hat{z}_3)dz_2 + 2(\hat{z}_1 + \hat{z}_2)dz_3 = 0,$$

$$dC_1 = \hat{z}_2\hat{z}_3 dz_1 + \hat{z}_1\hat{z}_3 dz_2 + \hat{z}_1\hat{z}_2 dz_3 = 0.$$

Upon solving the second for $dz_3$ and then using the result to eliminate $dz_3$ from the first, we obtain

$$\hat{z}_2\left(1 - \frac{\hat{z}_3}{\hat{z}_1}\right)dz_1 + \hat{z}_1\left(1 - \frac{\hat{z}_3}{\hat{z}_2}\right)dz_2 = 0.$$

This requires $\alpha_1(\hat{\mathbf{z}}) = \hat{z}_2(1 - \hat{z}_3/\hat{z}_1) = 0$ and $\alpha_2(\hat{\mathbf{z}}) = \hat{z}_1(1 - \hat{z}_3/\hat{z}_2) = 0$. For a rectangular parallelopiped with edges of finite length, the unique solution for these two conditions is again $\hat{z}_1 = \hat{z}_2 = \hat{z}_3 = \gamma_1^{1/3}$.

By working with $dJ$ and $\{dC_k\}$, we replaced the task of solving nonlinear equations by solving a linear system instead. However, the $m$ increments $\{dz_{n-m+1},\ldots,dz_n\}$ often do not all appear in the $m$ equations (A.3). A proper choice of $m$ increments to be eliminated will have to be selected by examining (A.3) for each specific problem. It is, therefore, desirable to avoid the intermediate step of solving for $m$ of the increments (or $m$ unknowns in the first approach) and leave any solution of simultaneous equations to the end, especially when it has to be done numerically. This is accomplished with the help of Lagrange multipliers, one for each constraint. We will introduce the notion of a *Lagrange multiplier* first through problems with a single constraint.

Consider the problem of minimizing $J(z_1,z_2)$ subject to a single constraint $C(z_1,z_2)=\gamma$. We will keep $\gamma \neq 0$ for this special case to obtain an interpretation of the Lagrange multiplier. We assume $\widehat{C}_{,2} \equiv C_{,2}(\widehat{\mathbf{z}}) \neq 0$. If $\widehat{C}_{,2}=0$ and $\widehat{C}_{,1} \neq 0$, we simply relabel the variables. (The *abnormal* case with $\widehat{C}_{,k}=0$ for both $k=1$ and $k=2$ will be discussed later.) Imagine the relation $C(\mathbf{z})=\gamma$ is solved for $z_2$ in terms of $z_1$ to get $z_2=\varphi(z_1,\gamma)$. Upon expressing $z_2$ in terms of $z_1$ in $J(\mathbf{z})$ and $C(\mathbf{z})$, we have

$$\left.\frac{dJ}{dz_1}\right|_{z_1=\hat{z}_1}=\left[J_{,1}+J_{,2}\frac{d\varphi}{dz_1}\right]_{z_1=\hat{z}_1}=0 \tag{A.6a}$$

and

$$C_{,1}(\mathbf{z})+C_{,2}(\mathbf{z})\frac{d\varphi}{dz_1}=0 \quad \text{or} \quad \frac{d\varphi}{dz_1}=-\frac{C_{,1}(\mathbf{z})}{C_{,2}(\mathbf{z})}. \tag{A.6b}$$

After using (A.6b) to eliminate $d\varphi/dz_1$ from (A.6a), we obtain

$$\frac{\widehat{J}_{,1}}{\widehat{J}_{,2}}=\frac{\widehat{C}_{,1}}{\widehat{C}_{,2}} \tag{A.7a}$$

or

$$\frac{\widehat{J}_{,1}}{\widehat{C}_{,1}}=\frac{\widehat{J}_{,2}}{\widehat{C}_{,2}}. \tag{A.7b}$$

The ratio on the right-hand side of (A.7a) is the negative of the slope of the tangent of the curve $C(\mathbf{z})=\gamma$ at $\widehat{\mathbf{z}}$. The ratio on the left side is the negative of the slope of the tangent of the level curve $J(z_1,z_2)=J(\mathbf{z})$ at $\mathbf{z}=\widehat{\mathbf{z}}$. The condition of stationarity stated in the form (A.7a) says that, given the constraint, the minimum (or maximum) attainable level curve of $J(\mathbf{z})$ is tangent to the constraining curve at the stationary point (see Fig. 10.2).

If we denote the common value of the two ratios in (A.7b) by the constant $\lambda$, the condition of stationarity (A.7b) itself can be stated as the two relations

$$\widehat{J}_{,k}-\lambda\widehat{C}_{,k}=0 \quad (k=1,2). \tag{A.8}$$

The two conditions (A.8) and the constraint $C(\widehat{\mathbf{z}})=\gamma$ determine $\widehat{z}_1$, $\widehat{z}_2$, and $\lambda$. [If there are restrictions on the range of $\mathbf{z}$ and $\widehat{\mathbf{z}}$ so obtained lies outside the

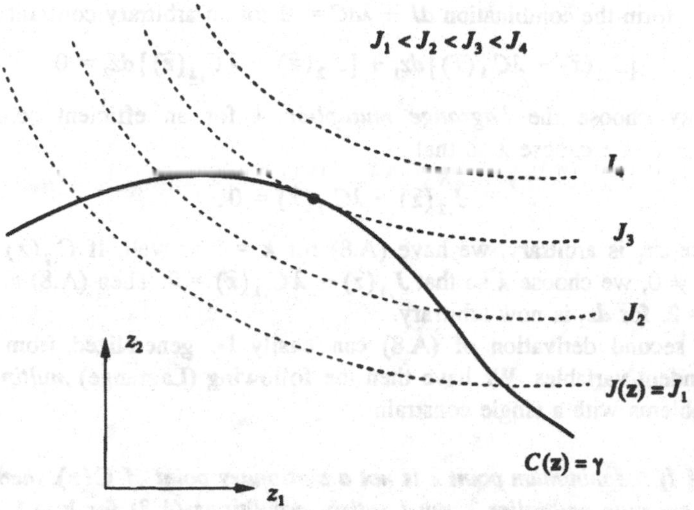


Figure 10.2

allowable range, the actual minimum point must lie on the boundary of this domain with (A.6a) replaced by $dJ/dz_1 > 0$ at the minimum point $\widehat{\mathbf{z}}$.]

EXAMPLE B: $J = z_1 - z_2$, with $C(z) \equiv z_1^2 + z_2^2 = \gamma (> 0)$.

Conditions (10.8) for this case are

$$1 - 2\lambda \widehat{z}_1 = 0, \quad -1 - 2\lambda \widehat{z}_2 = 0,$$

which requires $\lambda \neq 0$. These two conditions can be combined to give

$$2\lambda (\widehat{z}_1 + \widehat{z}_2) = 0.$$

Because $\lambda \neq 0$, we have $\widehat{z}_2 = -\widehat{z}_1$. The constraint then requires $2\widehat{z}_1^2 = \gamma$ and therewith $\widehat{z}_1 = -\widehat{z}_2 = -\sqrt{\gamma/2}$. [By the second-order necessary conditions, the other possible solution for $\widehat{z}_1^2 = \gamma/2$ corresponds to a maximum $J(\mathbf{z})$]. Either condition in (A.8) then gives $\lambda = -1/\sqrt{2\gamma}$.

Note that we would have obtained the same result (A.8) if we have $C_{,1}(\widehat{\mathbf{z}}) \neq 0$ instead of $C_{,2}(\widehat{\mathbf{z}}) \neq 0$. In fact, the symmetric appearance of the necessary conditions (A.8) is more readily seen from the following alternate derivation. From the two conditions

$$dJ = J_{,1}(\widehat{\mathbf{z}})\,dz_1 + J_{,2}(\widehat{\mathbf{z}})\,dz_2 = 0$$

and

$$dC = C_{,1}(\widehat{\mathbf{z}})\,dz_1 + C_{,2}(\widehat{\mathbf{z}})\,dz_2 = 0,$$

we may form the combination $dJ - \lambda dC = 0$ for an arbitrary constant $\lambda$ to get

$$[J_{,1}(\hat{\mathbf{z}}) - \lambda C_{,1}(\hat{\mathbf{z}})]\,dz_1 + [J_{,2}(\hat{\mathbf{z}}) - \lambda C_{,2}(\hat{\mathbf{z}})]\,dz_2 = 0. \tag{A.9}$$

We may choose the *Lagrange multiplier* $\lambda$ for an efficient solution. If $C_{,2}(\hat{\mathbf{z}}) \neq 0$, we choose $\lambda$ so that

$$J_{,2}(\hat{\mathbf{z}}) - \lambda C_{,2}(\hat{\mathbf{z}}) = 0.$$

Because $dz_1$ is arbitrary, we have (A.8) for $k = 1$ as well. If $C_{,2}(\hat{\mathbf{z}}) = 0$ but $C_{,1}(\hat{\mathbf{z}}) \neq 0$, we choose $\lambda$ so that $J_{,1}(\hat{\mathbf{z}}) - \lambda C_{,1}(\hat{\mathbf{z}}) = 0$. Then (A.8) also holds for $k = 2$, for $dz_2$ is now arbitrary.

The second derivation of (A.8) can easily be generalized from 2 to $n$ independent variables. We have then the following (Lagrange) *multiplier rule* for problems with a single constraint:

**[X.A]** *If the minimum point $\hat{\mathbf{z}}$ is not a stationary point of $C(\mathbf{z})$, then $\hat{\mathbf{z}}$ and the Lagrange multiplier $\lambda$ must satisfy conditions (A.8) for $k = 1, 2, \ldots, n$ and the constraint $C(\hat{\mathbf{z}}) = \gamma$.*

The advantage of the method of Lagrange multipliers should now be evident. We do not have to use $C(\mathbf{z})$ to eliminate anyone of the parameters $z_i$ from $J(\mathbf{z})$ or anyone of the increments $dz_i$ from $dJ = 0$. The necessary conditions in (A.8) for a minimum $J$ are easily formed from the partial derivatives of $J$ and $C$. The solution of the $n$ conditions (A.8) and the constraint $C(\mathbf{z}) = 0$ for $\hat{\mathbf{z}}$ and $\lambda$ can be done numerically as the final step of the solution process. We are not required to solve any other equations along the way to this final step but may take advantage of any possible intermediate simplifications. For some problems, the multiplier is of no particular interest and needs not be obtained.

### b. Abnormality and Sensitivity

The multiplier rule (A.8) may be treated as the conditions of stationarity of the function

$$I(\mathbf{z}, \lambda) = J(\mathbf{z}) + \lambda[\gamma - C(\mathbf{z})] \tag{A.10}$$

with respect to the independent variables $\{z_1, \ldots, z_n\}$ without any reference to the constraint. Because the multiplier $\lambda$ is also an unknown parameter, stationarity with respect to $\lambda$ formally reproduces the constraint $C(\mathbf{z}) = \gamma$ as another necessary condition to be satisfied by $\hat{\mathbf{z}}$ and $\lambda$. The same process applies to problems of maximization.

If $C_{,i}(\hat{\mathbf{z}}) = 0$ for all $i = 1, 2, \ldots, n$, the derivation of (A.8) breaks down because $\lambda$ cannot be chosen for (A.8) to hold for any $k$. Conditions (A.8) themselves are generally not operative. The following variation of Example B illustrates this observation.

**EXAMPLE C:** $J(\mathbf{z}) = z_1 - z_2$ with $C(\mathbf{z}) \equiv z_1^2 + z_2^2 = 0$.

For $\gamma = 0$, the previous results specialize to $\widehat{z}_1 = \widehat{z}_2 = 0$. In that case, neither of the two conditions from (A.8),

$$1 - 2\lambda\widehat{z}_1 = 0, \qquad -1 - 2\lambda\widehat{z}_2 = 0,$$

is satisfied at the "minimum" point. It is inappropriate to conclude that the problem has no solution, as the necessary conditions (A.8) themselves should not be applied to this problem. In fact, the only (real) pair of $(z_1, z_2)$ which satisfies the constraint $z_1^2 + z_2^2 = 0$ is the point $(0,0)$. The minimum of $J(\mathbf{z})$ subject to the constraint must be $J(\mathbf{0}) = 0$; there is no other choice.

The important point brought out by the example above is that the performance index $J(\mathbf{z})$ sometimes plays no role in the solution process for a given problem. To have a *multiplier rule* valid even for the exceptional case $C_{,i}(\mathbf{z}) = 0$ for all $i$, we introduce an additional multiplier $\lambda_0$ and form the linear combination

$$\lambda_0 dJ - \lambda dC = (\lambda_0 \widehat{J}_{,1} - \lambda \widehat{C}_{,1}) dz_1 + \ldots + (\lambda_0 \widehat{J}_{,n} - \lambda \widehat{C}_{,n}) dz_n = 0 \quad \text{(A.11)}$$

[instead of the $n$-variable version of (A.9)]. For the *normal* case where $\widehat{C}_{,k} \neq 0$ for some $k$, $\lambda_0$ should not vanish, otherwise we would have $\lambda = 0$. If $\lambda_0 \neq 0$, we can divide (A.11) through by $\lambda_0$ to recover (A.8) or (A.9) and we proceed as before. In other words, we can set $\lambda_0 = 1$ if $\widehat{\mathbf{z}}$ is not a stationary point of $C(\mathbf{z})$ so that the solution process described in [X.A] applies.

For the *abnormal* case with $C_{,i}(\widehat{\mathbf{z}}) = 0$ for all $i = 1, \ldots, n$, relation (A.11) requires $\lambda_0 J_{,i}(\widehat{\mathbf{z}}) = 0$ for all $i$. If $J_{,i}(\widehat{\mathbf{z}})$ does not vanish simultaneously for all $i$, the necessary condition (A.11) can only be satisfied by $\lambda_0 = 0$, corresponding to a situation where $J(\mathbf{z})$ plays no role in the solution process (as in Example B with $\gamma = 0$ discussed above). In this abnormal case, the remaining unknown multiplier $\lambda$ is not uniquely determined.

In the more general form (A.11), both the normal and the abnormal case require

$$\lambda_0 J_{,k} - \lambda C_{,k} = 0 \quad (k = 1, \ldots, n) \qquad \text{(A.12)}$$

at a minimum point $\widehat{\mathbf{z}}$. These necessary conditions may be viewed as stationary conditions for the function

$$I_0(\mathbf{z}, \lambda) = \lambda_0 J(\mathbf{z}) + \lambda[\gamma - C(\mathbf{z})] \qquad \text{(A.13)}$$

with respect to $\mathbf{z}$ without any reference to the constraint $C(\mathbf{z}) = 0$. Again, stationarity of $I_0(\mathbf{z}, \lambda)$ with respect to $\lambda$ formally requires the constraint as an additional necessary condition to be satisfied by $\widehat{\mathbf{z}}$ and $\lambda$. If the minimum point

$\hat{\mathbf{z}}$ of $J(\mathbf{z})$ subject to the constraint $C(\mathbf{z}) = \gamma$ is a stationary point of the constraint function $C(\mathbf{z})$, then we have an abnormal case. For abnormal problems, we take $\lambda_0 = 0$ unless $\hat{\mathbf{z}}$ is also a stationary point of $J(\mathbf{z})$. The result is summarized in the following extension of the multiplier rule [X.A]:

**[X.B]** *If $\hat{\mathbf{z}}$ minimizes $J(\mathbf{z})$ locally subject to the constraint $C(\mathbf{z}) = \gamma$, then there are two real multipliers $\lambda_0$ and $\lambda$, not both zero, which satisfy the n conditions (A.12) at $\hat{\mathbf{z}}$. For the normal case where $\hat{\mathbf{z}}$ is not a stationary point of $C(\mathbf{z})$, then we may set $\lambda_0$ equal to 1 and [X.A] applies. For the abnormal case, we take $\lambda_0 = 0$ unless $\hat{\mathbf{z}}$ is also a stationary point of $J(\mathbf{z})$. In the latter case, neither multiplier is specified.*

For the abnormal case, it is important to keep in mind that for a stationary point of $C(\mathbf{z})$ to be a solution of the problem, it must also satisfy the constraint $C(\mathbf{z}) = \gamma$. Even if it does, the stationarity conditions and the constraint may not completely determine the solution as seen from the following (somewhat contrived) example.

EXAMPLE D: $J(\mathbf{z}) = (z_1 - 1)^2 + z_2^2$ with

$$C(\mathbf{z}) \equiv \sin^2(z_1 - z_2) + (z_1 - z_2)^2 = 0.$$

The stationary conditions for $I_0(\mathbf{z}, \lambda)$ are

$$\lambda_0(\hat{z}_1 - 1) + \lambda[\cos(\hat{z}_1 - \hat{z}_2)\sin(\hat{z}_1 - \hat{z}_2) + (\hat{z}_1 - \hat{z}_2)] = 0,$$
$$\lambda_0\hat{z}_2 - \lambda[\cos(\hat{z}_1 - \hat{z}_2)\sin(\hat{z}_1 - \hat{z}_2) + (\hat{z}_1 - \hat{z}_2)] = 0.$$

The constraint requires $z_2 = z_1$. But any pair $(\hat{z}_1, \hat{z}_2) = (x, x)$ is a stationary point of $C(\mathbf{z})$ so that we should set $\lambda_0 = 0$ by [X.B]. This is also evident upon setting $\hat{z}_2 = \hat{z}_1$ in the two stationary conditions to get

$$\lambda_0(\hat{z}_1 - 1) = 0, \qquad \lambda_0\hat{z}_2 = 0.$$

Unless $\lambda_0 = 0$, we would have an inconsistency by obtaining $\hat{z}_1 (= \hat{z}_2) = 1$ from the first and $\hat{z}_1 (= \hat{z}_2) = 0$ from the second. With $\lambda_0 = 0$, the two stationary conditions for $I_0$ are satisfied without any additional restrictions on $\hat{\mathbf{z}}$ and $\lambda$. All that we have from the multiplier rule is $\hat{z}_2 = \hat{z}_1$.

On the other hand, it is straightforward to use the constraint (*in the form* $z_2 = z_1$) to eliminate $z_2$ from $J(\mathbf{z})$ to get

$$J \equiv \bar{J} = (z_1 - 1)^2 + z_1^2.$$

The minimum $J$ is attained at the unique stationary point $\hat{z}_1 = 1/2$ because $d^2\bar{J}/dz_1^2 = 4 > 0$. The minimum point $(\hat{z}_1, \hat{z}_2) = (1/2, 1/2)$ of the original problem is seen to satisfy (A.12) with $\lambda_0 = 0$ and $\lambda \neq 0$ (but otherwise unspecified) as well as the constraint, as required by [X.B]. However, these same conditions with $\lambda_0 = 0$ do not by themselves determine $\hat{\mathbf{z}}$ (or $\lambda$) in this abnormal case.

The Lagrange multiplier $\lambda$ has a useful interpretation. To obtain this interpretation, we note that the minimum point $\hat{\mathbf{z}}$ and the multiplier $\lambda$ both depend on the parameter $\gamma$ and write $\bar{J}(\gamma) \equiv J(\hat{\mathbf{z}})$. For the $n = 2$ and $m = 1$ case, we have

$$\bar{J}(\gamma) \equiv J(\hat{\mathbf{z}}) - \lambda[C(\hat{\mathbf{z}}) - \gamma]$$

with $\hat{\mathbf{z}} = \mathbf{Z}(\gamma) \equiv (Z_1(\gamma), Z_2(\gamma))^T$ and $\lambda = \lambda(\gamma)$. We can differentiate both sides of this expression with respect to $\gamma$ to get

$$\frac{d\bar{J}}{d\gamma} = [\hat{J}_{,1} - \lambda\hat{C}_{,1}]\frac{dZ_1}{d\gamma} + [\hat{J}_{,2} - \lambda\hat{C}_{,2}]\frac{dZ_2}{d\gamma} - [\hat{C} - \gamma]\frac{d\lambda}{d\gamma} + \lambda.$$

But (A.8) and the constraint $C(\mathbf{z}) = \gamma$ are satisfied at the stationary point $\hat{\mathbf{z}}$. This reduces the above relation to

$$\frac{d\bar{J}}{d\gamma} = \lambda. \tag{A.14}$$

Thus, we have the following result:

**[X.C]** ***The Lagrange multiplier $\lambda$, which depends on $\gamma$, is the rate at which the minimum value of the performance index $J(\mathbf{z})$ changes with a change of the value of the parameter $\gamma$ on the right side of the equality constraint $C(\mathbf{z}) = \gamma$.***

A multiplier is sometimes called the *marginal valuation* or the *shadow price* of the constraint. It tells us that an extra unit of $\gamma$ would increase the minimum $J(\mathbf{z})$ by an amount equal to $\lambda$ (which may be negative), i.e., the *sensitivity* of $\hat{J}$ to $\gamma$.

A less formal and simpler way to arrive at the same interpretation of $\lambda$ is to calculate the change in $J$ from its minimum value corresponding to a small change from $\hat{\mathbf{z}}$ to $\hat{\mathbf{z}} + d\mathbf{z}$:

$$dJ = \hat{J}_{,1}dz_1 + \hat{J}_{,2}dz_2 = \lambda(\hat{C}_{,1}dz_1 + \hat{C}_{,2}dz_2) = \lambda dC$$

where we have used (A.8) to eliminate $J_{,i}(\hat{\mathbf{z}})$. If the change in $\hat{\mathbf{z}}$ is the result of a change $d\gamma$ from $\gamma$, then we have from the constraint $dC = d\gamma$ so that $dJ = \lambda d\gamma$ as we expect.

From the expression for $dJ/dz_1$ in (A.6a), we obtain

$$\frac{d^2J}{dz_1^2} = \frac{d}{dz_1}\left[J_{,1} + J_{,2}\frac{d\varphi}{dz_1}\right]$$

$$= J_{,11} + 2J_{,12}\frac{d\varphi}{dz_1} + J_{,2}\frac{d^2\varphi}{dz_1^2} + J_{,22}\left(\frac{d\varphi}{dz_1}\right)^2. \tag{A.15}$$

Relation (A.6b) may be used to write $d\varphi / dz_1 = -C_{,1}/C_{,2}$ so that

$$\frac{d^2\varphi}{dz_1^2} = \frac{d}{dz_1}\left[-\frac{C_{,1}}{C_{,2}}\right]$$

$$= -\frac{1}{C_{,2}^2}\left\{C_{,2}\left[C_{,11} + C_{,12}\frac{d\varphi}{dz_1}\right] - C_{,1}\left[C_{,21} + C_{,22}\frac{d\varphi}{dz_1}\right]\right\}. \qquad \text{(A.16)}$$

We may now use (A.16) and (A.6b) to eliminate $d^2\varphi / dz_1^2$ and $d\varphi / dz_1$, respectively, from the expression (A.15) for $d^2J / dz_1^2$. The sufficient condition $d^2J / dz_1^2 > 0$ at the stationary point $\hat{\mathbf{z}}$ which guarantees $J(\hat{\mathbf{z}})$ to be a local minimum may then be taken in the form

$$\hat{C}_{,2}^2\hat{J}_{,11} - 2\hat{C}_{,1}\hat{C}_{,2}\hat{J}_{,12} + \hat{C}_{,1}^2\hat{J}_{,22} + 2\hat{C}_{,1}\hat{C}_{,12}\hat{J}_{,2} - \hat{C}_{,1}\hat{C}_{,22}\hat{J}_{,1}$$
$$- \hat{C}_{,2}\hat{C}_{,11}\hat{J}_{,2} > 0 \qquad \text{(A.17)}$$

where we have used (A.7a) to make (A.17) more symmetric in its appearance. Condition (A.17) may also be written as

$$\det[H] < 0 \qquad \text{(A.18)}$$

where the *bordered Hessian matrix* $H$ is defined by

$$H = \begin{bmatrix} \hat{I}_{,11} & \hat{I}_{,12} & \hat{C}_{,1} \\ \hat{I}_{,21} & \hat{I}_{,22} & \hat{C}_{,2} \\ \hat{C}_{,1} & \hat{C}_{,2} & 0 \end{bmatrix} \qquad \text{(A.19)}$$

with $I(\mathbf{z}, \lambda)$ given in (A.10) and $\lambda$ to be eliminated by (A.8).

### c. The Multiplier Rule

To generalize [X.A] and [X.B] to problems with more than one constraints, we return to the general problem (A.1). For $\hat{\mathbf{z}}$ to be a local minimum point for the problem, the stationary condition

$$dJ = J_{,1}(\hat{\mathbf{z}})\,dz_1 + \ldots + J_{,n}(\hat{\mathbf{z}})\,dz_n = 0 \qquad \text{(A.20)}$$

must hold for any incremental change $d\mathbf{z}$ from $\hat{\mathbf{z}}$ subject to the constraints (A.3). The $n$-tuple $(J_{,1}, \ldots, J_{,n})$ is the gradient row vector of $J(\mathbf{z})$ in the $n$-dimensional $\mathbf{z}$-space:

$$(J_{,1}, \ldots, J_{,n}) \equiv \nabla J. \qquad \text{(A.21a)}$$

Condition (A.20) can, therefore, be interpreted as the gradient vector $\nabla J$ at $\hat{\mathbf{z}}$ being orthogonal to the vector $d\mathbf{z}$.

Similarly, we have

$$(C_{i,1}, C_{i,2}, \ldots, C_{i,n}) \equiv \nabla C_i \quad (i = 1, 2, \ldots, m). \tag{A.21b}$$

In terms of the $n \times m$ matrix

$$D(\mathbf{z}) = \begin{bmatrix} \nabla C_1 \\ \vdots \\ \nabla C_m \end{bmatrix} = \begin{bmatrix} C_{1,1}(\mathbf{z}) & \ldots & C_{1,n}(\mathbf{z}) \\ \vdots & & \vdots \\ C_{m,1}(\mathbf{z}) & \ldots & C_{m,n}(\mathbf{z}) \end{bmatrix} \tag{A.21c}$$

conditions (A.20) and (A.3) may be written as

$$\nabla J(\widehat{\mathbf{z}})\, dz = 0 \tag{A.22a}$$

and

$$\mathrm{D}(\widehat{\mathbf{z}})\, dz = 0, \tag{A.22b}$$

respectively. That is, the (column) vector $dz$ is orthogonal to (the row vector) $\nabla\widehat{J}$ as well as to each of the $m$ rows of the matrix $\widehat{D} \equiv D(\widehat{\mathbf{z}})$.

Suppose for the moment that the rows of $\widehat{D}$ are linearly independent (so that $\widehat{D}$ is of rank $m$). It follows that $dz$ must lie in the *orthogonal complement* to the $m$-dimensional subspace (of the $n$-dimensional space) spanned by the row of $\widehat{D}$. Because $\nabla J(\widehat{\mathbf{z}})$ is also orthogonal to $dz$, it is in the row space of $\widehat{D}$. In other words, $\nabla\widehat{J}$ is a linear combination of the rows of the $m \times n$ matrix $\widehat{D}$:

$$\nabla J(\widehat{\mathbf{z}}) = \lambda_1 \nabla C_1(\widehat{\mathbf{z}}) + \ldots + \lambda_m \nabla C_m(\widehat{\mathbf{z}}) \tag{A.23a}$$

or

$$J_{,i}(\widehat{\mathbf{z}}) - \sum_{j=1}^{m} \lambda_j C_{j,i}(\widehat{\mathbf{z}}) = 0 \quad (i = 1, 2, \ldots, n) \tag{A.23b}$$

for a set of $m$ constants $\{\lambda_i\}$, not all zero. For $m = 1$ and $n = 2$, this set of $n$ conditions reduce to (A.8). The present derivation of (A.23) offers a new (linear algebra) interpretation of the previously derived multiplier rule for that special case. The multiplier rule for $n$ variables and $m$ ($< n$) constraints is summarized as follows:

**[X.D]** ***If $\widehat{\mathbf{z}}$ is a (local) minimum point for the general problem (A.1) and if the rank of the* $\mathrm{m} \times \mathrm{n}$ *matrix* $\widehat{\mathrm{D}} \equiv \mathrm{D}(\widehat{\mathbf{z}})$ *as defined in (A.21c) is not less than the rank of the augmented matrix***

$$\widehat{\mathrm{D}}_J = \begin{bmatrix} \widehat{\mathrm{D}} \\ \nabla\widehat{\mathrm{J}} \end{bmatrix},$$

***then there is a set of* m *Lagrange multipliers* $\{\lambda_i\}$, *not all zero, which satisfies the* n *conditions of (A.23b) for the particular* $\widehat{\mathbf{z}}$.**

If the rank of $\widehat{D}$ (and $\widehat{D}_J$) is $m$ as we have assumed up to now, then the set of multipliers is uniquely determined by (10.23) for a given $\widehat{\mathbf{z}}$. If the rank of $\widehat{D}$ is less than $m$ (so that the row vectors of $\widehat{D}$ are not linearly independent), the fact that $\widehat{D}$ and $\widehat{D}_J$ both have the same rank still ensure the existence of a solution for (A.23) as a linear system for the $m$ multipliers $\{\lambda_1, \ldots, \lambda_m\}$. But now the multipliers may not be unique.

A casual reading of [X.D] gives the impression that it serves no useful purpose. [If we have the minimum point $\widehat{\mathbf{z}}$, the problem is solved; the multipliers themselves are of no particular interest in many problems.] We will show by the following two examples that this is not the case; [X.D] is, in fact, useful for the solution process for $\widehat{\mathbf{z}}$.

EXAMPLE E: $J(\mathbf{z}) = z_1^2 + z_2^2 + z_3^2$, with

$$C_1(\mathbf{z}) \equiv z_1 + z_2 + z_3 - 1 = 0, \qquad C_2(\mathbf{z}) \equiv z_1 z_2 z_3 - \frac{1}{27} = 0.$$

To apply [X.D], we first use (A.23b) to get

$$2\widehat{z}_i - \lambda_1 - \lambda_2 \widehat{z}_j \widehat{z}_k = 0 \quad (i \neq j \neq k, i = 1,2,3) \tag{A.24a}$$

with

$$\widehat{D}_J = \begin{bmatrix} 1 & 1 & 1 \\ \widehat{z}_2\widehat{z}_3 & \widehat{z}_3\widehat{z}_1 & \widehat{z}_1\widehat{z}_2 \\ 2\widehat{z}_1 & 2\widehat{z}_2 & 2\widehat{z}_3 \end{bmatrix}.$$

One way to proceed is to solve (A.24a) and the two constraints to determine $\widehat{\mathbf{z}}$, $\lambda_1$, and $\lambda_2$. To this end, we use the second constraints in the form $z_j z_k = 1/27 z_i$ for $i \neq j \neq k$ to transform (A.24a) into

$$2\widehat{z}_i - \lambda_1 - \frac{\lambda_2}{27\widehat{z}_i} = 0 \quad (i = 1,2,3).$$

This is a quadratic equation for $\widehat{z}_i$ which gives

$$\widehat{z}_i = \frac{1}{4}\left[\lambda_1 \pm \sqrt{\frac{8\lambda_2}{27}}\right].$$

With $z_1 = z_2 = z_3$ as a stationary point, the first constraint then gives $z_i = 1/3$, $i = 1,2,3$, and the stationary conditions (A.24a) now reduce to the single relation

$$\frac{2}{3} - \lambda_1 - \frac{\lambda_2}{9} = 0. \tag{A.24b}$$

There are no other restrictions on the multipliers.

For the stationary point $\hat{\mathbf{z}} = (1,1,1,)^T/3$, the matrix $D_J$ is given by

$$\widehat{D}_J = \begin{bmatrix} 1 & 1 & 1 \\ \frac{1}{9} & \frac{1}{9} & \frac{1}{9} \\ \frac{2}{3} & \frac{2}{3} & \frac{2}{3} \end{bmatrix}.$$

Evidently, both $\widehat{D}$ and $\widehat{D}_J$ are of rank 1 ($< m = 2$). As implied by [X.D], a set of multipliers exists as long as the rank of $\widehat{D}$ and the rank of $\widehat{D}_J$ are the same. That $\widehat{D}$ is not of full rank only affects the uniqueness of the multipliers. All [X.D] did in this solution process was to assure us that the multipliers are not unique and there is no other conditions on $\lambda_1$ and $\lambda_2$.

For a more useful application of [X.D], we note that it is generally not so straightforward to solve (A.23) and the constraints for $\hat{\mathbf{z}}$ and $\boldsymbol{\lambda}$ as we did above. As an alternative method of solution, we reduce the transpose of $\widehat{D}_J$ to echelon form:

$$D_J^{\mathrm{T}} \to \begin{bmatrix} 1 & z_2 z_3 & 2z_1 \\ 0 & z_3(z_1 - z_2) & 2(z_2 - z_1) \\ 0 & z_2(z_1 - z_3) & 2(z_3 - z_1) \end{bmatrix} \to \begin{bmatrix} 1 & z_2 z_3 & 2z_1 \\ 0 & 1 & -2/z_3 \\ 0 & 1 & -2/z_2 \end{bmatrix}.$$

By [X.D], we need $\widehat{D}_J$ to be of the same rank as $\widehat{D}$. This requires $\hat{z}_2 = \hat{z}_3$; the first constraint then gives

$$\hat{z}_1 = 1 - \hat{z}_2 - \hat{z}_3 = 1 - 2\hat{z}_3.$$

Upon using these results to eliminate $\hat{z}_1$ and $\hat{z}_2$ from the second constraint, we obtain

$$2\hat{z}_3^3 - \hat{z}_3^2 + \frac{1}{27} = 0.$$

The (double) root $\hat{z}_3 = 1/3$ of this equation is for the minimum point of the problem. Correspondingly, we have $\hat{z}_2 (= \hat{z}_3) = 1/3$ and $\hat{z}_1 = 1 - \hat{z}_2 - \hat{z}_3 = 1/3$. For the minimum point $\hat{\mathbf{z}} = (1,1,1)^T/3$, the matrices $\widehat{D}_J$ and $\widehat{D}$ are both of rank 1 with $\lambda_1$ and $\lambda_2$ required to satisfy (A.24b) as we found earlier.

It is important to note that $\nabla J$ is generally not in the row space of $D$. [X.D] requires that it be so at $\hat{\mathbf{z}}$. This requirement effectively (and efficiently) distills $n - m$ essential features of the minimum point $\hat{\mathbf{z}}$ from the $n$ stationary conditions (A.23). They supplement the $m$ constraints to provide $n$ conditions for the determination of the $n$ components of the stationary points. The unused information in (A.23) are for the determination of the multipliers $\{\lambda_i\}$.

EXAMPLE F: $J(z) = 2z_1z_2 + 3z_1z_3$ with

$$C_1(\mathbf{z}) \equiv z_1^2 + z_2^2 - 3 = 0, \qquad C_2(\mathbf{z}) \equiv z_1z_3 - 2 = 0.$$

For this problem, the three stationary conditions (A.23b) take the form

$$\begin{bmatrix} 2\hat{z}_1 & \hat{z}_3 \\ \hat{z}_2 & 0 \\ 0 & \hat{z}_1 \end{bmatrix} \begin{bmatrix} \lambda_1 \\ \lambda_2 \end{bmatrix} = \begin{bmatrix} 2\hat{z}_2 + 3\hat{z}_3 \\ \hat{z}_1 \\ 3\hat{z}_1 \end{bmatrix}.$$

Instead of solving these relations and the two constraints for the five unknowns (as in the previous example), we apply [X.D] to obtain one condition on $\hat{\mathbf{z}}$ alone.

To see the rank of $\hat{D}$ and $\hat{D}_J$ for this problem, we row-reduce $\hat{D}_J^T$ to get (for $\hat{z}_1\hat{z}_2\hat{z}_3 \neq 0$)

$$\hat{D}_J^T \to \begin{bmatrix} 2\hat{z}_1 & \hat{z}_3 & 2\hat{z}_2 + 3\hat{z}_3 \\ \hat{z}_2 & 0 & \hat{z}_1 \\ 0 & \hat{z}_1 & 3\hat{z}_1 \end{bmatrix} \to \dots \to \begin{bmatrix} 1 & \hat{z}_3/2\hat{z}_1 & (2\hat{z}_2 + 3\hat{z}_3)/2\hat{z}_1 \\ 0 & 1 & [2\hat{z}_2^2 + 3\hat{z}_3\hat{z}_2 - 2\hat{z}_1^2]/\hat{z}_2\hat{z}_3 \\ 0 & 0 & 2(\hat{z}_2^2 - \hat{z}_1^2)/\hat{z}_2\hat{z}_3 \end{bmatrix}.$$

For $\hat{D}_J$ to have the same rank as $\hat{D}$ (which is 2), we need the lower right-hand corner entry to vanish. This is accomplished by $\hat{z}_2 = \pm\hat{z}_1$. It follows from the first constraint that $\hat{z}_1 = \pm\sqrt{3/2}$ and from the second, $\hat{z}_3 = \pm\sqrt{8/3}$. Hence, the equal rank condition from [X.D] and the two constraints completely determine all stationary points. Two of the stationary conditions then give $\lambda_2 = 3$ and $\lambda_1 = \pm 1$ for $\hat{z}_2 = \pm\hat{z}_1$. Because $\hat{z}_3$ and $\hat{z}_1$ have the same sign, $\hat{z}_2$ and $\hat{z}_1$ must be of opposite sign for $J(\mathbf{z})$ to be a minimum.

It should be evident from the two examples above that [X.D] does help to facilitate the determination of the stationary points. It really does not require a knowledge of the minimum point in order to apply the theorem as it would appear from a superficial reading.

EXAMPLE G: $J(\mathbf{z}) = z_1 - z_2$ with $C_1(\mathbf{z}) \equiv z_1^2 + z_2^2 = 0.$

The matrix $D_J$ for this problem is

$$D_J = \begin{bmatrix} 2z_1 & 2z_2 \\ 1 & -1 \end{bmatrix} \rightarrow \begin{bmatrix} 2z_1 & 2z_2 \\ 0 & 2(z_1 + z_2) \end{bmatrix}$$

The last row can be made $(0,0)$ by restricting $z_2$ to be $-z_1$. But, the only solution of the constraining equation $z_1^2 + z_2^2 = 0$ is $\hat{z} = (0,0)$. For this point, we have

$$\hat{D} = [0,0], \quad \hat{D}_J = \begin{bmatrix} 0 & 0 \\ 1 & -1 \end{bmatrix}$$

so that rank $(\hat{D}) <$ rank $(\hat{D}_J)$. Hence, we have an *abnormal* problem as there does not exist a multiplier $\lambda_1$ for which the stationary conditions (A.23)

$$1 - 2\lambda_1\hat{z}_1 = 0, \quad -1 - 2\lambda_1\hat{z}_2 = 0, \tag{A.25}$$

are satisfied by $\hat{z}$ and $\lambda_1$.

It is possible to extend [X.B] to cover abnormal problems with more than one constraint. Instead of insisting $\nabla J(\hat{z})$ to be a linear combination of the $m$ vectors $\{\nabla C_i(\hat{z})\}$, we merely require that these $(m+1)$ vectors be linearly dependent, which is all that (A.22a) and (A.22b) imply. In other words, there is a set of $(m+1)$ constants $\{\lambda_0, \lambda_1, \ldots, \lambda_m\}$ such that

$$\lambda_0 \nabla\hat{J} - [\lambda_1 \nabla\hat{C} + \ldots + \lambda_m \nabla\hat{C}_m] = 0 \tag{10.26a}$$

or

$$\lambda_0 \hat{J}_{,i} - [\lambda_1 \hat{C}_{1,i} + \ldots + \lambda_m \hat{C}_{m,i}] = 0 \quad (i = 1, \ldots, n). \tag{10.26b}$$

If the rows of $\hat{D}$ are linearly independent, then $\nabla\hat{J}$ lies in the row space of $\hat{D}$ so that $\lambda_0$ may be taken to be unity and (A.23) holds.

Even if it is not possible to choose $\hat{z}$ so that (A.23) holds, (A.26) is still valid for it can be made to hold by choosing $\lambda_0 = 0$. That is, if the rank of $D_J$ is not equal to the rank of $D$ for any $\hat{z}$, we can make it so by setting $\lambda_0 = 0$ so that the entries of the last column of $D_J$ are all zeros. On the other hand, if $\hat{D}$ and $\hat{D}_J$ are of equal rank, then $\lambda_0$ is not zero (and may be set equal to 1) and [X.D] is applicable. Altogether, we have the following multiplier rule:

**[X.E]** *If* $J(\hat{z})$ *is a local minimum subject to the* m *equality constraints* $C_i(z) = 0$, i, ..., m, *then there are* $(m+1)$ *real multipliers* $\{\lambda_0, \lambda_1, \ldots, \lambda_m\}$, *not all zeros, such that the* n *conditions of (A.26b) are satisfied by these multipliers at* $\hat{z}$. *If the rank of* $n \times m$ *matrix* $\hat{D} \equiv [\hat{D}_{ij}] = [C_{i,j}(\hat{z})]$ *is less than the rank of the augmented matrix* $\hat{D}_J$ *we set* $\lambda_0 = 0$ *for this abnormal case. Otherwise, set* $\lambda_0 = 1$ *and* [X.D] *applies.*

When the rank of $\widehat{D}$ is less than $m$, the set of $\{\lambda_1, \lambda_2, \ldots, \lambda_m\}$ is not unique. In Example C for instance, we have now

$$D_J = \begin{bmatrix} 2z_1 & 2z_2 \\ \lambda_0 & -\lambda_0 \end{bmatrix}, \qquad \widehat{D}_J = \begin{bmatrix} 0 & 0 \\ \lambda_0 & -\lambda_0 \end{bmatrix}.$$

Because rank $(\widehat{D}) <$ rank $(\widehat{D}_J)$, $\lambda_0$ is set to zero according to [X.E] and the multiplier $\lambda_1$ is completely arbitrary. Example E shows that the nonuniqueness of multipliers is a consequence of the rank of $\widehat{D}$ being less than $m$ and not whether it is equal to the rank of $\widehat{D}_J$ (before setting $\lambda_0 = 0$). For abnormal problems, the multiplier method may not completely determine the minimum point $\widehat{\mathbf{z}}$ (see Example D).

As in the one constraint case, the multiplier rule (A.26) may be formally treated as the necessary conditions for $\widehat{\mathbf{z}}$ to be a stationary point of

$$I(\mathbf{z}; \boldsymbol{\lambda}) = \lambda_0 J(\mathbf{z}) - \sum_{k=1}^{m} \lambda_k C_k(\mathbf{z}) \tag{A.27}$$

without any reference to the constraints. The constraints $\{C_k(\mathbf{z}) = 0\}$ themselves may be formally thought of as the necessary conditions for stationarity of $I$ with respect to $\{\lambda_1, \ldots, \lambda_m\}$.

# 11

# Pointwise Constraints on Extremals

## 1. Pointwise Equality Constraints

In Chapter 10, the basic problem of the calculus of variations was modified by imposing equality constraints on auxiliary integrals of functions of $x$, $\mathbf{y}$, and $\mathbf{y}'$, i.e., isoperimetric constraints. Other problems in the calculus of variations involve equality or inequality constraints on functions of $x$, $\mathbf{y}$ and $\mathbf{y}'$ at all points in $(a, b)$. When a pointwise equality constraint involves only the vector unknowns $\mathbf{y} = (y_1(x), y_2(x), \ldots, y_n(x))^T$ (and possibly also $x$ explicitly), it is called a ***holonomic constraint***; otherwise it is called a ***nonholonomic constraint***. We have already encountered some problems with holonomic constraints in Chapter 7. We shall see in the next chapter that the isoperimetric constraints of Chapter 10 are, in fact, equivalent to a certain kind of nonholonomic constraints.

An example of a variational problem with a holonomic (equality) constraint is finding a geodesic curve between two given points on a surface. In a parametric representation, a curve in space is given by $\{x = y_1(t), y = y_2(t), z = y_3(t),\ a \le t \le b\}$. The length of a curve with prescribed end points $\mathbf{y}(a) = \mathbf{A}$ and $\mathbf{y}(b) = \mathbf{B}$ is given by

$$J[\mathbf{y}] = \int_a^b \sqrt{\dot{y}_1^2 + \dot{y}_2^2 + \dot{y}_3^2}\, dt, \qquad (\dot{\ }) \equiv \frac{d(\ )}{dt}. \tag{11.1}$$

The curve of minimum length is to lie on a surface which we take to be given by $\varphi(x, y, z) = 0$; hence, we have the constraint on $\mathbf{y}(t)$

$$\varphi(y_1, y_2, y_3) \equiv \varphi(\mathbf{y}(t)) = 0 \quad (a \le t \le b). \tag{11.2}$$

On the surface of a unit sphere centered at the origin for instance, we have

$$\varphi(\mathbf{y}) = y_1^2 + y_2^2 + y_3^2 - 1 = 0 \tag{11.2'}$$

In this section, we develop a *multiplier method* of solution for variational problems with equality constraints such as the geodesic problem above. To simplify the development, we will first focus our attention on the case of two scalar unknowns, $\mathbf{y} = (y_1, y_2)^T$, and a single equality constraint. The method will then be extended to problems with more unknowns and more constraints as well as inequality constraints.

In variational problems with holonomic constraints, we typically wish to minimize an integral

$$J[\mathbf{y}] = \int_a^b F(x, \mathbf{y}, \mathbf{y}')dx = \int_a^b F(x, y_1, y_2, y_1', y_2')\, dx. \tag{11.3}$$

For simplicity, we take the minimization to be over all $C^2$ functions $y_1$ and $y_2$ which satisfy the end conditions

$$\mathbf{y}(a) = \mathbf{A}, \qquad \mathbf{y}(b) = \mathbf{B} \tag{11.4}$$

as well as the constraint

$$\varphi(x, \mathbf{y}) \equiv \varphi(x, y_1(x), y_2(x)) = 0 \qquad (a < x < b) \tag{11.5}$$

where $\varphi$ is at least $C^1$ in all of its arguments. Condition (11.5) is a relation between the two unknowns $y_1$ and $y_2$ at each point $x$ of the interval $(a, b)$, consistent with (11.4) so that $\varphi(a, \mathbf{A}) = \varphi(b, \mathbf{B}) = 0$.

One approach we might use for this problem would be to solve (11.5) for one of the two unknowns in terms of the other; the result can then be used to eliminate one component of $\mathbf{y}$ from (11.3) to get a conventional one-unknown basic problem (with no holonomic constraints). A more general version of this approach has been described and applied in Chapter 7. Such an approach is not always feasible. Even when it is, the resulting basic problem for the remaining unknown would most likely require the solution of one or more very difficult Euler DEs. Our discussion of parameter optimization problems with equality constraints in the Appendix of Chapter 10 suggests that we should explore the possibility of extending the method of Lagrange multipliers to variational problems with pointwise constraints.

To develop a multiplier rule for our problem, recall that for $J[\hat{\mathbf{y}}]$ to be a minimum, the corresponding first variation

$$\delta J = \int_a^b \{[\hat{F}_{,y_1} - (\hat{F}_{,y_1'})']\delta y_1 + [\hat{F}_{,y_2} - (\hat{F}_{,y_2'})']\delta y_2\}dx \tag{11.6}$$

must vanish. Because of the holonomic constraint (11.5), $\delta y_1$ and $\delta y_2$ are not independent variations; they are related by

$$\delta\varphi = [\varphi_{,y_1}\delta y_1 + \varphi_{,y_2}\delta y_2] = 0 \tag{11.7}$$

for any pair of admissible $y_1$ and $y_2$ and every point in $[a,b]$. In particular, (11.7) holds for $\mathbf{y} = \widehat{\mathbf{y}}$ so that

$$\int_a^b \lambda(x)[\widehat{\varphi}_{,y_1}\delta y_1 + \widehat{\varphi}_{,y_2}\delta y_2]\,dx = 0 \tag{11.8}$$

for an arbitrary function $\lambda(x)$. It follows from (11.8) and $\delta J = 0$ that

$$\int_a^b \{[\widehat{F}_{,y_1} - (\widehat{F}_{,y_1'})' - \lambda\widehat{\varphi}_{,y_1}]\delta y_1 + [\widehat{F}_{,y_2} - (\widehat{F}_{,y_2'})' - \lambda\widehat{\varphi}_{,y_2}]\delta y_2\}\,dx = 0 \tag{11.9}$$

So far, we have not specified $\lambda(x)$; we now choose it so that combination of terms inside the brackets multiplying $\delta y_2$ vanishes, i.e.,

$$\widehat{F}_{,y_2} - (\widehat{F}_{,y_2'})' - \lambda\widehat{\varphi}_{,y_2} = 0 \quad (a < x < b). \tag{11.10}$$

[Of course, we could have chosen $\lambda$ so that the combination of terms multiplying $\delta y_1$ vanishes instead and would have to do so if $\widehat{\varphi}_{,y_2} = 0$. We will return to the question of the existence of the needed multiplier(s) in the next section.] The remaining variation $\delta y_1$ is (smooth but otherwise) arbitrary except at the end points where it must be zero. It follows from the basic lemma [I.1] that its coefficient must vanish as well so that

$$\widehat{F}_{,y_k} - (\widehat{F}_{,y_k'})' - \lambda\widehat{\varphi}_{,y_k} = 0 \quad (k = 1, 2). \tag{11.11}$$

The two differential equations (11.11), together with the holonomic constraint (11.5), are three simultaneous equations for $\widehat{y}_1$, $\widehat{y}_2$, and $\lambda(x)$. By introducing the Lagrange multiplier $\lambda(x)$, we are able to avoid solving for one of the two unknowns in terms of the other unknown as an intermediate step of the solution process. The price we pay is having to solve for three unknowns instead of two. The substantive gain is a postponement of the solution of any algebraic or differential equation to the end of the solution process when it can be done numerically if necessary. As a bonus, there is usually a useful interpretation of the Lagrange multiplier(s). We illustrate the method of Lagrange multiplier for pointwise equality constraints with the geodesic problem on the unit sphere described earlier.

**EXAMPLE 1:** Minimize $J[\mathbf{y}] = \displaystyle\int_0^1 \sqrt{\dot{\mathbf{y}}\cdot\dot{\mathbf{y}}}\,dt$ subject to

$$\mathbf{y}\cdot\mathbf{y} = 1, \qquad \mathbf{y}(0) = (0,1,2)/\sqrt{5}, \qquad \mathbf{y}(1) = (1,2,3)/\sqrt{14}.$$

For this problem of three unknowns, $y_1$, $y_2$ and $y_3$, and one constraint, the multiplier method leads to the following three differential equations:

$$\left(\frac{\dot{y}_k}{F_0}\right)^{\bullet} + \lambda\varphi_{,y_k} = 0 \quad (k = 1,2,3), \tag{11.12}$$

where $F_0 = \sqrt{\dot{y}_1^2 + \dot{y}_2^2 + \dot{y}_3^2}$ and $\varphi = y_1^2 + y_2^2 + y_3^2 - 1 = 0$ with $(\ )^{\bullet}$ and $(\dot{\ })$ both indicating differentiation with respect to the independent variable $t$. To keep the notation simple, we have omitted the caret $\hat{}$ over $y_k$, $F_0$, and $\varphi$. Upon carrying out the various differentiations, the differential relations for this problem corresponding to (11.12) may be written as

$$\frac{\ddot{y}_k F_0 - \dot{y}_k \dot{F}_0}{y_k F_0^2} = -2\lambda \quad (k = 1, 2, 3). \tag{11.13a}$$

The three equations of (11.13a) corresponding to $k = 1, 2$, and 3 imply

$$\frac{\ddot{y}_i}{y_i} - \frac{\dot{y}_i}{y_i}\frac{\dot{F}_0}{F_0} = \frac{\ddot{y}_k}{y_k} - \frac{\dot{y}_k}{y_k}\frac{\dot{F}_0}{F_0} \quad (i \neq k)$$

or

$$\frac{\ddot{y}_i y_k - \ddot{y}_k y_i}{\dot{y}_i y_k - \dot{y}_k y_i} = \frac{(\dot{y}_i y_k - \dot{y}_k y_i)^{\bullet}}{\dot{y}_i y_k - \dot{y}_k y_i} = \frac{\dot{F}_0}{F_0} \quad (i \neq k). \tag{11.13b}$$

We can integrate both sides of the last equality of (11.13b) with respect to $t$ and take the result in the form

$$\dot{y}_i y_k - \dot{y}_k y_i = c_{ik} F_0 \quad (i \neq k), \tag{11.14}$$

where $c_{ik}$ is a constant of integration. By setting $(i, k)$ equal to $(1, 2)$ and $(2, 3)$, we obtain from (11.14)

$$\dot{y}_1 y_2 - \dot{y}_2 y_1 = c_{12} F_0 \quad \text{and} \quad \dot{y}_2 y_3 - \dot{y}_3 y_2 = c_{23} F_0.$$

When we use one of these two relations to eliminate $F_0$ from the other, we are left with the relation

$$\dot{y}_1 y_2 - \dot{y}_2 y_1 = c_0 (\dot{y}_2 y_3 - \dot{y}_3 y_2), \qquad c_0 = \frac{c_{12}}{c_{23}},$$

or

$$y_2(\dot{y}_1 + c_0 \dot{y}_3) = \dot{y}_2 (y_1 + c_0 y_3).$$

We write the above as a relation between two logarithmic derivatives:

$$\frac{\dot{y}_2}{y_2} = \frac{(y_1 + c_0 y_3)^{\bullet}}{(y_1 + c_0 y_3)}.$$

From this, we get

$$y_2 = c_1 y_1 + c_3 y_3, \tag{11.15}$$

where the two arbitrary constants of integration $c_1$ and $c_3$ are fixed by the end conditions (11.4).

For the end constraints with $\mathbf{A} = (0,1,2)/\sqrt{5}$ and $\mathbf{B} = (1,2,3)/\sqrt{14}$, we have from (11.15)

$$1 = 2c_3, \qquad 2 = c_1 + 3c_3,$$

giving $c_1 = c_3 = 1/2$ so that (11.15) becomes

$$y_2 = \frac{1}{2}(y_1 + y_3). \tag{11.15'}$$

The constraint (11.2′) provides the second relation for a complete characterization of the geodesic curve between the two prescribed points on the surface. For the specified **A** and **B**, we have

$$y_1^2 + y_2^2 + y_3^2 = y_1^2 + \frac{1}{4}(y_3 + y_1)^2 + y_3^2 = 1. \tag{11.16}$$

Relations (11.15′) and (11.16) together with $y_1 = t/\sqrt{14}$, $0 \le t \le 1$, may be solved for $y_1, y_2$ and $y_3$ in terms of $t$ to provide the following parametric representation of the geodesic curve sought:

$$y_1 = \frac{t}{\sqrt{14}}, \qquad y_2 = \frac{1}{5}\left\{\frac{2t}{\sqrt{14}} + \sqrt{5 - \frac{6t^2}{14}}\right\},$$

$$y_3 = \frac{1}{5}\left\{-\frac{t}{\sqrt{14}} + 2\sqrt{5 - \frac{6t^2}{14}}\right\}.$$

## 2. The Multiplier Rule for Equality Constraints

It is not difficult to extend the multiplier rule developed in the last section for one holonomic constraint to more (unknowns and more) holonomic constraints. For the more general problem, we wish to minimize

$$J[\mathbf{y}] = \int_a^b F(x, \mathbf{y}, \mathbf{y}')\,dx \equiv \int_a^b F(x, y_1, \ldots, y_n, y_1', \ldots, y_n')\,dx \tag{11.17a}$$

subject to

$$\mathbf{y}(a) = \mathbf{A}, \qquad \mathbf{y}(b) = \mathbf{B}, \tag{11.17b}$$

and

$$\varphi_i(x, \mathbf{y}) = 0 \qquad (i = 1, 2, \ldots, m) \tag{11.18}$$

for $m < n$. To obtain the admissible extremal(s) of this variational problem, we can again append the integral of $\lambda_1(x)\delta\varphi_1 + \ldots + \lambda_m(x)\delta\varphi_m = 0$ to $\delta J = 0$ as before and formally choose $\{\lambda_i\}$ to eliminate the coefficients of $m$ of the variations $\delta y_i$, leaving us with $n - m$ Euler DEs for the determination of the extremal(s) of the problem. The $m$ relations for $\{\lambda_i\}$ and the $n - m$ Euler DEs are all of the same form:

$$\widehat{F}_{,y_k} - (\widehat{F}_{,y_k'})' - \sum_{i=1}^{m} \lambda_i \widehat{\varphi}_{i,y_k} = 0 \quad (k = 1, \ldots, n). \tag{11.19}$$

The $n$ relations (11.19) and the $m$ constraints (11.18) determine $\{\lambda_i(x)\}$ and $\{\widehat{y}_k(x)\}$. Often they are conveniently taken to be the Euler DE of

$$\begin{aligned} I[\mathbf{y}, \boldsymbol{\lambda}] &\equiv \int_a^b \{F(x, \mathbf{y}, \mathbf{y}') - \sum_{i=1}^{m} \lambda_i(x) \varphi_i(x, \mathbf{y})\} dx \\ &= \int_a^b \{F(x, \mathbf{y}, \mathbf{y}') - \boldsymbol{\lambda}(x)^T \boldsymbol{\varphi}(x, \mathbf{y})\} dx, \end{aligned} \tag{11.20}$$

where $\mathbf{y}$ satisfies the end conditions (11.17b) and we have set $\boldsymbol{\lambda} = (\lambda_1, \ldots, \lambda_m)^T$, $\boldsymbol{\varphi} = (\varphi_1, \varphi_2, \ldots, \varphi_m)^T$. In other words, we formally seek the admissible extremal(s) of (11.20) by treating $(\mathbf{y}, \boldsymbol{\lambda})$ as $n + m$ unknowns to be varied independently. The $(n + m)$ Euler DEs together with the end conditions (11.18) are identical to the relations obtained by the method of the previous section for the minimizing function $\widehat{\mathbf{y}}$ of the original problem and the Lagrange multipliers $\{\lambda_i\}$. Evidently, the $m$ Euler DEs associated with $\delta\lambda_i$ are just the $m$ given constraints (11.18). The remaining $n$ Euler DEs associated with $\delta y_k$ are the $n$ conditions (11.19) satisfied by $\widehat{\mathbf{y}}(x)$ and $\boldsymbol{\lambda}(x)$.

As we indicated in the last chapter, we want to know the answer to the following important question for our constrained variational problem: Under what conditions is it possible to find a set of $\{\lambda_i(x)\}$, defined on $[a, b]$ and not all identically zero, for which the above multiplier rule holds? That it is not always possible is shown by the following example:

**EXAMPLE 2:** Minimize

$$J[\mathbf{y}] = \int_0^1 (y_1'^2 + y_2'^2) dt, \tag{11.21a}$$

subject to

$$\mathbf{y}(0) = (0, 0), \qquad \mathbf{y}(1) = (1, 0), \tag{11.21b}$$

and

$$y_2'^2 + 2(y_1 - t) = 0. \tag{11.22}$$

The two DEs corresponding to (11.19) for this problem are

$$\ddot{y}_1 = -\lambda, \qquad \ddot{y}_2 + \lambda y_2 = 0$$

where again we have omitted the caret over $y_i$. The unknown multiplier $\lambda$ is generally a function of time $t$; hence, the second ODE above is nonlinear. The solution of nonlinear ODE is generally difficult. For a constant $\lambda$, the ODE for $y_2$ and the prescribed end conditions $y_2(0) = y_2(1) = 0$ form a homogeneous BVP. It has a nontrivial solution if $\lambda$ is an eigenvalue, $n^2\pi^2$. For such a multiplier, the ODE and prescribed end conditions for $y_1$, in turn, require that $y_1$ be quadratic in $t$. Altogether, we have

$$y_1 = \left(1 + \frac{\lambda}{2}\right)t - \frac{\lambda}{2}t^2, \qquad y_2 = c_2 \sin(\sqrt{\lambda t}), \quad \lambda = (n\pi)^2 \equiv \lambda_n \tag{11.23}$$

for any constant $c_2$. The results given by (11.23) also include the trivial solution $\lambda = 0$ and $y_2 = 0$. The corresponding solution for $y_1$ is $y_1(t) = t$.

For the solution (11.23) to satisfy constraint (11.22) we must have

$$\frac{1}{2}c_2^2 \sin^2(\sqrt{\lambda_n t}) + \frac{\lambda_n}{2}(t - t^2) = 0$$

which is not possible for any choice of $c_2$ unless $n = 0$ so that $\lambda(t) \equiv 0$. This reduces (11.23) to $y_1 = t$ and $y_2 = 0$ which formally satisfy (11.19) [and the pointwise constraint (11.22)] of our problem. It can be shown that $\{\hat{y}_1 = t, \hat{y}_2 = 0, \text{ and } \lambda = 0\}$ is, in fact, the unique admissible extremal of $I[\mathbf{y}, \lambda]$ of (11.20) (see Exercise 6 at the end of the chapter). We will use a different argument below to show that $J[\hat{\mathbf{y}}]$ is the desired minimum.

For a given $y_1(t)$, the admissible comparison function $y_2(t) = 0$ makes $J[\mathbf{y}]$ of (11.21a) a minimum. With $y_2 = 0$ and ignoring constraint (11.22), we have the shortest path problem between two points in the $x, y_1$-plane. The solution is the straight line $y_1 = t$ which also satisfies constraint (11.22) automatically (given $y_2 = 0$).

The example above calls attention to the fact that the minimizing function $\hat{\mathbf{y}}$ for $J[\mathbf{y}]$ (or the extremals of $I[\mathbf{y}, \lambda]$) sometimes satisfies one or more Euler DE of $J[\mathbf{y}]$ as it does in our example. In that case, we may need $\lambda = 0$ (as $\hat{\varphi}_{,y_k}$ is usually not zero) which is at variance with the essence of the multiplier method we have began to develop. The following example shows what can happen when $\hat{\mathbf{y}}$ is an extremal of $\varphi$ instead.

EXAMPLE 3: Minimize $J[\mathbf{y}] = \int_0^1 (\dot{y}_1^2 + 2y_1 + \dot{y}_2^2)\,dt$ subject to (11.21b) and

$$y_2^2 + (y_1 - t)^2 = 0. \tag{11.24a}$$

In this case, the relevant Euler DEs are

$$\ddot{y}_1 + \lambda y_1 = \lambda t + 1, \qquad \ddot{y}_2 + \lambda y_2 = 0. \tag{11.24b}$$

Now, the only real-valued pair of $y_1$ and $y_2$ which satisfies constraint (11.24a) is $y_1 = t$ and $y_2 \equiv 0$. However, the two DEs (11.24b) are not satisfied by this pair. The difficulty lies in the fact that $(t, 0)$ is an extremal for $\varphi = y_2^2 + (y_1 - t)^2$ and not for the Lagrangian $F = \dot{y}_1^2 + 2y_1 + \dot{y}_2^2$.

To circumvent the difficulty within the framework of the multiplier method, we introduce an additional multiplier $\lambda_0$ as a multiplicative factor for $F(x, \mathbf{y}, \mathbf{y}')$, similar to what we did for isoperimetric problems, and form

$$I[\mathbf{y}, \boldsymbol{\lambda}, \lambda_0] = \int_a^b [\lambda_0 F(x, \mathbf{y}, \mathbf{y}') - \boldsymbol{\lambda}(x)^T \boldsymbol{\varphi}(x, \mathbf{y})]\, dx \tag{11.25}$$

instead of (11.20). When $F$ plays no role in the optimization process, we now have the option of setting $\lambda_0 = 0$. In Example 3, $\hat{\mathbf{y}} = (t, 0)$ is an extremal of $\varphi(x, \mathbf{y})$ and is the only pair which satisfies the constraint. In this case, we still have a multiplier rule by setting $\lambda_0 = 0$ (whereas $\lambda_1 = \lambda$ is unspecified).

Similar to the abnormal cases of isoperimetric problems discussed in Chapter 10, we have an *abnormal* variational problem of $n$ unknowns with $m$ holonomic constraints if the minimizing $\hat{\mathbf{y}}(x)$ satisfies more than $n - m$ Euler–Lagrange equations of $\boldsymbol{\lambda}(x) \cdot \boldsymbol{\varphi}(x, \mathbf{y}) = \lambda_1(x)\varphi_1(x, \mathbf{y}) + \ldots + \lambda_m(x)\varphi_m(x, \mathbf{y}) \equiv \Phi(x, \mathbf{y}, \boldsymbol{\lambda})$, where the multipliers $\{\lambda_1(x), \ldots, \lambda_m(x)\}$ are at least continuous functions of $x$ in $[a, b]$ and not all equal to zero. The notion of abnormality relates solely to the constraining functions $\{\varphi_i\}$; an abnormal solution curve is abnormal whatever the original Lagrangian $F(x, \mathbf{y}, \mathbf{y}')$ may be. A problem with holonomic constraints which is not abnormal is said to be *normal*. If the problem is normal (so that $\lambda_0 \neq 0$), we can set $\lambda_0 = 1$ as we did in Chapter 10. We summarize our observations as the following *multiplier rule* for the problem (11.17) and (11.18):

**[XI.1]** *If $\hat{\mathbf{y}}(\mathbf{x})$ minimizes (11.17a) subject to (11.17b) and (11.18), then $\hat{\mathbf{y}}$ is an extremal of* I *given by (11.25) for a nontrivial set of multipliers $\{\lambda_0, \lambda_1(\mathbf{x}), \ldots, \lambda_m(\mathbf{x})\}$, which are at least continuous on $[\mathrm{a}, \mathrm{b}]$. If the problem is normal, we may set $\lambda_0 = 1$.*

If the problem is normal, then we can choose $\boldsymbol{\lambda}$ to eliminate $m$ components of $\delta\mathbf{y}$ from

$$\delta I = \int_a^b \sum_{i=1}^n \{\lambda_0 [\hat{F}_{,y_i} - (\hat{F}_{,y_i'})'] - \boldsymbol{\lambda} \cdot \hat{\boldsymbol{\varphi}}_{,y_i}\} \delta y_i \, dx .$$

The Lagrangian $F$ is involved in the optimization process in that case and $\lambda_0$ is, therefore, not zero. A rigorous proof of the existence of the multipliers is usually done for the more general case of nonholonomic constraints to be discussed in Chapters 12-14.

## 3. Inequality Constraints on the Unknowns

In many applications, we have to solve the usual basic problem with a pointwise auxiliary constraint on the magnitude of the minimizing extremal in the form $y(x) \geq \varphi(x)$ in $[a, b]$ for some prescribed continuous function $\varphi(x)$. To be consistent, we should have $A \geq \varphi(a)$ and $B \geq \varphi(b)$, where $A$ and $B$ are the end values of $y$ at $x = a$ and $x = b$, respectively. In the minimum cost production problem of Section 2 of Chapter 1, the accumulated inventory cannot be negative so that we must have $y(t) \geq 0$. (If goods are not perishable and cannot be destroyed, we also have $y'(t) \geq 0$. Nonholonomic inequality constraints will be discussed in the next chapter.) In the minimal surface problem of Section 2 of Chapter 1, we also want $y(x) > 0$ to give us a conventional surface of revolution. Until now, we have not had to deal with inequality constraints explicitly. In these and other previous examples with inequality constraints (explicitly stated or not), conditions were assumed to be favorable so that the admissible extremals for the corresponding unconstrained problems automatically satisfy the relevant inequality constraints.

Unfortunately, inequality constraints are not always satisfied by the solution of the related unconstrained problem as seen from the following example:

**EXAMPLE 4:** Minimize

$$J[y] = \int_0^5 [(y')^2 + 4y]\, dx, \tag{11.26a}$$

subject to

$$y(0) = 10, \tag{11.26b}$$

$$y(5) = 0, \tag{11.26c}$$

$$y \geq 6 - 2x \equiv \varphi(x). \tag{11.27}$$

The solution of this problem without the inequality constraint (11.27) is $\hat{y}_f(x) = x^2 - 7x + 10$ with $J[\hat{y}_f] = 235/3$. Unfortunately, $\hat{y}_f(x)$ attains its minimum value at $x_m = 7/2$ with $\hat{y}_f(x_m) = -9/4$ which is less than $\varphi(x_m) = -1$. In fact, we have $\hat{y}_f(x) \leq \varphi(x)$ for $1 < x < 4$ (Fig. 11.1).

For the solution of our problem, it is tempting to take the minimizing $\hat{y}(x)$ to be $\hat{y}_f(x)$ over the portion of $[a, b]$ where the inequality constraint is satisfied (and therefore does not need to be enforced) and to be $\varphi(x)$ where it is not satisfied. We denote this "solution" by $\hat{y}_r(x)$ with

$$\hat{y}_r(x) = \begin{cases} \hat{y}_f(x) & (0 \leq x \leq 1) \\ \varphi(x) & (1 \leq x \leq 4) \\ \hat{y}_f(x) & (4 \leq x \leq 5). \end{cases} \tag{11.28}$$

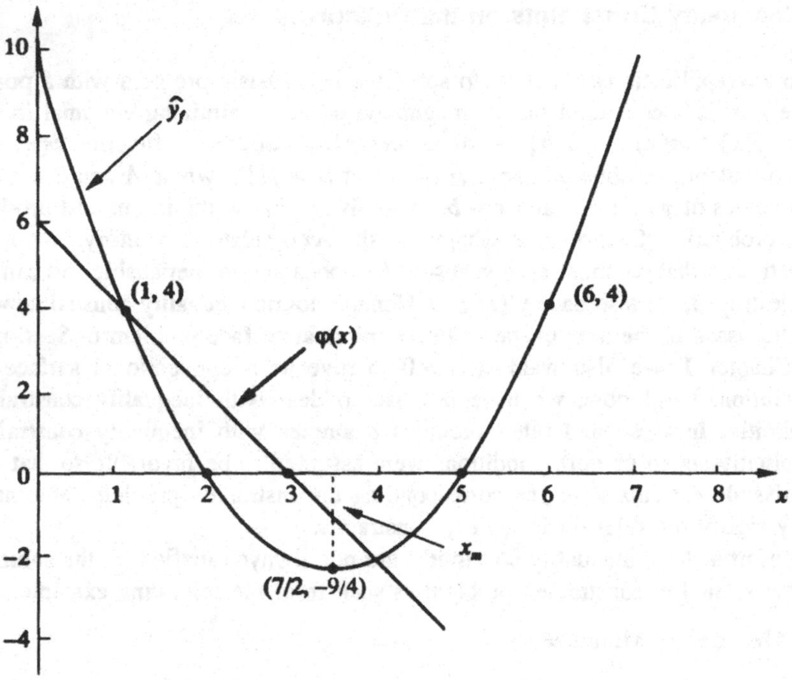


Figure 11.1

It can be verified that $J[\hat{y}_r]$ is 262/3 which is greater than $J[\hat{y}_f]$ as we would expect. It turns out, however, that $\hat{y}_r(x)$ does not minimize $J[y]$ in comparison with other admissible comparison functions. For example, we have $J[\hat{y}_c] = 248/3$ for

$$\hat{y}_c(x) = \begin{cases} x^2 - 6x + 10 & (0 \le x \le 2) \\ 6 - 2x & (2 \le x \le 3) \\ x^2 - 8x + 15 & (3 \le x \le 5). \end{cases} \tag{11.29}$$

and $\hat{y}_c(x)$ satisfies the inequality constraint (11.27) as well as the end conditions (11.26b) and (11.26c) (see Fig. 11.2). Thus, the true solution of the problem is *not* simply $\hat{y}_f$ with portions of it replaced by $\varphi(x)$ in intervals where the constraint is not satisfied. The portions of $\hat{y}_f$ which satisfy the inequality constraint are generally no longer optimal for the constrained problem; the optimality of $\hat{y}_f$ is usually altered by the changes in the unused portion.

What then is an appropriate strategy for problems with inequality constraints on $y(x)$? The answer is provided in part by the following:

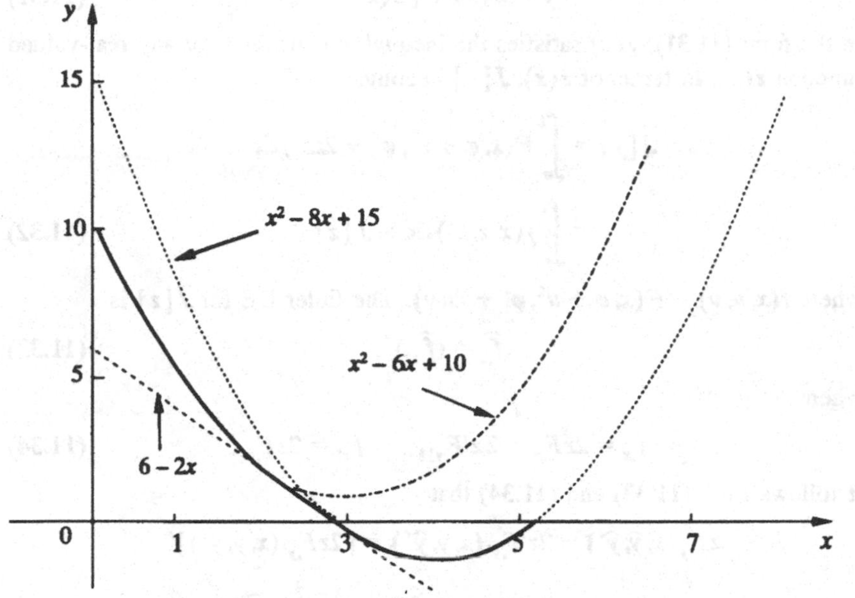


**Figure 11.2**

**[XI.2]** ***If*** $\hat{y}(x)$ ***minimizes*** $J[y]$ ***of the basic problem with the inequality constraint*** $y(x) \geq \varphi(x)$ ***in*** $[a,b]$, ***then*** $\hat{y}(x)$ ***consists of segments of the extremals of*** $J[y]$ ***and segments of*** $\hat{y}(x) = \varphi(x)$. ***At the switch points which join the two different types of segments,*** $\hat{y}(x)$ ***is continuous.***

Suppose the inequality constraint is not binding in the subinterval $(c,d)$ of $[a,b]$. Then we have $\hat{y}(c) = \varphi(c)$ and $\hat{y}(d) = \varphi(d)$ by the continuity of $\hat{y}(x)$, because $\hat{y}$ is required to be PWS. It follows that $\hat{y}$ in $[c,d]$ must be the extremal of the basic problem

$$\min_{y} \left\{ \int_c^d F(x,y,y')\,dx \,\middle|\, y(c) = \varphi(c), y(d) = \varphi(d) \right\} \qquad (11.30)$$

which has the same extremal as $J[y]$. Thus, $\hat{y}(x)$ is an extremal of $J[y]$ in $[c,d]$. This argument (and conclusion) applies to the case of more than one inequality constraint, including $y(x) \geq \varphi(x)$ and $y(x) \leq \psi(x)$, i.e., with both an upper and a lower bound. Hence, wherever the constraints are not effective, the solution must be a free extremal of $J[y]$ with known end values there.

For a more systematic proof of [XI.2], we write $y$ in terms of a *slack function* $z$ by setting

$$y = \varphi(x) + [z(x)]^2. \tag{11.31}$$

In the form (11.31), $y(x)$ satisfies the inequality constraint for any real-valued function $z(x)$. In terms of $z(x)$, $J[y]$ becomes

$$\begin{aligned} J[y] &= \int_a^b F(x, \varphi + z^2, \varphi' + 2zz')\,dx \\ &= \int_a^b f(x, z, z')\,dx \equiv \bar{J}[z], \end{aligned} \tag{11.32}$$

where $f(x, u, v) = F(x, \varphi + u^2, \varphi' + 2uv)$. The Euler DE for $\bar{J}[z]$ is

$$\widehat{f}_{,z} = (\widehat{f}_{,z'})', \tag{11.33}$$

where

$$f_{,z} = 2zF_{,y} + 2z'F_{,y'}, \qquad f_{,z'} = 2zF_{,y'}. \tag{11.34}$$

It follows from (11.33) and (11.34) that

$$\begin{aligned} 2zF_{,y}(x, \widehat{y}, \widehat{y}') + 2z'F_{,y'}(x, \widehat{y}, \widehat{y}') &= [2zF_{,y'}(x, \widehat{y}, \widehat{y}')]' \\ &= 2z'F_{,y'}(x, \widehat{y}, \widehat{y}') + 2z\frac{d}{dx}[\widehat{F}_{,y'}] \end{aligned}$$

or

$$2z\left[\frac{d}{dx}(\widehat{F}_{,y'}) - \widehat{F}_{,y}\right] = 0. \tag{11.35}$$

Thus, either $z$ vanishes [and, hence, $\widehat{y} = \varphi$ by (11.31)] or $\widehat{y}$ is an extremal of the Lagrangian $F$ free of any constraint.

It is now evident that (11.28) is not (likely to be) the correct solution of Example 4 as the switch points are specified in advance and not a part of the solution as implied by [XI.2]. The important question at hand is how we decide on the locations of the switch points. We will answer this question in the next section.

## 4. Binding Inequality Constraints

Suppose an (admissible) extremal of the given problem without the inequality constraints $\widehat{y}_f(x)$ violates one or more inequality constraints over a subinterval $(\bar{c}, \bar{d})$. We must have $\widehat{y}(x) = \varphi(x)$ in some portion of this subinterval for [XI.2] requires that it be an extremal of $J[y]$ otherwise. It is important to observe that the correct solution does not have to be equal to $\varphi(x)$ in the entire subinterval $(\bar{c}, \bar{d})$. By requiring the solution to be equal to $\varphi$ in $(c, d)$, $\bar{c} < c < d < \bar{d}$, and continuous at switching points $c$ and $d$, the corresponding admissible extremal outside $(c, d)$ may now satisfy the inequality constraint in a larger subinterval of $[a, b]$. We summarize this important corollary of [XI.2] as

**[XI.2a]** *If* $\bar{y}_f(x)$ *violates the inequality constraint* $y(x) \geq \varphi(x)$ *in a subinterval* $(\bar{c}, \bar{d})$ *of* $[a, b]$, *then the true solution* $\hat{y}(x)$ *must be equal to* $\varphi(x)$ *in* $(c, d)$ *with* $\bar{c} \leq c < d \leq \bar{d}$.

A method is still needed for finding the proper switch points of one or more subintervals in which the inequality constraint is enforced. For simplicity, suppose (the minimizing extremal of the unconstrained problem) $\bar{y}_f(x)$ is less than or equal to $\varphi(x)$ in $\bar{c} \leq x \leq b$ with strict inequality applied in the subinterval of $(\bar{c}, b)$ and $\varphi(b) = B$. Suppose the actual solution $\hat{y}(x)$ is identical to $\varphi(x)$ only in the interval $c \leq x \leq b$ *for some* $c > a$. (The inequality constraint is said to be *tight* or *binding* whenever strict equality holds for the actual solution $\hat{y}$). By [XI.2a], we should allow $c \geq \bar{c}$ so that an extremal of the given Lagrangian may be used inside $(\bar{c}, c)$ as well. Note that a similar flexibility is not available at the other end because the end condition requires $\hat{y}(b) = \varphi(b) = B$.

We now make use of the observation above to develop a method of solution for $\hat{y}$. Let $c$ be the boundary point which separates the interval $[a, c]$, where the constraint is not binding, from the interval $[c, b]$ where it is. In that case, we have

$$J[y] = \int_a^c F(x, y, y')\,dx + \int_c^b F(x, \varphi, \varphi')\,dx. \tag{11.36}$$

The main idea of our method for the optimal solution $\hat{y}(x)$ (which follows from [XI.2] and [XI.2a]) is to let $c$ be determined by the solution process (and not to assume that it be $\bar{c}$). The optimal solution consists now of the appropriate boundary point $\hat{c}$ and the minimizing extremal $\hat{y}(x)$ in $[a, \hat{c}]$; it must be equal to $\varphi(x)$ in $[\hat{c}, b]$. To find $\hat{y}(x)$ and $\hat{c}$, it suffices to recognize that each of the two integrals in (11.36) is of the type treated in Section 3 of Chapter 3. In the second integral, however, $y$ is equal to $\varphi(x)$ and $y'$ equal to $\psi(x) \equiv \varphi'(x)$; they do not vary.

As in Chapter 3, we work with the parametric representation (3.16) where now $x(\beta) = b$ is fixed, but, for a fixed $\gamma$,

$$y(\gamma) = c \quad (\alpha < \gamma < \beta) \tag{11.37}$$

is an unknown parameter. The integral $J[y]$ is taken in the form

$$J = \int_\alpha^\gamma F(x, y, \frac{y^\bullet}{x^\bullet}) x^\bullet\, dt + \int_\alpha^\beta F(x, \overline{\varphi}, \overline{\psi}) x^\bullet\, dt, \tag{11.38}$$

where $\overline{\varphi}(t) \equiv \varphi(x(t))$ and $\overline{\psi}(t) \equiv \varphi'(x(t))$ are known functions. For a stationary value of $J$, we need

$$\delta J = \int_\alpha^\gamma \{\widehat{F}_{,x} \widehat{x}^\bullet \delta x + \left(\widehat{F} - \frac{\widehat{y}^\bullet}{\widehat{x}^\bullet} \widehat{F}_{,y'}\right) \delta x^\bullet + \widehat{x}^\bullet \widehat{F}_{,y} \delta y + \widehat{F}_{,y'} \delta y^\bullet\}\, dt$$

$$+ \int_\gamma^\beta \{\widehat{x}^\bullet F_{,x}(\widehat{x}, \overline{\varphi}, \overline{\psi}) \delta x + F(\widehat{x}, \overline{\varphi}, \overline{\psi}) \delta x^\bullet\}\, dt = 0. \tag{11.39}$$

After integration by parts and observing the Euler–Lagrange equations of [II.3] as well as the end conditions $\delta x = \delta y = 0$ at $\alpha$ and $\beta$, relation (11.39) simplifies to

$$[(\widehat{F} - \widehat{y}' \widehat{F}_{,y'}) \delta x + \widehat{F}_{,y'} \delta y]_{t=\gamma_-} = [F(\widehat{x}, \overline{\varphi}, \overline{\psi}) \delta x]_{t=\gamma_+} \tag{11.40}$$

with $\widehat{x}(\gamma) = \widehat{c}$. At $t = \gamma$, we have $\delta y = \delta\varphi = \varphi'(\widehat{c}) \delta x(\gamma)$ so that (11.40) requires

$$\widehat{F}(\widehat{c}) - \Phi(\widehat{c}) - [\widehat{y}'(\widehat{c}) - \varphi'(\widehat{c})] \widehat{F}_{,y'}(\widehat{c}) = 0 \tag{11.41}$$

as the transversality condition for $\widehat{c}$ with

$$\Phi(\widehat{c}) = F(\widehat{c}, \varphi(\widehat{c}), \varphi'(\widehat{c})). \tag{11.42}$$

We have then

**[XI.3]** ***An optimal switch point $\widehat{c}$ between $\varphi(x)$ and an extremal of $J[y]$ is determined by (11.41) and the continuity requirement $\widehat{y}(\widehat{c}) = \varphi(\widehat{c})$.***

Equation (11.41) provides the condition we need to specify $\widehat{c}$. It should be clear from the development above that there is a similar optimality condition for every junction $\widehat{c}_i$ which separates an interval where an inequality constraint is binding from an interval where it is not binding.

To stress the continuity requirement $\widehat{y}(\widehat{c}) = \varphi(\widehat{c})$, it is often useful to set $f(z) \equiv F(\widehat{c}, \varphi(\widehat{c}), z) = F(\widehat{c}, \widehat{y}(c), z)$. We can then write the *optimal switching condition* (11.42) as

$$f(\varphi'(\widehat{c})) - f(\widehat{y}'(\widehat{c})) - [\varphi'(\widehat{c}) - \widehat{y}'(\widehat{c})] f^\bullet(\widehat{y}'(\widehat{c})) = 0, \tag{11.41'}$$

where $f^\bullet(z) = F_{,y'}(\widehat{c}, \varphi(\widehat{c}), z) = F_{,y'}(\widehat{c}, \widehat{y}(\widehat{c}), z)$. By Taylor's theorem, the left-hand side of (11.41′) is equal to $f^{\bullet\bullet}(\overline{z})[\varphi'(\widehat{c}) - \widehat{y}'(\widehat{c})]^2/2$ for some value $\overline{z}$ with $|\overline{z} - \widehat{y}'(\widehat{c})| \le |\varphi'(\widehat{c}) - \widehat{y}'(\widehat{c})|$. It follows from condition (11.41′) that we must have either $f^{\bullet\bullet}(\overline{z}) \equiv F_{,y'y'}(\widehat{c}, \widehat{y}(\widehat{c}), \overline{z}) = 0$ or $\varphi'(\widehat{c}) = \widehat{y}'(\widehat{c})$. For our example (11.26) and (11.27), we have $F_{,y'y'} = 2 \neq 0$; hence, the two segments

$\hat{y}(x)$ and $\varphi(x)$ join together smoothly (with the same slope as well as the same function value) at the switch point $\hat{c}$.

**[XI.4]** *If* $\hat{F}_{,y'y'}(\hat{c},\varphi(\hat{c}),z) > 0$ for $|z - \hat{y}'(\hat{c})| \le |\varphi'(\hat{c}) - \hat{y}'(\hat{c})|$, then $\hat{y}'(\hat{c}) = \varphi'(\hat{c})$.

The theorem still holds if $F_{,y'y'} > 0$ is replaced by $\hat{F}_{,y'y'} < 0$ so that it applies to problems of maximization as well.

**EXAMPLE 4 (continued):**

From the solution $\hat{y}_f(x)$ of the unconstrained problem and Figure 11.1, we see that the actual solution should be an extremal adjacent to the two end points with the constraint made binding in an interior subinterval $(\hat{x}_1, \hat{x}_2)$:

$$\hat{y}(x) = \begin{cases} x^2 + c_1 x + c_0 & (0 \le x \le \hat{x}_1) \\ \varphi(x) & (\hat{x}_1 \le x \le \hat{x}_2) \\ x^2 + c_3 x + c_2 & (\hat{x}_2 \le x \le 5) \end{cases}$$

The end condition $\hat{y}(0) = 10$, the continuity condition $\hat{y}(\hat{x}_1) = \varphi(\hat{x}_1)$, and the optimal switching condition (11.41) for $\hat{x}_1$ [or the simpler smoothness condition $\hat{y}'(\hat{x}_1) = \varphi'(\hat{x}_1)$ which applies in this case] require $c_0 = 10$, $c_1 = -6$, and $x_1 = 2$. Similarly, the end condition $\hat{y}(5) = 0$, the continuity condition $\hat{y}(\hat{x}_2) = \varphi(\hat{x}_2)$, and the optimal junction condition for $\hat{x}_2$ require $c_3 = -8$, $c_2 = 15$, and $x_2 = 3$. The final solution then is $\hat{y}_c(x)$ as given by (11.29) with $J[\hat{y}] = 248/3 < 262/3 = J[\hat{y}_r]$ (see also Fig. 11.2).

## 5. Brachistochrone with Limited Descent

As an application of [XI.2], [XI.3] and [XI.4] in particle dynamics, we consider a modified version of the Brachistochrone problem. Again, we wish to minimize

$$J[y] = \int_a^b \sqrt{\frac{1 + (y')^2}{2g(A - y)}}\, dx \qquad (a < x < b) \tag{11.43}$$

which is the time of descent of the bead along the curve $y(x)$ (with $y$ positive upward). The bead is initially at rest at point $(a,A)$ so that $y(a) = A$. Previously, we sought $y(x)$, which allows the bead to reach a lower point $(b,B)$ in the fastest time. We found the solution for that problem to be the cycloid

$$x = a + \sigma(\theta - \sin\theta), \qquad y = A - \sigma(1 - \cos\theta) \quad (0 < \theta < \theta_1), \tag{11.44}$$

where $\theta_1$ and $\sigma$ are determined by the conditions $x(\theta_1) = b$ and $y(\theta_1) = B$.

The new problem we wish to consider here is one with $b$ *fixed and* $y(b)$ unspecified but $y(x)$ restricted to be not less than $G(<A)$, i.e., $y(x) \geq G$ for all $x$ in $[a,b]$. Here, $G$ may be the ground below which the bead cannot go (Fig. 11.3). Without the inequality constraint, the problem is solved with the Euler boundary condition $\widehat{F}_{,y'}(b) = 0$ derived in Chapter 3. With the inequality constraint, the cycloid solution may still apply if $A - G$ is sufficiently large. Given that $A - y = \sigma(1 - \cos\theta)$ is maximum at $\theta = \pi$ (for which $x - a = \pi\sigma$), we need

$$A - G \geq \frac{2(b-a)}{\pi} \tag{11.45}$$

for the lowest point $y(b)$ of the cycloid not to be below ground level.

If condition (11.45) is not satisfied, then we expect the solution $\widehat{y}(x)$ to be an extremal of $J[y]$ for $a \leq x \leq c$ and $\widehat{y} = G$ for $c \leq x \leq b$ given that the extremals for the problem are generally monotone decreasing. It remains to determine the switch point $c$ of the composite minimizing $\widehat{y}$:

$$\widehat{y} = \begin{cases} Y(x) & (a \leq x \leq c) \\ G & (c \leq x \leq b), \end{cases} \tag{11.46}$$

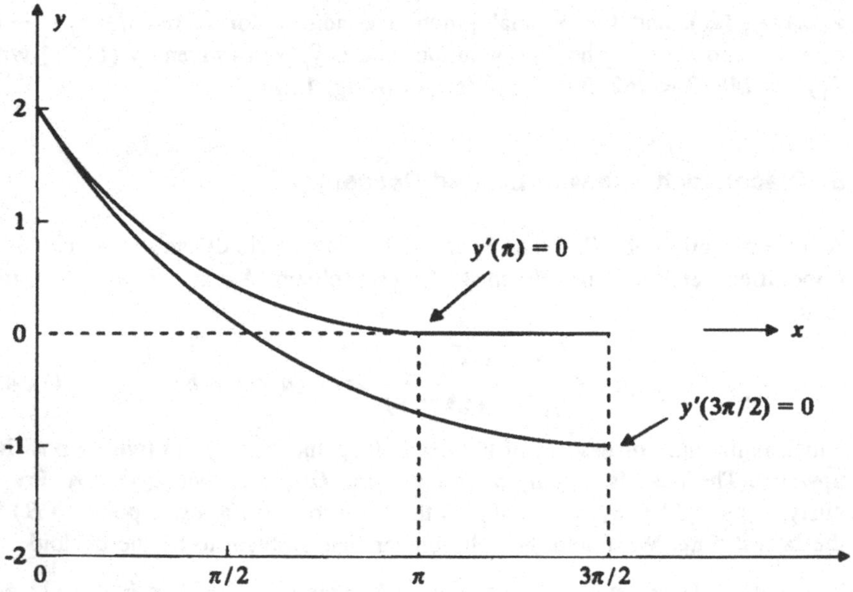


Figure 11.3

where $Y(x)$ is the cycloid in (11.44) with $\theta_1$ and $\sigma$ determined by $y(\theta_1) = G$, $x(\theta_1) = c$. By [XI.4], the optimal switch point $c = \widehat{c}$ is determined by

$$Y'(\widehat{c}) = \varphi'(\widehat{c}) = 0 \tag{11.47}$$

because we have $\widehat{F}_{,y'y'} > 0$ for our problem. With

$$\frac{dy}{d\theta} = -\sigma \sin\theta, \qquad \frac{dx}{d\theta} = \sigma(1 - \cos\theta),$$

condition (11.47) requires $\theta_1 = \pi$ (with $dx/d\theta = 2\sigma$ there) so that

$$G = A - 2\sigma \quad \text{or} \quad \sigma = \frac{1}{2}(A - G).$$

It follows that the curve $Y(x)$ in (11.46) is given parametrically by

$$\begin{cases} x(\theta) = a + \dfrac{1}{2}(A - G)(\theta - \sin\theta) \\ y(\theta) = A - \dfrac{1}{2}(A - G)(1 - \cos\theta) \end{cases} \qquad (0 < \theta < \pi). \tag{11.48}$$

The switch point $\widehat{c}$ is, in turn, given by

$$\widehat{c} = x(\pi) = a + \frac{1}{2}(A - G)\pi. \tag{11.49}$$

The time of descent $J[\widehat{y}]$ can be calculated by integrating $\widehat{F}$ with respect to $\theta$ (instead of $x$) in the range $[a, \widehat{c}]$:

$$\begin{aligned} J[\widehat{y}] &= \int_0^\pi \sqrt{\frac{(x^\bullet)^2 + (y^\bullet)^2}{2g(A - y)}}\, d\theta + \int_{\hat{c}}^b \frac{dx}{\sqrt{2g(A - G)}} \\ &= \pi \sqrt{\frac{A - G}{2g}} + \frac{b - \hat{c}}{\sqrt{2g(A - G)}} \\ &= \frac{\pi}{2}\sqrt{\frac{A - G}{2g}} + \frac{b - a}{\sqrt{2g(A - G)}}, \end{aligned} \tag{11.50}$$

where $(\ )^\bullet \equiv d(\ )/d\theta$ and use has been made of (11.49) to express $\widehat{c}$ in terms of known parameters. The value of $J[\widehat{y}]$ should be compared with the minimum time $T_{\min} = \sqrt{\pi(b - a)/g}$ for the unconstrained problem with an unspecified end value $y(b)$. (Note that $T_{\min}$ is independent of the initial position of the bead.) We expect $T_{\min} < J[\widehat{y}]$ and it can be shown to hold with the help of the known inequality $\sqrt{1 + 2\alpha^2} = \sqrt{(1 + \alpha^2)^2 - \alpha^4} < 1 + \alpha^2$ (for any real $\alpha$) and the relation $(b - a) > \pi(A - G)/2$. In particular, $J[\widehat{y}]$ reduces to $T_{\min}$ for

$G = A - 2(b-a)/\pi$ which is the value of $\widehat{y}(b)$ for the unconstrained problem (with an unspecified terminal value).

## 6. Inequality Constraints on an End Point

Inequality constraints may also be imposed on an end point of the minimizing function. For example, the end $b$ of the interval $[a,b]$ may be freely chosen as long as it does not exceed a given upper bound $\overline{b}$ so that $b \leq \overline{b}$. This situation arises frequently in problems where $x$ is time and the time-dependent process $y(x)$ may be terminated at some optimal time with a maximum cutoff for maximum gain or minimum loss. Another type of inequality constraints on end points is on an end value of $y(x)$. Typically, $y(b)$ is not specified but it cannot be below a specified lower limit $\overline{B}$ so that $y(b) \geq \overline{B}$. The lower bound $\overline{B}$ is often zero.

To develop a method of solution for these problems, we again consider the minimization of

$$J[y] = \int_a^b F(x,y,y')dx, \qquad y(a) = A, \tag{11.51}$$

with

$$y(b) \geq \overline{B} \tag{11.52a}$$

and/or

$$b \leq \overline{b}. \tag{11.52b}$$

Suppose $\widehat{y}(x)$ minimizes $J[y]$ over $[a,\widehat{b}]$ and $\widehat{y}(\widehat{b}) = \widehat{B}$. Because of the constraints (11.52a) and/or (11.52b), $\widehat{J} \equiv J[\widehat{y}]$ may not be a stationary value. All that we can say is $\delta J \geq 0$. To derive the necessary conditions for a (weak local) minimum for this free boundary problem, we will use the parametric form (3.18b) of the solution curve as in Chapter 3. By a straightforward calculation similar to that of Section 4 of that chapter, we obtain

$$\delta J = \int_\alpha^\beta \{\widehat{x}^\bullet \widehat{F}_{,x}\delta x + [\widehat{F} - \widehat{y}'\widehat{F}_{,y'}]\delta x^\bullet + \widehat{x}^\bullet \widehat{F}_{,y}\delta y + \widehat{F}_{,y'}\delta y^\bullet\}dt$$

$$= \ldots = [\{\widehat{F} - \widehat{y}'\widehat{F}_{,y'}\}\delta x + \widehat{F}_{,y'}\delta y]_{t=\beta} \geq 0.$$

By writing $\widehat{x}(\beta) = \widehat{b}, \widehat{y}(\beta) = \widehat{B}, \delta x(\beta) = db$, and $\delta y(\beta) = dB$, the inequality above becomes

$$[\widehat{F}(\widehat{b}) - \widehat{y}'(\widehat{b})\widehat{F}_{,y'}(\widehat{b})]db + \widehat{F}_{,y'}(\widehat{b})dB \geq 0. \tag{11.53}$$

We now consider the comparisons with $db = 0$. For this case, we have from (11.53)

$$\widehat{F}_{,y'}(\widehat{b})\,dB \geq 0. \tag{11.54}$$

We are interested here in the case where the terminal value of $y$ is neither fixed [otherwise $dB = 0$ and no further requirement results from (11.54)] nor freely varying [otherwise we recover the Euler boundary condition $\widehat{F}_{,y'}(\widehat{b}) = 0$]. Instead, we have the constraint (11.52a) which requires that $\widehat{B} > \overline{B}$ or $\widehat{B} = \overline{B}$. In the first case, $B = \widehat{B} + dB$ may be above or below $\widehat{B}$ so that $dB$ may be of either sign. Condition (11.54) then requires the coefficient of $dB$ to vanish for $\delta J \geq 0$, i.e.,

$$\widehat{F}_{,y'}(\widehat{b}) = 0 \quad (\widehat{B} > \overline{B}). \tag{11.55a}$$

In this case, the inequality constraint is not active (not binding, not tight, etc.) and the result is the same as if there were no constraint on $B$. In the second case, $\widehat{B} = \overline{B}$, admissible modifications of $\widehat{B}$ cannot be lower so that $dB \geq 0$. For $\delta J \geq 0$, expression (11.54) requires for this case

$$\widehat{F}_{,y'}(\widehat{b}) \geq 0 \quad (\widehat{B} = \overline{B}). \tag{11.55b}$$

We now combine (11.55a) and (11.55b) to get the equivalent requirement:

**[XI.6]** ***Let*** $\widehat{y}(x)$ ***minimize (11.51) for a fixed*** b ***subject to*** $y(a) = A$ ***and (11.52a). Then*** $\widehat{y}(x)$ ***must satisfy the three conditions***

$$\widehat{B} \geq \overline{B}, \qquad \widehat{F}_{,y'}(\widehat{b}) \geq 0, \qquad (\widehat{B} - \overline{B})\widehat{F}_{,y'}(\widehat{b}) = 0, \tag{11.56}$$

***where*** $\widehat{B} = \widehat{y}(\widehat{b})$.

**EXAMPLE 5:** Find $\widehat{y}(x)$ which minimizes

$$J[y] = \int_0^5 [(y')^2 + 4y]\,dx, \quad y(0) = 10$$

with $y(5) = B$ unspecified except that $B \geq \overline{B}$.

The extremals of $J[y]$ are $\widehat{y} = x^2 + c_1 x + c_0$. The end condition, $y(0) = 10$, requires $c_0 = 10$. If $\widehat{B} > \overline{B}$, then (11.56) requires $\widehat{F}_{,y'}(5) = 2\widehat{y}'(5) = 2[2(5) + c_1] = 0$ or $c_1 = -10$. In that case, $\widehat{B} = \widehat{y}(5) = -15$.

(*i*) If $\overline{B} = -20$, then $\widehat{B} > \overline{B}$ and the minimizing extremal is $\widehat{y} = x^2 - 10x + 10$.

(*ii*) If $\overline{B} = -10 > -15$, then we have to set $\widehat{y}(5) = \overline{B} = -10$ which requires $c_1 = -9$ and therewith $\widehat{y} = x^2 - 9x + 10$. For this admissible extremal, we have $\widehat{y}'(5) = 1$ so that $\widehat{F}_{,y'}(5) = 2 > 0$ as required by [XI.6]

We now return to the general problem (11.51) and (11.52) for which we have (11.53) as a necessary condition for a minimum. Suppose the comparison functions are those with a prescribed $B$ so that $dB = 0$ instead. Then, we have from (11.53)

$$[\hat{F}(\hat{b}) - \hat{y}'(\hat{b})\hat{F}_{,y'}(\hat{b})]\,db \geq 0. \tag{11.57a}$$

We are interested in the case where $b$ is neither fixed [otherwise we have $db = 0$ and no further requirement results from (11.56)] nor freely varying (otherwise we get [III.3]). Instead, we require that the inequality constraint (11.52b) be satisfied so that $\hat{b} < \bar{b}$ or $\hat{b} = \bar{b}$. In the first case, we have that $b = \hat{b} + db$ may be to the left or to the right of $\hat{b}$ so that $db$ may be of either sign. In that case, condition (11.53) or (11.57a) holds only if we have

$$\hat{F}(\hat{b}) - \hat{y}'(\hat{b})\hat{F}_{,y'}(\hat{b}) = 0 \qquad (\hat{b} < \bar{b}). \tag{11.57b}$$

When the inequality constraint is not binding, the result is the same as [III.3] for the case with no inequality constraint on $b$. In the second case ($\hat{b} = \bar{b}$), admissible comparison values for $\hat{b}$ cannot be larger so that $db < 0$. For $\delta J \geq 0$ to hold, expression (11.56) requires

$$\hat{F}(\hat{b}) - \hat{y}'(\hat{b})\hat{F}_{,y'}(\hat{b}) < 0 \qquad (\hat{b} = \bar{b}). \tag{11.57c}$$

We combine (11.57b) and (11.57c) to get

**[XI.7]** ***Let $\hat{y}(x)$ and $\hat{b}$ minimize (11.51) subject to fixed terminal values at both the fixed end a and the unspecified (but constrained) end*** **b**. ***Then $\hat{y}(x)$ and $\hat{b}$ must satisfy the three conditions***

$$\hat{b}\left\{\begin{matrix}<\\=\end{matrix}\right\}\bar{b}, \qquad [\hat{F}(\hat{b}) - \hat{y}'(\hat{b})\hat{F}_{,y'}(\hat{b})]\left\{\begin{matrix}=\\<\end{matrix}\right\}0,$$
$$(\bar{b} - \hat{b})[\hat{F}(\hat{b}) - \hat{y}'(\hat{b})\hat{F}_{,y'}(\hat{b})] = 0. \tag{11.58}$$

An application of this result will be given in the next section.

## 7. Land Use in a Long and Narrow City

A business district of prescribed total area $A$ is to be built within a rectangular area of a fixed width $W$ and maximum length $\bar{b}$ with $W\bar{b} > A$. The remaining land within the rectangular region is to be used for roads needed for the movement of traffic generated by the business district (see Fig. 11.4). Each square foot of business area generates $T$ tons of traffic (per unit time) with destination uniformly distributed over the remaining business area. The cost for moving one ton of traffic one foot lengthwise is taken to be proportional to $(v/w)^k$, where $v(x)$ is the total volume of traffic passing through the point $x$

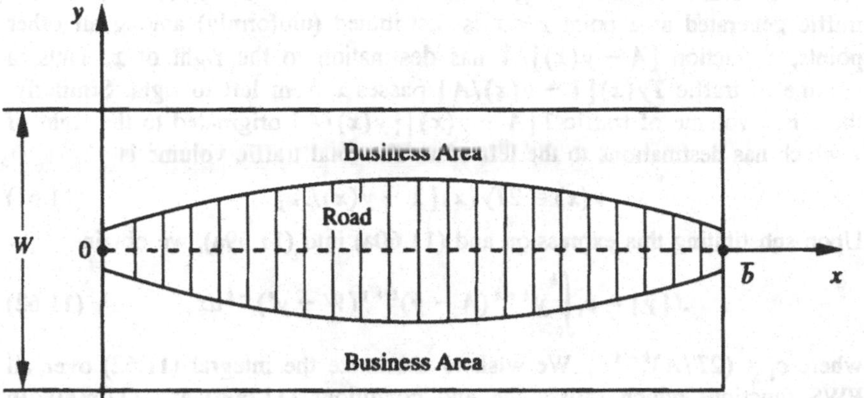


**Figure 11.4**

of the lengthwise coordinate (per unit time) and $w(x)$ is the road width at $x$. Hence, the total transport cost in the city (per unit time) is

$$J = c_0 \int_0^b v(x)[v(x)/w(x)]^k \, dx, \tag{11.59a}$$

where $c_0$ is the constant of proportionality in the unit traffic moving cost function $(= c_0[v/w]^k)$. With the volume of traffic $v(x)$ expected to depend on the road width $w(x)$, we want to choose $w(x)$ and the length of the city $b$ to minimize $J$ subject to the inequality end point constraint

$$b \leq \bar{b}. \tag{11.59b}$$

To relate $v$ to $w$, we introduce a new variable $y(x)$ by

$$y' = W - w(x), \tag{11.60a}$$

$$y(0) = 0 \tag{11.60b}$$

so that the total business area to the left of $x$ is

$$y(x) = \int_0^x [W - w(t)] \, dt, \tag{11.60c}$$

with the total business area to the left of $b$ being the prescribed area $A$:

$$y(b) = A. \tag{11.60d}$$

Now the volume of traffic originated to the left of $x$ is $Ty(x)$. Because the traffic generated at a point $z < x$ is distributed (uniformly) among all other points, a fraction $[A - y(x)]/A$ has destination to the right of $x$. Thus, a volume of traffic $Ty(x)[1 - y(x)/A]$ passes $x$ from left to right. Similarly, there is a volume of traffic $T[A - y(x)][y(x)/A]$ originated to the right of $x$ which has destinations to the left of $x$. The total traffic volume is

$$v(x) = 2Ty(x)[1 - y(x)/A]. \tag{11.61}$$

Upon substituting this expression and (11.60a) into (11.59a), we obtain

$$J[y] = c_1 \int_0^b y^{k+1}(A - y)^{k+1}(W - y')^{-k}\,dx, \tag{11.62}$$

where $c_1 = (2T/A)^{k+1}c_0$. We wish to minimize the integral (11.62) over all PWS functions which satisfy the end conditions (11.60d) and (11.60b). In addition, we are still free to choose the length parameter $b$ in the optimization process subject to the inequality end constraint (11.59b) and the nonholonomic (inequality) constraint $0 \le y' \le W$ on the road width throughout the length of the city. The integrand of (11.62) is real only if $W \ge y'$ and $w(x) \le W$; hence, we expect the inequality constraint on $y'$ to be automatically satisfied by our solution and, therefore, will not be binding.

Because the integrand of (11.62) does not depend on $x$ explicitly, the corresponding Euler DE has a first integral

$$\widehat{F} - \widehat{y}'\widehat{F}_{,y'} = \left[\frac{\widehat{y}(A - \widehat{y})}{W - \widehat{y}'}\right]^{k+1}[W - (k+1)\widehat{y}'] = C \tag{11.63}$$

for some constant of integration $C$. By [XI.7], we should investigate the two cases $\widehat{b} < \overline{b}$ and $\widehat{b} = \overline{b}$ separately. In either case $\widehat{y}'$ is symmetric about $b/2$ (see Exercise 10).

For $\widehat{b} < \overline{b}$, the conditions in (11.58) require $\widehat{F}(\widehat{b}) - \widehat{y}'(\widehat{b})\widehat{F}_{,y'}(\widehat{b}) = 0$. This implies $C = 0$ in (11.63). Because $\widehat{y}(x)$ is neither zero nor $A$ for all $x$ in $[0, \widehat{b}]$, we must have $(k+1)\widehat{y}'(x) = W$ or in view of (11.60b)

$$\widehat{y}(x) = \frac{Wx}{k+1}. \tag{11.64}$$

The end condition (11.60d) then requires

$$\frac{W\widehat{b}}{k+1} = A \tag{11.65a}$$

or

$$\widehat{b} = \frac{(k+1)A}{W}. \tag{11.65b}$$

From (11.60a) and (11.64), we see that

***The road width is uniform throughout the length of the city with***

$$\mathrm{w(x)} = \frac{\mathrm{kW}}{\mathrm{k+1}} \quad (0 \le \mathrm{x} \le \hat{\mathrm{b}} < \bar{\mathrm{b}}) \tag{11.66}$$

***if*** $(\mathrm{k}+1)\mathrm{A}/\mathrm{W}$ ***is less than*** $\bar{\mathrm{b}}$.

If $(k+1)A/W$ is not less than $\bar{b}$, then the solution (11.64) does not apply. For now we must have $\hat{b} = \bar{b}$ and the (11.57c) part of [XI.7] requires that $[\hat{F}(\hat{b}) - \hat{y}'(\hat{b})\hat{F}_{y'}(\hat{b})] < 0$ or, because $C = 0$ is no longer feasible, $C \equiv -D^{2(k+1)} < 0$ for some (positive) real number $D$. In that case, (11.63) requires that $W - (k+1)\hat{y}' < 0$ or

$$w(x) < \frac{kW}{k+1}. \tag{11.67}$$

Hence:

> ***When the available land is not long enough, the width of the road is everywhere narrower than the optimal width of (11.66) for the unconstrained configuration.***

It is possible to obtain an exact solution of the first-order ODE (11.63). To do so, we first solve (11.63) for $\hat{y}$ in terms of $\hat{y}'$ to get

$$\hat{y} = \frac{A}{2}\left\{1 \pm \sqrt{1 - \frac{4D^2(W-\hat{y}')}{A^2[(k+1)\hat{y}' - W]^{1/(k+1)}}}\right\}. \tag{11.68}$$

For $\hat{y}(0)$ to be zero as required by (11.60b), we need

$$\hat{y}'(0) = W \tag{11.69}$$

and the negative sign in (11.68). Condition (11.69) and the ODE for $y$ (11.60a) tell us that:

> ***If the end constraint (11.59b) is binding, the width of the road narrows to zero at the end*** $\mathrm{x} = 0$ ***(and by symmetry also at the end*** $\mathrm{x} = \bar{\mathrm{b}}$ ***as well).***

With the negative square roots, the ODE (11.68) is of the form

$$\hat{y} = Q(p; D^2) \equiv \frac{A}{2}\left\{1 - \sqrt{1 - \frac{4D^2(W-p)}{A^2[(k+1)p - W]^{1/(k+1)}}}\right\}. \tag{11.70}$$

where $p = \hat{y}'$. An exact solution of this type of equation is always possible by differentiating both sides of $y = Q(p; D^2)$ with respect to $x$ to get $\hat{y}' = (dQ/dp)p'$ or

$$p = q(p;D^2)p', \qquad q(p;D^2) \equiv \frac{dQ}{dp}. \tag{11.71}$$

The ODE (11.71) is separable; hence, we can always get $x(p)$ as a quadrature:

$$x = -\int_p^W \frac{q(p;D^2)\,dp}{p} \equiv R(p;W,D^2) \tag{11.72}$$

where we have used condition (11.69) to fix the constant of integration. Relations (11.70) and (11.72) provide a parametric representation of the function $\widehat{y}(x)$ with $p$ as the parameter. For our problem, the integrand of the integral in (11.72) is

$$-\frac{q(p;D^2)}{p} = \frac{kD^2/A}{[(k+1)p - W]^{(k+2)/(k+1)}} \times \left[1 - \frac{4D^2}{A^2}\frac{W-p}{\{(k+1)p - W\}^{1/(k+1)}}\right]^{-1/2}. \tag{11.73}$$

The remaining unknown constant $D^2$ may be determined by condition (11.60d) which, by symmetry, will be used in the form $\widehat{y}(\overline{b}/2) = A/2$. By (11.68), this condition requires

$$\frac{4D^2[W - p_m]}{A^2[(k+1)p_m - W]^{1/(k+1)}} = 1 \tag{11.74}$$

which gives $D^2$ in terms of $p_m \equiv p(\overline{b}/2)$. The unknown constant $p_m$ is then determined by (11.72) evaluated at $x = \overline{b}/2$:

$$\frac{\overline{b}}{2} = -\int_{p_m}^W \frac{q(p;D^2)}{p}\,dp = R(p_m;W,D^2) \equiv \overline{R}(p_m;W) \tag{11.75}$$

after expressing $D^2$ in terms of $p_m$ by (11.74).

Although the actual shape of the road for the case $(k+1)A/W \geq \overline{b}$ can only be found after we evaluate (11.75) and (11.72) (numerically or asymptotically), some qualitative feature of the road width can be deduced without evaluating the relevant integrals. From (11.71) and (11.73), we have

$$\begin{aligned} y'' = p' &= \frac{p}{q} \\ &= -\frac{A}{kD^2}[(k+1)p - W]^{(k+2)/(k+1)} \\ &\quad \times \left[1 - \frac{4D^2}{A^2}\frac{W-p}{\{(k+1)p - W\}^{1/(k+1)}}\right]^{1/2}. \end{aligned} \tag{11.76}$$

By (11.74) and (11.69), we have

$$y''(\overline{b}/2) = 0, \;\; y''(0) = -A\,(kW)^{(k+2)/(k+1)}/kD^2 < 0$$

and $y''(x)$ does not change the sign in $0 \le x < \overline{b}/2$. With $w'(x) = -\widehat{y}''$, we see that:

> *If* $(k+1)A/W > \overline{b}$, *the road width is no longer uniform. Instead, it is monotone increasing and concave, from (zero at) both ends of the city to a maximum width at the city center* $x = \overline{b}/2$.

Thus, Figure 11.4 (with the road width narrowed to a point at both ends) gives a qualitatively accurate description for this case.

## 8. Exercises

1. Use spherical coordinates $(r, \theta, \varphi)$ to obtain the geodesic between two points on the surface of a right circular cone with a semi vertex angle $\alpha$ (so that $ds^2 = dr^2 + r^2 \sin^2\alpha\, d\theta^2$).

2. Use cylindrical coordinates $(r, \theta, z)$ to obtain the geodesic between two points on a surface of revolution described by $z = f(r)$ (and, of course, $x = r\cos\theta$ and $y = r\sin\theta$ so that $ds^2 = [1 + (f')^2]\,dr^2 + r^2 d\theta^2$). Leave your solution in the form of an integral.

3. A particle moves on the surface $\varphi(x, y, z) = 0$ from $(x_1, y_1, z_1)$ to $(x_2, y_2, z_2)$ in the time period $[0, T]$. It moves in such a way that total kinetic energy is a minimum, i.e.,

$$\min_{(x,\,y,\,z)} \left[ J = \int_0^T \frac{1}{2}(\dot{x}^2 + \dot{y}^2 + \dot{z}^2)\,dt \,\middle|\, \varphi(x, y, z) = 0, x(0) = x_1, \ldots, z(T) = z_2 \right].$$

   (a) Show that $\ddot{x}/\varphi_{,x} = \ddot{y}/\varphi_{,y} = \ddot{z}/\varphi_{,z}$.

   (b) For a unit sphere, we have $\varphi = x^2 + y^2 + z^2 - 1 = 0$. Suppose the particle moves from $(0, 0, 1)$ to $(0, 0, -1)$ in time $T$. Show that $r = \sqrt{x^2 + y^2} = \sin(n\pi t/T)$, $\theta = \tan^{-1}(y/x) = \theta_0$ and $z = \cos(n\pi t/T)$, where $n$ is an odd integer.

   (c) Show that $J$ is a minimum when $n = 1$ and is equal to $\pi^2/2T$.

4. Consider the simple pulley system of Figure 11.5 where $y_1$ and $y_2$ are the vertical distances of the masses $m_1$ and $m_2$ from the pulley center pivot, respectively. Because the length $\ell$ of the cord connecting the two masses is fixed, we have $y_1 + y_2 = \ell$. The potential energy associated with the

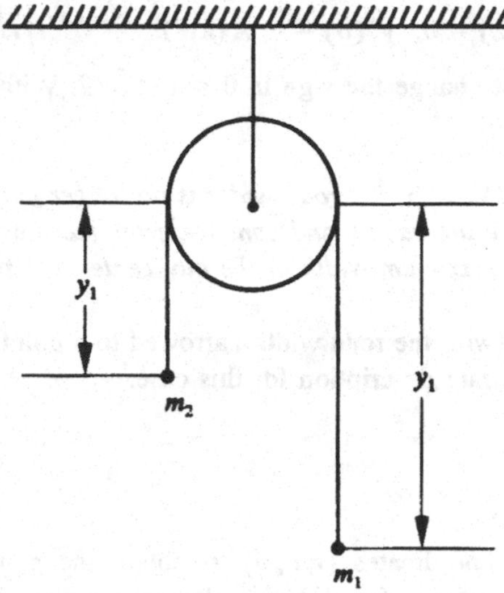


**Figure 11.5**

masses is due to gravity alone. By Hamilton's principle, the motion of the masses is governed by the minimization of the energy differential $J[y]$ given by

$$J[y] = \int_{t_1}^{t_2} \left\{ \frac{1}{2}[m_1(\dot{y}_1^2) + m_2(\dot{y}_2)^2] + mgy_1 + m_2 g y_2 \right\} dt.$$

(a) Deduce the Euler DE for $J[y]$ subject to the holonomic constraint $y_1 + y_2 = \ell$.

(b) Obtain an expression for the relevant Lagrange multiplier and deduce the physical interpretation by its appearance in the Euler DE.

(c) Obtain a single equation of motion for $y_1$ alone.

5. Consider the rolling of a right cylinder of mass $m$ and radius $R_1$ on another cylinder of the same mass but of a different radius $R_2$ (see Fig. (11.6)). The rolling motion of the points $P$ and $P'$ (of equal arclength from the point of contact $Q$ so that $\widehat{QP} = \widehat{QP'}$) on the two cylinders is again subject to Hamilton's principle. With $\theta_1$ and $\theta_2$ as the two primary angular displacement variables, Hamilton's principle requires the motion to minimize

$$J[r, \theta_1, \theta_2] = \int_{t_1}^{t_2} \left\{ \frac{1}{2} m \left( \dot{r}^2 + r^2 \dot{\theta}_1^2 + \frac{1}{2} R_2^2 \dot{\theta}_2^2 \right) - mgr\cos\theta_1 \right\} dt$$

subject to the holonomic constraints

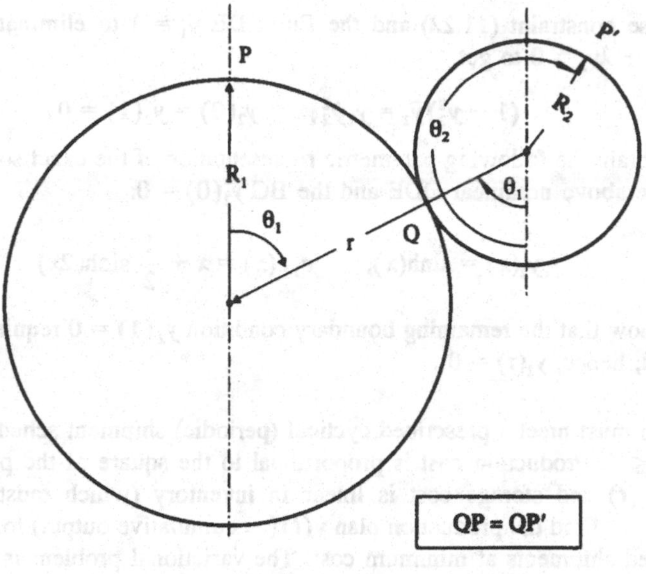


Figure 11.6

$$R_1 + R_2 = r, \qquad R_1\theta_1 = R_2(\theta_2 - \theta_1).$$

The first constraint is geometrically evident. The second of the constraints is a requirement imposed by frictional contact.

(a) For rigid cylinders, $r$ is a constant (because $R_1$ and $R_2$ remain unchanged). Use this directly to eliminate the $\dot{r}^2$ term in $J$ and to think of $J$ as $J[\theta_1, \theta_2]$. Obtain the Euler DE for $J[\theta_1, \theta_2]$ subject to the remaining constraint.

(b) Interpret the Lagrange multiplier.

(c) Obtain a single equation of motion for $\theta_1$ alone in the form

$$\ddot{\theta}_1 = \frac{2g}{3r}\sin\theta_1 .$$

(d) Do not make use of the fact that $r = R_1 + R_2$ is a constant directly but introduce both constraints by two multipliers instead. Obtain three Euler DEs for the new problem.

(e) Use the fact that $r$ is a constant in the new Euler DE of part (*d*) to obtain a physical interpretation of the additional multiplier. What happens physically when the multiplier vanishes?

6. (a) Use constraint (11.22) and the Euler DE $\ddot{y}_1 = \lambda$ to eliminate $\lambda$ from $\ddot{y}_2 + \lambda y_2 = 0$ to get

$$(1 - y_2^2)\ddot{y}_2 = y_2\dot{y}_2^2, \qquad y_2(0) = y_2(1) = 0.$$

(b) Obtain the following parametric representation of the exact solution for the above nonlinear ODE and the BC $y_2(0) = 0$:

$$y_2(x) = \sinh(x), \qquad c_1 t(x) = x + \frac{1}{2}\sinh(2x).$$

(c) Show that the remaining boundary condition $y_2(1) = 0$ requires $c_1 = 0$ nd, hence, $y_2(t) = 0$.

7. A firm must meet a prescribed cyclical (periodic) shipment schedule $S'(t)$, $0 \le t \le T$. Production cost is proportional to the square of the production rate $y'(t)$ and storage cost is linear in inventory (which must be non-negative). Find the production plan $y(t)$ ($\equiv$ cumulative output) to make the required shipments at minimum cost. The variational problem is

$$\min_y \left\{ J[y] \equiv \int_0^T \{c_p[y'(t)]^2 + c_s[y(t) - S(t)]\}dt \mid y(0) = y_0, \right.$$

$$\left. y(T) = y_T, \ \ y(t) \ge S(t) \right\}.$$

(a) Suppose $\hat{y}(t)$ satisfies the constraint over the interval $t_1 \le t \le t_2$. Determine $\hat{y}(t)$ in this interval by setting up the four conditions for finding $t_1, t_2$, and the two constants of integration in the solution of the Euler DE.

(b) Sketch qualitatively $\hat{y}'(t)$ and $S'(t)$ on the same graph.

(c) What can you say about the optimal plan if $S''(t) < c_s/2c_p$?

8. Find the shortest path between $(a,A)$ and $(b,B)$ in the $x, y$-plane subject to the inequality constraint $y \le x^2$. (The end points satisfy the constraint so that $A \le a^2$ and $B \le b^2$.) Discuss specifically the following cases:

(a) $(a,A) = (0,0)$ and $(b,B)$ in the first quadrant with $B \le b^2$.

(b) $a < 0, A < a^2, b > 0$, and $B \le b^2$.

9. Solve the brachistochrone problem of Section 5 of Chapter 1 with the additional inequality constraint $y \le x\tan\theta + h$, where $\theta$ and $h$ are given constants and $y$ is positive downward (see Fig. 11.7).

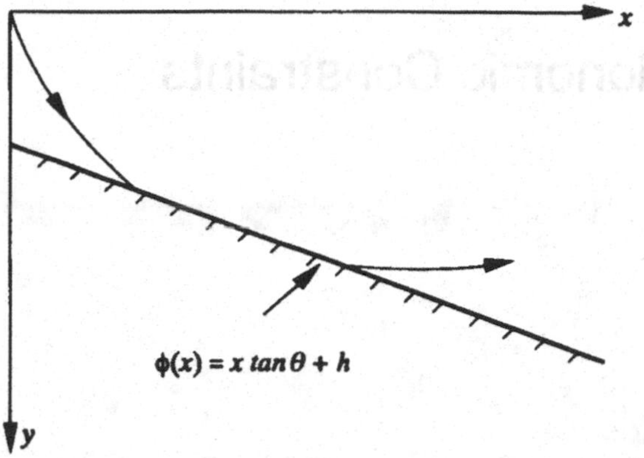


**Figure 11.7**

10. If $\widehat{y} = \widehat{y}_1(x) \equiv g(x)$ satisfies (11.63) for some function $g(\cdot)$, show that $\widehat{y} = \widehat{y}_2(x) \equiv A - g(b - x)$ also satisfies the same equation.

11. Consider basic problems of three unknowns, $(y_1, y_2, y_3)^T = \mathbf{y}$ with one holonomic equality constraint $\varphi(x, \mathbf{y}) = 0$.
    (a) Obtain the equivalent of (11.9) for these problems.
    (b) What minimum restriction(s) must we impose on $\varphi$ to be able to choose $\lambda$ to eliminate one component of $\delta\mathbf{y}$?

# 12

# Nonholonomic Constraints

## 1. Equality Constraints Involving Derivatives

Consider again the basic problem (11.3) and (11.4) for the two unknowns $y_1$ and $y_2$, but now with (11.5) replaced by the *nonholonomic* constraint

$$\varphi(x, y_1, y_2, y_1', y_2') = 0. \tag{12.1}$$

Several applications of this problem will be described in Examples 1-3. It is generally more difficult now to use (12.1) to eliminate one of the unknowns from the Lagrangian $F(x, \mathbf{y}, \mathbf{y}')$ and the method of Lagrange multipliers is even more indispensable. To take advantage of this method, we calculate the first variation of (12.1). The variation $\delta\varphi$ vanishes for all admissible $y_1$ and $y_2$ so that

$$\delta\varphi = \widehat{\varphi}_{,y_1}\delta y_1 + \widehat{\varphi}_{,y_2}\delta y_2 + \widehat{\varphi}_{,y_1'}\delta y_1' + \widehat{\varphi}_{,y_2'}\delta y_2' = 0. \tag{12.2}$$

We multiply (12.2) by $-\lambda(x)$, integrate over $[a, b]$, and add the result to $\delta J = 0$ to obtain

$$\delta I = \int_a^b \{[\widehat{F}_{,y_1}\delta y_1 + \widehat{F}_{,y_2}\delta y_2 + \widehat{F}_{,y_1'}\delta y_1' + \widehat{F}_{,y_2'}\delta y_2']$$
$$- \lambda[\widehat{\varphi}_{,y_1}\delta y_1 + \widehat{\varphi}_{,y_2}\delta y_2 + \widehat{\varphi}_{,y_1'}\delta y_1' + \widehat{\varphi}_{,y_2'}\delta y_2']\}\, dx = 0.$$

Integrating by parts and observing $\delta y_k = 0$ at $x = a$ and $x = b$ reduce the integral above to

$$\delta I = \int_a^b \{[(\widehat{F}_{,y_1} - \lambda\widehat{\varphi}_{,y_1}) - (\widehat{F}_{,y_1'} - \lambda\widehat{\varphi}_{,y_1'})']\delta y_1$$
$$+ [(\widehat{F}_{,y_2} - \lambda\widehat{\varphi}_{,y_2}) - (\widehat{F}_{,y_2'} - \lambda\widehat{\varphi}_{,y_2'})']\delta y_2\}\, dx = 0. \tag{12.3a}$$

Because of the constraint (12.1), $\delta y_1$ and $\delta y_2$ are not independent variations. The simplest way to avoid this complication would be to choose $\lambda$ so that

$$(\widehat{F}_{,y_2'} - \lambda\widehat{\varphi}_{,y_2'})' = (\widehat{F}_{,y_2} - \lambda\widehat{\varphi}_{,y_2}).$$

This choice of $\lambda$ eliminates the term involving $\delta y_2$. Because $\delta y_1$ may take either sign, it follows from [I.2] that the coefficient of $\delta y_1$ in (12.3a) must vanish as well so that

$$(\widehat{F}_{,y_k'} - \lambda\widehat{\varphi}_{,y_k'})' = (\widehat{F}_{,y_k} - \lambda\widehat{\varphi}_{,y_k}) \quad (k = 1,2). \tag{12.3b}$$

The two ODEs in (12.3b) along with (12.1) determine $\widehat{y}_1, \widehat{y}_2$, and $\lambda$.

For some problems, we may have to choose $\lambda$ to eliminate the term involving $\delta y_1$ instead, as it may not be possible to find a $\lambda$ to eliminate the $\delta y_2$ term. The end results are, of course, the same two ODEs in (12.3b). For PWS extremals, (12.3b) is replaced by the corresponding Euler–Lagrange equations.

Variational problems with nonholonomic constraints similar to (12.1) arise naturally in many applications but particularly in conjunction with optimal control problems to be discussed in the next two chapters. They also arise naturally when we try to transform a basic problem involving higher-order derivatives to one for many unknowns and their first derivatives. We illustrate this with the following problem of the transverse bending of an elastic beam on an elastic foundation.

**EXAMPLE 1:** $J[y] = \int_a^b \frac{1}{2}[(y'')^2 + ky^2 - 2py]\,dx.$ (12.4)

The total energy of a deformed elastic beam, whose underformed central axis spans the interval $[a,b]$ on the $x$-axis and sits on an elastic foundation, is known from Chapter 3, Section 5 to be given by (12.4). In this integral, $y(x)$ is the infinitesimally small transverse displacement of the central beam axis, $k$ is the dimensionless spring constant of the elastic foundation, and $p$ is the known dimensionless distributed transverse load along the length of the beam. We expect $J[y]$ to be a minimum when the deformed beam is in equilibrium. With all the results already available for Lagrangians of the form $F(x, \mathbf{y}, \mathbf{y}')$, it is natural to set $y = y_1$ and $y' = y_2$ and write (12.4) as

$$J[y_1, y_2] \equiv J[\mathbf{y}] = \int_a^b \frac{1}{2}[(y_2')^2 + ky_1^2 - 2py_1]\,dx. \tag{12.5}$$

But the definitions of $y_1$ and $y_2$ require the constraining relation

$$\varphi \equiv y_1' - y_2 = 0 \tag{12.6}$$

to hold for all points in $(a,b)$. For a beam clamped at both ends, the end conditions are

$$y_1(a) = y_1(b) = 0, \qquad y_2(a) = y_2(b) = 0. \tag{12.7}$$

A minimizing function $\hat{y}(x)$ for the problem is necessarily a solution of the system (12.3b) and (12.6). For the present problem, the two ODEs and the constraint specialize to

$$(-\lambda)' = ky_1 - p, \qquad (y_2')' = \lambda, \qquad y_2 = y_1'.$$

The last two equations may be combined to give $\lambda = y_1'''$, and upon substituting this result into the first, we obtain

$$y_1'''' + ky_1 = p \tag{12.8}$$

which is known to be the equation of transverse equilibrium for a beam on an elastic foundation. Although the same equation can be obtained directly as the Euler DE of (12.4) without any pointwise constraints, the introduction of the auxiliary variable $y_2$ through the constraint (12.6) offers us an opportunity to illustrate the mechanics of the Lagrange multiplier method. It also avoids the necessity of developing a complete theory for Lagrangians involving second or higher derivatives of the unknowns. For Lagrangians with higher derivatives, the introduction of new unknowns for unwanted derivatives will require the use of more than one multipliers.

Problems with isoperimetric constraints may also be transformed into problems with nonholonomic equality constraints. We illustrate this transformation with the following example.

**EXAMPLE 2: The isoperimetric problem (10.20) with $m = 1$.**

To transform the problem into one with pointwise constraints, we set $y_1 = y$ and introduce a second unknown $y_2$ by the first-order ODE

$$\varphi(x,y,y') \equiv G(x,y_1,y_1') - y_2' = 0$$

and the initial condition $y_2(a) = 0$. We recover the isoperimetric constraint when we require $y_2(b) = \gamma_1$. As a problem with a single nonholonomic constraint, we need to satisfy the two DEs in (12.3b) which for the present problem specialize to

$$(\lambda \cdot 1)' = 0, \qquad (\hat{F}_{,y_1'} - \lambda\hat{\varphi}_{,y_1'})' = (\hat{F}_{,y_1} - \lambda\hat{\varphi}_{,y_1}).$$

With the first condition implying that $\lambda$ is a constant, we recover the result of [X.4] keeping in mind that the ODE and initial condition defining $y_2$ give

$$y_2(x) = \int_a^x G(x,y_1,y_1')\,dx$$

and the auxiliary terminal condition $y_2(b) = \gamma_1$ reproduces the isoperimetric constraint (10.20c) with $m = 1$.

Many problems with nonholonomic constraints arise directly in application areas. The following optimal growth problem appears naturally in neoclassical economic growth theory.

**EXAMPLE 3:** In the simplest setting, an endowment of capital $k$ at time $t$ is used for production. A portion of the goods produced is consumed at the rate $c(t)$ and the rest is used to re-stock the depleted capital and possibly to increase the capital endowment for future production. If capital depreciation is proportional to $k$ with a constant of proportionality $\mu$, we have

$$f(k) = \frac{dk}{dt} + \mu k + c, \tag{12.9}$$

where $f(k)$ is the known production rate of goods in our simple economy. Individual satisfaction with the allocation of goods for consumption at any given instance is measured by a utility function of the consumption rate $U(c(t))$ which is a non-negative, monotone increasing, concave function. The aggregate social welfare over a planning period $[0, T]$ is given by an integral over $[0, T]$ of $U(c)$ discounted by $e^{-rt}$ to reflect a preference for more immediate consumption:

$$J[c] = \int_0^T e^{-rt} U(c(t))\, dt, \tag{12.10}$$

where $r$ is the constant discount rate. In a centrally planned economy, the central planning board may wish to choose $c(t)$ to maximize $J[c]$. In its optimization process, the board has to live with the technological constraint (12.9) on the limitation of production rate and with the initially endowed stock of capital $k_0$ (leftover from the previous planning period), i.e.,

$$k(0) = k_0. \tag{12.11a}$$

In turn, the planning board may wish to leave a prescribed endowment of capital stock $k_T$ for the next planning period so that

$$k(T) = k_T. \tag{12.11b}$$

Otherwise, a program which maximizes $J[c]$ may deplete the entire capital stock at the end of the planning period to achieve a higher consumption.

We will discuss the solution of the above single-sector economic growth problem (more) thoroughly in a later chapter. At this time, we merely note that the solution for the special case with $f(k) = ak$ is determined by the two ODEs (12.3b). When specialized to the present problem, we have

$$e^{-rt} \frac{dU}{dc} - \lambda = 0, \tag{12.12a}$$

$$\frac{d\lambda}{dt} = \lambda(\mu - a). \tag{12.12b}$$

Equation (12.12b) can be solved immediately to give

$$\lambda(t) = \Lambda_0 \exp[(\mu - a)t].$$

Equation (12.12a) then gives $dU/dc \equiv V(c) = \Lambda_0 e^{(\mu + r - a)t}$ which can be solved to get $c$ as a function of $\Lambda_0 \exp[(\mu + r - a)t]$, once we know the function $U(\cdot)$ and, therefore, its derivative (with respect to $c$) $V(c)$. The remaining unknown $k(t)$ is determined by the constraint (12.9) which now takes the form

$$\frac{dk}{dt} = (a - \mu)k - c, \tag{12.9'}$$

where $c$ is known up to the arbitrary constant of integration $\Lambda_0$. This equation along with the initial condition (12.11a) determines $k(t)$:

$$k(t) = e^{(a-\mu)t}\left[k_0 - \int_0^t e^{-(a-r-\mu)\tau} c(\tau)\, d\tau\right]. \tag{12.13}$$

The unknown constant $\Lambda_0$ is then determined by the end condition $k(T) = k_T$. Depending on the form of $U(\cdot)$, the solution for $\Lambda_0$ may have to be obtained some numerical method.

As $c(t)$ must be non-negative, $k(t)$ cannot be greater than $k_0$ for all $t > 0$ if $a < \mu$. In fact, $k(t)exp[(\mu - a)t]$ is monotone decreasing for $\mu > a$. It follows that for $k_T > k_0$, there is no value we can assign to the constant $\lambda_0$ in the expression for $c(t)$ to get $k(T) = k_T$. For other production functions $f(k)$, the variational problem may or may not have a solution depending on the actual form of $f(k)$ and the constants $\mu, k_0$, and $k_T$.

It is important to note that when the problem fails to have a solution for the situation described; it does so independent of $J[c]$. As long as $c$ is non-negative and $a < \mu$, $k(t)$ is forced to decrease monotonically from $k_0$ by the technical constraint (12.9) [or (12.9') for the problem with $f(k) = ak$], whatever $J[c]$ may be. In general, whether the solution trajectory $k(t)$ of the ODE (12.9) and the initial condition (12.11a) can be steered by an admissible comparison function $c(t)$ toward a particular target value in finite time is a question of *controllability* in control theory. A more thorough discussion of controllability will be postponed until Chapters 13 and 14.

## 2. The Multiplier Rule

In addition to the question of controllability illustrated by a specific example at the end of the last section, other complications are encountered in conjunction with the method of multipliers for variational problems for $\mathbf{y} = (y_1, ..., y_n)^T$ with nonholonomic constraints $\varphi_i(x, \mathbf{y}, \mathbf{y}') = 0$, $i = 1, 2, ..., m(\leq n)$. One complication arises when the minimizing function $\widehat{\mathbf{y}}(t)$ satisfies more than $(n - m)$ of the scalar Euler–Lagrange equations of

$$\Phi(x, \mathbf{y}, \mathbf{y}'; \boldsymbol{\lambda}) \equiv \lambda_1(x)\varphi_1(x, \mathbf{y}, \mathbf{y}') + ... + \lambda_m(x)\varphi_m(x, \mathbf{y}, \mathbf{y}') \equiv \boldsymbol{\lambda} \cdot \boldsymbol{\varphi}$$

for a set of functions $\{\lambda_i(x)\}$, not all zero. Such a problem is seen to be *abnormal*. A problem is *normal* if it is not abnormal. A few examples will help to demonstrate the complications involved in abnormal problems.

For problems with two unknowns and a single constraint (12.1), the method of multipliers which leads to the pair of conditions (12.3b) fails when we cannot choose $\lambda$ to eliminate either $\delta y_1$ or $\delta y_2$ in (12.3a). This occurs if terms involving $\lambda$ cancel out in (12.3b). The following two examples illustrate this situation:

**EXAMPLE 4:** $\varphi \equiv (y_1')^2 + (y_2')^2 = 0.$

For a real solution of this problem, the constraint can only be satisfied by $y_1' = 0$ and $y_2' = 0$ or $\widehat{\mathbf{y}}(x)$ is a constant vector:

$$(\widehat{y}_1, \widehat{y}_2) = (c_1, c_2) = \mathbf{c}^T.$$

With $\varphi_{,y_k} = 0$ and $\widehat{\varphi}_{,y_k'} = 2\widehat{y}_k' = 0$, the integral (12.3a) becomes

$$\delta I = \int_a^b \{[\widehat{F}_{,y_1} - (\widehat{F}_{,y_1'})']\delta y_1 + [\widehat{F}_{,y_2} - (\widehat{F}_{,y_2'})']\delta y_2\, dx = 0.$$

Because $\lambda$ does not appear in the integrand, there is no way we can choose it to eliminate either $\delta y_1$ or $\delta y_2$. As it stands, the multiplier method described in Section 1 is not applicable to this problem. Furthermore, the variational problem has no solution unless we have $\mathbf{B} = \mathbf{A}$. Whether the problem has a solution for $\mathbf{B} = \mathbf{A}$ depends on the structure of the Lagrangian $F$.

**EXAMPLE 5:** $F(x, \mathbf{y}, \mathbf{y}') = e^{y_1} + y_1 - \sin y_2$ and $\varphi = (y_1 - y_2)^2 + (y_1')^2 + (y_2')^2$.

Again the single constraint $\varphi = 0$ requires $y_1' = y_2' = 0$ so that $\widehat{\mathbf{y}}$ is a constant vector. In addition, it also requires $y_1 = y_2$ so that we have

$$(\widehat{y}_1(x), \widehat{y}_2(x)) = Y(1, 1),$$

where $Y$ is a constant determined by an end condition. [Needless to say, we must have $\mathbf{y}(a) = \mathbf{y}(b) = A_0(1, 1)$ for the problem to have a solution]. With this restricted choice of $\widehat{\mathbf{y}}(x)$, relation (12.3a) reduces to

$$\delta I = \int_a^b \{\widehat{F}_{,y_1}\delta y_1 + \widehat{F}_{,y_2}\delta y_2\}\, dx = \int_a^b \{\widehat{F}_{,y_1} + \widehat{F}_{,y_2}\}\delta y_1\, dx = 0,$$

given that we have $y_1 = y_2$ so that $\delta y_2 = \delta y_2$. For the given Lagrangian $F$, the simplified expression for $\delta I$ above becomes

$$\int_a^b (e^{\hat{y}_1} + 1 - \cos\hat{y}_2)\,\delta y_1\,dx = \int_a^b (e^Y + 1 - \cos Y)\,\delta y_1\,dx = 0$$

which is not possible for arbitrary variation $\delta y_1$ because $e^Y + 1 > 1$ and $\cos Y \leq 1$.

Given that there is a unique admissible comparison function $\hat{\mathbf{y}}(x) = A_0(1,1)^T$ for this problem, the minimum value of $J[\mathbf{y}]$ has to be

$$J[\hat{\mathbf{y}}] = \int_a^b (e^{A_0} + A_0 - \sin A_0)\,dx = (e^{A_0} + A_0 - \sin A_0)(b-a).$$

The Lagrangian $F$ plays no role in the determination of $\hat{\mathbf{y}}(x)$. To allow for this type of situation in the framework of the multiplier method, we should again work with an augmented Lagrangian of the form $\lambda_0 F + \lambda\varphi$ with the option of $\lambda_0 = 0$ when $F$ is not involved in the solution process.

EXAMPLE 6: $F(x,\mathbf{y},\mathbf{y}') \equiv F(\mathbf{y}')$ and $\varphi(x,\mathbf{y},\mathbf{y}') = \varphi(\mathbf{y}')$.

We can, in principle, solve $\varphi(\mathbf{y}') = \varphi(y_1', y_2') = 0$ for $y_2'$ in terms of $y_1'$ and use the result $y_2' = \psi(y_1')$ to eliminate $y_2'$ from $F(\mathbf{y}')$ to get

$$J = \int_a^b F(y_1', y_2')\,dx = \int_a^b \bar{F}(y_1')\,dx \equiv \bar{J}[y_1].$$

This result allows us to minimize $\bar{J}[y_1]$ as a basic problem free of constraints (except for the end conditions). We have $\hat{y}_1 = c_1 x + d_1$ for the smooth extremals of that problem and correspondingly $\hat{y}_2 = c_2 x + d_2$ with

$$c_2 = \psi(c_1) \quad \text{or} \quad \varphi(c_1, c_2) = 0.$$

In that case, relations (12.3b) of Section 1 reduce to

$$\hat{F}_{,y_1'} - \lambda\hat{\varphi}_{,y_1'} = F_{,y_1'}(c_1,c_2) - \lambda\varphi_{,y_1'}(c_1,c_2) = 0,$$
$$\hat{F}_{,y_2'} - \lambda\hat{\varphi}_{,y_2'} = F_{,y_2'}(c_1,c_2) - \lambda\varphi_{,y_2'}(c_1,c_2) = 0.$$

which imply $\lambda(x)$ is a constant and terms involving $\lambda$ disappear from the integrand of (12.3a) leaving us with

$$\delta J = \int_a^b [(\hat{F}_{,y_1'})'\,\delta y_1 + (\hat{F}_{,y_2'})'\,\delta y_2]\,dx = 0.$$

But $\hat{F}_{,y_1'}$ and $\hat{F}_{,y_2'}$ *are both constants* so that $\delta J = 0$ is satisfied even when we cannot find an appropriate $\lambda$ to eliminate $\delta y_1$ or $\delta y_2$. In other words, the presence of the Lagrangian does not interfere with the solution process, unlike

**the situation in Example 5. This will generally be the case whenever $\hat{y}$ is also an extremal of $F$.**

**The three examples above show that it is not always possible to choose scalar multipliers to eliminate the desired number of the $\delta y_i$ from the expression for $\delta J$. For other problems, we also need to have the option of not involving $F$ in the optimization process. Not surprisingly, a multiplier rule for problems with $m$ nonholonomic equality constraints can be constructed to allow for this possibility in a way similar to the rules for holonomic and isoperimetric constraints:**

**[XII.1]** ***If $\hat{\mathbf{y}}(\mathbf{x})$ minimizes $\mathbf{J}[\mathbf{y}]$ of a basic problem with m nonholonomic equality constraints, then $\hat{\mathbf{y}}$ is an admissible extremal of the unconstrained variational problem for the Lagrangian***

$$\begin{aligned} \mathbf{I}(\mathbf{x},\mathbf{y},\mathbf{y}') &\equiv \lambda_0 \mathbf{F}(\mathbf{x},\mathbf{y},\mathbf{y}') - \sum_{i=1}^{m} \lambda_i(\mathbf{x})\varphi_i(\mathbf{x},\mathbf{y},\mathbf{y}') \\ &= \lambda_0 \mathbf{F}(\mathbf{x},\mathbf{y},\mathbf{y}') - \boldsymbol{\lambda}(\mathbf{x})\cdot\boldsymbol{\varphi}(\mathbf{x},\mathbf{y},\mathbf{y}') \end{aligned} \tag{12.14}$$

***for a suitable set of multipliers $\{\lambda_0, \boldsymbol{\lambda}(\mathbf{x})\}$, not all zero.***

**As we will clarify later in Section 4, the multipliers $\{\lambda_1(x),\ldots,\lambda_m(x)\}$ may be PWS with a corner at a corner of $\hat{\mathbf{y}}$. Also, we will have to work with the Euler–Lagrange equations instead of Euler differential equations if the extremal is not smooth.**

**A problem which requires $\lambda_0$ to be set equal to zero is called abnormal. Otherwise, we have a normal case and we can set $\lambda_0 = 1$ without affecting the solution. The following summarizes the normal case:**

**[XII.2]** ***For a normal problem, $\lambda_0$ in (12.14) may be taken to be 1. The remaining multipliers and $\hat{\mathbf{y}}$ satisfy the stationary conditions of I, the m nonholonomic constraints, and the end conditions.***

**Taking $\lambda_0 = 1$ is equivalent to dividing the corresponding performance index $\mathbf{I}(\mathbf{y},\boldsymbol{\lambda})$ by the constant $\lambda_0$. The stationary conditions are not affected by a re-scaling of $\mathbf{I}[\mathbf{y}]$. A detailed proof for the existence of the multipliers can be found in Pars (1962); the same results may also be taken as a special case of those for optimal control problems (Pontryagin et al., 1962).**

## 3. Brachistochrone in a Resisting Medium

**Suppose that in the statement of the brachistochrone problem, the descent of the bead is resisted by a drag force per unit mass denoted by $R$. This drag force**

normally is a known function of the bead speed $v$. In Section 5 of Chapter 1, we formulated the basic variational problem for the case of no drag resistance. In the parametric form $x = x(\theta)$ and $y = y(\theta)$, $\theta_a \le \theta \le \theta_b$, the integral to be minimized is

$$J[x, y, v] = \int_{\theta_a}^{\theta_b} \frac{\sqrt{(\dot{x})^2 + (\dot{y})^2}}{v}\, d\theta, \tag{12.15a}$$

where $J$ denotes the total time needed for the bead's descent. When there is no air resistance, $v$ was related to the vertical coordinate $y$ by $dv/dt = g \sin \alpha = g\, dy/ds$ where $\alpha$ is the angle made by the downward slanted tangent of the curved path of the bead and the $x$-axis (see Fig. 12.1) and $ds = \sqrt{\dot{x}^2 + \dot{y}^2}\, d\theta = \sqrt{dx^2 + dy^2}$ is the incremental arclength. In the presence of a drag force $R$, the equation of motion for the bead is modified to read $dv/dt = g\, dy/ds - R$. With $dv/dt = (dv/ds)(ds/dt) = v(dv/ds)$, we have he following relation between $v$, $x$ and $y$:

$$v\frac{dv}{ds} = g\frac{dy}{ds} - R \quad \text{or} \quad v\dot{v} = g\dot{y} - R\sqrt{\dot{x}^2 + \dot{y}^2}, \tag{12.15b}$$

where $(\dot{\ }) = d(\ )/d\theta$ and $R$ is normally a function of $v^2$ (or $|v|$). Therefore the minimiziation of $J[x, y, v]$ in (12.15a) is *subject to nonholonomic constraint (12.15b)* and the end conditions

$$x(\theta_a) = a, \quad y(\theta_a) = A, \quad v(\theta_a) = v_a, \quad x(\theta_b) = b, \quad y(\theta_b) = B. \tag{12.15c}$$

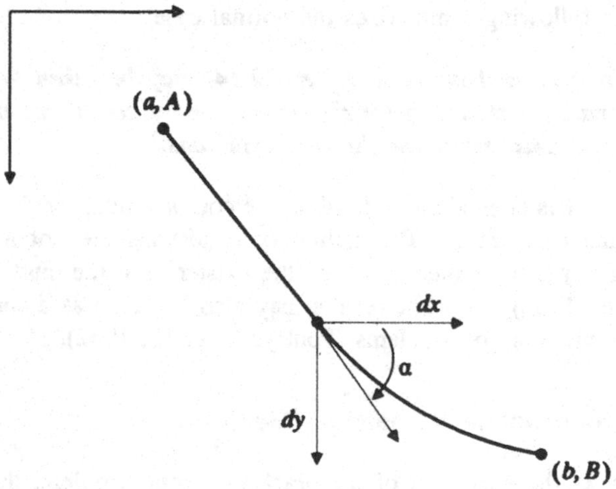


Figure 12.1

Note that in the absence of the drag term, (12.15b) can be integrated to give $v^2 = 2g(y - A) + v_a^2$ where $v_a$ is the known speed of the bead at the starting point. This relation is what allowed us to eliminate $v$ from $J$ in Chapter 1. For convenience, we will take $a = A = 0$ and the $y$-axis positive downward so that $B > 0$.

Because we do not have $\widehat{y}(x)$ in advance to check normality, we will proceed as if the problem is normal until we discover otherwise. The multiplier rule suggests that we seek the stationary point(s) of

$$I[x, y, v, \lambda] = \int_{\theta_a}^{\theta_b} \{\sqrt{\dot{x}^2 + \dot{y}^2}\left(\frac{1}{v} + \lambda R\right) + \lambda v\dot{v} - \lambda g\dot{y}\} \, d\theta.$$

The independent variable $\theta$ does not appear in the integrand explicitly, nor do $y$ and $x$. Two of the Euler differential equations can be integrated once to get

$$\left(\frac{1}{v} + \lambda R\right)\frac{dx}{ds} = c_1, \tag{12.16a}$$

$$\left(\frac{1}{v} + \lambda R\right)\frac{dy}{ds} = \lambda g + c_2, \tag{12.16b}$$

where we have set $ds = \sqrt{\dot{x}^2 + \dot{y}^2}\, d\theta$ and omit the usual caret $\frown$ for an extremal. The remaining conditions for the determination of the solution are

$$v\frac{d\lambda}{ds} = \left(\frac{1}{v} + \lambda R\right)_{,v} \tag{12.16c}$$

and (the constraint) (12.15b). The speed $v$ is one of the unknowns in $I[x, y, v, \lambda]$ and its terminal value at the point $(x, y) = (b, B)$ is not known until we have solved the problem. In that case, the term $\lambda v\, \delta v$ resulting from integrating the term $\lambda v\, \delta\dot{v}$ in the integrand of $\delta I$ by parts does not automatically vanish at $\theta_b$ as $\delta v$ does not need to vanish there. It follows that there is also an Euler boundary condition $\lambda v = 0$ at $\theta = \theta_b$. The physics of the problem tells us that the bead does not arrive at the terminal point with zero speed (for it would otherwise take infinite time to get there). Hence, we have as our Euler boundary condition

$$\lambda(\theta_b) = 0. \tag{12.15d}$$

The six boundary conditions (12.15c) and (12.15d) allow us to determine the six constants of integration of the problem: $c_1$, $c_2$, and the four constants of integration associated with the fourth-order system (12.15b), (12.16a), (12.16b), and (12.16c).

For the solution of our problem (first obtained by Euler), we square both sides of (12.16a) and (12.16b) and add. With $(dx)^2 + (dy)^2 = ds^2$, we get from this combination the algebraic relation

$$\varphi^2 = c_1^2 + (\lambda g + c^2)^2, \tag{12.17a}$$

$$\varphi(v,\lambda) \equiv \frac{1}{v} + \lambda R(v). \tag{12.17b}$$

The fact that $\lambda = 0$ at the terminal point gives

$$v^2(\theta_b) \equiv v_b^2 = \frac{1}{c_1^2 + c_2^2}. \tag{12.17c}$$

Because the independent variable $s$ does not appear explicitly in (12.16a)-(12.16c), we eliminate $s$ by forming the following two ODEs from these three relations:

$$\frac{dx}{d\lambda} = \frac{c_1 v}{\varphi\varphi_{,v}}, \tag{12.18a}$$

$$\frac{dy}{d\lambda} = \frac{(\lambda g + c_2)v}{\varphi\varphi_{,v}}. \tag{12.18b}$$

The right-hand side of both equations are functions of $\lambda$ and $v$ [by the definition of $\varphi$ in (12.17b)]. The algebraic relation (12.17a) enables us to express $v$ in terms of $\lambda$ (or vice versa). We can, therefore, integrate both sides of (12.18a) and (12.18b) with respect to $\lambda$ to get $x$ and $y$ in terms of $\lambda$. The (new) constants of integration are fixed by the conditions that $x = b$ and $y = B$ at $\lambda = 0$ [see (12.15d)]. The other terminal value of $\lambda$, denoted by $\lambda_a$, is determined in terms of $c_1$, and $c_2$ by (12.17) and the known parameter $v(\theta_a) = v_a$ [see (12.15c)]. The two conditions $x = a$ and $y = A$ at $\lambda = \lambda_a$ give two nonlinear equations for $c_1$ and $c_2$. Altogether then, we can, in principle, completely determine the solution for $x$, $y$, and $v$ as a function of $\lambda$ and the known parameters $a, b, A, B, v_a$ and $g$ once the function $R(v)$ is prescribed. We may, if we wish, integrate the separable first-order ODE (12.16c) to get $\lambda$ as a function of $s$ with the new constant of integration fixed by $\lambda(\theta_a) = \lambda_a$. With this result, the other three quantities, $x$, $y$ and $v$, can also be given as functions of $s$ and known parameters.

The above solution process is not attractive for we need to solve (12.17) for $v$ in terms of $\lambda$. Depending on the prescribed function $R(v)$, this may or may not be feasible. To circumvent this difficulty, we note that the combination $\varphi\varphi_{,v}$ which appears in the denominator of the right-hand side of (12.18a) and (12.18b) can be calculated by differentiating both sides of (12.17a) with respect to $\lambda$:

$$2\varphi[\varphi_{,v}\frac{dv}{d\lambda} + \varphi_{,\lambda}] = 2g(\lambda g + c_2)$$

which may be rearranged to read

$$\frac{d\lambda}{\varphi\varphi_{,v}} = \frac{dv}{\lambda(g^2 - R^2) + (c_2 g - R/v)}. \tag{12.19a}$$

Remarkably, the denominator of the right-hand member can be expressed in terms of $v$ alone for any $R(v)$! This is accomplished by multiplying both sides of $\varphi^2 = c_1^2 + (c_2 + \lambda g)^2$ [see (12.17a)] by $c_1^2 g^2$ and rearranging the result to read

$$(g^2 - R^2)\lambda + \left(c_2 g - \frac{R}{v}\right) = \psi(v), \tag{12.19b}$$

where

$$\frac{1}{v_b^2}\psi^2(v) \equiv \left[\frac{R(v)}{v_b^2} - \frac{c_2 g}{v}\right]^2 + c_1^2 g^2 \left[\frac{1}{v^2} - \frac{1}{v_b^2}\right] \tag{12.19c}$$

with $1/v_b^2 = c_1^2 + c_2^2$ as in (12.17c). [The validity of the identity (12.19b) can be confirmed by squaring both sides of (12.19b), canceling out terms, and recombining the remaining expressions.] It follows from (12.18) and (12.19) that

$$dx = \frac{c_1 v\,dv}{\psi(v)}, \qquad dy = \frac{(\lambda g + c_2) v\,dv}{\psi(v)} \tag{12.20a, b}$$

where $\lambda$ is now easily obtained in terms of $v$ from (12.19b). The two equations above may be integrated immediately to obtain

$$x - a = c_1 \int_{v_a}^{v} \frac{v}{\psi(v)}\,dv, \tag{12.21a}$$

$$y - A = c_1 \int_{v_a}^{v} \frac{[\lambda(v) g + c_2] v}{\psi(v)}\,dv. \tag{12.21b}$$

The two terminal conditions for $x$ and $y$,

$$b - a = c_1 \int_{v_a}^{v_b} \frac{v}{\psi(v)}\,dv, \tag{12.22a}$$

$$B - A = c_1 \int_{v_a}^{v_b} \frac{[\lambda(v) g + c_2] v}{\psi(v)}\,dv, \tag{12.22b}$$

determine $c_1$ and $c_2$, keeping in mind that $v_b$, $\lambda$, and $\psi$ all depend on the parameters of $c_1$ and $c_2$. It remains to solve (12.16c) with $\lambda(0) = \lambda_a$ for $\lambda$ as a function of arc length $s$.

Lest we get mired in the details of our illustrative example, it is worth noting that the following general result from Chapter 3 has again been useful in the solution process for this example. *Whenever an unknown is not prescribed at*

***a boundary point, a relevant Euler boundary condition should be obtained at that end point.***

## 4. Inequality Constraints

In many applications, the constraints involved are inequality relations among the unknowns and their derivatives of the form $\varphi(x, \mathbf{y}, \mathbf{y}') \leq 0$. It is customary to set $\mathbf{u} = \mathbf{y}'$ and to write a nonholonomic inequality constraint as $\varphi(x, \mathbf{y}, \mathbf{u}) \leq 0$. This device allows us to interpret the components of the auxiliary variable $\mathbf{u}$ as control variables and thereby make use of techniques in control theory for the problem. Whereas the inequality constraint does not involve the derivatives of $\mathbf{y}$ and $\mathbf{u}$, the differential equation $\mathbf{y}' = \mathbf{u}$ which defines $\mathbf{u}$ must now be included as a pointwise (vector) constraint for the problem. In this chapter, we will limit our discussion of variational problems with nonholonomic constraints to a number of special solution techniques which are not within the framework of optimal control theory. A more general treatment of these problems will be postponed until Chapters 13 and 14.

In this section, we briefly treat the special case where $\varphi$ does not depend on $\mathbf{y}$ explicitly. We will describe the method of solution for a basic problem involving a *scalar* unknown and one inequality constraint $\varphi(x, y') \leq 0$. As indicated above, we set

$$y' = u \quad (a < x < b) \tag{12.23a}$$

and write $J[y]$ and the inequality constraint as

$$J[y] = \int_a^b F(x, y, u)\,dx, \tag{12.23b}$$

$$\varphi(x, u) \leq 0. \tag{12.23c}$$

The equality constraint (12.23a) will be incorporated by the multiplier rule [XII.2] through

$$I = \int_a^b [F(x, y, u) - \lambda(y' - u)]\,dx$$

with

$$\delta I = \int_a^b \{\hat{F}_{,y}\delta y - \lambda\delta y' + (\hat{F}_{,y'} + \lambda)\delta u\}\,dx$$

$$= \int_a^b \{(\hat{F}_{,y} + \lambda')\delta y + (\hat{F}_{,y'} + \lambda)\delta u\}\,dx.$$

We choose $\lambda$ so that

$$\lambda' = -\widehat{F}_{,y} \tag{12.24}$$

for the particular minimizing function $\widehat{y}(x)$ to eliminate the term involving $\delta y$. This leaves us with

$$\delta I = \int_a^b (\widehat{F}_{,y'} + \lambda)\,\delta u\, dx. \tag{12.25}$$

With the inequality constraint (12.23c), $\delta u$ may be restricted so that we may not conclude from (12.25) the usual Euler DE $\widehat{F}_{,y'} + \lambda = 0$. For example, we have $\delta u \leq 0$ if $\varphi(x, u) = u$ and $\varphi(x, \widehat{u}) = 0$. Similarly, we have $\delta u \geq 0$ if $\varphi(x, u) = -u$ and $\varphi(x, \widehat{u}) = 0$. For $\widehat{y}$ to be minimizing when $\delta u$ must be of one sign, it is only necessary to have $\delta I \geq 0$; we do not need $\delta I = 0$. It follows from this that $\widehat{F}_{,y'} + \lambda = 0$ is no longer a necessary condition for $\widehat{y}$ to be a minimum point with $\varphi(x, \widehat{y}') = 0$. The following weaker necessary condition on $\widehat{y}$ takes account of a binding (inequality) constraint:

**[XII.3]** ***A minimizing $\widehat{y}(x)$ for a basic problem with $\varphi(x, y') \leq 0$ satisfies the inequality***

$$(\widehat{F}_{,y'} + \lambda)\,\delta u \geq 0, \quad (a \leq x \leq b) \tag{12.26a}$$

***with (i) $y' = u$, (ii) $\lambda$ determined by (12.24), and (iii) $\delta u$ restricted by***

$$\delta\varphi = \varphi_{,y'}(x, \widehat{u})\,\delta u \leq 0, \quad (a \leq x \leq b) \tag{12.26b}$$

***whenever $\varphi(x, \widehat{u}) = 0$.***

If $\varphi(x, \widehat{u}) < 0$, then there is no restriction on $\delta u$ as $\widehat{\varphi}_{,y'}\delta u$ can be either positive or negative. In that case, we must have

$$\widehat{F}_{,y'} + \lambda = 0. \tag{12.26c}$$

On the other hand, if $\varphi(x, \widehat{u}) = 0$, $\delta u$ must be such that

$$\varphi(x, u) \simeq \varphi(x, \widehat{u}) + \widehat{\varphi}_{,y'}\delta u = \varphi_{,y'}(x, \widehat{u})\,\delta u \leq 0$$

in view of (12.23c); hence, we have (12.26b).

How we may use the necessary conditions (12.24) and (12.26) together with the constraints (12.23a) and (12.23c) for a particular problem will be clear from the example below. In general, the minimizing function $\widehat{y}(x)$ is composed of one or more extremals of $F$ and one or more functions specified by the binding constraint $\varphi(x, u) = 0$ with the composite function $\widehat{y}(x)$ satisfying all the required necessary conditions for a minimum. The main difference between the results here and those of Chapter 11 is that $u = y'$ *may not be continuous.* More specifically, a switch point for problems with nonholonomic inequality constraints may be a corner point of the minimizing $\widehat{y}(x)$. If $u$ is discontinuous across a switch point, then $\lambda$ is PWS by (12.24). This confirms the remark following theorem [XII.1].

**EXAMPLE 7:**

$$J[y] = \frac{1}{2}\alpha^2[y(b)]^2 + \frac{1}{2}\int_a^b \left[\frac{y'}{\sigma(x)}\right]^2 dx \tag{12.27a}$$

with

$$\left|\frac{y'}{\sigma(x)}\right| \leq 1, \tag{12.27b}$$

$$y(a) = A, \tag{12.27c}$$

where $\sigma(x)$ is a given function and $\alpha$ is a given constant.

To solve this problem, we set (instead of $y' = u$)

$$y' = \sigma(x)u \tag{12.28a}$$

and write (12.27) as

$$J[y] = \frac{1}{2}\alpha^2[y(b)]^2 + \frac{1}{2}\int_a^b u^2 dx, \tag{12.28b}$$

$$|u| \leq 1. \tag{12.28c}$$

It follows from (12.24) that $\lambda' = 0$ or $\lambda(x) = \bar{\lambda}$. Because $y(b)$ is not prescribed, there is an Euler boundary condition

$$[\alpha^2\widehat{y}(b) - \lambda(b)] = 0 \quad \text{or} \quad \bar{\lambda} = \alpha^2\widehat{y}(b) \tag{12.28d}$$

which relates $\bar{\lambda}$ to $\widehat{y}(b)$. Because of the way we introduced $u$, (12.26a) is replaced by

$$(\widehat{u} + \bar{\lambda}\sigma)\,\delta u \geq 0 \tag{12.29}$$

and (12.26b) is replaced by $sgn(\widehat{u})\,\delta u \leq 0$ for $\widehat{u} = \pm 1$.

We treat three possible ranges of $u$ values separately:

(a) For $-1 < \widehat{u} < 1$, $\delta u$ can be of either sign; therefore, we must have

$$u + \lambda\sigma = \widehat{u} + \alpha^2\widehat{y}(b)\sigma = 0 \quad (|\widehat{u}| < 1)$$

or

$$\widehat{u}(x) = -\alpha^2\widehat{y}(b)\sigma(x) \quad (|\alpha^2\widehat{y}(b)\sigma(x)| < 1). \tag{12.30a}$$

(b) For $\widehat{u} = -1$, $\delta u$ must be non-negative; (12.29) then requires

$$\hat{u} + \lambda\sigma = \hat{u} + \alpha^2\hat{y}(b)\sigma \geq 0 \quad (\hat{u} = -1). \tag{12.30b}$$

A more useful form of this conclusion is that we must have $\hat{u} = -1$ if $\alpha^2\hat{y}(b)\sigma(x) \geq 1$.

(c) For $\hat{u} = 1$, $\delta u$ must be nonpositive so that

$$\hat{u} + \lambda\sigma = \hat{u} + \alpha^2\hat{y}(b)\sigma \leq 0 \quad (\hat{u} = 1). \tag{12.30c}$$

Equivalently, we must have $\hat{u} = 1$ if $\alpha^2\hat{y}(b)\sigma(x) \leq -1$.

We summarize these results in terms of a saturation function $S(z)$ which is $sgn(z)$ if $|z| \geq 1$ and is $z$ if $|z| < 1$:

$$\hat{u} = -S[\alpha^2\hat{y}(b)\sigma(x)] \equiv \begin{cases} -\alpha^2 B\sigma(x) & (|\alpha^2 B\sigma(x)| < 1) \\ 1 & (\alpha^2 B\sigma(x) < -1) \\ -1 & (\alpha^2 B\sigma(x) > 1), \end{cases} \tag{12.31}$$

where $B \equiv \hat{y}(b)$ is still to be determined. Correspondingly, we have from (12.28a)

$$\hat{y}(x) = A - \int_a^x \sigma(t) S[\alpha^2 B\sigma(t)]\, dt \tag{12.32a}$$

with

$$B \equiv \hat{y}(b) = A - \int_a^b \sigma(t) S[\alpha^2 B\sigma(t)]\, dt. \tag{12.32b}$$

Condition (12.32b) determines $B \equiv \hat{y}(b)$ and then (12.31) determines $\hat{u}(x)$.

To be move specific in illustrating the solution process, we take $b = 1$, $a = 0$, $\alpha^2 = 1$, and $\sigma(x) = \sin(\pi x)$. In that case, we have from (12.32b)

$$B \equiv \hat{y}(1) = A - \int_0^1 \sin(\pi t) S[B\sin(\pi t)]\, dt. \tag{12.33a}$$

Assuming for the time being $|B\sin(\pi t)| < 1$, the above relation becomes

$$B = A - \int_0^1 B\sin^2(\pi t)\, dt = A - \frac{B}{2} \tag{12.33b}$$

or

$$B = \frac{2}{3}A. \tag{12.33c}$$

The argument of the saturation function $\alpha^2 B\sigma(x)$ is, therefore,

$$\alpha^2 B\sigma(x) = \frac{2}{3}A\sin(\pi x). \tag{12.33d}$$

(a) For $|A| < 3/2$, we have from (12.31)

$$\hat{u}(x) = -\frac{2A}{3}\sin(\pi x), \tag{12.34a}$$

$$\lambda = \frac{2A}{3}, \tag{12.34b}$$

and

$$\begin{aligned}\hat{y}(x) &= A - \frac{2}{3}A\int_0^x \sin^2(\pi t)\,dt \\ &= A\left\{1 - \frac{1}{3}x + \frac{1}{6\pi}\sin(2\pi x)\right\}.\end{aligned} \tag{12.34c}$$

(b) If $|A| > 3/2$, the preceding development does not apply. We know for this case $|\alpha^2 B\sigma(x)| = |B\sin(\pi x)| > 1$ in the interval $x_1 < x < 1 - x_1$ for some $x_1$ and $|B\sin(\pi x)| < 1$ in the remaining portion of $[0,1]$. To be concrete, we consider the case $A > 3/2$. In that case, (12.32b) [or (12.33a)] becomes

$$\begin{aligned}B &= A - B\int_0^{x_1}\sin^2(\pi t)\,dt - \int_{x_1}^{1-x_1}\sin(\pi t)[1]\,dt - B\int_{1-x_1}^{1}\sin^2(\pi t)\,dt \\ &= A - \frac{2}{\pi}\cos(\pi x_1) - B\left\{x_1 - \frac{1}{2\pi}\sin(2\pi x_1)\right\}\end{aligned}$$

or

$$B = \frac{2[\pi A - 2\cos(\pi x_1)]}{2\pi(1 + x_1) - \sin(2\pi x_1)}. \tag{12.35}$$

The switch point $x_1$ is then determined by the condition

$$B\sin(\pi x_1) = \frac{2\sin(\pi x_1)[\pi A - 2\cos(\pi x_1)]}{2\pi(1 + x_1) - \sin(2\pi x_1)} = 1 \tag{12.36}$$

which follows from (12.31) and always has a solution in $(0, 1/2)$ for $A > 3/2$. Correspondingly, the minimizing function is given by

$$\hat{y}(x) = A + \int_a^x \sin(\pi t)\,\hat{u}(t)\,dt. \tag{11.37a}$$

with

$$\hat{u}(x) = \begin{cases} -B\sin(\pi x) & (0 \le x \le x_1) \\ -1 & (x_1 \le x \le 1 - x_1) \\ -B\sin(\pi x) & (1 - x_1 \le x \le 1) \end{cases} \tag{12.37b}$$

and

$$\lambda(x) = B. \tag{12.37c}$$

## 5. Singular Solutions

A class of variational problems important in applications involves a Lagrangian $F(x, y, y')$ which is linear in $y'$. An example of such a problem arises in the determination of the optimal harvesting of a homogeneous fish population described in Section 2, Chapter 1. Here, $y(t)$ is the total fish population (measured in units of biomass) at time $t$. Left alone, the population grows at a rate which depends on the size of the population so that $dy/dt \equiv y^{\bullet} = f(y)$. However, the fish population is to be harvested at the rate $h(t)$ so that the rate of increase of the population over a period $[0, T]$ is, in fact,

$$y^{\bullet} = f(y) - h(t) \quad (0 \le t \le T). \tag{12.38a}$$

The harvested fish is sold at a given constant price per unit biomass $p$. The cost of harvesting per unit biomass is $c$ which may depend on the size of the fish population; the larger the population, the easier it is to catch the fish. The profit from fishing over the period $[0, T]$ is given by

$$P = \int_0^T e^{-rt}[p - c(y)]h(t)\,dt \tag{12.38b}$$

where we have discounted future revenue at a constant discount rate $r$ to reflect a preference for more immediate returns from the fixed cost investments (such as fishing vessels and processing equipment). Given $f(\cdot), c(\cdot), p, r$, and the initial fish population

$$y(0) = A, \tag{12.38c}$$

what is the harvesting rate $\hat{h}(t)$ which maximizes total discounted profit $P$?

Without some kind of salvage value or government regulation, it would be unrealistic to expect the fishery to target a particular positive terminal fish population to be left at the end of the planning period. A larger profit may be had by harvesting some of the remaining fish. On the other hand, not all fish would be harvested; the cost of harvesting the last few fish may be too high to be worth the effort. Thus, we have a problem where the terminal value of the unknown $y(t)$ is unspecified and is to be determined by the process of optimization.

This problem in optimal control may be restated as a problem in the calculus of variations. We simply use (12.38a) to eliminate $h(t)$ from (12.38b) to get

$$P = \int_0^T e^{-rt}[p - c(y)][f(y) - y^{\bullet}]\,dt. \tag{12.38d}$$

The problem now is to choose $y$ to maximize (12.38d) subject to the initial condition (12.38c). The Euler DE for this problem is

$$\frac{df}{dy} - \frac{f\dfrac{dc}{dy} - p^{\bullet}}{p - c} = \delta \tag{12.39}$$

where we have again omitted $\widehat{\ }$ for the extremal. This equation does not contain any derivative of $y$ and may be solved for $y$ as a function of $t$. We denote the solution by $\widehat{y}_s(t)$. For a constant $p$, $\widehat{y}_s$ is a constant! In any case, $\widehat{y}_s(0)$ is generally not equal to the known initial fish population $A$. It would appear that the problem does not have a solution which is somewhat contrary to our intuition and expectation.

The fishery problem described above is typical of a general class of variational problems with the Lagrangian $F$ being linear in $y'$. For the general problem, we may write $F = F_0(x, y) + F_1(x, y)y'$ so that we have

$$J[y] = \int_a^b [F_0(x, y) + F_1(x, y)y']\,dx. \tag{12.40a}$$

For the basic problem, we also have the fixed end points

$$y(a) = A, \qquad y(b) = B. \tag{12.40b}$$

It can be shown (see Exercise 7) that

**[XII.4]** ***The Euler DE of (12.40a) is***

$$\frac{\partial F_0}{\partial y} = \frac{\partial F_1}{\partial x}. \tag{12.41}$$

Equation (12.41) is *not* a differential equation for $y(x)$ as it does not contain any derivative of $y$. It generally has one or more solutions giving $y$ as a function of $x$. A solution $\widehat{y}_s(x)$ of (12.41) normally does not satisfy either of the end conditions in (12.40b) [or, as in the fishery problem, an end condition $y(a) = A$ and an Euler BC $\widehat{F}_{,y'}(b) = F_1(b, \widehat{y}(b)) = 0$]. It does not contain any constants of integration which could be adjusted to meet either condition. (For example, we may have $\widehat{y}_s$ given by the curve $N \rightarrow W$ in Figure 12.2 while the prescribed end points are $P$ and $T$.) For this reason, the special case of a

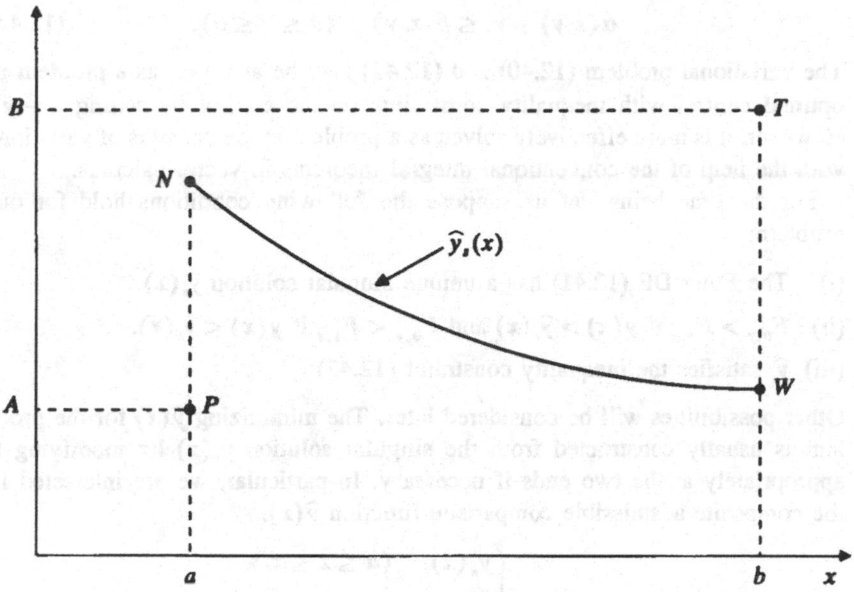


Figure 12.2

Lagrangian linear in $y'$ is called the *singular case*, and a solution of (12.41) is called a *singular solution*. It is rarely discussed in introductory courses in the calculus of variations.

## 6. The Most Rapid Approach

In this section, we provide a resolution to the difficulty posed by singular solutions. The key to the resolution lies in the fact that, for many problems of this type which arise naturally in science and engineering, there are usually some constraints on $y'$. For the fishery problem, the harvesting rate $h(t)$ is not completely arbitrary; it certainly cannot be negative so that we must have $h(t) \geq 0$. More importantly, $h(t)$ has to be bounded above by some finite number $\bar{h} > 0$; there is only so much harvesting capability in the fishery (in terms of fishermen, fishing boats, etc.) so that we cannot harvest any faster than some maximum rate $\bar{h}$. In terms of $y^{\bullet}(t)$, we have from (12.38a) that

$$f(y) - \bar{h} \leq y^{\bullet} \leq f(y).$$

For the more general problem (12.40), $y'$ is usually subject to an inequality constraint of the form

$$\alpha(x,y) \le y' \le \beta(x,y) \qquad (a \le x \le b). \tag{12.42}$$

The variational problem (12.40) and (12.42) may be analyzed as a problem in optimal control with inequality constraints on the control by setting $u = y'$. However, it is more effectively solved as a problem in the calculus of variations with the help of the conventional integral theorems in vector calculus.

For the time being, let us suppose the following conditions hold for our problem:

(i) The Euler DE (12.41) has a unique singular solution $\hat{y}_s(x)$.

(ii) $F_{0,y} > F_{1,x}$ if $y(x) > \hat{y}_s(x)$ and $F_{0,y} < F_{1,x}$ if $y(x) < \hat{y}_s(x)$.

(iii) $\hat{y}_s$ satisfies the inequality constraint (12.42).

Other possibilities will be considered later. The minimizing $\hat{y}(x)$ for the problem is usually constructed from the singular solution $\hat{y}_s(x)$ by modifying it appropriately at the two ends if necessary. In particular, we are interested in the composite adsmissible comparison function $\hat{y}(x)$:

$$\hat{y}(x) = \begin{cases} y_a(x) & (a \le x \le x_1) \\ \hat{y}_s(x) & (x_1 \le x \le x_2) \\ y_b(x) & (x_2 \le x \le b), \end{cases} \tag{12.43}$$

where $y_a(x)$, $y_b(x)$, and the switch points $x_1$ and $x_2$ are described below.

If $\hat{y}_s(a) < A$, then $y_a(x)$ is taken to be the "most rapid descent" curve from $(a,A)$ to a point on $\hat{y}_s(x)$:

$$y_a' = \alpha(x, y_a), \quad y_a(a) = A \qquad (a \le x \le x_1). \tag{12.44}$$

Assuming $\alpha(x,y) < 0$, $y_a(x)$ is the most rapid approach to $y_s(x)$ possible. For the fishery problem, it corresponds to harvesting at maximum rate $\bar{h}$ when the initial fish population $A$ is higher than the optimal level corresponding to $y_s(0)$. When the natural growth rate $f(y)$ for $\hat{y}_s(0) \le y \le A$ is slower than $\bar{h}$, then $y_a'$ is negative and the fish population decreases gradually toward $\hat{y}_s(x)$.

Let $x_1$ be the first point $y_a(x)$ which reaches $\hat{y}_s(x)$, i.e.,

$$y_a(x_1) = y_s(x_1).$$

Either $x_1$ exists and is less than $b$, or $\hat{y}_s(x) < y_a(x)$ for all $x$ in $[a,b]$. The latter would be the case if $\alpha \ge \hat{y}_s'(x)$ for example, and we would set $x_1 = b$.

Similarly, if $B > \hat{y}_s(b)$, $y_b(x)$ is defined by

$$y_b' = \beta(x, y_b), \quad y_b(b) = B, \qquad (x_2 \le x \le b) \tag{12.45}$$

and $x_2$ is the largest value for which $y_b(x_2) = \hat{y}_s(x_2)$. Evidently, $y_b(x)$ is the most rapid ascent curve to the point $(b,B)$ from the singular solution curve. If $x_2$ does not exist or if it is smaller than $a$, we take $x_2 = a$.

In most cases, we have $a < x_1 < x_2 < b$ and (12.43) defines $\widehat{y}(x)$. For the unusual situation where $x_2 < x_1$, the segment $(x_1 < x < x_2)$ involving $\widehat{y}_s(x)$ is absent and $\widehat{y}(x)$ is defined to be the maximum of $y_a(x)$ and $y_b(x)$. In the extreme case with $y_a(x) > y_b(x)$ or $y_a(x) < y_b(x)$ for the entire interval $[a, b]$, the problem has no admissible comparison function and, hence, no solution (in the class of PWS functions). The construction of $\widehat{y}(x)$ is similar if $A < \widehat{y}_s(a)$ or $B < \widehat{y}_s(b)$. In the first case, we replace $\alpha(x, y_a)$ by $\beta(x, y_a)$ in (12.44); in the second case, we replace $\beta(x, y_b)$ by $\alpha(x, y_b)$ in (12.45).

We are now ready for the solution of our basic problem (12.40) with the nonholonomic inequality constraint (12.42). When the constraint (12.42) is satisfied by $\widehat{y}_s(x)$ in $x_1 \le x \le x_2$, we have the following theorem:

**[XII.5]** ***The composite function $\widehat{y}(x)$ defined in (12.43) minimizes $J[y]$. In fact $J[\widehat{y}]$ is a global minimum.***

We need to show $J[y] \ge J[\widehat{y}]$ for all admissible comparison functions $y$ which satisfy the inequality constraint (12.42).

Suppose $\widehat{y}$ is as shown in Figure 12.3 given by the PWS curve indicated by $P \equiv (a, A) \to Q \to R \to S \to T \equiv (b, B)$ and $\widehat{y}_s$ is the curve indicated by $N \to Q \to R \to S \to W$. Suppose $y(x)$ is any other admissible comparison function indicated by $P \to U \to R \to V \to T$. The point $R \equiv (c, C)$ is the intersection

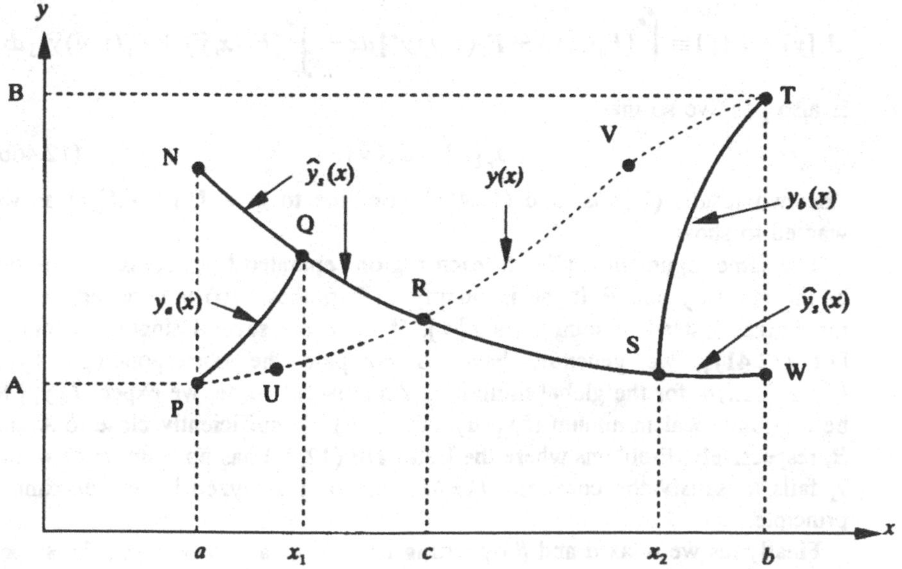


Figure 12.3

of $y(x)$ and $\hat{y}(x)$ with $y(c) = \hat{y}(c) = C$. Consider the following difference of the two integrals of $F(x, y, y')$ from $x = a$ to $x = c$ which may be written as a line integral over the closed curve $\Gamma$ from $P$ through $U, R, Q$ back to $P$:

$$\begin{aligned} J_\ell[y] - J_\ell[\hat{y}] &\equiv \int_a^c [F_0(x,y)dx + F_1(x,y)y']dx - \int_a^c [F_0(x,\hat{y}) + F_1(x,\hat{y})\hat{y}'dx] \\ &= \int_{P,U}^R [F_0(x,y)dx + F_1(x,y)dy] - \int_{P,Q}^R [F_0(x,y)dx + F_1(x,y)dy] \\ &= \oint_\Gamma [F_0(x,y)dx + F_1(x,y)dy], \end{aligned}$$

where $P,X$ indicates a curve going through the point $X$ on its way from $P$ to $R$. The loop integral along $\Gamma$ on the right-hand side may be written as a double integral over the area $A_\ell$ enclosed by the closed curve $\Gamma$ so that

$$J_\ell[y] - J_\ell[\hat{y}] = \iint_{A_\ell} \left[\frac{\partial F_1}{\partial x} - \frac{\partial F_0}{\partial y}\right] dx\, dy .$$

In $A_\ell$, we have $y(x) < \hat{y}(x) \leq \hat{y}_s(x)$ and, by hypothesis (ii), $F_{1,x} - F_{0,y} > 0$. Hence, we conclude that

$$J_\ell[y] > J_\ell[\hat{y}]. \tag{12.46a}$$

A similar calculation shows

$$J_r[y] - J_r[\hat{y}] \equiv \int_c^b [F_0(x,y) + F_1(x,y)y']dx - \int_c^b [F_0(x,\hat{y}) + F_1(x,\hat{y})\hat{y}']dx$$

is also positive so that

$$J_r[y] > J_r[\hat{y}]. \tag{12.46b}$$

The two results (12.46a) and (12.46b) combine to give $J[y] > J[\hat{y}]$ as we wanted to show.

The same argument applies to each region separated by successive crossing of the curves $y$ and $\hat{y}$. If the inequality in hypothesis (ii) is reversed, then $\hat{y}$ maximizes instead of minimizes $J[y]$. If there are several singular solutions [of (12.41)], we generally have to compare the corresponding $J[\hat{y}_k]$, $k = 1, 2, \ldots, n$, for the global minimum. As a useful guide, we expect $J[\hat{y}_1]$ to be a strong local minimum if $\hat{y}_1(a)$ and $\hat{y}_1(b)$ are sufficiently close to $A$ and $B$, respectively. Problems where the Euler DE (12.41) has no solution or where $\hat{y}_s$ fails to satisfy the constraint (12.42), are best analyzed by the maximum principle.

Finally, as we relax $\alpha$ and $\beta$ by letting $\alpha \to -\infty$ and $\beta \to -\infty$, the slopes of the curves $y_a(x)$ and $y_b(x)$ became more and more steep (and vertical in the limit) with $x_1$ and $x_2$ closer and closer to $a$ and $b$, respectively. In terms of the fishery problem, we are allowed to harvest the difference between the initial fish population $A$ and $\widehat{y}_s(0)$ instantaneously to bring the population immediately down to an optimal level. In that case, we can get on the optimal solution trajectory immediately and stay there until an adjustment is needed to meet the appropriate (Euler boundary) condition at $t = T$, namely,

$$e^{-\delta T}[p - c(y(T))] = 0. \tag{12.47}$$

Thus, the variational problem (12.40) with no inequality constraint is generally expected to have a solution but not a PWS solution.

## 7. The Hamilton–Jacobi Inequality

For problems with nonholonomic inequality constraints, we have so far only offered methods of solution applicable to two special classes of problems. In Section 4, we discussed techniques for constraints which involve only $y'$ and $x$ (and not the unknown $y$). In Section 6, the constraints were taken to involve $x, y$ and $y'$ but $y'$ entered the Lagrangian linearly. In this section, we describe a method of solution which in principle applies to all problems. It is limited only by our ability to implement the solution process. Specifically, we will consider here the basic problem but now with $y(x)$ and $y'(x)$ constrained rather generally by stipulating that the points $\{y(x), y'(x)\}$ for all $x$ in $[a, b]$ lie inside an open region $U$ in the $(y, v)$-plane. We call $U$ the *constraint set.*

For all $(y, y')$ defined on $[a, b]$ and belonging to the constraint set $U$, we define

$$J_X(y) \equiv \int_a^X F(x, y, y')\,dx \tag{12.48a}$$

and a *value function* $V(X, Y)$ for $a < X < b$ and $Y$ in $U$ by

$$V(X, Y) = \inf_{y \in U}\{J_X[y] \mid y(a) = A,\ y(X) = Y\} \tag{12.48b}$$

where the infimum is taken over all $(y, y')$ in $U$. Suppose that, for all such pairs of $X$ and $Y$, the infimum defining $V(X, Y)$ is attained and is a $C^1$ function of $X$ and $Y$. For a given pair $(X, Y)$, let the infimum be attained at an admissible comparison function $\widehat{y}(x)$ so that $V(X, Y) = J_X[\widehat{y}]$. For $x = X + \Delta X > X$ and for a sufficiently small $\Delta X$, we define $\widehat{y}(x) = Y + v(x - X) \equiv Y + v\Delta X$ for an arbitrary $v$ which is in $U$. Then, we have

$$V(X + \Delta X, Y + v\Delta X) \le \int_a^X F(x, \widehat{y}, \widehat{y}')\,dx + \int_X^{X+\Delta X} F(X + t, Y + tv, v)\,dt \tag{12.49}$$

because the extended $\hat{y}(x)$ generally does not minimize $J_{X+\Delta X}[y]$ with $y(X+\Delta X) = Y + v\Delta X$.

Inequality (12.49) can be rearranged to read

$$\frac{V(X+\Delta X, Y+v\Delta X) - V(X,Y)}{\Delta X} \leq \frac{1}{\Delta X}\int_X^{X+\Delta X} F(x, Y+v(x-X))\,dx.$$

In the limit as $\Delta X$ tends to zero, we have

$$\frac{dV}{dX} = V_{,X}(X,Y) + V_{,Y}(X,Y)v \leq F(X,Y,v). \tag{12.50}$$

Moreover, with

$$V(x,\hat{y}(x)) = \int_a^x F(t,\hat{y}(t),\hat{y}'(t))\,dt \equiv \hat{V}(x)$$

for all $x < X$ [because $\hat{y}$ must be minimizing from $(a,A)$ to $(x,\hat{y}(x))$ by the principle of optimality], we have

$$\frac{d\hat{V}}{dx} = V_{,x}(x,\hat{y}(x)) + V_{,Y}(x,\hat{y}(x))\hat{y}'_{,x}.$$

The properties satisfied by the *value function* $V(X,Y)$ are summarized in the following:

**[XII.6]** *Suppose* $V(X,Y)$ *is well defined by (12.48) and continuously differentiable for* $a < X < b$ *and* $Y$ *in* $U$. *Then* $V$ *satisfies the* **Hamilton–Jacobi inequality**

$$V_{,X}(X,Y) + V_{,Y}(X,Y)w \leq F(X,Y,w) \tag{12.51}$$

*for all* $w$. *Moreover, if* $\hat{y}(x)$ *is an PWS function for which (12.48b) holds so that* $V(X,Y) = J_X[\hat{y}]$, *then*

$$V_{,X}(x,\hat{y}(x)) + V_{,Y}(x,\hat{y}(x))\hat{y}'(x) = F(x,\hat{y}(x),\hat{y}'(x)). \tag{12.52}$$

The usefulness of the value function and its properties lies in the following converse of [XII.6]

**[XII.7]** *Suppose* $S(X,Y)$ *is a* $C^1$ *scalar function which satisfies the Hamilton–Jacobi inequality*

$$S_{,X}(x,y) + S_{,Y}(x,y)v \leq F(x,y,v) \tag{12.53}$$

*for all* $a \leq x \leq b$ *and* $(y,v)$ *in the constraint set* $U$. *Suppose* $\hat{y}(x)$ *is an admissible comparison function for the basic problem with* $(\hat{y}(x),\hat{y}'(x))$ *belonging to* $U$ *for all* $x$ *in* $[a,b]$ *and*

$$S_{,x}(x, \hat{y}(x)) + S_{,Y}(x, \hat{y}(x))\hat{y}'(x) = \hat{F}(x) \qquad (12.54)$$

*for each* x. *Then* $\hat{y}(x)$ *minimizes* J *over all admissible* y [*with* $(y(x), y'(x))$ *in* U].

The validity of the [XII.7] follows immediately from

$$\begin{aligned} J[y] - J[\hat{y}] &= \int_a^b F(x, y, y')dx - J[\hat{y}] \\ &\geq \int_a^b [S_{,x}(x, y(x)) + S_{,Y}(x, y(x))y'(x)]\,dx - J[\hat{y}] \\ &= \int_a^b \frac{d}{dx}[S(x, y(x))]\,dx - J[\hat{y}] \\ &= [S(x, y(x)) - S(x, \hat{y}(x))]_a^b = 0 \end{aligned}$$

because (12.54) and the prescribed end conditions (so that $\hat{y}(a) = y(a) = A$ and $\hat{y}(b) = y(b) = B$).

The results in [XII.6] and [XII.7] are to be applied in the following way. By working with the Euler DE (and the inequality constraints), we usually have some idea about the form of the minimizing function $\hat{y}(x)$ for any terminal point $(X, Y)$. For example, a part of it is likely to be an admissible extremal. For any assumed form of $\hat{y}(x)$, we calculate the corresponding $C^1$ value function $V(X, Y)$. If this value function satisfies the Hamilton–Jacobi inequality for any triplet $(x, y, v)$ and equality holds for $(x, \hat{y}, \hat{y}'(x))$, then [XII.7] assures us that $\hat{y}(x)$ is the solution of our problem. We illustrate this solution process with the following example:

**EXAMPLE 8:** $J[y] = \int_0^{3\pi/4} [(y')^2 - y^2]\,dx$,

Find $\hat{y}(x)$ which minimizes $J[y]$ subject to $y(0) = 0$, $y(3\pi/4) = 1$, and $y'(x) \geq 0$.

The solution of the unconstrained problem is $\hat{y}_f(x) = \sin(x)/\sin(3\pi/4)$, but it is not the solution of the constrained problem because $\hat{y}_f'(x) < 0$ for $\pi/2 < x \leq 3\pi/4$.

Fox $x \leq \pi/2$, the unconstrained solution is applicable. Take

$$\hat{y}(x) = Y\frac{\sin(x)}{\sin(X)} \quad \left(0 \leq x \leq X \leq \frac{\pi}{2}\right). \qquad (12.55)$$

Now, $\hat{y}$ is admissible and satisfies the inequality constraint $y'(x) \geq 0$ as well as conditions (I′), (II′), (III′), and (IV′). Hence, $J_X[\hat{y}]$ is a strong minimum for $J_X[y]$ [with $y(a) = A$, $y(X) = Y$, $y'(x) \geq 0$] and we have

$$V(X,Y) = J_X[\hat{y}] = \int_0^X \frac{Y^2}{\sin^2(X)}[\cos^2(x) - \sin^2(x)]\,dx$$
$$= Y^2 \cot(X) \quad \left(0 \le X \le \frac{\pi}{2}\right).$$

Observe that

$$V_{,x}(x,y) + V_{,Y}(x,y)v - F(x,y,v)$$
$$= -\frac{y^2}{\sin^2(x)} + 2vy\cot x - (v^2 - y^2) = -(v - y\cot x)^2 \le 0$$

and

$$V_{,x}(x,\hat{y}(x)) + V_{,Y}(x,\hat{y}(x))\hat{y}'(x) - \hat{F}(x) = [\hat{y}'(x) - \hat{y}(x)\cot x]^2$$
$$= -\left(\frac{Y}{\sin X}\cos x - \frac{Y}{\sin X}\sin x\,\cot x\right)^2 = 0$$

so that the hypotheses of [XII.7] holds for $V(X,Y)$ in the range $0 \le X \le \pi/2$.

For $X > \pi/2$, an educated guess would be

$$\hat{y}(x) = \begin{cases} Y\sin x & (0 \le x \le \pi/2) \\ Y & (\pi/2 \le x \le X) \end{cases} \tag{12.56}$$

because the constraint should be binding for $x > \pi/2$ [and the choice $\hat{y}(x) = Y$ for $x > \pi/2$ minimizes the contribution of the positive term $(y')^2$ in the integrand]. Now, we no longer have the assurance that $\hat{y}$ minimizes $J_X[y]$ in the range of $x$ of interest as $\hat{y}$ does not satisfy the Euler DE (in $x > \pi/2$). Nevertheless, we evaluate $J_X[\hat{y}]$ and set

$$V(X,Y) = J_X[\hat{y}] = \int_0^{\pi/2} Y^2(\cos^2 x - \sin^2 x)\,dx + \int_{\pi/2}^X (-Y^2)\,dx$$
$$= -Y^2\left(X - \frac{\pi}{2}\right) \quad \left(\frac{\pi}{2} < X \le \frac{3\pi}{4}\right).$$

Consider now $V(X,Y)$ defined by (12.55) for $X \le \pi/2$ and by (12.56) for $X \ge \pi/2$; we have only to test the Hamilton–Jacobi inequality for $\pi/2 < x \le 3\pi/4$. For this range of $x$, we have

$$V_{,x}(x,y) + V_{,Y}(x,y)v - F(x,y,v)$$
$$= -y^2 - 2vy\left(x - \frac{\pi}{2}\right) - [v^2 - y^2] = -v\left[v + 2y\left(x - \frac{\pi}{2}\right)\right] \le 0$$

because $v \ge 0$ (by constraint) and $y \ge 0$ [because $y' \ge 0$ and $y(0) = 0$]. Moreover, we have the following equality:

$$V_{,x}(x,\hat{y}) + V_{,Y}(x,\hat{y})\hat{y}' - \hat{F}(x) = -\hat{y}'\left[\hat{y}' + 2\hat{y}\left(x - \frac{\pi}{2}\right)\right] = 0$$

because $\hat{y}(x) = Y$ for $\pi/2 < x \leq 3\pi/4$ [and hence $\hat{y}'(x) \equiv 0$ there]. As $V(X, Y)$ satisfies the Hamilton–Jacobi inequality with equality holding for $y = \hat{y}(x)$ and $v = \hat{y}'(x)$, the admissible comparison function defined by (12.55) and (12.56) minimizes $J[y]$ by [XII.7].

## 8. Blocked Harvest of a Uniform Forest

In Chapter 10, we formulated and analyzed the problem of optimal harvesting of once-and-for-all forests for maximum total (discounted) profit. The present value of the total profit is given by

$$P = \int_{t_1}^{t_2} e^{-rt} vh\, dt, \tag{12.57}$$

where $h$ is the harvesting rate for the homogeneous forest and $v \equiv p(t) - c(t, h)$ is the net revenue per unit tree biomass. The optimization problem was to choose the harvest rate $h(t)$ and the time interval $(t_1, t_2)$ to maximize $P$ subject to the isoperimetric constraint

$$\int_{t_1}^{t_2} h(t)\, dt = W_0. \tag{12.58a}$$

In this section, we complement the discussion in Chapter 10 by developing a solution process for the special case where $v$ is a known function of $t$ and $h$ is restricted by

$$0 \leq h(t) \leq h_m. \tag{12.58b}$$

In other words, neither the unit price $p$ nor the unit harvesting cost $c$ depends on the harvesting rate. When the harvest rate is restricted, the problem is known as the blocked-harvest problem (see Clark (1976) and references therein).

To seek the solution within the framework of a calculus of variations for PWS comparison functions, we introduce a (PWS) function $y(t)$ by

$$y^{\bullet} = h, \tag{12.59a}$$

$$y(t_1) = 0, \tag{12.59b}$$

$$y(t_2) = W_0, \tag{12.59c}$$

where $(\ )^{\bullet} = d(\ )/dt$. It is not difficult to see that $y(t)$ is the accumulated harvest at time $t$. The initial condition $y(t_1) = 0$ indicates that no harvesting took place before $t_1$.

The terminal condition $y(t_2) = W_0$ indicates that the entire forest is harvested during the interval $(t_1, t_2)$. Evidently, the problem defined by (12.59) is completely equivalent to the isoperimetric constraint (12.58a). The performance index $P$ may also be written in terms of $y$ as

$$P[y] = \int_{t_1}^{t_2} e^{-rt} v(t) y^{\bullet}(t)\,dt \tag{12.60}$$

with the last two conditions in (12.59) providing the end conditions for the comparison function $y(t)$. The blocked-harvest constraints $0 \le h(t) \le h_m$ now serve as inequality constraints for $y^{\bullet}$

$$0 \le y^{\bullet}(t) \le h_m \tag{12.61}$$

as discussed in Section 4 of this chapter.

The "most rapid approach" technique is not applicable to the problem of *maximizing* (12.60) subject to (12.59) and (12.61); condition (12.41) does not determine a singular solution. The method based on [XII.3] is also not useful; condition (12.26a) does not involve $y$ or $y'$ and, therefore, cannot be used to determine $\hat{y}$ when the constraint on $h$ is not binding.

For the solution of our problem, we append the constraint (12.59a) to $P[y]$ to get

$$I \equiv \int_{t_1}^{t_2} e^{-rt} v(t) h\,dt - \int_{t_1}^{t_2} \lambda(t)[y^{\bullet} - h]\,dt$$

$$= [-\lambda y(t)]_{t_1}^{t_2} + \int_{t_1}^{t_2} \{[e^{-rt} v(t) + \lambda] h(t) + \lambda^{\bullet} y\}\,dt.$$

We choose

$$\lambda^{\bullet} = 0 \quad \text{or} \quad \lambda(t) = \bar{\lambda} \tag{12.62}$$

to eliminate the $\lambda^{\bullet} y$ term in the integrand of $I$. This leaves us with

$$I = \int_{t_1}^{t_2} [e^{-rt} v(t) + \bar{\lambda}] h(t)\,dt - \bar{\lambda} W_0. \tag{12.63}$$

With (12.59a), the portion of $I$ involving $y$ and $y^{\bullet} = u$ is of the form

$$J = \int_a^b f(t) u(t)\,dt \tag{12.64}$$

where a PWC function $u(t)$ is to be chosen to maximize $J$ subject to $u_{min} \le u(t) \le u_{max}$. For fixed end points $a$ and $b$, and a prescribed function $f(t)$, we have immediately

**[XII.8]** *J is maximized by*

$$\hat{u}(t) = \begin{cases} u_{max} & \text{if } f(t) > 0 \\ u_{min} & \text{if } f(t) < 0. \end{cases} \tag{12.65}$$

For the actual problem, we are free to choose the interval $[t_1, t_2]$ as well. In this case, we let

$$\tau = \frac{t - t_1}{t_2 - t_1} \quad \text{or} \quad t = t_1 + (t_2 - t_1)\tau$$

so that $0 \le \tau \le 1$ for $t_1 \le t \le t_2$. The augmented performance index (12.63) can be written as

$$I = \int_0^1 f(\tau) h\, dt - \bar{\lambda} W_0;$$

where

$$f(\tau) = [e^{-rt} v(t) + \bar{\lambda}](t_2 - t_1).$$

Because $(t_2 - t_1)$ is positive and $\bar{\lambda} W_0$ is fixed for a given $\bar{\lambda}$, [XII.8] requires

$$\hat{y}'(t) = \hat{h}(t) = \begin{cases} h_m & \text{if } e^{-rt} v(t) > -\bar{\lambda} \\ 0 & \text{if } e^{-rt} v(t) < -\bar{\lambda}. \end{cases} \tag{12.66}$$

It remains to determine $t_1$ and $t_2$ to complete the solution of the problem.

For the free end points $t_1$ and $t_2$, we have (from Chapter 3) the two transversality conditions

$$[e^{-rt_i} v(t_i) + \bar{\lambda}]\hat{h}(t_i) = 0 \quad (i = 1, 2).$$

These conditions require that $e^{-rt_i} v(t_i) + \bar{\lambda}$ vanishes for $i = 1$ and $i = 2$. If $e^{-rt_1} v(t_1) + \bar{\lambda}$ should be negative, then, by continuity, it would be negative also for $t_1 < t < t_1 + \delta$ and harvesting would start later than $t_1$. On the other hand, if $e^{-rt_1} v(t_1) + \bar{\lambda}$ should be positive, harvesting would start earlier. Similar arguments also apply at $t_2$. Hence, we have

$$e^{-rt_i} v(t_i) = -\bar{\lambda} > 0 \quad (i = 1, 2) \tag{12.67}$$

with $h(t_i)$ still unspecified [and $v(t_i) > 0$ for harvesting to take place]. The net worth of a unit of lumber, $v(t)$, is known to be generally *unimodal* with a single maximum. There is a unique pair of $t_1$ and $t_2$ which satisfies (12.67) for a given $\bar{\lambda}$ (see Fig. 12.4). Condition (12.58a), which now becomes

$$h_m[t_2(\bar{\lambda}) - t_1(\bar{\lambda})] = W_0, \tag{12.68}$$

determines $\bar{\lambda}$ and therewith $t_1$ and $t_2$.

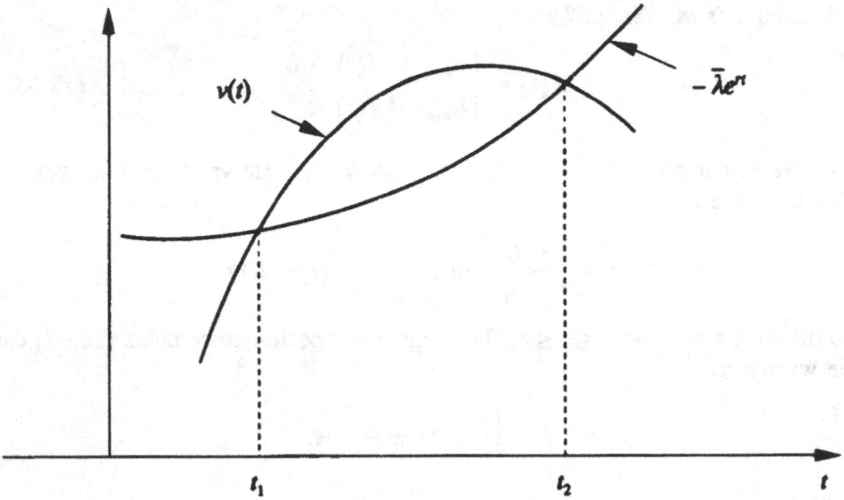


**Figure 12.4**

In practice, it is simpler to use (12.68) to express $t_2$ in terms of $t_1$:

$$t_2 = t_1 + \frac{W_0}{h_m} \tag{12.68'}$$

and combine the two conditions of (12.67) into

$$e^{-rt_2}v(t_2) = e^{-rt_1}v(t_1). \tag{12.69}$$

With (12.68′), condition (12.69) is effectively a condition on the starting time $t_1$ for the optimal blocked-harvest schedule. The unknown $t_2$ is then obtained from (12.68′) with $\bar{\lambda} = -e^{-rt_1}v(t_1)$.

A previous result [X.5] tells us that $-\bar{\lambda}$ is the shadow price of a unit of unharvested lumber. The optimal strategy (12.66), therefore, has the following interpretation:

> *Harvest as quickly as possible (at the maximum rate* $h_m$*) if the discounted net worth of an extra unit of unharvested lumber is higher than its shadow price; otherwise, do not harvest.*

When we do harvest, condition (12.69) requires us to:

> *Start logging (at maximum rate) at the instance* $t_1$ *when the present value (or the shadow price) of the first tree cut equals that of the last tree cut.*

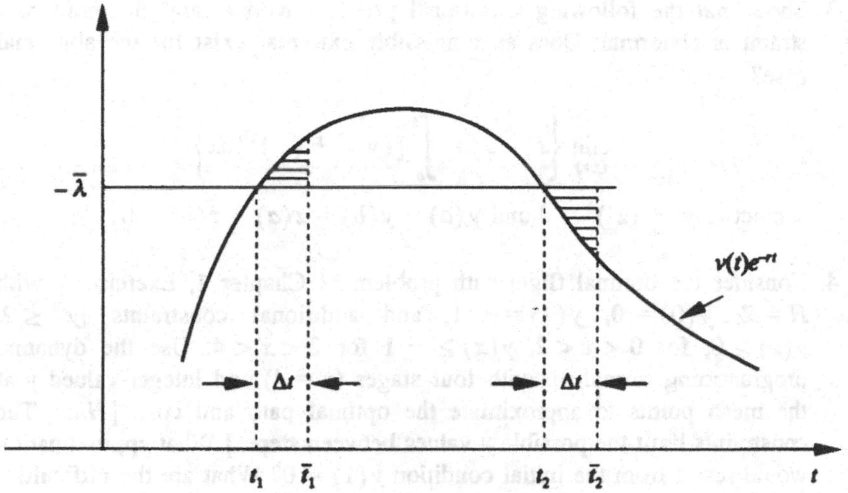


**Figure 12.5**

If the above optimal strategy for the maximum discounted total profit for the blocked harvest of the homogeneous forest is not followed, we can always do better. Suppose harvest is planned for $v(\bar{t}_1)e^{-r\bar{t}_1} > v(\bar{t}_2)e^{-r\bar{t}_2}$ for example. We see from Figure 12.5 that $P$ is less than the $\widehat{P}$ in this case because the discounted profits from the shaded areas are lost.

## 9. Exercises

1. The Problem of S.A. Chaplygin: An airplane flying with constant natural horizontal speed $v_0$ (relative to still air) is to encircle the largest area in a fixed time $T$.

   (a) Find the shape of the encircling curve and the area encircled by the airplane in still air.

   (b) Obtain $\widehat{y}(t)$ and $\widehat{x}(t)$ as well as the area encircled for the case where there is a constant wind speed $w_0$ in the positive $x$ direction.

2. Obtain an admissible extremal of the constrained basic problem

$$\min_{x,y,z}\left[J[x,y,z] \equiv \int_0^{\pi/2} x^2\,dt\right]$$

with $x(0) = 1$, $y(0) = 0$, $z(0) = 0$, $x(\pi/2) = 0$, $y(\pi/2) = 1$, $z(\pi/2) = 0$, $y' = x$, and $x' = [(y - z)^2 - 1]y$.

3. Show that the following variational problem with a nonholonomic constraint is abnormal. Does an admissible extremal exist for the abnormal case?

$$\min_{(y,z)} \left\{ J[y,z] = \int_a^b [(y')^2 + (z')^2]\, dx \right\}$$

subject to $y' + (z')^3 = 0$ and $y(a) = y(b) = z(a) = z(b) = 0$.

4. Consider the optimal flight-path problem of Chapter 1, Exercise 3, with $H = 2$, $y(0) = 0$, $y(4) = -1$, and additional constraints $|y'| \leq 2$, $y(x) \geq 0$, for $0 < x < 2$, $y(x) \geq -1$ for $2 < x < 4$. Use the dynamic programming formalism with four stages ($n = 4$) and integer-valued $y$ at the mesh points to approximate the optimal path and cost. [*Hint:* The constraints limit the possible $y$ values between steps.] What approximation would result from the initial condition $y(1) = 0$? What are the difficulties in applying the trial function method to this problem?

5. (a) Suppose $F(x, y, v)$ is (strictly) convex in $v$ for each pair of $(x, y)$ and the value function $V$ is $C^1$. If the infimum defining $V$ is always attained, prove that V satisfies the Hamilton–Jacobi equation, i.e.,

$$H(x, y, V_{,y}) + V_{,x} = 0$$

(b) Use the result of part (a) to find a solution of the Hamilton–Jacobi equation for $H(x, y, p) = -kyp + p^2/4$. [*Hint*: What is $F(x, y, y')$ in this case?]

6. Consider the problem of minimizing

$$\int_0^X e^{yy'}\, dx$$

subject to $y(0) = 1$, $y(X) = Y$, $y(x) > 0$, $y'(x) < 0$, where $X > 0$ and $0 < Y < 1$. By considering the extremals, formulate a conjecture regarding the solution to this problem for any $X, Y$ in the given range. Use this conjecture to calculate provisionally the value function $V$; then prove that your conjecture is correct.

7. If $F(x, y, y') = G(x, y) + H(x, y)y'$, prove that the corresponding Euler DE is

$$\frac{\partial G}{\partial y} = \frac{\partial H}{\partial x}.$$

8. A uniform flexible string of prescribed length $l$ is held by the ends at $(a,A)$ and $(b,B)$. Physicists assert that the string lies in a (vertical) plane containing the end points and the direction of gravity and its form is such that the potential energy is a minimum. (An equivalent requirement is that the center of gravity of the string must be as low as possible.) Mathematically, this gives rise to a variational problem with $F(x,y,y') = y$ up to a multiplicative factor) subject to the fixed length constraint.

   (a) Formulate this problem as an isoperimetric problem.

   (b) Formulate the same problem with arclength $s$ as the independent variable to get a problem with the pointwise nonholonomic constraint $(x')^2 + (y')^2 = 1$.

   (c) Obtain the admissible extremal of part (b) in parametric form.

   (d) Obtain $\widehat{y}(x)$.

9. Let $R(y(t))$ (with $dR/dy > 0$ and $d^2R/dy^2 < 0$) be the maximum revenue a firm can earn with $y(t)$ amount of "good will." Good will increases with advertising $I$ and decays at a constant proportional rate $\gamma$ so that $y^\bullet = I(t) - \gamma y$. The spending rate for advertising is constrained by $0 \le I \le \bar{I}$, where $\bar{I}$ is some upper bound on permissible spending. The firm is to choose advertising spending $I(t)$ to maximize the present value of the stream of profit given by

$$J[y] = \int_0^\infty e^{-rt}[R(y) - I]\,dt,$$

   starting with an initial amount of good will $y(0) = y_0$.

   (a) Reformulate the problem as a variational problem with nonholonomic inequality constraints.

   (b) Show that there is a unique singular solution given by $\widehat{y}_s(t) = y_{s0}$ (a constant).

   (c) Construct the most rapid approach solution for the case $y_0 < y_{s0}$.

   (d) Construct the most rapid approach solution for $y_0 > y_{s0}$.

10. Let

$$\varphi(x, \mathbf{y}, \mathbf{y}') = d\Phi(x, \mathbf{y})/dx = \Phi_{,x} + \Phi_{,y_1}y_1' + \Phi_{,y_2}y_2' + \dots + \Phi_{,y_n}y_n'.$$

   Show that any smooth $\widehat{\mathbf{y}}$ is a smooth extremal of $\varphi(x, \mathbf{y}, \mathbf{y}')$.

# 13

# Optimal Control with Linear Dynamics

## 1. Optimal Control

Many times throughout this volume, we have encountered problems concerning the dynamics of mass particles. In the Newtonian theory, the second law of motion requires the acceleration of a particle with a *fixed* mass $m$ to be proportional to the resultant force acting on the particle: $y^{\bullet\bullet} = F/m$. (A dot indicates a derivative with respect to time in this and the next chapter.) More than one kind of force may act on the particle. In a mass-spring and dashpot system, the mass is subjected to the restoring force of the spring $-ky$ and the frictional force of the dashpot $-cy^{\bullet}$. If the system is affected by the gravity field of the Earth, the gravitational force $-mg$ must also be considered. The oscillation of the mass may also be driven by time-varying forces supplied externally by motors or other mechanical or electrical devices and summarized by $F_m(t)$. The resulting vertical acceleration of the (lumped) mass of this system is given by

$$y^{\bullet\bullet} = -\frac{c}{m}y^{\bullet} - \frac{k}{m}y - g + \frac{1}{m}F_m(t). \tag{13.1}$$

For another example, consider the deceleration of a spacecraft reentering the Earth's atmosphere. The net force experienced by such a spacecraft is the sum of the drag force $c(y^{\bullet})^2$, the gravitational attraction $mg$ (which may be negligible) and the thrust $T(t)$ provided by the spacecraft's own power source. The coefficient $c$ associated with drag force is proportional to the air density, which is a decreasing function of height above ground. If $y$ measures the distance traveled by the spacecraft along a slanted straight line which makes an angle $\alpha$ with the horizon, we have

$$y^{\bullet\bullet} = \frac{c_0}{m}\rho_0 e^{-y/H}(y^{\bullet})^2 - \frac{g}{\sin\alpha} + \frac{T}{m}, \tag{13.2}$$

where $c_0$, $\rho_0$, and $H$ are known constants.

Evidently, applications of Newton's second law invariably lead to the study of differential equations such as (13.1) and (13.2). For the mathematical analysis of these equations of motion, it is often more convenient to rewrite them as a first order system by setting $y_1 = y$ and $y_2 = y^{\bullet}$. In that case, (13.1) becomes

$$y_1^{\bullet} = y_2, \quad y_2^{\bullet} = -\frac{k}{m}y_1 - \frac{c}{m}y_2 - g + u_2(t) \qquad (13.1')$$

where $u_2(t) = F_m(t)/m$, whereas (13.2) becomes

$$y_1^{\bullet} = y_2, \quad y_2^{\bullet} = c_1 e^{-y_1/H} y_2^2 - \frac{g}{\sin\alpha} + u_2(t) \qquad (13.2')$$

where now $u_2(t) = T(t)/m$. Both systems can be thought of as a vector equations for $\mathbf{y} = (y_1, y_2)^T$ with

$$\mathbf{y}^{\bullet} = \mathbf{f}(t, \mathbf{y}, \mathbf{u}), \qquad (13.3)$$

where $\mathbf{u} \equiv (u_1, u_2)^T = (0, u_2)^T$,

$$\mathbf{f} = \begin{pmatrix} y_2 \\ -\frac{k}{m}y_1 - \frac{c}{m}y_2 - g + u_2(t) \end{pmatrix}$$

$$= \begin{bmatrix} 0 & 1 \\ -\frac{k}{m} & -\frac{c}{m} \end{bmatrix}\mathbf{y} + \begin{bmatrix} 0 & 0 \\ 0 & 1 \end{bmatrix}\mathbf{u} + \begin{pmatrix} 0 \\ g \end{pmatrix} \equiv A\mathbf{y} + B\mathbf{u} + \mathbf{f}_0 \qquad (13.1'')$$

for (13.1′) and

$$\mathbf{f} = \begin{pmatrix} y_2 \\ c_1 e^{-y_1/H} y_2^2 \end{pmatrix} + \begin{pmatrix} 0 \\ \frac{g}{\sin\alpha} \end{pmatrix} + \begin{pmatrix} 0 \\ 1 \end{pmatrix} u_2 \qquad (13.2'')$$

for (13.2′). Moreover, the system (13.1′) is linear with $\mathbf{f}(t, \mathbf{y})$ given by $A\mathbf{y} + B\mathbf{u} + \mathbf{f}_0(t)$ where $\mathbf{f}_0(t)$ is a known vector function of $t$. On the other hand, the vector $\mathbf{f}$ in (13.2″) is nonlinear in $y_1$ and $y_2$ and cannot be so expressed.

When the coupled one-dimensional motion of $n$ particles is studied, the dynamics of this system of particles can still be represented in the form (13.3). The only difference is that $\mathbf{y}$ is now a $2n$ vector with two components corresponding to the position and velocity of each particle is one spatial direction.

Natural phenomena whose temporal evolution is modeled by (13.3) are called *dynamical systems*. The motion of a system of mass particles is an obvious example. An evolving economy and a growing fish population are two other examples. A single sector neoclassical economic growth model was discussed in Example 3 of Chapter 12 [see also E. Burmeister and A.R. Dobell (1970)]. In that model, the growth of the capital stock $k(t)$ is governed by

$$k^{\bullet} = f(k) - \mu k - c, \tag{13.4}$$

where $c(t)$ is the consumption rate, the constant $\mu$ is the unit capital stock depreciation rate, and $f(k)$ is the (normalized) output rate of consumption goods. The ordinary differential equation (13.4) which determines capital accumulation is a scalar equation of the form (13.3). So is

$$y^{\bullet} = f(y) - h \tag{13.5}$$

of Section 2 in Chapter 1, which governs the single-specie fish population $y$ being harvested at the harvest rate $h(t)$.

It should be evident from (13.1) and (13.2) that the "forces" acting on a dynamical system may be divided into two groups: those which depend on the *state* $y$ of the system and those which do not. The restoring spring force, the frictional force of the dashpot, the drag force of the atmosphere, etc., belong to the first group and gravity belongs to the second. The second group may also be separated into those forces which can be changed at will by an agent outside the system (e.g., rocket thrust) and those which cannot be (e.g., gravity). By judiciously and continually adjusting the forces which can be regulated, we often can get the system to perform in a way consistent with a specific objective. A typical objective may be to reach a particular target location by a certain time in a changing environment. The process of adjusting the forces at our disposal to achieve a specific goal is known as the *control process* and the controlling mechanisms are called the *control variables* or simply the *controls*. They will be denoted by the vector function $\mathbf{u} = (u_1, \ldots, u_m)^T$ in this and the next chapter.

For some problems, there may be more than one way to reach a particular target or to maintain a certain trajectory. For these problems, we may wish to select the control(s) to achieve our goal in an "optimal" way. In attempting to reach a particular target, we may want to do so in the fastest time. While maintaining a desired trajectory we may want to do so with a minimum fuel expenditure. Problems in which we select available controls to optimize the performance of a dynamical system are known as *optimal control* problems. The development of a mathematical theory for modern day optimal control problems was initially stimulated by interest in rocketry and space flight problems. The general mathematical results obtained were soon exploited to study optimal control problems in other areas of science and engineering. In this and the next chapter, we will present the basic features of optimal control theory. Some specific problems will be analyzed in detail to illustrate the solution process and its applications.

## 2. Statement of the Problem

Whatever the particular application, an optimal control problem typically includes the following features:

(i) a vector differential *equation of state*

$$\mathbf{y}^{\bullet}(t) = \mathbf{f}(t, \mathbf{y}(t), \mathbf{u}(t)) \quad (t > t_0) \tag{13.6}$$

for a PWS (vector) *state variable* $\mathbf{y} = (y_1, \ldots, y_n)^T$;

(ii) the (usually) known initial condition of the state variable

$$\mathbf{y}(t_0) = \mathbf{y}^0; \tag{13.7}$$

(iii) some constraints on the controls, usually of the form $\mathbf{u}(t)$ belonging to some prescribed constraint set $U$ of points in $m$-dimensional $\mathbf{u}$-space:

$$\mathbf{u}(t) \quad \text{in } U \tag{13.8}$$

for any fixed $t$ in the solution domain;

(iv) some equality and inequality constraints on the state variable $\mathbf{y}$:

$$\boldsymbol{\varphi}^e(t, \mathbf{y}) = \mathbf{0}, \tag{13.9}$$

$$\boldsymbol{\varphi}^i(t, \mathbf{y}) \le \mathbf{0} \tag{13.10}$$

for $\boldsymbol{\varphi}^e = (\varphi_1^e, \ldots, \varphi_m^e)^T$ and $\boldsymbol{\varphi}^i = (\varphi_1^i, \ldots, \varphi_k^i)^T$.

To illustrate feature (iii) of a general control problem, we note that the harvesting rate in the fishery problem must be non-negative and bounded above as there are only a limited number of fishing boats and fishermen available to a given fishery. On the other hand, the rocket thrust available to maneuver a spacecraft can be negative (reversed thrust); but there is a limit to the power the engine can supply. In such cases, the rather general (and abstract) condition (13.8) can be written more specifically as

$$u_m \le u \le u_M \tag{13.8a}$$

for two given numbers $u_m$ and $u_M$. In other cases, the constraints may depend on the state of the system in a way described by the functional relation

$$C(t, \mathbf{y}, \mathbf{u}) \le 0. \tag{13.8b}$$

In still other cases, the constraints cannot be described by any functional relation and description (13.8) is appropriate.

For each choice of admissible $\mathbf{u}(t)$, the initial-value problem (13.6) and (13.7) generates a solution $\mathbf{y}(t)$. In some cases, we merely seek an appropriate $\mathbf{u}(t)$ which guides $\mathbf{y}(t)$ toward a terminal target value $\mathbf{y}^*$ at time $t = T$:

$$\mathbf{y}(T) = \mathbf{y}^*. \tag{13.11}$$

In other cases, only some components of $\mathbf{y}(T)$ are prescribed. We are interested here in *optimal control* problems where $\mathbf{u}$ is chosen to minimize

$$J[\mathbf{u}] = \int_{t_0}^{T} F(t, \mathbf{y}, \mathbf{u})\, dt + \psi(T, \mathbf{y}(T)) \tag{13.12}$$

where $F$ and $\psi$ are given functions. The minimization may or may not be subject to the terminal condition (13.11). Note that the explicit dependence of $\psi$ on $T$ is relevant in problems which allow for an optimal choice of $T$, and the dependence on $\mathbf{y}(T)$ is relevant for components of $\mathbf{y}$ not prescribed at terminal time.

It should be evident that there is a definite relation between optimal control problems and problems in the calculus of variations. If, in the Lagrangian of (1.1), we set

$$y' = u, \tag{13.13}$$

and let the constraint set $U$ be the whole real line, then the performance index of (1.1) can be written as

$$J = \int_{a}^{b} F(x, y, u)\, dx \tag{13.14}$$

and the variational problem is transformed into an optimal control problem with (13.13) as the equation of state. Conversely, if we could solve the equation of state (13.6) for $\mathbf{u}$ in terms of $t$, $\mathbf{y}$, and $\mathbf{y}'$, the result could be used to eliminate $\mathbf{u}$ from (13.12) and transform the problem of optimal control into a problem in the calculus of variations. A more elaborate method of transformation to be described in chapter 14 is available when it is not practical to solve (13.6) for $\mathbf{u}$.

Whether or not we can transform one problem into the other, there is a significant conceptual difference between the two types of problems. The emphasis in the optimal control problem is on the controlling process. The adjustment available to optimize these problems is the control $\mathbf{u}$ and not the state variable $\mathbf{y}$, although the latter will be changed by changing $\mathbf{u}$. The minimal cost production problem of Section 2 in Chapter 1, is, strictly speaking, a genuine optimal control problem because what we can control (more directly) is the rate of production $y'$ ($= u$) and not $y$ itself. The accumulated goods produced $y(t)$ would change according to our choice of production rate but is not regulated directly. In contrast, what we can physically adjust in the minimal surface problem (Section 2, Chapter 1) or the brachistochrone problem (Section 5, Chapter 1) is the unknown shape (state) function **y** itself and not **y**′ (which corresponds to **u** in an optimal control formulation). What distinguishes optimal control problems from those of the calculus of variations is the fact that we can only regulate the state of the system by way of the controls and not directly. As such, optimal control is not the calculus of variations in disguise, although many problems can be stated in either form.

It has been said also that the difference between the two classes of problems mirrors the difference between science and engineering. The calculus of variations was developed to explain phenomena where nature has already done the controlling and optimization and all that is left is for us to find the optimal state. In contrast, optimal control was developed to achieve engineering and social objectives with controlling devices regulated externally by engineers and managers to guide the system toward those objectives in an optimal way.

In this chapter, we will limit our discussion of optimal control theory to linear systems so that

$$\mathbf{y}^{\bullet} = F(t, \mathbf{y}, \mathbf{u}) = A\mathbf{y} + B\mathbf{u}, \tag{13.15}$$

where $A$ is an $n \times n$ matrix and $B$ is an $n \times m$ matrix. External input such as gravity are lumped as components of $\mathbf{u}(t)$. We have no control over these components; hence, $U$ is a proper subspace of the $m$-dimensional space in this case. For systems with linear dynamics, the theory of optimal control is well developed for several special types of $J[\mathbf{u}]$. We will discuss only a few of these in later sections of this chapter. Consistent with the terminology of optimal control, the quantity $J[\mathbf{u}]$ will continue to be called the *performance index* of the optimal control problem unless the content of the problem suggests otherwise.

Constraints (13.9) and (13.10) on the state variable **y** can be handled in the solution process for optimal control problem by techniques similar to those already discussed in Chapters 11 and 12. Unless specifically indicated to the contrary, we will not include them in most of our discussion of optimal control problems in this and the next chapter. That is, most problems to be discussed will not involve these types of state constraints.

## 3. Controllability of Linear Autonomous Systems

The optimal control problem is more difficult if the terminal state is prescribed so that (13.11) holds. It is not always obvious that the system can be steered from the initial point $\mathbf{y}^0$ to a prescribed terminal state $\mathbf{y}^*$ in a prescribed time interval $[0, T]$ by some choice of control $\mathbf{u}(t)$ from the constraint set U. For example, we saw in Section 1, Chapter 12, that the capital stock of the simple economy described there cannot increase when the production rate (linear in $y$) is slower than the capital depreciation rate.

A more subtle situation is found in the double pendulum of Figure 13.1 which is governed by the following two coupled equations of motion for the angular displacement variables $\theta_1$ and $\theta_2$:

$$m\ell^2\theta_1^{\bullet\bullet} = -ka^2(\theta_1 - \theta_2) - mg\ell\theta_1 + u_1, \tag{13.16}$$

$$m\ell^2\theta_2^{\bullet\bullet} = -ka^2(\theta_2 - \theta_1) - mg\ell\theta_2 + u_2, \tag{13.17}$$

These two equations can, of course, be written as a fourth-order linear system of four first-order ordinary differential equations for further analysis. Instead, let us simply combine the two equations to obtain

$$m\ell^2(\theta_1 + \theta_2)^{\bullet\bullet} = -mg\ell(\theta_1 + \theta_2) + (u_1 + u_2). \tag{13.18}$$

This is a single equation which determines $(\theta_1 + \theta_2)$. Generally, we may, by an appropriate choice of $u_1 + u_2$, steer $\theta_1 + \theta_2$ to a target value at time $t = T$ from some initial configuration at $t = t_0$. An exception occurs when $\mathbf{u}$ is restricted by $u_2 = -u_1$. In this case, the forcing terms in (13.18) cancel out and $\theta_1 + \theta_2$ is completely determined by the initial conditions on $\theta_1$ and $\theta_2$. It is not possible to steer $\theta_1 + \theta_2$ to any other target.

The following question thus arises: How can we tell whether, for any pair of $\mathbf{y}_0$ and $\mathbf{y}^*$, there is a PWC $\mathbf{u}(t)$ for which the solution of the initial-value problem (13.6) and (13.7) satisfies $\mathbf{y}(T) = \mathbf{y}^*$ for a prescribed terminal time or any $T > t_0$? When we can find such a PWC function $\mathbf{u}$, the system characterized by $\mathbf{y}$ is said to be *controllable*. We will answer the question of *controllability* for linear systems in several ways. Linear systems (13.15) are clearly

Figure 13.1

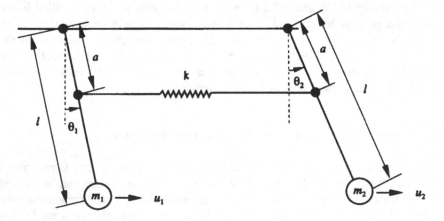

not controllable when $B = 0$. Such systems are not tied to controls and depend only on initial conditions. For $B$ not vanishing identically, we consider first the case of an autonomous system (also called a time-invariant system) for which $A$ and $B$ are constant matrices. Given that a change of variable $\tau = t - t_0$ does not change the form of an autonomous system of ordinary differential equations, it is evident that controllability of a linear autonomous system is independent of the initial time $t_0$. We will take $t_0 = 0$ whenever it is convenient to do so.

If $A$ is *nondefective* so that it has a full set of eigenvectors, we arrange these linearly independent vectors as columns of a matrix $M$, called the *modal matrix* of $A$. We then set $\mathbf{z} = M^{-1}\mathbf{y}$ and write the linear system as

$$\mathbf{z}^{\bullet} = \Lambda\mathbf{z} + M^{-1}B\mathbf{u} \equiv \Lambda\mathbf{z} + B_n\mathbf{u}, \tag{13.19}$$

where $M^{-1}AM = \Lambda$ is the diagonal matrix with the eigenvalues of $A$ as the diagonal elements and $B_n = M^{-1}B$. We have immediately from (13.19)

**[XIII.1]** *If* **A** *is nondefective, then the dynamical system is not controllable if* $\mathbf{B}_n \equiv \mathbf{M}^{-1}\mathbf{B}$ *has a zero row.*

If the $k$th row is a zero row, then we have from (13.19)

$$z_k^{\bullet} = \lambda_k z_k \quad \text{(no sum in } k) \tag{13.20}$$

so that $z_k(t)$ is completely determined by the initial data $\mathbf{y}^0$ and, therefore, not controllable. Actually, a stronger statement can be made:

**[XIII.2]** *If* **A** *is nondefective, a necessary and sufficient condition for the dynamical system to be controllable is that* $\mathbf{B}_n = \mathbf{M}^{-1}\mathbf{B}$ *has no zero rows.*

The proof of this result makes use of the fact that the initial-value problem for a linear autonomous dynamical system has an explicit exact solution which may be taken in the form

$$\mathbf{y}(t) = e^{A(t-t_0)}\mathbf{y}^0 + \int_{t_0}^{t} e^{A(t-\tau)}B\mathbf{u}(\tau)\,d\tau, \tag{13.21}$$

where, for a nondefective matrix $A$ with eigenvalues $\{\lambda_1, \ldots, \lambda_n\}$,

$$e^{At} = M^{-1}e^{\Lambda t}M = M^{-1}\begin{bmatrix} e^{\lambda_1 t} & & 0 \\ & \ddots & \\ 0 & & e^{\lambda_n t} \end{bmatrix}M. \tag{13.22}$$

[A derivation of (13.21) will be given in the next section.]

The sufficiency claim in [XIII.2] is far from obvious. This can be seen from the special case of time-invariant admissible controls (and $t_0 = 0$). For this case, we can write (13.21) as

$$\begin{aligned} \mathbf{z}(T) &= e^{AT}\mathbf{z}^0 + \int_0^T e^{A(T-\tau)}\mathbf{v}\,d\tau \\ &= e^{AT}\mathbf{z}^0 + (e^{AT} - I)A^{-1}\mathbf{v}, \end{aligned}$$

where $\mathbf{v} = B_n\mathbf{u} = M^{-1}B\mathbf{u}$ is a constant vector. Upon solving this for $\mathbf{v}$, we get

$$\mathbf{v} = M^{-1}B\mathbf{u} = A[I - e^{-AT}]^{-1}[e^{-AT}\mathbf{z}(T) - \mathbf{z}^0]$$

or

$$B\mathbf{u} = MA[I - e^{-AT}]^{-1}[e^{-AT}\mathbf{z}(T) - \mathbf{z}^0] \equiv \mathbf{w}.$$

Without exploiting the special properties of $\mathbf{w}$, we can expect to find the corresponding (constant) control $\mathbf{u}$ only if the $m \times m$ matrix $B^TB$ is invertible.

We do not pursue a proof of the sufficiency part of [XIII.2] here as the criterion for controllability in [XIII.1] and [XIII.2] involves the construction of the modal matrix $M$ and its inverse. As such, it is computationally inefficient and should not be used in practice except for the simplest problems. The following condition on the controllability of a general constant matrix $A$ (which may be defective so that it does not have a full set of eigenvectors) is more useful:

**[XIII.3]** ***A linear autonomous system is controllable (at*** $t_0$***) if and only if the*** **($n \times nm$)** ***matrix*** $C = [B, AB, A^2B, \ldots, A^{n-1}B]$ ***is of rank*** $n$**.**

If $C$ is of rank $r < n$, then $\mathbf{v}^TC = 0$ for some nonzero $\mathbf{v}$. This, in turn, implies $\mathbf{v}^TB = \mathbf{v}^TAB = \mathbf{v}^TA^2B = \ldots = \mathbf{v}^TA^{n-1}B = 0$. By the Cayley–Hamilton theorem ($A^n = c_0I + c_1A + \ldots + c_{n-1}A^{n-1}$), we have $\mathbf{v}^TA^kB = 0$ for all $k = 0, 1, 2, \ldots$. It follows from this result and the definition of the matrix exponential,

$$e^{Ax} \equiv I + Ax + \frac{1}{2!}A^2x^2 + \ldots + \frac{1}{n!}A^nx^n + \ldots, \tag{13.23}$$

that

$$\mathbf{v}^Te^{Ax}B = 0 \tag{13.24}$$

for all $x$. We now form $\mathbf{v} \cdot \mathbf{y}$ and use (13.21) and (13.24) to get

$$\mathbf{v}^T\mathbf{y}(t) = \mathbf{v}^Te^{A(t-t_0)}\mathbf{y}^0$$

for the same nonzero vector $\mathbf{v}$. In particular, we have $\mathbf{v} \cdot \mathbf{y} = 0$ for $\mathbf{y}^0 = \mathbf{0}$. Therefore, the collection of all possible solutions (corresponding to different controls $\mathbf{u}$) for the initial data $\mathbf{y}^0 = \mathbf{0}$ does not span $R^n$. For example, the point $\mathbf{v}$ cannot be reached using any control. The linear autonomous system is, therefore, not controllable if $C$ is of rank less than $n$.

Now suppose $C$ is of rank $n$. We want to show that the system is controllable. Consider again the case $\mathbf{y}^0 = \mathbf{0}$. If the system were not controllable, then there would be a nonzero vector $\mathbf{v}$ which could not be reached via any control at time $t = 1$, i.e.,

$$\mathbf{v} \cdot \int_0^1 e^{A(1-s)} B\, \mathbf{u}(s)\, ds = 0$$

or, for PWC controls,

$$\mathbf{v}^T e^{A(1-s)} B = \mathbf{0} \qquad (0 \leq s \leq 1) \tag{13.25}$$

by a DuBois–Reymond type argument (see Chapter 2). For $s = 1$, we would have $\mathbf{v}^T B = \mathbf{0}$. Upon repeatedly differentiating (13.25) with respect to $s$ and then setting $s = 1$, we would get also

$$\mathbf{v}^T AB = \mathbf{v}^T A^2 B = \ldots = \mathbf{v}^T A^{n-1} B = 0. \tag{13.26}$$

This result contradicts the hypothesis that $C$ is of rank $n$. Hence, the system is controllable for $\mathbf{y}^0 = \mathbf{0}$. For any fixed $\mathbf{y}^0 \neq \mathbf{0}$, the above argument applies to $\mathbf{z}(t) \equiv \mathbf{y}(t) - e^{A(t-t_0)}\mathbf{y}^0$ (because of linearity).

**EXAMPLE 1:**

$$A = \begin{bmatrix} -1 & 0 \\ 0 & -2 \end{bmatrix}, \qquad B = \begin{bmatrix} 1 & 0 \\ 1 & 1 \end{bmatrix}.$$

For this system we have

$$C = [B, AB] = \begin{bmatrix} 1 & 0 & -1 & 0 \\ 1 & 1 & -2 & -2 \end{bmatrix} \to \cdots \to \begin{bmatrix} 1 & 0 & -1 & 0 \\ 0 & 1 & -1 & -2 \end{bmatrix}$$

which is of rank 2. The system is therefore controllable.

**EXAMPLE 2:**

$$A = \begin{bmatrix} 0 & 1 & 0 & 0 \\ -(\alpha+\beta) & 0 & \alpha & 0 \\ 0 & 0 & 0 & 1 \\ \alpha & 0 & -(\alpha+\beta) & 0 \end{bmatrix}, \qquad B = \begin{bmatrix} 0 \\ 1 \\ 0 \\ -1 \end{bmatrix}.$$

This is just the double pendulum (13.17) with $u_2 = -u_1$ written as a first-order system with $\alpha = ka^2/m\ell^2$ and $\beta = g/\ell$ and $u_1$ rescaled by $m\ell^2$. For this problem, we have

$$C = [B, AB, A^2B, A^3B] = \begin{bmatrix} 1 & 0 & -2\alpha-\beta & 0 \\ 0 & -2\alpha-\beta & 0 & (2\alpha+\beta)^2 \\ -1 & 0 & 2\alpha+\beta & 0 \\ 0 & 2\alpha+\beta & 0 & -(2\alpha+\beta)^2 \end{bmatrix}$$

$$\rightarrow \cdots \rightarrow \begin{bmatrix} 1 & 0 & -(2\alpha+\beta) & 0 \\ 0 & -(2\alpha+\beta) & 0 & (2\alpha+\beta)^2 \\ 0 & 0 & 0 & 0 \\ 0 & 0 & 0 & 0 \end{bmatrix}.$$

The system is, therefore, not controllable.

## 4. Nonautonomous Linear Systems

The key element in the exact solution (13.21) for a linear autonomous system is the *principal matrix solution* or the *matrix resolvent* $R(t, t_0) \equiv e^{A(t-t_0)}$ which satisfies the homogeneous ordinary differential equation $R^{\bullet} = AR$ and the initial condition $R(t_0, t_0) = I$. For a *nondefective* matrix $A$, we have

$$[e^{At}]^{-1} = [Me^{\Lambda t}M^{-1}]^{-1} = M[e^{\Lambda t}]^{-1}M^{-1}$$
$$= Me^{-\Lambda t}M^{-1} = e^{-At}$$

because $e^{-\Lambda t}$ is a diagonal matrix with diagonal elements $\{e^{-\lambda_i t}\}$. It follows that $R^{-1} = e^{-A(t-t_0)}$ and

$$[R^{-1}]^{\bullet} = -e^{-A(t-t_0)}A = -R^{-1}A.$$

The same results hold for a general (constant) matrix $A$ because we can differentiate the right-hand side of (13.23) term by term. We now differentiate $R^{-1}\mathbf{y} = e^{-A(t-t_0)}\mathbf{y}$ to obtain

$$[e^{-A(t-t_0)}\mathbf{y}]^{\bullet} = -e^{-A(t-t_0)}A\mathbf{y} + e^{-A(t-t_0)}\mathbf{y}^{\bullet}$$
$$= -e^{-A(t-t_0)}A\mathbf{y} + e^{-A(t-t_0)}(A\mathbf{y} + B\mathbf{u}) = e^{-A(t-t_0)}B\mathbf{u}.$$

Upon integrating both sides with respect to $t$, we obtain

$$e^{-A(t-t_0)}\mathbf{y}(t) = \mathbf{y}^0 + \int_0^t e^{-A(\tau-t_0)}B\mathbf{u}(\tau)d\tau$$

which is just the exact solution given in (13.21).

For the initial value problem of a nonautonomous system,

$$\mathbf{y}^{\bullet} = A(t)\mathbf{y} + B(t)\mathbf{u}, \qquad \mathbf{y}(t_0) = \mathbf{y}^0, \tag{13.27}$$

an analogous solution process would seek the (matrix) resolvent $R(t, t_0)$ which satisfies the homogeneous equation $R^{\bullet} = A(t)R$ and the initial condition $R(t_0, t_0) = I$. Differentiating $[R^{-1}\mathbf{y}]$ gives

$$(R^{-1}\mathbf{y})^{\bullet} = (R^{-1})^{\bullet}\mathbf{y} + R^{-1}\mathbf{y}^{\bullet} = (R^{-1})^{\bullet}\mathbf{y} + R^{-1}(A\mathbf{y} + B\mathbf{u}).$$

With $RR^{-1} = I$, we have the identity $(RR^{-1})^{\bullet} = R^{\bullet}R^{-1} + R(R^{-1})^{\bullet} = 0$ or $(R^{-1})^{\bullet} = -R^{-1}R^{\bullet}R^{-1} = -R^{-1}(AR)R^{-1}$ leading to the following result for the resolvent of a nonautonomous linear system:

$$(R^{-1})^{\bullet} = -R^{-1}A, \tag{13.28}$$

which is formally identical to the result for the autonomous case. Hence, the expression for $(R^{-1}y)^{\bullet}$ can be written as

$$(R^{-1}\mathbf{y})^{\bullet} = -R^{-1}A\mathbf{y} + R^{-1}(A\mathbf{y} + B\mathbf{u}) = R^{-1}B\mathbf{u}$$

so that

$$[R^{-1}\mathbf{y}(t) - \mathbf{y}^0] = \int_0^t R^{-1}(\tau, t_0)B(\tau)\mathbf{u}(\tau)d\tau. \tag{13.29}$$

When an autonomous system is not controllable, the expression (13.21) [or (13.29) with $R(t, t_0) = e^{-A(t-t_0)}$] for the exact solution of the initial-value problem has enabled us to show that there must be a nonzero (constant) vector $\mathbf{v}$ for which

$$\mathbf{v} \cdot \int_0^{\bar{t}} R^{-1}(\tau, t_0)B(\tau)\mathbf{u}(\tau)d\tau = 0 \tag{13.30}$$

for any fixed time $\bar{t}$. A DuBois–Reymond type argument then requires

$$\mathbf{v}^T R^{-1}B = \mathbf{v}^T e^{-A(t-t_0)}B = \mathbf{0}. \tag{13.31}$$

The condition on the rank of $C = [B, AB, \ldots, A^{n-1}B]$ follows upon repeated differentiation of the left-hand side (13.31) and evaluating the resulting expressions at $t = t_0$ [keeping in mind $R^{-1}(t_0, t_0) = I$].

For the nonautonomous case with both $A$ and $B$ being matrix functions of $t$, a similar argument applied to the exact solution (13.29) leads to

**[XIII.4]** *If the nonautonomous system (13.27) is not controllable at* $t_0$, *then there is a nonzero (constant) vector* $\mathbf{v}$ *for which*

$$\mathbf{v}^T \mathrm{R}^{-1}(t, t_0)\mathrm{B}(t) = 0 \tag{13.32}$$

*for all* $t \geq t_0$.

The requirement (13.32) implies that all derivatives (with respect to $t$) of the left-hand side remain zero:

$$\begin{aligned}[\mathbf{v}^T R^{-1} B]^{\bullet} &= \mathbf{v}^T[(R^{-1})^{\bullet} B + R^{-1} B^{\bullet}] \\ &= \mathbf{v}^T[-R^{-1}AB + R^{-1}B^{\bullet}] \equiv \mathbf{v}^T R^{-1} B_1 = 0,\end{aligned}$$

where $B_1(t) = B^{\bullet} - AB$. Assuming $A(t)$ and $B(t)$ to be sufficiently differentiable, we continue the process to get

$$\begin{aligned}[\mathbf{v}^T R^{-1} B]^{\bullet\bullet} &= [\mathbf{v}^T R^{-1} B_1]^{\bullet} = \mathbf{v}^T[(R^{-1})^{\bullet} B_1 + R^{-1} B_1^{\bullet}] \\ &= \mathbf{v}^T[R^{-1}(B_1^{\bullet} - AB_1)] \equiv \mathbf{v}^T R^{-1} B_2 = 0, \\ &\ \ \vdots \\ [\mathbf{v}^T R^{-1} B]^{(k)} &= [\mathbf{v}^T R^{-1} B_{k-1}]^{\bullet} = \dots \\ &= \mathbf{v}^T R^{-1}(t, t_0) B_k(t) = 0,\end{aligned} \tag{13.33}$$

where $B_k = B_{k-1}^{\bullet} - AB_{k-1}$ with $B_0 \equiv B$. In particular, we have

$$[\mathbf{v}^T R^{-1} B_k]_{t=t_0} = \mathbf{v}^T B_k(t_0) = 0 \quad (k = 0, 1, 2, \dots).$$

The above calculation suggests the following conclusion:

**[XIII.5]** *Let*

$$\mathrm{C}(t) = [\mathrm{B}_0(t), \mathrm{B}_1(t), \dots, \mathrm{B}_k(t)].$$

*Then the* n*th-order nonautonomous linear system* $\mathbf{y}^{\bullet} = \mathrm{A}\mathbf{y} + \mathrm{B}\mathbf{u}$ *is controllable at* $t_0$ *if the rank of* $\mathrm{C}(t)$ *is equal to* n *for some positive integer* k *and some* $t \geq t_0$.

If the system were not controllable at $t_0$, we concluded earlier that there must be a nonzero vector $\mathbf{v}$ and $t_1 \geq t_0$ for which (13.32) holds and, hence, (13.33) also holds for all $k \geq 0$ and all $t$ in $[t_0, t_1)$. It follows that $\mathbf{v}^T R^{-1}(t, t_0)$ is orthogonal to the columns of $C(t)$ for any $k \geq 0$ and $\mathbf{v}^T R^{-1}(t, t_0)$ is not a zero vector. But this conclusion contradicts the fact that $C(t)$ is of full rank.

EXAMPLE 3: $A(t) = \begin{bmatrix} 0 & 0 \\ 0 & t \end{bmatrix}, \quad B(t) = \begin{bmatrix} 0 \\ 1 \end{bmatrix}.$

For this system we have

$$B_0(t) = B(t) = (0,1)^T \quad \text{with } B_0(0) = (0,1)^T,$$

$$B_1(t) = B^\bullet - AB = (1,t)^T \quad \text{with } B_1(0) = (1,0)^T$$

and therefore, for $k = 1$,

$$C(0) = [B_0(0), B_1(0)] = \begin{bmatrix} 0 & 1 \\ 1 & 0 \end{bmatrix}.$$

Because $C(0)$ is of rank 2 (which is equal to $n$ for this example), the nonautonomous linear system is controllable at $t_0 = 0$ (and, in fact, any $t_0 \geq 0$).

Unlike the autonomous case, the sufficient condition of [XIII.5] is not known to be necessary. The fact that $C(t)$ is not of full rank for all $k$ and a particular $t$ does not imply (13.33) for (all $k$ and) $t \geq t_0$ [and, hence, not (13.32) for all $t \geq t_0$].

## 5. Controllability with Constrained Controls

In discussing controllability of linear systems, we have so far allowed the vector control to be unrestricted. In most applications, the controls are restricted to a certain range of values or, more generally, to take on only values in a control set $U$ (in $R^m$). A controllable system may no longer be controllable when the control is subject to constraints. For example, the scalar system

$$y^\bullet = u, \qquad y(0) = 0$$

is controllable when $u$ is not restricted. But if $u$ can only be positive, then $y(t)$ cannot reach any negative value for all $t > 0$. It is instructive to note that if the terminal time is not prescribed we need only to allow $u$ to take on both positive and negative values, however small in magnitude, for this simple system to become controllable again even with control constraints. Suppose that the control is restricted to the range $|u| < \varepsilon$. Take $u = -\varepsilon + \delta$ for any $\delta$ in the range $0 < \delta < \varepsilon$. We get

$$y = (-\varepsilon + \delta)t$$

and $y$ can get to any negative value by taking $t$ sufficiently large. Similarly, by taking $u = \varepsilon - \delta$, we can make $y(t)$ as large as we wish by waiting long enough. This observation suggests that any controllable linear system remains controllable if $u$ is constrained to be inside a sphere of radius $r > 0$ centered at the origin of $R^m$ (or just to include the origin in the interior). Instead of pursuing a more thorough discussion along this line, we will consider controllability from another perspective which is more relevant to applications.

For many problems in engineering and social sciences, we want to know whether it is possible to steer a dynamical system from some initial state to a given target state (and, if possible, to find the optimal way to accomplish it). For convenience, we take the target point to be the origin of $R_n$. Let $N$ be the set of all initial points $\mathbf{y}^0$ which can be steered to $\mathbf{0}$ in finite time by some control $\mathbf{u}$ via a given *linear* system. The set $N$ is called the *region of controllability*. Clearly $\mathbf{y}^0 = \mathbf{0}$ is in $N$ if $\mathbf{u} = \mathbf{0}$ is in the control set $U$ [for then $\mathbf{y}(t) \equiv \mathbf{0}$ is a solution]. A constant matrix $A$ is called a *stability matrix* if for any initial state $\mathbf{y}^0$, the solution of the initial-value problem defined by $\{\mathbf{y}^{\boldsymbol{\cdot}} = A\mathbf{y}, \mathbf{y}(t_0) = \mathbf{y}^0\}$ goes to zero as $t \to \infty$. A constant matrix with all eigenvalues having a negative real part is a stability matrix; this follows directly from (13.32). Intuitively, we expect $N$ to be the whole space $R^n$ if $A$ is a stability matrix and $\mathbf{u} = \mathbf{0}$ is in the control set $U$. We know that $\mathbf{y}$ will tend to zero eventually; but the following important result will be useful for proving that we can get there in finite time:

**[XIII.6]** *Suppose the control set* U *contains an entire sphere* $\|\mathbf{u}\| = u_1^2 + \ldots + u_m^2 < r$ *for some positive* r. *Then* N *contains an entire sphere* $\|\mathbf{y}\| = y_1^2 + \ldots + y_n^2 < \rho$ *for some positive* $\rho$ *if and only if* $C \equiv [B, AB, \ldots, A^{n-1}B]$ *is of rank* n.

Its proof is not particularly relevant to our discussion of control theory and will not be given here [see F. Clarke (1980)]. One important consequence is the following result for linear autonomous systems:

**[XIII.7]** *Suppose*

(*i*) *the control set* U *contains a sphere of some positive radius about the origin (so that* $\|\mathbf{u}\| < r$ *is in* U);

(*ii*) $C = [B, AB, \ldots, A^{n-1}B]$ *is of full rank;*

(*iii*) *the matrix* A *is a stability matrix.*

*Then* N *is the entire* $R^n$ *(and any* $\mathbf{y}^0$ *can be steered to the origin in finite time).*

Because $A$ is a stability matrix, we can use $\mathbf{u} = \mathbf{0}$ for as long a (finite) period of time as we need to get the system from $\mathbf{y}^0$ to a point $\bar{\mathbf{y}}$ inside the sphere $\|\mathbf{y}\| < \rho$. For a sufficiently small $\rho$, we can apply [XIII.6] and steer the system from $\bar{\mathbf{y}}$ to $\mathbf{0}$ in finite time by an appropriate choice of $\mathbf{u}$.

When a dynamical system is controllable (including the special case where no terminal state is prescribed), there may exist many suitable controls. In that case, we have the option of choosing a control $\hat{\mathbf{u}}(t)$ which minimizes (or maximizes) a certain performance index. In this chapter, we restrict our attention to linear dynamical systems. It would be natural to consider optimal control problems for which the integrand (Lagrangian) of the performance index $J[\mathbf{u}]$ is also *linear*:

$$J[\mathbf{u}] = \boldsymbol{\gamma}\cdot\mathbf{y}(T) + \int_0^T [\sigma(t) + \mathbf{c}(t)\cdot\mathbf{y} + \mathbf{d}(t)\cdot\mathbf{u}]\,dt, \qquad (13.34)$$

where $\boldsymbol{\gamma}$ is a known constant vector and $\mathbf{c}(t)$, $\mathbf{d}(t)$ and $\sigma(t)$ are known functions. Without any restriction on $\mathbf{u}$, these problems are not very interesting as $\mathbf{u}$ can usually be eliminated so that

$$\begin{aligned}\int_0^T &[\sigma(t) + \mathbf{c}\cdot\mathbf{y} + \mathbf{d}\cdot\mathbf{u}]\,dt \\ &= \int_0^T [\sigma + (\mathbf{d}^T(B^TB)^{-1}B^T)\mathbf{y}^{\bullet} + \{\mathbf{c}^T - \mathbf{d}^T(B^TB)^{-1}(B^TA)\}\mathbf{y}]\,dt \\ &= [S(t) + \{\mathbf{d}^T(B^TB)^{-1}B^T\}\mathbf{y}]_0^T + \int_0^T \mathbf{v}_0^T(t)\cdot\mathbf{y}(t)\,dt,\end{aligned}$$

where $S^{\bullet} = \sigma$ and $\mathbf{v}_0^T = \mathbf{c}^T - [\mathbf{d}^T(B^TB)^{-1}B^T]^{\bullet} - \mathbf{d}^T(B^TB)^{-1}(B^TA)$. The performance index $J[\mathbf{u}]$ is, therefore, of type *d* in the classification in Section 4, Chapter 1, and the discussion there applies.

The optimal control problem for the performance index (13.34) becomes more interesting when $\mathbf{u}$ is constrained to be in a control set $U$. We will consider special cases of this class of problems in the next few sections. The (seemingly trivial) special case with $\sigma = 1$, $\boldsymbol{\gamma} = \mathbf{0}$, $\mathbf{c} = \mathbf{0}$, and $\mathbf{d} = \mathbf{0}$ corresponds to the important *time optimal problem* (TOP) in engineering. A typical application is for a spacecraft of mass $m_0$ in steady motion controlled by thrust of limited magnitude so that the single scalar (normalized) control $u(t)$ is restricted to the range $|u| \le 1$. The objective is to steer the spacecraft to a target point in a minimum time interval. The simplest dynamical system of this type is

$$\begin{bmatrix} y_1 \\ y_2 \end{bmatrix}^{\bullet} = \begin{bmatrix} 0 & 1 \\ 0 & 0 \end{bmatrix}\begin{bmatrix} y_1 \\ y_2 \end{bmatrix} + \begin{bmatrix} 0 \\ 1/m_0 \end{bmatrix} u(t) \quad (t > 0) \qquad (13.35)$$

with given initial position and velocity at $t = 0$:

$$y_1(0) = y_1^0, \quad y_2(0) = y_2^0. \qquad (13.36)$$

We want $y_1(T) = 0$ and $y_2(T) = 0$ so that the spacecraft is brought to rest at the origin (for a soft landing) by an appropriate choice of control $u(t)$ over the time interval $0 \le t \le T$. We want to accomplish the task in minimum time so that the performance index is

$$J[u] = T = \int_0^T dt. \tag{13.37}$$

The mathematical problem is to find $u(t)$ to minimize the performance index $J[u]$ given by (13.37) subject to the dynamics (the *equation of state*) (13.35), the initial state (13.36), and the constraint on the control $|u| \le 1$.

In the model (13.35), we have obviously neglected a number of features of the real problem which may or may not be important. If gravity is a factor and fuel constitutes a sizable fraction of the mass of the spacecraft, we will have to modify the equation for $y_2$ to read $(m_0 y_2)^\bullet = u - gm_0$ and a third equation on the rate of mass loss, $m_0^\bullet = -ku$, will have to be included (to reflect the fact that as fuel decreases so will the mass). However, the modified problem no longer has linear dynamics of the form (13.15).

Other examples of the optimal control problem with the performance index (13.34), linear dynamics, and control constraints will be discussed in the next two sections. A general method of solution will be developed in Section 8. A more intricate example which allows for a free boundary will be analyzed in Section 9.

## 6. An Inventory Control Model

Let $y_1(t)$ be the (weight) inventory of goods produced at the rate $u_1(t) \equiv u(t)$. Let $y_2(t)$ be the rate of (weight) sales of the goods. Evidently, the rate of change of the inventory is proportional to the difference between the production rate and the sale rate:

$$y_1^\bullet = u - y_2 \tag{13.38a}$$

where we have normalized time so that we can take the constant of proportionality to be unity. The sales rate may, of course, be taken from past record. However, it may be argued that the company would make a bigger effort to sell more if the inventory builds up. We take for the present study

$$y_2^\bullet = \alpha^2 y_1, \tag{13.38b}$$

where $\alpha$ is a known real positive constant. Because of the limited production facilities, the production rate is constrained to be in the range

$$0 \le u_m \le u \le u_M. \tag{13.39}$$

Suppose $p$ is the price per unit (weight) sale and $c_p$ the unit (weight) production cost. The total revenue over the planning period $[0, T]$ is

$$J[u] = \int_0^T [py_2 - c_p u - c_h y_1]\,dt \equiv \int_0^T F(t, \mathbf{y}, \mathbf{u})\,dt, \tag{13.40}$$

where $c_h$ is the positive constant storage cost for a unit weight of goods per unit time. We wish to maximize this total revenue given

$$y_1(0) = y_1^0, \quad y_2(0) = y_2^0. \tag{13.41}$$

We assume the problem to be normal (so that the Lagrangian is involved in the process of optimization) and append the two equations of state (13.38) to $J[u]$ by way of Lagrange multipliers $\lambda_1$ and $\lambda_2$ just as in Chapter 12:

$$\begin{aligned} I &= \int_0^T [F - \lambda_1(y_1^{\bullet} + y_2 - u) - \lambda_2(y_2^{\bullet} - \alpha^2 y_1)]\,dt \\ &= -[\lambda_1 y_1 + \lambda_2 y_2]_0^T \\ &\quad + \int_0^T [F + (\lambda_1^{\bullet} + \alpha^2\lambda_2)y_1 + (\lambda_2^{\bullet} - \lambda_1)y_2 + \lambda_1 u]\,dt. \end{aligned} \tag{13.42a}$$

Because of the constraints on $u$, the optimal program may not be a stationary solution. Hence, the first variation $\delta I$ is only nonincreasing for a maximum $J$. The condition $\delta I \le 0$ requires

$$\begin{aligned} -[\lambda_1\delta y_1 + \lambda_2\delta y_2]_{t=T} + \int_0^T [(\lambda_1^{\bullet} + \alpha^2\lambda_2 - c_h)\delta y_1 + (\lambda_2^{\bullet} - \lambda_1 + p)\delta y_2 \\ + (\lambda_1 - c_p)\delta u]\,dt \le 0 \end{aligned} \tag{13.42b}$$

where we have made use of the fact $\delta y_k(0) = 0$ because $y_k(0)$, $k = 1, 2$, are prescribed. To avoid having to determine the complex relations among $\delta y_k$ and $\delta u$, we choose the multipliers (or *adjoint variables* in the terminology of optimal control theory) $\lambda_1$ and $\lambda_2$ to eliminate terms involving $\delta y_1$ and $\delta y_2$. For this purpose, we need

$$\begin{aligned} \lambda_1^{\bullet} &= -\alpha^2\lambda_2 + c_h, \qquad & \lambda_1(T) &= 0, \\ \lambda_2^{\bullet} &= \lambda_1 - p, & \lambda_2(T) &= 0 \end{aligned} \tag{13.43}$$

which reduce (13.42b) to

$$\delta I = \int_0^T (\lambda_1 - c_p)\,\delta u\,dt \le 0. \tag{13.44}$$

The Euler boundary conditions $\lambda_1(T) = \lambda_2(T) = 0$ are necessitated by the fact that $y_1$ and $y_2$ are not prescribed at $t = T$. Together with the two ordinary differential equations in (13.43), they completely determine $\lambda_1(t)$ and $\lambda_2(t)$ uncoupled from the equations of state (13.38).

$$\lambda_1 = p\{1 - \cos(\alpha[T-t])\} - \frac{c_h}{\alpha}\sin(\alpha[T-t]),$$

$$\lambda_2 = \frac{c_h}{\alpha^2}\{1 - \cos(\alpha[T-t])\} + \frac{p}{\alpha}\sin(\alpha[T-t]). \tag{13.45}$$

(We shall see later that this uncoupling does not always occur for more general problems.) With $\lambda_1$ completely determined, it is generally not possible to have $\lambda_1(t) - c_p = 0$ (resulting in $\delta I = 0$) and certainly not for a constant $c_p$.

To select the proper production rate without the help of $\delta I = 0$, we return to (13.42a) which has been simplified by our choice of $\lambda_1(t)$ and $\lambda_2(t)$ to

$$I = \int_0^T (\lambda_1 - c_p)u\,dt - [\lambda_1(0)y_1^0 + \lambda_2(0)y_2^0]. \tag{13.46}$$

For a maximum $J[u]$, it is evident that we should take $u$ as large as possible if $\lambda_1(t) - c_p$ is positive and as small as possible if $\lambda_1(t) - c_p$ is negative. Given $\lambda_1(T) = 0$, we have $\lambda_1(T) - c_p < 0$ and, by continuity,

$$\lambda_1(t) \le c_p \quad (\bar{t} \le t \le T) \tag{13.47}$$

for some $\bar{t} \ge 0$. Hence, only two scenarios are possible:

(i) $\lambda_1 - c_p \le 0$ for all $0 \le t \le T$: For this case $(\lambda_1 - c_p)u$ is maximized by taking $u = u_m$. Production and storage costs are too high for any sale to be profitable.

(ii) $\lambda_1 - c_p$ changes sign one or more times in $(0, T)$: Suppose it changes from positive to negative at some switching time $t_s$, $(0 \le)\ t_1 < t_s < t_2$ $(\le T)$ as $t$ increases from $t_1$ to $t_2$. For maximum $J$, we should take

$$u(t) = \hat{u}(t) \equiv \begin{cases} u_M & (t_1 \le t \le t_s) \\ u_m & (t_s \le t \le t_2). \end{cases} \tag{13.48}$$

Note that for $\alpha T \le \pi/2$, $\lambda_1 - c_p$ has at most one sign change (with increasing $t$) from positive to negative. Hence, $u(t)$ switches from one extreme value to the other at most once in the interval $[0, T]$. An optimal control program which alternates between extreme values of the control is known as ***bang-bang control***.

For $\alpha T > \pi/2$, $\lambda_1 - c_p$ may change signs more than once and must be dealt with accordingly. By this, we mean more than just arranging for $u(t)$ to switch from one extreme value to the other when a change of sign occurs in $\lambda_1(t) - c_p$. For example, the equations of state for a constant $u(t)$ and the initial conditions give

$$y_1(t) = y_1^0\cos(\alpha t) - \frac{1}{\alpha}(y_2^0 - u)\sin(\alpha t),$$
$$y_2(t) = \alpha y_1^0\sin(\alpha t) + (y_2^0 - u)\cos(\alpha t) + 1. \tag{13.49}$$

For $\alpha T > \pi/2$, $y_1(t)$ or $y_2(t)$ may become negative for some later time. Neither is physically permissible, which suggests that we should stipulate the additional constraints

$$y_1(t) \geq 0, \qquad y_2(t) \geq 0 \tag{13.50}$$

not required up to now. It is clear from (13.40) that for a positive $p$, $J[u]$ would be larger if $y_1(t)$ could be negative. Without the non-negative constraints (13.50) on the state variable, the optimal program would unrealistically drive the inventory below zero while it continues to sell at a higher rate than the production rates.

Some methods of treating inequality constraints have already been described in Chapters 11 and 12. Others can be found in Bryson and Ho (1969). Instead of pursuing what may be considered technical details on such constraints, it is more important to point out a significant limitation in our argument leading to the conclusion that, for a maximum $J[u]$, it suffices to choose $u$ to maximize $(\lambda_1 - c_p)u$ for all $0 \leq t \leq T$. For other problems with a prescribed terminal state, there are no Euler boundary conditions for $\{\lambda_i\}$ alone; the conditions $\lambda_i(T) = 0$ are inappropriate for a basic problem. The determination of $\{\lambda_i(t)\}$ in this case is coupled to that of $\{\widehat{y}_k(t)\}$ through the end conditions on $\mathbf{y}$. Because $\mathbf{y}$ depends on $\mathbf{u}$, it would seem that the optimal controls $\{\widehat{u}_j\}$ should be determined concurrently with $\{\widehat{y}_k(t)\}$ and $\{\lambda_i(t)\}$. A similar situation also occurs for a performance index or system dynamics which is nonlinear in $\mathbf{y}$. Fortunately, Pontryagin's *maximum principle* assures us that the solution process is not so complex. We will give a more precise statement of this principle after looking at another specific example.

## 7. A Wheat-Trading Problem

A firm buys and sells wheat. At time $t$, the firm has an amount of cash $y_1(t)$ and an amount of wheat $y_2(t)$. The unit (bushel) price of wheat, denoted by $p(t)$, is known for the period $[0, T]$. Let $u(t)$ be the rate at which the firm buys $[u(t) > 0]$ or sells $[u(t) < 0]$ wheat. Then, we have

$$y_1^{\bullet} = -\alpha y_2 - pu,$$
$$y_2^{\bullet} = u, \tag{13.51}$$

where $\alpha$ is the cost of storing a unit of wheat per unit time. Physical and regulatory constraints limit $u$ to be in the range

$$-u_m \le u \le u_m, \tag{13.52}$$

where $u_m > 0$. Given the initial assets of the firm,

$$y_1(0) = y_1^0, \qquad y_2(0) = y_2^0, \tag{13.53}$$

we want to find a program $\hat{u}(t)$ which maximizes the firm's total assets at the terminal time $T$,

$$\max_{|u| \le u_m} \{J \equiv y_1(T) + p(T) y_2(T)\}, \tag{13.54}$$

subject to (13.51) and (13.52).

The equations of state and the performance index $J$ of the above problem are linear in both $\mathbf{y}$ and $\mathbf{u}$. The performance index is different from others considered in this volume in that $J$ depends only on the terminal values of $y_1$ and $y_2$. We again append the constraints (13.51) to $J$ by the multipliers $\lambda_1(t)$ and $\lambda_2(t)$

$$\begin{aligned} I \equiv J &- \int_0^T [\lambda_1(y_1^{\bullet} + \alpha y_2 + pu) + \lambda_2(y_2^{\bullet} - u)]\,dt \\ &= y_1(T)[1 - \lambda_1(T)] + y_2(T)[p(T) - \lambda_2(T)] + [\lambda_1(0) y_1^0 + \lambda_2(0) y_2^0] \\ &\quad + \int_0^T [\lambda_1^{\bullet} y_1 + (\lambda_2^{\bullet} - \alpha \lambda_1) y_2 + (\lambda_2 - p\lambda_1) u]\,dt. \end{aligned} \tag{13.55}$$

As in the last section, we choose $\lambda_1(t)$ and $\lambda_2(t)$ to eliminate the coefficients of $\delta y_i$ in $\delta I$. This requires

$$\begin{aligned} \lambda_1^{\bullet} &= 0, & \lambda_1(T) &= 1, \\ \lambda_2^{\bullet} &= \alpha \lambda_1, & \lambda_2(T) &= p(T). \end{aligned} \tag{13.56}$$

The final-value problem (13.56) completely determines $\lambda_1(t)$ and $\lambda_2(t)$:

$$\lambda_1(t) = 1, \qquad \lambda_2(t) = \alpha(t - T) + p(T), \tag{13.57}$$

again uncoupled from the solution for the state variables.

The choice of multipliers (13.57) reduces (13.55) to

$$I = \int_0^T (\lambda_2 - \lambda_1 p) u\,dt - [y_1^0 + \{p(T) - \alpha T\} y_2^0]. \tag{13.58}$$

The terms in brackets involving end values cannot be changed. To maximize $J$, we can only choose $u(t)$ to make $(\lambda_2 - \lambda_1 p)u$ as large as possible. As before, this requires $u(t)$ to assume one or the other extreme value allowable and to switch from one to the other when the switching function

$$s(t) \equiv \lambda_2(t) - p(t)\lambda_1(t)$$
$$= \alpha(t - T) + p(T) - p(t) \tag{13.59}$$

changes sign. More specifically, we should take [ see (13.52) for the range of $u$ ]

$$u(t) = \hat{u}(t) \equiv \begin{cases} u_m > 0 & (s(t) > 0) \\ -u_m > 0 & (s(t) > 0). \end{cases} \tag{13.60}$$

Hence, the optimal control program is again bang-bang.

It is of some interest to note that a switch point $t_s$ of $\hat{u}(t)$ is an intersection of the graph of the wheat price function $p(t)$ and the straight line connecting the points $(0, p(T) - \alpha T)$ and $(T, p(T))$ (Fig. 13.2). For the particular $p(t)$ in Figure 13.2, the optimal program requires selling wheat for a long period of time with the wheat stock given by

$$y_2(t) = -u_m t + y_2^0. \tag{13.61}$$

If the firm starts with a low wheat stock, $y_2(t)$ may become negative. The firm must then be allowed to "sell short" or we must modify our model to include a non-negative state constraint

$$y_2(t) \geq 0. \tag{13.62}$$

A similar non-negative restriction may also have to be imposed on $y_1(t)$ to improve the mathematical model of the actual problem.

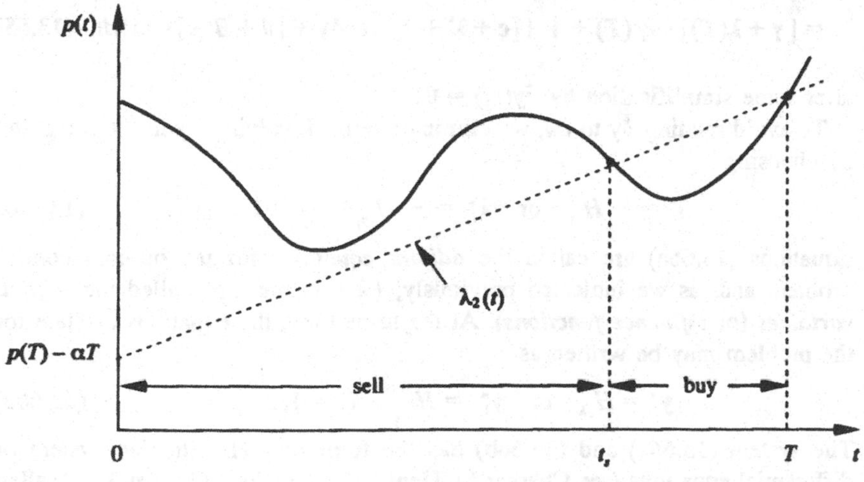


Figure 13.2

## 8. The Hamiltonian

From the development of the last two sections, it should be evident that the incorporation of the equations of state into the performance index $J$ and the elimination of the term involving $\delta \mathbf{y}$ in $\delta I$ are standard parts of the solution process for optimal control problems. For this reason, it is the general practice of the field to have these carried out automatically.

In this chapter, we limit our discussion to the case of a linear Lagrangian with linear dynamics. The optimal control problem here is to minimize (or maximize) (13.34) subject to (13.15). The use of multipliers then leads to

$$I = J - \int_{t_0}^{T} \boldsymbol{\lambda} \cdot [\mathbf{y}^{\bullet} - \mathbf{f}(t, \mathbf{y}, \mathbf{u})]\, dt$$

$$= \boldsymbol{\gamma} \cdot \mathbf{y}(T) + [\boldsymbol{\lambda} \cdot \mathbf{y}]_{t_0}^{T} + \int_{t_0}^{T} [H(t, \mathbf{y}, \mathbf{u}, \boldsymbol{\lambda}) + \boldsymbol{\lambda}^{\bullet} \cdot \mathbf{y}]\, dt, \qquad (13.63)$$

where

$$H(t, \mathbf{y}, \mathbf{u}, \boldsymbol{\lambda}) \equiv F(t, \mathbf{y}, \mathbf{u}) + \boldsymbol{\lambda} \cdot \mathbf{f}(t, \mathbf{y}, \mathbf{u})$$

$$= \sigma(t) + \mathbf{c}(t) \cdot \mathbf{y} + \mathbf{d}(t) \cdot \mathbf{u} + \boldsymbol{\lambda} \cdot [A\mathbf{y} + B\mathbf{u}]. \qquad (13.64)$$

The first variation of $I$ is

$$\delta I = [\boldsymbol{\gamma} + \boldsymbol{\lambda}(T)] \cdot \delta \mathbf{y}(T) + \int_{t_0}^{T} [(H_{,\mathbf{y}} + \boldsymbol{\lambda}^{\bullet}) \cdot \delta \mathbf{y} + H_{,\mathbf{u}} \cdot \delta \mathbf{u}]\, dt$$

$$= [\boldsymbol{\gamma} + \boldsymbol{\lambda}(T)] \cdot \delta \mathbf{y}(T) + \int_{t_0}^{T} \{[\mathbf{c} + \boldsymbol{\lambda}^{\bullet} + A^T \boldsymbol{\lambda}] \cdot \delta \mathbf{y} + [\mathbf{d} + B^T \boldsymbol{\lambda}] \cdot \delta \mathbf{u}\}\, dt \qquad (13.65)$$

after some simplification by $\delta \mathbf{y}(t_0) = \mathbf{0}$.

To avoid relating $\delta \mathbf{y}$ to $\delta \mathbf{u}$, we eliminate terms involving $\delta \mathbf{y}$ in the integrand by choosing

$$\boldsymbol{\lambda}^{\bullet} = -H_{,\mathbf{y}} \quad \text{or} \quad \lambda_i^{\bullet} = -H_{,y_i} \quad (i = 1, \ldots, n). \qquad (13.66a)$$

Equations (13.66a) are called the *adjoint equations* for the optimal control problem and, as we indicated previously, $\{\lambda_i(t)\}$ are also called the *adjoint variables* (or *influence functions*). At the same time, the equations of state for the problem may be written as

$$\mathbf{y}^{\bullet} = H_{,\boldsymbol{\lambda}} \quad \text{or} \quad y_i^{\bullet} = H_{,\lambda_i} \quad (i = 1, 2, \ldots, n). \qquad (13.66b)$$

The system (13.66a) and (13.66b) has the form of a Hamiltonian system of differential equations (see Chapter 6). Hence, the function $H(t, \mathbf{y}, \mathbf{u}, \boldsymbol{\lambda})$ is called the *Hamiltonian* for the optimal control problem. (To allow for abnormal problems, $H$ is more appropriately defined with another scalar multiplier $\lambda_0$ as a multiplicative factor of the first term of $H$. We will only be concerned with normal problems here so that the Lagrangian is to be involved in the optimization process and $\lambda_0$ may be assigned the value unity.) This Hamiltonian structure persists for a general Lagrangian and nonlinear system dynamics.

For the linear Lagrangian and linear dynamics considered in this chapter, (13.66a) and (13.66b) become

$$\boldsymbol{\lambda}^{\bullet} = -A^T\boldsymbol{\lambda} - \mathbf{c} \tag{13.67}$$

and (13.15), respectively. Special cases of these have been discussed in the last two sections. As we saw in the development there, we also need

$$\boldsymbol{\lambda}(T) = -\boldsymbol{\gamma} \tag{13.68}$$

for a minimum $J[\mathbf{u}]$ when $\mathbf{y}(T)$ is not prescribed. If $\mathbf{y}(T)$ (or any of its components) is prescribed so that $\delta\mathbf{y}(T) = \mathbf{0}$, (13.68) does not apply and we do not have any Euler boundary condition on $\mathbf{y}(T)$ (or the relevant component) in this case.

Either by $\delta\mathbf{y}(T) = \mathbf{0}$ or by our choice of $\boldsymbol{\lambda}(T)$, the expression for $\delta I$ in (13.65) simplifies to

$$\delta I = \int_{t_0}^{T} [H_{,\mathbf{u}} \cdot \delta\mathbf{u}]\, dt = \int_{t_0}^{T} [\mathbf{d} + B^T\boldsymbol{\lambda}] \cdot \delta\mathbf{u}\, dt. \tag{13.69a}$$

Because $[\mathbf{d} + B^T\boldsymbol{\lambda}]$ does not depend on $\mathbf{u}$, it is not possible to choose $\mathbf{u}$ so that $\delta I = 0$ (to make $I$ a stationary value). Moreover, $\mathbf{u}$ is usually constrained to lie in some control set so that we can only expect our optimal choice $\hat{\mathbf{u}}$ (to minimize $J$) to be nonimproving:

$$\delta I = \int_{t_0}^{T} [H_{,\mathbf{u}} \cdot \delta\mathbf{u}]\, dt \geq 0 \tag{13.69b}$$

for all admissible $\delta\mathbf{u}$ with the Hamiltonian given by (13.64).

For problems with no prescribed terminal conditions on $\mathbf{y}(t)$, $\boldsymbol{\lambda}(t)$ is completely determined by (13.67) and (13.68). The terminal-value problem for $\boldsymbol{\lambda}(t)$ may be used to eliminate $\delta\mathbf{y}(t)$ from $\delta I$. Expression (13.63) for $I$ then becomes

$$I = -\boldsymbol{\lambda}(t_0)\cdot\mathbf{y}(t_0) + \int_{t_0}^{T} \{\sigma(t) + [\mathbf{d}(t) + B^T\boldsymbol{\lambda}(t)]\cdot\mathbf{u}\}\, dt. \tag{13.70}$$

It is evident from (13.70) that a minimum $J[\mathbf{u}]$ is attained if each $u_i$ assumes an extreme value and switches from one extreme value to another whenever the *switching function* $s_i(t) \equiv d_i + (B^T\boldsymbol{\lambda})_i$ changes sign. Hence, *the optimal control is bang-bang.*

For problems with prescribed terminal conditions for some components of $\mathbf{y}$, $\lambda(t)$, in principle, depends on $\mathbf{u}$ through the initial-value problem for $\mathbf{y}$. Consequently, $\lambda(t)$ cannot be determined without determining $\mathbf{u}(t)$ simultaneously. Rather remarkably, the *Maximum Principle* of Pontryagin [see L.S. Pontryagin, et al., (1962) and L.D. Berkovitz (1974) for example] suceeds in giving us a useful characterization of $\mathbf{u}(t)$ without knowing $\lambda(t)$ explicitly:

**[XIII.8]** *Suppose* $\hat{\mathbf{u}}(t)$ *is the (PWC) optimal control function on the prescribed interval* $[t_0, T]$ *for the given problem under the previously stipulated conditions on* $\mathbf{f}$, $F$ *and* $\mathbf{u}$ *and suppose* $\hat{\mathbf{y}}(t)$ *is the corresponding optimal state determined by the equation of state (13.6), the prescribed initial conditions (13.7), and possibly some prescribed terminal condition* $y_i(T) = y_i^*$. *Then there is a (nonvanishing) vector multiplier* $\lambda(t)$ *for which the system (13.66), the given end conditions, and the Euler boundary conditions* $\lambda_k(T) = -\psi_{,y_k}(T, \hat{\mathbf{y}}(T))$ *[whenever* $y_k(T)$ *is not prescribed] hold with* $\mathbf{u} = \hat{\mathbf{u}}$ *and* $\mathbf{y} = \hat{\mathbf{y}}$. *Furthermore, we have for this pair of* $\{\hat{\mathbf{y}}(t), \lambda(t)\}$,

$$H(t, \hat{\mathbf{y}}(t), \hat{\mathbf{u}}(t), \lambda(t)) = \min_{\mathbf{u} \in U} [H(t, \hat{\mathbf{y}}(t), \mathbf{u}, \lambda(t))] \qquad (13.71)$$

*for all* $t_0 \le t \le T$.

For the problem of minimizing (13.34) subject to (13.15), we see from (13.70) that $\mathbf{u}$ cannot be chosen to make $\mathbf{d} + B^T\lambda = \mathbf{0}$ for $t_0 \le t \le T$ whether or not we have the terminal condition (13.11). If it could be done, then we have $\lambda = -(BB^T)^{-1}\mathbf{d}$; but such a $\lambda(t)$ generally would not satisfy (13.67). The maximum principle now enables us to specify some feature (such as the bang-bang nature) of $\hat{\mathbf{u}}$ when $H_{,\mathbf{u}}$ does not depend on $\mathbf{u}$ without knowing much about $\lambda(t)$.

The fundamental result of the optimal control [XIII.8] can be extended to problems with free boundaries by an argument similar to that for the calculus of variations case. For the case of an unspecified $T$, the independent variation of $T$ and $\mathbf{u}$ gives an additional *transversality condition* needed for the additional unknown:

**[XIII.9]** *If* $T$ *is unspecified, then the following transversality condition also holds*:

$$\hat{H}(T) = 0, \qquad (13.72)$$

*where* $\hat{H}(t) = H(t, \hat{\mathbf{y}}(t), \hat{\mathbf{u}}(t), \lambda(t))$.

This result will be useful in the next section.

## 9. The Linear Time Optimal Problem

We want to find the solution to the more difficult time optimal problem (TOP) defined in Section 5 in which $J$ does not involve the state variable $\mathbf{y}(t)$ or the control variable $\mathbf{u}(t)$. This unusual optimal control problem is further complicated by the fact that the terminal point $T$ is an unknown so that we have a free boundary problem previously discussed in Chapter 3. Recall that we want to minimize the terminal time $T$ taken in the form (13.37) subject to the general linear dynamics (13.15), the end conditions

$$\mathbf{y}(0) = \mathbf{y}^0, \quad \mathbf{y}(T) = \mathbf{0}, \tag{13.73}$$

and the control constraints

$$m_i \leq u_i \leq M_i \quad (i = 1, 2, \ldots, m), \tag{13.74}$$

where $m_i$ *and* $M_i$ (preferred to the more clumsy notations $u_{m_i}$ and $u_{M_i}$) may be negative. The Hamiltonian for this problem is

$$H = 1 + \boldsymbol{\lambda} \cdot [A\mathbf{y} + B\mathbf{u}], \tag{13.75}$$

where $\boldsymbol{\lambda} = \{\lambda_1(t), \ldots, \lambda_n(t)\}^T$ is an $(n \times 1)$ vector multiplier. From (13.66a), $\boldsymbol{\lambda}$ is the solution of

$$\boldsymbol{\lambda}^{\bullet} = -A^T\boldsymbol{\lambda} \tag{13.76}$$

which is a special case of (13.67a).

The system (13.76) determines $\boldsymbol{\lambda}$ up to $n$ constants of integration. With $\mathbf{y}(0)$ and $\mathbf{y}(T)$ specified, there are no Euler boundary conditions to supplement the differential system for $\boldsymbol{\lambda}(t)$. Although there is no coupling between the *adjoint equations* (13.76) and the equation state (13.15), there is a coupling in the determination of the state variables $\{y_k(t)\}$ and the multipliers $\{\lambda_i(t)\}$ through the end conditions (13.73). This appears to complicate the solution process for $\hat{\mathbf{u}}(t)$ because $\mathbf{y}(t)$ depends on $\mathbf{u}$. However, the maximum principle [XIII.8] assures us that we can choose, $\mathbf{u}$ to minimize $H$ as given by (13.75) with $\mathbf{y}$ and $\boldsymbol{\lambda}$ assuming the values corresponding to the optimal control $\hat{\mathbf{u}}(t)$. For the case $m_i < 0$ and $M_i > 0$, $i = 1, 2, \ldots, m$, this is accomplished by taking $u_i$ to be of greatest magnitude possible with a sign opposite to that of $i$th component of $\boldsymbol{\lambda}^T B$. If all $\{u_i\}$ are restricted by $|u_i| \leq 1$, this means

$$u_i = \hat{u}_i \equiv -\operatorname{sgn}[(\boldsymbol{\lambda}^T B)_i] \quad (i = 1, 2, \ldots, m). \tag{13.77}$$

Each control component can take on only the two extreme values of the control allowed. Moreover, if $(\boldsymbol{\lambda}^T B)_i$ changes sign over the interval $(0, T)$, $u_i(t)$ should switch from $+1$ to $-1$ or vice versa. The optimal control program is again bang-bang.

For our time optimal problem, the terminal time $T$ is not specified and allowed to vary. We need the transversality condition of [XIII.9] for the additional unknown $T$. The general method of solution described above will be illustrated by the problem of the *double integrator* defined by (13.35). For simplicity, we take the mass to be unit.

EXAMPLE 4: $A = \begin{bmatrix} 0 & 1 \\ 0 & 0 \end{bmatrix}, \quad B = \begin{bmatrix} 0 \\ 1 \end{bmatrix}, \quad |u| \le 1.$

For this problem, we have

$$\boldsymbol{\lambda}^{\bullet} = -A^T\boldsymbol{\lambda} = -\begin{bmatrix} 0 & 0 \\ 1 & 0 \end{bmatrix}\boldsymbol{\lambda} \quad \text{or} \quad \begin{bmatrix} \lambda_1 \\ \lambda_2 \end{bmatrix} = \begin{bmatrix} c_1 \\ c_1(T-t)+c_2 \end{bmatrix}. \tag{13.78}$$

Unlike the problems in the last two sections, there are no boundary conditions on $\boldsymbol{\lambda}(T)$ to fix $c_1$ and $c_2$. The Hamiltonian for this problem is $H = 1 + \lambda_1 y_2 + \lambda_2 u$. We see from this expression [or from the more general result in (13.77)] that we should take

$$\hat{u} = -\operatorname{sgn}[\lambda_2(t)]. \tag{13.79}$$

But $\lambda_2(t)$ is linear in $t$; hence, it changes sign at most once in $[0, T]$. The problem, therefore, involves a control with only one switch. The determination of the switch point is more intricate because $\lambda_2(t)$ is not completely determined. We will obtain this switch point by studying separately the consequence of $u = 1$ and $u = -1$.

For $u = +1$, we integrate the equation of state to get

$$y_2 = y_2^0 + t \quad \text{and} \quad y_1 = \left[y_1^0 - \frac{1}{2}(y_2^0)^2\right] + \frac{1}{2}(y_2^0 + t)^2. \tag{13.80}$$

Note that solution (13.80) satisfies the initial conditions $y_i(0) = y_i^0$ , $i = 1, 2$. We can eliminate $t$ from the two expressions to get

$$y_1 = \left[y_1^0 - \frac{1}{2}(y_2^0)^2\right] + \frac{1}{2}y_2^2 \quad (u = 1) \tag{13.81}$$

which is a parabola opening to the right (see Fig. 13.3). If the initial data are related by $y_1^0 = (y_2^0)^2/2$, then the parabola passes through the origin, entering from below ($y_2 < 0$) and leaving through the top ($y_2 > 0$). Clearly, if $y_2^0 < 0$ and $y_1^0 = (y_2^0)^2/2$, then we should take $u = 1$ which steers the initial state to the origin. No switching of control is necessary. To determine the corresponding terminal time $\hat{T}$ for this case, we use $\mathbf{y}(T) = \mathbf{0}$; this requires $\hat{T} = -y_2^0$

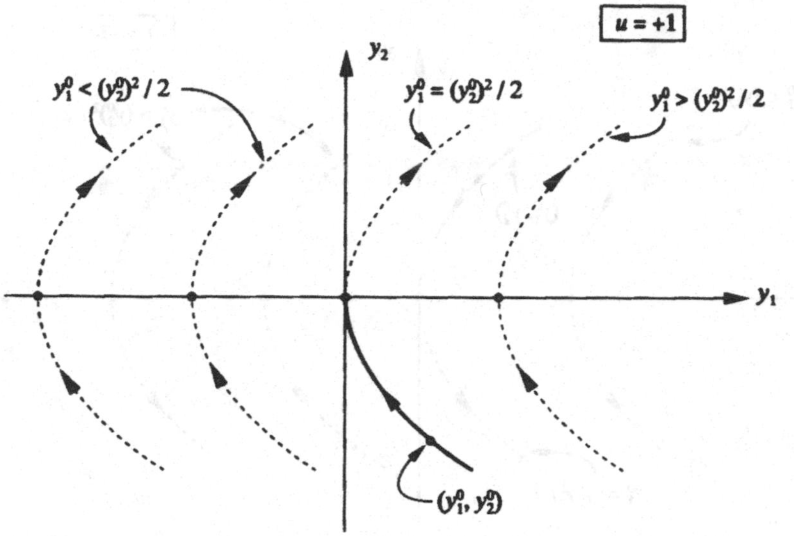


**Figure 13.3**

which is positive because $y_2^0 < 0$. The situation for $y_1^0 \neq (y_2^0)^2/2$ will be discussed later.

For $u = -1$, we have similarly

$$y_2 = y_2^0 - t, \qquad y_1 = \left[ y_1^0 + \frac{1}{2}(y_2^0)^2 \right] - \frac{1}{2}(y_2^0 - t)^2 \tag{13.82}$$

or, after eliminating $t$,

$$y_1 = \left[ y_1^0 + \frac{1}{2}(y_2^0)^2 \right] - \frac{1}{2}y_2^2, \tag{13.83}$$

which correspond to a parabola opening to the left (see Fig. (13.4). If the initial data are related by $y_1^0 = -(y_2^0)^2/2$, then the parabola passes through the origin, entering from the top ($y_2 > 0$) and leaving through the bottom ($y_2 < 0$). Hence, if $y_2^0 > 0$ and $y_1^0 = -(y_2^0)^2/2$, we should take $u = -1$ as it steers the initial state to the origin. Again no switching of control is necessary. To determine $\widehat{T}$ for this case, we again use $\mathbf{y}(T) = \mathbf{0}$ to get $\widehat{T} = y_2^0\ (>0)$.

Let $C_s^+$ be the half-parabola $y_1 = (y_2)^2/2$ for $y_2 < 0$ and $C_s^-$ be the half-parabola $y_1 = -(y_2)^2/2$ for $y_2 > 0$. The composite curve

$$C_s = \begin{cases} C_s^+ & (y_1 > 0) \\ C_s^- & (y_1 < 0) \end{cases} \tag{13.84}$$

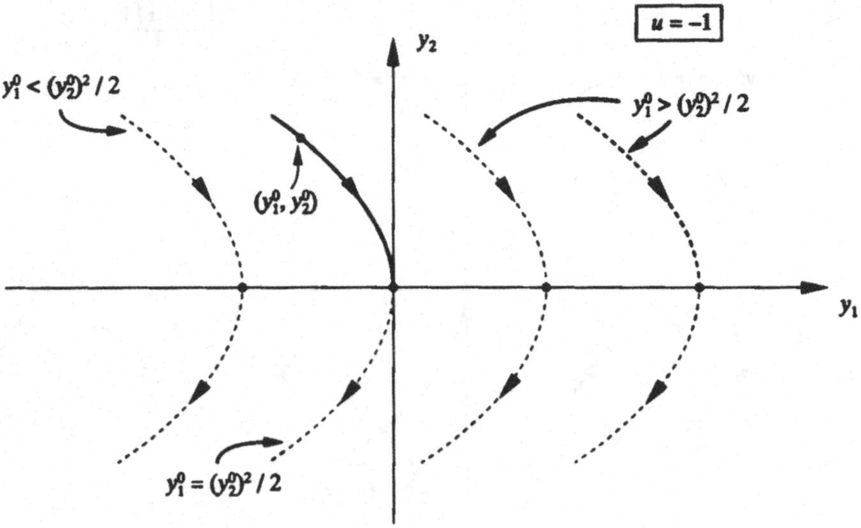


**Figure 13.4**

is a very special trajectory for our problem. If the prescribed initial condition $\mathbf{y}^0$ happens to be a point on this curve, the system (as noted previously) is driven to the origin *in minimum time* by taking $u(t) = 1$ for $\mathbf{y}^0$ on $C_s^+$ and $u(t) = -1$ for $\mathbf{y}^0$ on $C_s^-$. There should be no change in the control as a function of time.

For an initial point $(y_1^0, y_2^0)^T = \mathbf{y}^0$ not on $C_s$, the *maximum principle* requires the system to move along a parabolic trajectory $C_0$ determined by the initial data $\mathbf{y}^0$ at least for awhile (Fig. 13.5) For the system to get to the origin and still satisfy the maximum principle, we must eventually get the system onto the *switch curve* (or singular trajectory) $C_s$. To switch from $C_0$ to another admissible trajectory, $C_1$ can only take place when $\lambda_2(t)$ changes sign. Because $\lambda_2(t)$ is a linear function of $t$, this happens at most once. Hence, we must switch directly from $C_0$ to $C_s$. How this can be achieved depends on the location of the initial data $\mathbf{y}^0$:

(i) If $\mathbf{y}^0$ is above the switch curve $C_s$, the system can get on to $C_s$ only by following the left-facing parabola passing through $\mathbf{y}^0$, namely, (13.82) corresponding to $\hat{u} = -1$ for $t < t_s$. Once it reaches the switch curve $C_s$, we switch to $\hat{u} = 1$ for $t > t_s$ to head for the origin along $C_s^+$ (see Fig. 13.5).

(ii) If $\mathbf{y}^0$ is below the switch curve $C_s$, the system can get onto $C_s$ only by following the right-facing parabola passing through $\mathbf{y}^0$, namely, (13.81)

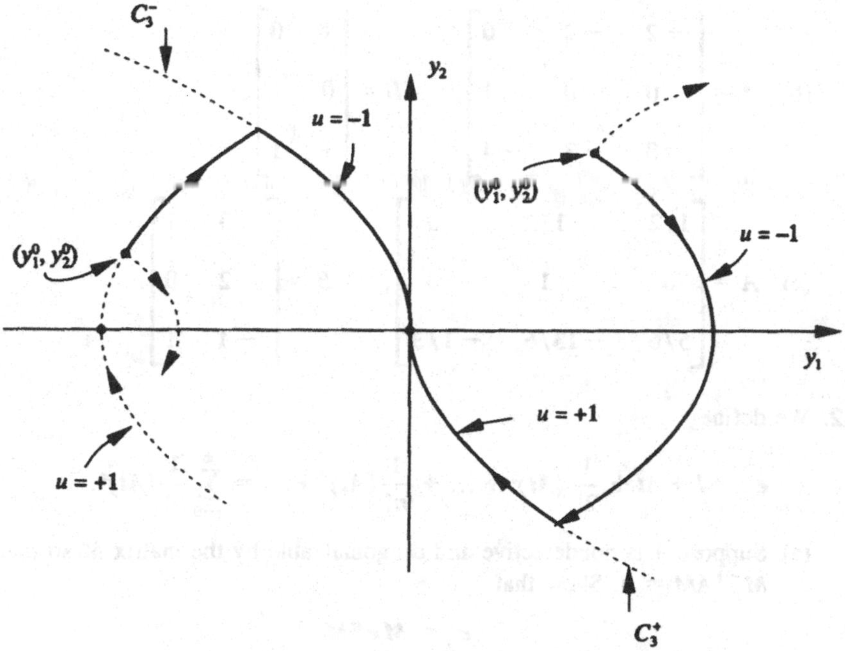


**Figure 13.5**

corresponding to $\hat{u} = 1$ for $t < t_s$, to head for $C_s$. Once it reaches $C_s$, we switch the control to $\hat{u} = -1$ for $t > t_s$ to head for the origin along $C_s^-$.

It remains to work out the details of the above optimal program. More specifically, we have to determine the switch time $t_s$, the switch point $(y_1^s, y_2^s) \equiv \mathbf{y}^s$, and the terminal time $\widehat{T}$ for the problem. We will leave the determination of these quantities as exercises.

## 10. Exercises

1. Decide on the controllability of $\mathbf{y}^{\bullet} = A\mathbf{y} + B\mathbf{u}$ for

(a) $A = \begin{bmatrix} 0 & 1 & 0 \\ 0 & 0 & 1 \\ -2 & -4 & -3 \end{bmatrix}, \quad B = \begin{bmatrix} 1 & 0 \\ 0 & 1 \\ -1 & 1 \end{bmatrix};$

(b) $A = \begin{bmatrix} -2 & -2 & 0 \\ 0 & 0 & 1 \\ 0 & -3 & -4 \end{bmatrix}, \quad B = \begin{bmatrix} 1 & 0 \\ 0 & 1 \\ 1 & 1 \end{bmatrix};$

(c) $A = \begin{bmatrix} 1/2 & 1/2 & 0 \\ 0 & 1 & 0 \\ 5/6 & -13/6 & -1/3 \end{bmatrix}, \quad B = \begin{bmatrix} 3 & 1 \\ 2 & 0 \\ -1 & 1 \end{bmatrix}.$

2. We define

$$e^{At} = I + At + \frac{1}{2!}(At)^2 + \dots + \frac{1}{n!}(At)^n + \dots = \sum_{n=0}^{\infty} \frac{1}{n!}(At)^n .$$

(a) Suppose $A$ is nondefective and diagonalizable by the matrix $M$ so that $M^{-1}AM = \Lambda$. Show that

$$e^{At} = M e^{\Lambda t} M^{-1}$$

with $\Lambda = [\Lambda_{ij}] = [\lambda_i \delta_{ij}]$ and $e^{\Lambda t} = [e^{\lambda_i t}\delta_{ij}]$, where $\{\lambda_1, \dots, \lambda_n\}$ are the eigenvalues of $A$.

(b) Whether or not $A$ is nondefective, show that

$$(e^{At})^{\cdot} = Ae^{At} = e^{At}A .$$

(c) Show that

$$e^{At}e^{-At} = e^{-At}e^{At} = I$$

whether or not $A$ is nondefective

3. Decide on the controllability of

$$\mathbf{y}^{\cdot} = \begin{bmatrix} t^3 & 0 \\ 0 & t^4 \end{bmatrix} \mathbf{y} + \begin{bmatrix} t^2 \\ t^3 \end{bmatrix} u .$$

4. Show that the $n$th-order linear ordinary differential equation

$$y^{(n)} + a_{n-1}(t)y^{(n-1)} + a_{n-2}(t)y^{(n-2)} + \dots + a_1(t)y^{\cdot} + a_0(t)y = u(t)$$

is controllable at any time $t_0 \geq 0$.

5. $$\mathbf{y}^{\bullet} = \begin{bmatrix} 0 & -1 \\ 1 & 0 \end{bmatrix} \mathbf{y} + \begin{bmatrix} \cos t \\ \sin t \end{bmatrix} u .$$

(a) Show that all solutions of the above system with $\mathbf{y}(0) = \mathbf{0}$ lie on the surface $y_1 \sin t - y_2 \cos t = 0$.

(b) Show that the system is not controllable at $t_0 = 0$.

6. Suppose the matrix $A$ in the linear system $\mathbf{y}^{\bullet} = A\mathbf{y} + Bu$ has multiple eigenvectors for one of its eigenvalues. Show that the system is not controllable. [Note: $u$ is a scalar.]

7. (a) Find the optimal control program which maximizes

$$J = \int_0^2 (2y - 3u)\,dt$$

subject to $y^{\bullet} = y + u, \quad 0 \le u \le 2, \quad y(0) = 5.$

(b) Repeat part (a) with "maximizes" replaced by "minimizes."

8. Find the optimal control program which minimizes

$$J = \int_0^1 (2 - 5t)u\,dt$$

subject to $y^{\bullet} = 2y + 4te^{2t}u, \quad y(0) = 0, \quad y(1) = e^2$, and $-1 \le u \le 1$.

9. Find the optimal control program which maximizes

$$J = \int_0^T (y - u)\,dt$$

subject to $\quad y^{\bullet} = u, \quad 0 \le u \le y, \quad y(0) = y_0.$

10. The revenue $r(t)$ that a machine earns at time $t$ is proportional to its quality $y$, i.e., $r(t) = py(t)$ for some positive constant $p$. The quality decays at a constant rate $a$ but can be enhanced by expenditure $u(t)$ on maintenance and upgrading. The machine will be sold at a predetermined time $T$; the unit sale price, $s$, is proportional to its quality. Find the optimal maintenance policy $\hat{u}(t)$, $0 \le t \le T$, which maximizes

$$J = \int_0^T e^{-\delta t}[py(t) - u(t)]\,dt + e^{-\delta T} s y(T)$$

subject to $y^{\bullet} = u - ay$, $0 \le u \le M$, $y(0) = y_0$ where $s < 1 < p/(\delta + a)$. Interpret this last inequality.

11. In the Time Optimal Problem for the double integrator, we know that the system should be on the switch curve $C_s$ for $t_s \le t \le T$. Suppose $y_1(t_s) = y_1^s$ and $y_2(t_s) = y_2^s$.
    (a) Determine $y_1(t)$ and $y_2(t)$ for $t \ge t_s$ in terms of $t_s$, $y_1^s$, and $y_2^s$ if $(y_1^s, y_2^s)$ is on $C_s^+$. Obtain also $y_1^s$ in terms of $y_2^s$.
    (b) Repeat part (a) for $(y_1^s, y_2^s)$ on $C_s^-$.
    (c) Use $y_2(T) = 0$ to determine $\widehat{T}$ in terms of $y_2^s$.

12. (a) Solve the adjoint equations for the double integrator. [Note that the solution contains two arbitrary constants of integration and applies to $(y_1^s, y_2^s)$ on either branch of $C_s$.]
    (b) Determine one of the constants by the transversality condition $H(\widehat{T}) = 0$.
    (c) Use $\lambda_2(t_s) = 0$ to determine the remaining constant in terms of $t_s$ and $y_2^s$.

13. (a) Solve the equations of state for the double integrator for $0 \le t \le t_s$. The solution will depend on the position of the given initial state $(y_1^0, y_2^0)$ relative to $C_s$.
    (b) At the switch time $t_s$, the system must be on the switch curve $C_s$ in order to be able to get to the origin in minimum time. Use this requirement to determine $y_2^s$ and $t_s$ in terms of $(y_1^0, y_2^0)$.

14. Solve the Time Optimal Problem with

$$y_1^{\bullet} = u_1, \qquad y_2^{\bullet} = u_2, \qquad \mathbf{y}(0) = \mathbf{y}^0, \qquad \mathbf{y}(T) = 0,$$

and $0 \le |u_1| + |u_2| \le 1$.

15. Solve the Time Optimal Problem with

$$y_1^{\bullet} = u_1, \quad y_2^{\bullet} = u_2, \quad \mathbf{y}(0) = \mathbf{y}^0, \quad \mathbf{y}(T) = 0,$$

and $|u_i| \le 1$.

# 14

# Optimal Control with General Lagrangians

## 1. The Maximum Principle

In this chapter, we consider optimal control problems with a general Lagrangian $F(t, \mathbf{y}, \mathbf{u})$ not necessarily linear in both the state variable $\mathbf{y}$ and the control variable $\mathbf{u}$. For the time being, we minimize the performance index (13.12) for a prescribed time interval $[t_0, t_1]$, subject only to the nonlinear equation of state (13.6) and the initial condition (13.7). Similar to problems in the calculus of variations, we append the first variation of the state equation to the first variation of the performance index by a vector Lagrange multiplier (or an *adjoint variable*) $\boldsymbol{\lambda}(t)$:

$$\begin{aligned}\delta I &= \delta J + \int_{t_0}^{t_1} \boldsymbol{\lambda}^T \delta[\mathbf{f}(t,\mathbf{y},\mathbf{u}) - \mathbf{y}^{\bullet}]\,dt \\ &= [\psi_{,\mathbf{y}}(\widehat{\mathbf{y}}(t_1)) - \boldsymbol{\lambda}^T(t_1)]\,\delta\mathbf{y}(t_1) \\ &\quad + \int_{t_0}^{t_1} [(\widehat{F}_{,\mathbf{u}} + \boldsymbol{\lambda}^T\widehat{\mathbf{f}}_{,\mathbf{u}})\delta\mathbf{u} + \{\widehat{F}_{,\mathbf{y}} + \boldsymbol{\lambda}^T\widehat{\mathbf{f}}_{,\mathbf{y}} + (\boldsymbol{\lambda}^T)^{\bullet}\}\delta\mathbf{y}]\,dt,\end{aligned}$$

where we have made use of $\delta\mathbf{y}(t_0) = 0$ because $\mathbf{y}(t_0)$ is prescribed. As usual, $\widehat{g}(t)$ denotes $g(t, \widehat{\mathbf{y}}(t), \widehat{\mathbf{u}}(t))$. As in Section 6 of Chapter 4, $g_{,\mathbf{x}}(x_1, \ldots, x_k)$ is the row vector $(g_{,x_1}, \ldots, g_{,x_k})$. Consistent with this, $\mathbf{f}_{,\mathbf{x}}$ is the $(n \times k)$ matrix

$$\mathbf{f}_{,\mathbf{x}} = \begin{pmatrix} f_{1,\mathbf{x}} \\ \vdots \\ f_{n,\mathbf{x}} \end{pmatrix} = \begin{bmatrix} f_{1,x_1} & \cdots & f_{1,x_k} \\ \vdots & & \vdots \\ f_{n,x_1} & \cdots & f_{n,x_k} \end{bmatrix}.$$

We may (assume normality and) choose the adjoint variable $\boldsymbol{\lambda}$ to eliminate terms involving $\delta\mathbf{y}$:

$$(\boldsymbol{\lambda}^{\bullet})^T = -\boldsymbol{\lambda}^T\widehat{\mathbf{f}}_{,\mathbf{y}} - \widehat{F}_{,\mathbf{y}}, \qquad \boldsymbol{\lambda}^T(t_1) = \psi_{,\mathbf{y}}(\widehat{\mathbf{y}}(t_1)). \tag{14.1}$$

The choice of $\boldsymbol{\lambda}(t)$ determined by (14.1) reduces $\delta I$ to

$$\delta I = \int_{t_0}^{t_1} (\widehat{F}_{,\mathbf{u}} + \boldsymbol{\lambda}^T\widehat{\mathbf{f}}_{,\mathbf{u}})\,\delta\mathbf{u}\,dt. \tag{14.2}$$

Because $\delta\mathbf{u}$ is arbitrary, we need

$$\widehat{H}_{,\mathbf{u}} = \mathbf{0}, \tag{14.3}$$

where $H = F + \boldsymbol{\lambda}^T\mathbf{f}$ is the *Hamiltonian* for the problem, to avoid the possibility $\delta I < 0$ for some choice of $\delta\mathbf{u}$ [in which case $(\widehat{\mathbf{y}}, \widehat{\mathbf{u}})$ would not minimize $J$].

If $F$, $\psi$ and $\mathbf{f}$ are all $C^1$ functions of their arguments and $\mathbf{u}$ is at least PWC, the maximum principle [XIII.8] also holds for the present more general problem. As there is no constraint on the control so that the control set $U$ is all of $R^m$, the minimizing condition in (13.71) implies the stationarity condition (14.3).

If $\mathbf{u}$ is restricted to some constraint set $U$ in $R^m$, the optimal solution which satisfies (14.3) may lie outside $U$. In that case, condition (13.71) locates the best choice of $\mathbf{u}$ which may be on the boundary of $U$. It is customary in the control literature to call this boundary solution a *corner solution*. For such an optimal solution $\widehat{\mathbf{u}}$, $\delta I$ can only be expected to be nonimproving for admissible choices of $\delta\mathbf{u}$:

$$\delta I = \int_{t_0}^{t_1} (\widehat{F}_{,\mathbf{u}} + \boldsymbol{\lambda}^T\widehat{\mathbf{f}}_{,\mathbf{u}})\,\delta\mathbf{u}\,dt \geq 0. \tag{14.4}$$

**EXAMPLE 1:** $\max_c \left\{J = \psi_T\, k(T) + \int_0^T c^{1-\sigma}\,dt\right\}$ subject to $k^{\bullet} = k^a - c$, $k(0) = k_0$, with $0 < \sigma, a < 1$, and $\psi_T > 0$.

This is a special case of Example 3 in Chapter 12. The Hamiltonian for the problem is

$$H = c^{1-\sigma} + \lambda(k^a - c).$$

The adjoint variable $\lambda$ is determined by

$$\lambda^{\bullet} = -\lambda a k^{a-1}, \qquad \lambda(T) = \psi_T.$$

The stationarity condition (14.3) becomes

$$(1-\sigma)c^{-\sigma} - \lambda = 0.$$

(Note that we have omitted the use of the for ⌢ the extremal solution and will continue to do when the omission is unlikely to cause any confusion.) We may use the above relation to eliminate $\lambda$ from the adjoint problem to obtain

$$c^{\bullet} = \frac{a}{\sigma} ck^{a-1}, \qquad c(T) = \left(\frac{1-\sigma}{\psi_T}\right)^{1/\sigma} \equiv c_T.$$

The equation of state $k^{\bullet} = k^a - c$ and the equation for $c$ above together wth $k(0) = k_0$ and $c(T) = c_T$ define a two-point boundary-value problem for $k(t)$ and $c(t)$.

Because the independent variable $t$ does not appear explicitly, we may combine the two first-order differential equations and obtain

$$\frac{dk}{dc} = \frac{\sigma}{a}\left[\frac{k}{c} - k^{1-a}\right].$$

The solution of this Bernoulli equation (see Boyce and Di Prima, 1976) can be obtained by rearranging it as

$$\frac{d}{dc}(k^a) = \frac{\sigma}{c} k^a - \sigma$$

which is a linear first-order equation in $k^a$. The exact solution of this equation is

$$k^a = a_0 c^{\sigma} - \frac{\sigma}{1-\sigma} c,$$

where $a_0$ is a constant of integration.

The relation between $c$ and $k$ above may be used to eliminate $k^{a-1}$ from the differential equation for $c$ to get

$$c^{\bullet} = \frac{a}{\sigma} c \left[a_0 c^{\sigma} - \frac{\sigma}{1-\sigma} c\right]^{\frac{a-1}{a}}.$$

This first-order equation and the terminal condition $c(T) = c_T$ determines $c(t)$ in terms of the parameter $a_0$. An explicit exact solution of this terminal-value problem as a function of $t$ is possible for $\sigma = a = ½$. The initial condition $k(0) = k_0$ then fixes $a_0$.

It should be emphasized that the maximum principle [XIII.8] gives only a set of necessary conditions for optimality. A solution obtained from these conditions corresponds only to an *extremal* in the calculus of variations and will also be so designated. Further documentation will be needed to show that such a solution is optimal. Also, capital stock and consumption rate are necessarily non-negative quantities (and often required to be positive). These requirements impose the non-negativity constraints $c(t) \geq 0$ and $k(t) \geq 0$ on the control and state variable. For the present problem, they are automatically satisfied by the solution obtained (for $T$ not too large). We will discuss in a later section the more general case where the constraints need to be enforced.

Although the reduction of the necessary conditions for optimality to a separable first-order differential equation for $c$ alone is attractive analytically, it turns out (as it is often the case) that a phase plane analysis of the coupled system

$$k^{\bullet} = k^{a} - c, \qquad c^{\bullet} = \frac{a}{\sigma} c k^{a-1}$$

is much more informative. This system has no critical point for $(0 <)\, a < 1$. The slope of the trajectories are horizontal at $c = 0$ [and therefore $c(t)$ cannot be negative given $c(T) > 0$] and vertical at $c = k^{a}$. The slopes must be headed in the NW direction for $c > k^{a}$ and in the NE direction for $c < k^{a}$. A qualitative phase portrait of the problem as shown in Figure 14.1 can easily be constructed.

In terms of the phase portrait, the solution process can be described as follows: For a given initial capital stock $k_0$, we locate an appropriate initial value $c(0) = c_0$ so that the trajectory that passes through $(k_0, c_0)$ reaches a point with ordinate $c_T$ at a later time $t = T$. Unless there is more than one such feasible solution, the trajectory found necessarily provides the optimal solution trajectory if a solution of the problem exists. However, the behavior of the

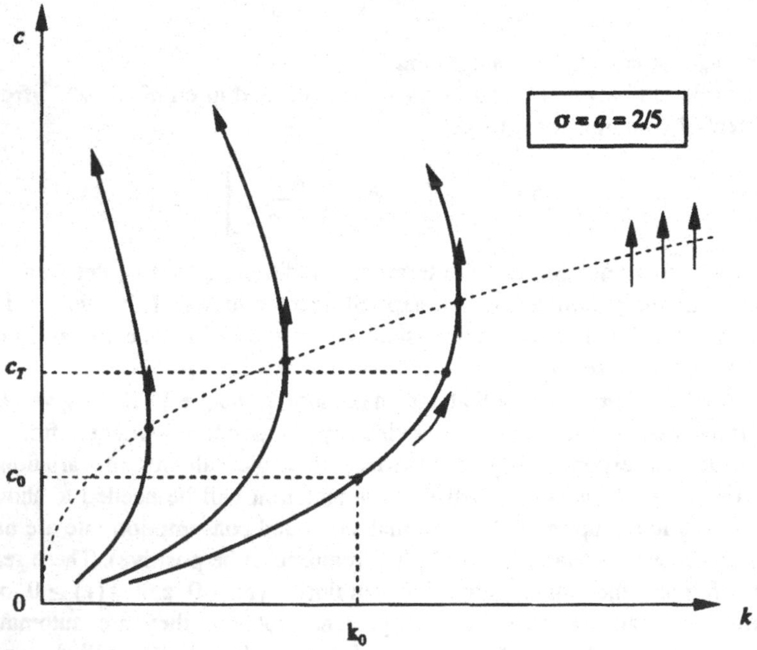


Figure 14.1

trajectories for large $c$ suggests that the model is not realistic (and should not be used) for large $t$.

**EXAMPLE 2:** Suppose production requires a natural resource flow $r(t)$, in addition to the capital stock input $k(t)$, so that the equation of state takes the form

$$k^{\cdot} = f(k, r) - \theta r - \mu k - c, \qquad k(0) = k_0,$$

where $\theta$ is the extraction cost for a unit of resource and $\mu$ is the constant rate of capital depreciation. Let $D(t)$ be total resource deposit remained at time $t$. Then we have

$$D^{\cdot} = -r, \qquad D(0) = D_0.$$

For social optimum and a fixed period planning, we want to choose $c(t)$ and $r(t)$ to maximize

$$J = \psi(k_T, D_T) + \int_0^T e^{-\delta t} U(c)\,dt$$

with $k_T \equiv k(T)$ and $D_T \equiv D(T)$, subject to the two equations of state and initial conditions for $k$ and $D$.

The Hamiltonian for the present problem is

$$H = e^{-\delta t} U + \lambda_1 (f - \theta r - \mu k - c) - \lambda_2 r.$$

The adjoint equations for $\lambda_1$ and $\lambda_2$ are

$$\lambda_1^{\cdot} = -H_{,k} = -\lambda_1 (f_{,k} - \mu), \qquad \lambda_2^{\cdot} = -H_{,D} = 0.$$

We will again assume that the non-negativity constraint on $c$, $r$, and $k$ are met by the extremals so that the stationarity conditions corresponding to (14.3),

$$H_{,c} = e^{-\delta t} U'(c) - \lambda_1 = 0, \qquad H_{,r} = \lambda_1 (f_{,r} - \theta) - \lambda_2 = 0,$$

are applicable. The prime in the first condition indicates differentiation with respect to the argument of the utility function $U(c)$. Together with the two equations of state, the prescribed initial conditions, and the Euler boundary conditions

$$\lambda_1 = \psi_{,k_T}(k_T, D_T), \qquad \lambda_2(T) = \psi_{,D_T}(k_T, D_T),$$

these necessary conditions for optimality are expected to determine the six unknown functions $k, r, c, D, \lambda_1$, and $\lambda_2$ once $f(\cdot, \cdot)$ and $U(\cdot)$ are prescribed. A typical *production function* is the Cobb–Douglas function $k^a r^b$ where $0 < b < a < a + b \leq 1$; a typical *utility function* $U(c)$ is $c^{1-\sigma}$ with $0 < \sigma < 1$.

Useful qualitative conclusions on the economy can still be drawn from the necessary conditions for optimality even if only certain features of the two functions $f$ and $U$ (such as monotonicity and concavity), and not the actual functional form, are known [see Wan (1989)]. We will not pursue a discussion of the solution of this problem here. It should be noted, however, that there is no guarantee that our houndary-value problem for the six unknowns has a solution. The maximum principle only asserts that we can always find a solution of the terminal-value problem for the adjoint variables $\boldsymbol{\lambda} = (\lambda_1, \ldots, \lambda_n)^T$ *if an optimal control exists.*

The stationarity conditions (14.3) are algebraic equations relating the unknown state, control, and adjoint variables. For the two examples above, they are simple relations which allow us to eliminate the adjoint variables from the boundary-value problem for the determination of the extremals. However, the stationarity conditions generally cannot be solved for some of the unknowns in terms of other unknowns, and a numerical solution is necessary for the unreduced problem which involves a combination of differential equations and algebraic equations. As such, boundary-value problems involving *differential-algebraic systems* arise naturally in optimal control problems. Numerical methods for such systems constitute an active area of current research in differential equations, scientific computing, and numerical softwares.

## 2. Controllability of Nonlinear Systems

When the terminal values of some or all of the components of the state variable $\mathbf{y} = (y_1, y_2, \ldots, y_n)^T$ are also prescribed, a particular optimal control problem may not have a solution. This could happen for different reasons. It may be that the system dynamics is incapable of driving the system to the desired terminal target whatever the control may be. The capital stock of the simple economy in Section 1, Chapter 12 simply cannot increase to a higher level than its initial state if the production rate is slower than the capital depreciation rate. System dynamics limitation is also the reason why the following optimal control problem has no solution:

EXAMPLE 3: $\max_u \left\{ J = \int_0^1 u(t)\,dt \right\}$ subject to $y^\bullet = y + u^2$, $y(0) = 1$, and $y(1) = 0$.

The exact solution of the initial-value problem for $y$ is

$$y(t) = 1 + \int_0^t e^{t-\tau} u^2(\tau)\,d\tau.$$

Because the integral is non-negative for any real-valued control $u(t)$, it is not possible to choose $u(t)$ so that $y(1) = 0$. Hence, the problem (is not controllable and) has no solution.

In these two examples, the optimal control problem has no solution whatever the performance index $J$ may be. The following optimal control problem also has no solution but for a different reason.

EXAMPLE 4: $\max_u \left\{ J = \int_0^1 u(t)\,dt \right\}$ subject to $y^\bullet = u^2$, $y(0) = 0$, and $y(1) = 0$.

The exact solution of the initial-value problem for $y$ is

$$y(t) = \int_0^t u^2(\tau)\,d\tau.$$

The only real-valued control $u(t)$ which allows $y(t)$ to satisfy the terminal condition $y(1) = 0$ is $u(t) = 0$ for $0 \le t \le 1$. But with $H = u + \lambda u^2$, the maximum of $H$ is attained at $2\lambda u + 1 = 0$ which cannot be met by $u = 0$.

In contrast to Example 3, there is a *feasible* solution (i.e., an admissible comparison function) for Example 4. Because $u(t) = 0$ is the only feasible solution, it should be the solution of the optimal control problem whatever the performance index may be. In other words, the actual performance index is irrelevant in this case. The optimal control problem would have a solution if the multiplier rule had allowed for this possibility. As in calculus of variations problems, Example 4 is a *controllable* optimal control problem which is *abnormal*. Its solution is straightforward if we work with a more general multiplier rule by taking

$$H = \lambda_0 F(t, \mathbf{y}, \mathbf{u}) + \boldsymbol{\lambda}^T \mathbf{f} \tag{14.5}$$

and having the option of setting $\lambda_0 = 0$.

The more general multiplier rule above resolves the kind of difficulty typified by Example 4 but does nothing for problems such as Example 3 where the difficulty is inherent in the system dynamics. When an exact expression for the solution of the initial-value problem is not available, we need a way to decide whether the system can be steered to the prescribed target by some admissible control. In other words, we need a way to decide on the *controllability* of (nonlinear) dynamical systems not treated in Chapter 13. To decide whether there is at least one feasible control to steer $\mathbf{y}$ to the prescribed target, usually taken to be the origin, is a difficult current research problem. If the origin is an asymptotically stable equilibrium state of the dynamical system, controllability may be analyzed with the help of the results already obtained in Chapter 13.

Because the origin is an equilibrium state, we have $\mathbf{f}(t, 0, 0) = \mathbf{0}$. We assume that the constraint set $U$ contains a neighborhood of $\mathbf{u} = 0$ and set

$$A(t) = [A_{ij}(t)] \equiv \mathbf{f}_{,\mathbf{y}}(t, 0, 0), \qquad B(t) = [B_{ij}(t)] \equiv \mathbf{f}_{,\mathbf{u}}(t, 0, 0) \tag{14.6}$$

so that $A_{ij}$ is $\partial f_i / \partial y_j$ evaluated at $(\mathbf{y}, \mathbf{u}) = (\mathbf{0}, \mathbf{0})$ and $B_{ij}(t)$ is $\partial f_i / \partial u_j$ evaluated at $(\mathbf{0}, \mathbf{0})$, the system behavior is adequately modeled by

$$\mathbf{z}^{\bullet} = A(t)\mathbf{z} + B(t)\mathbf{u}. \tag{14.7}$$

The controllabity of (14.7) is known from the results of Chapter 13. The conclusions can be shown to apply to the original nonlinear system if $\mathbf{y} = \mathbf{0}$ is a global, asymptotically stable equilibrium point. Without any control ($\mathbf{u} = \mathbf{0}$), the nonlinear system should reach inside a small neighborhood of the origin of the state space in finite time. At that point, the system behavior can be approximated by (14.7) and [XIII.7] (or its counterpart for nonautonomous linear systems) applies so that the system can be steered to the origin, again in finite time.

Clearly, the target state will not always be an asymptotically stable equilibrium state of the nonlinear system (not to mention global asymptotic stability). On the other hand, we would only need $\mathbf{y}(T)$ to be near a globally asymptotically stable point of the system. If $\mathbf{y}(T)$ lies inside the region of controllability $N$ of such an equilibrium state, we should be able to steer the system from the asymptotically stable equilibrium point to $\mathbf{y}(T)$ in finite time. The following example, however, shows that these results are not always useful for specific problems in applications.

**EXAMPLE 5:** Suppose $f(k, r)$ in Example 2 is the Cobb–Douglas production function $k^a r^b$ and there is a less expensive substitute for the resource input at the end of the planning period $(0, T)$. In that case, we should exhaust the entire resource deposit $D_0$ by $t = T$ for maximum benefit so that

$$D(T) = 0.$$

For this modified version of Example 2, we have

$$\mathbf{f}(\mathbf{y}, \mathbf{u}) = (k^a r^b - \theta r - \mu k - c, - r)^T$$

and $\mathbf{f}(\mathbf{0}, \mathbf{0}) = \mathbf{0}$. However, with

$$\mathbf{f}_{,r} = (b k^a r^{b-1} - \theta, - 1)^T$$

and $0 < b < 1$, it is not possible to linearize $\mathbf{f}$ about the origin. Hence, the conclusions reached by way of the linear system (14.7) are not useful here.

In the rest of this chapter, we will assume that the system of interest is *controllable* and *normal*. Abnormal situations are usually self-evident and can be resolved by ignoring the performance index.

## 3. Sustained Consumption with a Finite Resource Deposit

It is of some interest to note that Example 5 illustrates another aspect of optimal control problem concerning terminal conditions. In many applications, terminal conditions are prescribed for only some (but not all) of the state variables. Because the simple economy of Example 5 will continue beyond time $T$ (with only the resource flow replaced), it is *not* appropriate to require $k(T) = 0$ as well. On the other hand, in planning for the entire future without a substitute for the essential resource, we also cannot assume $k \to 0$ as $t \to \infty$. The following example shows that an optimal program may increase the capital stock without bound in some cases and deplete it completely in others.

EXAMPLE 6: Consider Example 2 with $f = k^a r^b$, and $U(c) = c^{1-\sigma}/(1-\sigma)$ for $\sigma > 0$ and with the single planning period being the entire future, $T = \infty$. In that case, there is no specific salvage value for $k(\infty)$ and, obviously, the natural resource deposit should eventually be used up so that $D(\infty) = 0$. [If $D(\infty) > 0$, we can increase $r(t)$ and consume a little more to get a larger $J$.] Hence, the optimal control problem now is to choose $r(t)$ and $c(t)$ to maximize

$$J = \int_0^\infty e^{-\delta t} \frac{c^{1-\sigma}}{1-\sigma}\, dt$$

subject to the equations of state

$$k^\bullet = k^a r^b - \theta r - \mu k - c, \qquad D^\bullet = -r \tag{14.8}$$

with

$$k(0) = k_0, \qquad D(0) = D_0, \qquad D(\infty) = 0. \tag{14.9}$$

With $f(k, 0) = 0$, the resource is essential for production. It is far from clear then that a finite consumption rate can be sustained for the indefinite future given only a finite resource deposit $D_0$. The Cobb–Douglas production function at least offer the possibility of different capital and resource flow mixes for the same output. With $r(t) \to 0$ as $t \to \infty$, it would appear that we need $k(t) \to \infty$ as $t \to \infty$ to sustain the economy.

Unfortunately, the actual (limiting) terminal value of $k(t)$ varies with the values for the model parameters as we shall see from the extremals for special cases. To get these extremals, we set

$$H = \frac{e^{-\delta t}}{1-\sigma} c^{1-\sigma} + \lambda_1 (k^a r^b - \theta r - \mu k - c) + \lambda_2(-r) \qquad (14.10)$$

and, as before, obtain the adjoint differential equations

$$\lambda_1^{\bullet} = -\lambda_1 (ak^{a-1} r^b - \mu), \qquad \lambda_2^{\bullet} = 0 \qquad (14.11)$$

so that $\lambda_2$ is a constant, $\lambda_2 = \bar{\lambda}_2$. The stationarity conditions (14.3) become

$$H_{,c} = e^{-\delta t} c^{-\sigma} - \lambda_1 = 0, \qquad H_{,r} = \lambda_1 (bk^a r^{b-1} - \theta) - \lambda_2 = 0. \qquad (14.12)$$

These two stationarity conditions together with the two equations of states (14.8) and the two equations (14.11) for the adjoint variables form a fourth-order system of six equations for the six unknowns $k$, $r$, $c$, $D$, $\lambda_1$ and $\lambda_2$. For this system, we have the three prescribed boundary conditions (14.9) on $k$ and $D$ and the Euler boundary condition

$$\lim_{t \to \infty} \lambda_1(t) = 0. \qquad (14.13)$$

Because $\lambda_2$ is a constant, we obtain from $(H_{,r})^{\bullet} = 0$ and the first adjoint differential equation:

$$\frac{(bk^a r^{b-1} - \theta)^{\bullet}}{(bk^a r^{b-1} - \theta)} = -(ak^{a-1} r^b - \mu). \qquad (14.14)$$

This so-called *Hotelling's rule* is an equation for only two unknowns: $k$ and $r$. To get another equation for the same two unknowns, we note that the two conditions $H_{,c} = 0$ and $H_{,r} = 0$ may be combined to express $c(t)$ in terms of $k$ and $r$ and the unknown parameter $\bar{\lambda}_2$:

$$c^{\sigma} = \frac{e^{-\delta t}}{\bar{\lambda}_2} (bk^a r^{b-1} - \theta). \qquad (14.15)$$

This result may be used to eliminate $c$ from the equation of state to obtain a second differential equation for $k$ and $r$ (with $\bar{\lambda}_2$ as an unknown parameter). The boundary-value problem for the extremals is thereby reduced to one for a pair of first-order differential equations.

Exact solutions for special cases of the boundary-value problem for $k$ and $r$ are possible (see the exercises at the end of this chapter). With $\theta = \mu = 0$ and $b = 1 - a$, it is not difficult to show

$$c(t) = (c_0^{b\sigma/a} + F_0 t)^{a/b\sigma} e^{-\delta t/\sigma}, \qquad (14.16)$$

where $F_0 = (b/\bar{\lambda}_2^{b})^{1/a}$ and $c_0^{b\sigma/a} = (b/\bar{\lambda}_2)^{b/a} (k_0/r_0)^b$. The two unknown constants $r_0$ [$= r(0)$] and $\bar{\lambda}_2$ are to be determined by

$$D(\infty) = \int_0^\infty r(t)\,dt = D_0, \qquad r(\infty) = 0 \tag{14.17}$$

with the latter being a consequence of $r(t) \geq 0$ and finiteness of the total resource deposit.

Expression (14.16) for $c(t)$ shows that the optimal economic growth program is very sensitive to the discount rate $\delta$. For $\delta > 0$, $c(t)$ clearly tends to zero as $t \to \infty$. Preference for current consumption eventually drives the economy to below subsistence level and, hence, extinction in finite time. On the other hand, $c(t)$ becomes unbounded as $t \to \infty$ if $\delta = 0$. (For $\delta = 0$, we take $\sigma > 1$ for the problem to have a solution.) This knife-edge dependence on $\delta$ and the possibility for extinction are obviously unacceptable and a different kind of optimality should be explored for a more robust economic growth model.

With the tendency of each generation to discount the future and the unborn (and possibly destined to become defunct) generations not represented in the planning process, social fairness would dictate that our performing index should have a built-in intergenerational equity. One possible way to ensure equity would be to require $c(t)$ to be the same for all time, i.e., $c(t) = c_0 \equiv c(0)$, $0 \leq t < \infty$ (Solow, 1972, Wan, 1989 and references therein). This stipulation can be formulated as an additional state equation

$$c^{\bullet} = 0 \tag{14.18}$$

without any end condition because we still want to maximize the uniform consumption rate. The following optimal control problem now offers a more egalitarian alternative to the conventional social optimum of Example 6.

**EXAMPLE 7:** $\max_r \{c(\infty)\}$ subject to $k^{\bullet} = k^a r^b - c$, $D^{\bullet} = -r$, and $c^{\bullet} = 0$ with $0 < b < a < a + b \leq 1$, $k(0) = k_0$, $D(0) = D_0$, and $D(\infty) = 0$. [The case $\theta \neq 0$ will be treated in the Exercises. This example is unusual in that neither an initial nor a terminal condition is prescribed for the state variable $c(t)$.]

In the usual notation for the performance index $J$ [see (13.12)], we have $F = 0$ and $\psi = c(\infty)$. The corresponding Hamiltonian is

$$H = \lambda_1[k^a r^b - c] + \lambda_2[-r] + \lambda_3[0].$$

It follows that the adjoint differential equations are

$$\lambda_1^{\bullet} = -\lambda_1 a k^{a-1} r^b, \qquad \lambda_2^{\bullet} = 0, \qquad \lambda_3^{\bullet} = \lambda_1 \tag{14.19}$$

with the Euler boundary conditions

$$\lambda_1(\infty) = 0, \qquad \lambda_3(0) = 0, \qquad \lambda_3(\infty) = 1. \tag{14.20}$$

Assuming the non-negativity constraints on the state and control variables are not binding, the stationarity condition

$$H_{,r} = \lambda_1[bk^a r^{b-1}] - \bar{\lambda}_2 = 0 \tag{14.21}$$

applies, with $\lambda_2(t) = \bar{\lambda}_2$ (a constant) being a consequence of $\lambda_2^{\bullet} = 0$.

Condition (14.21) may be used to eliminate $\lambda_1$ from the first adjoint differential equation to obtain $(k^a r^{b-1})^{\bullet}/k^a r^{b-1} = ak^{a-1}r^b$. With the help of the state equation of capital accumulation (14.8a) for $\theta = \mu = 0$, this Hotelling's rule may be written as

$$r^{\bullet} = -\frac{ac_0 r}{k(1-b)}, \tag{14.22}$$

with $c(t) = c_0$. For the solution of the boundary-value problem for $k$ and $r$ (with $c_0$ as a parameter), we form

$$\frac{dk}{dr} = \frac{k^{\bullet}}{r^{\bullet}} = -\frac{(1-b)}{ac_0}\frac{k}{r}(k^a r^b - c_0) \tag{14.23}$$

from the equation of capital accumulation and (14.22). The Bernoulli equation (14.23) may be rearranged to read (Boyce and Di Prima, 1976)

$$\frac{d}{dr}(k^{-a}) - \frac{1-b}{r}(k^{-a}) = \frac{1-b}{c_0} r^{b-1}.$$

The solution for this linear equations for $k^{-a}$ is

$$k^a r^b = \frac{r}{A_0 + [(1-b)/c_0]r}.$$

We need $r(t) \to 0$ as $t \to \infty$ for $D(t)$ not to become negative for sufficiently large $T$. At the same time, we do not want $k^a r^b$ to tend to zero as $t \to \infty$. [Otherwise, we have $k^{\bullet} < 0$ for $t > T_0$ for some finite $T_0 > 0$ and $k(t)$ becomes negative eventually for any finite $c_0 > 0$.] It follows from these two requirements that we must have $A_0 = 0$ so that

$$k^a r^b = \frac{c_0}{1-b}. \tag{14.24}$$

Upon using (14.24) to eliminate $k^a r^b$, the state equation for capital accumulation (14.8a) with $\theta = \mu = 0$ becomes $k^{\bullet} = bc_0/(1-b)$ or

$$k(t) = k_0 + \frac{bc_0 t}{1-b} \tag{14.25}$$

and therewith

$$r = \left(\frac{1-b}{c_0}\right)^{-1/b}\left(k_0 + \frac{bc_0 t}{1-b}\right)^{-a/b}. \tag{14.24'}$$

The boundary-value problem for $D(t)$ then gives

$$D_0 = \int_0^\infty r(t)\,dt = \frac{1}{a-b}\left(\frac{c_0}{1-b}\right)^{\frac{(1-b)}{b}} k_0^{\frac{(b-a)}{b}}. \tag{14.25}$$

This result determines $c(0) = c_0$ in terms of known quantities:

$$c_0 = (1-b)(a-b)D_0^{\frac{b}{(1-b)}} k_0^{\frac{(a-b)}{(1-b)}}. \tag{14.26}$$

Rather remarkably, it is possible to sustain a uniform consumption rate indefinitely (Solow, 1974). To do so requires an ever expanding stock of capital. [To stipulate $k(t) \to 0$ as $t \to \infty$ would be inappropriate for the present problem!] On the other hand, the dependence of $c_0$ on the initial endowment of capital and resource, $D_0$ and $k_0$, imposes a limitation on the uniform consumption rate. The maximum sustainable value possible may be below the subsistence level and, therefore, unacceptable. In the less extreme situation where $c_0$ is above subsistence, egalitarianism still has its price. A poor society stays poor!

Instead of expressing the optimal control $r$ of Example 7 explicitly as a function of time (called open-loop control) as in (14.24′), expression (14.24) gives $r(t)$ in terms of the state variable $k(t)$ [and the initial endowment of capital and resource stocks through (14.26)]. This second form of the optimal control solution is called a optimal *feedback control* or *closed-loop control*. It is actually more useful as it allows for an automatic adjustment in response to an unexpected (small) perturbation of the state of the system (after a recalculation of $c_0$ using the current values of the capital stock and resource deposit). In contrast, a preprogrammed open-loop control would not lead to an optimal solution if it should be allowed to continue after an unexpected perturbation of the state of the system.

## 4. The Linear–Quadratic Problem and Feedback Control

There is one class of optimal control problems for which the optimal control can always be given in feedback form. These are problems which involves linear system dynamics and a Lagrangian quadratic in both state and control variables. We consider in this section this linear-quadratic regulator problem (often called LQP in the optimal control literature) of minimizing

$$J = \frac{1}{2}\mathbf{y}^T(t_1)\Psi\mathbf{y}(t_1) + \frac{1}{2}\int_{t_0}^{t_1}[\mathbf{y}^TQ(t)\mathbf{y} + \mathbf{u}^TR(t)\mathbf{u}]\,dt \tag{14.27a}$$

subject to the (linear) equation of state

$$\mathbf{y}^{\bullet} = A(t)\mathbf{y} + B(t)\mathbf{u}, \tag{14.27b}$$

and the initial condition

$$\mathbf{y}(t_0) = \mathbf{y}^0. \tag{14.27c}$$

For simplicity, we will consider only the case where there is no terminal conditions on $\mathbf{y}$ and both $t_0$ and $t_1$ are prescribed. The $(n \times n)$ matrices $A$ and $Q$, the $(n \times m)$ matrix $B$, and the $(m \times m)$ matrix $R$ are all $C^1$ functions in $[t_0, t_1]$. $R(t)$ is (symmetric) positive definite and $Q$ is (symmetric) positive semidefinite for all $t$ in $[t_0, t_1]$, whereas $\Psi$ is an $(n \times n)$ symmetric constant matrix.

For the problem above, the necessary conditions for optimality are

$$\boldsymbol{\lambda}^{\bullet} = -A^T\boldsymbol{\lambda} - Q\mathbf{y}, \tag{14.28a}$$

$$\boldsymbol{\lambda}(t_1) = \Psi\mathbf{y}(t_1) \tag{14.28b}$$

and

$$R\mathbf{u} + B^T\boldsymbol{\lambda} = 0 \qquad \text{or} \qquad \mathbf{u} = -R^{-1}B^T\boldsymbol{\lambda}. \tag{14.28c}$$

We may use (14.28c) to eliminate $\mathbf{u}$ from (14.27b) to obtain

$$\mathbf{y}^{\bullet} = A\mathbf{y} - BR^{-1}B^T\boldsymbol{\lambda}. \tag{14.27'}$$

The system (14.28a) may be written as a single equation for the $2n$-component vector $(\mathbf{y}^T, \boldsymbol{\lambda}^T)^T \equiv \mathbf{z}$:

$$\mathbf{z}^{\bullet} = \begin{bmatrix} A & -BR^{-1}B^T \\ -Q & -A^T \end{bmatrix}\mathbf{z} \equiv C(t)\mathbf{z}. \tag{14.29}$$

Let $Z(t, t_0)$ be the *resolvent matrix* of (14.29) as defined in Section 4, Chapter 13, so that $Z^{\bullet} = C(t)Z$ with $Z(t_0, t_0) = I_{2n}$, where $I_{2n}$ is the $(2n \times 2n)$ identity matrix. In terms of $Z(\xi, \eta)$, we have

$$\mathbf{z}(t) = Z(t, t_0)\begin{bmatrix} \mathbf{y}^0 \\ \boldsymbol{\lambda}^0 \end{bmatrix} \quad \text{and} \quad \mathbf{z}(t_1) = Z(t_1, t)\mathbf{z}(t), \tag{14.30a}$$

where $\boldsymbol{\lambda}^0$ is the unknown vector $\boldsymbol{\lambda}(t_0)$, and

$$\mathbf{z}(t_1) = \begin{bmatrix} \mathbf{y}(t_1) \\ \Psi \mathbf{y}(t_1) \end{bmatrix} = Z(t_1, t)\mathbf{z}(t). \tag{14.30b}$$

By partitioning $Z(t_1, t)$ as

$$Z(t_1, t) = \begin{bmatrix} Z_{11}(t_1, t) & Z_{12}(t_1, t) \\ Z_{21}(t_1, t) & Z_{22}(t_1, t) \end{bmatrix},$$

we may write (14.30b) as

$$\mathbf{y}(t_1) = Z_{11}(t_1, t)\mathbf{y}(t) + Z_{12}(t_1, t)\boldsymbol{\lambda}(t),$$
$$\Psi\mathbf{y}(t_1) = Z_{21}(t_1, t)\mathbf{y}(t) + Z_{22}(t_1, t)\boldsymbol{\lambda}(t). \tag{14.30c}$$

The system (14.30c) can be solved for $\mathbf{y}(t_1)$ and $\boldsymbol{\lambda}(t)$ to obtain

$$\boldsymbol{\lambda}(t) = \Lambda(t, t_1)\mathbf{y}(t), \tag{14.31a}$$
$$\mathbf{y}(t_1) = Y(t_1, t)\mathbf{y}(t), \tag{14.31b}$$

where

$$\Lambda(t, t_1) = [Z_{22}(t_1, t) - \Psi Z_{12}(t, t_1)]^{-1}[\Psi Z_{11}(t_1, t) - Z_{21}(t_1, t)], \tag{14.31c}$$
$$Y(t_1, t) = [I_n - Z_{12}(t_1, t)Z_{22}^{-1}(t_1, t)\Psi]^{-1}[Z_{11}(t_1, t) - Z_{12}(t_1, t)Z_{22}^{-1}(t_1, t)Z_{21}(t_1, t)\Psi]. \tag{14.31d}$$

Note that $Z_{22}(t_1, t_1) = I_n$ and $Z_{12}(t_1, t_1) = 0$. The matrices $Z_{22}$, $(Z_{22} - \Psi Z_{12})$ and $(I_n - Z_{12}Z_{22}^{-1}\Psi)^{-1}$ are invertible in $[t_0, t_1]$. It follows from (14.28c) and (14.31a) that the extremal control may be written in the feedback form

$$\mathbf{u}(t) = -R^{-1}(t)B^T(t)\Lambda(t, t_1)\mathbf{y}(t). \tag{14.32}$$

Hence, we have the following result:

**[XIV.1]** ***If $\hat{\mathbf{u}}(t)$ is the optimal control for the LQP and $\hat{\mathbf{y}}(t)$ the corresponding optimal state trajectory, then $\hat{\mathbf{u}}$ and $\hat{\mathbf{y}}$ are related by the feedback control law (14.32).***

Instead of solving the initial-value problem $Z^{\bullet} = C(t)Z$, $Z(t_0, t_0) = I$ for $Z(t, t_0)$, it is often more efficient to solve for the matrix $\Lambda(t, t_1)$ directly by way of a terminal-value problem. From (14.31a), we have $\boldsymbol{\lambda}^{\bullet} = \Lambda^{\bullet}\mathbf{y} + \Lambda\mathbf{y}^{\bullet}$ or, after using (14.27b) and (14.32) to eliminate $\mathbf{y}^{\bullet}$ and $\mathbf{u}$,

$$\boldsymbol{\lambda}^{\bullet} = [\Lambda^{\bullet} + \Lambda A - \Lambda B R^{-1} B^T \Lambda]\mathbf{y}.$$

The left-hand side can be eliminated by (14.28a) and (14.31a) to leave

$$[\Lambda^{\bullet} + \Lambda A + A^T\Lambda - \Lambda BR^{-1}B^T\Lambda + Q]\mathbf{y} = 0.$$

The solution $\mathbf{y}$ depends on $\mathbf{y}^0$ and $t_0$; however the equation above must hold for any $\mathbf{y}^0$ at any $t_0 < T_1$. The quantity in brackets, which is independent of $t_0$, must, therefore, be a zero matrix so that

$$\Lambda^{\bullet} + \Lambda A + A^T\Lambda - \Lambda BR^{-1}B^T\Lambda + Q = 0. \tag{14.33a}$$

The first-order matrix (Riccati) equation (14.33a) for $\Lambda(t, t_1)$ is supplemented by the terminal condition

$$\Lambda(t_1, t_1) = \Psi \tag{14.33b}$$

which follows from (14.31c) with

$$Z_{22}(t_1, t_1) = Z_{11}(t_1, t_1) = I_n, \quad Z_{12}(t_1, t_1) = Z_{21}(t_1, t_1) = 0.$$

It is not difficult to show that $\Lambda(t, t_1)$ is a symmetric matrix. Hence, the terminal-value problem (14.33a) and (14.33b) is for $n(n+1)/2$ (instead of $n^2$) scalar differential equations.

Linear-quadratic problems often occur naturally in applications. They also occur when we seek a small perturbation from the optimal path associated with a small perturbation $\delta\mathbf{y}^0$ of the initial state. On expanding the augmented performance index about the optimal solution, we have (for problems with no terminal conditions)

$$I = \hat{I} + \frac{1}{2}[\delta\mathbf{y}(t_1) \cdot \hat{\Psi}_{,\mathbf{y}\mathbf{y}}\delta\mathbf{y}(t_1)]$$

$$+ \frac{1}{2}\int_{t_0}^{t_1} (\delta\mathbf{y}^T, \delta\mathbf{u}^T) \begin{bmatrix} \hat{H}_{,\mathbf{y}\mathbf{y}} & \hat{H}_{,\mathbf{y}\mathbf{u}} \\ \hat{H}_{,\mathbf{u}\mathbf{y}} & \hat{H}_{,\mathbf{u}\mathbf{u}} \end{bmatrix} \begin{pmatrix} \delta\mathbf{y} \\ \delta\mathbf{u} \end{pmatrix} dt$$

$$+ \text{higher order variations} \tag{14.34a}$$

subject to

$$\delta\mathbf{y}^{\bullet} = \hat{\mathbf{f}}_{,\mathbf{y}}\delta\mathbf{y} + \hat{\mathbf{f}}_{,\mathbf{u}}\delta\mathbf{u}, \tag{14.34b}$$

$$\delta\mathbf{y}(t_0) = \delta\mathbf{y}^0, \tag{14.34c}$$

where $\delta\mathbf{y}$ and $\delta\mathbf{u}$ are the perturbation of the state and control vectors, respectively, induced by $\delta\mathbf{y}^0$. For an extremal control of the perturbed problem in the neighborhood of the optimal control of the unperturbed problem, we want to determine $\delta\hat{\mathbf{u}}$ which minimizes the second variation of $I$ subject to (14.34b) and (14.34c). This problem is of the linear-quadratic type; all the matrices involved are evaluated at the optimal solution of the unperturbed problem and are, therefore, known quantities.

In view of the frequent occurrence of LQP in applications, it is of considerable importance to investigate whether the extremal control for the problem is, in fact, optimal. This leads to the question of sufficient conditions for optimality for general optimal control problems. We will discuss this in the next section.

## 5. A Sufficient Condition for Optimality

The Lagrangian $F = (\mathbf{y}^T Q \mathbf{y} + \mathbf{u}^T R \mathbf{u})/2$ is convex in $(\mathbf{y}, \mathbf{u})$ because both symmetric matrices $Q$ and $R$ have been restricted to be at least non-negative definite (see Section 4). For such a Lagrangian, we have from (4.22)

$$2\,[F(t,\mathbf{y},\mathbf{u}) - F(t,\hat{\mathbf{y}},\hat{\mathbf{u}})] \geq (\mathbf{y}-\hat{\mathbf{y}})^T Q \hat{\mathbf{y}} + (\mathbf{u}-\hat{\mathbf{u}})^T R \hat{\mathbf{u}}.$$

Similarly, we have for a non-negative definite (symmetric) matrix $\Psi$ a similar inequality

$$\Psi(\mathbf{y}(t_1)) - \Psi(\hat{\mathbf{y}}(t_1)) \geq [\mathbf{y}(t_1) - \hat{\mathbf{y}}(t_1)]^T \Psi \hat{\mathbf{y}}(t_1).$$

With these two inequalities, the difference between $J[\mathbf{y},\mathbf{u}]$ and $J[\hat{\mathbf{y}},\hat{\mathbf{u}}] \equiv \hat{J}$ satisfies the inequality

$$J - \hat{J} \geq \Delta\mathbf{y}^T(t_1)\Psi\hat{\mathbf{y}}(t_1) + \int_{t_0}^{t_1} [\Delta\mathbf{y}^T Q \hat{\mathbf{y}} + \Delta\mathbf{u}^T R \hat{\mathbf{u}}]\,dt,$$

where $\Delta\mathbf{g} \equiv \mathbf{g} - \hat{\mathbf{g}}$. We now use (14.28a) and (14.28c) to express $Q\hat{\mathbf{y}}$ and $R\hat{\mathbf{u}}$ in terms of $\boldsymbol{\lambda}$. Integration by parts then allows us to write the result as

$$J - \hat{J} \geq \Delta\mathbf{y}^T(t_1)[\Psi\hat{\mathbf{y}}(t_1) - \boldsymbol{\lambda}(t_1)] + \Delta\mathbf{y}^T(t_0)\boldsymbol{\lambda}(t_0)$$
$$+ \int_{t_0}^{t_1} [(\Delta\mathbf{y}^T)^{\cdot} - \Delta\mathbf{y}^T A^T - \Delta\mathbf{u}^T B^T]\,\boldsymbol{\lambda}\,dt. \qquad (14.35a)$$

The first term of (14.35a) vanishes because $\boldsymbol{\lambda}(t_1)$ satisfies (14.28b). The second term vanishes because $\mathbf{y}(t_0) = \hat{\mathbf{y}}(t_0) = \mathbf{y}^0$. The integrand of the third term vanishes by the equation of state (14.27b), leaving us with

$$J - \hat{J} \geq 0. \qquad (14.35b)$$

Thus, we have the following result:

**[XIV.2]** *If $\Psi$, $\mathrm{Q(t)}$ and $\mathrm{R(t)}$ are all symmetric non-negative definite matrices in $[t_0, t_1]$, the extremal control $\hat{\mathbf{u}}$ determined by (14.28c) for the LQP (with no control or state constraints) minimizes J globally.*

The convexity of the Lagrangian is assured by the non-negative definiteness of the matrices $Q$ and $R$ for the Linear–Quadratic Problem. That optimality follows as a consequence of convexity is not surprising given a similar result for the basic problem of the calculus of variations in Chapter 4. The latter suggests that we should be able to extend [XIV.2] to more general classes of optimal control problems, at least to those with a general convex Lagrangian and nonlinear dynamics.

To explore the possibility of an extension, suppose $\psi(\mathbf{y}(t_1))$ is convex in $\mathbf{y}$ and $F$ is jointly convex in both $\mathbf{y}$ and $\mathbf{u}$. Then we have for the general performance index (13.12):

$$J - \hat{J} \geq \psi_{,\mathbf{y}}(\mathbf{y}(t_1))\Delta\mathbf{y}(t_1) + \int_{t_0}^{t_1} [\hat{F}_{,\mathbf{y}}\Delta\mathbf{y} + \hat{F}_{,\mathbf{u}}\Delta\mathbf{u}]\,dt$$

$$= \psi_{,\mathbf{y}}(\hat{\mathbf{y}}(t_1))\Delta\mathbf{y}(t_1) - \int_{t_0}^{t_1} [(\boldsymbol{\lambda}^{\boldsymbol{\cdot}} + \hat{\mathbf{f}}_{,\mathbf{y}}^T\boldsymbol{\lambda})^T\Delta\mathbf{y} + \boldsymbol{\lambda}^T\hat{\mathbf{f}}_{,\mathbf{u}}\Delta\mathbf{u}]\,dt$$

$$= [\psi_{,\mathbf{y}}(\hat{\mathbf{y}}(t_1)) - \boldsymbol{\lambda}^T(t_1)]\Delta\mathbf{y}(t_1) + \boldsymbol{\lambda}^T(t_0)\Delta\mathbf{y}(t_0)$$

$$+ \int_{t_0}^{t_1} \boldsymbol{\lambda}^T[\Delta\mathbf{y}^{\boldsymbol{\cdot}} - \hat{\mathbf{f}}_{,\mathbf{y}}\Delta\mathbf{y} - \hat{\mathbf{f}}_{,\mathbf{u}}\Delta\mathbf{u}]\,dt.$$

The first term of the right-hand side above vanishes because of the second (Euler boundary) condition in (14.1). The second term vanishes because $\mathbf{y}(t_0) = \hat{\mathbf{y}}(t_0) = \mathbf{y}^0$. The $\Delta\mathbf{y}^{\boldsymbol{\cdot}}$ term in the integrand can be eliminated in favor of $\Delta\hat{\mathbf{f}}$ by the equation of state, leaving us with

$$J - \hat{J} \geq \int_{t_0}^{t_1} \boldsymbol{\lambda}^T[\Delta\hat{\mathbf{f}} - \hat{\mathbf{f}}_{,\mathbf{y}}\Delta\mathbf{y} - \hat{\mathbf{f}}_{,\mathbf{u}}\Delta\mathbf{u}]\,dt. \tag{14.36}$$

The following result is an immediate consequence of (14.36):

**[XIV.3]** *If* $\psi$ *is convex/concave in* $\mathbf{y}$ *and* F *is convex/concave in* $\mathbf{y}$ *and* $\mathbf{u}$ *then an extremal control* $\hat{\mathbf{u}}(t)$ *of (13.12), (13.6), and (13.7) minimizes/maximizes* J *if for all* t *in* $[t_0, t_1]$ *either* (i) $\mathbf{f}$ *is convex/concave in* $\mathbf{y}$ *and* $\mathbf{u}$ *and* $\boldsymbol{\lambda}(t) \geq \mathbf{0}$, or (ii) $\mathbf{f}$ *is concave/convex in* $\mathbf{y}$ *and* $\mathbf{u}$ *and* $\boldsymbol{\lambda}(t) \leq \mathbf{0}$

To illustrate the application of the result above, we consider the optimality of the extremal solution of Example 7.

**EXAMPLE 7 (continued): Optimality of extremal found previously.**

Having found $k(t)$ and $r(t)$ in (14.25) and (14.24′), respectively, it is straightforward to obtain from the first and third equation of (14.19)

$$\lambda_1 = \bar{\lambda}_1\left(k_0 + \frac{bc_0 t}{1-b}\right)^{-a/b},$$

$$\lambda_3 = \frac{\bar{\lambda}_1(1-b)}{c_0(a-b)}\left[k_0^{(b-a)/b} - \left(k_0 + \frac{bc_0 t}{1-b}\right)^{(b-a)/b}\right],$$

where $\bar{\lambda}_1$ is a constant of integration. These expressions satisfy the first two boundary conditions $\lambda_3(0) = 0$ and $\lambda_1(\infty) = 0$ of (14.20) because $a > b > 0$. The last condition of (14.20) then determines $\bar{\lambda}_1$ to be

$$\bar{\lambda}_1 = \frac{a-b}{1-b}\, c_0 k_0^{(a-b)/b}$$

which is positive. It follows from (14.21) that

$$\lambda_2(t) = \bar{\lambda}_2 = b(a-b)\left(\frac{c_0}{1-b}\right)^{(2b-1)/b} k_0^{(a-b)/b}$$

which is also positive.

The only nontrivial Hessian matrix of $\mathbf{f} = (k^a r^b - c, -r, 0)^T$,

$$\begin{bmatrix} f_{1,kk} & f_{1,kc} & f_{1,kr} \\ f_{1,ck} & f_{1,cc} & f_{1,cr} \\ f_{1,rk} & f_{1,rc} & f_{1,rr} \end{bmatrix} = \begin{bmatrix} a(a-1)k^{a-2}r^b & 0 & abk^{a-1}r^{b-1} \\ 0 & 0 & 0 \\ abk^{a-1}r^{b-1} & 0 & b(b-1)k^a r^{b-2} \end{bmatrix},$$

is nonpositive definite; hence, the components of $\mathbf{f}$ are jointly concave in $\mathbf{y} = (k, c)^T$ and $\mathbf{u} = (r)$. Because $\lambda_1$, $\lambda_2$ and $\lambda_3$ are all non-negative for all $t \geq 0$, [XIV.3] assures us that the extremal control (14.24′) maximizes the uniform consumption rate $c(\infty)\ (= c_0)$.

Note that if the system dynamics is linear, the integrand of the integral in (14.36) vanishes. The theorem therefore holds with no condition on $\mathbf{f}$ or $\boldsymbol{\lambda}$. This is certainly the case for the basic problem in the calculus of variations for which $\mathbf{u} = \mathbf{y}^{\bullet}$. More generally, we have as a corollary to [XIV.3]:

**[XIV.4]** ***If $\boldsymbol{\psi}$ is convex/concave in $\mathbf{y}$ and $\mathbf{F}$ is jointly convex/concave in $\mathbf{y}$ and $\mathbf{u}$, then an extremal control $\hat{\mathbf{u}}(t)$ for (13.12), (14.27b), and (14.27c) minimizes/maximizes J.***

Unfortunately, many optimal control problems do not have the convexity needed for the application of the sufficiency theorems of this section. As they stand, these results also may not apply to problems with equality and/or inequality constraints on the control and/or state variables. We will briefly discuss these additional features later in this chapter.

## 6. Household Optimum and Locational Equilibrium

Contrary to the previous examples in this chapter, most nonlinear problems in optimal control do not admit an exact solution in terms of simple or special functions. Similar to problems in the calculus of variations, approximate solutions by numerical methods are often necessary. However, it is usually tedious and often challenging to extract insight from numerical solutions. Whenever possible, it is still desirable to obtain accurate approximate solutions in simple analytic form. Perturbation and asymptotic methods offer this type of results; there is often a small or large dimensionless parameter lurking around in most problems (with the Reynolds' number in fluid mechanics as a prime example). In this section and the next, we describe a simple model of residential land economics formulated by R.M. Solow [1972, 1973; also Solow and Vickery (1971)] and studied by a number of other urban land economists [see Kanemoto (1980) and references therein]. We will show in Section 8 the benefits of a perturbation analysis for the problem.

In the conventional urban economic model for land use problems, the abstract city is taken to be circular with a circular Central Business District (CBD) of radius $R_i$ and an annular residential area extending from $X = R_i$ to $X = R_0$ where $X$ is the radial distance from the city center. For simplicity, the city is divided into pie-shaped sectors. In one of these sectors with a sectoral angle of $\theta$ radian ($\theta \leq 2\pi$), the residential area is inhabited by $N_0$ identical households each sending a worker traveling to and from the CBD by a single road of the sector (Fig. 14.2). Each household has the same annual income to be used for housing, for consumption of a composite good, and for transportation. We take the unit price, $p$, of the composite good to be unity. Then a household at distance $X$ from the city center has as its budget equation

$$c + rs + t = y, \tag{14.37a}$$

where $c(X)$ and $s(X)$ are the amount of consumption goods and residential land for that household per annum, respectively, and where $r(X)$ and $t(X)$ are the per annum unit land rent and total transportation cost for the household, respectively. Upon introducing a dimensionless after-transportation-income function $w(X)$, the budget equation (14.37a) may be written as

$$c + rs + = y\left(1 - \frac{t}{y}\right) \equiv yw(X). \tag{14.37b}$$

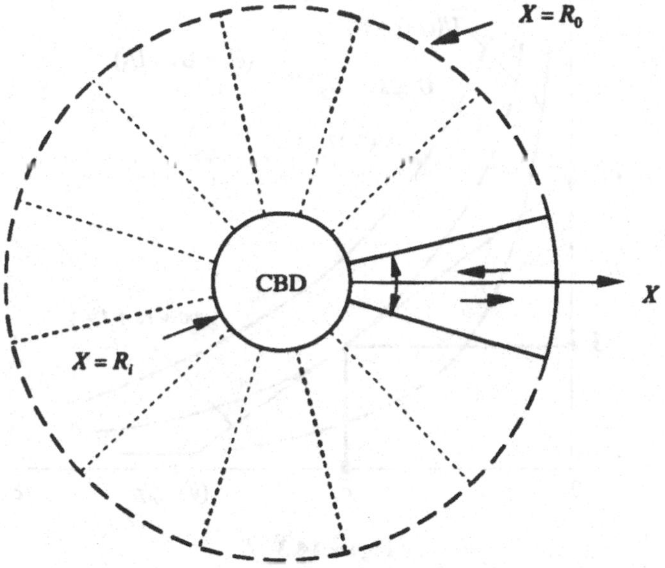


**Figure 14.2**

Each household in our model city derives the same satisfaction from a given amount of goods and space and, therefore, has the same utility function which will be taken to be

$$U(c, s) = U(\xi / \xi_0), \qquad \xi = c^{1-\sigma} s^{\sigma} \quad (0 < \sigma < 1), \tag{14.38}$$

where $\xi_0$ is a parameter having the same unit as $\xi$ so that $\xi / \xi_0$ is dimensionless (e.g., $\xi_0$ is the value of $\xi$ at the edge of CBD). $U(\cdot)$ is assumed to be monotone increasing and strictly concave. We have not allowed $U$ to depend on $X$ explicitly, but we expect it to depend on $X$ indirectly through $c$ and $s$ which will be seen to vary with the household location. Each household chooses $c$ and $s$ to maximize $U$ subject to the budget constraint (14.37). The maximum $U$ is attained at the stationary point $(\hat{c}, \hat{s})$ (see Fig. (14.3):

$$\hat{c} = (1 - \sigma)(y - t) = (1 - \sigma) yw, \tag{14.39a}$$

$$r\hat{s} = \sigma (y - t) = \sigma yw \tag{14.39b}$$

with $w = (1 - t/y)$. The corresponding stationary value of the utility function is

$$\hat{U} \equiv U(\hat{c}, \hat{s}) = U(\sigma^{\sigma}(1 - \sigma)^{1-\sigma}(y - t)/\xi_0 r^{\sigma}). \tag{14.39c}$$

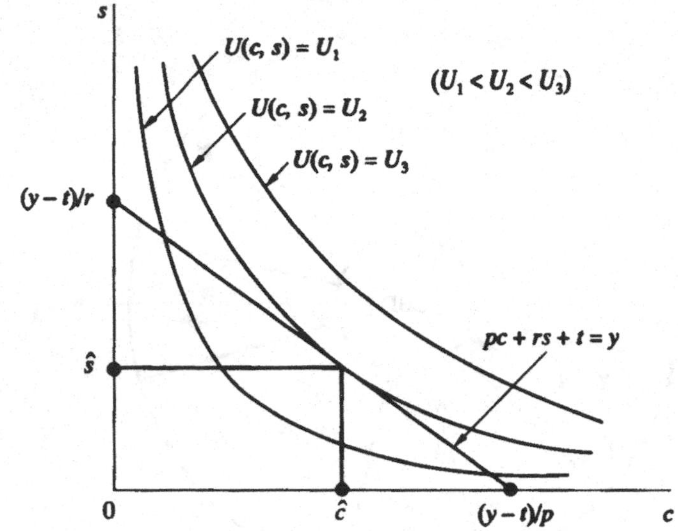


**Figure 14.3**

The monotone increasing and strictly concave properties of $U$ ensure that the stationary value $\widehat{U}$ is a maximum. From (14.39), we see that $\sigma$ and $(1-\sigma)$ are the fractions of after-travel-cost household income spent on housing and consumption goods, respectively.

In *locational equilibrium*, all household attain the same utility so that $\widehat{U}$ must be independent of $X$. In particular, $\widehat{U}$ at an arbitrary location $X$ must be the same as $\widehat{U}$ at the edge of the CBD:

$$U(\sigma^{\sigma}(1-\sigma)^{1-\sigma}(y-t)/\xi_0 r^{\sigma}) = U(\sigma^{\sigma}(1-\sigma)^{1-\sigma}(y-t_i)/\xi_0 r_i^{\sigma}) \equiv U_i, \quad (14.40)$$

where $t_i \equiv t(R_i)$ and $r_i \equiv r(R_i)$. We assume henceforth that transportation is free within the CBD so that $t_i = 0$. Because $U(\cdot)$ is monotone increasing, it follows from (14.40a) that

$$\frac{y-t}{r^{\sigma}} = \frac{y}{r_i^{\sigma}} \quad \text{or} \quad r = r_i w^{\alpha+1}, \qquad \alpha + 1 = \frac{1}{\sigma}. \quad (14.40b)$$

With (14.40b), the consumer's behavior is completely determined once the per annum transportation cost $t(X)$ and the fraction of land $b(X)$ within the sector allocated for housing are known.

In an annular sector of the residential area extending from $X$ to $X + dX$, a fraction $b(X)$ of the land is used for housing and the remaining land area $(= [1 - b(X)]\,\theta X\,dX)$ is for roads. The total per annum transportation cost $t(X)$ of a household at location $X$ is taken to be the sum of a *distance cost*

$t_d(X)$ with $t_d(R_i) = 0$ (because travel within the CBD is assumed free) and a congestion cost $t_c(X)$ which depends on the traffic density. We take $t$ in the form (Solow, 1972)

$$t(X) = t_d(X) + t_c(X) = \int_{R_i}^{X} \left[ \tau + a_0 \left\{ \frac{N(z)}{\theta z[1 - b(z)]} \right\}^k \right] dz, \qquad (14.41a)$$

where $N(X)$ is the number of households located outside the ring $X$, $\tau(X) = d(t_d)/dX$ is the known fixed unit distance travel cost within the residential area (per household per annum), and $k$ and $a_0$ are prescribed positive constants. We will work with the differential form of (14.41a):

$$\frac{dt}{dX} = \tau + a_0 \left\{ \frac{N(X)}{\theta X\,[1 - b(X)]} \right\}^k = -y\,\frac{dw}{dX}. \qquad (14.41b)$$

Because $N(X)$ is also an unknown, we need another condition to determine $w$ and $N$. This condition comes from a *conservation law for space*:

> *The amount of space occupied by the households in the annular sector (of area* $\theta X\,dX$*) must equal the total amount of space in the sector allocated for housing so that* $-\,s\,dN = b\theta X\,dX$.

We may write this conditions as

$$\frac{dN}{dX} = -\frac{\theta X b(X)}{\hat{s}(X)} = -\frac{\theta r_i\,(\alpha + 1)}{y} X b w^{\alpha} \equiv -\,n(X), \qquad (14.42)$$

where $n(X)$ is the *population density* in the incremental sector at $X$.

For the two first-order differential equations (14.41b) and (14.42), we have the initial conditions

$$w(R_i) = 1, \qquad (14.43b)$$

$$N(R_i) = N_0, \qquad (14.43b)$$

expressing the fact that households located at the edge of the CBD pay no travel expenses and that all households are located outside the CBD. For a given distribution of the fraction of land for housing $b(X)$, the initial-value problem determines $w(X)$ and $N(X)$ up to the unknown parameter $r_i$. Finally, the condition that there is no household located outside the city limit,

$$N(R_0) = 0, \qquad (14.43c)$$

determines $r_i$ [and therewith the common household utility $U_i$ from (14.39c) and (14.40a)]. Note that we have

$$0 < b(X) < 1, \qquad 0 \le w(X) \le 1, \qquad 0 \le N(X) \le N_0, \tag{14.44}$$

from the definition of the three quantities.

To have the simplest setting for the description of the perturbation method of solution, we will confine ourselves here to *closed* cities with a prescribed outer boundary $R_0$ and with an *absentee landlord* to whom all rents are paid. For such cities, $R_0$ and $y$ are known constants with the latter being the same wage earned by all households. These and other simplifying assumptions such as a homogeneous population may be removed [see Wan (1989)].

## 7. The Second Best Residential Land Allocation

With the choice of $b(X)$ still at our disposal, we may use it as a policy (control) variable to attain a certain social or economic objective. A typical residential land allocation policy is the *cost-benefit criterion* with $b(X)$ chosen by the condition of *marginal benefit from more land for roads* (leading to a reduction of the congestion cost of transportation) *equal to the marginal cost of land rent lost* [see the Appendix of this chapter as well as Kanemoto (1980)]:

$$\alpha k \left[ \frac{N}{\theta X(1-b)} \right]^{k+1} = r(X). \tag{14.45}$$

Because the social costs of traffic congestion (such as the costs in delay, aggravation, and wears and tears inflicted on other drivers and their vehicles) have not been priced by some kind of congestion tolls, the market land price in our model does not reflect the true social value of the land. Hence, the cost-benefit allocation policy generally leads to an excessive amount of land for roads compared to a socially optimal policy of maximizing the total utility of all the households (without the stipulation of locational equilibrium).

On the other hand, congestion tolls are politically unpopular and practically infeasible in most settings. Furthermore, social optimum is known to require unequal utility among households in different locations (Kanemoto, 1980) and, therefore, must be maintained by fiat. In a society where households are free to move, a more relevant issue is whether the cost-benefit criterion (using market rent without congestion tolls) allocates more lands for roads, resulting in less (common) household utility than a second best allocation which maximizes the equilibrium common household utility $U_i$, also without congestion tolls.

The uniform annual income of each household is known; the dependence of $U_i$ on $b(X)$ is, therefore, only through $r_i = r(R_i)$ in this case (see (14.40). Because this dependence is specified by the conditions (14.41) and (14.43), these conditions assume the role of equations of state and the relevant end conditions in the optimal control problem of

$$\max_{b} \{J = U_i \equiv U(\sigma^\sigma(1-\sigma)^{1-\sigma} y/\xi_0 r_i^\sigma)\}.$$

As it turns out, the inequality constraints (14.44) on $w$, $N$, and $b$ are automatically satisfied by the (interior) extremal control for the problem for the range of parameter values considered herein. To cast this optimal control problem in the form treated in this and the last chapter, we will think of $r_i$ as a function of $X$ but require it to be a constant by the additional equation of state

$$\frac{dr_i}{dX} = 0 \tag{14.46}$$

without any initial or terminal conditions. To avoid the tempting (but incorrect) identification of $r_i(X)$ with $r(X)$, we will regard $J$ as $U(\sigma^\sigma(1-\sigma)^{1-\sigma}y/\xi_0 r_i^\sigma(R_0))$.

The Hamiltonian of the problem is

$$H = -\lambda_1\left[\frac{r}{y} + \frac{a_0}{y}\left\{\frac{N}{\theta X(1-b)}\right\}^k\right] - \lambda_2\left[\frac{1}{y}\theta r_i(a+1)Xbw^a\right] + \lambda_3[0].$$

The differential equations and Euler boundary conditions for the adjoint variables $\lambda_1$, $\lambda_2$, and $\lambda_3$ associated with the three equations of state are, therefore,

$$\frac{d\lambda_1}{dX} = \frac{1}{y}a(a+1)\theta r_i Xbw^{a-1}\lambda_2, \tag{14.47a}$$

$$\lambda_1(R_0) = 0, \tag{14.47b}$$

$$\frac{d\lambda_2}{dX} = \frac{a_0 k\lambda_1}{yN}\left[\frac{N}{\theta X(1-b)}\right]^k, \tag{14.47c}$$

$$\frac{d\lambda_3}{dX} = \frac{1}{y}\lambda_2\theta(a+1)Xbw^a, \tag{14.47d}$$

$$\lambda_3(R_i) = 0, \tag{14.47e}$$

$$\lambda_3(R_0) + \frac{y\sigma^{\sigma+1}(1-\sigma)^{1-\sigma}}{\xi_0 r_i^{\sigma+1}(R_0)}U'\left(\frac{\sigma^\sigma(1-\sigma)^{1-\sigma}y}{\xi_0 r_i^\sigma(R_0)}\right) = 0, \tag{14.47f}$$

where $U'(\beta) = dU(\beta)/d\beta$. The stationarity condition $H_{,b} = 0$ for an interior extremal solution (when the inequality constraints are not binding) is

$$(a+1)\theta r_i X\lambda_2 w^a - \frac{\lambda_1 a_0 k}{1-b}\left[\frac{N}{\theta X(1-b)}\right]^k = 0. \tag{14.47g}$$

We may eliminate (14.46) and the boundary-value problem for $\lambda_3$ from further consideration by integrating (14.47d) and observing (14.47e) to obtain

$$\lambda_3(X) = \frac{\theta(\alpha+1)}{y}\int_{R_i}^{X} \lambda_2 X b w^{\alpha}\, dX. \tag{14.47h}$$

The boundary condition (14.47f) requires

$$\frac{y\sigma^{\sigma+2}(1-\sigma)^{1-\sigma}}{\theta\xi_0 r_i^{\sigma+1}}\, U'\left(\frac{\sigma^{\sigma}(1-\sigma)^{1-\sigma}y}{\xi_0 r_i^{\sigma}}\right) = -\int_{R_i}^{R_0} \lambda_2 X b w^{\alpha}\, dX \tag{14.47i}$$

where we have regarded $r_i$ as a constant $[\,= r_i(R_0) = r_i(R_i)\,]$ as required by (14.46).

The remaining coupled fourth-order differential-algebraic system (14.41b), (14.42), (14.47a), (14.47c), (14.47g), and the four boundary conditions (14.43) and (14.47b) define a two-point boundary-value problem for the five unknown (extremal) functions $w$, $N$, $\lambda_1$, $\lambda_2$ and $b$ with the constant $r_i$ as a parameter. The auxiliary condition (14.47i) then determines $r_i$.

For the purpose of obtaining perturbation solutions for the extremals of the optimal control problem, we need a dimensionless form of the nonlinear boundary-value problem for $w, N, \lambda_1, \lambda_2$, and $b$. For this, we introduce the following dimensionless variables in addition to $w$ as defined in (14.37b):

$$\tau(X) = \tau_0 T(x), \qquad u = \frac{N}{N_0}, \qquad \Lambda_2 = N_0\lambda_2, \qquad \bar{v} = \frac{r_i\,\theta R_i^2}{\sigma y N_0}, \tag{14.48a}$$

$$\varepsilon_i^{k+1}\eta = \frac{\tau_0 R_i}{y} \equiv \frac{\bar{t}_d}{y}, \qquad \varepsilon_i^{k+1}(1-\eta) = \frac{a_0 R_i}{y}\left(\frac{N_0}{\theta R_i}\right)^{k} \equiv \frac{\bar{t}_c}{y}, \tag{14.48b}$$

with $\varepsilon_i^{k+1} = (\bar{t}_d + \bar{t}_c)/y$ giving the order of magnitude of the fraction of income for the total annual household travel cost. We write the fourth-order boundary-value problem and the integrated condition in terms of these new variables:

$$w^{\bullet} = -\varepsilon_i^{k+1}\left\{\eta T(x) + (1-\eta)\left[\frac{u}{x(1-b)}\right]^{k}\right\}, \qquad w(1) = 1, \tag{14.49a,b}$$

$$u^{\bullet} = -\bar{v}xbw^{\alpha}, \quad u(1) = 1, \quad u(R) = 1, \tag{14.49c,d,e,}$$

$$\lambda_1^{\bullet} = -\alpha\bar{v}xb\Lambda_2 w^{\alpha-1}, \quad \lambda_1(R) = 0, \tag{14.49f,g}$$

$$\Lambda_2^{\bullet} = -\varepsilon_i^{k+1}(1-\eta)\frac{k\lambda_1}{u}\left[\frac{u}{x(1-b)}\right]^{k}, \tag{14.49h}$$

$$\bar{v}x(1-b)w^{\bullet}\Lambda_2 - \varepsilon_t^{k+1}(1-\eta)k\lambda_1\left[\frac{u}{x(1-b)}\right]^k = 0, \tag{14.49i}$$

$$\frac{\mu(\bar{v})}{\bar{v}} + \int_1^R (\alpha+1)xb(x)w^{\bullet}\Lambda_2\,dx = 0, \tag{14.49j}$$

where $(\ )^{\bullet} = d(\ )/dx$ and

$$\mu(\bar{v}) = \frac{\sigma^{\sigma}(1-\sigma)^{1-\sigma}y}{\xi_0 r_i^{\sigma}}\, U'\!\left(\frac{\sigma^{\sigma}(1-\sigma)^{1-\sigma}y}{\xi_0 r_i^{\sigma}}\right). \tag{14.49k}$$

The following three expressions for $\mu(\bar{v})$ correspond to utility functions used in existing literature:

(*i*) $\mu(\bar{v}) = 1$ if $U(z) = \ell n(z)$;

(*ii*) $\mu(\bar{v}) = \left[\dfrac{(\alpha+1)\theta R_i^2 y}{2N_0\xi_0^2}\right]\left(\dfrac{1}{\bar{v}}\right)$ if $U(z) = z^2$ and $\sigma = \dfrac{1}{2}$;

(*iii*) $\mu(\bar{v}) = \sigma^{\sigma}(1-\sigma)^{1-\sigma}\left(\dfrac{y}{\xi_0}\right)\left[\dfrac{(\alpha+1)\theta R_i^2 y}{yN_0}\right]^{\sigma}\left(\dfrac{1}{\bar{v}^{\sigma}}\right)$ if $U(z) = z$.

In all cases, $\mu$ is a dimensionless quantity for the definition $\xi_0$ given previously.

It should be observed that, for a fixed $\bar{v}$, the form of the two-point boundary-value problem (14.49) does not change for different utility functions of the form (14.38). However, this observation is of little practical value unless we can obtain the solution of the boundary-value problem explicitly in terms of elementary or special functions because the solution depends on $\bar{v}$ which varies with $U(\cdot)$ according to (14.49j) and (14.49k). In general, it is not possible to obtain such an explicit solution, and a numerical solution of the problem is necessary. Although an efficient numerical solution scheme is possible (Wan, 1993), it is of considerable interest to note that a perturbation solution of the problem suggests itself once a certain small parameter of the problem is identified. The first few terms of the perturbation solution for the various unknowns will be seen to be (essentially) polynomial or rational in the dimensionless distance variable $x$ and that they provide an adequate approximation of the exact solution. As such, a suitably truncated perturbation solution more efficiently delineates the structure of the solution of the optimization problem than any accurate numerical solution.

## 8. Perturbation Solution

The solution of the nonlinear boundary-value problem (14.49) depends on the value of $\varepsilon_t$ parametrically. For an interpretation of $\varepsilon_t$, we see from (14.31a) that the total transportation cost per annum $t$ for each household consists of a distance cost component and a congestion cost component. With $R_0 - R_i = O(R_i)$ for cases of interest, these two cost components are of the order of magnitude of $\bar{t}_d \equiv \tau_0 R_i$ and $\bar{t}_c \equiv a_0 R_i (N_0/\theta R_i)^k$, respectively. On the other hand, the two equations in (14.48b) defining $\varepsilon_t$ and $\eta$ may be combined to yield

$$\varepsilon_t^{k+1} = \varepsilon_t^{k+1}\eta + \varepsilon_t^{k+1}(1-\eta) = \frac{\bar{t}_d}{y} + \frac{\bar{t}_c}{y} = O\left(\frac{t}{y}\right). \tag{14.50}$$

The quantity $\varepsilon_t^{k+1}$ is, therefore, of the order of magnitude of the fraction of the household income allocated for transportation costs; this fraction is typically very small compared to unity, $0 < \varepsilon_t^{k+1} \ll 1$. We will take advantage of this observation for a simple but accurate approximate solution for our problem.

The algebraic equation (14.49i) rearranged as

$$x(1-b) = \varepsilon_t \left[\frac{(1-\eta)k\lambda_1 u^k}{\bar{v}\Lambda_2 w^a}\right]^{1/(k+1)}$$

or

$$b = 1 - \frac{\varepsilon_t}{x}\left[\frac{(1-\eta)k\lambda_1 u^k}{\bar{v}\Lambda_2 w^a}\right]^{1/(k+1)} \tag{14,51}$$

may be used to eliminate $b(x;\varepsilon_t)$ from (14.49a), (14.49c), (14.49f), and (14.49h). The form of the resulting four coupled first-order differential equations for $w$, $u$, $\lambda_1$, and $\Lambda_2$ together with the boundary conditions (14.49b), (14.49d), (14.49c), and (14.49g) indicates that parametric expansions for the four unknown functions of $x$ and the unknown constant $\bar{v}$ in powers of $\varepsilon_t$ are appropriate:

$$\{w, u, \lambda_1, \Lambda_2, \bar{v}\} = \sum_{m=0}^{\infty} \{\bar{w}_m(x), \bar{u}_m(x), \bar{\lambda}_m(x), \bar{\Lambda}_m(x), \bar{v}_m\}\varepsilon_t^m. \tag{14.52a}$$

Correspondingly, (14.51) suggests that we take an expansion for $b(x;\varepsilon_t)$ in the form

$$b(x;\varepsilon_t) = 1 - \frac{\varepsilon_t}{x}\,[\beta_1(x) + \beta_2(x)\varepsilon_t + \beta_3(x)\varepsilon_t^2 + \dots]. \tag{14.52b}$$

Upon substituting (14.52) into (14.49) and requiring the resulting equations to be satisfied identically in $\varepsilon_t$, we get a sequence of linear boundary-value problems for the determination of the coefficients of the various expansions. For illustration, we list here the first two problems of this sequence and their solution *for $k = 1$ and a logarithmic-additive utility function $U(\cdot) = \ell n(\cdot)$*:

*a. The O(1) Problem*

$$\begin{cases} \overline{w}_0^{\bullet} = 0, & \overline{w}_0(1) = 1, \quad (14.53a,b) \\ \overline{u}_0^{\bullet} + \overline{v}_0\, x \overline{w}_0^{a} = 0, & \overline{u}_0(1) = 1, \quad \overline{u}_0(R) = 0, \quad (14.53c,d,e) \\ \overline{\Lambda}_0^{\bullet} = 0, & (14.53f) \\ \overline{\lambda}_0^{\bullet} + a\overline{v}_0\, x\overline{\Lambda}_0 \overline{w}_0^{a-1} = 0, & \overline{\lambda}_0(R) = 0, \quad (14.53g,h) \end{cases}$$

and

$$\overline{v}_0 \int_1^R x\overline{\Lambda}_0 \overline{w}_0^{a} dx = 1. \qquad (14.53i)$$

The solution of this problem is

$$\overline{w}_0(x) = \overline{\Lambda}_0(x) = 1, \qquad \overline{v}_0 = \frac{2}{R^2 - 1}, \qquad (14.54a,b)$$

$$\overline{u}_0(x) = \frac{R^2 - x^2}{R^2 - 1}, \qquad \overline{\lambda}_0(x) = a\,\frac{R^2 - x^2}{R^2 - 1}. \qquad (14.54c,d)$$

*b. The O(ε₁) Problem*

$$\begin{cases} \beta_1 = \left[\dfrac{(1-\eta)\overline{\lambda}_0 \overline{u}_0}{\overline{v}_0 \overline{\Lambda}_0 \overline{w}_0^{a}}\right]^{1/2}, & (14.55a) \\ \overline{w}_1^{\bullet} + \dfrac{1-\eta}{\beta_1}\overline{u}_0 = 0, \qquad \overline{w}_1(1) = 0, & (14.55b,c) \\ \overline{u}_1^{\bullet} + \overline{v}_0 \overline{w}_0^{a}\left[\left(\dfrac{\overline{v}_1}{\overline{v}_0} + a\dfrac{\overline{w}_1}{\overline{w}_0}\right)x - \beta_1\right] = 0, \quad \overline{u}_1(1) = \overline{u}_1(R) = 0, & (14.55d,e) \\ \overline{u}_1^{\bullet} + (1-\eta)\dfrac{\overline{\lambda}_0}{\beta_1} = 0, & (14.55f) \\ \overline{\lambda}_1^{\bullet} + a\overline{v}_0 x\overline{\Lambda}_0 \overline{w}_0^{a-1}\left[\dfrac{\overline{v}_1}{\overline{v}_0} + (a-1)\dfrac{\overline{w}_1}{\overline{w}_0} + \dfrac{\overline{\Lambda}_1}{\overline{\Lambda}_0} - \dfrac{\beta_1}{x}\right] = 0, & (14.55g) \\ \overline{\lambda}_1(R) = 0, & (14.55h) \end{cases}$$

and

$$\overline{v}_1 \int_1^R x\overline{\Lambda}_0 \overline{w}_0^{a} dx = \overline{v}_0 \int_1^R x\overline{\Lambda}_0 \overline{w}_0^{a}\left[\frac{\beta_1}{x} - \frac{\overline{\Lambda}_1}{\overline{\Lambda}_0} - a\,\frac{\overline{w}_1}{\overline{w}_0}\right] dx. \qquad (14.55i)$$

With $\overline{w}_0, \overline{u}_0, \overline{\Lambda}_0, \overline{\lambda}_0$, and $\overline{v}_0$ known from (14.54), the solution of this problem is found to be

$$\beta_1(x) = \frac{\gamma_0}{\bar{v}_0}\frac{R^2 - x^2}{R^2 - 1}, \qquad \frac{\bar{v}_1}{\bar{v}_0^2} = \frac{\gamma_0}{3}(2R^3 - 3R^2 + 1),$$

$$\bar{w}_1(x) = -\frac{\gamma_0}{\alpha}(x - 1), \qquad \bar{\Lambda}_1 = \frac{2\gamma_0}{3}\left\{\frac{R^3 - 1}{R^2 - 1} - \frac{3}{2}x\right\}, \tag{14.56}$$

$$\bar{u}_1(x) = \frac{\gamma_0\bar{v}_0}{6}\left\{(x^3 - 1) - (4R^3 - 3R^2 - 1)\frac{x^2 - 1}{R^2 - 1} + 3R^2(x - 1)\right\},$$

$$\bar{\lambda}_1(x) = \alpha\bar{u}_1(x) - \bar{v}_0\gamma_0\left\{\left[\frac{1}{2}(R^2 - 1) - \frac{\alpha}{3}(R^3 - 1)\right]\frac{R^2 - x^2}{R^2 - 1} + \frac{\alpha - 1}{3}(R^3 - x^3)\right\},$$

where $\gamma_0 = [\bar{v}_0\alpha(1 - \eta)]^{1/2}$.

We see from (14.54) and (14.56) that $w$, $u$, $\lambda_1$, and $\Lambda_2$ are all polynomial in $x$, at least for the first two terms in their respective perturbation expansion. On the other hand, we have

$$b(x;\varepsilon_t) = 1 - \frac{\varepsilon_t}{x}\beta_1(x) - \frac{\varepsilon_t^2}{x}\beta_2(x) + O(\varepsilon_t^3)$$

$$= 1 - \varepsilon_t\left[\left(\frac{R^2 - x^2}{x}\right)\sqrt{\frac{\alpha(1 - \eta)}{2(R^2 - 1)}}\right] + O(\varepsilon_t^2), \tag{14.57a}$$

so that to order $\varepsilon_t$, the road width fraction $(1 - b)$ is a simple rational function of $x$. The corresponding two-term perturbation solution for the market allocation, denoted by $b_m(x;\varepsilon_t)$, is given by (A.15) in the Appendix of this chapter. The ratio

$$\frac{1 - b}{1 - b_m} = \sqrt{\frac{\alpha}{\alpha + 1}}\,[1 + O(\varepsilon_t)] \tag{14.57b}$$

implies the following:

*To order $\varepsilon_t$, the cost-benefit criterion based on market land price (14.45) allocates more land for roads than the second best allocation.*

Note that with (14.56), we can obtain $\beta_2(x)$ without solving another boundary-value problem because

$$\beta_2(x) = \frac{1}{2}\beta_1(x)\left[\frac{\bar{u}_1}{\bar{u}_0} + \frac{\bar{\lambda}_1}{\bar{\lambda}_0} - \frac{\bar{v}_1}{\bar{v}_0} - \frac{\bar{\Lambda}_1}{\bar{\Lambda}_0} - \alpha\frac{\bar{w}_1}{\bar{w}_0}\right]. \tag{14.58a}$$

The two-term perturbation solution for the net income fraction after transportation cost is

$$w(x;\varepsilon_t) = 1 - \varepsilon_t \left[ (x-1)\sqrt{\frac{2(1-\eta)}{\alpha(R^2-1)}} \right] + O(\varepsilon_t^2). \quad (14.58b)$$

The corresponding solution for the market allocation, denoted by $w_m(x;\varepsilon_t)$, is given in the Appendix of this chapter. The ratio

$$\frac{1-w}{1-w_m} = \sqrt{\frac{\alpha+1}{\alpha}}\,[1+O(\varepsilon_t)] \quad (14.58c)$$

implies the following:

> ***The annual transportation cost for a household at a given location is smaller in a market city (because there is more land for roads), giving it a larger after-transportation income for housing and consumption goods.***

The second best unit land rent at the edge of the CBD is

$$r_i = \frac{yN_0\bar{v}}{(\alpha+1)\theta R_i^2}, \quad (14.59a)$$

or

$$\frac{r_i}{r_0} = 1 + \varepsilon_t \left[ \frac{1}{3}(2R^3 - 3R^2 + 1)\sqrt{\alpha(1-\eta)\left(\frac{2}{R^2-1}\right)^3} \right] + O(\varepsilon_t^2) \quad (14.59b)$$

with

$$r_0 = \frac{2yN_0}{(\alpha+1)\theta(R_0^2 - R_i^2)}. \quad (14.59c)$$

From (A.14), we have

$$\frac{(r_i/r_0)-1}{(r_m/r_0)-1} = \sqrt{\frac{\alpha^2+\alpha}{\alpha^2+\alpha+\frac{1}{4}}}\,[1+O(\varepsilon_t)], \quad (14.60)$$

where $r_m$ is the value $r_i$ from the market allocation. To the extent that the common household utility increases as $r_i$ decreases, we conclude from the ratio (14.60) that

> ***The second best land allocation gives a larger individual utility.***

Evidently, better and cheaper housing more than makes up for the higher transportation cost of the second best allocation.

In the course of obtaining the perturbation solution for the second best allocation, we have ignored the inequality constraints (3.7). Accurate numerical solutions for the same problem show that the constraints (3.7) are automatically satisfied in all cases of interest.

## 9. Inequality Constraints

Inequality constraints on control and state variables have already appeared in several examples in this chapter. However, they were satisfied by the (interior) extremal solutions and, therefore, played no role in the solution process. Binding inequality constraints appeared in Chapter 13; but for the linear Lagrangians and linear dynamics of these problems, the optimal controls were readily seen to be bang-bang. In this section, we discuss the treatment of optimal controls (subject to inequality constraints) which are neither bang-bang nor met by "interior" extremals for entire solution domain. For these problems, the concept of a *slack (excess) variable* introduced in Section 3 of Chapter 11 to prove [XI.2] may be exploited for a systematic solution process.

Suppose inequality constraints of the form

$$\varphi_i(t, \mathbf{y}, \mathbf{u}) \leq 0 \quad (i = 1, 2, \ldots, \ell), \tag{14.61a}$$

with $\ell < m$, are imposed on the optimal control problem defined by (13.12), (13.6), (13.7), and possibly terminal conditions on some of the components of $\mathbf{y}$. As before, we assume that controllability is not an issue. By introducing slack variables $z_i(t)$, $i = 1, \ldots, \ell$, we can write (14.61a) as

$$\Phi_i(t, \mathbf{y}, \mathbf{u}, z_i) = \varphi_i(t, \mathbf{y}, \mathbf{u}) + z_i^2 = 0 \quad (i = 1, 2, \ldots, \ell) \tag{14.61b}$$

with the first variation of $\Phi_i(t, \mathbf{y}, \mathbf{u}, z_i)$ vanishing in the neighborhood of an extremal:

$$\widehat{\varphi}_{i,\mathbf{y}}\delta\mathbf{y} + \widehat{\varphi}_{i,\mathbf{u}}\delta\mathbf{u} + 2\widehat{z}_i\,\delta z_i = 0 \quad (i = 1, 2, \ldots, \ell). \tag{14.61c}$$

As in problems of the calculus of variations, we append (14.61c) and other similar conditions on the first variations of the equations of state to $\delta J$. After integration by parts, we obtain for a prescribed terminal time $t_1$

$$\delta I = \delta J - \boldsymbol{\lambda}^T(t_1)\delta\mathbf{y}(t_1) + \int_{t_0}^{t_1} [(\boldsymbol{\lambda}^\bullet)^T\delta\mathbf{y} + \boldsymbol{\lambda}^T(\widehat{\mathbf{f}}_{,\mathbf{y}}\delta\mathbf{y} + \widehat{\mathbf{f}}_{,\mathbf{u}}\delta\mathbf{u})$$
$$+ \sum_{i=1}^{\ell} \Lambda_i\{\widehat{\varphi}_{i,\mathbf{y}}\delta\mathbf{y} + \widehat{\varphi}_{i,\mathbf{u}}\delta\mathbf{u} + 2z_i\,\delta z_i\}]\,dt, \tag{14.62}$$

keep in mind that we have taken the gradient of a scalar function to be a row vector.

As before, we choose the adjoint variable $\lambda$ to eliminate the terms involving $\delta y$:

$$(\lambda^T)^\cdot = -\lambda^T \widehat{\mathbf{f}}_{,\mathbf{y}} - \widehat{F}_{,\mathbf{y}} - \Lambda_i \widehat{\varphi}_{i,\mathbf{y}}, \qquad (14.63a)$$

$$\lambda(t_1) = \widehat{\psi}_{,\mathbf{y}}(\mathbf{y}(t_1)). \qquad (14.63b)$$

In addition, we choose the new adjoint variables $\{\Lambda_i\}$ associated with (14.61) to eliminate the $\delta z_i$ terms so that

$$2\Lambda_i z_i = 0 \quad (i = 1, 2, \ldots, \ell). \qquad (14.63c)$$

With (14.63), $\delta I$ is reduced to the familiar form

$$\delta I = \int_{t_0}^{t_1} [\widehat{F}_{,\mathbf{u}} + \lambda^T \widehat{\mathbf{f}}_{,\mathbf{u}} + \sum_{i=1}^{\ell} \Lambda_i \widehat{\varphi}_{i,\mathbf{u}}]\, \delta\mathbf{u}\, dt.$$

We expect $\delta I$ to be non-negative for the control $\widehat{\mathbf{u}}$ to be optimal and for $\widehat{J}$ to be a minimum. With nonvanishing slack variables $\{z_i\}$, there is formally no restriction on the variations $\{\delta u_j\}$; hence, we require

$$\widehat{F}_{,\mathbf{u}} + \lambda^T \widehat{\mathbf{f}}_{,\mathbf{u}} + \sum_{i=1}^{\ell} \Lambda_i \widehat{\varphi}_{i,\mathbf{u}} = \mathbf{0}. \qquad (14.63d)$$

The necessary conditions (14.63a)-(14.63d) along with the initial-value problem for the state variable $\mathbf{y}$ and the constraints (14.61b) serve to determine $\widehat{\mathbf{y}}$, $\widehat{\mathbf{u}}$, $\lambda$, $\Lambda = (\Lambda_1, \ldots, \Lambda_\ell)^T$, and $\widehat{\mathbf{z}} = (\widehat{z}_1, \ldots, z_\ell)^T$. If terminal conditions are prescribed for some components of $\mathbf{y}$, there will not be an Euler boundary conditions in (14.63b) for the corresponding adjoint variables.

For the solution of the problem above, we note that (14.63c) requires

$$\Lambda_i = 0 \quad \text{or} \quad z_i = 0. \qquad (14.63e)$$

If $z_i \neq 0$, then (14.61b) implies $\varphi_i(t, \widehat{\mathbf{y}}, \widehat{\mathbf{u}}) < 0$ and (14.63c) requires $\Lambda_i = 0$. Furthermore, (14.63d) reduces to the familiar stationarity condition

$$\widehat{F}_{,\mathbf{u}} + \lambda^T \mathbf{f}_{,\mathbf{u}} = \mathbf{0} \qquad (14.63d')$$

if $z_i \neq 0$ for all $1 \leq i \leq \ell$. With $\lambda = \mathbf{0}$ in this case, we have effectively an unconstrained extremal solution (none of the inequality constraints needs to be enforced). If, on the other hand, $z_k = 0$ for some $k$, then we have $\varphi_k(t, \widehat{\mathbf{y}}, \widehat{\mathbf{u}}) = 0$ so that one component of $\widehat{\mathbf{u}}$ (or $\widehat{\mathbf{y}}$) is determined. In the extreme case of $\widehat{\mathbf{z}} = \mathbf{0}$, $\ell\,(< m)$ of the $m + n$ components of $\widehat{\mathbf{y}}$ and $\widehat{\mathbf{u}}$ are formally determined in terms of the remaining unknowns. The critical step in the solution process is to determine whether it is $z_i = 0$ or $\Lambda_i = 0$ for each $i$. The proper choice is often decided (or suggested) by the phenomenon modeled by the mathematical problem. This is illustrated by the following example:

**EXAMPLE 8:** $\max_u \left\{ J = \int_0^T (u^2 + 4y)\,dt \right\}$ subject to $y^{\bullet} = u$, $y(0) = 0$, $y(T) = B$, and $u \geq 0$.

This is the problem of minimum cost production schedule first formulated in Section 1 of Chapter 1 and re-stated as an optimal control problem. The production rate $u = y^{\bullet}$ is to be adjusted to minimize the production cost $J$. However, as long as we do not include in $J$ the cost of destroying goods already produced, we should require $u \geq 0$ for the solution to be meaningful.

Without the inequality constraint, it is straightforward to show that the extremal solution is given by

$$y_f(t) = t^2 + (B - T^2)\frac{t}{T}, \qquad u_f(t) = 2t + \frac{1}{T}(B - T^2).$$

The unconstrainted extremal control $u_f$ satisfies the inequality constraint [and is, therefore, an admissable extremal $\widehat{u}(t)$ for our problem] as long as $B \geq T^2$. For $B < T^2$, $\widehat{u}_f$ is negative for $t \leq (T^2 - B)/2T \equiv t^*$. Certainly $y_f$ and $u_f$ are not applicable in some range $0 \leq t < t^*$. It does not necessarily follow that the optimal control is obtained by piecing together $u(t) = 0$ in $0 \leq t \leq t^*$ and an unconstrained solution with $y(t^*) = 0$:

$$u_p(t) = \begin{cases} 0 & (0 \leq t < t^*) \\ 2t + c_1 & (t^* < t \leq T), \end{cases}$$

$$y_p(t) = \begin{cases} 0 & (0 \leq t < t^*) \\ t^2 + c_1 t + c_0 & (t^* < t \leq T), \end{cases}$$

with $c_1 = [B - (T^2 - t^{*2})]/(T - t^*)$ and $c_0 = (T - t^*)(B - T^2)/[B - (T^2 - t^{*2})]$. (Generally, $\widehat{u}$ may be discontinuous at $t^*$.)

The proper solution process starts with setting $\widehat{\varphi}_1(t, y, u) = -u \leq 0$ so that $\Phi_1 \equiv z^2 - u = 0$. Evidently, we should have $\widehat{u} = 0$ for $0 \leq t < t_s$ for some $t_s > 0$ because the unconstrained extremal is not applicable near $t = 0$. Near the start, we have from (14.63) and the equation of state the following set of equations for this range of $t$:

$$\lambda_1^{\bullet} = -4, \qquad \lambda_1 = \Lambda_1, \qquad \widehat{y}^{\bullet} = 0,$$

where $\lambda_1$ and $\Lambda_1$ are the adjoint variables for $y^{\bullet} = u$ $(= 0)$ and $\Phi_1 = 0$. Integration and application of the initial condition for $y(t)$ give

$$\lambda_1 = -4t + \overline{\lambda}_1 = \Lambda_1, \quad \widehat{y} = 0 \quad (0 \leq t < t_s),$$

where $\bar{\lambda}_1$ is an unknown constant of integration.

Eventually, we will have to produce. Hence, for $t_s < t \le T_s \le T$, we have $\hat{u} > 0$ so that $\Lambda_1 = 0$ and

$$\lambda_1^{\bullet} = -4, \qquad 2\hat{u} + \lambda_1 = 0, \qquad \hat{y}^{\bullet} = \hat{u} = -\frac{\lambda_1}{2}.$$

After integration, we obtain

$$\lambda_1 = -4t + \lambda_s, \qquad \hat{u} = -\frac{\lambda_1}{2}, \qquad \hat{y} = (t^2 - t_s^2) - \frac{\lambda_2}{2}(t - t_s) \quad (t > t_s)$$

where we have required $\hat{y}$ to be continuous at $t_s$. We need $u > 0$ and, therefore, $\lambda_1 < 0$ for $t > t_s$. This requirement is met if $\lambda_1(t_s) \le 0$. Because $J$ increases with (the integral of) $\hat{u}^2 = \lambda_1^2/4$, we should require $\lambda_1(t_s) = 0$ and therewith $\lambda_s = 4t_s$. Correspondingly, we have

$$\lambda_1 = 4(t_s - t), \qquad \hat{u} = 2(t - t_s), \qquad \hat{y} = (t - t_s)^2 \qquad (t > t_s).$$

Because $\hat{u} > 0$ for all $t > t_s$, we have $\lambda_1 = 0$ for $t_s < t \le T$ so that unconstrained extremal is applicable for all $t$ after the switch point $t_s$. The switch time $t_s$ is determined by the terminal condition $y(T) = (T - t_s)^2 = B$ which requires

$$t_s = T - \sqrt{B}.$$

With $t_s$ as a free boundary, the same conclusion also follows immediately from [III.3] with $u$ replaced by $y^{\bullet}$ in $J$. Note that $t_s \le 0$ for $T < \sqrt{B}$ so that the unconstrained extremal applies to the entire time interval as previously found for this case. The only (unessential) unknown left is the constant $\bar{\lambda}_1$ (in the $0 \le t < t_s$ range). It should be set equal to $4t_s$ for $\lambda_1$ and $\Lambda_1$ to be continuous at $t_s$.

## 10. Optimality Under Constraints

The sufficiency theorems of Section 4 in principle also apply to problems with inequality constraints for which the unconstrained extremal is not feasible. For example, [XIV.2]–[XIV.4] may be invoke to conclude that the extremal obtained for Example 8 minimizes the production cost of that problem. Unfortunately, the functions $F$ and $\mathbf{f}$ for some problems are not convex or concave. For these problems we need to establish other types of sufficient conditions. The following example from Leitman (1981) offers the possibility of another sufficient condition which does not involve convexity.

**EXAMPLE 9:** $\max_u \left\{ J = \int_0^T [r(t)y - c(t)u]\,dt \right\}$ subject to

$$y^{\bullet} = -\mu y + u - \frac{\gamma}{2}\frac{u^2}{Y - y}, \qquad y(0) = y^0,$$

and

$$0 \le u \le \frac{Y - y}{\gamma} \qquad (0 \le t \le T).$$

This problem models the selection of a maintenance policy for rental housing to maximize the net rental revenue. Here, $y(t)$ is the quality of housing at time $t$ with $Y$ being the quality of brand new (or top condition) housing; $r(t)y$ is taken to be the rent charged for the $y$-level housing. The maintenance effort denoted by $u(t)$ is the control with $c(t)u$ as the associated cost.

Without maintenance, housing quality deteriorates with time at the rate $\mu y$. Maintenance improves on the housing quality but with a diminishing return. Maximum rate of improvement is attained at $u_{\max}(t) \equiv [Y - y(t)]/\gamma$; a larger maintenance effort actually leads to a smaller rate of improvement. Hence, we stipulate

$$0 \le u \le \frac{1}{\gamma}[Y - y(t)].$$

We also expect $0 \le y(t) \le Y$; but this condition is automatically satisfied by the solution [see Leitman (1981) for a proof of this claim].

The Hamiltonian for the problem is

$$H = ry - cu + \lambda\left[-\mu y + u - \frac{\gamma u^2}{2(Y - y)}\right] + \Lambda_1[z_1^2 - u] + \Lambda_2\left[u + z_2^2 - \frac{1}{\gamma}(Y - y)\right].$$

The differential equations and boundary condition for the adjoint variables $\lambda$, $\Lambda_1$, and $\Lambda_2$ are

$$\lambda^{\bullet} = -r + \lambda\left[\mu + \frac{\gamma \hat{u}^2}{2(Y - \hat{y})^2}\right] + \frac{\Lambda_2}{\gamma}, \qquad \lambda(T) = 0,$$

$$2\Lambda_2\hat{z}_2 = 0, \qquad 2\Lambda_1\hat{z}_1 = 0,$$

whereas the stationary condition with respect to $u$ is

$$H_{,u} = -c + \lambda - \frac{\gamma\lambda\hat{u}}{Y - \hat{y}} = 0 \quad \text{or} \quad \hat{u} = \frac{Y - \hat{y}}{\gamma}\left(1 - \frac{c}{\lambda}\right)$$

because $\Lambda_1$ and $\Lambda_2$ necessarily vanish for an interior solution.

Evidently, we have an interior solution only if $\lambda \geq c(t)$. [Note that $\lambda$ cannot be negative for we would have $\lambda^{\bullet} < 0$ and it would not be possible to satisfy $\lambda(T) = 0$.] For $0 < \lambda \leq c$, we must enforce the constraint at the lower end so that $\widehat{u} = 0$ and therewith $\widehat{z}_1 = 0$ and $\Lambda_2 = 0$. The switching from one type of solution to the other takes place at $t_s$, determined by the free boundary condition from [III.3] which requires $\widehat{u}(t_s-) = 0$. This condition, in turn, implies

$$c(t_s) = \lambda(t_s).$$

Because $\lambda^{\bullet}$ is more positive for the unconstrained extremal, that solution must hold for $0 \leq t < t_s$. Hence, we have

$$\widehat{u} = \begin{cases} \dfrac{1}{\gamma}(Y - \widehat{y})\left(1 - \dfrac{c}{\lambda}\right) & (0 \leq t < t_s) \\ 0 & (t_s < t \leq T) \end{cases}$$

and, correspondingly,

$$\widehat{\lambda}^{\bullet} = \begin{cases} -r + \lambda\mu + \dfrac{\lambda}{2\gamma}\left(1 - \dfrac{c}{\lambda}\right)^2 & (0 \leq t < t_s) \\ -r + \lambda\mu & (t_s < t \leq T) \end{cases}$$

$$\widehat{y}^{\bullet} = \begin{cases} -\mu\widehat{y} + \dfrac{1}{\gamma}(Y - \widehat{y})\left(1 - \dfrac{c}{\lambda}\right)\left[1 - \dfrac{1}{2}\left(1 - \dfrac{c}{\lambda}\right)\right] & (0 \leq t < t_s) \\ -\mu\widehat{y} & (t_s < t \leq T). \end{cases}$$

With $\lambda(T) = 0$, the solution for $\lambda$ in the range $(t_s < t \leq T)$ is immediate:

$$\lambda = e^{\mu t}\int_t^T r(\tau)e^{-\mu\tau}\,d\tau.$$

The transversality condition $\lambda(t_s) = c(t_s)$ determines $t_s$:

$$\lambda(t_s) = e^{\mu t_s}\int_{t_s}^T r(\tau)e^{-\mu\tau}\,d\tau = c(t_s).$$

The continuity of $\lambda$ at $t_s$ [which follows from $\widehat{u}(t_s-) = 0$] provides the auxiliary condition for the first-order differential equation for $\lambda$ in the range $0 \leq t \leq t_s$. Except for constant $r$ and $c$, an exact solution for $\lambda$ is not available in simple form but can be obtained numerically. So can the solution for $\widehat{y}(t)$, with its continuity at $t_s$ providing the auxiliary condition needed to determine $y$ from the equation of state in the $t_s < t \leq T$ range.

Fortunately, the optimality of the extremal found above may be established without an explicit solution for $\lambda$. For any other admissible control $u$ [so that $0 \leq u \leq (Y - y)/\gamma$] and the corresponding system state $y$, consider the integral

$$\int_0^T K\,dt \equiv \int_0^T [\hat{F} - F + \lambda(\hat{f} - f) + \lambda^{\bullet}(\hat{y} - y)]\,dt$$
$$= \int_0^T (\hat{F} - F)\,dt + [\lambda(\hat{y} - y)]_0^T = \int_0^T (\hat{F} - F)\,dt,$$

keeping in mind both $\hat{y}$ and $y$ satisfy the same initial condition and $\lambda(T) = 0$.

Evidently, the extremal control maximizes $J$ if the integrand $K$ is non-negative. For $t > t_s$, we have $\hat{u} = 0$ and therewith

$$K \equiv \hat{F} - F + \lambda(\hat{f} - f) + \lambda^{\bullet}(\hat{y} - y)$$
$$= u\left[(c - \lambda) + \frac{\gamma \lambda u}{2(Y - y)}\right] \geq u(c - \lambda) \geq 0 \qquad (t_s < t \leq T)$$

because $0 \leq \lambda < c$ in this range of $t$. In the other range, $t < t_s$, the expression for $K$ is more complicated but can be readily simplified to

$$K = -(\lambda - c)u + \frac{\gamma \lambda u^2}{2(Y - y)} + (Y - y)\frac{\lambda}{2\gamma}\left(1 - \frac{c}{\lambda}\right)^2$$
$$= \frac{\gamma \lambda}{2(Y - y)}\left[u - \frac{Y - y}{\gamma}\left(1 - \frac{c}{\lambda}\right)\right]^2 \geq 0 \qquad (0 \leq t \leq t_s).$$

Altogether, we have $K \geq 0$ for $0 \leq t \leq T$ so that

$$\hat{J} - J = \int_0^T (\hat{F} - F)\,dt \geq 0.$$

The results for Example 9 suggest a new sufficient condition. Let

$$K_m \equiv -F(t, \mathbf{y}, \mathbf{u}) + \mathbf{p} \cdot \mathbf{f}(t, \mathbf{y}, \mathbf{u}). \tag{14.64}$$

It is then straightforward to prove the following:

**[XIV.5]** ***Suppose the pair*** $(\mathbf{u}, \hat{\mathbf{y}})$ ***extremizes J as in (13.12) subject to (13.6), (13.7), and (14.61a). If for some*** $C^1$ ***function*** $\mathbf{p}(t)$***, the inequality***

$$K_m(t, \hat{\mathbf{y}}, \hat{\mathbf{u}}, \mathbf{p}) - K_m(t, \mathbf{y}, \mathbf{u}, \mathbf{p}) + (\mathbf{p}^{\bullet})^T(\hat{\mathbf{y}} - \mathbf{y}) \geq 0 \tag{14.65}$$

***for any*** $(\mathbf{u}, \mathbf{y})$ ***pair which satisfies (13.6), (13.7), and (14.61a), then*** $(\hat{\mathbf{u}}, \hat{\mathbf{y}})$ ***minimizes J.***

For a maximum $\hat{J}$, [XIV.5] applies after a sign change in $K_m$, i.e., with $K_m$ replaced by $K_M = F + \mathbf{p} \cdot \mathbf{f}(t, \mathbf{y}, \mathbf{u})$. The function $\mathbf{p}$ may often be identified with $\boldsymbol{\lambda}$.

## 11. Methods of the Calculus of Variations

The few sufficient conditions for optimality developed in this chapter are applicable to only a small fraction of all optimal control problems. Other types of sufficiency theorems which require more elaborate proofs have also been established. A discussion of these theorems is beyond the scope of this volume; readers are referred to references in Bibliography. However, there is one easy way to enlarge the available set of sufficient conditions considerably. This is accomplished by transforming the optimal control problems into problems in the calculus of variations. The sufficiency theorems of the classical theory can then be invoked for the transformed problems whenever applicable.

A natural way to transform an optimal control problem into a problem in the calculus of variations is to solve $m$ of the $n$ scalar equations of state for the $m$ control variables $\mathbf{u} = (u_1, \ldots, u_m)$ and use the results to eliminate $\mathbf{u}$ from the Lagrangian and the remaining $n - m$ state equations. The transformed state equations will then serve as $n - m$ equality constraints and the problem can be treated by the methods discussed in Chapters 11 and 12. If $m = n$ and there are no other constraints on the state or control variables, the material in Chapters 4 and 5 are then applicable to the resulting calculus of variations problem.

Unfortunately, it is often not possible or practical to solve for $\mathbf{u}$ from the equations of state. A more reliable method of transformation consists of setting

$$\mathbf{y}^1 = \mathbf{y}, \tag{14.66a}$$

$$(\mathbf{y}^2)^\cdot = \mathbf{u}, \tag{14.66b}$$

$(\mathbf{y}^3)^\cdot$ = the adjoint variable associated with (13.6), $\boldsymbol{\lambda} = (\lambda_1, \ldots, \lambda_n)^T$,
$(\mathbf{y}^4)^\cdot$ = the adjoint variable, $\boldsymbol{\Lambda} = (\Lambda_1, \ldots, \Lambda_\ell)^T$, for $\boldsymbol{\varphi} = (\varphi_1, \ldots, \varphi_\ell)^T \le 0$ written as an equality constraint

$$\begin{aligned}\boldsymbol{\Phi} &= \boldsymbol{\varphi}(t, \mathbf{y}, \mathbf{u}) + [(\mathbf{y}^5)^\cdot]^2 \\ &= \boldsymbol{\varphi} + ([(y_1^5)^\cdot]^2, \ldots, [(y_\ell^5)^\cdot]^2)^T = \mathbf{0}\end{aligned} \tag{14.66c}$$

with the help of the slack variables

$$((y_1^5)^\cdot, (y_2^5)^\cdot, \ldots, (y_\ell^5)^\cdot)^T = (\mathbf{y}^5)^\cdot, \tag{14.66d}$$

and, to make $\{\mathbf{y}^k(t)\}$ unique,

$$\mathbf{y}^2(t_0) = \ldots = \mathbf{y}^5(t_0) = \mathbf{0}. \tag{14.66e}$$

In terms of

$$\mathbf{y} = \begin{pmatrix} \mathbf{y}^1 \\ \vdots \\ \mathbf{y}^5 \end{pmatrix}, \tag{14.67}$$

we write the augmented performance index over a prescribed interval $[t_0, t_1]$ as

$$I = \psi(\mathbf{y}^1(t_1)) + \int_{t_0}^{t_1} \Big\{ F(t, \mathbf{y}^1, (\mathbf{y}^2)^\bullet) + [(\mathbf{y}^3)^\bullet]^T [\mathbf{f} - (\mathbf{y}^1)^\bullet]$$
$$+ [(\mathbf{y}^4)^\bullet]^T [\boldsymbol{\Phi} + \{(\mathbf{y}^5)^\bullet\}^2] \Big\} dt$$
$$\equiv \psi(\mathbf{y}^1(t_1)) + \int_{t_0}^{t_1} G(t, \mathbf{y}, \mathbf{y}^\bullet) dt. \tag{14.68}$$

The single-vector Euler differential equation for $I$ is

$$[G_{,\mathbf{y}^\bullet}]^\bullet = [G_{,\mathbf{y}}]. \tag{14.69}$$

It is equivalent to the following five systems of differential equations:

$$[-(\mathbf{y}^3)^\bullet]^\bullet = [(\mathbf{y}^4)^\bullet]^T \boldsymbol{\varphi}_{,\mathbf{y}^1} + F_{,\mathbf{y}^1} + [(\mathbf{y}^3)^\bullet]^T \mathbf{f}_{,\mathbf{y}^1}, \tag{14.70a}$$

$$\{F_{,(\mathbf{y}^2)^\bullet} + [(\mathbf{y}^3)^\bullet]^T \mathbf{f}_{,(\mathbf{y}^2)^\bullet} + [(\mathbf{y}^4)^\bullet]^T \boldsymbol{\varphi}_{,(\mathbf{y}^2)^\bullet}\}^\bullet = 0, \tag{14.70b}$$

$$[\mathbf{f} - (\mathbf{y}^1)^\bullet]^\bullet = 0, \tag{14.70c}$$

$$\{\boldsymbol{\Phi} + [(\mathbf{y}^5)^\bullet]^2\}^\bullet = 0, \tag{14.70d}$$

$$2[(y_i^4)^\bullet (y_i^5)^\bullet]^\bullet = 0 \quad (i = 1, 2, \ldots, \ell). \tag{14.70e}$$

The Euler boundary conditions are

$$t = t_1: \qquad G_{,(\mathbf{y}^1)^\bullet} = \psi_{,\mathbf{y}^1}(\mathbf{y}^1(t_1)), \tag{14.71a}$$

$$G_{,(\mathbf{y}^i)^\bullet} = 0, \quad (i = 2, 3, \ldots, 5). \tag{14.71b}$$

We integrate (14.70) and make use of (14.71) to obtain

$$\boldsymbol{\lambda}^\bullet = -F_{,\mathbf{y}} - \boldsymbol{\lambda}^T \mathbf{f}_{,\mathbf{y}}, \tag{14.72a}$$

$$F_{,\mathbf{u}} + \boldsymbol{\lambda}^T \mathbf{f}_{,\mathbf{u}} + \boldsymbol{\Lambda}^T \boldsymbol{\varphi}_{,\mathbf{u}} = 0, \tag{14.72b}$$

$$\mathbf{y}^\bullet = \mathbf{f}, \qquad \boldsymbol{\varphi} + \mathbf{z}^2 = 0,$$

$$\Lambda_i z_i = 0 \quad (i = 1, 2, \ldots, \ell) \tag{14.72c}$$

after writing $\mathbf{y}, \mathbf{u}, \boldsymbol{\lambda}, \boldsymbol{\Lambda}$, and $\mathbf{z}$ for $\mathbf{y}^1, (\mathbf{y}^2)^\bullet, (\mathbf{y}^3)^\bullet, (\mathbf{y}^4)^\bullet$, and $(\mathbf{y}^5)^\bullet$, respectively with $\mathbf{z}^2 = (z_1^2, z_2^2, \ldots, z_\ell^2)^T$. Hence, we conclude

**[XIV.6]** ***The smooth extremals of (14.68), (13.7), and (14.66c) are also the extremals of the optimal control problem (13.12), (13.6), (13.7), and (14.61a) and conversely.***

To illustrate, we apply the general method to Example 2 of this chapter. We begin by setting $[t_0, t_1] = [0, T]$, $y_1^1 = k$, $y_2^1 = D$, $(y_1^2)^{\cdot} = c$, and $(y_2^2)^{\cdot} = r$. The inequality constraints, $k \geq 0$, $D \geq 0$, $r \geq 0$, and $c \geq 0$ may be written as

$$\varphi_1 \equiv -y_1^1 + [(y_1^5)^{\cdot}]^2 = 0, \tag{14.73a}$$

$$\varphi_2 \equiv -y_2^1 + [(y_2^5)^{\cdot}]^2 = 0, \tag{14.73b}$$

$$\varphi_3 \equiv -(y_1^2)^{\cdot} + [(y_3^5)^{\cdot}]^2 = 0, \tag{14.73c}$$

$$\varphi_4 \equiv -(y_2^2)^{\cdot} + [(y_4^5)^{\cdot}]^2 = 0. \tag{14.73d}$$

In that case, we have

$$\begin{aligned} G = {} & e^{-\delta t} U((y_1^2)^{\cdot}) + (y_1^3)^{\cdot}\{f - \theta(y_2^2)^{\cdot} - \mu y_1^1 - (y_1^2)^{\cdot} - (y_1^1)^{\cdot}\} \\ & + (y_2^3)^{\cdot}\{-(y_2^2)^{\cdot} - (y_2^1)^{\cdot}\} + (y_1^4)^{\cdot}\{-y_1^1 + [(y_1^5)^{\cdot}]^2\} \\ & + (y_2^4)^{\cdot}\{-y_2^1 + [(y_2^5)^{\cdot}]^2\} + (y_3^4)^{\cdot}\{-(y_1^2)^{\cdot} + [(y_3^5)^{\cdot}]^2\} \\ & + (y_4^4)^{\cdot}\{-(y_2^2)^{\cdot} + [(y_4^5)^{\cdot}]^2\}. \end{aligned} \tag{14.74}$$

The Euler differential equations for the problem are

$$-(y_1^3)^{\cdot\cdot} = (y_1^3)^{\cdot}(f_{,k} - \mu), \tag{14.75a}$$

$$-(y_2^3)^{\cdot\cdot} = 0, \tag{14.75b}$$

$$\{e^{-\delta t} U'((y_1^2)^{\cdot}) - (y_1^3) - (y_3^4)^{\cdot}\}^{\cdot} = 0, \tag{14.75c}$$

$$\{(y_1^3)^{\cdot}(f_{,r} - \theta) - (y_2^3)^{\cdot} - (y_4^4)^{\cdot}\}^{\cdot} = 0, \tag{14.75c}$$

$$\{f - \theta(y_2^2)^{\cdot} - \mu y_1^1 - (y_1^2)^{\cdot} - (y_1^1)^{\cdot}\}^{\cdot} = 0, \tag{14.75e}$$

$$\{-(y_2^2)^{\cdot} - (y_2^1)^{\cdot}\}^{\cdot} = 0, \tag{14.75f}$$

$$\{-y_k^1 + [(y_k^5)^{\cdot}]^2\}^{\cdot} = 0, \quad \{-(y_k^2) + [(y_k^5)^{\cdot}]^2\}^{\cdot} = 0 \quad (k = 1, 2), \tag{14.75g}$$

$$\{2(y_i^4)^{\cdot}(y_i^5)^{\cdot}\}^{\cdot} = 0 \quad (i = 1, 2, 3, 4). \tag{14.75i}$$

The Euler boundary conditions are

$$(y_1^3)^{\cdot}|_{t=T} = \psi_{,k_T}, \tag{14.76a}$$

$$(y_2^3)^{\cdot}|_{t=T} = \psi_{,D_T}, \tag{14.76b}$$

$$[e^{-\delta t} U'((y_1^2)^{\cdot}) - (y_1^3)^{\cdot} - (y_3^4)^{\cdot}]|_{t=T} = 0, \tag{14.76c}$$

$$[(y_1^3)^{\cdot}(f_{,r} - \theta) - (y_2^3)^{\cdot} - (y_4^4)^{\cdot}]|_{t=T} = 0, \tag{14.76d}$$

$$\{f - \theta (y_2^2)^{\bullet} - \mu y_1^1 - (y_1^2)^{\bullet} - (y_1^1)^{\bullet}\}\big|_{t=T} = 0, \tag{14.76e}$$

$$\{(y_2^2)^{\bullet} + (y_2^1)^{\bullet}\}\big|_{t=T} = 0, \tag{14.76f}$$

$$\{-y_j^k + [(y_j^5)^{\bullet}]^2\}\big|_{t=T} = 0 \quad (j = 1,2; k = 1,2), \tag{14.76g}$$

$$\{(y_i^4)^{\bullet}(y_i^5)^{\bullet}\}\big|_{,t=T} = 0 \quad (i = 1,2,3,4), \tag{14.76h}$$

where $k_T \equiv k(T)$ and $D_T \equiv D(T)$. We integrate (14.75c)-(14.75*i*) and apply (14.76c)-(14.76h) to set the constants of integration to zero. Upon writing the results in the more familiar notations, we have

$$e^{-\delta t} U'(c) = \lambda_1 + \Lambda_3, \quad \lambda_2 + \Lambda_4 = \lambda_1 (f_{,r} - \theta), \tag{14.77a,b}$$

$$k^{\bullet} = f - \theta r - \mu k - c, \quad D^{\bullet} = -r, \tag{14.77c,d}$$

$$-k + z_1^2 = 0, \quad -D + z_2^2 = 0, \tag{14.77e,f}$$

$$-c + z_3^2 = 0, \quad -r + z_4^2 = 0, \tag{14.77g,h}$$

$$\Lambda_i z_i = 0 \quad (i = 1,2,3,4). \tag{14.77i}$$

The remaining Euler equations (14.75a) and (14.75b) take the form

$$\lambda_1^{\bullet} = -\lambda_1 (f_{,k} - \mu), \tag{14.78a}$$

$$\lambda_2^{\bullet} = 0. \tag{14.78b}$$

They are supplemented by the Euler boundary conditions (14.76a) and (14.76b) written as

$$\lambda_1(T) = \psi_{,k_T}, \quad \lambda_2(T) = \psi_{,D_T}. \tag{14.79a,b}$$

When we discussed Example 2 in Section 1, we noted that the extremal solutions $\hat{k}(t)$, $\hat{D}(t)$, $\hat{c}(t)$, and $\hat{r}(t)$ are strictly positive. In that case, we must have $\Lambda_i(t) = 0$, $i = 1, \ldots, 4$, by (14.77i). Consequently, (14.77a) and (14.77b) simplify to

$$e^{-\delta t} U'(c) - \lambda_1 = 0, \quad \lambda_1 (f_{,r} - \theta) - \lambda_2 = 0. \tag{14.78c,d}$$

Equations (14.78a)-(14.78b) and (14.78c)-(14.78d) are the two adjoint equations and the two stationarity conditions obtained in Section 1. The two boundary conditions (14.79a) and (14.79b) at $t = T$ are identical to the two Euler boundary conditions for $\lambda_1$ and $\lambda_2$ found there.

With [XIV.6], we can now bring the theoretical results for classical problems of calculus of variations to bear on optimal control problems. Numerical methods such as those described in Chapters 8 and 9 for the classical problems can also be used for optimal control problems after a preliminary transformation as described above. More recently developed numerical techniques may be found in Gregory and Lin (1992) and references therein.

## 12. Exercises

1. Obtain the exact solution of Example 1 (in Section 1) for $\sigma = \alpha = 1/2$ as follow:

   (a) Obtain $t$ as a function of $c$ with $a_0$, $T$ and $c_T$ as parameters.

   (b) Write the condition $k(0) = k_0$ as an equation for the only unknown $a_0$ alone (with $T$, $c_T$ and $k_0$ being prescribed constants).

2. Consider Example 2 (in Section 1) with the more general equation of state

$$k^{\bullet} = f(k, r) - \theta r - \mu k - c,$$

   where $\theta$ and $\mu$ are known non-negative constants.

   (a) Obtain the Hotelling's rule

$$\frac{[f_{,r}]^{\bullet}}{f_{,r} - \theta} = f_{,k} - \mu$$

   from the various necessary conditions and state equations.

   (b) Specialize the Hotelling's rule above for $f(k, r) = k^a r^b$.

3. Consider Example 2 with $U(c) = c^{1-\sigma}/(1-\sigma)$, $f(k, r) = k^a r^{1-a}$ $(b = 1 - a)$, $T = \infty$, and $\psi(k_T, D_T) = \mu = \theta = 0$.

   (a) Show that $c(t) = e^{\delta t/a}[c_0^{b\sigma/a} + F_0 t]$.

   (b) Obtain also $r(t)$ and $k(t)$; give conditions for determining all unknown constants.

   (c) Discuss the long time behavior of $c(t)$ for $\delta > 0$ and $\delta = 0$ (with $0 < \sigma < 1$ and $a < \sigma$). Interpret the mathematical results.

4. Suppose $\mu \neq 0$ in Exercise 3.

   (a) Deduce

$$c(t) = e^{-\delta t/\sigma}\left\{\left[c_0^{\sigma b/a} - \frac{a}{\mu b}\left(\frac{b}{A^b}\right)^{1/a}\right]e^{-b\mu t/a} + \frac{a}{\mu b}\left(\frac{b}{A^b}\right)^{1/a}\right\}^{a/\sigma b}.$$

   (b) Discuss the long time behavior of $c(t)$ for $\delta > 0$ and $\delta = 0$ and interpret the mathematical results.

5. Repeat Example 7 of Section 3 for $\theta \neq 0$.

6. (a) Obtain the feedback control law for the following LQP:

$$\min_u \int_0^T [y^2 + c_0^2 u^2]\, dt$$

subject to $y^\bullet = u$, $y(0) = y_0(> 0)$ for some real constant $c_0$.

(b) Repeat part (a) with the additional terminal condition $y(T) = 0$.

7. (a) Repeat part (b) of Exercise 6 with the additional inequality constraints

$$\varphi_1 \equiv a_1 - b_1 t - y \leq 0, \qquad \varphi_2 \equiv y - a_2 + b_2 t \leq 0$$

with $a_i, b_i > 0$, $a_2 > y_0 > a_1$, and $a_2/b_2 > a_1/b_1$.

(b) Repeat part (a) of this problem by allowing an optimal choice of $T$.

8. Deduce (14.53) and (14.55).

9. Obtain (14.54) from (14.53).

10. Obtain (14.56) from (14.55).

11. Obtain the first two terms of a perturbation solution for the cost-benefit allocation (14.45) (see also the Appendix of this chapter).

12. Transform Example 9 of Section 10 into a problem of the calculus of variations.

13. Obtain the feedback control law for the following problems:

(a) Find $\hat{u}(t)$ to minimize

$$J = \frac{1}{2} c_0 [y(T)]^2 + \frac{1}{2} \int_0^T u^2\, dt \qquad (c_0 > 0)$$

subject to $y^\bullet = u$, $y(0) = y_0$. (Both $u$ and $y$ are scalar unknowns.)

(b) Minimize

$$J = \frac{1}{2} [c_1 y_1^2(T) + c_2 y_2^2(T)] + \frac{1}{2} \int_0^T u^2\, dt$$

subject to $y_1^\bullet = y_2$, $y_2^\bullet = u$, and given initial conditions.

(c) Minimize $J$ of part (a) with the equations of state changed to $y^\bullet = -ay + bu$.

(d) Minimize

$$J = \frac{1}{2} c_0 [y_1(T)^2] + \frac{1}{2} \int_0^T u^2 \, dt$$

subject to $y_1^{\bullet} = y_2$, $y_2^{\bullet} = -\omega^2 y_1 + u$, and the initial condition $y(0) = y^0$.

## Appendix: A Cost–Benefit Criterion

In terms of the dimensionless quantities introduced in (14.48), condition (14.45) becomes

$$\frac{u}{x(1-b)} = \frac{\zeta}{\varepsilon_t} w^{(\alpha+1)/(k+1)}, \quad \zeta = \left[ \frac{\bar{v}}{k(\alpha+1)(1-\eta)} \right]^{1/(k+1)}. \quad \text{(A.1)}$$

We use (A.1) to eliminate $u$ from (14.49a) to obtain

$$w' = -\varepsilon_t (1-\eta) \zeta^k w^{k(\alpha+1)/(k+1)} - \varepsilon_t^{k+1} \eta. \quad \text{(A.2)}$$

Rather remarkably, the housing land fraction $b$ does not appear in this first-order equation. Hence, the separable first-order ODE (A.2) along with the initial condition $w(1) = 1$ determines $w(x; \varepsilon_t)$.

Next, we use (A.1) to express $u$ in terms of $\ell_T \equiv x(1-b)$ and the known function $w(x; \varepsilon_t)$:

$$u = \frac{\zeta}{\varepsilon_t} \ell_T w^{(\alpha+1)/(k+1)}, \quad \text{(A.3)}$$

which is, in turn, used to eliminate $u$ form (14.49c), taken in the form

$$u^{\bullet} = \bar{v} \ell_T w^{\alpha} - \bar{v} x w^{\alpha} = -k \zeta^{k+1} (\alpha+1)(1-\eta)(x - \ell_T) w^{\alpha} \quad \text{(A.4)}$$

to get

$$[\ell_T w^{(\alpha+1)/(k+1)}]^{\bullet} = -k \varepsilon_t \zeta^k (\alpha+1)(1-\eta)(x - \ell_T) w^{\alpha}. \quad \text{(A.5)}$$

The boundary condition $u(R) = 0$ [see (14.49e)] and (A.3) imply

$$\ell_T (R; \varepsilon_t) = 0; \quad \text{(A.6)}$$

it serves as an auxiliary condition for (A.5). An exact solution of the "terminal-value" problem (A.5) and (A.6) for $\ell_T$ is also immediate because (A.5) is a first-order linear differential equation. However, this exact solution is in the form of a quadrature from which it is difficult to extract useful information without extensive numerical work.

Useful information can be more simply obtained by seeking a perturbation solution for the various initial and terminal-value problems as in Section 7. We omit the calculation and simply give the following results for the $k = 1$ case for comparison with the second-best solution:

$$w(x;\varepsilon_t) = 1 - \frac{t}{y} = 1 - \varepsilon_t\left[\sqrt{\frac{2(1-\eta)}{(R^2-1)(\alpha+1)}}\,(x-1)\right] + O(\varepsilon_t^2), \quad \text{(A.7)}$$

$$r_i = \frac{yN_0\bar{v}}{(\alpha+1)\theta R_i^2} \equiv r_0\bar{v}(\varepsilon_t, \sigma, R, \eta), \quad \text{(A.8)}$$

$$r_0 = \frac{2yN_0}{(\alpha+1)\theta(R_0^2 - R_i^2)}, \quad \text{(A.9)}$$

$$\bar{v} = 1 + \varepsilon_t\left[\frac{1}{6}(2R^3 - 3R^2 + 1)\left(\frac{2\alpha+1}{\alpha+1}\right)\sqrt{\left(\frac{2}{R^2-1}\right)^3(1-\eta)(\alpha+1)}\right]$$
$$+ O(\varepsilon_t^2), \quad \text{(A.10)}$$

$$b(x;\varepsilon_t) = 1 - \varepsilon_t\left[\frac{R^2-x^2}{x}\sqrt{\frac{(\alpha+1)(1-\eta)}{2(R^2-1)}}\right] + O(\varepsilon_t^2). \quad \text{(A.11)}$$

Although we have limited ourselves to the case $k = 1$ and a fixed city boundary, it is evident that the same technique can be used to obtain approximate solutions for other values of $k$ and/or an unknown outer city limit which is determined as a part of the solution process by the residual rent beyond.

# 15

# Several Independent Variables

## 1. The Plateau Problem

**Let $\Gamma$ be a simple closed curve in space whose projection in the $x,y$-plane is the simple closed curve $\Gamma_0$ characterized by $f_0(x,y) = 0$. To simplify our discussion, we take $\Gamma$ to be entirely above the $x,y$-plane and described by a well-defined function $Z(x,y)$ for all $(x,y)$ on $\Gamma_0$. Let $z(x,y)$ be a surface in space with $\Gamma$ as its boundary (see Fig. 15.1). Evidently, there are many such surface even if we limit ourselves to only smooth (simply-connected) surface whose projection on the $x,y$-plane is the entire interior of $\Gamma_0$. The surface area for each such surface is given by the double integral**

$$J[z] = \iint_A \sqrt{1 + (z_{,x})^2 + (z_{,y})^2}\, dx\, dy, \tag{15.1}$$

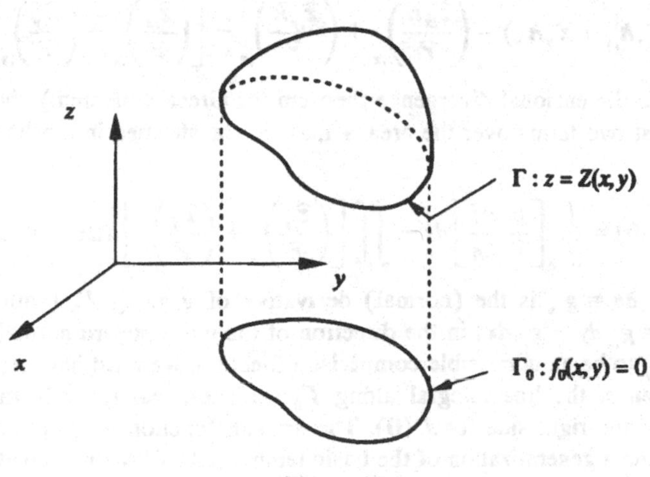


**Figure 15.1**

where $(\ )_{,t} \equiv \partial(\ )/\partial t$ (unless specifically indicated otherwise) and $A$ is the region of the $x, y$-plane enclosed by the simple smooth plane curve $\Gamma_0$. Suppose that among all surfaces $z(x,y)$ described above, the surface $\hat{z}(x,y)$ has the smallest surface area. In that case, $\hat{z}(x,y)$ minimizes the functional $J[z]$ and assumes a prescribed value at each boundary point of the solution domain:

$$\hat{z}(x,y) = Z(x,y) \quad \text{for} \quad f_0(x,y) = 0, \tag{15.2}$$

where both $f_0(x,y)$ and $Z(x,y)$ are prescribed functions. Finding the minimal surface $\hat{z}(x,y)$ is, therefore, a basic problem in the calculus of variations. The problem involves comparison functions of two variables not treated in our development to this point.

For smooth comparison surfaces, we may again reduce the search for $\hat{z}(x,y)$ to the solution of an Euler differential equation. To get this differential equation, we set $z(x,y) = \hat{z}(x,y) + \varepsilon h(x,y)$ so that $J[z] \equiv J(\varepsilon)$ with

$$J(\varepsilon) = \iint_A \sqrt{1 + [(\hat{z} + \varepsilon h)_{,x}]^2 + [(\hat{z} + \varepsilon h)_{,y}]^2}\, dx\, dy$$

and $J(0) = J[\hat{z}]$ being the minimum surface area. For $J(0)$ to be a stationary value, we need

$$J^{\cdot}(0) = \iint_A \frac{1}{\hat{r}} [\hat{z}_{,x} h_{,x} + \hat{z}_{,y} h_{,y}]\, dx\, dy = 0,$$

where $r \equiv (1 + z_{,x}^2 + z_{,y}^2)^{1/2}$. As in the single-variable case, we wish to integrate by parts to get the integrand into the form $(\dots)h(x,y)$. This is accomplished with the help of the identity

$$\frac{1}{\hat{r}}(\hat{z}_{,x} h_{,x} + \hat{z}_{,y} h_{,y}) = \left(\frac{\hat{z}_{,x} h}{\hat{r}}\right)_{,x} + \left(\frac{\hat{z}_{,y} h}{\hat{r}}\right)_{,y} - \left[\left(\frac{\hat{z}_{,x}}{\hat{r}}\right)_{,x} + \left(\frac{\hat{z}_{,y}}{\hat{r}}\right)_{,y}\right] h.$$

By the two-dimensional divergence theorem (or Green's theorem), the integral for the first two terms over the area $A$ may be transformed into a line integral so that

$$J^{\cdot}(0) = \oint_{\Gamma_0} \left[\frac{h}{\hat{r}} \frac{\partial \hat{z}}{\partial n}\right] ds - \iint_A \left[\left(\frac{\hat{z}_{,x}}{\hat{r}}\right)_{,x} + \left(\frac{\hat{z}_{,y}}{\hat{r}}\right)_{,y}\right] h\, dx\, dy = 0,$$

where $\partial g/\partial n \equiv g_{,n}$ is the (normal) derivative of $g$ along $\Gamma_0$ (with $g_{,n} ds \equiv \nabla g \cdot \mathbf{n}\, ds = g_{,x} dy - g_{,y} dx$) in the direction of the unit outward normal $\mathbf{n}$ of $\Gamma_0$. For $z(x,y)$ to be an admissible comparison function, we must have $h(x,y) = 0$ on $\Gamma_0$. Hence, the line integral along $\Gamma_0$ vanishes, leaving only the double integral on the right side for $J^{\cdot}(0)$. The smooth function $h(x,y)$ is otherwise arbitrary and a generalization of the basic lemma [I.1] allows us to deduce from $J^{\cdot}(0) = 0$ the following Euler differential equation for our (Plateau) problem:

$$\left(\frac{\hat{z}_{,x}}{\hat{r}}\right)_{,x} + \left(\frac{\hat{z}_{,y}}{\hat{r}}\right)_{,y} = 0 \tag{15.3a}$$

or, with $\hat{r} > 0$,

$$D[\hat{z}] \equiv (1 + \hat{z}_{,y}^2)\hat{z}_{,xx} - 2\hat{z}_{,x}\hat{z}_{,y}\hat{z}_{,xy} + (1 + \hat{z}_{,x}^2)\hat{z}_{,yy} = 0. \tag{15.3b}$$

This is a partial differential equation (PDE) to be solved for $\hat{z}$ in conjunction with the boundary condition (15.2). A solution of the Euler differential equation (15.3) is called a (smooth) *extremal (surface)* of the given Lagrangian $F(x, z, z_{,x}, z_{,y}) = r$. Those extremals which also satisfy the prescribed boundary condition are called *admissible extremals.*

**EXAMPLE 1:** $Z(x, y) = y + 1$ on $f_0(x, y) = x^2 + (y - 1)^2 - 1 = 0$.

By inspection, $\hat{z}(x, y) = y + 1$ satisfies the Euler differential equation (since $\hat{z}_{,xx} = \hat{z}_{,xy} = \hat{z}_{,yy} = 0$). It also satisfies the prescribed boundary condition. Hence, it is an admissible extremal. Intuitively, it is the minimal surface for our problem because the curve $\Gamma$ is really planar and the surface $\hat{z} = y + 1$ is the flat area enclosed by $\Gamma$ in its own plane. Any other surface over the interior of $\Gamma_0$ would have a larger surface area.

The observation above is further supported by the fact that the product of $\hat{r}^{-1}$ and the left-hand side of (15.3b), denoted by $M$, has a simple geometrical interpretation:

$$M = \frac{1}{r}[(1 + z_{,y}^2)z_{,xx} - 2z_{,x}z_{,y}z_{,xy} + (1 + z_{,x}^2)z_{,yy}] = \frac{1}{2}\left[\frac{1}{r_1} + \frac{1}{r_2}\right], \tag{15.4}$$

where $r_1$ and $r_2$ are the two principal radii of curvature of the surface $z(x, y)$ [see Struik (1961)]. The Euler differential equation (15.3a) effectively stipulates that the mean curvature of the surface must vanish for a minimal surface with a fixed boundary curve $\Gamma$. The flat extremal surface of Example 1 certainly has this property. In fact, its curvature is zero for all points inside $\Gamma_0$.

Still, the development above does not prove that an admissible (smooth) extremal $\hat{z}$ minimizes $J[z]$ for the present problem or a general Lagrangian $F$. The establishment of sufficient conditions for a minimum is similar to that for the one variable case and can be found in Forsyth (1960). We will not pursue a detailed discussion of sufficient conditions in this volume but will briefly comment on such conditions for a weak local minimum in a later section.

## 2. Euler Differential Equation and Boundary Conditions

The derivation of an Euler differential equation for the Plateau problem in Section 1 also applies to problems with a more general Lagrangian of the type

$F(\mathbf{x}, u, \nabla u) \equiv F(\mathbf{x}, u, u_{,1}, \ldots, u_{,n})$ involving one unknown of several variables and its first partial derivatives. Suppose we want to find a smooth function $\widehat{u}(\mathbf{x}) \equiv \widehat{u}(x_1, x_2, x_3)$ which minimizes

$$J[u] \equiv \iiint_R F(\mathbf{x}, u, \nabla u)\, dx_1\, dx_2\, dx_3$$

$$\equiv \iiint_R F(x_1, x_2, x_3, u, u_{,1}, u_{,2}, u_{,3})\, dx_1\, dx_2\, dx_3, \tag{15.5}$$

where $\nabla u = (u_{,1}, u_{,2}, u_{,3})$ is the gradient (row vector) of $u$. In addition, $u$ takes on prescribed values on the whole or portions of the boundary of the three-dimensional region $R$. As before, we set $u = \widehat{u} + \varepsilon h(\mathbf{x})$ with $h(\mathbf{x}) = 0$ on the boundary $S$ of $R$. It is straightforward to compute the derivative of $J[u] \equiv J(\varepsilon)$ with respect to $\varepsilon$ to get

$$J^{\boldsymbol{\cdot}}(0) = \iiint_R \{\widehat{F}_{,u} h + \widehat{F}_{,v_1} h_{,1} + \widehat{F}_{,v_2} h_{,2} + \widehat{F}_{,v_3} h_{,3}\}\, dx_1\, dx_2\, dx_3,$$

where $v_k = u_{,k}$, $F_{,v_k} \equiv \partial F / \partial v_k$ and $\widehat{f}(\mathbf{x}) \equiv f(\mathbf{x}, \widehat{u}(\mathbf{x}), \widehat{u}_{,1}(\mathbf{x}), \widehat{u}_{,2}(\mathbf{x}), \widehat{u}_{,3}(\mathbf{x}))$. As in Section 1, we write

$$\widehat{F}_{,v_k} h_{,k} = (\widehat{F}_{,v_k} h)_{,k} - (\widehat{F}_{,v_k})_{,k} h$$

and transform the volume integral of

$$(\widehat{F}_{,v_1} h)_{,1} + (\widehat{F}_{,v_2} h)_{,2} + (\widehat{F}_{,v_3} h)_{,3} = \sum_{k=1}^{3} (F_{,v_k} h)_{,k}$$

into a surface integral by the divergence theorem:

$$J^{\boldsymbol{\cdot}}(0) = \oint\!\!\oint_S B[u]\, h\, dS + \iiint_R [\widehat{F}_{,u} - \sum_{k=1}^{3} (\widehat{F}_{,v_k})_{,k}]\, h\, dx_1\, dx_2\, dx_3 = 0, \tag{15.6}$$

with

$$B[u] \equiv n_1 \widehat{F}_{,v_1} + n_2 \widehat{F}_{,v_2} + n_3 \widehat{F}_{,v_3} = \sum_{k=1}^{3} n_k \widehat{F}_{,v_k}, \tag{15.7}$$

where $n_1$, $n_2$, and $n_3$ are the components of the unit (positive outward) normal vector $\mathbf{n}$ of $S$ in the cartesian reference frame. For $u(\mathbf{x})$ to take on a prescribed value at each point on the surface(s) $S$, we must have $h(\mathbf{x}) = 0$ on $S$, and, therefore, the surface integral in (15.6) vanishes. Because $h$ is otherwise arbitrary and we assume that the combination inside the brackets of the volume integral is continuous, we must have

$$\widehat{F}_{,u} - \nabla \cdot (\nabla_v \widehat{F}) = \widehat{F}_{,u} - \sum_{k=1}^{m} (\widehat{F}_{,v_k})_{,k} = 0, \tag{15.8}$$

where $m = 3$, $\nabla_v(g) = (g_{,v_1}, g_{,v_2}, g_{,v_3})$ and $\nabla \cdot (\ )$ is the conventional divergence operation with $\nabla \cdot \mathbf{w} = w_{1,1} + w_{2,2} + w_{3,3}$.

It is straightforward to extend these results to the case of $m$ independent variables given that the divergence theorem holds in higher dimensions as well (Flander, 1963). Hence, we have the following more general result:

**[XV.1]** *For a smooth function $\hat{\mathbf{u}}(\mathbf{x})$ to minimize $J[\mathbf{u}]$ as given in (15.5) and to fit the prescribed data on S, it is necessary that $\hat{\mathbf{u}}$ satisfy the Euler differential equation (15.8) where m is the number of independent variables (and of course fit the prescribed data on S).*

EXAMPLE 2: $F = \dfrac{1}{2}(u_{,1}^2 + u_{,2}^2 + u_{,3}^2)$.

For this Lagrangian, we have $F_{,u} = 0$ and the Euler differential equation becomes

$$\sum_{k=1}^{3} (\hat{F}_{,v_k})_{,k} = \sum_{k=1}^{3} v_{k,k} = 0$$

(with $v_k = u_{,k}$) which is just Laplace's equation in three dimensions:

$$u_{,11} + u_{,22} + u_{,33} \equiv \nabla^2 u = 0.$$

This linear partial differential equation appears in many different applications. Among them are the flow of irrotational and incompressible fluids (for which the gradient of $u$ is the velocity vector of the fluid) and the steady-state temperature distribution in a heat-conducting body (for which $u$ is the steady-state temperature distribution in the homogeneous and isotropic body).

As an example of a region $R$ with more than one boundary, consider the minimal surface problem of the last section with the surface now bounded by two given nonintersecting closed curves in space. We already analyzed one such problem in Section 5 of Chapter 2. In that case, the two nonintersecting closed curves were concentric parallel circles of (generally) different radius. In the context of our present discussion, the resulting minimal surface is a catenoid, a surface of revolution generated by revolving a catenary about the $z$-axis (Fig. 15.2).

EXAMPLE 3: $\cosh\left(\dfrac{z}{c_0} + c_1\right) = \dfrac{1}{c_0}\sqrt{x^2 + y^2}$.

The two constants of integration $c_0$ and $c_1$ of the above minimal surface of revolution are to be determined by the position of two prescribed circles, one of radius $A$ at $z = a$ and the other of radius $B$ at $z = b$ (Fig. 15.2):

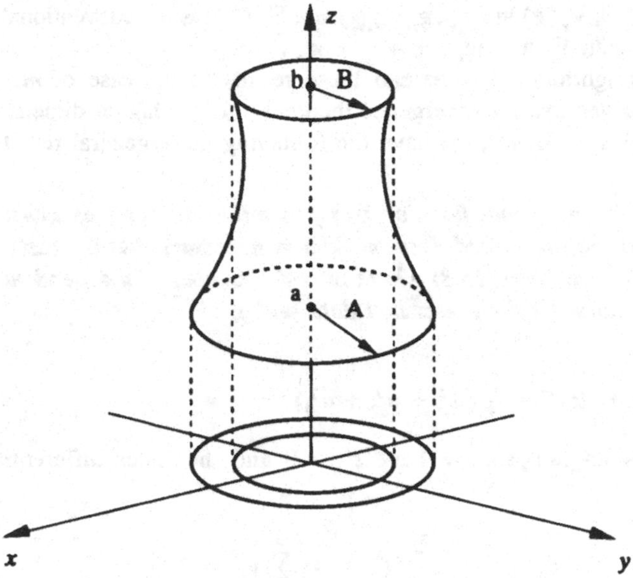


Figure 15.2

$$\cosh\left(\frac{a}{c_0}+c_1\right)=\frac{A}{c_0}, \qquad \cosh\left(\frac{b}{c_0}+c_1\right)=\frac{B}{c_0}.$$

For simplicity, consider the case $c_1 = 0$ and $c_0 = 1$ (which correspond to $a = -b$ and $A = B$). Upon differentiating the given equation for the surface with respect to $x$ and $y$, we obtain

$$x = z_{,x}\cosh z \sinh z, \qquad y = z_{,y}\cosh z \sinh z,$$

$$1 = z_{,xx}\cosh z \sinh z + (z_{,x})^2\cosh(2z),$$

$$0 = z_{,xy}\cosh z \sinh z + z_{,x}z_{,y}\cosh(2z),$$

$$1 = z_{,yy}\cosh z \sinh z + (z_{,y})^2\cosh(2z).$$

Form $\cosh^3 z \sinh z D[z]$, where $D[z]$ is the left-hand side of the Euler differential equation (15.3) for the minimal surface problem; we obtain, with the help of the relations above and some simplifications,

$$\begin{aligned}\cosh^3 z \sinh z D[z] &= z\cosh^2 z[1-(z_{,x}^2+z_{,y}^2)\sinh^2 z] \\ &= 2[\cosh^2 z - 2(x^2+y^2)] = 0.\end{aligned}$$

Hence, the catenoid $\cosh^2 z = x^2 + y^2$ in fact satisfies the general Euler differential equation for a minimal surface as it should.

If $u$ is not required to take on a prescribed value on a portion of the boundary of $R$, then $h$ needs not vanish on that portion of $S$. However, (15.6) must hold for those comparison functions with $h$ vanishing on $S$; hence, the Euler differential equation still is a necessary condition for the new minimization problem. With the elimination of the volume integral, the vanishing of $J'(0)$ for all smooth functions $h$ which need not vanish on $S$ requires that $\widehat{u}$ satisfies the *Euler boundary condition*

$$B[u] \equiv \sum_{k=1}^{m} n_k \widehat{F}_{,v_k} = 0 \tag{15.9}$$

for all points $\mathbf{x}$ on the portion of the boundary where $u$ is not prescribed. In (15.9), the quantities $n_k = \mathbf{i}_k \cdot \mathbf{n}$, $k = 1, 2, 3$, are the directional cosines of the unit normal of $S$. Hence,

**[XV.2]** *If* u *is not prescribed on (a portion of) the boundary* S *of the region* R*, it is necessary for the extremal* $\widehat{u}$ *to satisfy the Euler boundary condition (15.9) on (that portion of)* S *if* $\widehat{u}$ *is to render* J[u] *a minimum.*

EXAMPLE 4: $F = \dfrac{1}{2}(u_{,1}^2 + u_{,2}^2 + u_{,3}^2)$.

For this Lagrangian, the boundary term in (15.9),

$$B[u] = n_1 u_{,1} + n_2 u_{,2} + n_3 u_{,3} = \frac{\partial u}{\partial n},$$

is just the normal derivative of $u$ on $S$. The Euler boundary condition (15.9) requires that the normal derivative of $u$ vanish:

$$\frac{\partial u}{\partial n} = 0$$

for all $\mathbf{x}$ on that portion of $S$ where $u$ is not prescribed.

It is customary to call the problem of solving Laplace's equation with a prescribed value of $u$ on the boundary of the solution domain the *Dirichlet problem* and the boundary condition itself the *Dirichlet condition*. When the normal derivative of the unknown $u$ is prescribed on the boundary instead, the BVP for Laplace's equation is called the *Neumann problem* and the prescribed boundary condition the *Neumann condition*.

## 3. Sufficient Conditions

The development of conditions which assure us a minimum value for $J[u]$ defined in (15.5) is similar to that for the one-dimensional case discussed in

Chapters 4 and 5. The details can be found in Forsyth (1960) and will not be discussed here. We merely remark that for a weak local minimum, the sufficient conditions for the case of one unknown are analogous to those given by [IV.6]. We have already confined ourselves to *smooth* admissible extremals [which by definition satisfy the Euler differential equation (15.8)]. The strengthened Legendre condition $\hat{F}_{,y'y'} > 0$ naturally becomes a requirement that

**[II″]** *The Hessian matrix* $\mathrm{H}[F] \equiv [\hat{F}_{,v_i v_j}(\mathbf{x})]$ *be positive definite.*

The analogue of the Jacobi condition requires a clarification of the concept of conjugate points. Just as in the one-dimensional case, we can obtain an accessory equation for $h$ from $J^{\cdot\cdot}(\varepsilon)$. This equation then allows us to develop a theory of conjugate points (which are now curves or surfaces) analogous to the theory of conjugate points in Chapter 4. For the two-variable case, we can find curves "conjugate" to a boundary curve of the domain of integration $A$ of $J[u]$. As for one-dimensional problems, we can do this analytically or geometrically (by looking for possible formulation of an envelope for the extremals passing through the boundary). The result [IV.6] for the one-dimensional case suggests that we do not want such conjugate curves (or hypersurfaces in the case of higher dimensions) in the solution domain.

We state without proof the following result which guarantees a weak local minimum:

**[XV.3]** *If the domain of the extremals does not contain points (curves or surfaces) conjugate to its boundaries and if the Hessian* $[\hat{\mathrm{F}}_{,v_i v_j}]$ *(where* $\mathrm{v}_i \equiv \mathrm{u}_{,i}$) is positive definite, then the smooth admissible extremal $\hat{\mathrm{u}}$ *of the given Lagrangian renders* $\mathrm{J}[\mathrm{u}]$ *of (15.5) a weak local minimum if such a minimum exists for the given boundary data.*

**EXAMPLE 5:** For the minimal surface problem of Section 1, we have

$$F_{,v_1v_1} = \frac{1+v_2^2}{(1+v_1^2+v_2^2)^{3/2}} > 0, \qquad F_{,v_2v_2} = \frac{1+v_1^2}{(1+v_1^2+v_2^2)^{3/2}} > 0,$$

$$F_{,v_1v_2} = \frac{v_1 v_2}{(1+v_1^2+v_2^2)^{3/2}}.$$

In that case, the Hessian matrix

$$H[F] = \begin{bmatrix} F_{,v_1v_1} & F_{,v_1v_2} \\ F_{,v_2v_1} & F_{,v_2v_2} \end{bmatrix}$$

is positive definite as

$$\det|H[F]| = F_{,v_1v_1}F_{,v_2v_2} - F_{,v_1v_2}F_{,v_2v_1}$$
$$= \frac{(1+v_1^2)(1+v_2^2) - v_1^2v_2^2}{[1+v_1^2+v_2^2]^3} = \frac{1}{[1+v_1^2+v_2^2]^2} > 0$$

for any $\mathbf{v}$, including the special case $v_i = \hat{u}_{,i}(\mathbf{x})$.

The analysis of "conjugate points" (or, more properly, conjugate boundary) to a given boundary of the solution domain is much more complex and we do not wish to describe it here. For the minimal surface of revolution $\cosh^2 z = x^2 + y^2$ (as in Example 3), the intersection of this surface with the plane $z = z_1$ is a circle. The conjugate boundary (curve) of this circle is known to be another such circle $z = z_2$ with

$$\coth z_1 - z_1 = \coth z_2 - z_2 .$$

For the catenoid bounded by two parallel circles $z = c_1$ and $z = c_2$, we should, in principle, determine all the conjugate curves of $z = c_1$. If all of them lie outside the two bounding circles, then the surface area of the catenoid is a minimum among all surfaces with the same two circles as their boundaries. Fortunately, we have $z = z(r)$ in cylindrical coordinates so that the minimal surface problem for a surface of revolution is reduced to a problem of one (independent) variable and can be analyzed by the techniques of Chapters 1-5.

Though conceptually similar to the one-variable case, sufficient conditions for several-variable problems are generally much more difficult to verify. The determination of conjugate curves or surfaces is an example of these difficulties. Consequently, sufficient conditions in higher dimensions are generally not very useful in practice. Fortunately, for many problems in science and engineering the variational approach is used mainly to obtain approximate solutions of the Euler differential equation. Whether or not $J[\hat{u}]$ is the minimum value is less important or of no interest for these problems. This perspective will be reflected in the development of the remaining sections of this volume.

## 4. Dirichlet's Problem on a Unit Disk

There is another kind of difficulty associated with sufficiency for higher-dimensional problems not encountered in the one-variable case. Prescribed data on the boundary are now functions of position. Depending on their smoothness, the solution of a given variational problem may or may not exist in the prescribed class of comparison functions. Hence, a qualification pertaining to the boundary data (unnecessary for the one-variable case) was inserted at the end of [XV.3]. For instance, take the Dirichlet problem on a unit disc, with $u(1,\theta) = g(\theta)$. If $g(\theta)$ is a continuous function, it would seem reasonable that a solution should exist. Let us now examine this problem a little more carefully.

A formal solution of the boundary-value problem (BVP) can be obtained by "separation of variables" in the form

$$u(r,\theta) = a_0 + \sum_{n=1}^{\infty} r^n (a_n \cos n\theta + b_n \sin n\theta) \tag{15.10}$$

with the Fourier coefficients determined by

$$a_0 = \frac{1}{2\pi}\int_0^{2\pi} g(\theta)\,d\theta, \qquad \begin{pmatrix} a_n \\ b_n \end{pmatrix} = \frac{1}{\pi}\int_0^{2\pi} g(\theta)\begin{pmatrix} \cos n\theta \\ \sin n\theta \end{pmatrix} d\theta. \tag{15.11}$$

The series for $u(r,\theta)$ is convergent for all $0 \le r < 1$, $0 \le \theta \le 2\pi$. It tends to $g(\theta)$ uniformy as $r \to 1$. Therefore, $u(r,\theta)$ tends to $g(\theta_0)$ when the point $(r,\theta)$ approaches $(1,\theta_0)$ from inside the unit circle in any way.

Moreover, the solution found above is the only solution. If there were another, say $u_1(r,\theta)$, we could let $\bar{u} = u - u_1$ and $\bar{u}$ would satisfy Laplace's equation for $r < 1$ and $\bar{u}(1,\theta) = 0$, $0 \le \theta \le 2\pi$. In that case, we would have

$$\iint_{r\le 1} (\bar{u}\nabla^2\bar{u})\,dx\,dy = \iint_{r\le 1} [\nabla(\bar{u}\nabla\bar{u}) - |\nabla\bar{u}|^2]\,dx\,dy$$

$$= \oint_{r=1} (\bar{u}\nabla\bar{u})\cdot\mathbf{n}\,ds - \iint_{r\le 1} |\nabla\bar{u}|^2\,dx\,dy = 0.$$

But we know $\bar{u} = 0$ on $r = 1$ so that the line integral vanishes, leaving us with

$$\iint_{r\le 1} |\nabla\bar{u}|^2\,dx\,dy = 0.$$

This is only possible if $\nabla\bar{u} = 0$ inside the unit disc or $\bar{u} = \bar{u}_0$ (a constant). Because $\bar{u}(1,\theta) = 0$, we must have $u_0 = 0$ so that $\bar{u} = 0$ or $u_1 = u$.

Given that the integrand of the Dirichlet integral

$$J[u] = \iint_{r\le 1} [u_{,x}^2 + u_{,y}^2]\,dx\,dy = \iint_{r\le 1} \left[u_{,r}^2 + \left(\frac{1}{r}u_{,\theta}\right)^2\right] r\,dr\,d\theta \tag{15.12}$$

is non-negative, we know $J[u] \ge 0$. A minimum should exist and should not be smaller than zero. In addition, the Hessian matrix for $F(x, y, u, u_{,x}, u_{,y})$,

$$\begin{bmatrix} 2 & 0 \\ 0 & 2 \end{bmatrix},$$

is positive definite so that the strengthened Legendre condition is satisfied. It can be shown that the Jacobi test is also satisfied for typical boundary data. These facts suggest that the unique solution obtained above should minimize $J[u]$.

Now consider the case

$$g(\theta) = \sum_{m=1}^{\infty} \frac{1}{m^2} \sin(m!\theta).$$

For this $g(\theta)$, we have [from (15.10)] as the solution of the Dirichlet problem

$$\hat{u}(r,\theta) = \sum_{m=1}^{\infty} \frac{r^{m!}}{m^2} \sin(m!\theta),$$

which converges uniformly and absolutely for all $0 \leq r \leq 1$. Correspondingly, (15.12) becomes

$$J[\hat{u}] = \sum_{m=1}^{\infty} \frac{m!}{m^4}.$$

The series $J[\hat{u}]$ diverges; hence, $\hat{u}$ does not minimize $J[u]$ despite all indications suggesting that it should.

The development above illustrates the added mathematical complexity associated with problems in higher dimensions. Even what may be described as the simplest problem of this type turns out to require extreme care in its formulation. For an introduction to the calculus of variations, we will assume that the minimum exists for our problems and limit the scope of our discussion to material on necessary conditions which are useful in applications.

## 5. Several Unknowns

In many applications in science and engineering, the variational problems are for many unknowns which are functions of several independent variables. These include problems in important areas such as solid mechanics, fluid mechanics, and electromagnetics. We will discuss the classical theories of elastostatics and fluid dynamics in some detail in the next few chapters. Some essential features of the electromagnetic problem will be presented in Section 6 of this chapter.

Even without an excursion into these rich areas of application, the geometrical problem of minimal surfaces sometimes calls for a theory for several unknowns. For a closed surface or a large portion of such a surface, the description $z(x,y)$ used in Section 1 may not be adequate. For example, $z = (x^2 + y^2)^{1/2}$ cannot completely describe a portion of a sphere which contains an equator of the sphere in its interior. For such cases, the surface involved should be described in parametric form

$$x = x(\xi,\eta), \qquad y = y(\xi,\eta), \qquad z = z(\xi,\eta) \tag{15.13}$$

for $\alpha \le \xi \le \beta$ and $\gamma \le \eta \le \delta$. We know from vector calculus that if

$$\mathbf{R}(\xi,\eta) = x\mathbf{i} + y\mathbf{j} + z\mathbf{k} \tag{15.14}$$

is the position vector of a point on the surface, the total surface area is given by

$$J[\mathbf{R}] = \iint_A \left|\mathbf{R}' \times \mathbf{R}^{\bullet}\right| d\xi\, d\eta, \tag{15.15}$$

where $(\ )' \equiv (\ )_{,\xi}$, $(\ )^{\bullet} \equiv (\ )_{,\eta}$ and $A$ is the region spanned by the range of $\xi$ and $\eta$. By the identity $(\mathbf{A} \times \mathbf{B})\cdot(\mathbf{C} \times \mathbf{D}) = (\mathbf{A}\cdot\mathbf{C})(\mathbf{B}\cdot\mathbf{D}) - (\mathbf{A}\cdot\mathbf{D})(\mathbf{B}\cdot\mathbf{C})$, we have

$$\left|\mathbf{R}' \times \mathbf{R}^{\bullet}\right|^2 = (\mathbf{R}' \times \mathbf{R}^{\bullet})\cdot(\mathbf{R}' \times \mathbf{R}^{\bullet}) = g_{11}g_{22} - g_{12}^2, \tag{15.16}$$

where $g_{11} \equiv \left|\mathbf{R}'\right|^2 = (x')^2 + (y')^2 + (z')^2$, $g_{22} \equiv \left|\mathbf{R}^{\bullet}\right|^2 = (x^{\bullet})^2 + (y^{\bullet})^2 + (z^{\bullet})^2$, and $g_{12} \equiv (\mathbf{R}'\cdot\mathbf{R}^{\bullet}) = x'x^{\bullet} + y'y^{\bullet} + z'z^{\bullet} = g_{21}$. Hence, the problem of minimal surface becomes one of finding functions $\widehat{x}(\xi,\eta), \widehat{y}(\xi,\eta)$, and $\widehat{z}(\xi,\eta)$ which minimize the integral

$$J[x,y,z] \equiv \iint_A \sqrt{g_{11}g_{22} - g_{12}^2}\, d\xi\, d\eta \tag{15.17}$$

with $x, y$, and $z$ taking on prescribed values along the boundary of $A$. This problem involves three unknown functions of two independent variables.

More generally, we want to find $n$ unknown scalar functions $\widehat{\mathbf{u}}(\mathbf{x}) = (\widehat{u}_1, \ldots, \widehat{u}_n)^T$ of $m$ independent variables $\{x_1, \ldots, x_m\} = \mathbf{x}$ which minimize the integral

$$J[\mathbf{u}] = \int \cdots \int_{R_m} F(\mathbf{x}, \mathbf{u}, v_{ij})\, dx_1 \ldots dx_m, \tag{15.18}$$

where $v_{ij} = u_{i,j} \ne u_{j,i} = v_{ji}$ and $R_m$ is the domain of $\mathbf{u}(x)$ with $S_m$ as its boundary. For the basic problem, $\mathbf{u}$ is required to take on prescribed values on the boundary $S_m$.

**[XV.3]** *If the smooth function $\widehat{\mathbf{u}}$ minimizes $J[\mathbf{u}]$, then $\widehat{\mathbf{u}}$ satisfies the system of Euler differential equation*

$$\widehat{F}_{,u_i} - \sum_{j=1}^{m} (\widehat{F}_{,v_{ij}})_{,j} = 0 \tag{15.19}$$

*for* $i = 1, 2, \ldots, n$ *where* $\widehat{\mathbf{f}}(\mathbf{x}) \equiv \mathbf{f}(\mathbf{x}, \widehat{\mathbf{u}}(\mathbf{x}), \widehat{v}_{ij}(\mathbf{x}))$.

This result can be derived by the usual method after setting $u_i = \hat{u}_i + \varepsilon h_i(\mathbf{x})$, where $h_i$ vanishes on the boundary of $R_m$ if $u_i$ is prescribed there. For each $u_i$ not prescribed on $S_m$ there is an additional necessary condition in the form of an Euler boundary condition.

The general result [XV.3] will be applied to an important class of problems of mathematical physics in the next section. In particular, we will show that electromagnetic phenomena are consequences of a variational problem for an appropriate Lagrangian involving multiple unknowns.

## 6. Maxwell's Equations

Electromagnetic phenomena in a vacuum can be described by two vector fields: the electric field vector $\mathbf{E} = (E_1, E_2, E_3)^T$ and the magnetic field vector $\mathbf{B} = (B_1, B_2, B_3)^T$. Both $\mathbf{E}$ and $\mathbf{B}$ vary with location $\mathbf{x} = (x_1, x_2, x_3)^T$ in space and with time $t$. In the absence of electric charges, $\mathbf{E}$ and $\mathbf{B}$ are related by (the following normalized form of) Maxwell's equations:

$$\nabla \times \mathbf{E} = -\mathbf{B}_{,t}, \tag{15.20a}$$

$$\nabla \times \mathbf{B} = \mathbf{E}_{,t} \tag{15.20b}$$

with

$$\nabla \cdot \mathbf{E} = 0, \tag{15.21a}$$

$$\nabla \cdot \mathbf{B} = 0. \tag{15.21b}$$

The scalar partial differential equation (15.21b) is satisfied identically by setting (see Greenspan and Benney, 1973).

$$\mathbf{B} = \nabla \times \mathbf{A} \tag{15.22}$$

for some vector field $\mathbf{A} = (A_1, A_2, A_3)^T$. Upon substituting (15.22) into (15.20a), we get $\nabla \times (\mathbf{E} + \mathbf{A}_{,t}) = \mathbf{0}$, or

$$\mathbf{E} = \nabla A_0 - \mathbf{A}_{,t} \tag{15.23}$$

for some scalar potential $A_0(\mathbf{x}, t)$. In terms of $\mathbf{A}(\mathbf{x}, t)$ and $A_0(\mathbf{x}, t)$, (15.21a) and (15.20b) become

$$\nabla^2 A_0 - (\nabla \cdot \mathbf{A})_{,t} = 0 \tag{15.24}$$

and

$$\nabla \times (\nabla \times \mathbf{A}) = \nabla(\nabla \cdot \mathbf{A}) - \nabla^2 \mathbf{A} = \nabla A_{0,t} - \mathbf{A}_{,tt}, \tag{15.25}$$

respectively. If we require $\mathbf{A}$ and $A_0$ to satisfy the *Lorentz condition*

$$\nabla \cdot \mathbf{A} = A_{0,t}, \tag{15.26}$$

then (15.25) simplifies to a vector *wave equation* for **A**:

$$\nabla^2 \mathbf{A} = \mathbf{A}_{,tt} \tag{15.25'}$$

and (15.24) gives a scalar wave equation for $A_0$ after (15.26) is used to eliminate $\nabla \cdot \mathbf{A}$:

$$\nabla^2 A_0 = A_{0,tt}. \tag{15.24'}$$

Note that the expressions for **B** and **E** in (15.22) and (15.23) satisfy all Maxwell equations (15.20) and (15.21) if **A** and $A_0$ satisfy (15.24'), (15.25') and (15.26).

To appreciate the importance of the Lorentz condition, we observe that a term involving a scalar potential $\varphi(\mathbf{x}, t)$ can always be added to **A** and $A_0$ without changing **B** or **E**. More specifically, suppose we replace **A** by $\mathbf{A} + \nabla\varphi$ and $A_0$ by $A_0 + \varphi_{,t}$ (which is known as a *gauge transformation*). Given $\nabla \times \nabla f = 0$ for any scalar function $f$, we have

$$\mathbf{B} = \nabla \times (\mathbf{A} + \nabla\varphi) = \nabla \times \mathbf{A},$$

$$\mathbf{E} = \nabla(A_0 + \varphi_{,t}) - (\mathbf{A} + \nabla\varphi)_{,t} = \nabla A_0 - \mathbf{A}_{,t}.$$

Hence, a gauge transformation leaves **E** and **B** invariant. In other words, **A** and $A_0$ are not completely specified by **E** and **B** and are determined by them up to a scalar potential. The Lorentz condition is precisely what is needed to remove the ambiguity because it requires

$$\nabla \cdot (\mathbf{A} + \nabla\varphi) = (A_0 + \varphi_{,t})_{,t}$$

or

$$\nabla^2 \varphi = \varphi_{,tt}.$$

The nonuniqueness is just another *wave function* and, therefore, introduces nothing new.

We can now show that the wave function description of electromagnetic phenomena is a consequence of extremizing the difference between electrical and magnetic energy in a region $R$ in space over the time interval $[0, T]$:

$$J[\mathbf{A}, A_0] \equiv \frac{1}{8\pi} \int_0^T \iiint_R \{\mathbf{E} \cdot \mathbf{E} - \mathbf{B} \cdot \mathbf{B}\} dV\, dt, \tag{15.27}$$

where **E** and **B** are given in terms of **A** and $A_0$ by (15.22) and (15.23). In particular, we have the following result for $J[\mathbf{A}, A_0]$:

**[XV.4]** ***With* E *and* B *given by (15.22) and (15.23) subject to the Lorentz condition (15.26), the energy differential* J *is rendered stationary if* $A_0$ *and* A *satisfy the wave equations (15.24') and (15.25').***

In terms of **A** and $A_0$, we have

$$J[\mathbf{A},A_0]=\frac{1}{8\pi}\int_0^T\iiint_R\left\{|\nabla A_0-\mathbf{A}_{,t}|^2-|\nabla\times\mathbf{A}|^2\right\}dV\,dt. \qquad (15.27')$$

The Euler differential equation for $\delta J=0$ are given by [XV.3] to be

$$\frac{\partial F}{\partial A_k}=\left(\frac{\partial F}{\partial A_{k,j}}\right)_{,j} \qquad (k=0,1,2,3), \qquad (15.28)$$

where $8\pi F=|\nabla A_0-\mathbf{A}_{,t}|^2-|\nabla\times\mathbf{A}|^2$ and $(\;)_{,0}\equiv(\;)_{,t}$. *For* $k=0$, we have from the expression for $F$

$$\frac{\partial F}{\partial A_0}=0, \qquad \frac{\partial F}{\partial(A_{0,t})}=0,$$

$$\frac{\partial F}{\partial(A_{0,j})}=\frac{1}{4\pi}(A_{0,j}-A_{j,t}) \qquad (j=1,2,3). \qquad (15.29)$$

The corresponding Euler differential equation (for $k=0$) can now be written out:

$$\begin{aligned}0&=\frac{1}{4\pi}\{(A_{0,j}-A_{j,t})_{,j}\}=\frac{1}{4\pi}\{\nabla^2A_0-(\nabla\cdot\mathbf{A})_{,t}\}\\&=\frac{1}{4\pi}\{\nabla^2A_0-A_{0,tt}\}\end{aligned} \qquad (15.30)$$

where the Lorentz condition was used in the last equality. The other three Euler differential equations for $k=1,2,3$ can also be similarly reduced to $\nabla^2A_k=A_{k,tt}$.

## 7. Higher Derivatives

Many variational problems for the behavior of continuous (solid, fluid, or gaseous) media involve Lagrangians which depend on partial derivatives higher than the first. An important example is the Germain–Kirchhoff theory for thin elastic plates. In this classical plate bending theory (to be derived in Chapter 17), we effectively treat a thin elastic flat plate before deformation as an area $A$ in the $x,y$-plane endowed with certain mechanical properties. When subject to transverse external loads (such as pressure) distributed over $A$, the plate will sag. We are concerned here only with loadings which give rise to a small transverse displacement field $w(x,y)$ throughout the plate area $A$. For simplicity, we will limit our discussion to plates which are *simply-connected* with an edge curve $C$.

The dependence of $w(x, y)$ on the distributed transverse surface load $p(x, y)$ in $A$ can be shown to result from minimizing the performance index

$$J[w] = \frac{1}{2}\iint_A \{D[(w_{,xx})^2 + (w_{,yy})^2 + 2\nu w_{,xx} w_{,yy} + 2(1-\nu)(w_{,xy})^2] - 2pw\}\,dx\,dy, \quad (15.31)$$

where $D$ is the known bending stiffness factor of the plate and $\nu$ is Poisson's ratio. The latter is a dimensionless number which characterizes the shortening in one ($y$) direction of the plate accompanying stretching in the other inplane ($x$) direction. That (15.31) is the appropriate functional to be minimized for thin plate problems will be justified in Chapter 17. It will be indicated there that $J[w]$ is the potential energy in the bent plate. In this section, we take both $D$ and $\nu$ to be uniform throughout the plate and limit further consideration to plates with clamped edge(s). In the context of the classical plate theory, a clamped edge requires

$$w = w_{,n} = 0, \quad (15.32)$$

along an edge curve $C$ of the midplane of the plate. Here, $w_{,n} \equiv \nabla w \cdot \mathbf{n}$ is the derivative of $w$ in the plane of, and normal to the edge curve of the plate.

For $J[w]$ to be a local minimum, we must have $\delta J = 0$. To deduce the Euler differential equation, we need to integrate by parts repeatedly by several applications of the two-dimensional divergence theorem:

$$\begin{aligned}
\delta J &= D\iint_A \{w_{,xx}\delta w_{,xx} + w_{,yy}\delta w_{,yy} + \nu(w_{,xx}\delta w_{,yy} + w_{,yy}\delta w_{,xx}) \\
&\qquad + 2(1-\nu)w_{,xy}\delta w_{,xy} - \frac{p}{D}\delta w\}\,dx\,dy \\
&= -\iint_A \{[M_{xx}\delta w_{,x} + M_{xy}\delta w_{,y}]_{,x} + [M_{xy}\delta w_{,x} + M_{yy}\delta w_{,y}]_{,y} \\
&\qquad - [Q_x\delta w]_{,x} - [Q_y\delta w]_{,y} - (D\nabla^2\nabla^2 w - p)\delta w\}\,dx\,dy \\
&= \iint_A (p - D\nabla^2\nabla^2 w)\delta w\,dx\,dy - \oint_C (M_{nn}\delta w_{,n} + M_{ns}\delta w_{,s} + Q_n\delta w)\,ds,
\end{aligned}$$

where

$$\begin{gathered}
M_{xx} = -D(w_{,xx} + \nu w_{,yy}), \qquad M_{yy} = -D(w_{,yy} + \nu w_{,xx}), \\
M_{xy} = M_{yx} = -D(1-\nu)w_{,xy}, \\
Q_x = -D(\nabla^2 w)_{,x}, \qquad Q_y = -D(\nabla^2 w)_{,y}, \\
\nabla^2(\ ) = (\ )_{,xx} + (\ )_{,yy}, \qquad Q_n = n_1 Q_x + n_2 Q_y, \qquad \text{etc.,} \quad (15.33)
\end{gathered}$$

and $(\ )_{,s}$ is a partial derivative of ( ) with respect to the arclength variable $s$. The subscripts $n$ and $t$ indicate the normal and tangential direction (in the plane of the region $A$) along the edge curve $C$, respectively. By the edge conditions (15.32), all admissible $w$ and their normal derivatives must vanish along $C$. (The tangential derivative $w_{,s}$ also vanishes whenever $w$ vanishes along $C$.) It follows that the line integral along $C$ vanishes. With $\delta w$ otherwise arbitrary, $\delta J = 0$ requires

$$D\nabla^2\nabla^2 w = p \quad (\text{in } A). \tag{15.34}$$

The partial differential operator $\nabla^2\nabla^2(\ ) \equiv \nabla^4(\ )$ is called the ***biharmonic*** (or ***bi-Lalpacion***) ***operator*** and the partial differential equation (15.34) is called the (inhomogeneous) biharmonic equation. This equation appears in many other important applications. Among them are the axisymmetric deformation of a three-dimensional elastic solid (Timosenko and Goodier, 1951) and the creeping motion of very viscous incompressible fluid flows in two dimensions (Lamb, 1945).

Evidently, an elementary exact solution of the fourth-order partial differential equation (15.34) which also satisfies the boundary conditions (15.32) is rarely possible even for a uniform pressure distribution for which $p$ is constant. On the other hand, a relatively simple approximate solution is possible for simple plate geometries. For a rectangular plate with edges at $x = 0, a$ $(0 \le y \le b)$ and $y = 0, b$ $(0 \le x \le a)$ we can take

$$w \approx c_0\left[1 - \cos\left(\frac{2\pi x}{a}\right)\right]\left[1 - \cos\left(\frac{2\pi y}{b}\right)\right] \equiv w_c \tag{15.35}$$

for some constant $c_0$. Note that the trial function $w_c$ satisfies the clamped edge conditions (15.32) at all four edges but not the partial differential equation (15.34). The variational formulation (15.31) leads us to pick an appropriate $c_0$ to minimize $J[w] \equiv J(c_0)$, giving the best approximation of the type (15.35) for the actual solution of the problem. Upon substituting (15.35) into (15.31), we obtain, after evaluation of the various integrals:

$$J[w_c] = Dc_0^2\left\{\frac{2\pi^4}{a^3b^3}\left[3(a^4 + b^4) + 2a^2b^2\right]\right\} - pc_0 ab \equiv J(c_0). \tag{15.36}$$

For a minimum $J[w_c]$, $c_0$ must be chosen so that $dJ/dc_0 \equiv J^{\bullet}(c_0) = 0$. This gives the following choice of $c_0$:

$$\widehat{c}_0 = \frac{pa^4b^4}{4\pi^4 D[3(a^4 + b^4) + 2a^2b^2]}. \tag{15.37}$$

$J(\widehat{c}_0)$ is, in fact, a minimum as we have $J^{\bullet\bullet}(\widehat{c}_0) > 0$.

The flat plate problem described above is one of the simplest yet scientifically (very) important problems in the mechanics of continuous media. However, a tractable exact solution for the problem is possible only for simple geometries and very special boundary conditions. This situation is typical for variational problems in several independent variables. For these problems, we are mainly concerned with approximate solutions for the Euler differential equation or the performance index $J[y]$ by direct (and semidirect) methods and with qualitative information about the solution. The most useful trial functions for a direct method of solution are generally acknowledged to be those constructed by the finite element method. We will sketch the basic idea of this solution technique for two-dimensional problems in the next two sections. The mathematics of the finite element method is discussed in Strang and Fix (1973).

## 8. Finite Elements in Two Dimensions

In higher dimensions, the method of finite elements offers flexibility and advantages beyond those we saw in the one-dimensional case. For one-dimensional problems, the localized coordinate functions are defined over subintervals $(a_i, a_i + 1)$, $i = 1, 2, \ldots, n$, of the solution domain $[a, b]$. The only flexibility available is in the location of the nodes $\{a_i\}$ or, what is the same, the subinterval lengths $a_{i+1} - a_i = h_i$. For two-dimensional problems over a region $R$ with a chosen set of nodes, we have the additional option of defining the local coordinate functions $\{\varphi_j(x, y)\}$ for the problem over triangular, rectangular, quadrilateral, or other shaped subdomains (including those with curved edges) with the given nodes as their vertices. For a given problem, different subdomains, called the *finite elements*, offer different advantages and disadvantages in computing an approximate solution. We will limit our discussion here to two-dimensional problems using triangular elements (such as those shown in Fig. (15.3)) and rectangular elements (such as those shown in Fig. 15.4).

### a. Linear Triangular Element

Suppose we introduce in the region $R$ a set of $N$ nodes with coordinates $(x_i, y_i)$, $i = 1, \ldots, N$, and connect the nodes so that, except for a narrow strip (called the skin) adjacent to the boundary, $R$ is subdivided into $M$ triangles of area $A_j, j = 1, \ldots, M$. As in the one-dimensional case, we define a *localized coordinate function* $\varphi_i(x, y)$ by requiring it to vanish in $R$ except in those triangles having the $i$th node as a vertex and to assume the value 1 at the node, i.e., $\varphi_i(x_i, y_i) = 1$. The simplest coordinate functions are linear inside each triangle having $(x_i, y_i)$ as a vertex; such a triangle is called a *linear triangular (finite)*

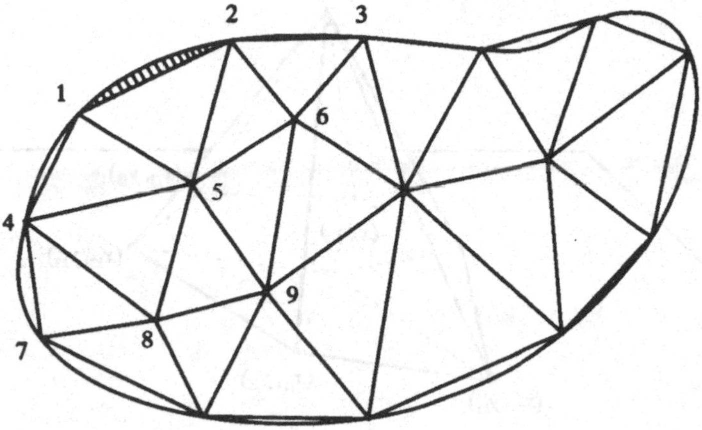


**Figure 15.3**

*element*. For $\varphi_i(x, y)$ to be a PWS function, it must be continuous across the common edges of these triangles and must vanish along edges opposite the $i$th node. Depending on the number of triangles having the $i$th node as a vertex, the graph of $\varphi_i(x, y)$ has a different geometrical configuration. Figures 15.5a and 15.5b show two possible configurations. The different triangularizations of $R$ available for the same set of nodes offers an analyst a new kind of flexibility not available in the one-dimensional finite element method.

Within the $m$th linear triangular element which has vertices at the $i$th, $j$th, and $k$th node (with $m$ being either $i$, $j$ or $k$), the trial function $u_N^{(m)}$ is linear in $x$ and $y$. The additional superscript $(m)$ is introduced here only to indicate the

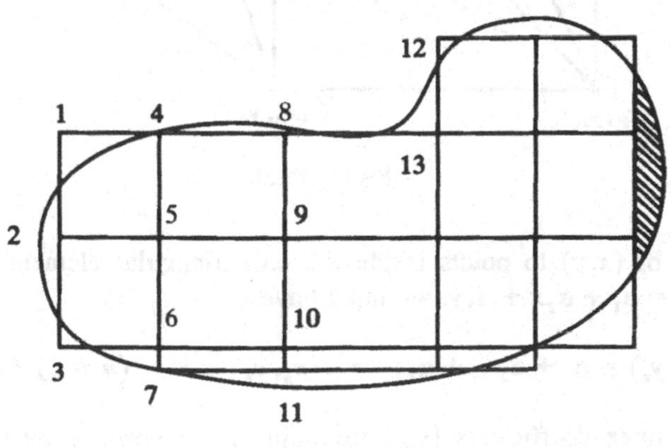


**Figure 15.4**

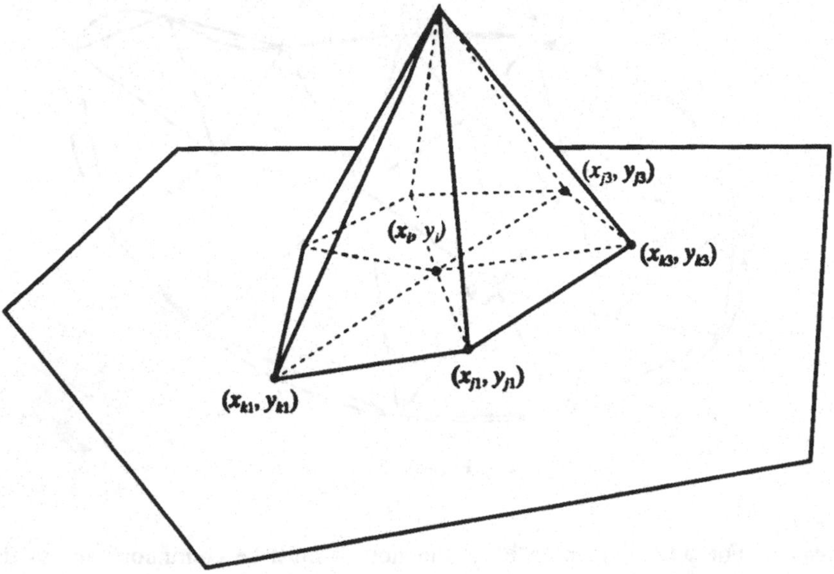


**Figure 15.5a**

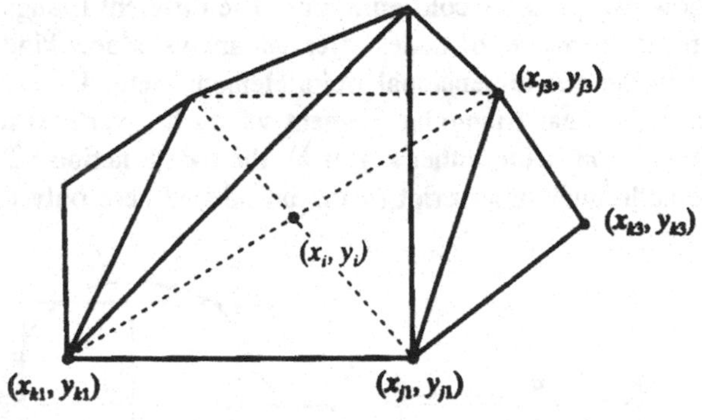


**Figure 15.5b**

restriction of $(x, y)$ to points inside the $m$th triangular element. If we write $u_N^{(m)}(x, y) = \alpha_1 + \alpha_2 x + \alpha_3 y$, we must have

$$u_N^{(m)}(x_n, y_n) = \alpha_1 + \alpha_2 x_n + \alpha_3 y_n = u(x_n, y_n) \equiv U_n \quad (n = i, j, k). \qquad (15.38)$$

Thus, the three coefficients $\{\alpha_\ell\}$ are uniquely determined by the values of $u(x, y)$ at the three vertices:

$$\alpha_1 = \frac{1}{2A_m}[(x_j y_k - x_k y_j)U_i + (x_k y_i - y_k x_i)U_j + (x_i y_j - y_i x_j)U_k],$$

$$\alpha_2 = \frac{1}{2A_m}[(y_j - y_k)U_i + (y_k - y_i)U_j + (y_i - y_j)U_k], \tag{15.39a}$$

$$\alpha_3 = \frac{1}{2A_m}[(x_k - x_j)U_i + (x_i - x_k)U_j + (x_j - x_i)U_k],$$

where

$$2A_m = \begin{vmatrix} 1 & x_i & y_i \\ 1 & x_j & y_j \\ 1 & x_k & y_k \end{vmatrix} = (x_j y_k - x_k y_j) - (x_i y_k - x_k y_i) + (x_i y_j - x_j y_i) \tag{15.39b}$$

is twice the area of the triangle. The trial function in the $m$th element can, therefore, be written

$$u_N^{(m)}(x,y) = U_i T_i^{(m)}(x,y) + U_j T_j^{(m)}(x,y) + U_k T_k^{(m)}(x,y) \tag{15.40}$$

where the *shape functions* $T_n^{(m)}(x,y)$, $n = i, j, k$, are defined by

$$T_n^{(m)}(x,y) = \frac{1}{2A_m}[a_n + b_n x + c_n y] \quad (n = i, j, k), \tag{15.41a}$$

where

$$\begin{aligned} a_i &= x_j y_k - x_k y_j, & b_i &= y_j - y_k, & c_i &= x_k - x_j; \\ a_j &= x_k y_i - x_i y_k, & b_j &= y_k - y_i, & c_j &= x_i - x_k; \\ a_k &= x_i y_j - x_j y_i, & b_k &= y_i - y_j, & c_k &= x_j - x_i. \end{aligned} \tag{15.41b}$$

Note that $T_n^{(m)}(x,y)$ is element specific as the nodes ($i$, $j$ and $k$) vary with the ($m$th) element.

At the $i$th node, we have

$$T_i^{(m)}(x_i, y_i) = \frac{1}{2A_m}[a_i + b_i x_i + c_i y_i] = 1, \tag{15.42a}$$

$$T_n^{(m)}(x_i, y_i) = \frac{1}{2A_m}[a_n + b_n x_i + c_n y_i] = 0 \quad (n = j, k). \tag{15.42b}$$

At the $j$th and $k$th nodes, we have also

$$T_i^{(m)}(x_p, y_p) = 0 \quad (p = j, k). \tag{15.42c}$$

Because $T_n^{(m)}(x, y)$ is linear in $x$ and $y$, (15.42c) implies $T_n^{(m)}(x, y) = 0$ along the side of the triangle opposite the $n$th node. It follows from the definition of a localized coordinate function that

$$\varphi_i(x, y) = \begin{cases} T_i^{(m)}(x, y) & \text{for } (x, y) \text{ in any } m\text{-th element} \\ & \text{with } i\text{th node as a vertex} \\ 0 & \text{elsewhere} \end{cases} \tag{15.43}$$

Two different coordinate functions are shown in Figures 15.5a and 15.5b. The trial function for the $N$ nodes triangularization of $R$ may be taken as a linear combination of the localized coordinate functions $\{\varphi_i(x, y)\}$:

$$u_N(x, y) = \sum_{i=1}^{N} U_i\, \varphi_i(x, y), \tag{15.44}$$

where the coefficient $U_i$ is the (approximate) value of $u(x, y)$ at the $i$th node. As we shall see later, it turns out to be more convenient in practice to work with the shape function representation (15.40) for $u_N(x, y)$ instead of the local coordinate function representation (15.44).

### *b. Bilinear Rectangular Element*

Suppose instead of triangularizing $R$, we subdivide $R$ into $M$ rectangles with areas $A_m$, $m = 1, \ldots, M$, and the shaded "skin" adjacent to the boundary, as in Figure 15.4. A typical rectangular element has the four nodes $i$, $j$, $k$ and $\ell$ as vertices and two sides of length $2a$ in the $x$-direction and two others of length $2b$ in the $y$-direction. It is conventional to assign the $i$th node to the vertex at the lower left-hand corner and the $\ell$th node diametrically opposite the $i$th node (see Fig. 15.6). Again, a set of localized coordinate functions $\{\varphi_i(x, y)\}$ is to be defined on $R$ for a given variational problem. The function $\varphi_i(x, y)$ associated with the $i$th node at $(x_i, y_i)$ is to vanish in $R$ except in those rectangles having that node as a vertex and to have the value 1 at the node, i.e., $\varphi_i(x_i, y_i) = 1$. The simplest form of $\varphi_i$ is one which is bilinear inside the four rectangles which share a vertex at the $i$th node. For $\varphi_i$ to be PWS, it must be continuous across the four common edges of these rectangles and it must vanish along those edges of these rectangles which do not have the $i$th node as an end point.

Inside the $m$th rectangle which has a vertex at the $i$th node, the trial function for the $N$ nodes subdivision can be taken as

$$u_N^{(m)}(x, y) = c_1 + c_2 s + c_3 t + c_4 st, \tag{15.45}$$

where $s$ and $t$ are coordinates in the $x$ and $y$ directions, respectively, measured from the $i$th node, i.e., $s = x - x_i$ and $t = y - y_i$. Along any line of constant

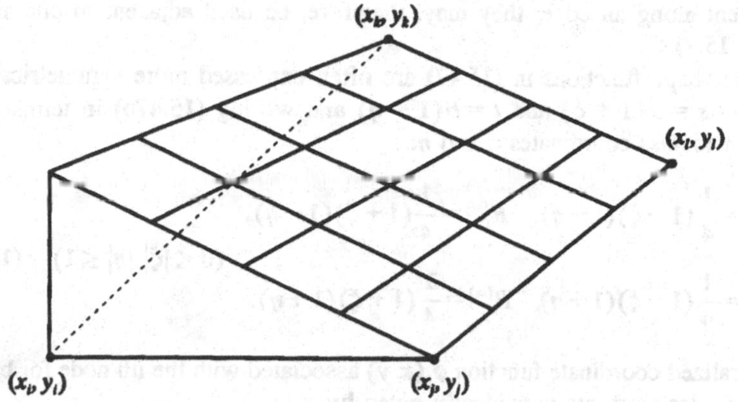


Figure 15.6

$t$, $u_N^{(m)}$ is linear in $s$. Similarly, along any line of constant $s$, $u_N^{(m)}$ is linear in $t$. Such an element is said to be *bilinear*.

For continuity, we need

$$u_N^{(m)}(s=0,t=0) = c_1 = u(x_i, y_i) \equiv U_i,$$
$$u_N^{(m)}(s=2a,t=0) = c_1 + 2ac_2 = u(x_j, y_j) \equiv U_j,$$
$$u_N^{(m)}(s=0,t=2b) = c_1 + 2bc_3 = u(x_k, y_k) \equiv U_k,$$
$$u_N^{(m)}(s=2a,t=2b) = c_1 + 2ac_2 + 2bc_3 + 4abc_4 \equiv u(x_\ell, y_\ell) \equiv U_\ell.$$

These four conditions determine $c_1, \ldots, c_4$. The resulting expressions for these coefficients, in turn, allow us to write

$$u_N^{(m)} = U_i R_i^{(m)}(x,y) + U_j R_j^{(m)}(x,y) + U_k R_k^{(m)}(x,y) + U_\ell R_\ell^{(m)}(x,y) \quad (15.47a)$$

where

$$R_i^{(m)}(x,y) = \left(1 - \frac{s}{2a}\right)\left(1 - \frac{t}{2b}\right), \qquad R_j^{(m)}(x,y) = \frac{s}{2a}\left(1 - \frac{t}{2b}\right),$$

$$R_k^{(m)}(x,y) = \frac{t}{2b}\left(1 - \frac{s}{2a}\right), \qquad R_\ell^{(m)}(x,y) = \frac{st}{4ab}. \quad (15.47b)$$

The quantities $R_n^{(m)}(x,y)$, $n = i, j, k, \ell$ are called *shape functions for bilinear rectangular elements* (Fig. 15.6). Note that $R_n^{(m)}$ is element specific as the nodes $(i, j, k, \ell)$ involved vary with $m$. For a fixed $n$, $R_n^{(m)}$ varies linearly along the two edges joining the $n$th node with its adjacent nodes in the rectangle. As such, a bilinear rectangular element is compatible with a linear triangular element along an edge; they may, therefore, be used adjacent to one another (Fig. 15.7).

The shape functions in (15.47) are often expressed more symmetrically by setting $s = a(1+\xi)$ and $t = b(1+\eta)$ and writing (15.47b) in terms of the dimensionless coordinates $\xi$ and $\eta$:

$$R_i^{(m)} = \frac{1}{4}(1-\xi)(1-\eta), \quad R_j^{(m)} = \frac{1}{4}(1+\xi)(1-\eta),$$
$$(0 \le |\xi|, |\eta| \le 1) \qquad (15.47c)$$
$$R_k^{(m)} = \frac{1}{4}(1-\xi)(1+\eta), \quad R_\ell^{(m)} = \frac{1}{4}(1+\xi)(1+\eta).$$

A localized coordinate function $\varphi_i(x, y)$ associated with the $i$th node for bilinear rectangular elements is evidently given by

$$\varphi_i(x,y) = \begin{cases} R_i^{(m)}(\xi,\eta) & \text{for } (x,y) \text{ in any } m\text{-th element with } i\text{th node as a vertex} \\ 0 & \text{elsewhere} \end{cases} \qquad (15.48)$$

with $\varphi_i(x_i, y_i) = R_i^{(m)}(\xi = -1, \eta = -1) = 1$.

Unlike the straight contour lines in a linear triangular element, a contour line in a bilinear rectangular element is generally curved. In fact, the nonvanishing part of $\varphi_i(x, y)$ has the appearance of the curved roof of a four-sided pagoda. Hence, the local basis functions of a bilinear rectangular element assembly are called *pagoda functions* [see Fig. 15.8].

It should now be evident that finite elements of other polygonal shapes or shapes with curved edges are also possible in both two and three dimensions. Localized coordinate functions of higher-degree polynomials may be used with

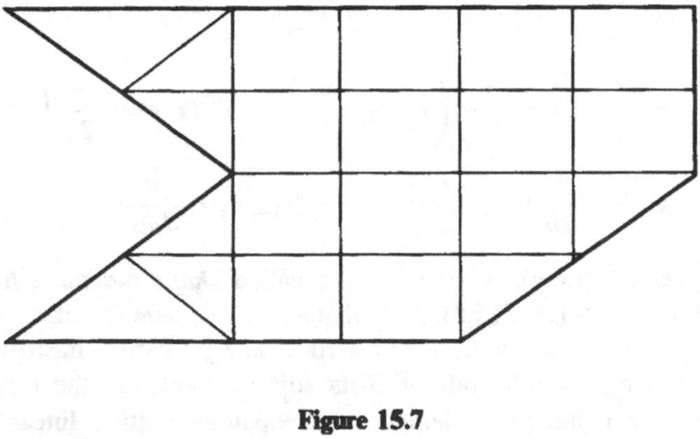

**Figure 15.7**

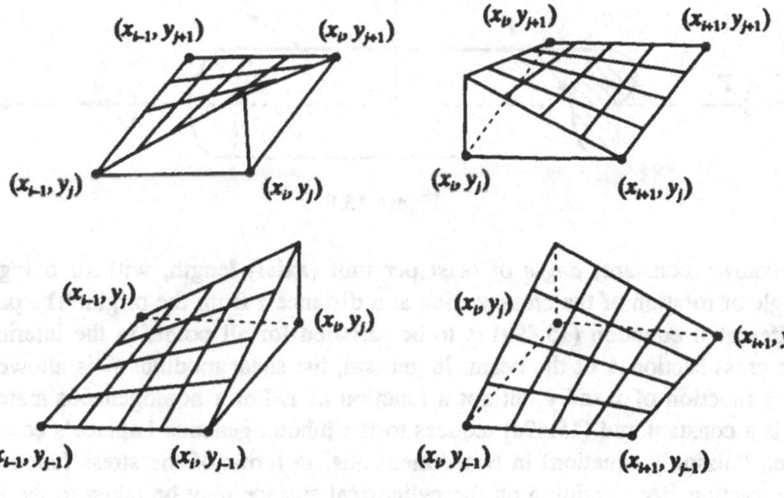


**Figure 15.8**

these elements as well. However, our detailed description of linear triangular elements and bilinear rectangular elements suffices to indicate how to construct other types of finite elements. We turn next to a discussion of how such finite elements may be used in conjunction with the direct method of the calculus of variations.

## 9. Torsion of an Elastic Bar

Consider a straight and slender elastic beam with the same cross section along its length (Fig. 15.9). The beam is free of internal loading and free of surface traction on its cylindrical surface. The two ends of the beam are subjected to equal and opposite torques $T$. We are interested here in the stresses of the beam away from the ends.

For an isotropic material with a shear modulus $G$, Saint-Venant's solution for this torsion problem (to be derived in Chapter 16) involves the solution of the following inhomogeneous second-order partial differential equation in two dimensions for a stress function $u(x, y)$

$$\left(\frac{u_{,x}}{G}\right)_{,x} + \left(\frac{u_{,y}}{G}\right)_{,y} = -2\theta, \tag{15.49a}$$

where $x$ and $y$ are Cartesian coordinates in the planes of the beam's (simply connected) cross sections (and $z$ is along the length of the beam) and $\theta$ is the

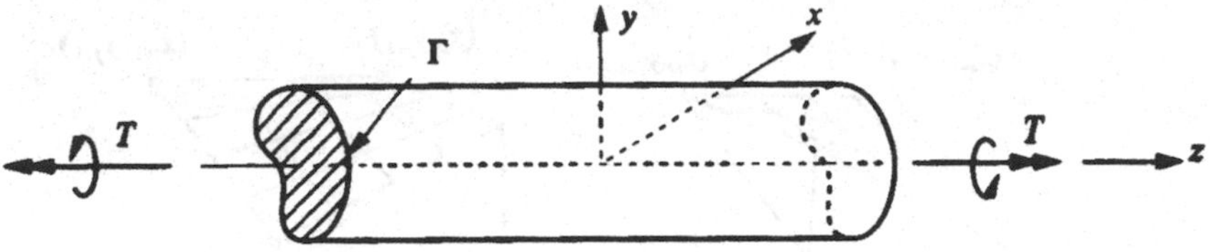


Figure 15.9

(unknown constant) *angle of twist* per unit (axial) length, with $\theta z$ being the angle of rotation of the cross section at a distance $z$ from the origin. The partial differential equation (15.49a) is to be satisfied for all points in the interior of the cross section $A$ of the beam. In general, the shear modulus $G$ is allowed to be a function of $x$ and $y$ but not a function of $z$. For a homogeneous material, $G$ is a constant and (15.49a) reduces to the inhomogeneous Laplace's equation (i.e., Poisson's equation) in two dimensions. In terms of the stress function $u$, the traction-free condition on the cylindrical surface may be taken in the form

$$u(x, y) = 0, \quad (x, y) \text{ on } \Gamma, \tag{15.49b}$$

where $\Gamma$ is the boundary of the (simply-connected) cross section $A$ of the beam. The boundary-value problem defined by (15.49a) and (15.49b) determines the stress function $u(x, y)$ uniquely. (Uniqueness is proved in the same way as for the Dirichlet problem sketched earlier.) Once we have $u(x, y)$, the stresses and deformation of the beam can be calculated from $u(x, y)$ by the process to be described in Chapter 16.

Except when $G$ is a constant, an exact solution of the boundary-value problem for $u(x, y)$ is generally not possible in terms of known functions. Even for a constant $G$, no explicit exact solution is possible if the shape of the cross section is not simple. In most cases, we have to settle for an approximate solution. We now apply the Rayleigh–Ritz method and construct an equivalent variational problem by a method analogous to the one introduced in Chapter 8 for one-dimensional problems. We first multiply (15.49a) by $\delta u$ and integrate over the cross section to obtain

$$\begin{aligned}
0 &= -\iint_A \left[\left(\frac{u_{,x}}{G}\right)_{,x} + \left(\frac{u_{,y}}{G}\right)_{,y} + 2\theta\right]\delta u \, dx\, dy \\
&= -\iint_A \left[\left(\delta u \frac{u_{,x}}{G}\right)_{,x} + \left(\delta u \frac{u_{,y}}{G}\right)_{,y} - \frac{u_{,x}\delta u_{,x} + u_{,y}\delta u_{,y}}{G} + 2\theta\delta u\right] dx\, dy \\
&= -\oint_\Gamma \left[\frac{1}{G} u_{,n} \delta u\right] ds + \delta \iint_A \left[\frac{1}{2G}(u_{,x}^2 + u_{,y}^2) - 2\theta u\right] dx\, dy
\end{aligned}$$

where the line integral is taken over the closed boundary $\Gamma$ of the cross section. Because $u$ is prescribed on the boundary, we have $\delta u = 0$ there. In that case, we conclude that

*The second-order partial differential equation (15.49a) is the Euler differential equation of*

$$J[u] = \frac{1}{2}\iint_A \left[\frac{1}{G}(u_{,x}^2 + u_{,y}^2) - 4\theta u\right] dx\, dy \qquad (15.50)$$

*subject to (15.49b).*

We now apply the Rayleigh–Ritz method to the variational problem above for the case of a *square cross section*. We will use an assembly of $N$ bilinear rectangular elements. With the $z$-axis through the centroid of the square, the solution $u(x, y)$ is symmetric about both the $x$-axis and the $y$-axis. We thus need only the solution for that part of the cross section in the first quadrant of the $x, y$-plane.

For the $m$th element of our finite element assembly, we have from (15.47a)

$$u_N^{(m)}(x,y) = U_i R_i^{(m)}(x,y) + U_j R_j^{(m)}(x,y) + U_k R_k^{(m)}(x,y) + U_\ell R_\ell^{(m)}(x,y) \qquad (15.51a)$$

$$= [R_i^{(m)}, R_j^{(m)}, R_k^{(m)}, R_\ell^{(m)}]\begin{bmatrix} U_i \\ U_j \\ U_k \\ U_\ell \end{bmatrix} \equiv \mathbf{R}^{(m)}\cdot\mathbf{U}^{(m)}$$

with

$$\frac{\partial u_N^{(m)}}{\partial x} = [R_{i,x}^{(m)}, R_{j,x}^{(m)}, R_{k,x}^{(m)}, R_{\ell,x}^{(m)}]\,\mathbf{U}^{(m)}$$

$$= \left[-\frac{1}{2a}\left(1-\frac{t}{2b}\right), \frac{1}{2a}\left(1-\frac{t}{2b}\right), -\frac{t}{4ab}, \frac{t}{4ab}\right]\mathbf{U}^{(m)} \equiv \mathbf{R}_x^{(m)}\cdot\mathbf{U}^{(m)}, \qquad (15.51b)$$

$$\frac{\partial u_N^{(m)}}{\partial y} = \left[-\frac{1}{2b}\left(1-\frac{s}{2a}\right), -\frac{s}{4ab}, \frac{1}{2b}\left(1-\frac{s}{2a}\right), \frac{s}{4ab}\right]\mathbf{U}^{(m)}$$

$$\equiv \mathbf{R}_y^{(m)}\cdot\mathbf{U}^{(m)}. \qquad (15.51c)$$

For $(x, y)$ outside the $m$th element, $u_N(x, y)$ does not involve $\mathbf{R}^{(m)}$. At the same time, $u_N(x, y)$ does not involve $\mathbf{R}^{(n)}$, $n \neq m$, for $(x, y)$ inside the $m$th element. It follows that the integral $J[u_N]$ is a sum of integrals of the type

$$\frac{1}{2}\iint_{\mathcal{A}_m}\left\{[\mathbf{U}^{(m)}]^T\frac{\mathbf{R}_x^{(m)}[\mathbf{R}_x^{(m)}]^T+\mathbf{R}_y^{(m)}[\mathbf{R}_y^{(m)}]^T}{G}\mathbf{U}^{(m)}-4\theta\mathbf{R}^{(m)}\cdot\mathbf{U}^{(m)}\right\}dx\,dy$$

$$\equiv\frac{1}{2}[\mathbf{U}^{(m)}]^T B^{(m)}\mathbf{U}^{(m)}-2\theta\mathbf{f}^{(m)}\cdot\mathbf{U}^{(m)},$$

where the symmetric matrix $B^{(m)}$ is given by

$$B^{(m)}=\iint_{\mathcal{A}_m}\frac{1}{G}\{\mathbf{R}_x^{(m)}[\mathbf{R}_x^{(m)}]^T+\mathbf{R}_y^{(m)}[\mathbf{R}_y^{(m)}]^T\}\,dx\,dy$$

$$=\begin{bmatrix} B_{ii}^{(m)} & B_{ij}^{(m)} & B_{ik}^{(m)} & B_{i\ell}^{(m)} \\ & B_{jj}^{(m)} & B_{jk}^{(m)} & B_{j\ell}^{(m)} \\ & & B_{kk}^{(m)} & B_{k\ell}^{(m)} \\ & & & B_{\ell\ell}^{(m)} \end{bmatrix}\tag{15.52a}$$

and

$$\mathbf{f}^{(m)}=\iint_{\mathcal{A}_m}\mathbf{R}^{(m)}\,dx\,dy,\tag{15.52b}$$

one integral for each element. Altogether, we have

$$J[u]=\sum_{m=1}^{M}\left[\frac{1}{2}\mathbf{U}^{(m)}\cdot\{B^{(m)}\mathbf{U}^{(m)}\}-2\theta\mathbf{f}^{(m)}\cdot\mathbf{U}^{(m)}\right]$$

$$\equiv J(U_1,U_2,\ldots,U_N)\tag{15.52c}$$

where the sum is taken over the $M$ elements of the finite element assembly. The four nodes involved in an element differ from element to element; two elements should have no more than two nodes in common.

In expression (15.52c) for $J[u]$, those $U_n$ for nodes on the boundary curve of the cross section are prescribed by (15.49b) and should be set to zero along $\Gamma$ (corresponding to the cylindrical surface of the beam). Let $\overline{N}\ (\leq N)$ be the number of the remaining free nodes at which the value of $u(x,y)$ is not known. For the Rayleigh–Ritz method, we choose the $\overline{N}$ remaining unknowns $\{U_n\}\equiv\overline{\mathbf{U}}$ to extremize $J(\overline{\mathbf{U}})$. This requires

$$\sum_{m=1}^{M}\overline{B}^{(m)}\overline{\mathbf{U}}^{(m)}=2\theta\sum_{m=1}^{M}\overline{\mathbf{f}}^{(m)},\tag{15.53a}$$

where $\overline{B}^{(m)}$ is $B^{(m)}$ with the rows and columns associated with the known $U_n$'s deleted and $\overline{f}^{(m)}$ is $f^{(m)}$ adjusted to include terms from known $U_n$'s. The vector equation (13.53a) may be written in the form

$$B\overline{\mathbf{U}} = 2\theta\overline{\mathbf{f}} = 2\theta\{\bar{f}_1, \ldots, \bar{f}_{\overline{N}}\}^T, \tag{15.53b}$$

where $B$ is now an $\overline{N} \times \overline{N}$ matrix and $\overline{\mathbf{U}}$ and $\overline{\mathbf{f}}$ are $\overline{N}$-vectors. The notation for the elements of the matrix $B^{(m)}$ used in (15.52a) was designed to facilitate the transition from (15.53a) to (15.53b). The latter is a linear system of $\overline{N}$ equation for the $\overline{N}$ unknowns $\{U_n\}$.

Note that we could have obtained the same system by proceeding as in the one-dimensional case in Chapter 8 by using

$$u_N(x, y) = \sum_{n=1}^{N} U_n \varphi_n(x, y) \tag{15.54}$$

[instead of $\{u_N^{(m)}\}$ (15.51a)] in the performance index (15.50) and carrying out the integration [which is possible when the integrand is polynomial in the unknown $u(x, y)$ and its various partial derivatives]. The result is again given by (15.53). However, when automating these calculations for more complex problems, it is much more convenient to work with the individual elements and shape functions instead of the coordinate functions as in (15.54).

To illustrate the actual steps leading to (15.53b), consider a square cross section with edges of length 2. We partition the quarter of this square cross section in the first quadrant into four congruent square elements with edges $2a = 2b = 1/2$ (Fig. 15.10). Though this choice of finite element assembly is not likely to produce an accurate approximate solution of our problem, it is adequate for illustrating the solution process.

Given $s = x - x_i$ and $t = y - y_i$, we have for a typical element and a constant $G$:

$$\iint_{A_m} \frac{1}{G} \mathbf{R}_x^{(m)} [\mathbf{R}_x^{(m)}]^T dx\, dy$$

$$= \int_0^{1/2}\int_0^{1/2} \frac{1}{G} \begin{bmatrix} 4(1-2t)^2 & -4(1-2t)^2 & 8t(1-2t) & -8(1-2t) \\ & 4(1-2t)^2 & -8t(1-2t) & 8(1-2t) \\ & & 16t^2 & -16t^2 \\ & & & 16t^2 \end{bmatrix} ds\, dt$$

$$= \frac{1}{6G} \begin{bmatrix} 2 & -2 & 1 & -1 \\ -2 & 2 & -1 & 1 \\ 1 & -1 & 2 & -2 \\ -1 & 1 & -2 & 2 \end{bmatrix}. \tag{15.55a}$$

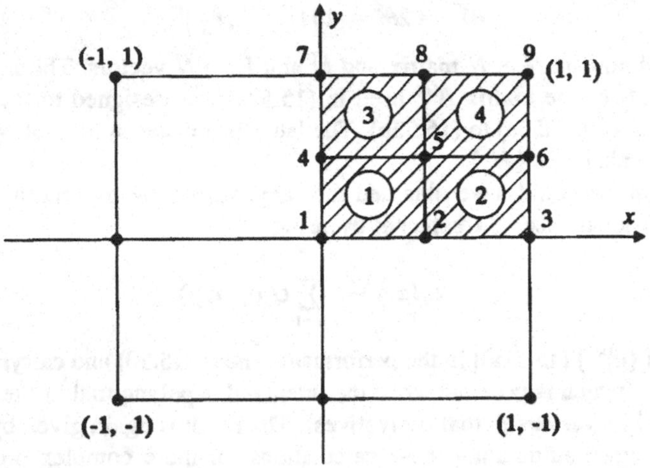


Figure 15.10

The integral of $\mathbf{R}_y^{(m)}[\mathbf{R}_y^{(m)}]^T/G$ is similarly obtained:

$$\iint_{\Omega_m} \frac{1}{G}\mathbf{R}_y^{(m)}[\mathbf{R}_y^{(m)}]^T dx\,dy = \frac{1}{6G}\begin{bmatrix} 2 & 1 & -2 & -1 \\ 1 & 2 & -1 & -2 \\ -2 & -1 & 2 & 1 \\ -1 & -2 & 1 & 2 \end{bmatrix}. \tag{15.55b}$$

for a constant $G$. The two integrals are then combined to give

$$B^{(m)} = \frac{1}{6G}\begin{bmatrix} 4 & -1 & -1 & -2 \\ -1 & 4 & -2 & -1 \\ -1 & -2 & 4 & -1 \\ -2 & -1 & -1 & 4 \end{bmatrix}. \tag{15.55c}$$

For the present problem, the element matrix $B^{(m)}$ is the same for all elements. However, the nodes involved are different. For element 1 [see Fig. 15.10], the nodes involved are 1, 2, 4, and 5. Correspondingly, we have

$$B^{(1)} = \begin{bmatrix} B_{11} & B_{12} & B_{14} & B_{15} \\ & B_{22} & B_{24} & B_{25} \\ & & B_{44} & B_{45} \\ & & & B_{55} \end{bmatrix}. \tag{15.55d}$$

On the other hand, we have for element 4 (with nodes 5, 6, 8, and 9),

$$B^{(4)} = \begin{bmatrix} B_{55} & B_{56} & B_{58} & B_{59} \\ & B_{66} & B_{68} & B_{69} \\ & & B_{88} & B_{89} \\ & & & B_{99} \end{bmatrix}. \tag{15.55d}$$

We also need

$$\mathbf{f}^{(m)} = \iint_{A_m} \mathbf{R}^{(m)}\,dx\,dy = \int_0^{1/2}\int_0^{1/2} \begin{bmatrix} (1-2s)(1-2t) \\ 2s(1-2t) \\ 2t(1-2s) \\ 4st \end{bmatrix} ds\,dt.$$

For the bilinear square element with $2a = 2b = 1/2$, we obtain

$$\mathbf{f}^{(m)} = \int_0^{1/2}\int_0^{1/2} \begin{bmatrix} (1-2s)(1-2t) \\ 2s(1-2t) \\ 2t(1-2s) \\ 4st \end{bmatrix} ds\,dt = \frac{1}{16}\begin{bmatrix} 1 \\ 1 \\ 1 \\ 1 \end{bmatrix}. \tag{15.56}$$

As well, the vector $\mathbf{f}^{(m)}$ is the same for all elements in the assembly, but the nodes involved differ from element to element. For instance, we have

$$\mathbf{f}^{(1)} = \begin{bmatrix} f_1 \\ f_2 \\ f_4 \\ f_5 \end{bmatrix}, \qquad \mathbf{f}^{(4)} = \begin{bmatrix} f_5 \\ f_6 \\ f_8 \\ f_9 \end{bmatrix}.$$

The next step is to assemble the $\overline{N} \times \overline{N}$ matrix $\overline{B}$ and the $\overline{N}$-vector $\overline{\mathbf{f}}$ for the linear system (15.53b). For our problem, there are four elements ($M = 4$) and a total of nine nodes ($N = 9$). Of these, nodes 3, 6, 7, 8, and 9 are on the boundary and hence prescribed (to be zero). This leaves us with four unknowns $U_n$ for $n = 1, 2, 4$, and 5 so that $\overline{N} = 4$ and $\overline{\mathbf{U}} = [U_1, U_2, U_4, U_5]^T$. The $4 \times 4$ matrix $\overline{B}$ is, therefore, a "sum" of all the element matrices $\overline{B}^{(m)}$ which involve these four nodes, all treated as $4 \times 4$ matrices:

$$
\begin{aligned}
\overline{B} &= \overline{B}^{(1)} + \overline{B}^{(2)} + \overline{B}^{(3)} + \overline{B}^{(4)} \\
&= \begin{bmatrix} B_{11}^{(1)} & B_{12}^{(1)} & B_{14}^{(1)} & B_{15}^{(1)} \\ & B_{22}^{(1)} & B_{24}^{(1)} & B_{25}^{(1)} \\ & & B_{44}^{(1)} & B_{45}^{(1)} \\ & & & B_{55}^{(1)} \end{bmatrix} + \begin{bmatrix} 0 & 0 & 0 & 0 \\ & B_{22}^{(2)} & 0 & B_{25}^{(2)} \\ & & 0 & 0 \\ & & & B_{55}^{(2)} \end{bmatrix} \\
&\quad + \begin{bmatrix} 0 & 0 & 0 & 0 \\ & 0 & 0 & 0 \\ & & B_{44}^{(3)} & B_{45}^{(3)} \\ & & & B_{55}^{(3)} \end{bmatrix} + \begin{bmatrix} 0 & 0 & 0 & 0 \\ 0 & 0 & 0 & 0 \\ 0 & 0 & 0 & 0 \\ 0 & 0 & 0 & B_{55}^{(4)} \end{bmatrix} \\
&= \frac{1}{6G}\begin{bmatrix} 4 & -1 & -1 & -2 \\ & 4 & -2 & -1 \\ & & 4 & -1 \\ & & & 4 \end{bmatrix} + \frac{1}{6G}\begin{bmatrix} 0 & 0 & 0 & 0 \\ & 4 & 0 & -1 \\ & & 0 & 0 \\ & & & 4 \end{bmatrix} \\
&\quad + \frac{1}{6G}\begin{bmatrix} 0 & 0 & 0 & 0 \\ & 0 & 0 & 0 \\ & & 4 & -1 \\ & & & 4 \end{bmatrix} + \frac{1}{6G}\begin{bmatrix} 0 & 0 & 0 & 0 \\ & 0 & 0 & 0 \\ & & 0 & 0 \\ & & & 4 \end{bmatrix} \\
&= \frac{1}{6G}\begin{bmatrix} 4 & -1 & -1 & -2 \\ & 8 & -2 & -2 \\ & & 8 & -2 \\ & & & 16 \end{bmatrix}.
\end{aligned}
\tag{15.57a}
$$

The $4 \times 1$ vector $\bar{\mathbf{f}}$ is a "sum" of all the element vectors $\bar{\mathbf{f}}^{(m)}$ which involve nodes 1, 2, 4, and 5, all treated as $4 \times 1$ vectors:

$$\bar{\mathbf{f}} = \bar{\mathbf{f}}^{(1)} + \bar{\mathbf{f}}^{(2)} + \bar{\mathbf{f}}^{(3)} + \bar{\mathbf{f}}^{(4)}$$

$$= \begin{bmatrix} f_1 \\ f_2 \\ f_4 \\ f_5 \end{bmatrix} + \begin{bmatrix} 0 \\ f_2 - B_{23}u_3 - B_{26}u_6 \\ 0 \\ f_5 - B_{53}u_3 - B_{56}u_6 \end{bmatrix} + \begin{bmatrix} 0 \\ 0 \\ f_4 - B_{47}u_7 - B_{48}u_8 \\ f_5 - B_{57}u_7 - B_{58}u_8 \end{bmatrix}$$

$$+ \begin{bmatrix} 0 \\ 0 \\ 0 \\ f_5 - B_{56}u_6 - B_{58}u_8 - B_{59}u_9 \end{bmatrix} = \frac{1}{16}\begin{bmatrix} 1 \\ 2 \\ 2 \\ 4 \end{bmatrix}. \tag{15.57b}$$

With (15.57a) and (15.57b), the linear system (15.53b) becomes

$$\begin{bmatrix} 4 & -1 & -1 & -2 \\ & 8 & -2 & -2 \\ & & 8 & -2 \\ & & & 16 \end{bmatrix} \overline{\mathbf{U}} = \frac{3G\theta}{4}\begin{bmatrix} 1 \\ 2 \\ 2 \\ 4 \end{bmatrix} \tag{15.53c}$$

which can be solved to give an approximate solution for $\overline{\mathbf{U}} = (U_1, U_2, U_4, U_5)^T$.

As we noted earlier, the approximate solution obtained above is not likely to give an accurate approximation of the exact solution. We expect, however, the accuracy of the finite element solution to improve with more (smaller) elements and/or higher-degree polynomial elements. This is confirmed in Strang and Fix (1973) and elsewhere.

## 10. Pointwise Equality Constraints

Our discussion in Section 7 of variational problems involving partial derivatives of higher order in the Lagrangian has been confined to specific applications. No effort will be made in this volume to obtain results for the general class of problems with higher derivatives. As with the one-variable case, a problem with higher derivatives can be reduced to an equivalent problem involving several unknowns but no higher partial derivative than first. By setting $w_{,x} = u$ and $w_{,y} = v$, we can write the Lagrangian of (15.31) as

$$F(w, w_{,x}, w_{,y}, w_{,xx}, w_{yy}, w_{,xy}) = F(w, u, v, u_{,x}, v_{,y}, u_{,y}, v_{,x}) \qquad (15.58)$$

in which $u_{,y} = w_{,xy} = v_{,x}$. With only $u, v, w$, and their first partial derivatives as its arguments, $F$ now belongs to the same class of Lagrangians for which we have [XV.3]. However, that result is not directly applicable as we learned from the corresponding one independent variable problem discussed in Chapter 3. The equations which define the new variables $u = w_{,x}$ and $v = w_{,y}$ must be incorporated into our analysis as equality constraints for the new problem. Hence, we need multiplier rules for problems of several variables.

Constraints on variational problems may also occur naturally and not as the result of eliminating higher derivatives. As an example, we recall the problem of electromagnetic fields in Section 5. Equations (15.22) and (15.23) which give **E** and **B** in terms of **A** and $A_0$ are really constraints for (15.27); so is the Lorentz condition (15.26). We are fortunate that (15.22) and (15.23) can be used to directly eliminate **E** and **B** from (15.27). The use of the Lorentz condition in (15.30) and (15.24) is also straightforward. Hence, multiplier rules were not necessary for that problem.

On the other hand, many important practical problems have similar constraints which cannot be incorporated so directly. We will see some of these in the next two chapters. Here, we will work out the multiplier rule for the simplest problem of this type. Multiplier rules for other situations can be developed by the same approach. The simplest problem in higher dimensions with constraints is to minimize

$$J[\mathbf{u}] = \iint_A F(\mathbf{x}, \mathbf{u}, \mathbf{u}_{,1}, \mathbf{u}_{,2})\, dx_1\, dx_2 \qquad (15.59)$$

subject to the single pointwise equality constraint

$$\varphi(\mathbf{x}, \mathbf{u}, \mathbf{u}_{,1}, \mathbf{u}_{,2}) = 0 \quad (\mathbf{x} \textit{ in } A) \qquad (15.60)$$

and the prescribed boundary condition on $\mathbf{u}$ along the boundary curve $C$ of $A$. In general, $\mathbf{u}$ has $n$ components so that we have $n$ unknown functions to be determined. The special case $n = 1$ is not particularly meaningful as the partial differential equation (15.60) and the boundary condition on $C$ generally specify $u(\mathbf{x})$ so that there is really no choice in $u$ for the minimization of $J[u]$. The simplest nontrivial case is, therefore, for $n = 2$.

For the case of two unknowns and two independent variables, we can proceed as in Chapter 11 and calculate the first variations of $J$ and $\varphi$. With $v_{kj} = u_{k,j} (\neq u_{j,k} = v_{jk})$ in $F$ and $\varphi$, the conditions $\delta J = 0$ and $\delta\varphi = 0$ require

$$\delta J = \iint_A \{F_{,u_1}\delta u_1 + F_{,v_{11}}\delta u_{1,1} + F_{,v_{12}}\delta u_{1,2} + F_{,u_2}\delta u_2 + F_{,v_{21}}\delta u_{2,1} + F_{,v_{22}}\delta u_{2,2}\}\,dx_1\,dx_2$$

$$= \iint_A \sum_{k=1}^{2}\left[F_{,u_k} - \sum_{j=1}^{2}(F_{,v_{kj}})_{,j}\right]\delta u_k\,dx_1\,dx_2 = 0, \qquad (15.61)$$

$$\delta\varphi = \sum_{k=1}^{2}[\varphi_{,u_k}\delta u_k + \varphi_{,v_{k1}}\delta u_{k,1} + \varphi_{,v_{k2}}\delta u_{k,2}] = 0 \qquad (15.62)$$

where we have "integrated by parts" and made use of $\delta\mathbf{u} = 0$ on the boundary curve to eliminate a line integral in (15.61). In $\delta J$, the two variations $\delta u_1$ and $\delta u_2$ are not completely independent; they are related by (15.62). As the latter holds for all $\mathbf{x}$ in $A$, we have

$$\iint_A \sum_{k=1}^{2}[\varphi_{,u_k}\delta u_k + \varphi_{,v_{k1}}\delta u_{k,1} + \varphi_{,v_{k2}}\delta u_{k,2}]\,\lambda(\mathbf{x})\,dx_1\,dx_2$$

$$= \iint_A \sum_{k=1}^{2}\{\lambda\varphi_{,u_k} - (\lambda\varphi_{,v_{k1}})_{,1} - (\lambda_{,v_{k2}})_{,2}\}\,\delta u_k\,dx_1\,dx_2 = 0,$$

where we have again "integrated by parts" and used $\delta\mathbf{u} = 0$ on the boundary to eliminate the line integral. We now add the above to (15.61) to obtain

$$\delta J + \iint_A \lambda\,\delta\varphi\,dx_1\,dx_2$$

$$= \iint_A \sum_{k=1}^{2}\left[(F_{,u_k} + \lambda\varphi_{,u_k}) - \sum_{j=1}^{2}(F_{,v_{kj}} + \lambda\varphi_{,v_{kj}})_{,j}\right]\delta u_k\,dx_1\,dx_2 = 0.$$

To avoid dealing with the interdependence of $\delta u_1$ and $\delta u_2$, we choose $\lambda$ so that the coefficient of $\delta u_2$ vanishes, i.e.,

$$(F_{,u_2} + \lambda\varphi_{,u_2}) - \sum_{j=1}^{2}(F_{,v_{2j}} + \lambda\varphi_{,v_{2j}})_{,j} = 0.$$

We can then vary $\delta u_1$ arbitrarily so that

$$(F_{,u_1} + \lambda\varphi_{,u_1}) - \sum_{j=1}^{2}(F_{,v_{1j}} + \lambda\varphi_{,v_{1j}})_{,j} = 0$$

as well.

The development above suggests the following multiplier rule:

**[XV.5]** *If the pair* $(\widehat{\mathbf{u}}_1, \widehat{\mathbf{u}}_2)$ *renders (15.59) stationary, subject to the constraint (15.60) and to the prescribed boundary condition on* $\mathbf{u}$, *and is not an extremal pair of* $\varphi(\mathbf{x}, \mathbf{u}, \mathbf{v}_{ij})$, *then it is necessarily an extremal pair for* $F - \lambda\varphi(\mathbf{x}, \mathbf{u}, \mathbf{v}_{ij})$ *for some (continuous) Lagrange multiplier* $\lambda(\mathbf{x})$ *i.e., the triplet* $(\widehat{\mathbf{u}}_1, \widehat{\mathbf{u}}_2, \lambda)$ *satisfies the two Euler differential equations*

$$(F + \lambda\varphi)_{,u_k} - \sum_{j=1}^{2} [(F + \lambda\varphi)_{,v_{kj}}]_{,j} = 0 \quad (k = 1, 2) \qquad (15.63)$$

*and the constraint* $\varphi = 0$.

If $\widehat{u}_1$ and $\widehat{u}_1$ constitute an extremal pair of an integral with $\varphi(\mathbf{x}, \mathbf{u}, v_{ij})$ as a Lagrangian, we have again an *abnormal case*. A more general multiplier rule which includes a multiplier $\lambda_0$ for $F$ is needed for such problems. Such a multiplier rule is similar to the corresponding situation in Chapters 10, 11, and 12. In particular, $\lambda_0$ may be set equal to one if it is not zero. The central mathematical issue (not discussed here) is again a proof of the existence of the required multiplier(s).

For more independent variables, the result [XV.5] applies for the normal case with the sum in (15.63) to range over the number of variables. For more unknowns, there will be more Euler differential equations of the form (15.63), one for each of the unknowns. Extension to more constraints of the same type is also straightforward. We simply introduce one more multiplier for each additional pointwise equality constraint and interpret $\lambda\varphi$ as the scalar product of two vector quantities. Applications of these results will be found in the next three chapters.

## 11. Isoperimetric Constraints

There is another class of constraints which cannot be incorporated directly into $J[\mathbf{u}]$ (without the use of a multiplier) to free the variational problem from constraints. Consider a simple closed curve $\Gamma_0$ in the $x, y$-plane and a surface $S$ given by $z(x, y)$ above the simply connected domain $A$ enclosed by $\Gamma_0$ with $z = 0$ along $\Gamma_0$. We know from (15.1) that the surface area of $S$ is given by

$$\Phi[z] \equiv \iint_A \sqrt{1 + z_{,x}^2 + z_{,y}^2}\, dx\, dy = S_0. \qquad (15.64)$$

Consider the variational problem of choosing $z(x, y)$ with a prescribed surface area $S_0$ to maximize the volume

$$V = \iint_A z\, dx\, dy \qquad (15.65)$$

*enclosed by the surface* and the planar area $A$. This is the higher-dimensional version of the isoperimetric problem of Chapter 10. It is simply not possible to incorporate the constraint (15.64) directly in (15.65) (as we did for the problem of electromagnetics). On the other hand, analogous rules similar to those in Chapter 10 can be established.

As the simplest class of isoperimetric problems in higher dimensions, we want to minimize

$$J[u] = \iint_A F(x_1, x_2, u, u_{,1}, u_{,2})\, dx_1\, dx_2 \tag{15.66}$$

subject to a single constraint

$$C[u] \equiv \iint_A G(x_1, x_2, u, u_{,1}, u_{,2})\, dx_1\, dx_2 - \gamma = 0 \tag{15.67}$$

and with $u$ prescribed on the boundary curve $\Gamma_0$ of the (simply-connected) region $A$ in the $x_1, x_2$-plane. The problem of maximizing the volume $V$ for a fixed surface area $S_0$ is of this form. For this class of problems, the following multiplier rule applies:

**[XV.6]** ***If*** $\mathbf{J[\hat{u}]}$ ***is an extremum and*** $\hat{\mathbf{u}}$ ***is not an extremal of*** $\mathbf{C[u]}$, ***then there is a nonzero constant*** $\boldsymbol{\lambda}$ ***for which***

$$\hat{L}_{,u} - \sum_{j=1}^{2} (\hat{L}_{,v_j})_{,j} = 0 \tag{15.68}$$

***where*** $\mathbf{L(x, u, v_1, v_2, \lambda) \equiv F(x, u, v_1, v_2) - \lambda G(x, u, v_1, v_2)}$ ***and*** $\hat{\mathbf{L}}(\mathbf{x}) = \mathbf{L(x, \hat{u}(x), \hat{u}_{,1}(x), \hat{u}_{,2}(x), \lambda)}$.

The existence of the multiplier $\lambda$ will not be proved here.

Extension of the [XV.6] to more independent variables is accomplished by letting the summation in (15.68) range over the number of independent variables. Extension to more constraints is accomplished by allowing $\lambda$ and $G$ be vector quantities and $\lambda G_{,v}$ be interpreted as a scalar product.

For the maximum volume problem, [XV.6] requires $\hat{z}$ and $\lambda$ to satisfy the Euler differential equation

$$1 + \lambda \left[\frac{\hat{z}_{,x}}{\hat{r}}\right]_{,x} + \lambda \left[\frac{\hat{z}_{,y}}{\hat{r}}\right]_{,y} = 0, \tag{15.69}$$

where $r = (1 + z_{,x}^2 + z_{,y}^2)^{1/2}$. The solution of the nonlinear partial differential equation (15.69) depends on the unknown constant $\lambda$. The constraint (15.64) provides the condition needed to determine this constant. For an arbitrary boundary $\Gamma_0$, (15.69) generally does not admit an exact solution in terms of known functions. Still, we can extract useful information from (15.69) itself. In terms of the mean curvature of the surface $S$ [see (15.4)], (15.69) for $\hat{z}(x, y)$ may be written as

$$\frac{1}{\rho_1} + \frac{1}{\rho_2} = -\frac{1}{\lambda}; \tag{15.70}$$

the extremal surface $\hat{z}$ must, therefore, be of constant mean curvature ($= -1/2\lambda$).

For the special case where the curve $\Gamma_0$ is a circle of radius $r_0$ centered at $(x_0, y_0)$, solution of the partial differential equations is

$$\hat{z} = [c_0^2 - (x - x_0)^2 - (y - y_0)^2]^{1/2} - [c_0^2 - r_0^2]^{1/2}. \tag{15.71}$$

Note that the above expression describes a spherical surface of radius $c_0$ with the center of the sphere at $(x_0, y_0, z_0)$ where $z_0 = -(c_0^2 - r_0^2)^{1/2}$. It can be verified by direct substitution that $\hat{z}$ satisfies the partial differential equation (15.69) if we take $c_0 = 2\lambda$. Along the boundary curve $\Gamma_0$, we have $(x - x_0)^2 + (y - y_0)^2 = r_0^2$ so that $\hat{z}$, as given by (15.71), satisfies the prescribed boundary condition along $\Gamma_0$, namely, $\hat{z} = 0$ there. The constraint (15.64) then gives a condition for $\lambda$ (or $c_0$):

$$\iint_A \frac{c_0\,dx\,dy}{\sqrt{c_0^2 - (x - x_0)^2 - (y - y_0)^2}} = S_0.$$

Upon evaluating the double integral over the region $(x - x_0)^2 + (y - y_0)^2 \le r_0^2$ (by doing the integration in polar coordinates), we obtain

$$c_0 = \frac{S_0/2\pi}{\sqrt{(S_0/\pi) - r_0^2}}, \qquad \lambda = \frac{c_0}{2} = \frac{S_0/4\pi}{\sqrt{(S_0/\pi) - r_0^2}}$$

and hence completely determine the admissible extremal (15.71).

## 12. Exercises

1. Find the minimal surface whose equation is expressible in the form $z(x, y) = f_1(x) + f_2(y)$.

2. $J[u] \equiv \iint_A F(x, y, u, u_{,x}, u_{,y})\,dx\,dy.$

   (a) With the help of Green's theorem and $u$ prescribed on the boundary curve $\Gamma_0$, obtain the following Euler differential equation of $J[u]$:

   $$(\hat{F}_{,v_x})_{,x} + (\hat{F}_{,v_y})_{,y} = \hat{F}_{,u},$$

   where $v_k \equiv u_{,k}$.

(b) If $u$ is not prescribed on $\Gamma_0$, obtain the Euler boundary condition for $u$.

3. (a) Specialize the two results above in the case

$$F = a(x,y)u_{,x}^2 + b(x,y)u_{,y}^2 - c(x,y)u^2$$

and further reduce the Euler BC for the case $a(x,y) = b(x,y)$.

(b) Specialize the Euler DE for $F = (1+x^2)u_{,x}^2 + 3u_{,y}^2 - e^y(u^2+u^3)$.

(c) Specialize the Euler DE for $F = u^2 u_{,x} + u_{,y}^2 + xu_{,x}u_{,y}$.

4. $$J[u] = \int_0^{2\pi}\int_a^b \left[u_{,r}^2 + \frac{1}{r^2}u_{,\theta}^2\right] r\,dr\,d\theta.$$

(a) Obtain the Euler DE for $J[u]$.

(b) Obtain the Euler BC along the two circular edges $r = a$ and $r = b$.

(c) Obtain the Euler BC along $r = b$ if a boundary integral $J_b[u]$ is added to $J[u]$ where

$$J_b[u] = -\int_0^{2\pi} \varphi(\theta,u)\,d\theta.$$

(d) Specialize the result of part (c) for $\varphi = f(\theta)u(b,\theta)$.

5. $$J[u] = \int_0^T\int_0^\ell \{[u_{,t}^2 - u_{,x}^2] + a(x,t)u\}\,dx\,dt.$$

(a) Obtain the Euler DE for $u(0,t) = u(\ell,t) = 0$, $u(x,0) = f(x)$, and $u(x,T) = g(x)$.

(b) Obtain the Euler BC if only $u(0,t) = 0$ and $u(x,0) = f(x)$ are prescribed.

6. (a) Different Lagrangian may give rise to the same Euler DE. We saw from (13.31) of Section 7 that the Lagrangian

$$F = (w_{,xx})^2 + (w_{,yy})^2 + 2\nu w_{,xx}w_{,yy} + 2(1-\nu)w_{,xy}^2$$

has as its Euler DE $\nabla^2\nabla^2 w = 0$ for any value of (constant) parameter $\nu$. Verify that the same biharmonic equation is also the Euler DE of

$$F = (w_{,xx} + w_{,yy})^2$$

(which corresponds to the special case $\nu = 0$).

(b) The two different Lagrangians above which give the same Euler DE lead to different Euler BC. Obtain the Euler BC for $F = (w_{,xx} + w_{,yy})^2$ if only $w$ is prescribed on $\Gamma_0$ and compare them to those implied in the expression for $\delta J$ in Section 7.

(c) Compare the single Euler BC for the two cases if $w_{,n}$ is prescribed along $\Gamma_0$.

7. (a) The transverse displacement $w(x, y)$ of a flat (stretched) membrane fixed at its edge(s) and subject to a normalized distributed pressure loading $p(x, y)$ is governed by the PDE $\nabla^2 w = p$. Formulate a basic problem which has this PDE as the Euler DE.

   (b) For a triangular membrane with vertices at $(0,0)$, $(1,0)$, and $(0,1)$ and with fixed edges so that $w = 0$ along the edges, use $w_1 = c_1 xy(x + y - 1)$ to get an approximate solution by the direct method. Determine $c_1$ for $p(x, y) = p_0$ (a constant).

8. Obtain the Euler DE for

$$J[u] = \iint_A [u_{,x}^2 + u_{,y}^2]\, dx\, dy$$

subject to

$$\iint_A u^2\, dx\, dy = 1, \qquad \iint_A u\, v\, dx\, dy = 0$$

for some prescribed function $v(x, y)$.

9. $J[u, v] = \iint_A \{u_{,x} + v_{,y}\}^2\, dx\, dy.$

   (a) Obtain the Euler DE for $J[u, v]$ subject to the constraint $u_{,y} = v_{,x}$ in $A$ and $u$, $v$ both specified on the boundary curve $\Gamma_0$.

   (b) Formulate an equivalent variational problem without the constraint $u_{,y} = v_{,x}$. [Hint: The condition $u_{,y} = v_{,x}$ is satisfied by $u = \psi_{,x}$ and $v = \psi_{,y}$.

10. For a triangular membrane with vertices at $(0,0)$, $(1,0)$, and $(0,1)$ and with all edges fixed so that $w = 0$ along the edges, we can apply the trial function method to the appropriate Rayleigh quotient to obtain approximate values of the lowest natural frequency of a vibrating membrane:

    (a) With $\nabla^2 w + \lambda w = 0$ [where $\lambda = \rho\omega^2 a^2 / T$ and $\nabla^2(\ ) = (\ )_{,xx} + (\ )_{,yy}$], deduce the Rayleigh quotient for the fixed edge case:

$$R[w] = \frac{\iint_A [w_{,x}^2 + w_{,y}^2]\,dx\,dy}{\iint_A w^2\,dx\,dy}.$$

(b) With $A$ being the area inside the triangle, it would be appropriate to take

$$w_n(x,y) = xy(x+y-1)(c_1 + c_2x + c_3y + c_4x^2 + c_5y^2 + c_6xy + \ldots).$$

Obtain $\lambda_1$ by minimizing $R[w_1]$ [Note: $\lambda_1 = 56$.]

(c) Repeat (b) for $R[w_3]$.

# 16

# Linear Theory of Elasticity

## 1. Continuum Mechanics and Elasticity Theory

A body of substance is generally made of a large number of molecules bound together by intermolecular forces. Under external influence (such as gravitational forces, electromagnetic forces, heat, etc.), these molecules will redistribute themselves to reach a state of equilibrium. This redistribution generally results in a *deformation* of the body which involves changes in the position and shape of the body. In principle, it is possible to determine this redistribution of molecular units by the laws of mechanics, electromagnetics, and thermodynamics. However, interaction among a large number of molecules is too complex for a detailed analysis of the activity and the final equilibrium position of individual molecules.

If we are only interested in the macroscopic behavior of the body at the level of our daily life experience, a much less detailed description and analysis of the substance often suffices. The less detailed (macroscopic) approach idealizes the substance as a mathematical continuum $V$ (a closed connected set of points $V$ in the three-dimensional physical space $E^3$) endowed with certain physical properties. In the absence of cracks and tears, the body is assumed to remain a continuum $V^*$ after deformation. As such, the deformation of a body of substance may be thought of as a mapping of $V$ in $E^3$. For most problems of interest and certainly for those studied in this course, we stipulate that the map be one-to-one.

For a given body acted on by external influence (also called *external loads* or *loading*), we would like to know the deformed configuration $V^*$ as well as the "internal mechanical reaction" (to be defined later as *stresses*) of the body to this deformation. In other words, we want to know the relevant mapping which sends $V$ into $V^*$. In general, this mapping, along with the material properties of the substance, tells us whether the strength requirements of a structural design are met or whether the intermolecular ties would be broken by the external loading leading to some kind of structural failure.

Starting from the macroscopic idealization of a body of substance, we can use the laws of physics and the macroscopic properties intrinsic to the particular substance along with mathematical methods to analyze the influence and predict the outcome of the action of the external loads on the body. Such an endeavor is called *Continuum Mechanics* or the *Mechanics of Deformable Bodies*. Occasionally, new mathematics may have to be developed to cope with a new problem. In fact, fluid and solid mechanics have been a major source of stimuli for the development of mathematics over the years.

The *Theory of Elasticity* is a branch of Continuum Mechanics dealing with deformable *solid* bodies having the physical property called *elasticity*. Roughly, a body is *elastic* if, when the external disturbances are removed, the deformed body will return to its original (undeformed) shape. A more specific description of elasticity will be given in Section 4. To varying degrees, this property is shared by a very wide range of substances, especially commonly used metals.

To understand the behavior of an elastic body under external loads, we can attack each problem ab initio and in an ad hoc manner. This approach has been and is still being employed by experienced applied scientists and engineers. However, its success often depends on the insight and experience of the analyst. An alternative approach would be to develop a general theory which is applicable to all problems involving elastic solids. In principle, this second approach makes the solution of a specific problem strictly an application of mathematical methods to a special case of the general theory. It is an approach generally favored by applied mathematicians and the first correct general continuum theory of elasticity was, in fact, developed by A.L. Cauchy, widely known in the mathematical community for his contributions to the theory of functions of complex variables. In the history of science, however, Cauchy's contributions to continuum mechanics are no less monumental. In this chapter, we outline the fundamentals of the classical *linear theory of elastostatics* as developed by Cauchy and Navier [see A.E.H. Love (1944) and J.R. Barber (1992) for example] as well as several variational formulations of this theory.

## 2. Components of Displacement and Strain

When subject to external loads, a body of substance will deform and (for elasostatics) reach a new equilibrium configuration. To extract quantitative information about this deformation, it is not enough to talk abstractly about mappings or transformations of a point set in Euclidean space. Ideally, we would like to know the mapping itself explicitly in terms of simple functions. As a first step toward the explicit determination of the specific mapping associated with a given set of external loads, let us describe a point in the body $V$ before deformation by a *position vector* $\mathbf{r} = x_1\mathbf{i}_1 + x_2\mathbf{i}_2 + x_3\mathbf{i}_3 \equiv x_k\mathbf{i}_k$ with reference to a set of Cartesian axes with directional unit vectors $\{\mathbf{i}_1, \mathbf{i}_2, \mathbf{i}_3\}$. In this chapter, we adopt the convention that expressions with *repeated indices will be summed over the range of the indices* and the conventional summation sign is therefore omitted. After deformation, the same point is now in a new position with a new position vector $\mathbf{r}^* = x_k^* \mathbf{i}_k$. For different points in $V$, the coordinates $\{x_1, x_2, x_3\}$ take on different numerical values. The corresponding coordinates of the new equilibrium position $\{x_1^*, x_2^*, x_3^*\}$ are evidently scalar functions of $\mathbf{x} = (x_1, x_2, x_3)$, i.e., $x_k^* = X_k^*(x_1, x_2, x_3)$ (and also time $t$ if we should be interested in the evolution of the deformation).

We now define a *displacement vector* $\mathbf{u}$ by

$$\mathbf{u} = u_k \mathbf{i}_k = \mathbf{r}^* - \mathbf{r} \tag{16.1}$$

so that $u_k = x_k^* - x_k$ (Fig. 16.1a). Evidently, we know all about the deformed configuration $V^*$ if we know $\mathbf{u}$ (because we know $\mathbf{r}$). For the special case $\mathbf{u} = \mathbf{u}_0$, where $\mathbf{u}_0$ is a constant vector, the body has simply undergone a rigid translation moving from one location to another. It has not deformed in the ordinary sense of the word and the elastic property of the body played no role in this change of position. A body experiences a genuine deformation and exhibits its elasticity only if there is a nonuniform or relative displacement of points in the body, i.e., only if the body changes its size or shape. This leads us to the more useful concept of deformation gradient or *strain*. Very crudely, strain may be thought of as a relative displacement per unit distance between any two points in a body. Evidently, the body suffers no strain if it experiences only a rigid motion, either translation or rotation.

To describe relative displacements quantitatively, we consider two points $P_1$ and $P_2$ in the undeformed body with position vector $\mathbf{r}$ and $\mathbf{r} + d\mathbf{r}$, respectively.

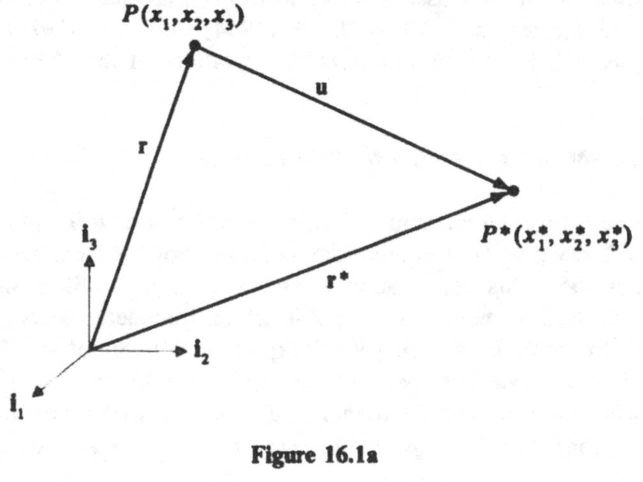


Figure 16.1a

After deformation, the two points have moved to new locations $P_1^*$ and $P_2^*$ with new position vectors $\mathbf{r}^*$ and $\mathbf{r}^* + d\mathbf{r}^*$, respectively (Fig. 16.1b). The distance between $P_1$ and $P_2$ is $ds = |d\mathbf{r}| = (d\mathbf{r} \cdot d\mathbf{r})^{1/2}$ and the distance between $P_1^*$ and $P_2^*$ is $ds^* = |d\mathbf{r}^*| = (d\mathbf{r}^* \cdot d\mathbf{r}^*)^{1/2}$. The difference $ds^* - ds$ is clearly the change in distance between the points due to deformation. But the relative displacement of these two points involves more than just this change of distance between them. For instance, it involves the orientation of one point relative to the other before and after deformation. It turns out that a complete description of the relative displacement is contained in the quantity $(ds^*) - (ds)^2 = d\mathbf{r}^* \cdot d\mathbf{r}^* - d\mathbf{r} \cdot d\mathbf{r}$. This quantity is used to define the *components of finite strain* $E_{jk}$:

$$(ds^*)^2 - (ds)^2 \equiv 2E_{jk}\, dx_j\, dx_k. \tag{16.2}$$

We will now relate the strain components to the displacement components $u_i$ introduced in (16.1). We limit our discussion to external loads which give

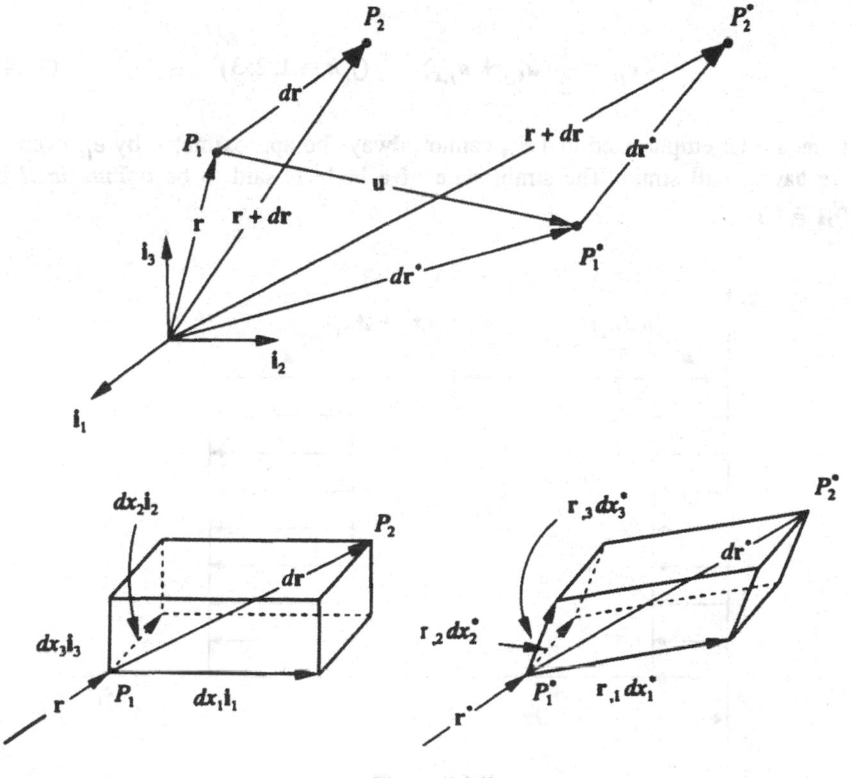


Figure 16.1b

rise only to (*continuous and differentiable*) *displacement fields* which have piecewise continuous second derivatives. In that case, we have

$$d\mathbf{r} = \mathbf{r}_{,j}dx_j = x_{k,j}dx_j\mathbf{i}_k = \delta_{kj}dx_j\,\mathbf{i}_k = dx_k\,\mathbf{i}_k$$

and

$$d\mathbf{r}^* = \mathbf{r}^*_{,j}dx_j = (\mathbf{r} + \mathbf{u})_{,j}dx_j = (\delta_{kj} + u_{k,j})dx_j\,\mathbf{i}_k .$$

where $(\ )_{,j} = \partial(\ )/\partial x_j$. It then follows from (16.2) that

$$2E_{jk} = u_{k,j} + u_{j,k} + u_{m,j}u_{m,k} \qquad (j,k = 1,2,3). \tag{16.3}$$

Note that $E_{jk}$ is *symmetric* in subscripts, i.e., $E_{jk} = E_{kj}$.

The strain in the body is small if $|E_{jk}| \ll 1$ for all $j$ and $k$. That portion of the expression for the components of finite strain which is linear in the derivatives of the displacement components is called the *linear strain components* and denoted by $e_{jk}$:

$$e_{jk} = \frac{1}{2}(u_{k,j} + u_{j,k}) \qquad (j,k = 1,2,3). \tag{16.4}$$

It should be emphasized that $E_{jk}$ cannot always be approximated by $e_{jk}$ even if we have small strain. The strain state of a body is said to be *infinitesimal* if $E_{jk} \cong e_{jk}$.

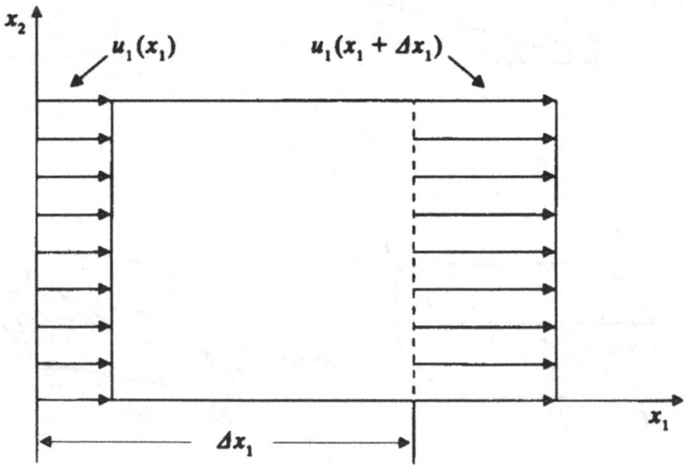


**Figure 16.2a**

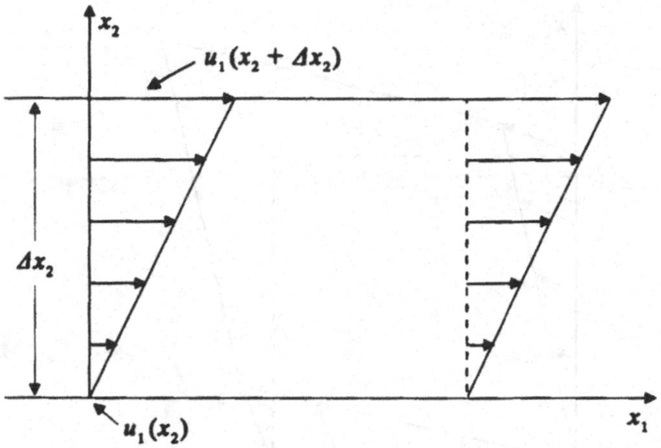


**Figure 16.2b**

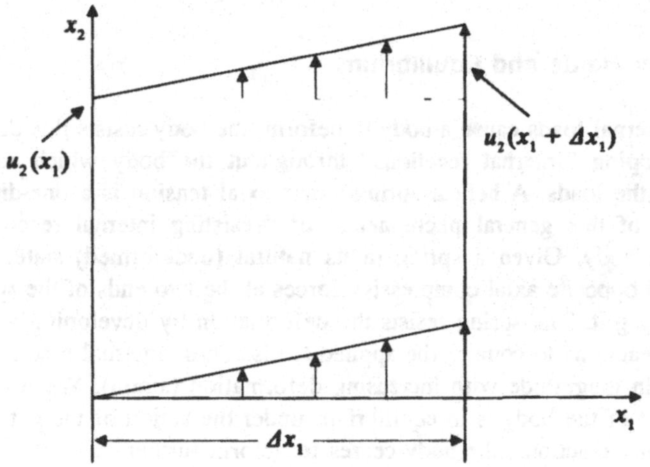


**Figure 16.2c**

An infinitesimal strain theory of elasticity has been found adequate for a large class of technical problems. In the engineering literature, $e_{11}, e_{22}$, and $e_{33}$ are called *normal strain* components whereas $2e_{jk}$ $(j \neq k)$ are known as *shear strain* components. We will not attempt to indicate the connotative nature of these expressions and refer interested readers to Fung (1967) (see also Fig. 16.2a–16.2d).

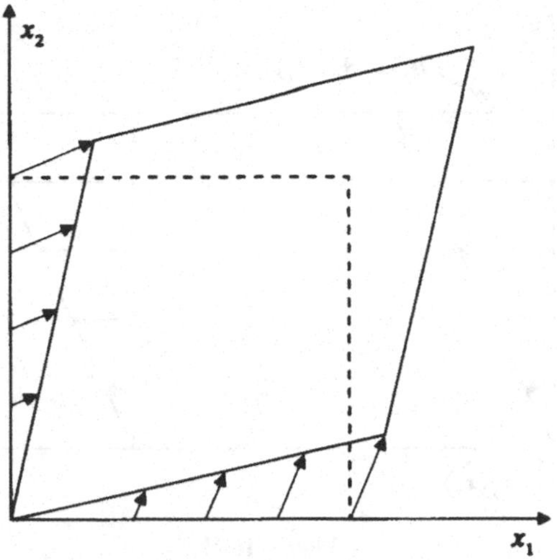


**Figure 16.2d**

## 3. Stress Fields and Equilibrium

When external loads cause a body to deform, the body resists this deformation by developing "internal reactions" throughout the body which counter the effect of the loads. A helical spring under axial tension is a one-dimensional analogue of this general phenomenon of "resisting internal reactions" in a deformed body. Given a spring in its natural (undeformed) state, we apply equal and opposite axial compressive forces at the two ends of the spring, thus compressing it. The spring resists the deformation by developing a system of internal reactions to counter the applied loads. These internal reactions usually increase in magnitude with increasing deformation (strain). When every (isolated) part of the body is in equilibrium under the action of the external loads and internal reactions, the body ceases to deform further.

To get a better feel for the internal reactions, we consider a body $V$ in equilibrium with external forces $\mathbf{F}_1, \ldots, \mathbf{F}_5$. Imagine that the body has been cut arbitrarily into two isolated part, denoting the new (fictitious) surface to the left of the cut by $S_+$ with unit (outward) normal $\mathbf{n}$, and the surface to the right of the cut by $S_-$, with unit normal $-\mathbf{n}$ (Fig. 16.3). Let $\mathbf{F}_+$ and $\mathbf{M}_+$ be the resultant internal force and moment acting on $S_+$, and $\mathbf{F}_-$ and $\mathbf{M}_-$ be acting on $S_-$. By Newton's third law, we have $\mathbf{F}_- = -\mathbf{F}_+$ and $\mathbf{M}_- = -\mathbf{M}_+$. Whatever the external loads may be, it is clear that $\mathbf{F}_+$ and $\mathbf{M}_+$ are the resultants of certain fields of force and moment intensity, respectively, distributed over the entire surface $S_+$. [For example, the internal reaction of the roof of a building to the

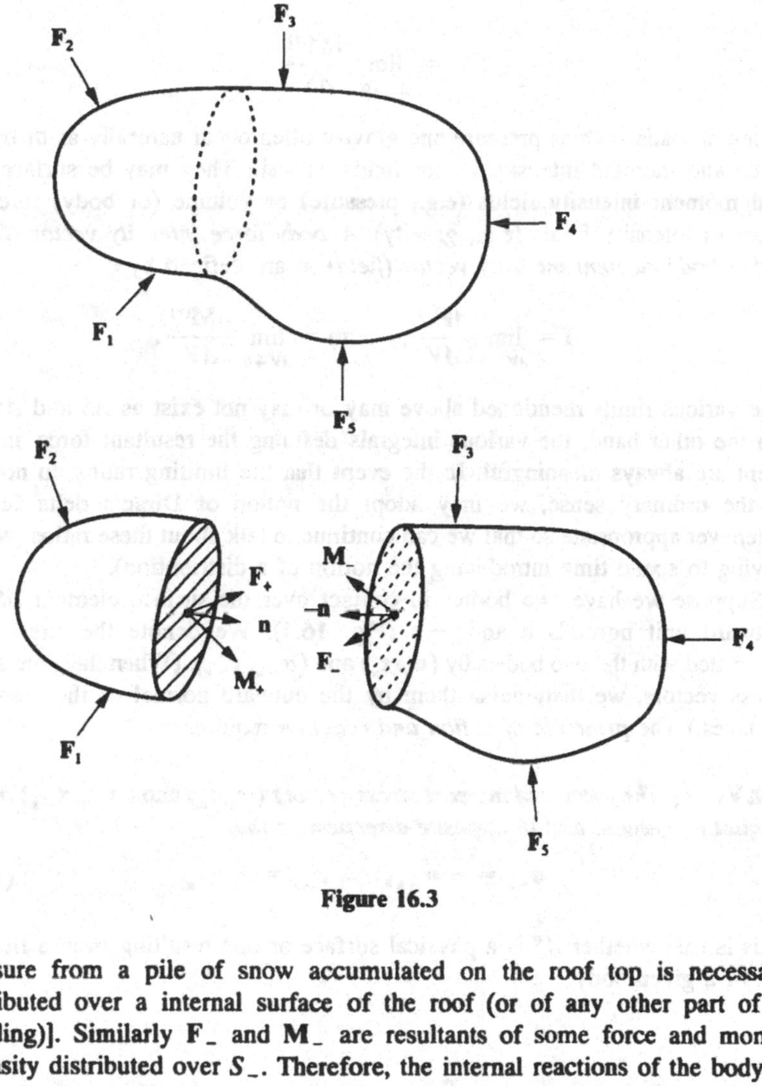


Figure 16.3

pressure from a pile of snow accumulated on the roof top is necessarily distributed over a internal surface of the roof (or of any other part of the building)]. Similarly $\mathbf{F}_-$ and $\mathbf{M}_-$ are resultants of some force and moment intensity distributed over $S_-$. Therefore, the internal reactions of the body are more fundamentally described by force and moment intensity *vector fields* defined for all points of the body.

We define a *force stress vector (field)* $\boldsymbol{\sigma}(\mathbf{x}, t)$ as the limit as $\Delta S \to 0$ of the ratio of the resultant surface force $\Delta \mathbf{F}^{(s)}$ acting over a surface of an elemental volume to the surface area $\Delta S$ of the elemental volume:

$$\boldsymbol{\sigma} = \lim_{\Delta S \to 0} \frac{\Delta \mathbf{F}^{(s)}}{\Delta S}.$$

Analogously, we define a *moment* (or *couple*) *stress vector* (*field*) $\boldsymbol{\tau}(\mathbf{x}, t)$ by

$$\boldsymbol{\tau} = \lim_{\Delta S \to 0} \frac{\Delta \mathbf{M}^{(s)}}{\Delta S}.$$

External loads such as pressure and gravity often occur naturally as distributed force and moment intensity vector fields as well. They may be surface force and moment intensity fields (e.g., pressure) or volume (or body) force and moment intensity fields (e.g., gravity). *A body force intensity vector (field)* $\mathbf{f}$ and *a body moment intensity vector (field)* $\mathbf{m}$ are defined by

$$\mathbf{f} = \lim_{\Delta V \to 0} \frac{\Delta \mathbf{F}^{(v)}}{\Delta V}, \qquad \mathbf{m} = \lim_{\Delta V \to 0} \frac{\Delta \mathbf{M}^{(v)}}{\Delta V}.$$

The various limits mentioned above may or may not exist as $\Delta S$ and $\Delta V \to 0$. On the other hand, the various integrals defining the resultant force and moment are always meaningful. In the event that the limiting ratios do not exist in the ordinary sense, we may adopt the notion of Dirac's delta function whenever appropriate so that we can continue to talk about these ratios (without having to spend time introducing the notion of a distribution).

Suppose we have two bodies in contact over the surface element $\Delta S$ with outward unit normals $\mathbf{n}$ and $-\mathbf{n}$ (Fig. 16.4). We denote the stress fields associated with the two bodies by $(\boldsymbol{\sigma}_n, \boldsymbol{\tau}_n)$ and $(\boldsymbol{\sigma}_{-n}, \boldsymbol{\tau}_{-n})$. (When there are several stress vectors, we distinguish them by the outward normal of the associated surfaces.) The *principle of action and reaction* requires:

**[XVI.1a]** ***The force and moment stress vectors*** $(\boldsymbol{\sigma}_n, \boldsymbol{\tau}_n)$ **and** $(\boldsymbol{\sigma}_{-n}, \boldsymbol{\tau}_{-n})$ ***are of equal magnitude and in opposite direction so that***

$$\boldsymbol{\sigma}_{-n} = -\boldsymbol{\sigma}_{-n}, \qquad \boldsymbol{\tau}_{-n} = -\boldsymbol{\tau}_{-n}. \tag{16.5}$$

This is true whether $\Delta S$ is a physical surface or one resulting from a fictitious cut of a given body.

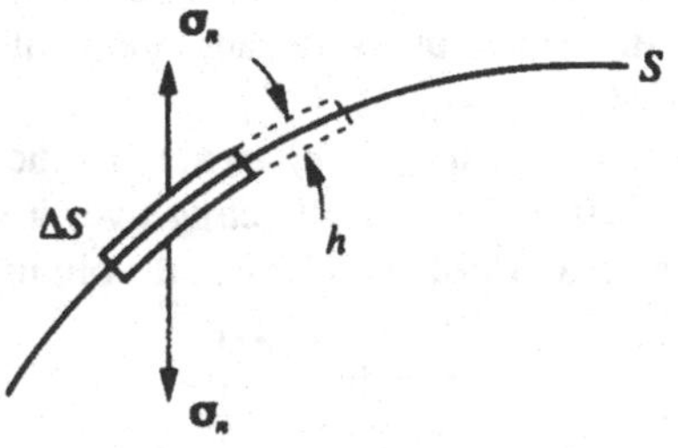


**Figure 16.4**

For problems treated in the classical theory of elasticity, the internal moment stress vector fields $\boldsymbol{\tau}$ and $\mathbf{m}$ are negligible and assumed to be absent in that theory. We will henceforth omit them from our discussion and refer to force stress vectors as *stress vectors* without any ambiguity. Furthermore, we will consider only static problems so that time and motion are not involved in the subsequent development. The fundamental postulate of the classical theory of elastostatics is:

> *On any imaginary closed surface* $S_0$ *in the interior of a continuum, a stress vector field can be defined in such a way that its action on the material occupying the space interior to* $S_0$ *is equipollent to the action of the exterior material on the material inside* $S_0$.

At any point $P$ of a body there is an infinity of different surface elements with $P$ as their centroid. Hence, associated with the point $P$, there are uncountably many stress vectors. Fortunately, not all stress vectors are independent of each other. Given any three elemental surfaces with centroid $P$ and with unit normals $\mathbf{n}_1, \mathbf{n}_2, \mathbf{n}_3$ which are not coplanar, the stress vector associated with a fourth elemental surface with unit normal $\mathbf{n}$ (and also with its centroid at $P$) can be expressed in terms of the three stress vectors associated with the first three elemental surfaces. The relation among the four stress vectors is given by Cauchy's formula. [A proof of this result is based on the condition of force equilibrium and can be found in most texts of elasticity such as Fung (1967)]. This formula is most often used for the special case where $\mathbf{n}_1, \mathbf{n}_2$, and $\mathbf{n}_3$ are simply the Cartesian base vectors $\mathbf{i}_1, \mathbf{i}_2$, and $\mathbf{i}_3$, respectively. For this case, we have the following form of Cauchy's formula:

**[XVI.1b]** ***Let*** $\boldsymbol{\sigma}_k$ ***be the stress vector associated with a unit normal in the*** $\mathbf{i}_k$ ***direction and*** $\mathbf{v}$ ***be any unit normal of a surface with same centroid. Then, we have***

$$\boldsymbol{\sigma}_\nu = \boldsymbol{\sigma}_\nu \nu_k, \qquad \nu_k = \mathbf{i}_k \cdot \mathbf{v}. \tag{16.6}$$

The formula (16.6) holds also for any orthonormal set of $\{\mathbf{n}_1, \mathbf{n}_2, \mathbf{n}_3\}$ where $\boldsymbol{\sigma}_k$ denotes $\boldsymbol{\sigma}_\nu$ for $\mathbf{v} = \mathbf{n}_k$.

The simplest representation of the state of stress at a point in terms of scalar quantities is to resolve the stress vector $\boldsymbol{\sigma}_j$ with respect to the Cartesian reference frame:

$$\boldsymbol{\sigma}_j = \sigma_{jk}\mathbf{i}_k. \tag{16.7}$$

It is customary to call the components $\sigma_{jj}$ (no sum) the *normal stress components* and the components $\sigma_{jk}$ with $j \neq k$ the *shearing stress components*. With the help of Cauchy's formula, we see that the nine scalar stress components $\sigma_{ij}$, $i,j = 1,2,3$, completely describe the state of stress at any point in the body. We will show presently that the condition of moment equilibrium (with $\boldsymbol{\tau} = 0$ and $\mathbf{m} = 0$) requires that

$$\sigma_{ij} = \sigma_{ji} \qquad (i,j = 1,2,3) \tag{16.8}$$

so that in the classical theory of elasticity, *there are only six distinct stress components.*

The classical theory of elasticity also postulates, as *conditions of equilibrium* for elastic bodies, that the following two equations of *vanishing resultant force and vanishing resultant moment,*

$$\oiint_{S_0} \boldsymbol{\sigma}_n \, dS + \iiint_{V_0} \mathbf{f} \, dV = 0, \tag{16.9a}$$

$$\oiint_{V_0} (\mathbf{r} \times \boldsymbol{\sigma}_n) \, dS + \iiint_{V_0} (\mathbf{r} \times \mathbf{f}) \, dV = 0 \tag{16.9b}$$

must be *valid for the entire body as well as any of its (fictitiously) isolated parts* (denoted by $V_0$ with $S_0$ as its boundary).

Using Cauchy's formula for $\boldsymbol{\sigma}_j$ in the surface integral of the force equilibrium equation, we obtain

$$\oiint_{S_0} \boldsymbol{\sigma}_j n_j \, dS + \iiint_{V_0} \mathbf{f} \, dV = 0 .$$

For piecewise smooth stress vectors $\boldsymbol{\sigma}_j$, we may apply the divergence theorem to the surface integral to get

$$\iiint_{V_0} (\boldsymbol{\sigma}_{j,j} + \mathbf{f}) \, dV = 0 .$$

Because the last integral holds for arbitrary $V_0$, it can be shown that the integrand must vanish if the integrand is continuous. We have then:

**[XVI.1c]** ***The differential equation of force equilibrium***

$$\boldsymbol{\sigma}_{j,j} + \mathbf{f} = 0 \tag{16.10}$$

***must hold for all points in the interior of the body*** $\mathbf{V}_0$.

A similar consideration of the integral moment equilibrium condition requires that

**[XVI.1d]** ***The equation of moment equilibrium***

$$\mathbf{r}_{,j} \times \boldsymbol{\sigma}_j = \mathbf{0} \tag{16.11}$$

***must hold for all interior points of the body*** $\mathbf{V}_0$.

The two vector equations of equilibrium (16.10) and (16.11) are each equivalent to three scalar equations. With the component representation (16.7) for $\boldsymbol{\sigma}_j$ and the corresponding component representation for $\mathbf{f}$,

$$\mathbf{f} = f_k \mathbf{i}_k, \tag{16.12}$$

we get from the vector force equilibrium equation

$$\sigma_{jk,j} + f_k = 0 \quad (k = 1,2,3). \tag{16.13}$$

The moment equilibrium equation implies $\sigma_{jk} - \sigma_{kj} = 0$ $(j \neq k)$ which are the symmetry conditions (16.8) stated earlier, reducing the number of distinct stress components from nine to six. The symmetry conditions cease to be valid if there are either moment stresses (from $\boldsymbol{\tau}_j$) or body moment intensities (from $\mathbf{m}$) present in the body.

## 4. Elasticity and Isotropy

In the absence of thermal, electromagnetic and chemical effects, experimental evidence indicates that, within a certain allowable limit of deformation, most materials encountered in our daily life exhibit the following properties to varying degrees:

1. If it is not under the influence of any external disturbance, a body of material is free of any internal stress and can remain in this "unstressed" or "natural" state indefinitely.
2. When subject to external loads, the state of stress at each point in the body depends only on the state of strain at the same point and conversely.
3. The body returns to the unstressed state once the external loads are removed.

We call such a body of material an *elastic body* and properties 1-3 *elasticity*. To indicate other possible modes of behavior, we note, for instance, that the stress state at a point of a body may depend on the time history of the strain state at that same point (viscoelasticity) or on the strain state of all points in some neighborhood of the given point (nonlocal theory). We are interested here only in elastic bodies. Because the state of strain and stress at each point of such a body is described by the six distinct components of the strain tensor $E_{ij}$

and the stress tensor $\sigma_{ij}$, respectively, it appears that the elasticity of a body should be adequately described by six (elastic) stress-strain relations of the form

$$\sigma_{ij} = f_{ij}(E_{11}, E_{12}, E_{13}, E_{22}, E_{23}, E_{33}) \quad (i, j = 1, 2, 3) \tag{16.14}$$

with $f_{ij} = f_{ji}$ and $f_{ij}(0, 0, \ldots, 0) = 0$. [Strictly speaking, the relations in (16.14) are not meaningful because of the definitions of $E_{ij}$ and $\sigma_{mn}$. However, they can be shown to be appropriate for the linear theory of elasticity developed in this chapter.]

For sufficiently "small" external loads, experimental results indicate that the strain components of an elastic medium will be small compared to unity and that the elastic stress-strain relations are effectively linear. In this chapter, we will be interested only in this particular range of elasticity. In this range, the stress-strain relations are essentially a generalization of Hooke's original observation and are often referred to as *generalized Hooke's laws*, and the body is said to be *linearly elastic*. In the subsequent development, we will be concerned only with *infinitesimal* strains characterized by $E_{ij} \cong e_{ij}$. In that case, the most general form of the linear stress-strain relations is

$$\sigma_{ij} = C_{ijk\ell} e_{k\ell} \quad (i, j = 1, 2, 3), \tag{16.15}$$

where the quantities $C_{ijk\ell}$ are known functions of position and are determined by suitable experiments. An elastic body is *homogeneous* if the elastic moduli $C_{ijk\ell}$ assume the same value throughout the body. It is inhomogeneous if they are functions of position. In the absence of initial stresses, relations (16.15) may be thought of as the first (nonzero) terms of the Taylor series expansion of relations (16.14) for infinitesimal strains, i.e., $E_{ij} \cong e_{ij}$.

A few observations on the linear stress-strain relation (16.15) are useful in subsequent development:

(*i*) The 81 elastic moduli $C_{ijk\ell}$ are not completely independent of each other. Because $\sigma_{ij} = \sigma_{ji}$ and $e_{ij} = e_{ji}$, we have

$$C_{ijk\ell} = C_{jik\ell} = C_{ij\ell k}. \tag{16.16}$$

These symmetry conditions reduce the number of distinct $C_{ijk\ell}$ to 36.

(*ii*) For $e_{mn}$ to be completely described by $\sigma_{ij}$, we must be able to invert these relations to get

$$e_{k\ell} = c_{k\ell ij} \sigma_{ij} \quad (k, \ell = 1, 2, 3) \tag{16.17}$$

with $c_{k\ell ij} = c_{\ell kij} = c_{k\ell ji}$.

(*iii*) The stress-strain relations for a certain class of (linearly or nonlinearly) elastic bodies can be expressed in terms of a scalar function $S$ of the strain components. For infinitesimal strains, we have $S = S(e_{11}, \ldots, e_{33})$ and

$$\sigma_{ij} = \frac{\partial S}{\partial e_{ij}} \quad (i,j = 1,2,3). \tag{16.18}$$

Such an elastic body is said to be ***hyperelastic***. As long as there is no dissipative mechanism within the body, it can be shown by thermodynamical considerations that such a scalar functional $S$ exists. It is called the ***strain energy (density) function*** in the engineering literature. We will occasionally refer to $S$ more descriptively as a ***stress potential*** (in strain space).

(*iv*) For the linear relations (16.15), it can be shown (see the Exercises) that the existence of a stress potential imposes the additional requirement

$$C_{ijk\ell} = C_{k\ell ij} \tag{16.19}$$

on the elastic moduli. In that case, the number of distinct $C_{ijk\ell}$ is further reduced to 21, and the strain energy function is given by

$$S = \frac{1}{2} C_{ijk\ell} e_{ij} e_{k\ell}. \tag{16.20}$$

In a crystalline body, the elastic coefficients $C_{ijk\ell}$ at a given point in the body depend on the orientation of the Cartesian frame relative to the body. For this reason, the elastic coefficients are, in general, different for different choices of the Cartesian reference frame subject only to the symmetry conditions (16.16) and possibly (16.19). However, many metals have a sufficiently small and randomly oriented crystal grain structure so that the elastic properties averaged over a few grains are essentially independent of direction. An elastic medium is called elastically ***isotropic*** if there are no preferred directions insofar as relations between stress and strain are concerned.

For the linear relations (16.15), it can be shown (C.E. Pearson, 1959) that this lack of preferred directions drastically reduces the number of independent elastic modulij to 2. We may take the elastic moduli to be the so-called ***Lamé parameters*** $\lambda$ and $\mu$ with

$$C_{ijkm} = \lambda \delta_{ij} \delta_{km} + 2\mu \delta_{ik} \delta_{jm}. \tag{16.21}$$

The parameters $\lambda$ and $\mu$ may be functions of position in space. Correspondingly, the stress-strain relations (16.15) become

$$\sigma_{ij} = \lambda \delta_{ij} e_{kk} + 2\mu e_{ij}. \tag{16.22}$$

The corresponding strain energy density function $S$ may be constructed without additional constraints imposed on the elastic coefficients $\lambda$ and $\mu$:

$$S = \frac{1}{2}\lambda(e_{kk})^2 + \mu e_{ij}e_{ij}. \tag{16.23}$$

As such, isotropic media form a subclass of hyperelastic media, at least in the linear range. From (16.23), we have immediately

**[XVI.2a]** ***The quadratic form S is positive definite if μ and λ are positive.***

The inverted linear isotropic stress-strain relations are (see the Exercises)

$$2\mu e_{ij} = \sigma_{ij} - \frac{\lambda}{2\mu + 3\lambda}\delta_{ij}\sigma_{kk} \qquad (i,j = 1,2,3) \tag{16.24}$$

provided that $\mu$ and $2\mu + 3\lambda$ do not vanish. These relations may also be expressed in terms of a strain potential (in stress space) or ***complementary energy (density) function*** by

$$C = \frac{1}{4\mu}\left[\sigma_{ij}\sigma_{ij} - \frac{\lambda}{2\mu + 3\lambda}(\sigma_{kk})^2\right] \tag{16.25a}$$

$$e_{ij} = \frac{\partial C}{\partial \sigma_{ij}} \qquad (i,j = 1,2,3). \tag{16.25b}$$

For a given material, the values of the two Lamé parameters, $\mu$ and $\lambda$ are to be determined experimentally. A typical experiment shown in Figure 16.5 is for a prismatic bar with a rectangular cross section subject only to equal and opposite uniform axial tension at the two ends (whose unit surface normals are taken to be $\mathbf{i}_1$ and $-\mathbf{i}_1$, respectively). For this configuration, the stress state in the bar is known from the solution of the stress BVP to be $\sigma_{11} = T_0$ and $\sigma_{ij} = 0$ for $i, j \neq 1$ (see the Exercises).

If the bar material is isotropic and homogeneous, the strain components are related to the stress components by (16.24). For the particular loading above, we have

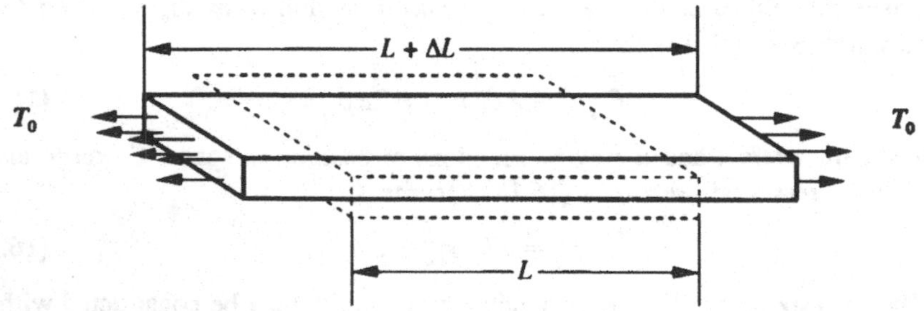


**Figure 16.5**

$$2\mu e_{11} = \frac{2(\mu+\lambda)}{2\mu+3\lambda}\sigma_{11} = \frac{2(\mu+\lambda)}{2\mu+3\lambda}T_0$$

or

$$\frac{T_0}{e_{11}} = \frac{\mu(2\mu+3\lambda)}{\mu+\lambda} \equiv E. \tag{16.26}$$

The quantity $E$ defined by (16.26) is called *Young's modulus* (or *modulus of elasticity*) and it can be determined by measuring $e_{11} = \frac{\Delta L}{L}$ and $T_0$.

In addition, we have

$$2\mu e_{22} = -\frac{\lambda}{2\mu+3\lambda}T_0, \qquad 2\mu e_{33} = -\frac{\lambda}{2\mu+3\lambda}T_0$$

so that

$$\frac{e_{33}}{T_0} = \frac{e_{22}}{T_0} = -\frac{\lambda}{2\mu(2\mu+3\lambda)}$$

or

$$-\frac{e_{33}}{e_{11}} = -\frac{e_{22}}{e_{11}} = \frac{\lambda}{2(\mu+\lambda)} \equiv \nu. \tag{16.27}$$

The quantity $\nu$ defined by (16.27) is called *Poisson's ratio*. If $\nu > 0$, then $\nu$ is a measure of the lateral contraction (in the $\mathbf{i}_2$ and $\mathbf{i}_3$ directions) accompanying the axial extension (in the $\mathbf{i}_1$ direction). Evidently $\nu$ can also be determined from measurements.

To get a feel for the magnitude of these two important new elastic parameters, we note that $E \cong 10^7$ psi for aluminum and $\nu \cong 0.3$ for most metal alloys. From the definition of Young's modulus and Poisson's ratio, we can obtain the values of $\lambda$ and $\mu$ from those of $E$ and $\nu$. In particular, we get from

$$E = 2\mu + \frac{\lambda\mu}{\mu+\lambda} = 2\mu(1+\nu)$$

the expression

$$\mu = \frac{E}{2(1+\nu)}. \tag{16.28}$$

The definition of $\nu$ implies $\lambda = 2\nu(\mu+\lambda)$ so that

$$\lambda = \frac{\nu E}{(1+\nu)(1-2\nu)}. \tag{16.29}$$

Evidently, $\lambda$ and $\mu$ are also determined once $E$ and $\nu$ are, provided that $\nu \neq -1$ and $1/2$. In fact, we see from expression (16.23) for the stress potential $S(e_{ij})$ of a linearly elastic and isotropic material that $\mu$ must be positive (and, therefore, $\nu > -1$ given $E > 0$) for $S$ to be positive definite. Otherwise, $S$ would be negative for some pure shear strain state. For $\lambda > 0$, we also need $\nu < 1/2$ [see (16.24)] and, hence, we must have $-1 < \nu < 1/2$ for $S$ to be positive definite.

Among other experiments which can be performed to obtain the Lamé parameters or to check the results obtained by the tensile test, we have the case of a body subject to uniform normal pressure. The stress state of such a body is given by $\sigma_{ij} = -p_0 \delta_{ij}$, where $p_0$ is the uniform normal pressure (see the Exercises). With in the framework of linear elasticity theory, the change of volume per unit volume (of the undeformed body) is

$$e_{jj} = -\frac{3p_0}{2\mu + 3\lambda} \equiv -\frac{p_0}{K}.$$

We can determine the *bulk modulus K by measuring the change of volume of the body after deformation.* $K$ is related to $E$ and $\nu$ by

$$K = \frac{2}{3}\mu + \lambda = \frac{E}{3(1-2\nu)}.$$

Note that we have $Ee_{jj} = (1-2\nu)\sigma_{kk}$. Therefore, an elastic material with $\nu = 1/2$ is called an *incompressible* medium (no change in volume).

It may be verified directly that expression (16.23) for the stress potential is equivalent to

$$S(e_{ij}) = \frac{1}{2}K(e_{kk})^2 + \mu \tilde{e}_{ij}\tilde{e}_{ij}$$

where

$$\tilde{e}_{ij} \equiv e_{ij} - \frac{1}{3}e_{kk}\delta_{ij}$$

is called the *strain deviation* tensor. If $e_{ij} = 0$ for $i \neq j$ and $e_{11} = e_{22} = e_{33} = a$, then

$$\tilde{e}_{ij}\tilde{e}_{ij} = e_{ij}e_{ij} - \frac{1}{3}(e_{kk})^2 = 3a^2 - \frac{1}{3}(3a)^2 = 0.$$

It is, therefore, necessary and sufficient to have $K > 0$ and $\mu > 0$ for $S$ to be positive definite. From the expressions for $\mu$ and $K$ in terms of $E$ and $\nu$, we have the following theorem which we obtained earlier in terms of $\lambda$ and $\mu$:

**[XVI.2b]** ***The stress potential of a linearly elastic and isotropic material is positive definite if and only if*** $E > 0$ ***and*** $-1 < \nu < 1/2$.

The inverted stress-strain relations (16.24) for an isotropic medium can be written in terms of $E$ and $\nu$ as

$$e_{11} = \frac{1}{E}\left[\sigma_{11} - \nu(\sigma_{22} + \sigma_{33})\right], \qquad e_{22} = \frac{1}{E}\left[\sigma_{22} - \nu(\sigma_{11} + \sigma_{33})\right],$$

$$e_{33} = \dots, \qquad e_{ij} = \frac{1}{2G}\sigma_{ij} \quad (i \neq j), \tag{16.30}$$

where

$$G = \frac{E}{2(1+\nu)} = \mu \tag{16.31}$$

is connotatively called the *shear modulus* (identical to $\mu$ but used universally in the engineering literature). Correspondingly, the complementary energy (density) function $C$ given in (16.25a) may be written as

$$C = \frac{1}{4G}\sigma_{ij}\sigma_{ij} - \frac{\nu}{2E}(\sigma_{kk})^2. \tag{16.32}$$

We will occasionally refer to it (by the more descriptive term) as the *strain potential* (in stress space).

For (16.25b) to be identical to (16.24), whether we use (16.25a) or (16.32), $\sigma_{ij}$ and $\sigma_{ji}$ should be treated as distinct quantities in the expression for $C$. In engineering literature, it is customary to work with engineering shear strain components $\bar{e}_{ij} = e_{ij} + e_{ji} = u_{i,j} + u_{j,i}$ $(i \neq j)$. Correspondingly, we should replace all $\sigma_{ji}$ by $\sigma_{ij}$ in $C$ and be concerned only with $\bar{e}_{ij} = \partial C / \partial \sigma_{ij}$ (and not $\bar{e}_{ji}$).

## 5. Navier's Reduction

The governing equations for the infinitesimal strain theory of linearly elastic solids are now seen to consist of the strain-displacement relations (16.4), the stress-strain relations (16.15), and the equilibrium equations (16.13) (with $\sigma_{ij} = \sigma_{ji}$) where all subscripts range from 1 to 3. For isotropic media, relations (16.15) reduce to (16.22). In terms of $E$ and $\nu$, the corresponding inverted relations (16.24) may be written as

$$e_{ij} = \frac{1+\nu}{E}\sigma_{ij} - \frac{\nu}{E}\delta_{ij}\sigma_{kk}. \tag{16.33}$$

[which summarizes (16.30)].

Equations (16.4), (16.15), and (16.13) may be reduced to three second-order linear partial differential equations for $u_1, u_2$, and $u_3$. (We assume here that the body force intensities $f_k$ are either known functions or dependent on no higher than the second derivatives of the displacement components.) For an isotropic and homogeneous medium, we accomplish this reduction by combining the strain-displacement relations and the stress-strain relations to get

$$\sigma_{ij} = \mu(u_{i,j} + u_{j,i}) + \lambda\delta_{ij}u_{k,k} \tag{16.34}$$

and then by substituting (16.34) into the equilibrium equations (16.13) to get

$$(\mu + \lambda)u_{k,kj} + \mu u_{j,ii} + f_j = 0 \qquad (j = 1, 2, 3) \tag{16.35}$$

with $u_{j,ii} = \nabla^2 u_j$.

To completely determine the stress, strain, and displacement fields of an elastic body, the equations of linear elasticity theory must be supplemented by appropriate conditions on the boundary surface(s) of the body. For example, $\boldsymbol{\sigma}_n$ should be equal to $-p_0\mathbf{n}$ on the boundary in the case of a body subject to uniform (normal) surface pressure $p_0 > 0$. Among many realizable sets of boundary conditions on a bounding surface $S$, defined by $g_s(x_1, x_2, x_3) = 0$, we mention here the case of a *prescribed displacement* (vector) field and the case of a *prescribed stress* vector field.

For the prescribed displacement case, we have

$$\mathbf{u} = \mathbf{U}(\mathbf{x}) \quad \text{on } S, \tag{16.36}$$

where $\mathbf{U}(\mathbf{x})$ is a given function of position on $S$. If only the displacement field is prescribed on the boundary, the problem of determining the solution of the equations of elasticity theory (subject to these prescribed displacement conditions) is called the *first fundamental problem of elastostatics*.

For the case of prescribed surface traction, we have

$$\boldsymbol{\sigma}_n = \mathbf{T}(\mathbf{x}) \quad \text{on } S, \tag{16.37}$$

where $\mathbf{T}(\mathbf{x})$ is a prescribed vector function of position on $S$. Note that this is equivalent to prescribing combinations of the derivatives of the displacement components. Such a boundary-value problem is called the *second fundamental problem of elastostatics*.

Other kinds of boundary conditions are also admissible. For example, the displacement field may be prescribed on a part $S_d$ of the closed surface $S$, whereas the stress field may be prescribed on the rest of the surface $S_\sigma$. A different type of mixed condition occurs when the body is bounded by two or more closed surface (e.g., two concentric spherical surfaces), with the displacement field prescribed on one of these surfaces and the stress field prescribed on the other(s).

Still another kind of mixed condition is to prescribe a combination of stress and displacement at every point on the surface. A typical example of this is when the boundary has an an elastic support characterized by $\beta u_i + \alpha\sigma_{ni} = 0$, $i = 1, 2, 3$, where $\alpha$ and $\beta$ are known quantities.

Evidently, the first fundamental problem is analogous to the Dirichlet problem for Laplace's equation (because only the unknowns $u_i$ are prescribed), whereas the second fundamental problem is akin to the Neumann problem. From these analogies, it is not difficult to imagine that prescribing both the stress and displacement field at the same point(s) of the boundary is not permissible in the sense that the problem would be overdetermined. This observation will be supported by the results of variational considerations to be discussed in the remaining sections of this chapter.

## 6. Minimum Potential Energy

We learned from Hamilton's principle (in Chapter 6) that the mechanics of particles and rigid bodies is the consequence of minimizing some kind of (action or) energy integral of these dynamical systems. For elastostatics, kinetic energy is not involved. However, in addition to the potential energy associated with the work done by the external body and surface loads, there is now for an elastically deformable body another kind of potential energy associated with straining of the body. This new kind of potential energy is called the *strain energy* and can be shown to be the volume integral of the stress potential $S(e_{ij})$ introduced in Section 4. We limit discussion here to external body and surface loads which can be expressed in terms of load potentials $L_b(\mathbf{u})$ and $L_s(\mathbf{u})$ by

$$f_i = -\frac{\partial L_b}{\partial u_i}, \qquad T_i = -\frac{\partial L_s}{\partial u_i} \qquad (i = 1, 2, 3). \qquad (16.38,39)$$

For the case of fixed external loads, we have

$$L_b = -\mathbf{f}\cdot\mathbf{u}, \qquad L_s = -\mathbf{T}\cdot\mathbf{u}. \qquad (16.40a,b)$$

The total energy $P$ of the deformed body in that case is given by

$$P \equiv \iiint_V [S(e_{ij}) + L_b(\mathbf{u})]\,dV + \iint_{S_\sigma} L_s(\mathbf{u})\,dS, \qquad (16.41)$$

where surface load is prescribed on a portion $S_\sigma$ of $S$. On the remaining portion of $S$, denoted by $S_d$, $\mathbf{u}$ is prescribed. The potential energy $P$ evidently varies with $e_{ij}$ and $u_k$.

With $e_{ij}$ defined in terms of $u_k$ by the six strain-displacement relations, only the three displacement components can vary independently. The first variation of $P$ is

$$\delta P = \iiint_V [\frac{1}{2}\sigma_{ij}(\delta u_{i,j} + \delta u_{j,i}) - f_i\,\delta u_i\,]\,dV - \iint_{S_\sigma} \mathbf{T}\cdot\delta\mathbf{u}\,dS, \quad (16.42)$$

where $\sigma_{ij}$ is an abbreviation for $\partial S/\partial e_{ij}$. By the symmetry condition $\sigma_{ij} = \sigma_{ji}$, we have

$$\frac{1}{2}\sigma_{ij}(\delta u_{i,j} + \delta u_{j,i}) = \frac{1}{2}(\sigma_{ji}\delta u_{i,j} + \sigma_{ij}\delta u_{j,i}) = \sigma_{ji}\delta u_{i,j}$$

$$= (\sigma_{ji}\delta u_i)_{,j} - \sigma_{ji,j}\delta u_i\,. \quad (16.43)$$

The divergence theorem then enables us to write $\delta P$ as

$$\delta P = -\iiint_V (\sigma_{ji,j} + f_i)\,\delta u_i\,dV + \iint_{S_\sigma} (\boldsymbol{\sigma}_n - \mathbf{T})\cdot\delta\mathbf{u}\,dS \quad (16.44)$$

$$+ \iint_{S_d} \boldsymbol{\sigma}_n\cdot\delta\mathbf{u}\,dS,$$

where $S_d$ is the remaining portion of $S$ on which $\mathbf{u}$ is prescribed. Now, the comparison functions $\mathbf{u}$ are restricted to those which satisfy the displacement condition $\mathbf{u} = \mathbf{U}(\mathbf{x})$ on $S_d$. For these admissible comparison functions, we have $\delta\mathbf{u} = \mathbf{0}$ there so that the surface integral over $S_d$ vanishes. Thus, the Euler DE for $\delta P = 0$ are the three scalar equilibrium equations (16.13) and the Euler BC is the stress boundary condition (16.39) on $S_\sigma$. In other words:

**[XVI.3]** ***An admissible comparison displacement field*** $\{\hat{\mathbf{u}}\}$ ***which renders the potential energy P stationary must leave the deformed body in static equilibrium and therefore must satisfy (16.13) and (16.39) (with*** $\hat{\sigma}_{ij}$ ***given in terms of*** $\{\hat{u}_j\}$ ***by the stress-strain relations and the strain-displacement relations).***

For a positive definite quadratic strain energy density function, the stationary value $P[\hat{\mathbf{u}}]$ is actually a minimum value. This is seen from the fact that

$$P[\hat{u}_j + \delta u_j\,] - P[\hat{u}_j\,] = \iiint_V \frac{\partial^2 S}{\partial e_{ij}\partial e_{k\ell}}\,\delta e_{ij}\delta e_{k\ell}\,dV \equiv \delta^2 P \quad (16.45)$$

with $\partial^2 S/\partial e_{ij}\partial e_{k\ell}$ independent of the strain components in such a way that

$$P[\hat{u}_j + \delta u_j\,] - P[\hat{u}_j\,] = \iiint_V 2S(\delta e_{mn})\,dV \geq 0\,. \quad (16.46)$$

Strict inequality holds unless $\delta e_{mn} = 0$ for all $m$, $n$, and $\mathbf{x}$. We summarize the results in the following *theorem of minimum potential energy*:

**[XVI.4]** ***Of all linear elastostatic displacement fields*** $\{\mathbf{u}_j\}$ ***satisfying the given displacement boundary conditions on*** $\mathbf{S}_d$***, the actual displacement field*** $\{\hat{\mathbf{u}}_j\}$ ***of the deformed body satisfies all requirements of equilibrium in*** $\mathbf{V}$ ***and on*** $\mathbf{S}_\sigma$ ***and renders*** $\mathbf{P}[\mathbf{u}]$ ***a minimum.***

For a positive definite $S(e_{ij})$, the extremal $\{\hat{u}_j\}$ can be shown to be unique (see the Exercises) and $\hat{P}$ is the global weak minimum. However, the linear theory of elastostatics is usually not adequate for an accurate description of the behavior of an elastic body except in a small neighborhood of the unstrained state. Therefore, the conclusion of a global minimum on the basis of this linear theory must be understood in this context.

## 7. Reissner's Variational Principle

As in the case of a single independent variable in Chapter 6, it is possible (and often preferable) to work with the Hamiltonian instead of the Lagrangian of a given problem. For elasticity theory, it is customary to introduce the Legendre transformation

$$C(\sigma_{ij}) = \sigma_{ij}e_{ij} - S(e_{ij}) \tag{16.47}$$

[analogous to the defining relation (6.1) for the one-variable case] where $\sigma_{ij} = \partial S/\partial e_{ij}$. For the quadratic strain energy density (16.23), $C$ is the complementary energy density function introduced in Section 4 with $C(\sigma_{ij}) = S(e_{ij})$. More generally, relation (16.47) defines the complementary density function $C(\sigma_{ij})$ by using the inverted stress-strain relations to express $e_{ij}$ in terms of $\sigma_{mn}$.

We now write the expression for the potential energy $P$ in terms of $C$ with the help of (16.47) to get

$$P = \iiint_V \{\sigma_{ij}e_{ij} - C(\sigma_{mn}) + L_b(u_k)\}\,dV + \iint_{S_\sigma} L_s(u_i)\,dS.$$

In the theorem of minimum potential energy, the admissible comparison displacement functions are restricted to those which meet the prescribed displacement boundary conditions (16.38) on $S_d$. We incorporate this constraint into the functional to be extremized with the help of a (vector) Lagrange multiplier which can be identified with the traction vector $\boldsymbol{\sigma}_n$ on $S_d$. Let

$$I[u_j, \sigma_{ik}] = \iiint_V \{\sigma_{ij}e_{ij} - C(\sigma_{mn}) + L_b(u_k)\}\,dV + \iint_{S_\sigma} L_s(u_i)\,dS + \iint_{S_d} \boldsymbol{\sigma}_n \cdot (\mathbf{U} - \mathbf{u})\,dS. \tag{16.48}$$

As in the single-variable case, we can no longer expect $I$ to have a minimum (or maximum) value. We have instead the following variational principle of E. Reissner (1950):

**[XVI.5]** ***With $\sigma_{ij}$ and $u_k$ allowed to vary independently (and $e_{ij}$ defined in terms of $u_k$ by the strain-displacement relations) and no boundary conditions imposed on these comparison functions, the Euler differential equation for a stationary value of $I[u_i, \sigma_{jk}]$ are the three scalar differential equations of equilibrium (16.13) and six stress-strain relations (in the inverted form) (16.25b), whereas Euler boundary conditions are the stress boundary conditions (16.39) on $S_\sigma$ and the displacement boundary conditions (16.38) on $S_d$***

For approximate solutions of BVP of linear elastostatics by the direct method of calculus of variations (or the semidirect method to be described in Section 9), Reissner's variational principle does not require the trial functions for stress and displacement fields to be related (by the stress-strain relations) or restricted on the boundary by the displacement boundary conditions. Unlike the theorem of minimum potential energy, it does not favor one set of field variables (displacement) over another (stress).

As long as we are allowing more field variables to vary independently, the question of whether we can let $\delta e_{ij}$ be independent variations naturally suggests itself. The Hu–Washizu variational principle provides an affirmative answer to this question (see K. Washizu (1975) and references therein). The relevant functional $J[u_i, e_{mn}, \sigma_{jk}]$ of this stationary principle is the potential energy $P$ with all the constraints in $V$ and on $S$ appended by Lagrange multipliers:

$$J[u_i, e_{mn}, \sigma_{jk}] = \iiint_V \left\{ \sigma_{ij} \left[ \frac{1}{2}(u_{i,j} + u_{j,i}) - e_{ij} \right] + S + L_b(u_j) \right\} dV + \iint_{S_\sigma} L_s[u_m]\, dS + \iint_{S_d} \boldsymbol{\sigma}_n \cdot (\mathbf{U} - \mathbf{u})\, dS. \qquad (16.49)$$

where we have taken the relevant Lagrange multipliers to be $\{\sigma_{ij}\}$ in $V$ and $\boldsymbol{\sigma}_n$ on $S_d$ to eliminate the few nonessential Euler differential equations. It is then straightforward to verify that:

**[XVI.6]** ***The Euler differential equations for a stationary value of J, with $\delta u_i$, $\delta e_{mn}$, and $\delta\sigma_{jk}$ varying independently in V (and not required to vanish on the boundary surface(s) of V), are the 15 equations of linear elastostatics, (16.4), (16.18), and (16.13), and the Euler boundary conditions are the stress condition (16.37) on $S_\sigma$ and the displacement condition (16.36) on $S_d$.***

There are still other (different) variational formulations of the boundary-value problems of linear elastostatics. Some of the more recent developments can be found in E. Reissner (1987). While we do not wish to pursue an exhaustive discussion of these formulations, we will consider in the next section one further development of the Reissner variational principle which, as we shall see, will be useful in the next chapter.

## 8. Minimum Complementary Energy

The first term in the Reissner functional $I[u_i, \sigma_{jk}]$ defined in (16.48) may be written in terms of the displacement components by steps indicated in (16.43):

$$\begin{aligned}\iiint_V \sigma_{ij} e_{ij}\,dV &= \iiint_V \frac{1}{2}\sigma_{ij}(u_{i,j}+u_{j,i})\,dV = \iiint_V \sigma_{ij} u_{j,i}\,dV \\ &= \iiint_V [(\sigma_{ij}u_j)_{,i} - (\sigma_{ij,i}u_j)]\,dV \\ &= \oint\!\!\oint_S (\boldsymbol{\sigma}_n \cdot \mathbf{u})\,dS - \iiint_V \sigma_{ij,i} u_j\,dV. \end{aligned} \tag{16.50}$$

It follows that for the case of fixed external loads given by (16.40′), the functional $I$ itself may be transformed into

$$I = -\iiint_V \{C + [\sigma_{ij,i} + f_j]u_j\}\,dV + \iint_{S_\sigma} (\boldsymbol{\sigma}_n - \mathbf{T})\cdot\mathbf{u}\,dS + \iint_{S_d} \mathbf{U}\cdot\boldsymbol{\sigma}_n\,dS. \tag{16.51}$$

We now require that the admissible comparison stress fields $\sigma_{ij}$ satisfy the equilibrium equations (16.13) and the prescribed stress boundary conditions (16.39) on $S_\sigma$. If we denote the resulting specialized $I[u_i, \sigma_{mn}]$ of (16.51) by $I_s[\sigma_{ij}]$, we have

$$I_s[\sigma_{ij}] = -\left\{\iiint_V C(\sigma_{ij})\,dV - \iint_{S_d} \mathbf{U}\cdot\boldsymbol{\sigma}_n\,dS\right\} \equiv -\mathscr{P}[\sigma_{ij}]. \tag{16.52}$$

As suggested by the notation $I_s[\sigma_{ij}]$, the new functional (16.52) depends only on $\sigma_{ij}$ and not $u_k$. Furthermore, the stress components are restricted by the requirement that they satisfy equilibrium in $V$ and on $S_\sigma$. In particular, the variations $\delta\sigma_{ij}$ must satisfy the three homogeneous equilibrium equations (as $f_i$ do not vary):

$$\delta\sigma_{ij,i} = 0 \quad (\text{in } V) \tag{16.53a}$$

and

$$\delta\boldsymbol{\sigma}_n = 0 \quad (\text{on } S_\sigma). \tag{16.53b}$$

Suppose $\widehat{\sigma}_{ij}$ extremizes $\mathcal{P}[\sigma_{ij}] \equiv -I_s[\sigma_{ij}]$ (subject to the equilibrium constraints on $\sigma_{ij}$). Then for a complementary energy density $C[\sigma_{ij}]$ quadratic in its arguments, we have

$$\mathcal{P}[\widehat{\sigma}_{ij} + \delta\sigma_{ij}] - \mathcal{P}[\widehat{\sigma}_{ij}] = \delta\mathcal{P} + \delta^2\mathcal{P}, \tag{16.54a}$$

where

$$\delta\mathcal{P} = \iiint_V \left[\frac{\partial C}{\partial \sigma_{ij}}\right]_{\hat{\sigma}_{mn}} \delta\sigma_{ij}\, dV - \iint_{S_u} \mathbf{U}\cdot\delta\sigma_n\, dS, \tag{16.54b}$$

$$\delta^2\mathcal{P} = \frac{1}{2}\iiint_V \left[\frac{\partial^2 C}{\partial\sigma_{ij}\,\partial\sigma_{mn}}\right]_{\hat{\sigma}_{mn}} \delta\sigma_{ij}\,\delta\sigma_{mn}\, dV. \tag{16.54c}$$

Now $\delta\mathcal{P}$ must vanish as $\{\widehat{\sigma}_{ij}\}$ renders $\mathcal{P}$ stationary and $C$ is quadratic so that

$$\frac{\partial^2 C}{\partial\sigma_{ij}\,\partial\sigma_{mn}}\delta\sigma_{ij}\,\delta\sigma_{mn} = 2C(\delta\sigma_{kl}); \tag{16.55}$$

therefore, equation (16.54a) may be written as

$$\mathcal{P}[\widehat{\sigma}_{ij} + \delta\sigma_{ij}] - \mathcal{P}[\widehat{\sigma}_{ij}] = \delta^2\mathcal{P} = \iiint_V C(\delta\sigma_{ij})\, dV. \tag{16.56}$$

From (16.56) follows the *theorem of minimum complementary energy*:

**[XVI.7]** ***Of all linear elastostatic stress fields in equilibrium with the external loads in V and on $S_\sigma$, the stress field $\{\widehat{\sigma}_{ij}\}$ of the deformed body which renders $\mathcal{P}[\sigma_{ij}]$ stationary actually minimizes $\mathcal{P}$ if C is (quadratic and) positive definite.***

The minimum is actually a global minimum; however, we must keep in mind again that the linear theory of elastostatics is only appropriate for a limited range of the deformed body's strain field.

With $I_s[\sigma_{ij}] = -\mathcal{P}[\sigma_{ij}]$ and $I_s[\widehat{\sigma}_{ij}] = I[\widehat{\sigma}_{ij}] \equiv \widehat{I}$, we have

$$-\mathcal{P}[\sigma_{ij}] = I_s[\sigma_{mn}] \leq \widehat{I}.$$

But the theorem of minimum potential energy implies

$$\widehat{I} = P[\widehat{u}_j] \leq P[u_j].$$

It follows that the two minimum principles provide an upper and lower bound for the stationary value of the Reissner functional

$$\widehat{I} = I[\widehat{u}_j, \widehat{\sigma}_{mn}].$$

More specifically, we have

$$\text{[XVI.8]} \quad -\mathscr{P}[\sigma_{ij}] \le I[\widehat{u}_j, \widehat{\sigma}_{mn}] \le P[u_k]. \tag{16.57}$$

We now return to the condition of stationarity to see what necessary conditions (Euler differential equations and boundary conditions) follow from $\delta\mathscr{P} = 0$. To get these, we note that the general solution for (16.53a) is given in terms of (the first variations of) six stress functions $\delta\varphi_{ij} = \delta\varphi_{ji}$, $i, j = 1, 2, 3$ (by B. Finzi, 1934) with

$$\delta\sigma_{ij} = e_{imr} e_{jns} \delta\varphi_{rs,mn}, \tag{16.58}$$

where $e_{ijk}$ is the usual *alternator* with $e_{ijk} = 0$ if two indices are the same, $e_{123} = e_{231} = e_{312} = 1$ and $e_{ijk} = -1$ otherwise. The stress function representation for typical components of $\delta\sigma_{ij}$ are

$$\delta\sigma_{11} = \delta\varphi_{22,33} + \delta\varphi_{33,22} - 2\delta\varphi_{23,23},$$

$$\delta\sigma_{12} = \delta\varphi_{31,32} + \delta\varphi_{32,31} - \delta\varphi_{33,12} - \delta\varphi_{12,33},$$

etc. Because the three constraints in (16.53a) are satisfied with no restrictions on the six stress functions, $\delta\varphi_{ij}$ can vary independently. For brevity, we set $\widehat{e}_{ij}$ to be $\partial C/\partial\sigma_{ij}$ evaluated at the extremal $\{\widehat{\sigma}_{mn}\}$. Then (16.54b) may be written as

$$\delta\mathscr{P} = \iiint_V \widehat{e}_{ij} e_{imr} e_{jns} \delta\varphi_{rs,mn} dV - \iint_{S_d} \mathbf{U} \cdot \delta\boldsymbol{\sigma}_n dS$$

$$= \iiint_V (\widehat{e}_{ij} e_{imr} e_{jns})_{,mn} \delta\varphi_{rs} dV - \oiint_S (\ldots) dS. \tag{16.59}$$

Independent variations of $\delta\varphi_{rs}$ give the six Euler differential equations

$$e_{imr} e_{jns} \widehat{e}_{ij,mn} = 0 \qquad (r, s = 1, 2, 3). \tag{16.60}$$

These are seen to be six *compatibility equations* for the six strain components. Recall that the strain components are defined in terms of three displacement components and, therefore, cannot be completely independent. For example, we see from

$$e_{11} = u_{1,1}, \qquad e_{22} = u_{2,2}, \qquad 2e_{12} = u_{1,2} + u_{2,1}$$

that $e_{11}$, $e_{22}$, and $e_{12}$ are related by

$$e_{11,22} + e_{22,11} - 2e_{12,12} = 0$$

which is one of the six equations in (16.60), namely, when $r = s = 3$. The remaining five relations can also be read off from the six strain-displacement relations with no difficulty.

In the complementary energy formulation, nothing is said about the strain-displacement relations. The strain components are given in terms of the stress components by $e_{ij} = \partial C / \partial \sigma_{ij}$ and the six compatibility equations among these strain components will have to be satisfied in order to be consistent with their actual definition (16.4). These six compatibility equations provide the six conditions needed for the determination of the six unknown stress functions $\varphi_{ij}(= \varphi_{ji})$.

The relevant Euler boundary conditions can also be obtained from the surface integral after writing

$$\delta\boldsymbol{\sigma}_n = \delta\sigma_{nn}\mathbf{n} + \delta\sigma_{n1}\mathbf{t}_1 + \delta\sigma_{n2}\mathbf{t}_2,$$

where $\mathbf{t}_1$ and $\mathbf{t}_2$ are unit tangent vectors on the surface $S$ with $\mathbf{t}_1 \cdot \mathbf{t}_2 = 0$ and $\mathbf{t}_1 \times \mathbf{t}_2 = \mathbf{n}$. We will not pursue a discussion of these Euler boundary conditions here as they will not be needed in the subsequent development.

## 9. Semidirect Method

We indicated in Chapter 15 that direct methods for one-dimensional problems such as those discussed in Chapter 8 can be extended in a natural way to problems in higher dimensions. Trial functions are now specified as a linear combination of coordinate functions of several variables with a number of unknown constants to be determined in the extremizing process. For the two-variable case, we typically take

$$u_N(x, y) = \varphi_0(x, y) + c_1\varphi_1(x, y) + c_2\varphi_2(x, y) + \dots + c_N\varphi_N(x, y) \tag{16.61}$$

where the coordinate functions $\{\varphi_1, \varphi_2, \dots, \varphi_N\}$ are specified in such a way that they satisfy the *homogeneous* counterpart of the prescribed boundary condition(s) of the problem, and $\varphi_0$ is any function which satisfies the prescribed boundary conditions. With $J[u_N] \equiv J(c_1, \dots, c_N)$, we now have a classical parameter optimization problem of choosing $(c_1, \dots, c_N)$ to extremize $J(c_1, \dots, c_N)$.

With two or more independent variables, some novel variations of the above technique may be adopted to take advantage of special features of a given problem. For a two-variable problem, suppose we have a good estimate of the dependence of the solution on one of the variables, say $x$, but not on the other variable $y$. In that case, we may take as our trial functions

$$u_N(x, y) = \varphi_0(x, y) + c_1\psi_1(x)f_1(y) + c_2\psi_2(x)f_2(y) + \dots + c_N\psi_N(x)f_N(y). \tag{16.62}$$

Here, $\varphi_0$ is as previously described and the functions $\{\psi_k(x)\}$ are suitably chosen (prescribed) functions of $x$ alone and the functions $\{f_m(y)\}$ are unknown functions of $y$ alone. For example, if we expect $u(x,y)$ to be sinusoidal in $x$, we might take $\psi_m(x) = \sin(m\pi x)$, $m = 1, 2, \ldots$. In the form (16.62), the trial functions are only partially specified and the variational method will determine the unknown functions $\{f_1, f_2, \ldots\}$. Typically, the integration with respect to $x$ can be carried out in $J[\mathbf{u}]$, reducing it from a double integral to a one-dimensional integral. The Euler differential equations for $\{f_k(y)\}$ will be ordinary differential equations instead of partial differential equations, and ordinary differential equations are generally much easier to analyze. In this section, we will indicate the essential features of this *semidirect method* using a simple membrane problem as an illustrative example. The same method will then be used for genuine elasticity problems in section 10 and in Chapter 17.

Consider a flat elastic membrane lying inside the closed curve $C$ in the $x,y$-plane, in static equilibrium under edge tension $T$ per unit arc length. The membrane is subject to an external transverse load intensity (force per unit undeformed area) $p(x,y)$ in the $z$ direction while the condition at the edge of the membrane remains unchanged. We are interested in the small-amplitude transverse deflection of the membrane.

We limit our discussion to problems with negligible in-plane shear and negligible deformation parallel to the $x, y$-plane. The small-amplitude transverse displacement $w(x,y)$ at a point $(x,y)$ of such a membrane is governed by the second-order partial differential equation [see Wan, (1989)]

$$(T_x w_{,x})_{,x} + (T_y w_{,y})_{,y} + p = 0, \tag{16.63}$$

where $T_\xi(x,y)$ is the tension in the $\xi$ direction of the stretched flat membrane known from in-plane equilibrium calculation. For a small deflection theory, these components of tension are taken to be the same value $T(x,y)$. The edge may be *fixed* (in which case we have

$$w(x,y) = 0 \tag{16.64a}$$

there) or *free* from restraints in the transverse direction (in which case we have

$$Tw_{,n}(x,y) = 0 \tag{16.64b}$$

there). Either (16.64a) or (16.64b) serves as the boundary condition for (16.63) along $C$*.

The boundary-value problem for $w(x,y)$ generally does not have an explicit exact solution especially when the tension in the membrane is not uniform. An

* For small-amplitude transverse deflections, it is consistent to regard the boundary of the membrane and the tension needed to maintain equilibrium along the edge to remain $C$ and $T$, respectively, after deformation.

approximate solution using a variational formulation of the problem is of interest for these problems. To construct the appropriate functional $J[w]$, we form

$$0 = -\iint_R [(Tw_{,x})_{,x} + (Tw_{,y})_{,y} + p]\,\delta w\,dx\,dy$$
$$= \iint_R [Tw_{,x}\delta w_{,x} + Tw_{,y}\delta w_{,y} - p\,\delta w]\,dx\,dy - \oint_C Tw_{,n}\delta w\,ds. \quad (16.65)$$

The line integral vanishes for either $\delta w = 0$ for that portion of $C$ where the membrane is fixed (against transverse displacement) or $Tw_{,n} = 0$ along the free portion of the edge. With $p$ and $T$ prescribed (and not function of $w$), we may write the double integral in (16.65) as $\delta J = 0$ with

$$J[w] = \frac{1}{2}\iint_R [T(w_{,x})^2 + T(w_{,y})^2 - 2pw]\,dx\,dy. \quad (16.66)$$

The integral $J[w]$ has the dimension of work (force × length) and is, in fact, the potential energy of the deformed membrane. Evidently, the partial differential equation for $w$ is the Euler differential equation of $J[w]$.

Consider the problem of a rectangular membrane spanning the region $\{|x| \le a, |y| \le b\}$ with three of the four edges fixed so that

$$w(-a, y) = w(a, y) = w(x, -b) = 0 \quad (16.67a,b,c)$$

and the remaining edge free of transverse tractions or constraints so that (for $T \neq 0$ along $C$)

$$w_{,y}(x, b) = 0. \quad (16.67d)$$

The loading on the membrane is quadratic in $x$:

$$p(x, y) = -p_0(a^2 - x^2). \quad (16.68)$$

The solution of this boundary-value problem is nontrivial even for the case of uniform tension, $T = T_0$ (a constant). However, we may expect the transverse displacement $w$ to be approximately parabolic in $x$ in view of the prescribed load distribution. If we do not know much about the $y$ dependence of $w(x, y)$, we may seek a solution of the form

$$w(x, y) = (a^2 - x^2)f(y). \quad (16.69)$$

Note that $w(\pm a, y) = 0$ so that (16.67a) and (16.67b) are satisfied by (16.69). Unfortunately, such a solution does not satisfy the partial differential equation (16.63) (with $T_x = T_y = T_0$) for any $f(y)$. One approach to obtaining an approximate solution of the form (16.69) would be to extremize $J[w]$ relative to this restricted class of comparison functions with $f(-b)=0$. Upon substitution of (16.69) into (16.66), we obtain

$$J[w]=\int_{-b}^{b}\int_{-a}^{a}\left\{\frac{T_0}{2}\{-2xf\}^2+\{(a^2-x^2)f'\}^2+p_0(a^2-x^2)^2f\right\}dx\,dy$$

$$=\frac{T_0}{2}\int_{-b}^{b}\left\{\frac{8a^3}{3}f^2+\frac{16}{15}a^5(f')^2+p_0\frac{16}{15}a^5f\right\}dy.$$

The condition

$$\delta J=T_0\int_{-b}^{b}\left\{\frac{8a^3}{3}f\delta f+\frac{16}{15}a^5f'\delta f'+\frac{p_0}{T_0}\frac{16}{15}a^5\delta f\right\}dy$$

$$=\frac{16}{15}a^5T_0[f'\delta f]_{-b}^{b}+T_0\int_{-b}^{b}\left\{\frac{8a^3}{3}f-\frac{16}{15}a^5f''+\frac{p_0}{T_0}\frac{16}{15}a^5\right\}\delta f dy=0$$

requires that $f$ satisfy the Euler differential equation

$$f''-\frac{5}{2a^2}f=\frac{p_0}{T_0} \tag{16.70a}$$

and the Euler boundary condition

$$f'(b)=0. \tag{16.70b}$$

The approximate solution $(a^2-x^2)f(y)$ must also satisfy the end condition (16.67c), $w(x,-b)=0$, so that all comparison functions $f(y)$ must satisfy the end constraint

$$f(-b)=0. \tag{16.70c}$$

The boundary-value problem (16.70a)-(16.70c) determines $f(y)$ and, hence, the approximate extremal (16.69).

It should be evident that the key to our approximate solution process is the construction of the variational problem of (16.66) from the boundary-value problem for the stretched membrane. The variational formulation of the same problem makes it possible for us to apply the semidirect method for a simpler determination of approximate extremals. Note that the original boundary conditions (16.67a) and (16.67c) on the unknown $w(x,y)$ continue to be boundary conditions for the variational problem. However, the boundary condition of the original boundary-value problem on the normal derivative of $w$ turns out to be an Euler boundary condition and is, therefore, not a part of the prescribed end conditions for the variational problem. The situation is analogous to that for one-dimensional problems discussed in Chapter 3.

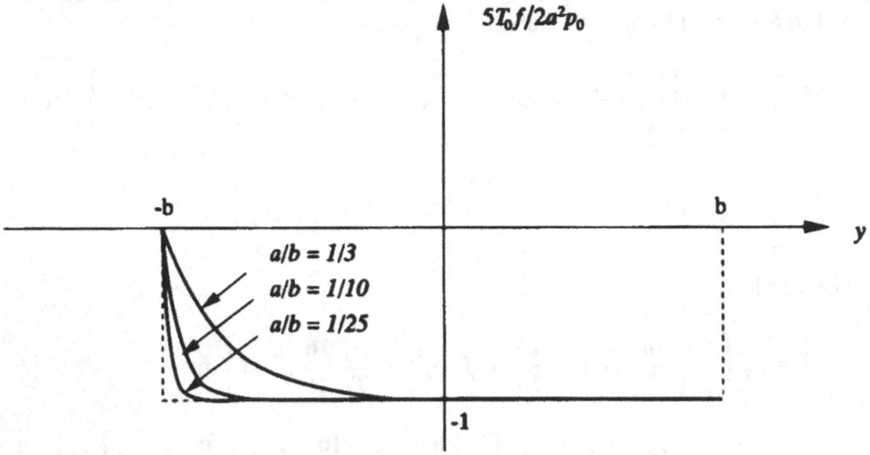


**Figure 16.6**

It is of some interest to note that the exact solution of the boundary-value problem (16.70a)-(16.70c) is

$$f(y) = \frac{2a^2 p_0}{5T_0}\left\{\frac{\cosh(\sqrt{5/2}\,(b-y)/a)}{\cosh(\sqrt{10}\,b/a)} - 1\right\}. \tag{16.71}$$

For a narrow rectangular membrane so that $a/b \ll 1$, we have for $y + b > \gamma a$ $(\gamma \geq 2)$

$$f(y) \approx \frac{2a^2 p_0}{5T_0}\{e^{-[\sqrt{5/2}\,(3b+y)/a]} - 1\} \tag{16.71'}$$

except for (exponentially small) terms of the order of $e^{(-\sqrt{10}\,b/a)}$. For $y + b > 2a$, the exponential term is negligibly small compared to 1. Therefore, for $a/b \ll 1$, the solution for $f(y)$ is effectively a constant, $-2a^2p_0/5T_0$, corrected abruptly near $y = -b$ to satisfy the boundary condition $f(-b) = 0$ (see Fig. 16.6). It would take a large number of terms for trial functions in the form of a power or trigonometric series to provide an adequate approximation of this solution with such a sharp gradient near an end point ($y = -b$).

## 10. Saint–Venant Torsion

In most engineering applications, the actual stress distributions applied over the boundary surface(s) of a body are not known in detail, but only their resultant forces and moments. A typical example is the torsion of a straight prismatic beam whose uniform, simply-connected cross section is bounded by the closed curve $C$ described by $f(x_1,x_2) = 0$. The cylindrical surface of the beam is free of surface traction. The two ends, $x_3 = 0, L$, are subject to external surface loads resulting in equal and opposite prescribed axial torques but no other resultant forces and moments. The distribution of the end tractions is not known in detail. Because many different stress distributions have the same resultant force and moment, the known boundary data are incomplete for the purpose of specifying the response of the elastic body to the external loading. Many different solutions are, therefore, possible for the problem of a prismatic beam subject to equilibrating axial end torques. In what follows, we obtain one such solution by the semidirect method described in the last section and discuss the significance of this so-called Saint–Venant torsion solution.

Away from its ends, the cross sections of an isotropic elastic beam under equal and opposite axial end torques are known to rotate more or less rigidly about a point in the plane of the cross section (known as the *center of twist*). At the same time, the out-of-plane warping of the cross section changes very little from cross section to cross section. On the basis of these observations, we seek Rayleigh-Ritz solution for the elastostatic boundary-value problem of the beam using the following trial solution in the theorem of minimum potential energy:

$$u_1 = \theta x_2 x_3, \qquad u_2 = -\theta x_1 x_3, \qquad u_3 = \theta w(x_1,x_2). \tag{16.72}$$

In the expressions for $u_k(x_1,x_2,x_3)$, $\theta$ is an unknown constant parameter and $w(x_1,x_2)$ is the unknown *warping function* [To minimize new nomenclatures, we will not distinguish the trial solution from the actual and will understand that (16.72) and its consequences are generally approximations.]

For the trial solution (16.72), we have the following expressions for the infinitesimal strain components:

$$\begin{aligned} &e_{11} = e_{22} = e_{33} = e_{12} = e_{21} = 0, \\ &e_{13} = e_{31} = \theta(w_{,1} + x_2), \qquad e_{23} = e_{32} = \theta(w_{,1} - x_1). \end{aligned} \tag{16.73}$$

The expression for the potential energy $P$ in (16.41) then becomes

$$\begin{aligned} P &= \theta^2 L \iint_A G[(w_{,1} + x_2)^2 + (w_{,2} - x_1)^2]\,dx_1\,dx_2 \\ &\quad - \theta \iint_A [\sigma_{13}^* x_2 x_3 - \sigma_{23}^* x_1 x_3]_{x_3=0}^{L}\,dx_1\,dx_2 \\ &= \theta^2 L \iint_A G[(w_{,1} + x_2)^2 + (w_{,2} - x_1)^2]\,dx_1\,dx_2 - \theta L T, \end{aligned} \tag{16.74}$$

where $A$ is the cross-sectional area of the beam and

$$T = \iint_A [\sigma_{13}^* x_2 - \sigma_{23}^* x_1]\, dx_1\, dx_2 \tag{16.75}$$

is the known axial end torque (and by overall equilibrium, equal to the axial torque over any cross section).

The Euler differential equations associated with $\delta P = 0$ is seen to be

$$(Gw_{,1})_{,1} + (Gw_{,2})_{,2} = (Gx_1)_{,2} - (Gx_2)_{,1}. \tag{16.76}$$

For a homogeneous beam, $G$ is a constant and the partial differential equation (16.76) reduces to the Laplace's equation

$$\nabla^2 w = 0. \tag{16.77}$$

The Euler boundary condition resulting from $\delta P = 0$ is

$$(w_{,1} + x_2)\nu_1 + (w_{,2} - x_1)\nu_2 = 0 \tag{16.78}$$

or

$$w_{,n} = -\left(x_2 \frac{dx_2}{ds} + x_1 \frac{dx_1}{ds}\right) = -\frac{1}{2}\frac{d}{ds}(x_2^2 + x_1^2), \tag{16.78'}$$

where $w_{,n} = w_{,1}\nu_1 + w_{,2}\nu_2$ with $\nu_1 = dx_2/ds$ and $\nu_2 = -dx_1/ds$ being the directional cosines of the outward unit normal of the edge curve $C$ in its own plane. In addition to the Euler differential equation (16.76) and the Euler boundary condition (16.68), the independent variation of the unknown parameter $\theta$ requires also that the following integral condition be satisfied by $w(x_1, x_2)$ and $\theta$:

$$\theta = \frac{T}{\iint_A G[(w_{,1} + x_2)^2 + (w_{,2} - x_1)^2]\, dx_1\, dx_2} \tag{16.79}$$

With $w(x_1, x_2)$ determined by (16.76) and (16.78), this condition determines the unknown $\theta$ in terms of the prescribed axial end torque $T$.

After a preliminary transformation, the boundary-value problem (16.76) and (16.78) and the integral condition (16.79) [which determine the Rayleigh–Ritz solution (16.72)] all have a simple physical interpretation. Equation (16.76) is satisfied identically by setting

$$G(w_{,1} + x_2) = -\varphi_{,2}, \qquad G(w_{,2} - x_1) = \varphi_{,1} \tag{16.80}$$

for any $\varphi$. Relations (16.80) which define $\varphi$ themselves require a compatibility condition be satisfied by $\varphi$:

$$(w_{,1})_{,2} = -\left(\frac{\varphi_{,2}}{G} + x_2\right)_{,2} = \left(\frac{\varphi_{,1}}{G} + x_1\right)_{,1} = (w_{,2})_{,1}$$

or

$$\left(\frac{\varphi_{,1}}{G}\right)_{,1}+\left(\frac{\varphi_{,2}}{G}\right)_{,2}=-2. \tag{16.81}$$

The boundary condition for $\varphi$ which supplements the partial differential equation (16.81) follows upon substituting (16.80) into (16.78):

$$-\varphi_{,2}\nu_1+\varphi_{,1}\nu_2=-\frac{d\varphi}{ds}=0 \quad \text{for} \quad f(x_1,x_2)=0 \tag{16.82}$$

or $\varphi = C_0$ along the single boundary curve $C$ of the simply-connected cross section. For such a cross section, the constant $C_0$ may be set to zero as it plays no role in the actual solution of the physical problem as only $\varphi_{,1}$ and $\varphi_{,2}$ are involved in the expressions for stresses and displacements. We may, therefore, use, instead of (16.82), the simpler condition

$$\varphi = 0 \quad \text{for} \quad f(x_1,x_2)=0\,. \tag{16.82'}$$

The inhomogeneous boundary-value problem (16.81) and (16.82′) determines $\varphi$ uniquely. For a uniform beam, $G$ is a constant and (16.81) reduces to a Poisson's equation in two dimensions:

$$\nabla^2\varphi = 2G\,. \tag{16.81'}$$

In terms of $\varphi(x_1,x_2)$, the denominator of the integral condition (16.79) becomes

$$\iint_A \frac{1}{G}\left[\varphi_{,2}^2+\varphi_{,1}^2\right]dx_1\,dx_2$$

$$=\oint_C \frac{1}{G}\varphi\varphi_{,n}\,ds-\iint_A\left[\left(\frac{\varphi_{,2}}{G}\right)_{,2}+\left(\frac{\varphi_{,1}}{G}\right)_{,1}\right]\varphi\,dx_1\,dx_2=\iint_A 2\varphi\,dx_1\,dx_2$$

where we have used (16.82′) to eliminate the line integral and (16.81) to simplify the double integral. The integral condition itself becomes

$$\theta=\frac{T}{2\iint_A \varphi\,dx_1\,dx_2}\,. \tag{16.83}$$

It is evident from (16.80) that the shear stress components transverse to the axis of the beam are $\sigma_{13}=\sigma_{31}=-\theta\varphi_{,2}$ and $\sigma_{23}=\sigma_{32}=\theta\varphi_{,1}$. The function $\varphi$ is, therefore, known as a *stress function* for the problem. Given that $\varphi$ is completely determined by the boundary-value problem (16.81) and (16.82′) and $\theta$ by (16.83), the transverse shear stresses $\sigma_{31}$ and $\sigma_{32}$ cannot be arbitrarily prescribed at the two ends of the beam (and the axial stress $\sigma_{33}$ must vanish there).

Yet, there are many other end stress fields which have the same resultant axial torque (and vanishing resultant forces and bending moments). In what sense, then, do we have an adequate understanding of the elastostatic response of slender bodies to the external loads of which we know only their resultant forces and moments? As an answer to this critical question, Saint–Venant postulated that

> *the difference between two solutions for the present problem with the same resultant end force and moment is negligible at a distance away from the ends long compared to a representative span of the cross section.*

This conjecture has been successfully justified in a number of special cases (Horgan and Knowles, 1983). Although the issue involved is beyond the subject of variational calculus, it is worth noting that this so-called Saint–Venant principle is essentially concerned with boundary-layer phenomena and may, therefore, be treated by asymptotic analysis.

The boundary-value problem for $\varphi$ was solved by the finite element method in Chapter 15. The variational principle used there for the Rayleigh-Ritz solution is easily obtained by multiplying (16.81) by $\delta\varphi$ and integrating by parts to get

$$\iint_A \left[\left(\frac{\varphi_{,1}}{G}\right)_{,1} + \left(\frac{\varphi_{,2}}{G}\right)_{,2} + 2\right]\delta\varphi\, dx_1\, dx_2$$

$$= \oint_C \frac{1}{G}\,\varphi_{,n}\,\delta\varphi\, ds - \delta\iint_A \left[\left(\frac{1}{2G}\right)(\varphi_{,1}^2 + \varphi_{,2}^2 - 4G\varphi)\right] dx_1\, dx_2 .$$

Given that $\varphi$ must satisfy the boundary condition (16.82′) along $C$, the partial differential equation (16.81) is evidently the Euler differential equations of

$$J[\varphi] \equiv \iint_A \frac{1}{G}[\varphi_{,1}^2 + \varphi_{,2}^2 - 4G\varphi]\, dx_1\, dx_2 . \qquad (16.84)$$

With $u = \theta\varphi$, this result was used in Section 11 of Chapter 15 to illustrate the finite element method [see (15.50)].

## 11. Exercises

1. (a) Determine $E_{ij}$ and $e_{ij}$ for the displacement field

$$\mathbf{u} = \mathbf{c} + \boldsymbol{\omega} \times \mathbf{r},$$

where $\mathbf{c}$ and $\boldsymbol{\omega}$ are constant vectors. If a *rigid body displacement* is defined as a displacement field which gives rise to no strain in the body, so that $E_{ij} = 0$ $(i, j = 1, 2, 3)$ for a finite strain formulation, in what sense is the above $\mathbf{u}$ a rigid body displacement?

(b) Show that the vector $\mathbf{u}$ defined by

$$\mathbf{u} = (c_1 - x_1 + x_1 \cos\alpha - x_2 \sin\alpha)\mathbf{i}_1$$
$$+ (c_2 - x_2 + x_1 \sin\alpha - x_2 \cos\alpha)\mathbf{i}_2 + c_3\mathbf{i}_3$$

is a true rigid body displacement for arbitrary constants $c_k$ and $\alpha$.

2. Show that the area $dA_3$ of the face of the deformed box enclosed by the vectors $\mathbf{r}^*_{,1}dx_1$ and $\mathbf{r}^*_{,2}dx_2$ is given by

$$dA_3 = (1 + E_{11} + E_{22} + E_{11}E_{22} - E_{12}^2)^{1/2}\, dx_1\, dx_2.$$

Give also an expression for $dA_3$ if all nonlinear terms in the derivatives of the displacement components are neglected.

3. Show that the volume $dV$ of the deformed box is given by the formula

$$dV = \begin{vmatrix} 1 + u_{1,1} & u_{2,1} & u_{3,1} \\ u_{1,2} & 1 + u_{2,2} & u_{3,2} \\ u_{1,3} & u_{2,3} & 1 + u_{3,3} \end{vmatrix} dx_1\, dx_2\, dx_3$$

and that, except for nonlinear terms, the relative change in volume due to deformation of the rectangular box with edges $dx_k$ is given by $\nabla \cdot \mathbf{u}$.

4. (a) The notion of *physical strain* $\varepsilon$ is defined as the change in distance between two points per unit (undeformed) distance (Fig. 16.7a), i.e.,

$$\varepsilon = \frac{|d\boldsymbol{\rho}| - |d\mathbf{r}|}{|d\mathbf{r}|}.$$

Let $\varepsilon_1$ be $\varepsilon$ for two points $(x_1, x_2, x_3)$ and $(x_1 + dx_1, x_2, x_3)$, etc. Express $\varepsilon_k$ in terms of $E_{ij}$.

(b) Let $\gamma$ be the change of angle between two vectors $d\boldsymbol{\rho}^{(1)}$ and $d\boldsymbol{\rho}^{(2)}$ which correspond to two vectors $d\mathbf{r}^{(1)}$ and $d\mathbf{r}^{(2)}$ at right angles to each other before deformation (Fig. 16.7b). Let $\gamma_{23}$ be $\gamma$ for the case $d\mathbf{r}^{(1)} = dx_2\mathbf{i}_2$ anb $d\mathbf{r}^{(2)} = dx_3\mathbf{i}_3$, etc. Express $\gamma_{jk}$ in terms of $E_{mn}$.

5. Obtain the solution of the homogeneous equilibrium equations (with $f_k = 0$, $k = 1, 2, 3$) which satisfies the conditions

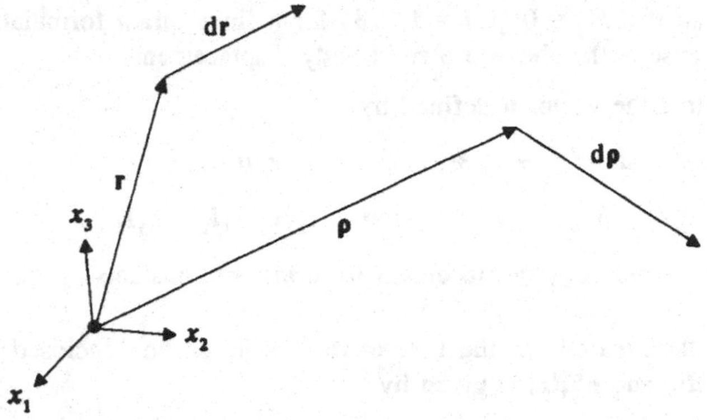


**Figure 16.7a**

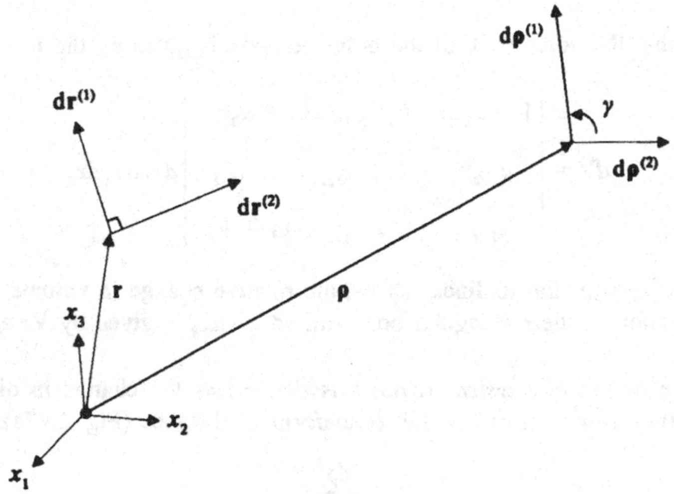


**Figure 16.7b**

(a) $\boldsymbol{\sigma}_n = -T_0\mathbf{n}$ (uniform pressure $T_0$ is a constant) on the surface of a body of arbitrary shape,

(b) $\boldsymbol{\sigma}_1 = \mp T_0\mathbf{i}_1$ at the end sections $x_1 = \pm L$ and $\boldsymbol{\sigma}_n = 0$ on the cylindrical surface of a cylindrical body (Fig. 16.5).
[*Hint*: You have only three equations for six unknowns, so do something obvious! ]

6. In a theory which allows for moment stresses, the vector moment equilibrium equation will be different from that of the classical theory.

   (a) Assume the validity of Cauchy's formula for the moment stress vectors and use it to derive the vector moment equilibrium equation for such a theory.

   (b) Use the component representation

$$\boldsymbol{\tau}_j = \tau_{jk}\mathbf{i}_k \quad \text{and} \quad \mathbf{m} = m_k\mathbf{i}_k$$

   to obtain the corresponding three scalar moment equilibrium equations.

   (c) If we now set $\tau_{ij} = 0$, $i, j = 1, 2, 3$, do these equilibrium equations imply $\sigma_{ij} = \sigma_{ji}$?

7. Recall that if $s_{jk}$ is any Cartesian tensor, then $I_1 = s_{kk}$ is an invariant with respect to a rotation of the Cartesian coordinate axes about the origin. With $I_1$ expressed as $\delta_{jk}s_{jk}$, prove the following *without the assumption of symmetric tensors*:

   (a) If $t_{ij}$ is also a Cartesian tensor, then $t_{ij}s_{ji}$ is an invariant.

   (b) $s_{ik}s_{ki}$ is an invariant.

   (c) $I_2 = [(s_{ij}s_{ji} - (s_{kk})^2]/2$ and is an invariant.

8. (a) If the stress potential $S$ is a function of the invariants $I_1, I_2$ and $I_3$ of the strain tensor $E_{ij}$, show that the relations

$$\sigma_{ij} = \frac{\partial S}{\partial E_{ij}}$$

   are invariant with respect to an orthogonal transformation of the Cartesian reference frame.

   (b) Show that there are no preferred elastic directions in the $x_1, x_2$-plane for a material with a strain potential of the form

$$C = \frac{\sigma_{11}^2 + \sigma_{22}^2 - 2\nu\sigma_{11}\sigma_{22} + (1+\nu)\sigma_{12}^2}{2E} + \frac{\sigma_{13}^2 + \sigma_{23}^2}{4G_3}$$

$$+ \frac{\sigma_{33}^2}{2E_3} - \frac{\nu_3\sigma_{33}(\sigma_{11} + \sigma_{22})}{E},$$

   where $E, \nu, E_3, G_3$, and $\nu_3$ are independent elastic parameters.

   (c) Obtain the stress-strain relations for the *transversely isotropic* material of part (b).

9. A medium is *incompressible* if $e_{kk} = 0$ (no change in volume in the context of an infinitesimal strain theory) for all loading conditions.

   (a) Show that a linearly isotropic medium is incompressible if and only if $\nu = 1/2$.

   (b) The corresponding stress-strain relations can be written as

$$e_{jk} = \frac{3}{2E}(\sigma_{jk} - \sigma_0 \delta_{jk}),$$

   where $\sigma_0$ is the *mean stress* $( = I_1/3)$.

10. Show that $I_2$ is invariant with respect to an orthogonal transformation.

11. Suppose a body is linearly elastic and its stress potential is positive definite.
    (a) Show that the solution of the stress BVP is unique up to a rigid body displacement field.

    (b) Show that the solution of the displacement BVP is unique.
    [*Hint*: In the absence of body forces, we have $\sigma_{ij,i}u_j = 0$. Its volume integral is still zero]

12. *The Bette–Rayleigh Reciprocal Theorem*. Suppose two different sets of body forces $\mathbf{f}^{(1)}$ and $\mathbf{f}^{(2)}$ and surface tractions $\mathbf{T}^{(1)}$ and $\mathbf{T}^{(2)}$ are applied to the same linearly elastic body resulting in two different static equilibrium states $\{\sigma_{ij}^{(k)}, e_{mn}^{(k)}, u_\ell^{(k)}\}$, $k = 1, 2$, respectively. Let

$$U_{mn} = \frac{1}{2}\iiint \sigma_j^{(m)} e_{ij}^{(n)}\, dV,$$

$$W_{ik} = \frac{1}{2}\iiint_V \mathbf{f}^{(i)} \cdot \mathbf{u}^{(k)}\, dV + \frac{1}{2}\oint\!\!\oint_S \mathbf{T}^{(i)} \cdot \mathbf{u}^{(k)}\, dS \quad (i \neq k).$$

    (a) Show that $U_{mn} = U_{nm}$.

    (b) Show that $W_{ki} = W_{ik}$. [*Hint*: Use $U_{mn} = U_{nm}$.]

13. Suppose we write in Exercise (12), $\mathbf{f}^{(2)} = \mathbf{f}^{(1)} + \delta\mathbf{f}$, $\mathbf{T}^{(2)} = \mathbf{T}^{(1)} + \delta\mathbf{T}$, $\mathbf{u}^{(2)} = \mathbf{u}^{(1)} + \delta\mathbf{u}$, etc.

    (a) Deduce from the results of Exercise (12) the following *principle of virtual work*:

$$\iiint_V \mathbf{f}^{(1)} \cdot \delta\mathbf{u}\, dV + \iint_{S_\sigma} \mathbf{T}^{(1)} \cdot \delta\mathbf{u}\, dS = \iiint_V \sigma_{ij}^{(1)} \delta e_{ij} \delta e_{ij}\, dV,$$

    where $S_\sigma$ is the portion of $S$ on which surface tractions are prescribed.

(b) Deduce the *principle of complementary virtual work*

$$\iiint_V \mathbf{u}^{(1)} \cdot \delta \mathbf{f}\, dV + \iint_{S_d} \mathbf{u}^{(1)} \cdot \delta \mathbf{T}\, dS = \iiint_V e_{ij} \delta \sigma_{ij}\, dV,$$

where $S_d$ is the portion of $S$ on which displacements are prescribed.

14. *Plane stress theory*: Suppose $\mathbf{f} = \mathbf{0}$ and $\sigma_{3k} = \sigma_{k3} = 0$.

(a) Show that the two nontrivial equilibrium equations are satisfied identically by

$$\sigma_{11} = F_{,22}, \quad \sigma_{22} = F_{,11}, \quad \sigma_{12} = \sigma_{21} = -F_{,12}$$

for the Airy stress function $F$ (still to be determined).

(b) Show that

$$e_{11,22} + e_{22,11} = 2e_{12,12}.$$

(c) Use the stress-strain relations to express $e_{ij}, i, j = 1, 2$, in terms of $F$ and obtain an equation for $F$.

15. *Generalized Plane Stress*: Consider a flat plate extending from $-h/2$ to $h/2$ in the $x_3$ direction. Suppose $\sigma_{31}$ and $\sigma_{32}$ are not assumed to vanish except on $x_3 = \pm h/2$ (but $\sigma_{33} = 0$). Integrate the first two equilibrium equations across the thickness and write

$$N_{ij} = \int_{-h/2}^{h/2} \sigma_{ij}\, dx_3.$$

Repeat parts (a)-(c) for the integrated equilibrium equations, stress-strain relations, and strain-displacement relations.

16. *Plane strain*: Suppose $\mathbf{f} = \mathbf{0}$, $u_3 = 0$, and all other quantities are uniform in the $x_3$ direction so that $(\ )_{,3} = 0$.

(a) Show that $e_{3k} = e_{k3} = 0$ and use $e_{33} = 0$ to eliminate $\sigma_{33}$ from $e_{11}$ and $e_{22}$.

(b) Show that a reduction similar to that of Exercise 14 can be carried out so that again we only have to solve the biharmonic equation in two dimensions.

17. (a) For the membrane problem with $p(x, y) = -p_0$ [instead of (16.68)] and the same boundary conditions as given by (16.67), assume again (16.69) and obtain the Euler DE for $f(y)$.

(b) Obtain the exact solution for $f(y)$ and its (asymptotic) behavior for $a/b \ll 1$.

18. Obtain the exact solution for the actual membrane problem (Exercise 17) as follows:

(a) Obtain an exact solution of the PDE in the form $w_p(x)$ with $w_p(\pm a) = 0$. [Note that this solution also satisfies the BC, $w_{,y}(x, b) = 0$ but not $w(x, -b) = 0$.]

(b) Obtain exact solutions $w_m(x, y)$ of the homogeneous PDE and the BC, $w_m(\pm a, y) = 0$, $w_{m,y}(x, b) = 0$ by separation of variables.

(c) By linearity, we may take

$$w(x,y) = w_p(x) + \sum_{m=1}^{\infty} a_m w_m(x, y).$$

Choose the Fourier coefficients $\{a_m\}$ so that $w(x, b) = 0$.

(d) Compare $w(0,0)$ of the present solution with that of Exercise (17).

19. (a) Verify that each of the following $F$ satisfies $\nabla^2\nabla^2 F = 0$, with arbitrary constants $a_i$, $b_i$, etc., unless specified otherwise and calculate the corresponding stress resultants $N_{ij}$:

(*i*) $F = \dfrac{1}{2!}[a_2 x_1^2 + 2b_2 x_1 x_2 + c_2 x_2^2]$,

(*ii*) $F = \dfrac{1}{3!}[a_3 x_1^3 + 3b_3 x_1^2 x_2 + 3c_3 x_1 x_2^2 + d_3 x_2^3]$,

(*iii*) $F = \dfrac{1}{4!}[a_4 x_1^4 + 4b_4 x_1^3 x_2 + 6c_4 x_1^2 x_2^2 + 4d_4 x_1 x_2^3 + e_4 x_2^4]$

with $e_4 = -(2c_4 + d_4)$.

(b) Choose $d_4$ and $b_4$ in $F = d_4 x_1 x_2^3 + b_2 x_1 x_2$ so that $N_{22} = N_{21} = 0$ at $x_2 = \pm \ell_2$ and

$$\int_{-\ell_2}^{\ell_2} N_{12}\, dx_2 = -P.$$

# 17

# Plate Theory

## 1. The Elastostatics of Flat Plates

The boundary-value problems of linear elastostatics formulated in the last chapter generally do not admit an exact solution in terms of elementary or special functions. An accurate numerical solution for any fully three-dimensional problem is usually not feasible or practical with the computing power available to many engineers and designers. In the presence of layer phenomena, even two-dimensional problems (resulting from some kind of symmetry inherent in the geometry and loading) require an unacceptaly high level of computing. Historically, this situation gave considerable impetus to the search for adequate approximate theories for special classes of problems. One approach takes advantage of the special geometrical features of the elastic body to be analyzed. Intuition and experience gained from exact solutions for specific problems suggest simplyfying approximations for broad classes of problems involving thin or slender bodies. Applications of these approximations lead to simpler boundary value problems in lower dimensions, such as the one-dimensional theory of beams and two-dimensional theories of plates and shells. In this chapter, we discuss three approximate theories for the three-dimensional elastostatics of thin flat bodies bounded by two parallel planes and one or more cylindrical edge surfaces (see Fig. 17.1). We show how the variational formulations of the three-dimensional elasticity theory can be used in conjunction with the semidirect method to derive these approximate theories. In the process, we also resolve a long-standing difficulty concerning the appropriate stress boundary conditions for thin plate theory.

We begin our discussion of plate theories with a derivation of the classical Germain–Kirchhoff theory for thin plates [see Bucciarelli and Dworsky, 1980, for a historical background of this theory]. This approximate theory was originally developed by adopting a set of assumptions about the behavior of thin elastic bodies, known as the *Kirchhoff hypotheses*, which simplify the three-dimensional boundary-value problem of elastostatics. Although they are physi-

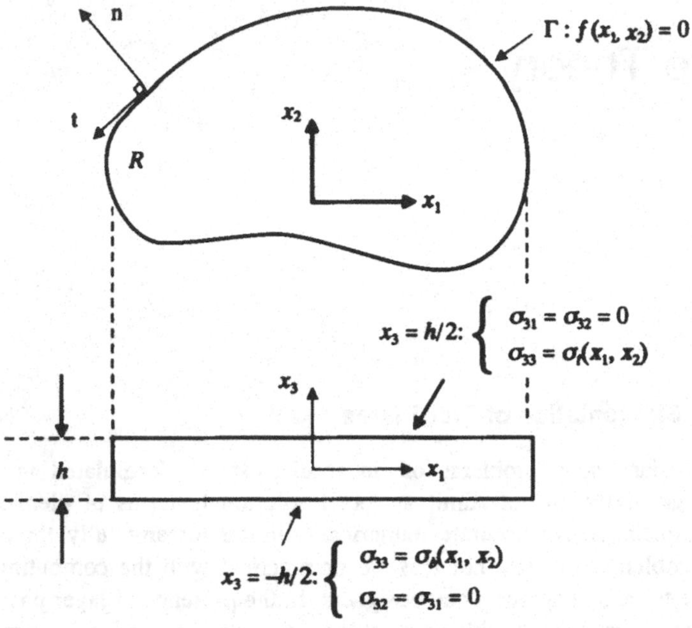


**Figure 17.1**

cally reasonable, these assumptions appeared internally inconsistent and required a great deal of rationalization by investigators trying to understand and apply the resulting attractive and useful theory. The inconsistencies in the Kirchhoff hypotheses were resolved by establishing the singular perturbation structure of the boundary-value problem. An appropriate asymptotic analysis shows that the classical thin plate theory is nothing more than the leading term of an *outer* (*asymptotic expansion* of the exact) *solution* of the problem (Friedrichs and Dressler, 1961). We will see in this chapter how we can avoid altogether the inconsistencies associated with the Kirchhoff hypotheses by the semidirect method of the calculus of variations.

We know from the technique of matched asymptotic expansions (Kevorkian and Cole, 1981) that the outer solution for a given problem generally cannot, by itself, satisfy all the given boundary conditions of the problem. This is generally the case for flat plate problems. Unfortunately, it is also rarely possible to obtain the leading term of the *inner* (*asymptotic expansion* of the exact) *solution* for most of these same problems. For these reasons, there has been considerable research effort in the last three decades directed toward removing the difficulty of fitting the appropriate (plate theory) solution to the prescribed boundary data. We will indicate in this chapter how the calculus of variations offers a different approach to this problem.

We limit our discussion of plate theory here mainly to flat plates bounded by two parallel planes and only one cylindrical surface normal to the planes. A Cartesian coordinate system is chosen so that the two bounding parallel planes are located at $x_3 = \pm h/2$ and the cylindrical surface is given by $\{f(x_1,x_2) = 0, |x_3| \leq h/2\}$. The bounding planes $x_3 = \pm h/2$ are called the *top and bottom* (or *upper and lower*) faces of the plate, respectively, and the cylindrical boundary is called the *edge* of the plate (Fig. 17.1). The boundary curve $\Gamma$ of the *middle plane* $x_3 = 0$, described by $f(x_1,x_2) = 0$, is assumed to have a minimum radius of curvature large compared to $h$ (except possibly for a few corners). A flat plate is said to be *thin* if the uniform plate *thickness* $h$ is negligibly small compared to the characteristic length (such as the minimum span) of the *midplane.*

For simplicity, we assume there are no body forces in the plate interior. The faces of the plate are subject only to surface traction normal to the midplane so that

$$\sigma_{31}(x_1,x_2, \pm h/2) = \sigma_{32}(x_1,x_2, \pm h/2) = 0 \qquad (17.1a,b)$$

$$\sigma_{33}(x_1,x_2, -h/2) = \sigma_b(x_1,x_2), \qquad (17.1c)$$

$$\sigma_{33}(x_1,x_2, h/2) = \sigma_t(x_1,x_2) \qquad (17.1d)$$

for all points $(x_1,x_2)$ in the region $R$ inside the boundary curve $\Gamma$. By linearity, the presence of body forces or applied shear stresses on the faces can always be eliminated by a particular solution of the boundary-value problem.

Along the edge of the plate, three appropriate boundary conditions are prescribed in terms of displacement and/or stress components. For prescribed edge displacements, we have from (16.38)

$$u_k(x_1,x_2,x_3) = U_k(x_1,x_2,x_3) \quad (k = 1,2,3) \qquad (17.2)$$

on the boundary $S \equiv \{f(x_1,x_2) = 0, |x_3| \leq h/2\}$. For prescribed edge tractions, we have from (16.39)

$$\sigma_{nk}(x_1,x_2,x_3) = T_k(x_1,x_2,x_3) \quad (k = 1,2,3) \qquad (17.3)$$

on $S$ where $\mathbf{n}$ is the unit normal to the curve $\Gamma$ in the $x_1,x_2$-plane. Other combinations of stress and displacement edge conditions can also be prescribed, but they will not be considered here.

Whatever the prescribed edge conditions, an exact solution of the boundary-value problem in terms of known functions is rarely possible. At the same time, accurate approximate solutions by numerical methods are often costly or cannot be found using available computing power. On the other hand, simplyfying approximations of the boundary-value problem based on the special geometry of the elastic body have proven effective for accurate approximate solutions.

## 2. The Germain–Kirchhoff Thin Plate Theory

**When a long, slender, and straight elastic body with a rectangular cross section is bent by equal and opposite moments at its ends, straight lines normal to its central axis (away from the two ends) are seen to remain straight and normal to the deformed central axis without any noticeable change in length (see Fig. 17.2). This observation was used along with several other approximating assumptions by L. Euler and D. Bernoulli to develop approximate solutions for the elastostatics of general beam bending problems involving transverse displacement of the central axis. G. Kirchhoff extended the Euler–Bernoulli hypotheses into an approximate method for analyzing the elastostatics of general plate bending problems, i.e., problems involving transverse displacement of the plate's midplane.**

**The Kirchhoff hypotheses may be taken in the form of two assumptions: (1) the plate experiences effectively a plane strain deformation so that $e_{3k} = e_{k3} = 0$, $k = 1, 2, 3$, and (2) the contributions of the transverse stress components $\sigma_{3k} = \sigma_{k3}$ are negligible (only) in the stress-strain relations. The stipulation of**

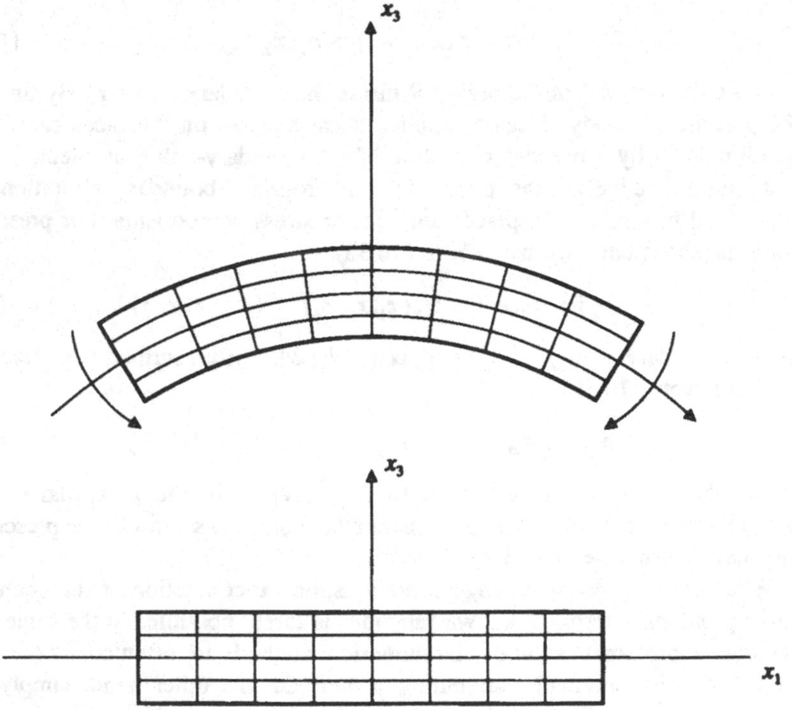


Figure 17.2

$e_{33}=0$ requires that the transverse displacement $u_3$ be uniform across the plate thickness, whereas $e_{31}=0$ and $e_{32}=0$ require that the *in-plane displacements* $u_1$ and $u_2$ be linear across the thickness so that

$$u_3(x_1,x_2,x_3)=w(x_1,x_2),$$
$$u_k(x_1,x_2,x_3)=-x_3 w_{,k}(x_1,x_2) \quad (k=1,2), \tag{17.4}$$

where $w(x_1,x_2)$ is still to be determined and we have omitted an unknown funciton of $(x_1,x_2)$ in $u_1$ and $u_2$ for problems without in-plane action. [Physically, assumption (17.4) corresponds to the requirement that a normal of the undeformed middle plane deforms without extension into a normal of the deformed middle plane]. The in-plane strain components are then given by

$$e_{ij}=\kappa_{ij}x_3 \quad (i,j=1,2) \tag{17.5}$$

with

$$\kappa_{ij}=-w_{,ij}. \tag{17.6}$$

At this point, we would normally invoke the second Kirchhoff assumption to simplify the stress-strain relations for the in-plane strain components to

$$e_{11}=\kappa_{11}x_3=\frac{\sigma_{11}-\nu\sigma_{22}}{E},\quad e_{22}=\kappa_{22}x_3=\frac{\sigma_{22}-\nu\sigma_{11}}{E},$$
$$e_{12}=\kappa_{12}x_3=\frac{1+\nu}{E}\sigma_{12}=\frac{1+\nu}{E}\sigma_{21}=\kappa_{21}x_3=e_{21} \tag{17.7}$$

(by neglecting the contribution of $\sigma_{33}$ in $e_{11}$ and $e_{22}$) as well as to maintain consistency in the two transverse shear stress-strain relations

$$e_{3j}=e_{j3}=0 \quad (j=1,2). \tag{17.8}$$

Upon inverting relations (17.7), we obtain

$$\sigma_{11}=x_3m_{11}=\frac{Ex_3}{1-\nu^2}(\kappa_{11}+\nu\kappa_{22}),\quad \sigma_{22}=x_3m_{22}=\frac{Ex_3}{1-\nu^2}(\kappa_{22}+\nu\kappa_{11}),$$
$$\sigma_{12}=x_3m_{12}=\frac{Ex_3}{1+\nu}\kappa_{12}=\frac{Ex_3}{1+\nu}\kappa_{21}=x_3m_{21}=\sigma_{21}. \tag{17.9}$$

For the stress components (17.9) to be a solution of the equations of linear elastostatics, they still have to satisfy equilibrium. The second Kirchhoff hypothesis requires that $\sigma_{k3}=\sigma_{3k}$ be negligible *only* in the stress-strain relations so that the three equilibrium equations

$$\sigma_{1j,1}+\sigma_{2j,2}+\sigma_{3j,3}=0 \quad (j=1,2,3), \tag{17.10}$$

in fact, determine the three transverse stress components in Kirchhoff's theory. We substitute the expressions (17.9) for $\sigma_{ij}$, i,j = 1,2, in the first two equilibrium equations

$$\sigma_{3k,3} = -(\sigma_{1k,1} + \sigma_{2k,2}), \quad (k = 1,2),$$

and integrate them with respect to $x_3$. For homogeneous and isotropic plates of uniform thickness without in-plane (stretching) deformation, the boundary conditions $\sigma_{3k} = 0$ $(k = 1,2)$ on $x_3 = \pm h/2$ then simplify the results to

$$\sigma_{3k} = q_k(x_1,x_2)(1 - z^2) \quad (k = 1,2), \tag{17.11}$$

where $z = x_3/(h/2)$ and [with $m_{ij}$ given in terms of $w$ by (17.9) and (17.6)]

$$q_k = \frac{h^2}{8}(m_{1k,1} + m_{2k,2}) = -\frac{Eh^2}{8(1-\nu^2)}(\nabla^2 w)_{,k} \quad (k = 1,2). \tag{17.12}$$

The remaining equilibrium equation gives

$$\sigma_{33} = \sigma_0(x_1,x_2) - \frac{h}{2}[q_{1,1} + q_{2,2}]\left(z - \frac{1}{3}z^3\right). \tag{17.13}$$

The boundary conditions for $\sigma_{33}$ on $x_3 = \pm h/2$, (17.1c) and (17.1d), then require the unknown function $\sigma_0(x_1,x_2)$ to be $(\sigma_t + \sigma_b)/2$. For the external loads to give rise to no in-plane stretching, we need

$$\sigma_0 = \frac{1}{2}(\sigma_t + \sigma_b) = 0. \tag{17.14a}$$

The same face conditions also require the expressions $q_1$ and $q_2$ to satisfy

$$q_{1,1} + q_{2,2} + \frac{3}{2h}p(x_1,x_2) = 0, \tag{17.14b}$$

where

$$p = (\sigma_t - \sigma_b), \tag{17.14c}$$

so that

$$\sigma_{33} = \frac{3}{4}(\sigma_t - \sigma_b)(z - \frac{1}{3}z^3) = \frac{3}{4}p(z - \frac{1}{3}z^3). \tag{17.15}$$

The requirement (17.14b) is effectively a condition on the only unknown $w(x_1,x_2)$ in our solution process. Upon substituting (17.12) into (17.14b), we obtain

$$D\nabla^2\nabla^2 w = p, \qquad D = \frac{Eh^3}{12(1-\nu^2)}. \tag{17.16}$$

This is an inhomogeneous fourth-order (biharmonic) partial differential equation to be supplemented by two boundary conditions at each edge of the plate. We will examine in the next section the implications of the condition (17.16) on the prescribed edge conditions of the plate.

With (17.15), $\sigma_{33}$ is completely determined by the face tractions specified for the plate. It is possible then to examine the validity of Kirchhoff's stipulation that $\sigma_{33}$ is negligible in the in-plane stress-strain relations

$$\sigma_{11} = \frac{Ex_3}{1-\nu^2}(\kappa_{11} + \nu\kappa_{22}) + \frac{\nu}{1-\nu}\sigma_{33},$$
$$\sigma_{22} = \frac{Ex_3}{1-\nu^2}(\kappa_{22} + \nu\kappa_{11}) + \frac{\nu}{1-\nu}\sigma_{33}. \tag{17.17}$$

Briefly, we have from the partial differential equation (17.16)

$$w = O(pL^4/D) = O(pL^4/Eh^3)$$

as an order-of-magnitude estimate. Correspondingly, we have

$$\kappa_{ij} = -w_{,ij} = O(pL^2/Eh^3).$$

With $\sigma_{33} = O(p)$ from (17.15), the contribution of the $\sigma_{33}$ terms in (17.17) is $O(h^2/L^2)$ compared to the contributions of the $\kappa_{ij}$ terms and is, therefore, negligible. This is consistent with taking the in-plane stresses (in terms of $w$) from (17.9) instead of (17.17). We already have the transverse stress components from (17.11), (17.12), and (17.15). Hence, all the stress and displacement components are known once we have $w(x_1, x_2)$ from the partial differential equation (17.16) and the relevant boundary conditions.

## 3. The Kirchhoff Contracted Stress Boundary Conditions

The two hypotheses which form the basis of the Germain–Kirchhoff theory for thin plates are internally inconsistent. The requirements $e_{31} = e_{32} = 0$ (which follow from the assumed displacement distributions in the thickness direction) and the stress-strain relations $\sigma_{3k} = 2\mu e_{3k}$, $k = 1, 2$ for an isotropic material require $\sigma_{31} = \sigma_{32} = 0$. These are at variance with (17.11) and (17.12) obtained from the equilibrium equations (and allowed by the second part of the hypotheses). Also, the stipulation that $\sigma_{33}$ be negligibly small in the in-plane stress-strain relations is incompatible with the remaining plane strain requirement $Ee_{33} = \sigma_{33} - \nu(\sigma_{11} + \sigma_{22}) = 0$. It follows from the latter that either we have $\sigma_{33} = 0$ and $\sigma_{11} + \sigma_{22} = 0$ or $\sigma_{33}$ is of the same order of magnitude as $\sigma_{11} + \sigma_{22}$; both are false in general. Given the usefulness of the Germain–Kirchhoff thin plate theory in engineering applications, a resolution of the inconsistencies of the Kirchhoff hypotheses became a fundamental issue in plate theory. We will see in the next section that one possible resolution is accomplished by the semidirect method of the calculus of variations.

Even with the difficulty of the Kirchhoff hypotheses resolved or avoided, there is still another difficulty associated with the appropriate boundary conditions for the biharmonic equation (17.16). This governing equation for the Germain–Kirchhoff classical thin plate theory is a fourth-order elliptic partial differential equation. For a well-posed boundary-value problem, *two* appropriate boundary conditions on $w$ and/or its derivatives are to be prescribed along each edge of the midplane $R$ of the plate. On the other hand, we have from the three-dimensional elasticity problem *three* independent (combinations of displacement and/or stress) boundary conditions on a cylindrical edge.

For the displacement field (17.4) to satisfy the three canonical prescribed displacement conditions along an edge, the across-the-thickness distributions of the prescribed edge data must have the form $U_3 = U_3(x_1,x_2), U_k = x_3\Phi_k(x_1,x_2)$, $k = 1,2$. Even if this requirement is met, the data must still be further restricted in their $x_1$ and $x_2$ dependence. For example, we must have $U_2 = -x_3U_{3,2}(x_1,x_2)$ (or $\Phi_2 = -U_{3,2}$) along an $x_1$ = constant edge to be compatible with (17.4). As such, only $w$ and its normal derivative can be prescribed along $\Gamma$. This situation is consistent with $w$ being a solution of the biharmonic equation. The following question now suggests itself: If the prescribed function $U_3$ is not uniform in $x_3$ and/or $U_1$ and $U_2$ are not linear in $x_3$, how should the data be fitted by the Kirchhoff solution (so that the latter is an accurate approximation of the exact solution at least away from the edge)? A complete resolution of this important issue (see R.D. Gregory and F.Y.M. Wan, 1985 and 1988) would lead us too far away from the calculus of variations and hence will not be pursued here. We will see in the next section how variational methods play a role in a partial resolution.

For prescribed stress data along the cylindrical edge of the plate, $\{f(x_1,x_2) = 0, |x_3| \le h/2\}$, the situation is even less favorable, whether or not the data $T_1$, $T_2$, and $T_3$ are consistent with the across-the-thickness distributions of $\sigma_{11}, \sigma_{12}$, and $\sigma_{13}$. If they are not, it is customary for the theory to require the data and the (approximate) solution to have the same edge resultants and edge couples. [ A plate theory version of Saint-Venant's principle is then invoked at this point to validate the correctness of the approximate solution away from the plate edge; see Love (1944) and Timoshenko and Goodier (1951).] For this purpose, we introduce the transverse shear resultants $Q_j$ and moment resultants (or stress couples) $M_{ij}$ by

$$Q_j = \int_{-h/2}^{h/2} \sigma_{j3}\,dx_3, \qquad M_{ij} = \int_{-h/2}^{h/2} \sigma_{ij3}x_3\,dx_3, \qquad (i,j = 1,2). \quad (17.18)$$

In terms of these quantities, we have from (17.12)

$$M_{1j,1} + M_{2j,2} + [x_3\sigma_{3j}]_{-h/2}^{h/2} - Q_j = 0$$

which simplifies to

$$M_{1j,1} + M_{2j,2} - Q_j = 0 \quad (j = 1,2) \tag{17.19a}$$

in view of the homogeneous face conditions (17.1a,b). Integration of (17.14b) gives

$$Q_{1,1} + Q_{2,2} + p = 0. \tag{17.19b}$$

The two conditions of (17.19a) are the first moments of the first two equilibrium equations of three-dimensional elastostatics and (17.19b) is the zeroth moment of the third. As such, they are the two-dimensional moment equilibrium equations and the transverse force equilibrium equation, respectively.

Along an $x_1$ = constant edge, it is expected from physical consideration (and confirmed later in Section 7 by a variational formulation) that we should be able to prescribe

$$M_{11} = M_{11}^*, \quad M_{12} = M_{12}^*, \quad Q_1 = Q_1^*, \tag{17.20}$$

where quantities with an asterisk are the various resultants of the prescribed edge tractions whether or not these edge stress data have the required polynomial distribution across the thickness as in (17.9) and (17.11). For the Germain–Kirchhoff approximate (thin plate) solution, we have from (17.9), (17.11), (17.12), and (17.18)

$$-D(w_{,11} + \nu w_{,22}) = M_{11}^*, \tag{17.21a}$$

$$-D(1-\nu) w_{,12} = M_{12}^*, \tag{17.21b}$$

$$-D(\nabla^2 w)_{,1} = Q_1^*. \tag{17.21c}$$

Unfortunately, (17.16) is a fourth-order PDE and, therefore, permits the specification of only two auxiliary conditions at each edge. How then do we extract two appropriate conditions from the three given conditions in (17.21) in order to have a well-defined boundary value problem for $w$?

It will be shown in the next section that the two appropriate stress boundary conditions along an $x_1$ = constant edge are (17.21a) and a second condition which is a contraction of the two remaining conditions to a single condition on the *effective transverse shear resultant* $Q_1^e \equiv Q_1 + M_{12,2}$ with

$$Q_1^e = -D[w_{,11} + (2-\nu) w_{,22}]_{,1} = Q_1^* + \frac{\partial M_{12}^*}{\partial x_2}. \tag{17.22a}$$

For an arbitrary (smooth) curved edge, the corresponding two appropriate stress boundary conditions are

$$M_{nn} = M^*_{nn}(s), \qquad Q^e_n = Q^*_n + \frac{\partial M^*_{nt}}{\partial s}, \tag{17.22b}$$

where $n$ and $t$ indicate directions in the $x_1,x_2$-plane normal and tangent to the edge curve $\Gamma$, respectively, and $s$ is the arclength along the edge curve measured relative to some reference point. The two quantities $M_{nn}$ and $Q^e_n = Q_n + M_{nt,s}$ are combinations of normal and tangential derivatives of $w$ up to second and third order, respectively [see (15.33)]. Hence, the two contracted stress edge conditions (17.22b) are compatible with the governing partial differential equation for $w$.

## 4. A Semidirect Method of Solution

As a first attempt to obtain a two-dimensional theory for thin elastic plates by the semidirect method of the calculus of variations, we use the displacement fields (17.4) as trial functions in the principle of minimum potential energy for our elastostatic plate problem. We anticipate that the Euler differential equation will be a single partial differential equation for the only unknown function $w(x_1,x_2)$ in this approximate method of solution.

Consider for the moment only problems with prescribed displacements along the cylindrical edge $\{f(x_1,x_2)=0, |x_3| \le h/2\}$ satisfied by the trial solution (17.4). With the strain components given by (17.5) and (17.6), the potential energy (16.41) of the isotropic, homogeneous, linearly elastic plate becomes

$$P = \iint_R [S_K(w) - pw]\, dx_1\, dx_2, \tag{17.23a}$$

where

$$\begin{aligned} S_K &= \frac{1}{2} D_0[\kappa_{11}^2 + \kappa_{22}^2 + 2\nu_0\kappa_{11}\kappa_{22} + (1-\nu_0)(\kappa_{12}^2 + \kappa_{21}^2)] \\ &= \frac{1}{2} D_0[w_{,11}^2 + w_{,22}^2 + 2\nu_0 w_{,11} w_{,22} + 2(1-\nu_0) w_{,12}^2] \end{aligned} \tag{17.23b}$$

with

$$D_0 = \frac{E_0 h^3}{12(1-\nu_0^2)}, \qquad E_0 = \frac{E}{1-\nu^2}, \qquad \nu_0 = \frac{\nu}{1-\nu}. \tag{17.23c}$$

As in Section 7 of Chapter 15, it is not difficult to show that the Euler differential equation of $\delta P = 0$ is the biharmonic equation (17.16) for the only unknown $w$ with $D_0$ instead of $D$ as the *bending stiffness factor*. For a solution $w$ of this Euler differential equation, i.e., for an extremal $\widehat{w}$ of the variational problem, $\delta P = 0$ simplifies to

$$\delta P = \oint_{\Gamma} (M_{nn}\delta w_{,n} + M_{nt}\delta w_{,s} - Q_n \delta w)\, ds$$

$$= \left[ M_{nt}\delta w \right]_{\bar{s}-}^{\bar{s}+} + \oint_{\Gamma} (M_{nn}\delta w_{,n} - Q_n^e \delta w)\, ds = 0. \qquad (17.24a)$$

For an edge curve $\Gamma$ with a continuously turning unit tangent vector $\mathbf{t}$, the qualities $M_{nn}, M_{nt}, Q_n$ and $Q_n^e = Q_n + M_{nt,s}$ are linear combinations of $w$ and its derivatives up to third order, as given in (15.33):

$$M_{11} = -D_0(w_{,11} + \nu_0 w_{,22}), \qquad M_{22} = -D_0(w_{,22} + \nu_0 w_{,11}),$$

$$M_{12} = M_{21} = -D_0(1 - \nu_0)\, w_{,12}, \qquad Q_j = -D_0(\nabla^2 w)_{,j} \quad (j = 1,2)$$

$$Q_n = n_1 Q_1 + n_2 Q_2, \qquad n_k = \mathbf{n}\cdot\mathbf{i}_k \quad (k = 1,2) \qquad (17.24b)$$

$$M_{nn} = \mathbf{M}_n \cdot \mathbf{n} = n_1^2 M_{11} + n_2^2 M_{22} + 2n_1 n_2 M_{12},$$

$$M_{nt} = \mathbf{M}_n \cdot \mathbf{t} = (n_1^2 - n_2^2) M_{12} + 2n_1 n_2 (M_{22} - M_{11})$$

with $\mathbf{n}\cdot\mathbf{t} = 0$ and $\mathbf{n}\times\mathbf{t} = \mathbf{i}_3$. The bracketed term in (17.24a) vanishes for continuous single-valued functions $M_{nt}$ and $\delta w$ along the edge $\Gamma$ (which has no corner). It follows that $\delta P$ vanishes if we have $\delta w = 0$ and $\delta w_{,n} = 0$. The two appropriate displacement edge conditions associated with the biharmonic equation (17.16) are, therefore, prescribed conditions on $w$ and $w_{,n}$.

For an edge with prescribed edge tractions, the expression for $P$ has an additional line integral for the work done by the edge stresses:

$$P = \iint_R [S_K(w) - pw]\, dx_1\, dx_2 + \oint_{\Gamma} (M_{nn}^* w_{,n} + M_{nt}^* w_{,s} - Q_n^* w)\, ds, \quad (17.25a)$$

where

$$\{M_{nn}^*, M_{nt}^*\} = \int_{-h/2}^{h/2} \{T_n, T_t\} x_3\, dx_3, \qquad Q_n^* = \int_{-h/2}^{h/2} T_3\, dx_3 \qquad (17.25b)$$

in which $T_n$, $T_t$ and $T_3$ are the prescribed edge tractions along the cylindrical edge. With $\nabla^2\nabla^2 w = p/D_0$, the vanishing of the first variation implies the line integral

$$\delta P = \oint_{\Gamma} (\delta w_{,n} \Delta M_{nn} + \delta w_{,s} \Delta M_{nt} - \delta w\, \Delta Q_n)\, ds$$

$$= [\delta w\, \Delta M_{nt}]_{\bar{s}-}^{\bar{s}+} + \oint_{\Gamma} (\delta w_{,n} \Delta M_{nn} - \delta w\, \Delta Q_n^e)\, ds = 0, \qquad (17.26)$$

where $M_{nn}, M_{nt}$, and $Q_n$ are combinations of $w$ and its derivatives as summarized in (17.24b). The bracketed term vanishes for single-valued continuous functions

$\delta w$ and $\Delta M_{nt}$. For an $x_1$ = constant edge, the two Euler boundary conditions reduce to (17.22b) with the left-hand side expressed in terms of $w$ and its derivatives. They are the two appropriate stress boundary conditions for the Euler differential equation (for $w$) associated with $\delta P = 0$.

The preceding semidirect method of solution for plate problems manages to circumvent the inconsistency of the transverse shear stresses $\sigma_{13}$ and $\sigma_{23}$ being neglected in the stress-strain relations but not in the equilibrium equations as well as the problem in the transverse normal stress-strain relation caused by $e_{33} = 0$. It also establishes the two appropriate boundary conditions on $w$ along $\Gamma$ for both the displacement boundary-value problem and the stress boundary-value problem. However, the method of minimum potential energy does not provide a way to incorporate the assumption of negligible transverse normal stress contribution to all stress-strain relations. This deficiency led to a replacement of $D$ and $\nu$ by $D_0$ and $\nu_0$ which we know to be incorrect from an asymptotic analysis of the plate problem (Friedrichs and Dressler, 1961, Gregory and Wan 1985, 1988).

Although we can remove the deficiency described above (which we will do later by way of Reissner's variational principle), there is also the challenge of seeking a properly formulated sixth-order plate theory which would allow for three natural and independently prescribed stress boundary conditions on the transverse shear resultant $Q_n$, the bending couple $M_{nn}$, and the twisting couple $M_{nt}$ [as in (17.20)]. These are the physically independent conditions expected for plate problems; it seems rather artificial to have to combine them just because our approximate (plate) theory happens to be fourth order. The challenge of formulating a sixth-order plate theory was finally met by an ingenius application of the semidirect method of the calculus of variations almost a century after Kirchhoff's resolution of the stress boundary conditions. We will outline E. Reissner's elegant solution to the problem of stress boundary conditions for plate theory in the next section (Reissner, 1944, 1945, 1947). As a limiting case of Reissner's more general theory, we recover the correct classical thin plate theory of Germain–Kirchhoff (with the correct bending stiffness factor $D$ and Poisson's ratio $\nu$) in a later section with no inconsistent assumptions. This specialization to the classical theory will allow us to deduce once more Kirchhoff's two contracted stress boundary conditions, now by a slightly different method.

## 5. Minimum Complementary Energy

Although the stresses given by (17.9) and (17.11)-(17.15) do not necessarily provide the (exact) solution of the boundary-value problem governing the linear elastostatics of flat plates, they satisfy the three differential equations of equilibrium and the stress boundary conditions on the upper and lower faces. As they stand, these expressions are too restrictive as they cannot be made to satisfy three independent stress boundary conditions such as (17.20) along an $x_1$ = constant edge. For a more general solution, we retain the $x_3$ dependence of these stress distributions and use the semidirect method of calculus of variations to obtain a more adequate approximate solution (in the form of a more refined plate theory). In this section, we consider the case where the prescribed stresses at the plate edge are satisfied exactly by the trial solution to be specified in (17.27). In this case, the theorem of minimum complementary energy is an appropriate vehicle for the semidirect method.

To get an approximate two-dimensional theory for a thin flat elastic plate similar to, but more refined than the Germain–Kirchhoff theory, by way of the minimum complementary energy theorem, we begin with the following expressions for the stresses:

$$\sigma_{ij}(x_1,x_2,x_3) = \frac{6}{h^2} M_{ij}(x_1,x_2)z,$$

$$\sigma_{3j}(x_1,x_2,x_3) = \frac{3}{2h} Q_j(x_1,x_2)(1-z^2), \tag{17.27}$$

$$\sigma_{33}(x_1,x_2,x_3) = \frac{3}{4} p(x_1,x_2)(z - \frac{1}{3} z^3)$$

with $z = x_3/(h/2)$ and $i,j = 1,2$. Whereas $M_{ij} = M_{ji}$ and $Q_j$ are known in terms of $w(x_1,x_2)$ in the Germain–Kirchhoff theory, they are now unknown functions of $x_1$ and $x_2$ to be determined by the minimum complementary energy theorem. In particular, they are not related to the midplane displacement $w(x_1,x_2)$ as were $m_{ij}$ and $q_i$. In fact, nothing will be said about the displacement fields here.

To apply of the minimum complementary energy theorem, we must ensure that the stresses are in equilibrium with the external loads. In the absence of body forces, the equilibrium equations, $\sigma_{ij,i} = 0$, require that the coefficients $M_{ij}$ and $Q_k$ be related by

$$Q_{k,k} + p = 0, \tag{17.28a}$$

$$M_{ij,i} - Q_j = 0 \quad (j = 1,2) \tag{17.28b}$$

inside $\Gamma$. Moreover, the stress boundary conditions

$$Q_n = Q_n^*(s), \qquad M_{nn} = M_{nn}^*(s), \qquad M_{nt} = M_{nt}^*(s) \tag{17.29}$$

must be satisfied on the portion $\Gamma_\sigma$ of the boundary curve $\Gamma$. In (17.29), quantities with an asterisk are prescribed functions on the edge curve $\Gamma$ as defined in (17.25b).

To obtain an approximate description for the linear elastostatics of a flat plate, we use the trial functions (17.27) and apply the semidirect method to the minimum complementary energy of the plate

$$\mathcal{P}[\sigma_{ij}] = \iint_R \int_{-h/2}^{h/2} C(\sigma_{ij})\, dx_3\, dx_1\, dx_2 - \int_{\Gamma_d} \int_{-h/2}^{h/2} \sigma_n \cdot \mathbf{U}\, dx_3\, ds \quad (17.30)$$

where $R$ is the region of the midplane inside $\Gamma$ (which has $\Gamma_a$ and $\Gamma_d$ as its two complementary parts) and $\mathbf{U}$ is the prescribed displacement vector field along the edge $\Gamma_d$. The surface integral over the plate edge is absent if stress boundary conditions are prescribed along the entire plate edge. For a little more generality, we will work with a complementary energy density function for transversely isotropic plates,

$$C(\sigma_{ij}) = \frac{1}{2E}[\sigma_{11}^2 + \sigma_{22}^2 - 2\nu\sigma_{11}\sigma_{22}] + \frac{1}{4G}(\sigma_{12}^2 + \sigma_{21}^2)$$

$$+ \frac{1}{4G_3}[\sigma_{13}^2 + \sigma_{31}^2 + \sigma_{23}^2 + \sigma_{32}^2] + \frac{1}{2E_3}\sigma_{33}^2 - \frac{\nu_3}{E}\sigma_{33}(\sigma_{11} + \sigma_{22}), \quad (17.31)$$

where $E, \nu, G_3$ and $\nu_3$ are independent parameters and $G = E/[2(1+\nu)]$. The expression (17.31) reduces to (16.33) for isotropic plates if we set $G_3 = G, \nu_3 = \nu$, and $E_3 = E$. For simplicity, we limit our discussion to homogeneous materials so that these parameters are constant in the whole plate.

The stress-strain relations corresponding to (17.31) are (obtained from $e_{ij} = \partial C / \partial \sigma_{ij}$):

$$e_{11} = \frac{1}{E}[\sigma_{11} - \nu\sigma_{22} - \nu_3\sigma_{33}], \qquad e_{22} = \frac{1}{E}[\sigma_{22} - \nu\sigma_{11} - \nu_3\sigma_{33}]$$

$$e_{33} = \frac{1}{E_3}\sigma_{33} - \frac{\nu_3}{E}(\sigma_{11} + \sigma_{22}), \quad (17.32)$$

$$e_{12} = \frac{1}{2G}\sigma_{12}, \qquad e_{j3} = \frac{1}{2G_3}\sigma_{j3} \quad (j = 1,2).$$

The relations above provide a new interpretation of Kirchhoff's hypotheses for the classical thin plate theory We see from (17.32) that these hypotheses are equivalent to requiring $\nu_3 = 1/E_3 = 1/G_3 = 0$. A material with such properties is evidently transversely rigid (or inelastic). The strain suffered by such materials under any loading condition will always be in the $x_1, x_2$-plane only, while $\sigma_{33}$ has no effect on the normal strain components. Moreover, the transverse stresses can only be computed from the equilibrium equations. In fact, it can be shown that Kirchhoff's thin plate theory is an exact consequence of the three-dimensional theory of elasticity if the material is transversely rigid (see the Exercises). There would not be any inconsistency whatsoever. (For an isotropic material, however, the removal of the apparent inconsistencies will still have to be accomplished by an appropriate asymptotic method, as mentioned previously.)

With the trial solution (17.27), integration with respect to $x_3$ [in (17.30)] gives

$$\int_{-h/2}^{h/2} \sigma_{ij}^2\,dx_3 = \frac{12}{h^3} M_{ij}^2, \qquad \int_{-h/2}^{h/2} \sigma_{11}\sigma_{22}\,dx_3 = \frac{12}{h^3} M_{11}M_{22},$$

$$\int_{-h/2}^{h/2} \sigma_{3j}^2\,dx_3 = \int_{-h/2}^{h/2} \sigma_{j3}^2\,dx_3 = \frac{6}{5h} Q_j^2, \qquad \int_{-h/2}^{h/2} \sigma_{33}^2\,dx_3 = \frac{17h}{140} p^2, \tag{17.33}$$

$$\int_{-h/2}^{h/2} \sigma_{33}(\sigma_{11} + \sigma_{22})\,dx_3 = \frac{6}{5h} p(M_{11} + M_{22}).$$

In that case, we have

$$\begin{aligned}\int_{-h/2}^{h/2} C(\sigma_{ij})\,dx_3 &= \frac{6}{Eh^3}[M_{11}^2 + M_{22}^2 - 2\nu M_{11}M_{22}] + \frac{3}{Gh^3}(M_{12}^2 + M_{12}^2)\\ &\quad + \frac{3}{5G_3h}(Q_1^2 + Q_2^2) - \frac{6\nu_3}{5Eh} p(M_{11} + M_{22}) + \frac{17h}{280E_3} p^2\\ &\equiv C_p(M_{ij}, Q_k)\end{aligned} \tag{17.34}$$

for all points $(x_1, x_2)$ in $R$ and

$$\int_{-h/2}^{h/2} \mathbf{U}\cdot\boldsymbol{\sigma}_n\,dx_3 = W^*Q_n + \Phi_n^*M_{nn} + \Phi_t^*M_{nt}, \tag{17.35}$$

where

$$\begin{aligned} W^*(x_1,x_2) &= \frac{3}{2h}\int_{-h/2}^{h/2} U_3(x_1,x_2,x_3)(1 - z^2)\,dx_3,\\ \Phi_n^*(x_1,x_2) &= \frac{12}{h^3}\int_{-h/2}^{h/2} U_n(x_1,x_2,x_3)x_3\,dx_3,\\ \Phi_t^*(x_1,x_2) &= \frac{12}{h^3}\int_{-h/2}^{h/2} U_t(x_1,x_2,x_3)x_3\,dx_3\end{aligned} \tag{17.36}$$

for $(x_1, x_2)$ on $\Gamma_d$ and $z = x_3/(h/2)$. Results (17.34)-(17.36) reduce (17.30) to

$$\begin{aligned}\mathcal{P}[M_{ij}, Q_k] &= \iint_R C_p(Q_i, M_{jk})\,dx_1\,dx_2\\ &\quad - \int_{\Gamma_d} (W^*Q_n + \Phi_n^*M_{nn} + \Phi_t^*M_{nt})\,ds.\end{aligned} \tag{17.37}$$

The variational problem is now to seek $\widehat{Q}_j$ and $\widehat{M}_{ik}$ which render $\mathscr{P}[M_{ij}, Q_k]$ stationary. The comparison functions must satisfy the two-dimensional equilibrium equations (17.28) and the three stress boundary conditions on $\Gamma_\sigma$ [(17.29)]. For our purpose, it suffices to find the Euler differential equations and Euler boundary conditions for this problem.

To avoid the nontrivial task of finding the general solution of (17.28) for the purpose of determining the appropriate independent variations, we use Lagrange multipliers $W, \Phi_1, \Phi_2$ and $\lambda$ to append the constraints to the functional $\mathscr{P}$. For a minimum $\mathscr{P}$, we have $\delta\overline{\mathscr{P}} = 0$ where

$$\delta\overline{\mathscr{P}} \equiv \delta\mathscr{P} + \iint_R [\Phi_j(\delta M_{ij,i} - \delta Q_j) + W\delta Q_{j,j} + \lambda(\delta M_{12} - \delta M_{21})]\, dx_1 dx_2. \tag{17.38}$$

The vanishing of the first variation $\delta\overline{\mathscr{P}}$ requires the Euler differential equations

$$\overline{\gamma}_j = \frac{\partial C_p}{\partial Q_j}, \qquad \overline{\kappa}_{ij} = \frac{\partial C_p}{\partial M_{ij}} \quad (i,j = 1,2) \tag{17.39a,b}$$

where [after eliminating the unessential multiplier $\lambda = (\Phi_{2,1} - \Phi_{1,2})/2$ which will not be needed henceforth]

$$\overline{\gamma}_j = W_{,j} + \Phi_j, \qquad \overline{\kappa}_{ij} = \frac{1}{2}(\Phi_{i,j} + \Phi_{j,i}) \quad (i,j = 1,2) \tag{17.40a,b}$$

and the Euler boundary conditions

$$W = W^*, \qquad \Phi_j = \Phi_j^* \quad \text{on } \Gamma_d. \tag{17.41}$$

Although we may interpret $W$ and $\Phi_j$ as weighted (across the thickness) averages of the relevant displacement fields, they are not to be identified with $w$ and $-w_{,j}$ of the Germain–Kirchhoff theory.

We summarize the results obtained above as follows:

**[XVII.1]** *For the stress components (17.27), which are in equilibrium with the external loads [in accordance with (17.28) in R and (17.29) on $\Gamma_\sigma$] to render the complementary energy $\mathscr{P}$ stationary, they must be the solution of the Euler differential equations (17.39) with $\overline{\gamma}_k$ and $\overline{\kappa}_{ij}$ defined in terms of the multipliers W and $\Phi_i$ by (17.40), and the multipliers themselves must satisfy the Euler boundary conditions (17.41) on $\Gamma_d$ where edge tractions are not prescribed.*

The equilibrium equations (17.28), the stress-strain relations (17.39) and the "strain-displacement relations" (17.40) constitute the governing differential equations for our new (two-dimensional) theory of the linear elastostatics of flat plates. They are appropriately supplemented by the prescribed stress boundary conditions (17.29) on $\Gamma_\sigma$ and the Euler boundary conditions (17.41) on $\Gamma_d$. We will see in the next section why this new approximate theory is an improvement over the Germain–Kirchhoff theory.

## 6. Reduction of Reissner's Plate Equations

For the plate strain potential $C_p(M_{ij}, Q_j)$ given in (17.34), the stress-strain relations (17.39) take the form

$$\bar{\kappa}_{11} = \frac{M_{11} - \nu M_{22}}{D(1-\nu^2)} - \frac{6\nu_3}{5Eh}p, \tag{17.42a}$$

$$\bar{\kappa}_{22} = \frac{M_{22} - \nu M_{11}}{D(1-\nu^2)} - \frac{6\nu_3}{5Eh}p, \tag{17.42b}$$

$$\bar{\kappa}_{12} = \bar{\kappa}_{21} = \frac{6M_{12}}{Gh^3} = \frac{M_{12}}{D(1-\nu)}, \tag{17.42c}$$

$$\bar{\gamma}_j = \frac{6}{5G_3h}Q_j \equiv BQ_j \quad (j = 1, 2). \tag{17.42d}$$

with $M_{21} = M_{12}$. Equations (17.40a) for $\bar{\gamma}_j$ may be combined with the stress-strain relations (17.42d) and rewritten as an expression for $\Phi_j$ in terms of $W$ and $Q_j$:

$$\Phi_j = BQ_j - W_{,j}, \tag{17.43a}$$

$$B = \frac{6}{5G_3h}. \tag{17.43b}$$

The relations for $\bar{\kappa}_{ij}$ in (17.40b) can then be written as

$$\bar{\kappa}_{ij} = \frac{1}{2}B(Q_{i,j} + Q_{j,i}) - W_{,ij}. \tag{17.44}$$

The stress-strain relations (17.42a)–(17.42c) are now inverted to give [with the help of (17.44)] $M_{ij}$ in terms of $W$ and $Q_j$:

$$M_{11} = -D[W_{,11} + \nu W_{,22}] + DB[Q_{1,1} + \nu Q_{2,2}] + \frac{3\nu_3 D}{5Gh}p, \tag{17.45a}$$

$$M_{22} = -D[W_{,22} + \nu W_{,11}] + DB[\nu Q_{1,1} + Q_{2,2}] + \frac{3\nu_3 D}{5Gh}p, \tag{17.45b}$$

$$M_{12} = M_{21} = -D(1-\nu)W_{,12} + \frac{1}{2}DB(1-\nu)[Q_{2,1} + Q_{1,2}]. \qquad (17.45c)$$

The two moment equilibrium equations (17.28b) are then used to express $Q_j$ in terms of $W$ and a stress function $\chi$

$$Q_1 = -D\nabla^2 W_{,1} + DB\nabla^2 Q_1 + \frac{1}{2}(1+\nu)DB\chi_{,2} + \frac{3\nu_3 D}{5Gh}p_{,1}, \qquad (17.46a)$$

$$Q_2 = -D\nabla^2 W_{,2} + DB\nabla^2 Q_2 - \frac{1}{2}(1+\nu)DB\chi_{,1} + \frac{3\nu_3 D}{5Gh}p_{,2}, \qquad (17.46b)$$

where

$$\chi = Q_{2,1} - Q_{1,2}. \qquad (17.47)$$

Upon substituting (17.46a) and (17.46b) into the transverse force equilibrium equation (17.28a), we obtain

$$-D\nabla^2\nabla^2 W + DB\nabla^2(Q_{1,1} + Q_{2,2}) + \frac{3\nu_3 D}{5Gh}\nabla^2 p + p = 0.$$

The term $Q_{1,1} + Q_{2,2}$ may again be eliminated by (17.28a), leaving us with an inhomogeneous biharmonic equation for $W$ alone:

$$D\nabla^2\nabla^2 W = [1 - D(B - A_3)\nabla^2]p, \qquad (17.48)$$

where $A_3 = 3\nu_3/5Gh$.

The combination $Q_{1,1} + Q_{2,2}$ used to obtain the partial differential equation (17.48) for $W$ only makes use of a portion of the content of (17.46). An independent combination of the same two expressions in (17.46) is $Q_{2,1} - Q_{1,2}$ which is just $\chi$ by (17.47). If we form this combination from (17.46), we obtain

$$\frac{1}{2}DB(1-\nu)\nabla^2\chi - \chi = 0 \qquad (17.49)$$

which is a Helmholtz-type partial differential equation for $\chi$.

For the special case $1/G_3 = 1/E_3 = \nu_3 = 0$ (so that $B = A_3 = 0$) corresponding to a transversely rigid plate, (17.49) simplifies to $\chi = 0$ and (17.48) reduces to (17.16). These results provide an independent confirmation of our previous observation that the Germain–Kirchhoff theory can be obtained without any approximation or inconsistencies by assuming the plate to be transversely rigid. For isotropic plates, $DB$ is proportional to $h^2$ so that the solution of (17.49) is of a boundary layer type. This is consistent with (but not a proof of) our previous remark that the classical plate theory for isotropic materials turns out to be the leading term of the outer asymptotic expansion of the exact solution.

The two partial differential equations (17.48) and (17.49) form an uncoupled sixth-order system, the general solution of which can satisfy *three* independent boundary conditions along an edge of the plate. As such, we no longer have "too many" boundary conditions for the governing partial differential equation as we did in the classical theory and the Kirchhoff contraction of the stress boundary conditions is, therefore, *not* necessary for the new theory of E. Reissner (1944). To satisfy the prescribed stress boundary conditions, we need to express $Q_n, M_{nn}$, and $M_{nt}$ in terms of $W$ and $\chi$. To obtain these expressions, we recall from (17.28a) and (17.47)

$$Q_{1,1} + Q_{2,2} = -p \qquad Q_{2,1} - Q_{1,2} = \frac{1}{B}\chi.$$

From these, we obtain

$$\nabla^2 Q_1 = -p_{,1} - \chi_{,2}, \qquad \nabla^2 Q_2 = -p_{,2} - \chi_{,1} \tag{17.50}$$

We then use (17.50) to eliminate $\nabla^2 Q_j$ from (17.46) to get

$$Q_1 = -D[\nabla^2 W + (B - A_3)p]_{,1} - \frac{1}{2} DB(1-\nu)\chi_{,2},$$

$$Q_2 = -D[\nabla^2 W + (B - A_3)p]_{,2} + \frac{1}{2} DB(1-\nu)\chi_{,1}.$$

The expressions for $Q_j$ in terms of $W$ and $\chi$ alone can now be used to eliminate $Q_{i,j}$ from (17.45) to get $M_{ij}$ in terms of $W$ and $\chi$ alone. A more attractive form of the results is obtained by first using the equilibrium equation $Q_{1,1} + Q_{2,2} = -p$ to eliminate $Q_{2,2}$ from (17.45a) and $Q_{1,1}$ from (17.45b), leaving us with

$$M_{,11} = -D[W_{,11} + \nu W_{,22}] + DB(1-\nu)Q_{1,1} + D(A_3 - \nu B)p,$$

$$M_{12} = M_{21} = -D(1-\nu)W_{,12} + \frac{1}{2}DB(1-\nu)[Q_{2,1} + Q_{1,2}]. \tag{17.45'}$$

Another form of $M_{ij}$ can be found in the Exercises.

We summarize these results as follows:

**[XVII.2]** ***The governing equations of Reissner's sixth-order plate theory are equivalent to the biharmonic equation (17.48) for W and the Helmholtz equation for $\chi$ (17.49) with the stress resultants and couples given in terms of W and $\chi$ by (17.51) and (17.45) [or (17.45')]. If needed (for displacement boundary conditions), the multipliers $\Phi_1$ and $\Phi_2$ can be obtained from (17.39a) and (17.40a) which can be combined to give***

$$\Phi_j = -W_{,j} + \frac{\partial C_p}{\partial Q_j} = -W_{,j} + \frac{6Q_j}{5G_3h}.$$

*The solution of the sixth-order system of partial differential equations for W and $\chi$ can be specialized to satisfy three independently prescribed boundary conditions [such as the stress conditions (17.29) or displacement conditions (17.41)] at an edge of the plate.*

## 7. A Variational Principle for Stresses and Displacements

The theorem of minimum complementary energy requires that all trial functions satisfy all stress boundary conditions. This requirement is generally not met by (17.27) because the prescribed edge stress data generally do not have the same simple $x_3$ dependence of the assumed trial stress distributions. Consequently, we need a variational formulation which does not require that stress edge conditions or displacement edge conditions be satisfied exactly. Reissner's variational principle for stresses and displacements offers one way to meet this need.

As before, we take the trial stress components to be (17.27). To use Reissner's variational principle, we also need trial functions for the displacement components. We take these to be

$$u_3(x_1,x_2,x_3) = w(x_1,x_2),$$

$$u_j(x_1,x_2,x_3) = x_3\varphi_j(x_1,x_2) \quad (j = 1,2) \tag{17.52}$$

which are similar to those assumed for the Germain–Kirchhoff theory except that $\varphi_1$ and $\varphi_2$ are new unknowns not restricted to be $-w_{,1}$ and $-w_{,2}$, respectively. The corresponding strain components are

$$e_{ij} = x_3\kappa_{ij} \equiv \frac{1}{2}x_3(\varphi_{i,j} + \varphi_{j,i}),$$

$$e_{3j} = \frac{1}{2}(w_{,j} + \varphi_j) \equiv \frac{1}{2}\gamma_j, \qquad e_{33} \equiv 0 \quad (i,j = 1,2). \tag{17.53}$$

We substitute (17.27), (17.52), and (17.53) into the performance index (16.48) with the complementary energy density function given by (17.31). The integration in $x_3$ can be carried out to give

$$I[u_j,\sigma_{ik}] = \iint_R \left\{ \sum_{j=1}^{2} \left[ \sum_{i=1}^{2} M_{ij}\kappa_{ij} + Q_j\gamma_j \right] - C_p(M_{mt}, Q_k) \right\} dx_1\, dx_2$$
$$- \iint_R pw\, dx_1\, dx_2 - \int_{\Gamma_\sigma} (\varphi_n M^*_{nn} + \varphi_t M^*_{nt} + wQ^*_n)\, ds \qquad (17.54)$$
$$+ \int_{\Gamma_u} (M_{nn}\Delta\varphi_n + M_{nt}\Delta\varphi_t + Q_n\Delta w)\, ds \equiv I_P[M_{ij}, Q_k, \varphi_m, w]$$

with $C_p$ given in (17.34). Note that we have $f_i = 0$ for our plate problem and we continue to use the notation $\Delta g \equiv g^* - g$. It is not difficult to show (see the Exercises):

**[XVII.3]** ***The Euler differential equations for a stationary value of*** $I_P$ ***are the equilibrium equations (17.28) and the stress-strain relations***

$$\kappa_{ij} = \frac{\partial C_p}{\partial M_{ij}}, \qquad \gamma_k = \frac{\partial C_p}{\partial Q_k} \quad (i,j,k = 1,2). \qquad (17.55a,b)$$

***The Euler boundary conditions are the three stress boundary conditions***

$$\Delta Q_n = \Delta M_{nn} = \Delta M_{nt} = 0 \qquad (17.56)$$

***on*** $\Gamma_\sigma$ ***and the three displacement boundary conditions***

$$\Delta W = \Delta\varphi_n = \Delta\varphi_t = 0 \qquad (17.57)$$

***on*** $\Gamma_d$.

Note that the constraint $M_{12} = M_{21}$ is observed throughout the derivation of [XVII.3]. Also, we may now identify the Lagrange multipliers $W$ and $\Phi_j$ of Section 6 with $w$ and $\varphi_j$ of (17.52).

For the special case of a transversely rigid plate with $\nu_3 = 1/G_3 = 1/E_3 = 0$, we have from the stress-strain relations (17.55b) $\gamma_k = 0$ or $\varphi_k = -w_{,k}$, $k = 1, 2$. With all Euler differential equations satisfied, we have from (17.54)

$$\delta I_P = -\oint_\Gamma (\Delta M_{nn}\delta\varphi_n + \Delta M_{nt}\delta\varphi_t + \Delta Q_n \delta w)\, ds = 0 \qquad (17.58)$$

if stress data are prescribed along the entire cylindrical edge so that $\Gamma_\sigma = \Gamma$. The three unit vectors **n**, **t**, and $\mathbf{i}_3$ form an orthonormal triad. Therefore, we have

$$\delta\varphi_n = -\delta w_{,n}, \qquad \delta\varphi_t = -\delta w_{,s}, \qquad (17.59)$$

where $(\ )_{,n}$ and $(\ )_{,s}$ are the normal and tangential derivatives, respectively, along $\Gamma$ with $s$ being the arclength parameter. The quantity $\delta w_{,s}$ is completely determined once we know $\delta w$ on $\Gamma$; hence, $\delta w$ and $\delta w_{,s}$ do not vary independently. In fact, upon substituting (17.59) into (17.58), we can integrate by parts to obtain for a smooth simple closed curve $\Gamma$

$$\delta I_P = \oint_\Gamma (\Delta M_{nn}\delta w_{,n} + \Delta M_{nt}\delta w_{,s} - \Delta Q_n \delta w)\,ds$$

$$= [\Delta M_{nt}\delta w]_0^{s_0} + \oint_\Gamma [\Delta M_{nn}\delta w_{,n} - \Delta Q_n^e \delta w]\,ds, \qquad (17.60)$$

where $Q_n^e = Q_n + M_{nt,s}$ is the effective transverse shear resultant and $s_0$ is the total arclength of the edge curve $\Gamma$. Both $M_{nt}$ and $w$ are expected to be single valued and continuous on $\Gamma$; hence, the bracketed term vanishes. With $\delta w_{,n}$ and $\delta w$ varying independently, we have

**[XVII.4]** ***The Euler stress boundary conditions for*** $\delta \mathrm{I}_P = 0$ ***with*** $\nu_3 = 1/\mathrm{G}_3 = 1/\mathrm{E}_3 = 0$ ***are***

$$\Delta \mathrm{M}_{nn} = 0, \qquad \Delta \mathrm{Q}_n^e = 0, \qquad (17.61)$$

*or*

$$\mathrm{M}_{nn} = \int_{-h/2}^{h/2} \mathrm{T}_n \mathrm{x}_3\, d\mathrm{x}_3, \qquad \mathrm{Q}_n^e = \int_{-h/2}^{h/2} [\mathrm{T}_3 + \mathrm{x}_3 \mathrm{T}_{t,s}]\, d\mathrm{x}_3. \qquad (17.62)$$

The two edge conditions (17.61) are just the Kirchhoff contracted stress boundary conditions (17.22) for the classical plate theory. They are now derived rationally as consequences of the Kirchhoff hypotheses (in the form of transverse rigidity) imposed on the sixth-order plate theory.

If $\Gamma$ is not smooth at a point $s = s_1$ on $\Gamma$ (e.g., at the corners of a rectangle), it is possible for $M_{nt}$ to be discontinuous (or even unbounded) there.

## 8. Twisting of a Rectangular Plate

In addition to allowing for the satisfaction of three independent edge conditions, Reissner's plate theory also offers more accurate approximate solutions to many plate problems in engineering applications. In this section, we discuss a simple example to illustrate this improvement.

Consider an isotropic and homogeneous rectangular plate of uniform thickness $h$ and with a midplane spanning the area $\{|x| \le L, |y| \le a/2\}$. The plate is free of distributed surface (and body) loads so that $p = 0$ throughout its midplane. The edges $y = \pm a/2$ are free of edge tractions so that

$$y = \pm\frac{a}{2}: \qquad M_{yy} = M_{yx} = Q_y = 0. \tag{17.63a,b,c}$$

The ends $x = \pm L$ are also free of normal stress but are rotated without distortion so that

$$x = \pm L: \qquad w = \pm\theta L y, \tag{17.64a}$$

$$\varphi_y = \mp\theta L, \tag{17.64b}$$

$$M_{xx} = 0, \tag{17.64c}$$

where $\mp\theta L$ is the angle of rotation of the end cross section at $x = \pm L$. The condition of distortionless rotation implies that line elements perpendicular to each other before deformation remain so after deformation. In other words, we have $\varphi_y = -w_{,y}$ along the rotated end sections and, hence, the second end condition $\varphi_y = \mp\theta L$ along $x = \pm L$.

The end conditions (17.64a) along $x = \mp L$ suggest that we seek a solution for $w(x, y)$ in the form

$$w(x, y) = \theta x y. \tag{17.65}$$

This solution satisfies the partial differential equation (17.48) (with $p = 0$). With $x_1 = x$ and $x_2 = y$, we have from (17.45′) and (17.46)

$$M_{xx} = DB(1 - \nu)Q_{x,x}, \tag{17.66a}$$

$$M_{yy} = DB(1 - \nu)Q_{y,y}, \tag{17.66b}$$

$$M_{xy} = M_{yx} = \frac{1}{2}DB(1 - \nu)[Q_{y,x} + Q_{x,y}] - D(1 - \nu)\theta, \tag{17.66c}$$

$$Q_x = \frac{1}{2}DB(1 - \nu)\nabla^2 Q_x = -\frac{1}{2}DB(1 - \nu)\chi_{,y}, \tag{17.67a}$$

$$Q_y = \frac{1}{2}DB(1 - \nu)\nabla^2 Q_y = \frac{1}{2}DB(1 - \nu)\chi_{,x}. \tag{17.67b}$$

The edge conditions (17.63) suggest that we seek a solution for $\chi$ for which $Q_y = 0$ and $Q_x$ independent of $x$. It follows that we should take $\chi$ to be a function of $y$ alone and the partial differential equation (17.49) for $\chi$ becomes

$$\frac{h^2}{10}\frac{d^2\chi}{dy^2} - \chi = 0 \tag{17.68}$$

for an isotropic plate, or

$$\chi = A_1 \sinh\left(\frac{\sqrt{10}}{h}y\right) + A_2 \cosh\left(\frac{\sqrt{10}}{h}y\right). \tag{17.69}$$

Correspondingly, we have from (17.66), (17.67), and the boundary condition $M_{yx}(x, \pm a/2) = 0$

$$Q_x = -\frac{h^2}{10}\chi_{,2} = \frac{h}{\sqrt{10}} A_2 \sinh\left(\frac{\sqrt{10}}{h} y\right), \tag{17.70}$$

$$M_{xy} = \frac{h^2}{10} Q_{x,y} - D(1-\nu)\theta$$

$$= -\frac{h^2}{10} A_2 \cosh\left(\frac{\sqrt{10}}{h} y\right) - D(1-\nu)\theta, \tag{17.71}$$

$$Q_y = M_{xx} = M_{yy} = 0. \tag{17.72}$$

It is evident from (17.65) and (17.72) that all boundary conditions in (17.64) are satisfied by our choice of $w$ and $\chi$, keeping in mind that the relation $\gamma_2 = \varphi_2 + w_{,2}$ in (17.53) requires

$$\varphi_y(\pm L, y) = [BQ_y - w_{,y}] = \mp \theta L. \tag{17.73}$$

Two of the three boundary conditions in (17.63) for $y = \pm a/2$ are also trivially satisfied. The remaining condition $M_{xy}(x, \pm a/2) = 0$ requires

$$\frac{h^2}{10} A_2 = -\frac{D(1-\nu)\theta}{\cosh(\sqrt{10}\, a/2h)} \tag{17.74}$$

so that

$$M_{xy} = D(1-\nu)\theta \left\{ \frac{\cosh(\sqrt{10}\, y/h)}{\cosh(\sqrt{10}\, a/2h)} - 1 \right\} \tag{17.75}$$

and

$$Q_x = \frac{\sqrt{10}}{h} D(1-\nu)\theta \left\{ \frac{\sinh(\sqrt{10}\, y/h)}{\cosh(\sqrt{10}\, a/2h)} \right\}. \tag{17.76}$$

Of interest in application is the torque required to produce the end rotation, given by

$$T = -\int_{-a/2}^{a/2} (M_{xy} - yQ_x)\,dy = -\int_{-a/2}^{a/2} \left( M_{xy} - y\frac{dM_{xy}}{dy} \right) dy \tag{17.77a}$$

$$= -[yM_{xy}]_{-a/2}^{a/2} - \int_{-a/2}^{a/2} 2M_{xy}\,dy = k_1 G h^3 a \theta,$$

where

$$k_1 = \frac{1}{3}\left[1 - \frac{2h}{a\sqrt{10}} \tanh\left(\frac{\sqrt{10}\,a}{2h}\right)\right]. \qquad (17.77b)$$

The values of $k_1$ for a range of width-to-thickness ratios are given in Table 17.1. The corresponding values of the exact three-dimensional elasticity solution (Timoshenko and Goodier, 1951) are also given in the same table for comparison.

Table 17.1. Dependence of $k_1$ on the width-to-thickness ratio

| $a/h$ | 1 | 2 | $\infty$ |
|---|---|---|---|
| $k_1$ (Reissner) | 0.139 | 0.228 | 0.333 |
| $k_1$ (Exact) | 0.1406 | 0.229 | 0.333 |
| $k_1$ (Classical) | 0.333 | 0.333 | 0.333 |

These values for $k_1$ should also be compared with the corresponding values obtained by the classical Germain–Kirchhoff plate theory. This simpler theory can be obtained from our results by formally setting $B = 0$. It follows from (17.49) that $\chi = 0$ so that all stress resultants and couples vanish except for $M_{xy} = M_{yx}$:

$$Q_x^c = Q_y^c = M_{xx}^c = M_{yy}^c \equiv 0, \qquad M_{xy}^c = -D(1-\nu)\theta. \qquad (17.78)$$

Consistent with the Kirchhoff contracted stress boundary conditions, the torque-twist relation (17.77a) is replaced by

$$T = \int_{-a/2}^{a/2} (Q_x^c + M_{xy,y}^c)\, y\, dy - [2yM_{xy}^c]_{-a/2}^{a/2} = 2D(1-\nu)\theta a. \quad (17.79)$$

The value $k_1$ from the classical theory is therefore $1/3$, independent of $a/h$. It differs from the exact value (obtained from the Saint–Venant torsion solution) by more than 45% for $a/h = 2$. In contrast, the solution by Reissner's theory differs from the exact solution by less than 0.5% for the same width-to-thickness ratio and still by less than 1.2% for $a/h = 1$.

## 9. A Finite Deflection Plate Theory

For a plate constrained from any displacement along its edges, any deformation in the transverse ($\mathbf{i}_3$) direction must be accompanied by some deformation in the lateral ($\mathbf{i}_1$ and $\mathbf{i}_2$) directions. For example, if we have $\mathbf{u} = \mathbf{0}$ along $x_1 = a$ and $x_1 = b$ as in Figure 17.3, a load applied to the upper face of the plate will cause the plate to sag and so it must also stretch both faces of the plate as well as any parallel plane. In particular, the horizontal line $x_3 = 0$ in an undeformed cross section $x_2$ = constant would become a (convex) curve. With its ends held fixed, this (deformed) curve is evidently longer than the corresponding underformed line. Hence, $u_1$ and $u_2$ are generally nonvanishing at the deformed midplane of the plate. Physical evidence indicates that the lateral deformation is usually small compared to the transverse deformation. It is much easier to bend a sheet of thin paper than to stretch it. Let $L$ be a representative span of the plate (such as the wavelength of the external surface loads $\sigma_t$ and $\sigma_b$ or the minimum plate span) and $u_0$ be the maximum magnitude of $u_3$. Then it is usually the case that $u_i = O(hu_0/L)$, $i = 1,2$, where the relation $a = O(\beta^n)$ has already been defined to be $a = c\beta^n$ for some constant $c$ independent of $\beta$.

All the plate theories encountered so far in this chapter are *infinitesimally small* deflection theories. In these theories, all the displacement components are assumed to be so small in magnitude so that the nonlinear terms in $E_{ij}$ may be neglected and the transverse bending (and twisting) action is effectively uncoupled from the in-plane stretching (and shearing) action as a consequence. In this section, we discuss a plate theory which applies to problems with moderate transverse displacements for which the accompanying stretching actions, as well as the nonlinear terms in the expressions for strain components, cannot be neglected. In other words, we want to consider plate problems for which the linear strain approximation $\{e_{ij}\}$ is not adequate and we need to work with the actual strain components $\{E_{ij}\}$ (which may, in fact, be small) or their nonlinear approximations. We expect $u_i/u_3 = O(h/L)$, $i = 1,2$, to hold for this class of problems. We also expect the expressions for $E_{ij}$ in terms of the displacement components to be adequately approximated by omitting terms which are $O(h^2/L^2)$ compared to unity so that

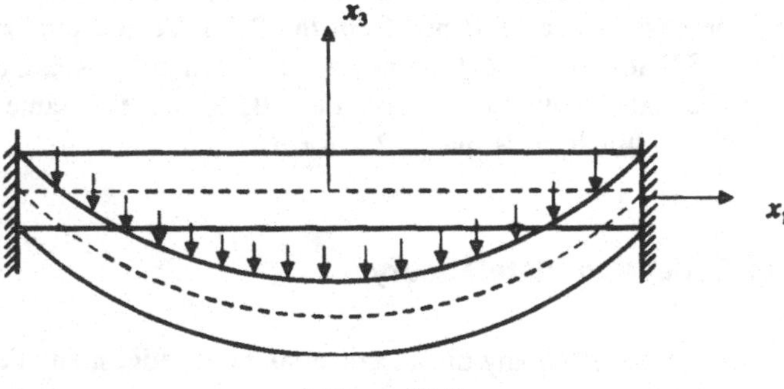


Figure 17.3

$$E_{11}=u_{1,1}+\frac{1}{2}(u_{1,1}^2+u_{2,1}^2+u_{3,1}^2)\cong u_{1,1}+\frac{1}{2}u^2_{3,1};$$

$$E_{22}\cong u_{2,2}+\frac{1}{2}u_{3,2}^2,\qquad E_{12}=E_{21}\cong\frac{1}{2}(u_{1,2}+u_{2,1}+u_{3,1}u_{3,2}),\qquad (17.80)$$

$$E_{3j}\cong\frac{1}{2}(u_{3,j}+u_{j,3}+u_{3,3}u_{3,j}),\qquad E_{33}\cong u_{3,3}+\frac{1}{2}u_{3,3}^2.$$

Implicit in the above approximations is the assumption that differentiation does not change order of magnitude. In other words, the new plate theory applies to problems in which the displacement fields away from the edge(s) do not have steep gradients over a distance of the order of the plate thichness. (It should not be used otherwise).

For a thin flat body to exhibit plate-like behavior, we want to retain the general features of the Germain–Kirchhoff theory, but now allow for midplane stretching and shearing action. We take for approximating displacement fields

$$u_3=w(x_1,x_2),\qquad u_j=v_j(x_1,x_2)+x_3\varphi_j(x_1,x_2)\qquad (17.81)$$

instead of (17.4) or (17.52). Correspondingly, the approximate strain-displacement relations (17.80) become

$$E_{ij}=\varepsilon_{ij}+x_3\kappa_{ij},\qquad E_{3j}=\frac{1}{2}\gamma_j,\qquad E_{33}\equiv 0\qquad (i,j=1,2),\qquad (17.82)$$

where

$$\varepsilon_{ij}=\frac{1}{2}(v_{i,j}+v_{j,i}+w_{,i}w_{,j}),$$

$$\kappa_{ij}=\frac{1}{2}(\varphi_{i,j}+\varphi_{j,i}),\qquad \gamma_j=w_{,j}+\varphi_j.\qquad (17.83)$$

Note that the expressions for $\kappa_{ij}$ and $\gamma_j$ are the same as those in (17.53).

Similar to the development of Section 7, we allow for separate approximations of stresses and displacements in our semidirect variational method of solution for the nonlinear plate theory. Just as we allowed for stretching and shearing plate action (in addition to bending and twisting) in the trial functions (17.81) for the three displacement components, we take the trial functions for the stress components in the form

$$\sigma_{ij}(x_1,x_2,x_3)=\frac{1}{h}N_{ij}(x_1,x_2)+\frac{6}{h^2}zM_{ij}(x_1,x_2)$$

$$\sigma_{3j}(x_1,x_2,x_3)=\frac{3}{2h}Q_j(x_1,x_2)(1-z^2)\qquad (i,j=1,2).\qquad (17.84)$$

$$\sigma_{33}(x_1,x_2,x_3)=\frac{3}{4}p(x_1,x_2)(z-\frac{1}{3}z^3)$$

with $z = x_3/(h/2)$, $N_{ij} = N_{ji}$ and $M_{ji} = M_{ji}$. Correspondingly, the variational formulation needed for the semidirect method is a variational principle for stresses-displacements appropriate for our finite deformation theory. By a suitable interpretation of $\sigma_{ij}$ [as components of Kirchhoff's stress tensor see Fung, 1967] together with the assumption of small strains (so that $|E_{ij}| \ll 1$), the appropriate variational principle is that for the functional $I[u_j, \sigma_{ik}]$ of (16.48) with $e_{ij}$ replaced by $E_{ij}$. We denote the modified performance index by $I_{NR}[u_i, \sigma_{jk}]$.

Upon substituting (17.81), (17.82), and (17.84) into $I_{NR}[u_i, \sigma_{jk}]$ with the complementary energy density $C(\sigma_{ij})$ given by (17.31), we obtain after integration with respect to $x_3$

$$I_{NR}[u_i,\sigma_{jk}] = \iint_R \left\{ \sum_{i=1}^{2} \left[ \sum_{j=1}^{2} (N_{ij}\varepsilon_{ij} + M_{ij}\kappa_{ij}) + Q_i\gamma_i \right] - pw - C_P \right\} dx_1\, dx_2$$

$$- \int_{\Gamma_u} (v_n N^*_{nn} + v_t N^*_{nt} + w V^*_n + \varphi_n M^*_{nn} + \varphi_t M^*_{nt})\, ds$$

$$+ \int_{\Gamma_d} (N_{nn}\Delta v_n + N_{nt}\Delta v_t + V_n \Delta w + M_{nn}\Delta\varphi_n + M_{nt}\Delta\varphi_t)\, ds$$

$$\equiv I_{NP}[v_i, w, N_{k\ell}, Q_m, M_{pq}], \tag{17.85}$$

where

$$C_P = C_p + \frac{1}{2Eh}(N_{11}^2 + N_{22}^2 - 2\nu N_{11}N_{22}) + \frac{1}{4Gh}(N_{12}^2 + N_{21}^2), \tag{17.86}$$

$$V_n = Q_n + w_{,n}N_{nn} + w_{,s}N_{nt}, \qquad V_n^* = \int_{-h/2}^{h/2} \mathbf{i}_3 \cdot \mathbf{T}\, dx_3, \tag{17.87a,b}$$

$$N^*_{nn} = \int_{-h/2}^{h/2} \mathbf{n} \cdot \mathbf{T}\, dx_3, \qquad N^*_{nt} = \int_{-h/2}^{h/2} \mathbf{t} \cdot \mathbf{T}\, dx_3, \tag{17.87c,d}$$

$$v_n^* = \int_{-h/2}^{h/2} \mathbf{n} \cdot \mathbf{U}\, dx_3, \qquad v_t^* = \int_{-h/2}^{h/2} \mathbf{t} \cdot \mathbf{U}\, dx_3. \tag{17.88a,b}$$

In relation (17.86) which defines $C_P$, the quantity $C_p$ is given by (17.34). With $\delta v_i$, $\delta w$, $\delta\varphi_j$, $\delta N_{m\ell}$, $\delta Q_k$, and $\delta M_{pq}$ varying independently and with (17.83) as defining relations for $\varepsilon_{ij}$, $\kappa_{ij}$, and $\gamma_i$, it is not difficult to establish (see the Exercises):

**[XVII.4]** ***The Euler differential equations of $\delta I_{NP} = 0$ are the stress-strain relations***

$$\varepsilon_{ij} = \frac{\partial C_P}{\partial N_{ij}}, \quad \kappa_{ij} = \frac{\partial C_P}{\partial M_{ij}}, \quad \gamma_i = \frac{\partial C_P}{\partial Q_i}, \quad (i, j = 1, 2), \quad (17.89)$$

***and the equilibrium equations***

$$N_{1j,1} + N_{2j,2} = 0, \qquad V_{1,1} + V_{2,2} + p = 0 \qquad (17.90a,b)$$

$$M_{1j,1} + M_{2j,2} - Q_j = 0, \qquad (17.90c)$$

***for*** $j = 1, 2$, ***where***

$$V_k = Q_k + w_{,1}N_{k1} + w_{,2}N_{k2} \quad (k = 1, 2). \qquad (17.90d)$$

***The Euler boundary conditions are the stress boundary conditions***

$$\Delta N_{nn} = \Delta N_{nt} = \Delta V_n = \Delta M_{nn} = \Delta M_{nt} = 0 \qquad (17.91)$$

***along*** $\Gamma_\sigma$ ***and***

$$\Delta v_n = \Delta v_t = \Delta w = \Delta \varphi_n = \Delta \varphi_t = 0 \qquad (17.92)$$

***along*** $\Gamma_d$.

The quantities $V_k, k = 1, 2$, in (17.90d) are the vertical shear resultants (in the $i_3$ direction) along a constant $x_k$ edge. For the theories (which assume infinitesimal strain and small displacement) considered in previous sections, they differ negligibly from the transverse shear resultants $Q_1$ and $Q_2$ which are normal to the middle surface of the deformed plate.

The five equilibrium equations (17.90), the ten stress-strain relations (17.89), and the ten strain-displacement relations (17.83) for all $(x_1, x_2)$ in R, together with the five boundary conditions [such as the stress booundary conditions (17.91) or the displacement boundary conditions (17.92)] along $\Gamma$, define a boundary-value problem in partial differential equations (in two independent variables) for the ten stress resultants and couples $(N_{ij}, Q_j, M_{ij})$, ten strain measures $(\varepsilon_{ij}, \gamma_j, \kappa_{ij})$, and five displacement components $(v_i, w, \varphi_j)$. If all nonlinear terms are omitted from the governing equations and boundary conditions, the tenth-order system of partial differential equations decouples into two unrelated smaller subsystems: a fourth-order system for plate stretching and shearing and a sixth-order system for plate bending and twisting. The latter is identical to Reissner's plate theory formulated in Section 5 of this chapter.

## 10. The von Kármán Plate Equations

For transversely rigid plates with $\nu_3 = 1/E_3 = 1/G_3 = 0$, we have from the stress-strain relations $\gamma_1 = \gamma_2 = 0$ (along with $E_{33} = 0$) as in the Germain–Kirchhoff theory. It follows from (17.83) that we have again

$$\varphi_j = -w_{,j}, \qquad \kappa_{ij} = -w_{,ij}. \tag{17.93}$$

The three stress-strain relations for $\kappa_{ij}$ in (17.89) become

$$M_{11} = -D(w_{,11} + \nu w_{,22}), \qquad M_{22} = -D(w_{,22} + \nu w_{,11}),$$
$$M_{12} = M_{21} = -D(1-\nu)w_{,12}. \tag{17.94}$$

The two equilibrium equations (17.90c) then give

$$Q_j = M_{1j,1} + M_{2j,2} = -(D\nabla^2 w)_{,j}. \tag{17.95}$$

In the Germain–Kirchhoff theory, we substitute (17.95) into the transverse force equilibrium equation to get a single equation for $w$. For the present nonlinear theory, the corresponding force equilibrium equation (17.90b) involves the new unknowns $N_{11}, N_{22}$, and $N_{12} = N_{21}$. After some simplification with the help of the in-plane equilibrium equations $N_{1j,1} + N_{2j,2} = 0$, this equation of vertical force equilibrium takes the form

$$Q_{1,1} + Q_{2,2} + w_{,11}N_{11} + 2w_{,12}N_{12} + w_{,22}N_{22} + p = 0. \tag{17.96}$$

Evidently, there is a coupling between the in-plane stretching (and shearing) action and the transverse bending (and twisting) action of the plate.

We may continue our reduction and express $N_{ij}$ in the three force equilibrium equations [in (17.90)] in terms of $v_1, v_2$ and $w$ by way of the stress-strain relations for $N_{ij}$ in (17.89) to obtain

$$N_{11} = \frac{Eh}{1-\nu^2}(\varepsilon_{11} + \nu\varepsilon_{22}) = \frac{Eh}{1-\nu^2}\left[v_{1,1} + \nu v_{2,2} + \frac{1}{2}(w_{,1}^2 + \nu w_{,2}^2)\right]$$

$$N_{22} = \ldots, \qquad N_{12} = N_{21} = \frac{Eh}{2(1+\nu)}\left[v_{1,2} + v_{2,1} + w_{,1}w_{,2}\right]. \tag{17.97}$$

In this way, we obtain three partial differential equations for $v_1, v_2$ and $w$, two second order and one fourth order. This reduction of the original problem to a boundary-value problem with three equations for the three displacement components $v_1, v_2$ and $w$ was accomplished by Kirchhoff more than a century ago. It will not be carried out here because there is a more attractive and widely used alternative reported by T. von Kármán in 1910. We will derive the von Kármán plate equations below.

The two in-plane equilibrium equations are for three unknowns. We can, in principle, solve for two of them in terms of the third. Alternatively, we may solve for all three in terms of a new unknown stress function $F$. One such stress function representation for $N_{ij}$ is

$$N_{11} = F_{,22}, \qquad N_{22} = F_{,11}, \qquad N_{12} = N_{21} = -F_{,12}. \tag{17.98}$$

With (17.98), we have effectively used up the two equilibrium equations for $N_{ij}$. For a plate with uniform properties and thickness, the transverse equilibrium equation (17.96) may now be written, with the help of (17.95) and (17.98), as

$$D\nabla^2\nabla^2 w = w_{,11}F_{,22} - 2w_{,12}F_{,12} + w_{,22}F_{,11} + p \tag{17.99}$$

where $\nabla^2(\ \ ) = (\ \ )_{,11} + (\ \ )_{,22}$ is the two-dimensional Laplace's operator. This is one (fourth-order) partial differential equations for two unknowns $w$ and $F$. We need a second equation for the same two unknowns to complete the reduction.

To get this second equation, we recall the derivation of the Euler differential equations in the principle of complementary energy for three-dimensional elasticity in Chapter 16 (see Section 8). It was shown there that when the equilibrium equations are satisfied identically by way of a stress function solution (representation), the Euler differential equation for that minimum principle are compatibility equations (16.60) for the strain components. Had nonlinear terms involving the product of $w_{,i}$ and $w_{,j}$ not been present, the expressions for $\varepsilon_{11}$, $\varepsilon_{22}$, and $\varepsilon_{12} = \varepsilon_{21}$ in terms of $v_1$ and $v_2$ would be identical in form to (16.61). Hence, these strain components (without the $w_{,i}w_{,j}$ terms) also satisfy the compatibility (16.62). In the presence of the nonlinear terms, it is not difficult to verify directly that the appropriate compatibility equation for $\varepsilon_{ij}$ is

$$\varepsilon_{11,22} + \varepsilon_{22,11} - 2\varepsilon_{12,12} = w_{,11}w_{,22} - (w_{,12})^2. \tag{17.100}$$

The above nonlinear "compatibility" equation may be transformed into an equation for $w$ and $F$ alone. To do this, we first invert the stress-strain relations for $N_{ij}$ to obtain

$$\varepsilon_{11} = A[N_{11} - \nu N_{22}], \qquad \varepsilon_{22} = A[N_{22} - \nu N_{11}],$$

$$\varepsilon_{12} = A(1+\nu)N_{12} = A(1+\nu)N_{21} = \varepsilon_{21}, \qquad A = \frac{1}{Eh}. \tag{17.101}$$

With $N_{ij}$ expressed in terms of $F$, we use these relations in (17.100) to eliminate $\varepsilon_{ij}$ and obtain the equation

$$A\nabla^2\nabla^2 F = w_{,11}w_{,22} - (w_{,12})^2 \tag{17.102}$$

for a plate of uniform properties and constant thickness. This is the second equation sought for $w$ and $F$.

It is evident from the von Kármán equations (17.99) and (17.102) that, in the absence of transverse shear deformation so that $\gamma_j = 0$ (and, hence, $\varphi_j = -w_{,j}$), the system of nonlinear plate equations is an eighth-order system. The number of (Euler) boundary conditions along an edge of the plate should, therefore, be reduced from 10 to 8. The four appropriate displacement edge conditions,

$$\Delta v_n = \Delta v_t = \Delta w = \varphi_n^* + w_{,n} = 0, \tag{17.103}$$

are evident from (17.92) and $\varphi_j = -w_{,j}$, $j = 1,2$; we cannot prescribe $\varphi_t^*$ arbitrarily once $w^*$ is given. A straightforward integration by parts for the term involving $\delta w_{,s}$ in the integral along $\Gamma_\sigma$ (similar to the linear case) shows that the four appropriate (Euler) stress boundary conditions are

$$\Delta N_{nn} = \Delta N_{nt} = \Delta M_{nn} = V_n^* + M_{nt,s}^* - V_n^c = 0, \tag{17.104}$$

where $V_n^c = V_n + M_{nt,s} = Q_n^c + w_{,n}N_{nn} + w_{,s}N_{nt}$.

When the deformation of the plate is infinitesimal so that the nonlinear terms may be omitted, the two von Kármán plate plate equations (17.99) and (17.102) uncouple. Equation (17.102) becomes a homogeneous biharmonic equation for the Airy stress function $F(x_1,x_2)$:

$$A\,\nabla^2\nabla^2 F = 0 \tag{17.105}$$

for all $(x_1,x_2)$ in $R$. This fourth-order linear partial differential equation for the stretching and shearing plate action is appropriately supplemented by two boundary conditions along an edge of the plate. If edge tractions are prescribed, the two edge conditions are

$$F_{,ss} = N_{nn}^*, \qquad -F_{,ns} = N_{nt}^*. \tag{17.106}$$

The other equation, (17.99), becomes an inhomogeneous biharmonic equation for the transverse displacement function $w(x_1,x_2)$:

$$D\,\nabla^2\nabla^2 w = p \tag{17.107}$$

for all $(x_1,x_2)$ in $R$. The two appropriate stress boundary conditions are now known to be Kirchhoff's contracted conditions $\Delta M_{nn} = \Delta Q_n^c = 0$.

## 11. Finite Twisting and Bending of Rectangular Plates

Consider the rectangular plate of Section 8 with the displacement conditions (17.64) replaced by prescribed resultant forces and moments. We are interested in particular in the case where the ends $x = \pm L$ are subject only to equal and opposite axial torques $\pm T$ and equal and opposite bending moments $\pm M$

(turning about the $y$ direction). From relations (17.87) and (17.104), these end conditions may be taken in the form

$$x = \pm L : \begin{cases} \displaystyle\int_{-a/2}^{a/2} \mathbf{N}_x^e dy - [2M_{xy}\mathbf{i}_z]_{-a/2}^{a/2} = 0, \\ \displaystyle\int_{-a/2}^{a/2} (\mathbf{M}_x^e + \mathbf{r}_0 \times \mathbf{N}_x^e) dy - [\mathbf{r}_0 \times (2yM_{xy}\mathbf{i}_z)]_{-a/2}^{a/2} = M\mathbf{i}_y + T\mathbf{i}_z, \end{cases} \tag{17.108}$$

where the *effective* stress resultant vector $\mathbf{N}_x^e$ and stress couple vector $\mathbf{M}_x^e$ are given by

$$\mathbf{N}_x^e = N_{xx}\mathbf{i}_x + N_{xy}\mathbf{i}_y + (Q_x^e + w_{,x}N_{xx} + w_{,y}N_{xy})\mathbf{i}_z,$$
$$\mathbf{M}_x^e = \mathbf{i}_z \times M_{xx}\mathbf{i}_x = M_{xx}\mathbf{i}_y.$$

and $\mathbf{r}_0 = x\mathbf{i}_x + y\mathbf{i}_y + w\mathbf{i}_z$. The bracketed terms in (17.108) are the effects of the (fictitions) corner forces associated with the introduction of the Kirchhoff hypotheses. In practice, we rarely know the actual distributions of the relevant stress resultants and stress couples along an edge of the plate. Hence, the overall conditions (17.108) are usually all that can be said about the conditions at an edge with prescribed edge loads.

The solution $w_L = \theta xy - k(x^2 - \nu y^2)/2$ for the corresponding linear problem (see Section 8 and Exercise 10) is no longer appropriate for the nonlinear theory. It does not satisfy the governing differential equations (17.99) and (17.102) for the new problem; they require a nontrivial solution for the stress function $F$ as well. Moreover, the possible presence of in-plane stress resultants necessitates a more elaborate prescription of boundary conditions for the free edges. According to (17.104) we should have

$$y = \pm\frac{a}{2}: \qquad N_{yy} = N_{yx} = V_y = M_{yy} = 0 \tag{17.110}$$

with $V_y = Q_y^e + w_{,x}N_{yx} + w_{,y}N_{yy}$. Conditions (17.110) must hold for all $|x| \leq L$; we may, therefore, attempt a solution for $F$ which is independent of $x$:

$$F(x, y) = f(y). \tag{17.111a}$$

Such a nontrivial stress function in turn requires that we modify the $y$-dependent portion of $w_L(x, y)$ and work with

$$w(x,y) = \theta xy - \frac{1}{2}kx^2 + W(y) \tag{17.111b}$$

instead with the second term being the expected pure bending behavior in respense to the moments $\pm M$ at the two ends.

Upon substituting (17.111) into (17.99) and (17.102), we obtain

$$DW'''' = -kf'', \qquad Af'''' = \theta^2 + kW'' \tag{17.112}$$

with $(\ )' \equiv d(\ )/dy$. The second equation may be integrated to obtain

$$Af'' = \frac{1}{2}\theta^2 y^2 + kW \tag{17.113}$$

where we have set the two constants of integration equal to zero. It is not difficult to show that they do not give rise to nontrivial stress fields and may be omitted from further consideration. We now use (17.113) to eliminate $f$ from the first equation of (17.112) to obtain

$$DAW'''' + k^2W = -\frac{1}{2}\theta^2 k^2 y^2. \tag{17.114}$$

An exact solution of (17.114) is immediate. Because $W(y)$ is associated with plate bending actions, we retain only terms of the solution even in $y$ so that

$$W(y) = c_1 \cosh(\beta y)\cos(\beta y) + c_2 \sinh(\beta y)\sin(\beta y) - \frac{1}{2k}\theta^2 y^2, \tag{17.115}$$

where $\beta^4 = k^2/4DA$.

The corresponding stress resultants and stress couples are

$$N_{yy} = N_{xy} = 0, \qquad N_{xx} = \frac{1}{2A}\theta^2 y^2 + \frac{k}{A}W,$$

$$M_{xx} = D[k^2 + \nu\frac{\theta^2}{k} + 2\nu\beta^2\{c_1 \sinh(\beta y)\sin(\beta y) - c_2 \cosh(\beta y)\cos(\beta y)\}],$$

$$M_{yy} = D\left[\frac{\theta^2}{k} + \nu k^2 + 2\beta^2\{c_1 \sinh(\beta y)\sin(\beta y) - c_2\cosh(\beta y)\cos(\beta y)\}\right], \tag{17.116}$$

$$M_{xy} = -(1-\nu)D\theta, \qquad V_x = (\theta y - kx)N_{xx},$$

$$V_y = Q_y = -DW''' = 2D\beta^3[c_1\{\cosh(\beta y)\sin(\beta y) + \sinh(\beta y)\cos(\beta y)\} + c_2\{\cosh(\beta y)\sin(\beta y) - \sinh(\beta y)\cos(\beta y)\}].$$

Two of the four free edge conditions in (17.110) are satisfied trivially. The other two conditions determine $c_1$ and $c_2$ in terms of $\theta$ and $k$:

$$\{c_1, c_2\} = \frac{1}{\beta^2}\left(\frac{\theta^2}{k} + \nu k\right)\{\xi_m, \xi_p\}, \tag{17.117a}$$

$$\begin{Bmatrix}\xi_m \\ \xi_p\end{Bmatrix} = \frac{\sinh\mu\cos\mu \pm \cosh\mu\sin\mu}{\sinh(2\mu) + \sin(2\mu)}, \qquad \mu = \frac{1}{2}a\beta. \tag{17.117b}$$

It remains to choose $k$ and $\theta$ to satisfy the integral conditions (17.108) which correspond to six scalar conditions:

$$\int_{-a/2}^{a/2} N_{xx}\,dy = 0, \qquad \int_{-a/2}^{a/2} N_{xy}\,dy = 0, \tag{17.118a,b}$$

$$\int_{-a/2}^{a/2} N_{xy}\,y\,dy = 0, \qquad \int_{-a/2}^{a/2} V_x^e\,dy - [2M_{xy}]_{-a/2}^{a/2} = 0, \tag{17.118c,d}$$

$$\int_{-a/2}^{a/2} V_x^e\,y\,dy - [2yM_{xy}]_{-a/2}^{a/2} = T, \qquad \int_{-a/2}^{a/2} (M_{xx} + wN_{xx})\,dy = M \tag{17.118e,f}$$

Of these six conditions, (17.118b) and (17.118c) are trivially satisfied whereas (17.118d) becomes

$$\int_{-a/2}^{a/2} (\theta y - kx)N_{xx}\,dy + [2(1-\nu)D\theta]_{-a/2}^{a/2} = -kx\int_{-a/2}^{a/2} N_{xx}\,dy$$

$$= -kx\int_{-a/2}^{a/2} f''\,dy = Dx\int_{-a/2}^{a/2} W''''\,dy$$

$$= Dx[W''']_{-a/2}^{a/2} = -x[V_y]_{-a/2}^{a/2} = 0$$

where we have used of the first ODE of (17.112) and the boundary condition $V_y = 0$ at $y = \pm a/2$. This shows that (17.118d) and, in the process, (17.118a) are both satisfied without any restriction on $k$ and $\theta$. The remaining two conditions, (17.118e) and (17.118f), give two relations for $\theta$ and $k$ in terms of $T$ and $M$. These relations may be taken in the form

$$T = 2(1-\nu)D\theta a\left[1 + \frac{\nu}{1-\nu}F_1(\mu) + \frac{4\lambda^4}{45(1-\nu)}F_2(\mu)\right], \tag{17.119a}$$

$$M = (1-\nu^2)Dka\left[1 + \frac{\nu^2}{1-\nu^2}G_1(\mu) + \frac{8\lambda^4}{45(1-\nu^2)}G_2(\mu) - \frac{32\lambda^8}{945(1-\nu^2)}G_3(\mu)\right], \tag{17.119b}$$

where

$$\mu^4 = \frac{3(1-\nu^2)k^2a^4}{16h^2}, \qquad \lambda^4 = \frac{3(1-\nu^2)\theta^2a^4}{16h^2}. \tag{17.120}$$

The quantities $\{F_i(\mu), G_j(\mu)\}$ can be found in Reissner (1957). For $0 \le \mu \ll 1$, we have

$$F_1 = \frac{4}{45}\mu^4 + O(\mu^8), \qquad F_2 = 1 + O(\mu^4), \tag{17.121a}$$

$$G_1 = \frac{8}{45}\mu^4 + O(\mu^8), \quad G_2 = 1 + O(\mu^4), \quad G_3 = 1 + O(\mu^4). \quad (17.121b)$$

Hence, we recover the results for the linear theory if $\mu^4$ and $\lambda^4$ are negligibly small. Note that the coupling between bending and twisting plate actions persists in the nonlinear theory if $\mu^2$ and $\lambda^4$ are not both negligibly small compared to unity.

## 12. Exercises

1. For the classical plate theory, an edge of the plate is "simply-supported" if $w = 0$ and $w_{,nn} = 0$ (where $w_{,n}$ is the normal derivative of $w$). Consider a rectangular plate spanning $\{0 \le x \le a, 0 \le y \le b\}$ with all four edges simply-supported. Use the results of Section 7 to obtain an approximate solution of the form

$$u \cong u_{mk}(x, y) = c_{mk} \sin\left(\frac{m\pi x}{a}\right) \sin\left(\frac{k\pi y}{b}\right),$$

where $m$ and $k$ are positive integers.

(a) Show that $w_{mk}(x, y)$ satisfies the simply-supported edge conditions along all four edges.

(b) Obtain an expression for $c_{mk}$ for a general $p(x,y)$.

(c) Evaluate the relevant integral(s) for $p(x, y) \equiv p_0$ (a constant).

(d) How would you obtain a better approximate solution for the problem (than $w_{mk}$)?

2. In Section 7 of Chapter 15, we obtained an approximate solution of the form

$$w \cong w_1(x, y) = c_1\left[1 - \cos\left(\frac{2\pi x}{a}\right)\right]\left[1 - \cos\left(\frac{2\pi y}{b}\right)\right] \equiv c_1 \overline{\varphi}_1(x, y)$$

for a rectangular plate spanning $\{0 \le x \le a, 0 \le y \le b\}$ with all four edges *clamped* so that both $w$ and $\partial w/\partial n$ vanish along all four edges. For a better approximate solution, we wish to take

$$w \cong c_1\overline{\varphi}_1(x, y) + c_2\overline{\varphi}_2(x, y),$$

where $\overline{\varphi}_1(x, y)$ is as given above. Without actually doing the calculations to get $c_1$ and $c_2$, what would you suggest that we take for $\overline{\varphi}_2$? Sketch the process for the determination of $c_1$ and $c_2$ in a Rayleigh–Ritz solution.

3. For a transversely rigid (and otherwise isotropic, linearly elastic) plate, we have $\nu_3 = 1/E_3 = 1/G_3 = 0$ in (17.32).
   (a) Obtain the six stress-strain relations.
   (b) Use $e_{j3} = 0$ to determine the dependence of $u_i$ $(i = 1,2,3)$ on $x_3$.
   (c) Use the remaining stress-strain relations to determine $\sigma_{ij}$, $i,j = 1,2$.
   (d) Use the equilibrium equations and the stress BCs on the faces to determine $\sigma_{3k} = \sigma_{k3}$, $k = 1,2,3$.

4. In cylindrical coordinates $(r, \theta, x_3)$, three relevant equilibrium equations for $Q_r, Q_\theta, M_{rr}, M_{\theta r}$ and $M_{\theta\theta}$ are

$$\frac{1}{r}[(rQ_r)_{,r} + Q_{\theta,\theta}] + p = 0,$$

$$\frac{1}{r}[(rM_{rr})_{,r} + M_{\theta r,\theta} - M_{\theta\theta}] = Q_r, \quad \frac{1}{r}[(rM_{r\theta})_{,r} + M_{\theta\theta,\theta} - M_{\theta r}] = Q_\theta.$$

With the same dependence on $x_3$ for stress components as those given in (17.27) for the Cartesian coordinates and the equilibrium equations above, derive the polar coordinate formulation of Reissner's plate theory.

5. Obtain the Euler DE and Euler BC of (17.37) subject to the constraint $M_{12} = M_{21}$. Eliminate the Lagrange multiplier for $M_{12} = M_{21}$ from the final results. (In other words, establish [XVII.1].)

6. Establish [XVII.3].

7. Establish [XVII.4].

8. In cylindrical coordinates, the expressions for $\varepsilon_{ij}, \kappa_{ij}$ and $\gamma_j$ (with $1 \leftrightarrow r$ and $2 \leftrightarrow \theta$) are given by

$$\varepsilon_{11} = v_1' + \frac{1}{2}(w')^2, \quad \varepsilon_{22} = \frac{1}{r}(v_2^\bullet + v_1) + \frac{1}{2}\left(\frac{w^\bullet}{r}\right)^2,$$

$$\varepsilon_{12} = \varepsilon_{21} = \frac{1}{2}\left[v_2' + \frac{1}{r}(v_1^\bullet - v_2) + \frac{1}{r}w^\bullet w'\right]$$

$$\gamma_1 = \varphi_1 + w', \quad \gamma_2 = \varphi_2 + \frac{1}{r}w^\bullet,$$

$$\kappa_{11} = \varphi_r' = -w'', \quad \kappa_{22} = \frac{1}{r}(\varphi_2^\bullet + \varphi_1) = -\left(\frac{1}{r}w' + \frac{1}{r^2}w^{\bullet\bullet}\right).$$

With the same dependence on $x_3$ for the displacement and stress components for Cartesian coordinates, derive the polar coordinates formulation of the finite deflection plate theory.

9. (a) Let $u = w + BD\nabla^2 w$. Show that expressions for $M_{ij}$ and $Q_k$ (in Cartesian coordinates) can be written in terms of $u$ and $\chi$ as

$$M_{11} = -D[u_{,11} + \nu u_{,22}] + (1-\nu)BD\chi_{,12},$$

$$M_{22} = -D[u_{,22} + \nu u_{,11}] - (1-\nu)BD\chi_{,12},$$

$$M_{12} = M_{21} = -(1-\nu)Du_{,12} + \frac{1}{2}(1-\nu)DB(\chi_{,22} - \chi_{,11}),$$

$$Q_1 = -D(\nabla^2 u)_{,1} + \chi_{,2}, \qquad Q_2 = -D(\nabla^2 u)_{,2} + \chi_{,1}.$$

(b) With the same definition of $u$, show that the corresponding expressions in polar coordinate are

$$M_{rr} = -D\left[u_{,rr} + \nu\left(\frac{u_{,r}}{r} + \frac{u_{,\theta\theta}}{r^2}\right)\right] + (1-\nu)BD\left(\frac{\chi_{,\theta}}{r}\right)_{,r},$$

$$M_{\theta\theta} = -D\left[\left(\frac{u_{,r}}{r} + \frac{u_{,\theta\theta}}{r^2}\right) + \nu u_{,rr}\right] - (1-\nu)BD\left(\frac{\chi_{,\theta}}{r}\right)_{,r},$$

$$M_{r\theta} = -(1-\nu)D\left(\frac{u_{,\theta}}{r}\right)_{,r} + \frac{1}{2}(1-\nu)BD\left[\left(\frac{\chi_{,r}}{r} + \frac{\chi_{,\theta\theta}}{r^2}\right) - \chi_{,rr}\right],$$

$$Q_r = -D(\nabla^2 u)_{,r} + \frac{\chi_{,\theta}}{r}, \qquad Q_r = -\frac{1}{r}D(\nabla^2 u)_{,\theta} - \chi_{,r}.$$

10. Consider the problem of transverse bending of a rectangular plate $\{|x| \le L, |y| \le a/2\}$ by equal and opposite end bending moments of magnitude $M$. The plate is free of traction at the two long edges $y = \pm b$ so that

$$y = \pm b: \qquad M_{yy} = Q_y^e = 0,$$

where $Q_y^e = Q_y + M_{xy,x}$. The two ends of the plate $x = \pm L$ are subject only to equal and opposite bending moments so that

$$x = \pm L: \begin{cases} \displaystyle\int_{-b}^{b} Q_y^e\, dy - [2M_{xy}]_{-b}^{b} = 0, \quad \int_{-b}^{b} M_{xx}\, dy = M, \\ \displaystyle\int_{-b}^{b} Q_y^e\, y\, dy - [2yM_{xy}]_{-b}^{b} = 0. \end{cases}$$

(a) We expect from elementary beam theory that the midplane transverse displacement $w$ should be quadratic in $x$ (so that the deformed plate has constant curvature in $x$ direction). Show that the boundary conditions of the problem requires $w(x, y) = -k(x^2 - \nu y^2)/2$ for some constant $k$.

(b) Show that $w(x, y)$ as found in part (a) satisfies the homogeneous biharmonic equation and all the boundary conditions at $y = \pm b$.

(c) Show that the overall end conditions at $x = \pm L$ are also satisfied if

$$M = 2(1 - \nu^2)Dbk.$$

# 18

# Fluid Mechanics

## 1. Mass and Entropy

To understand macroscopic phenomena involving motion of fluids (both liquids and gases), we work with mathematical models which characterize them by their kinematic and dynamic properties such as velocity, mass density, and pressure. In mathematical models for solid mechanics, the description of these properties (field variables) are for each material point of the body and their evolution over time. This type of description is known as the *Lagrangian representation*. In contrast, mathematical theories for fluid mechanics are more conventionally formulated in an *Eulerian description* instead, although Lagrangian representation is also used sometimes. In the Eulerian representation, the field variables are given as functions of position in space $\mathbf{x}$ and time $t$. In particular, $\mathbf{v}(\mathbf{x}, t)$ is the velocity vector of a fluid particle occupying the point $\mathbf{x}$ in space at time $t$; $\rho(\mathbf{x}, t)$ and $p(\mathbf{x}, t)$ are the corresponding mass density of a fluid element which happens to occupy $\mathbf{x}$ at time $t$ and the pressure acting on the element.

In the Eulerian representation, time rates of change are no longer simply derivatives with respect to time because a new fluid element has moved in to occupy $\mathbf{x}$ after the lapse of an incremental time $dt$. Instead, the acceleration of a fluid element in $\mathbf{x}$ at time $t$ is given by the *total derivative* of its velocity $\mathbf{v}(\mathbf{x}, t)$

$$\frac{D\mathbf{v}}{Dt} = \frac{\partial \mathbf{v}}{\partial t} + \mathbf{v} \cdot \operatorname{grad} \mathbf{v} = \mathbf{v}_{,t} + \mathbf{v} \cdot \nabla(\mathbf{v}). \tag{18.1}$$

(The terms "material derivative" and "substantial derivative" are also used in the literature.) More generally, the total derivative of ( ) is given by

$$\frac{D(\ )}{Dt} = \frac{\partial(\ )}{\partial t} + \mathbf{v} \cdot \nabla(\ ). \tag{18.2}$$

To stay with the fluid element as we calculate rates of change, we must include the effect of moving value of x (which is known to be) represented by the term $\mathbf{v}\cdot\nabla(\ \ )$.

In Lagrangian representation, the motion of a fluid element originally at the point $\mathbf{a}$ in space at time $t = 0$ is described by its position vector $\mathbf{x}(\mathbf{a},t)$, with $\mathbf{x}$ being the position of that fluid element at a later time $t$. Clearly, we have $\mathbf{x}(\mathbf{a},0) = \mathbf{a}$ and the velocity and acceleration at $t$ of the fluid element are given by

$$\mathbf{v} = \mathbf{x}_{,t}(\mathbf{a},t), \qquad \mathbf{v}_{,t} = \mathbf{x}_{,tt}(\mathbf{a},t). \tag{18.3}$$

In continuum models of fluid motion without shocks, the position vector $\mathbf{x}(\mathbf{a},t)$ is assumed to be at least a $C^2$ function in all its arguments. For a variational formulation of various fluid models, it is often convenient for us to work with the Lagrangian representation and transform the results obtained into the Euler formulation whenever needed.

Consider the mass $\Delta M$ in a small volume $\Delta V_x$ of the fluid surrounding the fluid element at $\mathbf{x}$. The mass density at $\mathbf{x}$ (and time $t$) is defined by $\rho(\mathbf{x},t)\Delta V_x = \Delta M$ in the limit as $\Delta V_x$ tends to zero. If $\Delta V_a$ is the volume about $\mathbf{a}$ which has evolved into $\Delta V_x$, then we must have

$$\rho(\mathbf{x},t)\Delta V_x = \rho(\mathbf{a},0)\Delta V_a. \tag{18.4}$$

As $\Delta V_x$ is given by $J\Delta V_a$ where $J = \partial(x_1,x_2,x_3)/\partial(a_1,a_2,a_3)$ is the relevant Jacobian, the conservation of mass (18.3) becomes

$$\rho(\mathbf{x},t)J = \rho(\mathbf{a},0) = \rho_0. \tag{18.5}$$

In particular, we must have $J = 1$ if the fluid is *incompressible*, that is, if the fluid density remains constant in time and position.

In Eulerian representation, we focus our attention on a small volume $\Delta V$ at a fixed $\mathbf{x}$ at time $t$. The rate of increase of mass $\Delta M$ in this volume must be equal to the fluid mass flowing in through the boundary surface $\Delta S$ of the volume:

$$\frac{d(\Delta M)}{dt} = \iiint_{\Delta V}\frac{\partial\rho}{\partial t}dV_x = \iint_{\Delta S}\rho\mathbf{v}\cdot(-\mathbf{n})\,dS,$$

where $dV_x \equiv dx_1dx_2dx_3 \equiv d\mathbf{x}$ and $\mathbf{n}$ is the unit surface normal, positive outward. By the divergence theorem, we have

$$\iiint_{\Delta V}[\rho_{,t} + \nabla\cdot(\rho\mathbf{v})]\,dV_x = 0. \tag{18.6}$$

Because $\Delta V$ is arbitrary (and the integrand is assumed to be continuous), this integral condition requires the integrand of (18.6) to vanish for all points in the body of fluid:

$$\rho_{,t} + \nabla\cdot(\rho\mathbf{v}) = 0. \tag{18.7}$$

We have altogether:

**[XVIII.1]** ***The conservation of fluid mass requires that (18.5) and (18.7) be satisfied by*** $\rho(\mathbf{x},t)$ ***and*** $\mathbf{v}(\mathbf{x},t)$.

If the fluid is incompressible so that $\rho = \rho_0$, the *equation of continuity* (18.7) reduces to $\nabla\cdot\mathbf{v} = 0$.

For a fluid with no dissipation of energy, its motion involves no change in *entropy*. Let $s(\mathbf{x},t)$ be the entropy per unit mass of the fluid at the point x and time $t$. Then, the conservation of entropy in a volume $\Delta V_x$ around x requires

$$\begin{aligned} s(\mathbf{x},t)\Delta M &= s(\mathbf{x},t)\rho(\mathbf{x},t)\Delta V_x \\ &= s(\mathbf{x},t)\rho(\mathbf{x},t)J\Delta V_a = s(\mathbf{a},0)\rho_0\Delta V_a \end{aligned}$$

so that we have $s(\mathbf{x},t)\rho(\mathbf{x},t)J = s(\mathbf{a},0)\rho_0$. But by (18.5), this reduces to

$$s(\mathbf{x},t) = s(\mathbf{a},0) = s_0. \tag{18.8}$$

By an argument similar to that leading up to (18.7), we can also obtain the corresponding requirement in the Eulerian representation:

$$(\rho s)_{,t} + \nabla\cdot(\rho s\mathbf{v}) = 0. \tag{18.9}$$

We may use (18.7) to simplify (18.9) to obtain:

**[XVIII.2]** ***The conservation of entropy requires that (18.8) and***

$$s_{,t} + \mathbf{v}\cdot\nabla s = 0 \tag{18.10}$$

***be satisfied by*** $s(\mathbf{x},t)$ ***and*** $\mathbf{v}(\mathbf{x},t)$.

## 2. A Lagrangian Variational Principle for Ideal Fluids

For a conservative dynamical system, Hamilton's principle of Chapter 6 requires the action integral of the system, i.e., the difference between the kinetic energy and potential energy, be stationary. It is natural to extend this principle to fluid motion. Because there is no consideration of thermodynamics in Hamilton's principle, its simplest extension to fluid dynamics will necessarily have to be restricted to fluids with no change in entropy, i.e., to *ideal fluids*.

As in the theory of elasticity, the rate of change of potential energy $\mathcal{P}$ in a deformable continuum is the sum of the load potential associated with external loads and the internal energy corresponding to the strain energy in elasticity theory:

$$\mathcal{P} = \iiint_V \rho(U+L)\,dV_x = \iiint_{V_0} \rho_0(U+L)\,dV_a \tag{18.11}$$

where $dV_a \equiv d\mathbf{a}$ and $V_0$ is the volume of fluid element which occupies $V_x$ at time $t$. The integral of load potential per unit mass $L(\mathbf{x},t)$ in (18.11) is similar to the load potential of Chapter 6. The internal energy density function $U$ is a function of mass density $\rho$ and entropy $s$ and does not depend on $\mathbf{x}$ and $t$ explicitly. The pressure $p$ and temperature $\tau$ of the fluid are obtained from $U$ by the relations

$$p(\mathbf{x},t) = \rho^2\frac{\partial U}{\partial \rho}, \qquad \tau = \frac{\partial U}{\partial s}. \tag{18.12}$$

With the kinetic energy of the fluid given by

$$T = \iiint_V \frac{1}{2}\rho\mathbf{x}_{,t}\cdot\mathbf{x}_{,t}\,dV_x = \iiint_{V_0} \frac{1}{2}\rho_0\mathbf{x}_{,t}\cdot\mathbf{x}_{,t}\,dV_a, \tag{18.13}$$

the analogue of Hamilton's principle for an ideal fluid is:

**The Generalized Hamilton's Principle:** *The actual motion of the fluid which evolves from a given initial state at* t = 0 *to a given final state at* t = t$_f$ *subject to the conservation of mass and entropy, renders stationary the action integral* I *given by*

$$\mathrm{I} = \int_0^{t_f} [T - \mathcal{P}]\,dt. \tag{18.14}$$

The stationary value is expected to be a minimum for a sufficiently short elapsed time.

By the method of Lagrangian multipliers, we consider the variational equation

$$\delta\left[I + \int_0^{t_f}\iiint_{V_0} \{\lambda_1(\rho J - \rho_0) + \lambda_2(s - s_0)\}\,dV_a\,dt\right] = 0 \tag{18.15}$$

in which the two constraints (18.5) and (18.8) are appended to $J$ by the multipliers $\lambda_1$ and $\lambda_2$. The independent variations of $\delta s$ and $\delta\rho$ give immediately $\lambda_2 = \rho_0 U_{,s}$ and $\lambda_1 J = \rho_0 U_{,\rho}$. By (18.12) and (18.5), these may be written as

$$\lambda_1 = \frac{p}{\rho}, \qquad \lambda_2 = \rho_0\tau. \tag{18.16}$$

Hence, the multipliers $\lambda_1$ and $\lambda_2$ are related to pressure and temperature, respectively.

The remaining Euler DEs associated with the independent variations of $\{\delta x_i\}$ are to be obtained from

$$\int_0^{t_1}\iiint_{V_0}\left\{\rho_0 x_{i,t}\delta x_{i,t} - \rho_0\frac{\partial L}{\partial x_i}\delta x_i + \lambda_1\rho\frac{\partial J}{\partial x_{i,j}}\delta x_{i,j}\right\}dV_a = 0 \qquad (18.17)$$

where, throughout this chapter, repeated indices are to be summed over 1, 2, and 3, and $(\ )_{,j} = \partial(\ )/\partial a_j$. After integrating by parts in both time and space variables (the latter with the help of the divergence theorem) and observing that $\delta x_i = 0$ on the boundary of $V_0$ as well as at the initial and terminal time, we obtain from (18.17)

$$\rho_0\frac{\partial^2 x_i}{\partial t^2} + \rho_0\frac{\partial L}{\partial x_i} + (\rho J_{ij})_{,j} = 0 \quad (i = 1,2,3), \qquad (18.18)$$

where $J_{ij} = \partial J/\partial x_{i,j}$ is the $(i,j)$-minor of the determinant $J$. The matrix identity $J_{ij,j} = 0$ allows us to write $(\rho J_{ij})_{,j} = \rho_{,j}J_{ij}$. With the help of a second matrix identity $x_{k,j}J_{ij} = J\delta_{ik}$, we have

$$\rho_{,j}J_{ij} = \frac{\partial\rho}{\partial a_j}J_{ij} = \frac{\partial\rho}{\partial x_k}\frac{\partial x_k}{\partial a_j}J_{ij} = \frac{\partial\rho}{\partial x_k}J\delta_{ik} = \frac{\partial\rho}{\partial x_i}J \qquad (18.19)$$

so that (18.18) becomes

$$\frac{\partial^2 x_i}{\partial t^2} + \frac{\partial L}{\partial x_i} + \frac{1}{\rho}\frac{\partial p}{\partial x_i} = 0 \quad (i = 1,2,3). \qquad (18.20)$$

Note that the entropy constraint $s = s_0$ has no role in the three equations of motion for the fluid. The same results would have been obtained if we had taken the internal energy $U$ to be a function of the mass density alone.

If we now set $v_i = x_{i,t}$ and $-L_{,x_i} = f_i$, we can write (18.20) as

$$\left(\frac{\partial v_i}{\partial t}\right)_{\mathbf{a}\ \text{fixed}} + \frac{1}{\rho}\frac{\partial p}{\partial x_i} = f_i \quad (i = 1,2,3). \qquad (18.21)$$

But with

$$\left(\frac{\partial v_i}{\partial t}\right)_{\text{fixed}\ \mathbf{a}} = \left(\frac{\partial v_i}{\partial t}\right)_{\text{fixed}\ \mathbf{x}} + \frac{\partial v_i}{\partial x_j}\left(\frac{\partial x_j}{\partial t}\right)_{\text{fixed}\ \mathbf{a}}$$
$$= \left(\frac{\partial v_i}{\partial t}\right)_{\text{fixed}\ \mathbf{x}} + v_j\frac{\partial v_i}{\partial x_j},$$

the three scalar equations of motion (18.21) may be combined into a single-vector equation

$$\frac{\partial \mathbf{v}}{\partial t} + \mathbf{v}\cdot\nabla\mathbf{v} + \frac{1}{\rho}\nabla p = \mathbf{f} \qquad (18.22\text{a})$$

or

$$\frac{D\mathbf{v}}{Dt} + \frac{1}{\rho}\nabla p = \mathbf{f}. \tag{18.22b}$$

With $\mathbf{f}$ being a known function of $\mathbf{x}$ and $U$ only a function of $\rho$, the vector equation of motion (18.22), the continuity equation (18.7), and the pressure-density relation in (18.12) form a system of five scalar equations for the determination of the five unknowns $v_i$, $\rho$ and $p$ as functions of $\mathbf{x}$ and $t$. As such, they constitute the Eulerian formulation of an ideal fluid model for which we should have a variational principle. Given in terms of gradients, they are applicable to other coordinate systems as well.

## 3. Ideal Fluid Motion Not Always Irrotational

We expect the generalized Hamilton's principle of Section 2 to apply to the Eulerian representation of fluid-flow problems as well. Because conservation of entropy plays no role in the resulting equations of motion, we will omit the constraint $s = s_0$ and take $U = U(\rho)$ to simplify our presentation.

In Eulerian representation, the action integral (18.14), with the mass-conservation constraint appended to it by a multiplier $\varphi$, becomes

$$\bar{I} = I + \int_0^{t_f}\iiint_V \varphi\,[\rho_{,t} + \nabla\cdot(\rho\mathbf{v})]\,dV_x\,dt \tag{18.23}$$

so that after integration by parts (with $\delta\rho$ and $\delta\mathbf{v}$ vanishing on the boundary $S$ of $V$ as well as at $t = 0$ and $t_f$)

$$\delta\bar{I} = \int_0^{t_f}\iiint_V\left\{\left[\frac{1}{2}|\mathbf{v}|^2 - \varphi_{,t} - L - (\rho U)_{,\rho} - \mathbf{v}\cdot\nabla\varphi\right]\delta\rho - [\rho\mathbf{v} - \rho\nabla\varphi]\cdot\delta\mathbf{v}\right\}dV_x\,dt. \tag{18.24}$$

The three scalar Euler DEs associated with $\delta\mathbf{v}$ are, therefore, given by the vector equation

$$\mathbf{v} = \nabla\varphi. \tag{18.25}$$

Upon substituting (18.25) into the remaining single Euler DE associated with $\delta\rho$, we obtain

$$\varphi_{,t} + \frac{1}{2}|\nabla\varphi|^2 + (\rho U)_{,\rho} = -L(\mathbf{x},t) \tag{18.26a}$$

or, in view of (18.12),

$$\varphi_{,t} + \frac{1}{2}|\mathbf{v}|^2 + \frac{p}{\rho} + U(\rho) = -L(\mathbf{x}, t). \tag{18.26b}$$

Together with the continuity equation, written as

$$\rho_{,t} + \nabla\cdot(\rho\nabla\varphi) = 0, \tag{18.27}$$

we have a pair of PDEs for $\rho$ and $\varphi$ as the load potential $L$, which depends only on $\mathbf{x}$ (and possibly $t$), is now a known function.

Equation (18.25) requires that the ideal fluid flow be irrotational because

$$\nabla \times \mathbf{v} = \nabla \times \nabla\varphi = 0.$$

This requirement is unnecessarily restrictive. (For example, the fluid must be irrotational initially.) Not all ideal fluid flows are irrotational (see also Section 5) and we must reexamine our variational formulation to eliminate this inappropriate restriction.

The difficulty was removed by C.C. Lin (1963). Lin observed that we need a device to distinguish the position vector $\mathbf{x}$ in space and the (initial) position vector $\mathbf{a}$ of the material point. Because $\mathbf{a}$ does not change with time, we have for $\mathbf{a}(\mathbf{x}, t)$

$$\left(\frac{\partial \mathbf{a}}{\partial t}\right)_{\text{fixed } \mathbf{x}} + \frac{\partial \mathbf{x}}{\partial t}\cdot\nabla\mathbf{a} = 0$$

so that

$$\frac{\partial \mathbf{a}}{\partial t} + \mathbf{v}\cdot\nabla\mathbf{a} = 0 \tag{18.28a}$$

or

$$\frac{da_i}{dt} = \left(\frac{\partial a_i}{\partial t}\right)_{\text{fixed } \mathbf{x}} + \frac{\partial a_i}{\partial x_j}\frac{\partial x_j}{\partial t} = 0 \quad (i = 1, 2, 3). \tag{18.28b}$$

Either (18.28b) or (18.28a) should be incorporated as a constraint in an Eulerian form of the variational principle for fluid flows which follows from the generalized Hamilton's principle (see also J. Serrin, 1959, for another report on Lin's contribution).

## 4. An Eulerian Variational Principle for Ideal Fluids

We now append the constraint (18.28a) to the integral (18.23) by a vector multiplier $\rho\mathbf{u}$ and consider the variational equation

$$\delta\{\bar{I} + \int_0^{t_f}\iiint_V \rho\mathbf{u}\cdot[\mathbf{a}_{,t} + \mathbf{v}\cdot\nabla\mathbf{a}]\,dV_x\,dt\} = 0. \tag{18.29}$$

The four Euler equations associated with $\delta\mathbf{v}$ and $\delta\rho$ now take the form

$$v_i - \frac{\partial\varphi}{\partial x_i} + u_k\frac{\partial a_k}{\partial x_i} = 0 \quad (i = 1,2,3), \tag{18.30}$$

$$\frac{1}{2}|\mathbf{v}|^2 - \frac{\partial}{\partial\rho}(\rho U) - L - \frac{\partial\varphi}{\partial t} - \mathbf{v}\cdot\nabla\varphi = 0. \tag{18.31}$$

In addition, there are three new Euler equations associated with $\{\delta a_k\}$:

$$\frac{\partial}{\partial t}(\rho u_j) + \frac{\partial}{\partial x_i}(\rho v_i u_j) = 0 \quad (j = 1,2,3)$$

which can be simplified by the continuity equation (18.7) to

$$\frac{\partial u_j}{\partial t} + \mathbf{v}\cdot\nabla u_j = 0 \quad (j = 1,2,3). \tag{18.32}$$

The Euler DEs (18.30)-(18.32) along with the continuity equation (18.7), the constraint equations (18.28), and the pressure-density relation in (18.12) form a system of 12 equations. Along with suitable initial and boundary conditions, this system determines the 12 unknowns $\{v_i, \rho, p, \varphi, u_j, a_k\}$. The question is how is this system related to the smaller system (18.22), (18.7), and the pressure-density relation previously found to characterize ideal fluid motion in Section 2.

To relate the two systems, we need to obtain equations involving accelerations from the larger new system. This can be accomplished by taking the time partial derivative of (18.30). The term involving $\varphi$ in the resulting equation can be eliminated by substracting from it the gradient of (18.31). The net result is

$$\frac{\partial v_i}{\partial t} + \frac{1}{\rho}\frac{\partial p}{\partial x_i}\left(\frac{1}{2}|\mathbf{v}|^2\right) + \frac{\partial}{\partial t}\left(u_j\frac{\partial a_j}{\partial x_i}\right) + \frac{\partial}{\partial x_i}\left(u_j v_k\frac{\partial a_j}{\partial x_k}\right) = f_i(\mathbf{x},t) \tag{18.33}$$

where we have made use of the relation $f_i = -\partial L/\partial x_i$ and the pressure-density relation $p = \rho^2 dU/d\rho$ so that

$$\frac{\partial}{\partial x_i}\frac{\partial}{\partial\rho}(\rho U) = \frac{\partial}{\partial x_i}\left[U(\rho) + \rho\frac{\partial U}{\partial\rho}\right] = \frac{dU}{d\rho}\frac{\partial\rho}{\partial x_i} + \frac{\partial}{\partial x_i}\left(\frac{p}{\rho}\right)$$

$$= \frac{p}{\rho^2}\frac{\partial\rho}{\partial x_i} + \frac{1}{\rho}\frac{\partial p}{\partial x_i} - \frac{p}{\rho^2}\frac{\partial\rho}{\partial x_i} = \frac{1}{\rho}\frac{\partial p}{\partial x_i} \tag{18.34}$$

By (18.28) and (18.32), the last two terms involving $a_j$ on the left side of (18.33) can be written as

$$u_j \frac{\partial}{\partial x_i}\left[\frac{\partial a_j}{\partial t} + v_k \frac{\partial a_j}{\partial x_k}\right] + \frac{\partial a_j}{\partial x_i}\frac{\partial u_j}{\partial t} + v_k \frac{\partial a_j}{\partial x_k}\frac{\partial u_j}{\partial x_i}$$

$$= -\frac{\partial a_j}{\partial x_i}\mathbf{v}\cdot\nabla u_j + \frac{\partial u_j}{\partial x_i}\mathbf{v}\cdot\nabla a_j . \tag{18.35}$$

We will show presently that the right-hand side of (18.35) is, in fact, the *i*th component of $-\mathbf{v}\times(\nabla\times\mathbf{v})$. With $\nabla(\frac{1}{2}|\mathbf{v}|^2) - \mathbf{v}\times\nabla\times\mathbf{v} = \mathbf{v}\cdot\nabla\mathbf{v}$, (18.33) then becomes

$$\frac{\partial \mathbf{v}}{\partial t} + \mathbf{v}\cdot\nabla\mathbf{v} + \frac{1}{\rho}\nabla p = \mathbf{f} \tag{18.22a}$$

which is just what we found to be a consequence of the generalized Hamilton principle in Section 2.

It remains to prove that the right-hand side of (18.35) is the *i*th component of $-\mathbf{v}\times(\nabla\times\mathbf{v})$ to complete the argument. We begin by writing the Euler DE (18.30) in vector form and take its curl to obtain

$$\nabla\times\mathbf{v} + \mathbf{w} = \mathbf{0}, \tag{18.36a}$$

where

$$\mathbf{w} = \nabla\times\left(\mathbf{u}\cdot\frac{\partial \mathbf{a}}{\partial x_k}\mathbf{i}_k\right). \tag{18.36b}$$

It is then straightforward to verify that the right-hand side of (18.35) is, in fact, the *i*th component of $\nabla\times\mathbf{w}$. The desired result then follows from (18.36a)

The analysis of this section has effectively established

**[XVIII.3]** ***In an Eulerian representation, the actual motion of an ideal fluid which evolves from a given initial state at t = 0 to a given final state at t = t$_p$, subject to the conservation of mass (i.e., the continuity equation) and the time-independent contraint of material point designation, renders the action integral (18.14) stationary.***

## 5. Incompressible Fluids

Liquids such as water are essentially *incompressible*; their mass density remains constant in both space and time under various types of external loads. The mathematical theory for the dynamical behavior of these fluids simplifies significantly by $\rho(\mathbf{x},t) = \rho_0$. Several consequences of this simplification will be discussed in this section in preparation for a description of variational principles for water waves in Section 6.

We noted in Section 3 that not all ideal fluids are irrotational. Although this remains the case for incompressible ideal fluids, they are more likely to be irrotational. More specifically, Helmholtz's theorem guarantees the following:

**[XVIII.4]** ***If the motion of an ideal, i.e., inviscid, fluid is irrotational at one instant in time, it will remain irrotational thereafter.***

The proof of this result involves the notion of *circulation* in a fluid.

Consider a closed curve $C$ which moves with the fluid, i.e., the curve consists always of the same fluid particles. The circulation $\Gamma(t)$ around $C$ is defined by

$$\Gamma(t) = \oint_C \mathbf{v}\cdot\mathbf{t}\,ds = \oint_C \mathbf{v}\cdot d\mathbf{x}, \tag{18.37}$$

where the unit tangent vector $\mathbf{t}$ of $C$ is given by the derivative of the position vector $\mathbf{x}(s)$ of points on $C$ with respect to the arclength variable $s$. Upon taking the time derivative of (18.37), we obtain

$$\begin{aligned}\frac{d\Gamma}{dt} &= \int_0^{s_0}\left[\frac{d\mathbf{v}}{dt}\cdot\frac{d\mathbf{x}}{ds} + \mathbf{v}\cdot\frac{d}{dt}\left(\frac{d\mathbf{x}}{ds}\right)\right]ds \\ &= \int_0^{s_0}\left[\frac{d\mathbf{v}}{dt}\cdot\mathbf{t} + \mathbf{v}\cdot\frac{d\mathbf{v}}{ds}\right]ds.\end{aligned} \tag{18.38}$$

We now use the vector equation of motion (18.22a) to express $d\mathbf{v}/dt$ as $-\rho^{-1}\nabla p - \nabla L$. Because the fluid is incompressible so that $\rho = \rho_0$, (18.38) may be written as

$$\begin{aligned}\frac{d\Gamma}{dt} &= \int_0^{s_0}\left[-\frac{1}{\rho_0}\frac{dp}{ds} - \frac{dL}{ds} + \frac{1}{2}\frac{d}{ds}(|\mathbf{v}|^2)\right]ds \\ &= \left[-\frac{1}{\rho_0}p - L + \frac{1}{2}|\mathbf{v}|^2\right]_0^{s_0} = 0,\end{aligned} \tag{18.39}$$

where we have made use of the fact that $s = 0$ and $s = s_0$ correspond to the same point on $C$, and $p$, $L$, and $\mathbf{v}$ are single-valued functions of position. Condition (18.39) implies

**[XVIII.5]** ***In an incompressible and inviscid fluid, the circulation around any closed curve consisting of the same fluid particles is constant in time.***

If the motion of an incompressible and inviscid fluid involves no non-vanishing circulation for all closed curves at a particular instant $t_0$ in time (e.g., the motion was initially at rest), then $\Gamma(t) = 0$ for all $C$ in the solution domain thereafter. By Stokes' theorem, we have

$$\Gamma = \oint_C \mathbf{v} \cdot \mathbf{t} ds = \iint_S (\nabla \times \mathbf{v}) \cdot \boldsymbol{\nu} dS_x \tag{18.40}$$

for any simply-connected surface $S$ with $C$ as its edge and $\boldsymbol{\nu}$ as its unit normal. Because $\Gamma(t) = 0$ for all curves $C$ in the solution domain, it follows from (18.40) that we must have $\nabla \times \mathbf{v} = \mathbf{0}$ throughout the solution domain. In other words, the motion of the fluid after the instant $t_0$ must continue to be irrotational. This confirms [XVIII.4].

Most fluid motions start from rest; hence, it is not at all restrictive to consider only irrotational motions for fluids which are inviscid (ideal) and incompressible. For such motions, we have $\mathbf{v} \cdot \nabla \mathbf{v} = \nabla(|\mathbf{v}|^2)/2$ [given the known vector identity $\nabla(\mathbf{u} \cdot \mathbf{v}) = \mathbf{u} \cdot \nabla \mathbf{v} + \mathbf{v} \cdot \nabla \mathbf{u} + \mathbf{u} \times (\nabla \times \mathbf{v}) + \mathbf{v} \times (\nabla \times \mathbf{u})$]. The vector equation of motion for an incompressible ideal fluid may then be written as

$$\mathbf{v}_{,t} + \frac{1}{2} \nabla(|\mathbf{v}|^2) + \nabla\left(\frac{p}{\rho_0}\right) + \nabla L = 0. \tag{18.41a}$$

For an irrotational flow, we have $\mathbf{v} = \nabla \varphi$ for some velocity potential $\varphi(\mathbf{x}, t)$ [see (18.25)] so that (18.41a) may be integrated to obtain

$$\varphi_{,t} + \frac{1}{2} |\nabla \varphi|^2 + \frac{1}{\rho_0} p + L = c_0(t) \tag{18.41b}$$

for some arbitrary function of time $c_0(t)$. This function has no effect on the flow velocity and may be set to zero. To see this, we write

$$\varphi = \Phi + \int^t c_0(\xi) d\xi$$

Evidently, we have $\nabla \Phi = \nabla \varphi$ and (18.41b) becomes

$$\Phi_{,t} + \frac{1}{2} |\nabla \Phi|^2 + \frac{1}{\rho_0} p + L = 0. \tag{18.41c}$$

The continuity equation (18.27) for the incompressible fluid reduces to

$$\nabla^2 \varphi = 0 \quad \text{or} \quad \nabla^2 \Phi = 0. \tag{18.42}$$

Thus, we have

**[XVIII.6]** ***The Bernoulli law (18.41c) and the potential equation (18.42) may be used in place of the equations of motion and continuity (18.22) and (18.7) to completely determine the motion of an incompressible ideal fluid initially at rest (or any incompressible ideal fluid which is also irrotational).***

## 6. A Surface Wave Problem

As a typical surface wave problem, consider a body of water initially at rest and filling the space $R$ (as shown in Fig. 18.1 and) defined by

$$R = \{ -h(x_1,x_2) \le x_3 \le 0,\ |x_2| < \infty,\ 0 \le x_1 < \infty \}.$$

For time $t > 0$, the free surface, initially located at $x_3 = 0$, is subject to some disturbance (by wind over a finite region of $x_3 = 0$ for example). We are interested in the subsequent motion of the water, particularly the evolution of the free surface $x_3 = z(x_1,x_2,t)$ with $z(x_1,x_2,0) = 0$. Because of its physical properties, water is effectively an incompressible ideal fluid (if we ignore

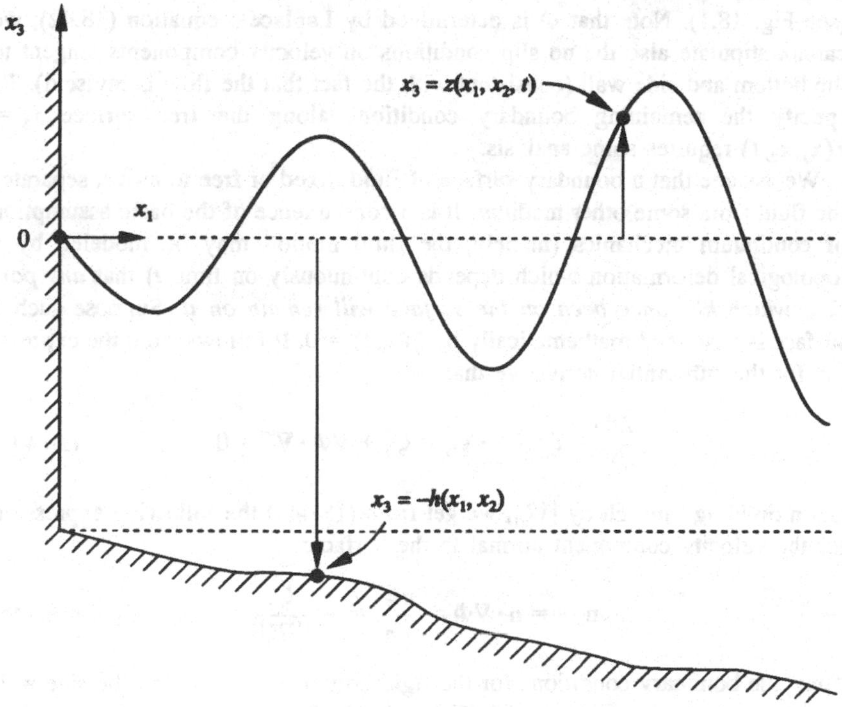


**Figure 18.1**

viscous effects primarily at the boundaries and surface tension). Because it is initially at rest, its subsequent motion will be irrotational by [XVIII.4] We can invoke [XVIII.6] to determine the motion caused by the disturbance over the free surface for $t > 0$ by equations (18.41c) and (18.42) supplemented by the initial (at rest) condition

$$z(x_1, x_2, 0) = 0, \qquad \mathbf{v}(\mathbf{x}, 0) = \nabla\Phi(\mathbf{x}, 0) = 0 \tag{18.43a}$$

and appropriate boundary conditions at the boundary surface of $R$.

For simplicity, we consider the boundaries $x_3 = -h$ and $x_1 = 0$ to be fixed, so that

$$\mathbf{n}\cdot\mathbf{v} = \frac{\partial\Phi}{\partial n} = 0 \quad \text{at} \quad x_3 = -h(x_1, x_2, t), \tag{18.43b}$$

$$\mathbf{n}\cdot\mathbf{v} = \frac{\partial\Phi}{\partial x_1} = 0 \quad \text{at} \quad x_1 = 0, \tag{18.43c}$$

where $\mathbf{n}$ is the unit normal of the surface involved. For example, $x_3 = -h$ may be the sloping bottom and $x_1 = 0$ may be a side wall of an ocean shore beach (see Fig. 18.1). Note that $\Phi$ is determined by Laplace's equation (18.42); we cannot stipulate also the no slip conditions on velocity components tangent to the bottom and side wall (consistent with the fact that the flow is inviscid). To specify the remaining boundary conditions along the free surface $x_3 = z(x_1, x_2, t)$ requires some analysis.

We assume that a boundary surface of fluid, fixed or free to move, separates the fluid from some other medium. It is a consequence of the basic assumption of continuum mechanics (namely, the fluid motion may be modeled by a topological deformation which depends continuously on time $t$) that *any particle which has once been on the surface will remain on it*. Suppose such a surface is described mathematically by $\zeta(\mathbf{x}, t) = 0$. It follows from the expression for the substantial derivative that

$$\frac{D\zeta}{Dt} = \zeta_{,t} + \mathbf{v}\cdot\nabla\zeta = \zeta_{,t} + \nabla\Phi\cdot\nabla\zeta = 0. \tag{18.44a}$$

Upon dividing through by $|\nabla\zeta|$, we get from (18.44a) the following expression for the velocity component normal to the surface:

$$\mathbf{n}\cdot\mathbf{v} = \mathbf{n}\cdot\nabla\Phi = \frac{\partial\Phi}{\partial n} = -\frac{\zeta_{,t}}{|\nabla\zeta|}. \tag{18.44b}$$

Thus, the boundary conditions for the rigid bottom $x_3 = -h$ and the side wall $x_1 = 0$ are properly given by (18.43b) and (18.43c), respectively, because $\zeta$ is independent of time there.

For the free surface initially at $x_3 = 0$ and denoted by $x_3 = z(x_1, x_2, t)$ for $t > 0$, we have $\zeta = x_3 - z(x_1, x_2, t)$. The *kinematic condition* (18.44a) becomes in this case

$$z_{,t} - \Phi_{,3} + (z_{,1}\Phi_{,1} + z_{,2}\Phi_{,2}) = 0 \quad \text{on} \quad x_3 = z(x_1, x_2, t). \qquad (18.44c)$$

In addition, the disturbance on the free surface may be taken as a known pressure distribution $p_0(\mathbf{x}, t)$ on $x_3 = z(x_1, x_2, t)$. Upon eliminating $x_3$ from $p_0$, we have

$$p(\mathbf{x}, t) = P(x_1, x_2, t) \quad \text{on} \quad x_3 = z(x_1, x_2, t), \qquad (18.45a)$$

where $P(x_1, x_2, t) = p_0(x_1, x_2, z(x_1, x_2, t), t)$. In the context of [XVIII.6], such a disturbance can only be introduced into the mathematical problem by the Bernoulli equation (18.41c) applied to the free surface. For most problems, the only other external loading is the weight of the water. In that case, we have $L(\mathbf{x}, t) = gx_3$ so that

$$f_1 = f_2 = 0, \qquad f_3 = -\frac{\partial L}{\partial x_3} = -g. \qquad (18.45b)$$

Altogether, the solution of the problem must satisfy the *dynamic condition* on the free surface:

$$\Phi_{,t} + \frac{1}{2}|\nabla\Phi|^2 + gx_3 + \frac{1}{\rho_0}p_0(\mathbf{x}, t) = 0 \quad \text{on} \quad x_3 = z(x_1, x_2, t) \qquad (18.46a)$$

or

$$\Phi_{,t} + \frac{1}{2}|\nabla\Phi|^2 + gz(x_1, x_2, t) + \frac{1}{\rho_0}P(x_1, x_2, t) = 0 \quad \text{on} \quad x_3 = z(x_1, x_2, t), \qquad (18.46b)$$

where $z(x_1, x_2, t)$, $P(x_1, x_2, t)$ and $\rho_0$ are known quantities.

The solution process for the free surface water wave problem consists of solving Laplace's equation (18.42) subject to the rigid wall boundary conditions (18.43b) and (18.43c) for $x_3 = -h(x_1, x_2, t)$ and $x_1 = 0$, respectively, the two free surface conditions (18.44c) and (18.46b), the initial conditions (18.43a), and some boundedness conditions for large $|x_2|$ and large positive $x_1$. This process determines $\Phi(\mathbf{x}, t)$ and $z(x_1, x_2, t)$. The Bernoulli law (18.41c) is then used to determine $p(\mathbf{x}, t)$ in the interior of $R$. The velocity components are calculated from $\Phi$ by $\mathbf{v} = \nabla\Phi$.

Although the governing PDE for the surface wave problem above is linear, the initial-boundary-value problem for the fluid motion is nonlinear because of the two boundary conditions (18.44c) and (18.46b) for the free surface. The generally nonlinear nature of the problem makes it difficult to obtain an exact solution in terms of elementary or special functions. Accurate approximate solutions are usually needed for most problems. One way to obtain an approximate solution is by the direct (or semidirect) method of the calculus of variations. For this approach a variational principle for the initial-boundary-value problem described above is a requisite. Such a variational principle has been formulated by Luke (1967). In what follows, we will give a variational principle for the *two-dimensional version* of this free surface water wave problem.

Suppose the motion of the water is uniform in the $x_2$ direction so that $v_2(\mathbf{x},t)=0$ and $(\ )_{,2}=0$. In that case, we may take $\Phi=\Phi(x_1,x_3,t)$, $z=z(x_1,t)$, and $p_0(\mathbf{x},t)=p_0(x_1,x_3,t)$ so that $P=P(x_1,t)$. The mathematical model for this water wave problem consists of

$$\Phi_{,11}+\Phi_{,33}=0 \quad \text{in } R \tag{18.47a}$$

with

$$\frac{\partial\Phi}{\partial x_3}=0 \quad \text{along } x_3=-h(x_1,t), \tag{18.47b}$$

$$\frac{\partial\Phi}{\partial x_1}=0 \quad \text{along } x_1=0, \tag{18.47c}$$

$$z_{,t}+z_{,1}\Phi_{,1}-\Phi_{,3}=0 \quad \text{on } x_3=z(x_1,t) \tag{18.47d}$$

$$\Phi_{,t}+\frac{1}{2}[\Phi_{,1}^2+\Phi_{,3}^2]+gz(x_1,t)+\frac{1}{\rho_0}P(x_1,t)=0 \quad \text{on } x_3=z(x_1,t), \tag{18.47e}$$

$$\Phi_{,1}(x_1,x_3,0)=\Phi_{,3}(x_1,x_3,0)=0, \qquad z(x_1,0)=0. \tag{18.47f}$$

With the Lagrangian given by

$$F\equiv\int_{-h}^{z(x_1,t)}\left[\frac{1}{2}(\Phi_{,1}^2+\Phi_{,3}^2)+\Phi_{,t}+gx_3+\frac{1}{\rho_0}p_0\right]dx_3, \tag{18.48a}$$

Luke worked with the integral

$$I[\Phi,z]=\int_0^{\bar{t}}\int_0^{\infty}F(\Phi_{,1},\Phi_{,3},\Phi_{,t},z,x_1,t)\,dx_1\,dt \tag{18.48b}$$

and established the following result:

**[XVIII.7]** *With the independent variation $\delta\Phi$ vanishing at $t=0$ and $t=\bar{t}$, the stationarity condition $\delta I=0$ has as its Euler differential equation the Laplace's equation (18.47a) for $\Phi(x_1,x_3,t)$ and as Euler boundary conditions the rigid wall conditions (18.47b) and (18.47c) and the free surface conditions (18.47d) and (18.47e).*

Verification of the result above requires some careful handling of details. With

$$\delta I = \int_0^{\bar t}\int_0^{\infty}\left\{\left[\frac{1}{2}\Phi_{,1}^2 + \frac{1}{2}\Phi_{,3}^2 + \Phi_{,t} + gx_3 + \frac{1}{\rho_0}p_0\right]_{x_3=z}\delta z + \int_{-h}^{z(x_1,t)}[\Phi_{,1}\delta\Phi_{,1} + \Phi_{,3}\delta\Phi_{,3} + \delta\Phi_{,t}]dx_3\right\}dx_1\,dt \qquad (18.49a)$$

and

$$\Phi_{,1}\delta\Phi_{,1} + \Phi_{,3}\delta\Phi_{,3} + \delta\Phi_{,t} = -(\Phi_{,11} + \Phi_{,33})\delta\Phi + [(\Phi_{,1}\delta\Phi)_{,1} + (\Phi_{,3}\delta\Phi)_{,3} + (\delta\Phi)_{,t}], \qquad (18.49b)$$

we apply the divergence theorem in the $(x_1, x_3, t)$-space to get for a finite but large positive $\bar x$

$$\int_0^{\bar t}\int_0^{\bar x}\int_{-h}^{z(x_1,t)}[\Phi_{,1}\delta\Phi_{,1} + \Phi_{,3}\delta\Phi_{,3} + \delta\Phi_{,t}]dx_3\,dx_1\,dt$$
$$= \iint_{S_{\bar x}}[v_1\Phi_{,1} + v_3\Phi_{,3} + v_t]\delta\Phi\,dS - \int_0^{\bar t}\int_0^{\bar x}\int_{-h}^{z(x_1,t)}(\Phi_{,11} + \Phi_{,33})\delta\Phi\,dx_3\,dx_1\,dt, \qquad (18.50)$$

where the parameters $\{v_a\}$ are the directional cosines of the unit normal of the surface $S_{\bar x}$ enclosing the region $\bar R = \{0 \le t \le \bar t, 0 \le x_1 \le \bar x, -h \le x_3 \le z\}$. It is clear that the Euler differential equation is Laplace's equation (18.47a). Because $\Phi$ is stipulated to satisfy the initial conditions (18.47f) and appropriate boundedness conditions as $\bar x \to \infty$, $\delta\Phi$ vanishes on these boundaries and we need only to consider the surface integrals on the three boundary surfaces $x_1 = 0, x_3 = 0$, and $x_3 = z(x_1, t)$.

The unit normal on $x_k = 0$ is $-\mathbf{i}_k$ whereas it is

$$\mathbf{v} = \frac{\zeta_{,1}\mathbf{i}_1 + \zeta_{,3}\mathbf{i}_3 + \zeta_{,t}\mathbf{i}_t}{\sqrt{\zeta_{,1}^2 + \zeta_{,3}^2 + \zeta_{,t}^2}} = \frac{-z_{,1}\mathbf{i}_1 + \mathbf{i}_3 - z_{,t}\mathbf{i}_t}{\sqrt{1 + z_{,1}^2 + z_{,t}^2}} \qquad (18.51)$$

on $x_3 = z(x_1, t)$ because $\zeta = x_3 - z(x_1, t) = 0$ (and therewith $\mathbf{v} = \nabla\zeta/|\nabla\zeta|$). It follows that

$$\iint_{S_{\bar x}}[v_1\Phi_{,1} + v_3\Phi_{,3} + v_t]\delta\Phi\,dS$$
$$= -\int_0^{\bar t}\int_{-h}^{z(x_1,t)}[\Phi_{,1}\delta\Phi]_{x_1=0}dx_3\,dt - \int_0^{\bar t}\int_0^{\bar x}[\Phi_{,3}\delta\Phi]_{x_3=-h}dx_1\,dt$$
$$+ \int_0^{\bar t}\int_0^{\bar x}[(-z_{,1}\Phi_{,1} + \Phi_{,3} - z_{,t})\delta\Phi]_{x_3=z}dx_1\,dt. \qquad (18.52)$$

Independent variation of $\delta\Phi$ and $\delta z$ on the boundary integrals in (18.52) and (18.49a) gives (in the limit as $\bar{x} \to \infty$) as the Euler BC the two homogeneous conditions (18.47b) and (18.47c) and the two inhomogeneous conditions (18.47d) and (18.47e).

## 7. Slow Dispersion of Wave Trains

A uniform periodic wave train is specified by certain characteristic parameters such as amplitude, wave number, etc. Many wave phenomena of interest occur in the form of nonuniform wave trains. Some appear to be essentially uniform periodic waves with their characteristic parameters varying slowly in space and time; changes over one wavelength or over one period are relatively small. This type of slow dispersion of (linear and nonlinear) wave trains can be observed in water waves governed by the surface wave theory of Section 6. Although an exact solution of the relevant surface wave problem for an appropriate description of the particular wave behavior is not likely, a useful approximate description can be obtained by the semidirect method of the calculus of variations (Whitham, 1965, 1967). The variational principle [XVIII.7] for this class of problems forms the basis of this approach to an approximate solution.

The free surface height of a uniform plane surface wave in the $x_1$ direction is sinusoidal in the variable $\theta = \kappa x_1 - \omega t$, e.g., $z = b + a \sin(\kappa x_1 - \omega t)$. For nonuniform wave trains (which are uniform in the $x_2$ direction), the velocity potential $\Phi(x_1, x_3, t)$ and the free surface height $z(x_1, t)$ may depend on $x_1$ and $t$ more generally in the form

$$\Phi(x_1,x_3,t) = \psi(x_1,t) + \Psi(\theta,x_3), \qquad z(x_1,t) = N(\theta), \qquad (18.53a)$$

where

$$\theta_{,1} = \kappa, \qquad \theta_{,t} = -\omega, \qquad \psi_{,1} = \beta, \qquad \psi_{,t} = -\gamma \qquad (18.53b)$$

with $\kappa, \omega, \beta, \gamma$ and the various amplitude factors for $z$ and $\Phi$ all taken to be slowly varying functions of $x_1$ and $t$. (For linear waves, we have $\theta = \kappa x_1 - \omega t$ and $\psi = \beta x_1 - \gamma t$.) The velocity potential $\Phi$ is not a physical quantity and, therefore, allows for the possibility of a $\psi(x_1, t)$ term; the corresponding velocity field $\mathbf{v} = \nabla\Phi = (\beta + \Phi_{,1}, 0, \Phi_{,3})$ is physically acceptable as long as $\beta$ is slowly varying and bounded in space and time. The equations governing these slowly varying functions may be derived from an averaged form of the variational principle [XVIII.7].

For nonuniform wave trains where the scale for nonuniformity is large compared with one wavelength, we may remove the fine structure of the solution at the shorter time/length scale by averaging the Lagrangian $F$ of (18.48a) over one wavelength:

$$\mathcal{L} = \frac{1}{2\pi}\int_0^{2\pi} F\,d\theta \tag{18.54a}$$

where $\mathcal{L}$ depends on $\kappa, \omega, \beta, \gamma$, etc. The averaged variational principle

$$\delta J \equiv \delta \int_0^T\int_0^X \mathcal{L}\,dx_1\,dt = 0 \tag{18.54b}$$

then allows us to focus on the more global nonuniformity of the wave phenomenon. For the actual application of (18.54), we take $[p_0(\mathbf{x},t) = 0,]$ $z$ to be periodic in $\theta$ and the corresponding potential $\Phi$ to satisfy (18.47a) in the fine scale (i.e., when $\kappa$ is treated as a constant):

$$\begin{aligned} z(x_1,t) &= N(\theta) = a_0 + a_1\cos\theta + \sum_{n=2}^{\infty} a_n \cos n\theta, \\ \Phi(x_1,x_3,t) &= \psi(x_1,t) + \psi(\theta,x_3) = \psi + \sum_{n=1}^{\infty}\frac{A_n}{n}\cosh(n\kappa[h+x_3])\sin n\theta. \end{aligned} \tag{18.54c}$$

Note that $\Phi$ also satisfies the condition of no normal velocity at the bottom $x_3 = -h$.

Upon substituting (18.54c) into (18.48a) and setting $p_0 = 0$ in $F$, we obtain

$$\begin{aligned} F &= \left(\frac{1}{2}\beta^2 - \gamma\right)H + \frac{1}{2}gN^2 - (\omega - \beta\kappa)\int_{-h}^{N(\theta)} \Psi_{,\theta}\,dx_3 \\ &\quad + \frac{1}{2}\int_{-h}^{N(\theta)} (\kappa^2\Psi_{,\theta}^2 + \Psi_{,3}^2)\,dx_3 \\ &= \left(\frac{1}{2}\beta^2 - \gamma\right)H + \frac{1}{2}gN^2 - (\omega - \beta\kappa)\sum_{n=1}^{\infty}\frac{A_n}{n}\sinh(n\kappa H)\cos n\theta \\ &\quad + \frac{1}{4}\kappa\sum_{n=2}^{\infty}\sum_{m=1}^{n-1} A_m A_{n-m}\left\{\frac{\sinh(n\kappa H)}{n}\cos(n-2m)\theta\right. \\ &\quad \left. + \frac{\sinh([n-2m]\kappa H)}{n-2m}\cos n\theta\right\} \end{aligned} \tag{18.55a}$$

with $H = h + a_0$ being the mean depth of the free surface. The mean value of $F$ over one wavelength is

$$\begin{aligned} \mathcal{L} &= \left(\frac{1}{2}\beta^2 - \gamma\right)H + \frac{1}{2}ga_0^2 + \frac{1}{4}ga_1^2 + \frac{1}{4}ga_2^2 - \frac{\omega - \beta\kappa}{\kappa}(\mu_1 A_1 + \mu_2 A_2) \\ &\quad + \kappa\left(\frac{1}{2}\mu_{11}A_1^2 + \mu_{12}A_1A_2 + \frac{1}{2}\mu_{22}A_2^2\right) + \dots, \end{aligned} \tag{18.55b}$$

where

$$\begin{aligned}
\mu_1 &= \frac{1}{2}\kappa a_1 \cosh(\kappa H) + \frac{1}{4}\kappa^2 a_1 a_2 \sinh(\kappa H) + \frac{1}{16}\kappa^3 a_1^3 \cosh(\kappa H),\\
\mu_2 &= \frac{1}{2}\kappa a_2 \cosh(2\kappa H) + \frac{1}{4}\kappa^2 a_1^2 \sinh(2\kappa H)\\
\mu_{11} &= \frac{1}{4}\sinh(2\kappa H) + \frac{1}{4}\kappa^2 a_1^2 \sinh(2\kappa H) + \frac{1}{4}\kappa a_2\\
\mu_{12} &= \frac{1}{4}\kappa a_1 \cosh(3\kappa H), \qquad \mu_{22} = \frac{1}{8}\sinh(4\kappa H).
\end{aligned} \tag{18.55c}$$

Euler equations of (18.54b) associated with the independent variations of $\delta A_j$ require $\{A_j\}$ to satisfy a linear system of equations. By retaining only two terms in the series for $\Psi$, this system reduces to the following two linear equations for $A_1$ and $A_2$:

$$\mu_{11}A_1 + \mu_{12}A_2 = \frac{\omega - \beta\kappa}{\kappa^2}\mu_1, \qquad \mu_{12}A_1 + \mu_{22}A_2 = \frac{\omega - \beta\kappa}{\kappa^2}\mu_2. \tag{18.55d}$$

It is known that $\kappa a_n$ and $\kappa A_n$ are $O(\varepsilon^n)$, where $\varepsilon = \kappa a_1 \ll 1$ (Whitham, 1967). Relations (18.55d) allow us to eliminate $A_1$ and $A_2$ from (18.55b) so that $\mathcal{L}$ becomes

$$\mathcal{L} = \underline{\left(\frac{1}{2}\beta^2 - \gamma\right)H + \frac{1}{2}ga_0^2} + \frac{1}{4}ga_1^2 + \frac{1}{4}ga_2^2$$
$$-\frac{1}{4}\frac{(\omega-\beta\kappa)^2}{\kappa T}\left\{a_1^2 - \frac{2T^2-1}{4T^2}\kappa^2 a_1^4 - \frac{3-T^2}{2T}\kappa a_1^2 a_2 + (1+T^2)a_2^2\right\} + O(\varepsilon^6) \tag{18.56}$$

where $T = \tanh(\kappa H)$.

For our small-amplitude wave phenomenon, the first two (underscored) terms in (18.56) are expected to be small of the order of the fourth term, $ga_2^2/4$, because they do not appear in a linear theory. (An exception may be the $\gamma h$ term which does not contribute to the final result.) To the *lowest order of approximation*, expression (18.56) may be further simplified to

$$\mathcal{L} = E\left(1 - \frac{\omega^2}{\omega_0^2}\right) + O(\varepsilon^2) \equiv G(\omega,\kappa)E + O(\varepsilon^2) \tag{18.57a}$$

with

$$E = \frac{1}{4} g a_1^2, \qquad \omega_0{}^2 = \kappa g \tanh(\kappa h). \tag{18.57b}$$

The Euler condition of $\delta J = 0$ associated with $\delta a_1$ (or $\delta E$) gives the dispersion relation $G(\omega, \kappa) = 0$ or

$$\omega^2 = \omega_0^2(\kappa) + O(\varepsilon^2) = g\kappa \tanh(\kappa h) + O(\varepsilon^2). \tag{18.57c}$$

The quantities $\omega$ and $\kappa$ are constrained by (18.53b); only $\theta$ may be varied independently. The dispersion of the linear wave trains is, therefore, governed by (18.57c), and the Euler condition of $\delta J = 0$ resulting from the independent variation of $\delta\theta$ requires

$$\left(\frac{\partial \mathcal{L}}{\partial \kappa}\right)_{,1} - \left(\frac{\partial \mathcal{L}}{\partial \omega}\right)_{,t} = 0, \tag{18.57d}$$

subject to the constraint on $\omega$ and $\kappa$ stipulated in (18.53b) and taken now more conveniently in the form

$$\kappa_{,t} + \omega_{,1} = 0. \tag{18.57e}$$

The three equations (18.57c), (18.57d), and (18.57e), supplemented by suitable auxiliary conditions, determine $a_1$, $\omega$ and $\kappa$ as (slowly varying) functions of $x_1$ and $t$. Upon using (18.57c) to eliminate $\omega$ from the remaining two equations, we need only to solve the two coupled first-order partial differential equtaion

$$E_{,t} + (C_0 E)_{,1} = 0, \tag{18.58a}$$

$$\kappa_{,t} + C_0 \kappa_{,1} = 0, \tag{18.58b}$$

where

$$C_0(\kappa) \equiv \frac{d\omega_0}{d\kappa} = \frac{\omega_0(\kappa)}{2\kappa}\left[1 + \frac{2\kappa h}{\sinh(2\kappa h)}\right]. \tag{18.58c}$$

We may use (18.57c) to simplify the approximate expression for $\mathcal{L}$ in (18.56) to get

$$\mathcal{L} = \left(\frac{1}{2}\beta^2 - \gamma\right)H + \frac{1}{2} g a_0^2 + \frac{1}{2} E\left\{1 - \frac{(\omega - \beta\kappa)^2}{g\kappa \tanh(\kappa H)}\right\} \tag{18.59}$$

$$+ \frac{g}{4}\left\{\frac{2T_0^2 - 1}{4T_0^2}\kappa^2 a_1^4 + \frac{3 - T_0^2}{2T_0}\kappa a_1^2 a_2 - T_0^2 a_2^2\right\} + O\left\{\varepsilon^6, \left(\frac{\omega^2}{\omega_0^2} - 1\right)\varepsilon^2\right\}$$

with $T_0 = \tanh(\kappa h)$. An approximate theory which includes all terms of the order $\varepsilon^4$ in $\mathcal{L}$ may now be formulated with the help of (18.59). The Euler equation of $\delta J = 0$ resulting from the independent variation of $\delta a_2$ requires

$$a_2 = \frac{3 - T_0^2}{4T_0^3}\kappa a_1^2. \tag{18.60a}$$

With the help of (18.60a), the Euler equation associated with $\delta a_1$ (or equivalently $\delta E$) gives the nonlinear dispersion relation

$$\frac{(\omega - \beta\kappa)^2}{g\kappa \tanh(\kappa H)} = 1 + \frac{2\kappa^2 D_0 E}{g \tanh(\kappa h)} + O(\varepsilon^4), \tag{18.60b}$$

where

$$D_0 = \frac{9T_0^4 - 10T_0^2 + 9}{8T_0^3}, \qquad H = h + a_0. \tag{18.60c}$$

Again, the dispersion relation determines $\omega$ but now in terms $\beta, a_1, a_0$ as well, not just in terms of $\kappa$ along. To the order of approximation made in (18.59), we may write (18.60b) as

$$\omega = \omega_0 + \frac{\kappa^3}{\omega_0} D_0 E + \frac{\kappa}{h} B_0 b + \kappa\beta + O(\varepsilon^4), \tag{18.60d}$$

where $B_0 = C_0 - \omega_0 / 2\kappa$.

Three other Euler equations of $\delta J = 0$ associated with the independent variations of $\delta a_0$, $\delta\theta$, and $\delta\psi$ require

$$\gamma = g a_0 + \frac{\kappa}{\omega_0 h_0} B_0 E + O(\varepsilon^4), \tag{18.61a}$$

$$\left(\frac{\partial \mathcal{L}}{\partial \kappa}\right)_{,1} = \left(\frac{\partial \mathcal{L}}{\partial \omega}\right)_{,t}, \qquad \left(\frac{\partial \mathcal{L}}{\partial \beta}\right)_{,1} = \left(\frac{\partial \mathcal{L}}{\partial \gamma}\right)_{,t}. \tag{18.61b,c}$$

Together with the two constraints for $\kappa, \omega, \beta$, and $\gamma$ implied by (18.53b), taken more conveniently in the form

$$\kappa_{,t} + \omega_{,1} = 0, \qquad \beta_{,t} + \gamma_{,1} = 0, \tag{18.61d,e}$$

they provide the five relations needed for the determination of $a_0, \beta, \gamma, \kappa$, and $a_1$. In these relations, $\mathcal{L}$ is to be taken from (18.59) and $\omega$ from (18.62c). Within the error magnitude in the various approximations, we have

$$\frac{\partial \mathcal{L}}{\partial \gamma} = -\frac{E}{\omega_0} + O(\varepsilon^4), \qquad \frac{\partial \mathcal{L}}{\partial \kappa} = C_0 \frac{E}{\omega_0} + O(\varepsilon^4),$$
$$\frac{\partial \mathcal{L}}{\partial \omega} = -H + O(\varepsilon^4), \qquad \frac{\partial \mathcal{L}}{\partial \beta} = \beta h + \frac{\kappa E}{\omega_0} + O(\varepsilon^4), \tag{18.62}$$

and further simplificaton of (18.61b) is possible.

The increased complexity of the more refined dispersion relation (18.60b) or (18.60d) is significant beyond the additional difficulty encountered in the solution process for determining $a_0, a_1, a_2, \kappa, \omega, \beta$, and $\gamma$ as functions of $x_1$ and $t$. The dependence of $\omega$ on $a_0, a_1$ and $\beta$ in addition to $\kappa$ associated with the effects of the small nonlinearity of the basic wave train does not merely give small corrections to the results of the lowest-order (linear wave) theory. A significant qualitative change in the predicted solution behavior may result from the modification of (18.57c). It can be shown (Whitham, 1967) that for $\kappa h > 1.36$, the system (18.61) is elliptic and the wave trains are generally unstable. In that case, a periodic wave train may grow in time without bound and ultimately break up into some other form.

## 8. Creeping Motion of an Incompressible Fluid

As we know from elasticity theory, there are tangential stresses acting on a fluid element, in addition to the normal stress characterized by the pressure distribution $p(\mathbf{x}, t)$. A theory which takes account of the tangential stresses may be formulated simply by replacing the pressure term in (18.22) by the total contribution from all the relevant stress components, analogous to the terms in the equilibrium equations of elasticity theory in Chapter 14. We obtain in this way

$$\frac{\partial v_i}{\partial t} + \mathbf{v} \cdot \nabla v_i - \frac{1}{\rho} \sigma_{ji,j} = f_i \quad (i = 1,2,3). \tag{18.63}$$

The sign change is appropriate because the pressure $p$ is taken to be positive inward. The stress components $\sigma_{ij} = \sigma_{ji}$ for an isotropic fluid are related to the strain rates by

$$\sigma_{ij} = -p\delta_{ij} + \mu(v_{i,j} + v_{j,i}) - \frac{2}{3}\mu \nabla \cdot \mathbf{v}\delta_{ij}. \tag{18.64}$$

Upon substituting (18.64) into (18.63), we obtain the following three equations of motion:

$$\frac{\partial v_i}{\partial t} + \mathbf{v} \cdot \nabla v_i = f_i(\mathbf{x}, t) - \frac{1}{\rho}\frac{\partial p}{\partial x_i} + \frac{\nu}{3}\frac{\partial}{\partial x_i}(\nabla \cdot \mathbf{v}) + \nu \nabla^2 v_i \quad (i = 1,2,3) \tag{18.65a}$$

or the single vector equation

$$\frac{D\mathbf{v}}{Dt} = \mathbf{f} - \frac{1}{\rho}\nabla p + \frac{\nu}{3}\nabla(\nabla \cdot \mathbf{v}) + \nu \nabla^2 \mathbf{v}. \tag{18.65b}$$

where $\nu = \mu/\rho$ is the *kinematic viscosity coefficient*. Together with the continuity equation, we have four scalar equations for five unknowns. It is tempting to close the system again by a pressure and density relation such as the first equation of (18.12). But the fluid motion is now dissipative because of the presence of the shear stress components; the internal energy may now involve other thermodynamic quantities.

For *incompressible* fluids, the density function is a constant in space and time. In that case we have $\rho(\mathbf{x}, t) = \rho_0$ and $D\rho/Dt = 0$ so that the equation of continuity reduces to

$$\nabla \cdot \mathbf{v} = 0. \tag{18.66}$$

Equation (18.65b), in turn, simplifies to

$$\frac{D\mathbf{v}}{Dt} = \mathbf{f} - \frac{1}{\rho}\nabla p + \nu\nabla^2\mathbf{v}. \tag{18.67}$$

Together, (18.66) and (18.67) form a system of four scalar partial differential equations for the four unknowns $\{v_i, p\}$.

We are interested here in two-dimensional motions of incompressible fluids in the $x_1, x_2$-plane. For such fluid motions, the continuity equation (18.66) becomes $v_{1,1} + v_{2,2} = 0$ where $(\ )_{,i} = \partial(\ )/\partial x_i$. This divergence free condition is satisfied identically by the stream function $\psi(x_1, x_2, t)$ with

$$v_1 = \psi_{,2}, \qquad v_2 = -\psi_{,1}. \tag{18.68}$$

The two relevant equations of motion for $v_1$ and $v_2$ can be written in terms of the stream function as

$$\frac{D\psi_{,2}}{Dt} = f_1 - \frac{1}{\rho_0}p_{,1} + \nu\nabla^2\psi_{,2}, \tag{18.69a}$$

$$-\frac{D\psi_{,1}}{Dt} = f_2 - \frac{1}{\rho_0}p_{,2} - \nu\nabla^2\psi_{,1}. \tag{18.69b}$$

A single equations for $\psi$ alone may be obtained by cross-differentiating and subtracting, leaving us with

$$\nabla^2\psi_{,t} + \psi_{,2}\nabla^2\psi_{,1} - \psi_{,1}\nabla^2\psi_{,2} = -\Omega + \nu\nabla^2\nabla^2\psi, \tag{18.70}$$

where $\Omega = f_{2,1} - f_{1,2}$ is a known function and we have assumed that the kinematic viscosity $\nu$ is a constant. The left-hand side of (18.70) is $D(\nabla^2\psi)/Dt$ with $\nabla^2\psi - (\nabla \times \mathbf{v})\cdot\mathbf{i}_3$. Hence, (18.70) itself gives the rate of change of the rotation of a fluid element due to friction and the external load.

In steady motion, we have $D(\nabla^2\psi)/Dt = 0$ so that the fluid behavior is completely determined by the inhomogeneous biharmonic equation

$$\nu\nabla^2\nabla^2\psi = \Omega \tag{18.71}$$

and the prescribed conditions on the normal and tangential velocity at the boundary of the flow:

$$-\psi_{,n} = u_t^*, \qquad \psi_{,s} = u_n^* \tag{18.72}$$

which are equivalent to prescribing $\psi$ and its normal derivative along the boundary. Fluid flows governed by (18.71) are called ***Stokes' flows.***

An exact solution for the boundary-value problem (18.71) and (18.72) is possible only for very simple boundaries; a Rayleigh–Ritz type solution is often useful such problems. For this purpose, we need to obtain a performance index $I[\psi]$ for which (18.71) and (18.72) are the Euler differential equation and boundary conditions. This is a relatively straightforward task for the present problem. We simply multiply (18.71) by $\psi$ and integrate by parts to get

$$\iint_A [\nu\nabla^2\nabla^2\psi - \Omega]\,\delta\psi\,dx_1\,dx_2 = \oint_C \{\nu(\nabla^2\psi)_{,n}\delta\psi - \nu\nabla^2\psi(\delta\psi)_{,n}\}\,ds$$
$$+\,\delta\iint_A\left\{\frac{\nu}{2}(\nabla^2\psi)^2 - \Omega\psi\right\}dx_1\,dx_2, \tag{18.73}$$

where $A$ is the fluid-flow region and $C$ its boundary curve(s). The prescribed conditions on $\psi$ and $\psi_{,n}$ require the comparison functions and their normal derivative to vanish along $C$. Hence, we have

**[XVIII.8]** ***The stream function for Stokes' flow renders stationary the integral***

$$I[\psi] = \iint_A\left[\frac{\nu}{2}(\nabla^2\psi)^2 - \Omega\psi\right]dx_1\,dx_2. \tag{18.74}$$

We should keep in mind, however, that more than one Lagrangian gives (18.71) as an Euler differential equation, a fact we learned in Section 7 of Chapter 15.

When the motion is not steady, $D(\nabla^2\psi)/Dt$ does not vanish identically. Nevertheless, the solution of the boundary-value problem (18.71) and (18.72) is still useful as an approximate solution for fluid motion for which the inertial terms are negligible compared to the viscous effect. Such an approximate solution has been found to agree with observed phenomena only for very slow motion flows. Let $V$ be a typical magnitude of the flow velocity (such as the flow velocity far away from any obstacles) and $L$ a typical length scale (such as the cross-section radius of a long cylindrical obstacle). The dimensionless quantity

$$R = \frac{VL}{\nu} \tag{18.75}$$

is called the *Reynolds' number* of the problem. The solution of (18.71) and (18.72) provides a good approximate description of the fluid flow only if $R \ll 1$. Such fluid flows are called *creeping motions*. Falling water particles in a cloud and rising air bubbles in syrup are examples of such creeping motion. In engineering applications, the boundary-value problem (18.71) and (18.72) is widely used in *lubrication theory*.

## 9. Oseen's Approximation

For many other problems involving slow fluid motion, the flow behavior is adequately described by Stokes' flow in one part of the flow region but not in another part. For a steady flow of an incompressible and highly viscous fluid moving slowly past a sphere for example, the Stokes' flow solution is adequate except at distances far away from the sphere. For this and other problems, the inertia term is not negligible because the convective acceleration term $\mathbf{v}\cdot\nabla\mathbf{v}$ in $D\mathbf{v}/Dt$ is not small compared to the viscous term $\nu\nabla^2\mathbf{v}$ in one or more parts of the flow region. Where $\mathbf{v}\cdot\nabla\mathbf{v}$ is significant, the flow is essentially the undisturbed main flow of the problem with flow velocity $V_k, k = 1, 2$. One way to make the problem more tractable is to approximate the coefficients of $\nabla\mathbf{v}$ in the convective part of $D\mathbf{v}/Dt$ by the known velocity components of the main flow so that (18.44) becomes

$$\nabla^2\psi_{,t} + V_1\nabla^2\psi_{,1} + V_2\nabla^2\psi_{,2} = -\Omega + \nu\nabla^2\nabla^2\psi, \qquad (18.76)$$

When the local acceleration term is negligible, (18.76) further simplifies to

$$\nu\nabla^2\nabla^2\psi - (V_1\nabla^2\psi_{,1} + V_2\nabla^2\psi_{,2}) - \Omega = 0 \qquad (18.77)$$

which is known as *Oseen's improvement of Stokes' theory*.

For a uniform main flow with constant velocity components $V_1$ and $V_2$, (18.76) may be regarded as the leading-term approximation of (18.70), linearized about the main flow. In the context of matched asymptotic expansions, Stokes and Oseen theories also follow from the Navier–Stokes equations formally as limits with $R \to 0$ in the outer and inner variables (Kevorkian and Cole, 1981).

For a Rayleigh–Ritz solution of the more accurate Oseen's theory for creeping motion, we again need a performance index for which (18.77) is its Euler DE. The usual procedure for solving the inverse problem fails in this case because

$$\iint_A \{\nu\nabla^2\nabla^2\psi - (V_1\nabla^2\psi_{,1} + V_2\nabla^2\psi_{,2}) - \Omega\}\delta\psi\, dx_1\, dx_2$$

$$= \delta\iint_A \left\{\frac{\nu}{2}(\nabla^2\psi)^2 - \Omega\psi\right\}dx_1\, dx_2 + \iint_A \nabla^2\psi[(V_1\delta\psi)_{,1} + (V_2\delta\psi)_{,2}]\, dx_1\, dx_2. \qquad (18.78)$$

The term $(V_1 \nabla^2 \psi_{,1} + V_2 \nabla^2 \psi_{,2})\delta\psi$ in (18.78) is not the first variation of any quantity. Hence, the relevant performance index (if it exists) cannot be obtained by this method. On the other hand, either side of the relation above can be used as the basis for a *weak solution*, in the sense described in Section 8 of Chapter 8, extended now to solutions for partial differential equations. In actual applications, it is preferable to use the right-hand side of (18.78) for a weak solution; less differentiability requirements are needed for the functions involved. This advantage is important for finite element methods.

## 10. Exercises

1. Let $J$ be the Jacobian $\partial(x_1,x_2,x_3)/\partial(a_1,a_2,a_3)$ and

$$J_{ij} = \frac{\partial J}{\partial x_{i,j}},$$

where $x_{i,j} = \partial x_i / \partial a_j$. Prove the following two identities:

(a) $\dfrac{\partial J_{ij}}{\partial a_j} = 0 \quad (i = 1,2,3);$

(b) $\dfrac{\partial x_k}{\partial a_j} J_{ij} = J\delta_{ik}.$

2. The governing equation for steady two-dimensional (supersonic) flow of inviscid perfect gas is given in terms of a dimensionless potential $u(x,y)$ by

$$\nabla^2 u[1 + \frac{1}{2}(\gamma - 1)M^2(1 - u_{,x}^2 - u_{,y}^2)] = M^2[u_{,x}^2 u_{,xx} + 2u_{,x}u_{,y}u_{,xy} + u_{,y}^2 u_{,yy}],$$

where the ratio of specific heat $\gamma(>1)$ and the Mach number $M(>1)$ are constants (Cole and Cook, 1986). Show that it is the Euler DE of

$$J[u] = \iint_A [\frac{1}{2}(\gamma - 1)M^2(u_{,x}^2 + u_{,y}^2) - \{1 + \frac{1}{2}(\gamma - 1)M^2\}]^{\gamma/(\gamma-1)}\, dx\, dy.$$

3. The governing equation for two-dimensional steady incompressible viscous flow in a region $A$ of the $x, y$-plane is

$$\nu \nabla^2 \nabla^2 \psi + \psi_{,x} \nabla^2 \psi_{,y} - \psi_{,y} \nabla^2 \psi_{,x} = 0$$

where $\psi$ is the stream function and $\nu$ is the coefficient of kinematic viscosity.

(a) Attempt to obtain the PDE as an Euler DE of an integral $I[\psi]$ by multiplying through by $\delta\psi$ and integrate by parts. What keeps you from finding $I[\psi]$?

(b) Do the best you can to get a weak form with an integrand which involves no derivative of $\psi$ higher than the second.

# Appendix

# Approximate Methods for Euler's Differential Equations

## 1. Two-Point Boundary-Value Problems

A pivotal step in the solution of the simplest problems (of one independent variable) in the calculus of variations is the determination of admissible extremals. These extremals are the solutions of a two-point boundary-value problem in ODE defined by the Euler DE

$$(\hat{F}_{,y'})' = \hat{F}_{,y} \quad (a < x < b)$$

and the prescribed or Euler boundary conditions at the two ends $x = a$ and $x = b$ of the solution domain. Even if the extremals are $C^2$ (and we will so assume in this appendix), it is often not possible to obtain an exact solution of this (BVP) boundary-value problem in terms of elementary or special functions. On the other hand, an accurate approximate solution usually suffices for most problems in applications. Chapters 8 and 9 describe two approaches to approximate solutions which implement the process of extremization inherent in the calculus of variations in two different approximate ways. In this appendix, we will discuss a third approach to approximate solutions for variational problems. The approach here is to construct a discrete analogue of the two-point boundary-value problem for the Euler DE and to develop efficient methods of solution for the resulting algebraic problem. The mathematics involved here no longer pertains to the calculus of variations proper or optimization in general, but exclusively to numerical linear algebra and numerical analysis. The description of this class of methods is, therefore, appropriately relegated to an appendix.

Boundary-value problems are mathematically much more complex than initial-value problems (IVP). Sometimes, the solution of a BVP does not exist even if the related IVP has a solution and it may not be unique even if the corresponding IVP has a unique solution. Therefore, approximate solutions should not be considered reliable without an appropriate theoretical justification.

As an illustration, we consider the one-dimensional model of heat conduction in a straight slender rod subject to a known distributed (time-independent) heat source $f(x)$ along the length of the rod, $0 < x < \ell$. The two ends of the rod, $x = 0$ and $x = \ell$, are maintained at two different constant temperatures $A$ and $B$. For this model, the *steady-state* temperature distribution $y(x)$ along the rod is determined by the ODE

$$\frac{d^2y}{dx^2} = f(x) \qquad (0 < x < \ell), \tag{A.1}$$

and the auxiliary conditions

$$y(0) = A, \qquad y(\ell) = B. \tag{A.2}$$

It is not difficult to verify that (A.1) is the Euler DE of a basic problem with the Lagrangian $F(x,y,y') = [(y')^2 + 2f(x)y]/2$.

The method of solution for the BVP above is straightforward. Integrate the ODE (A.1) twice to obtain

$$y(x) = c_0 + c_1 x + g(x), \qquad g(x) = \int_0^x \int_0^t f(z)\,dz\,dt.$$

The two boundary conditions (A.2) require

$$y(0) = c_0 + g(0) = A, \qquad y(\ell) = c_0 + c_1\ell + g(\ell) = B$$

which uniquely determine $c_0$ and $c_1$. With $g(0) = 0$, we have

$$c_0 = A, \qquad c_1 = \frac{1}{\ell}\{[B - A] - g(\ell)\}.$$

If the condition of prescribed end temperature $y(\ell) = B$ is replaced by a prescribed heat flux at that end so that $y'(\ell) = \beta$, we obtain instead $c_1 = \beta - g'(\ell)$ with $c_0 = A - g(0) = A$ as before. On the other hand, if the heat flux is prescribed at both ends, say $y'(0) = \alpha$ and $y'(\ell) = \beta$, we have

$$c_1 + g'(0) = \alpha \qquad c_1 + g'(\ell) = \beta.$$

It is not possible for both conditions to be satisfied by a single choice of $c_1$ unless $\alpha - g'(0) = \beta - g'(\ell)$. Therefore, the BVP does not have a solution in general. In the exceptional case where $\alpha - g'(0) = \beta - g'(\ell)$, we have $c_1 = \alpha - g'(0)$ and $c_0$ arbitrary; the solution is, therefore, not unique in this case.

Unusual features in a mathematical problem usually reflect a certain anomaly in the physical problem. It is not difficult to see that unless the net heat flux through the two ends balances the internal heat production by the heat source $f(x)$, the temperature of the rod cannot be in a steady state. If there is a balance, then the temperature distribution in the rod is determined up to an additive constant which depends on the condition of the rod prior to reaching steady state. We will assume in this chapter that the solution of our two-point BVP exists and is unique and concentrate on methods for obtaining an (accurate) approximate solution of the problem. One class of approximate solutions makes use of the conventional well-known numerical methods for solving initial-value problems in ODE. In the next section, we will review some useful background information on numerical solutions for IVP before we describe the so-called *shooting method* for BVP.

## 2. Numerical Solution of Initial-Value Problems

All modern texts for a first course in ODE at the level of Boyce and DiPrima (1976) discuss some numerical methods for ODE. For a first-order equation

$$y' = f(x,y), \qquad y(a) = A, \tag{A.3}$$

where the value of $y(x)$ is sought for the interval $(a, b)$, the simplest numerical method subdivides the interval $[a,b]$ into $N$ equal subintervals. The end points of these subintervals are denoted by $x_k = kh + a$, $k = 0,1,2,\ldots,N$, $h = (b-a)/N$ with $x_0 = a$ and $x_N = b$. We assume here that $f$ is continuously differentiable in both arguments. In that case $y'' = f_{,x} + f_{,y}y' = f_{,x} + f_{,y}f$ is continuous and we may use Taylor's theorem to write

$$\begin{aligned} y(x_{k+1}) &= y(x_k) + hy'(x_k) + \frac{1}{2}h^2y''(\xi_k) \\ &= y(x_k) + hf(x_k, y(x_k)) + \frac{1}{2}h^2y''(\xi_k) \end{aligned} \tag{A.4}$$

with $x_k \le \xi_k \le x_{k+1}$. Incidentally, if $y''$ is continuous over the bounded interval $[a,b]$, then it must be bounded in $[a,b]$, say $|y''| < M$. For a sufficiently small $h$, we may work with the following approximate version of (A.4) for an approximate solution of the IVP (A.3):

$$y_{k+1} = y_k + hf(x_k, y_k) \qquad (k = 0,1,2,\ldots). \tag{A.5}$$

Starting with $y_0 = A$, (A.5) gives successively

$$y_1 = A + hf(a, A), \quad y_2 = y_1 + hf(x_1, y_1),$$

$$\ldots, \quad y_N = y_{N-1} + hf(b, y_{N-1}).$$

Intuitively, the value $y_k$ generated by the (simple) Euler method (A.5) is an approximation of the actual solution $y(x_k)$. Furthermore, we expect the approximation to be more and more accurate as $h$ decreases. Roughly, an error proportional to $h^2$ is made in advancing a step of the solution process. To get from $x = a$ to $x = b$, $N$ such *local truncation errors* are made for a total accumulated (truncation) error proportional to $Nh^2 = (b - a)h$. We say the simple Euler method is *first-order accurate* because the global (*accumulated*) *error* after $N$ steps is proportional to the first power of the step (*mesh*) size $h$.

For a more precise statement about the accuracy of the simple Euler scheme for IVP in ODE, consider the accumulated error at the $k$th step

$$\varepsilon_k = y(x_k) - y_k \tag{A.6}$$

with $\varepsilon_0 = y(a) - A = 0$. To get an estimate on $\varepsilon_k$, we subtract (A.5) from (A.4) to obtain

$$\varepsilon_{k+1} = \varepsilon_k + h[f(x_k, y(x_k)) - f(x_k, y_k)] + \frac{h^2}{2} y''(\xi_k).$$

By the mean value theorem, we have $f(x_k, y(x_k)) - f(x_k, y_k) = f_{,y}(x_k, \bar{y}_k) \cdot [y(x_k) - y_k]$ with $|\bar{y}_k - y_k| \leq |y(x_k) - y_k|$. The expression for $\varepsilon_{k+1}$ can then be written as

$$\varepsilon_{k+1} = \varepsilon_k[1 + hf_{,y}(x_k, \bar{y}_k)] + \frac{h^2}{2} y''(\xi_k). \tag{A.7}$$

This formula allows us to conclude the following

**[A.1]** *Let* $|f_{,y}| < L$ *and* $|y''(x)| < M$ *for* $a \leq x \leq b$. *We have the following bound for the accumulated error after* $k$ *steps:*

$$|\varepsilon_k| \leq \frac{hM}{2L}(e^{khL} - 1). \tag{A.8}$$

Evidently, $|\varepsilon_k|$ tends to zero with $h$.

To prove (A.8), consider the first-order difference equation

$$r_{k+1} = r_k(1 + hL) + \frac{h^2}{2} M, \qquad r_0 = 0. \tag{A.9}$$

The solution of (A.9) is

$$r_k = \frac{hM}{2L}(1 + hL)^k - \frac{hM}{2L} \quad (h \geq 0). \tag{A.10}$$

With (A.7) giving

$$|\varepsilon_{k+1}| \leq |\varepsilon_k|(1 + hL) + \frac{h^2}{2} M, \tag{A.11}$$

it is not difficult to show by induction that $|\varepsilon_k| \le r_k$. Obviously $|\varepsilon_0| \le r_0$ because $|\varepsilon_0| = r_0 = 0$. Suppose $|\varepsilon_k| \le r_k$, then we have from (A.9) and (A.11)

$$r_{k+1} = (1 + hL)r_k + \frac{h^2}{2}M \ge (1 + hL)|\varepsilon_k| + \frac{h^2}{2}M \ge |\varepsilon_{k+1}|.$$

The result [A.1] follows from this and (A.10).

The hypothesis requiring $f$ to be continuously differentiable may be weakened considerably. The results may be extended to first-order systems for which $y = (y_1, y_2, \ldots, y_n)$. Algorithms with higher-order accuracy have also been developed. Most notable among these are the improved (or second-order) Euler method and the (fourth-order) Runge–Kutta method. The formulation and error analysis for these and other more sophisticated algorithms can be found in Henrici (1962) and will not be discussed here. Instead, we will show how the vector form of the simple Euler (or any other well-defined numerical) method can be used to solve the BVP for admissible extremals approximately.

## 3. Linear Boundary-Value Problems

If a BVP is linear, an approximate solution to the problem may be obtained by solving a few IVP numerically. We will illustrate this using the following second-order, linear homogeneous Euler DE:

$$y'' + p(x)y' + q(x)y = 0 \quad (a < x < b), \tag{A.12}$$

$$y(a) = A, \qquad y(b) = B. \tag{A.13}$$

For simplicity, we assume that the ODE (A.12) has no singularities in the solution domain, end points included, and take its two complementary solutions $\{\varphi_1(x), \varphi_2(x)\}$ to satisfy the following initial conditions:

$$\varphi_1(a) = 1, \qquad \varphi_1'(a) = 0, \tag{A.14}$$

$$\varphi_2(a) = 0, \qquad \varphi_2'(a) = 1, \tag{A.15}$$

Note that $\varphi_1(x)$ and $\varphi_2(x)$ so defined may be obtained numerically to a specified accuracy for $a < x \le b$ by any suitable method for IVP in ODE as indicated in the last section.

The solution of the linear ODE (A.12) may then be expressed as a linear combination of $\varphi_1$ and $\varphi_2$,

$$y(x) = y_0\varphi_1(x) + v_0\varphi_2(x), \tag{A.16}$$

where $y_0$ and $v_0$ are constants to be determined. The two boundary conditions (A.13) require

$$y(a) = y_0\varphi_1(a) + v_0\varphi_2(a) = y_0 \cdot 1 + v_0 \cdot 0 = A \tag{A.17}$$

$$y(b) = y_0\varphi_1(b) + v_0\varphi_2(b) = B. \tag{A.18}$$

Condition (A.17) gives immediately

$$y_0 = A \tag{A.19}$$

and, for $\varphi_2(b) \neq 0$, condition (A.18) gives

$$v_0 = [B - A\varphi_1(b)]/\varphi_2(b). \tag{A.20}$$

With (A.19) and (A.20), the solution of the BVP (A.12) and (A.13) is completely determined by (A.16) provided $\varphi_2(b) \neq 0$. We summarize the result as:

**[A.2]** *For the case $\varphi_2(b) \neq 0$, one numerical solution process for (A.12) and (A.13) generates $\varphi_1(x)$ and $\varphi_2(x)$ for $a < x \leq b$ by solving numerically two IVPs, namely, (A.12) with (A.14) and (A.12) with (A.15), respectively. Expression (A.16) then gives $y(x)$ with $y_0$ and $v_0$ taken from (A.19) and (A.20).*

If $\varphi_2(b) = 0$, we have one of two situations: either BVP has no solution if $B - A\varphi_1(b) \neq 0$ or it has a one-parameter family of solutions if $B - A\varphi_1(b) = 0$. In the latter case, the solution of the BVP is not unique.

The solution technique for BVP in linear ODE described above is easily modified to handle inhomogeneous equations as well as equations of higher order and (linear) systems. The development of these modifications will be left as exercises.

## 4. The Shooting Method

Numerical solutions of BVP in ODE are much more difficult to obtain if either the ODE or the boundary conditions involved are nonlinear in the unknown(s). To illustrate the nature of the difficulty, we note that the general solution of the second-order nonlinear ODE

$$xy'' + (2y + 1)y' = 0 \tag{A.21}$$

is

$$y = \frac{c_2}{2}\frac{x^{c_2} - c_1}{x^{c_2} + c_1}. \tag{A.22}$$

In contrast to solution (A.16) for a linear ODE, the constants of integration of a nonlinear ODE normally do not appear linearly in its general solution. It follows that the linear system of algebraic equations associated with the boundary conditions of a linear BVP for the determination of the constants of integration will generally be replaced by a nonlinear system of equations. Such a system must be solved by iterative methods, often using a generalization of Newton's method to several unknowns called the *Newton–Raphson method* in most references.

To get some idea of this more complex solution process, suppose the boundary conditions for (A.21) are

$$y(1) = 0 \quad \text{and} \quad y(2) = 5. \tag{A.23}$$

The first condition requires $c_1 = 1$; we discard the possibility $c_2 = 0$ in order to have a nontrivial solution. The second condition in (A.23) may then be written as

$$g(c_2) \equiv \frac{c_2}{2}\frac{2^{c_2}-1}{2^{c_2}+1} - 5 = 0. \tag{A.24}$$

The nonlinear equation (A.24) may be solved by Newton's method to get $c_2$. A reasonable initial guess would be $c_2^{(0)} = 10$. Subsequent iterates $c_2^{(k)}$ are calculated using Newton's iteration formula (*Greenspan* and *Benney*, 1973)

$$c_2^{(k+1)} = c_2^{(k)} - \frac{g(c_2^{(k)})}{g^{\bullet}(c_2^{(k)})} \qquad (k = 1, 2, 3, \ldots), \tag{A.25}$$

where $g^{\bullet} = dg/dc_2$.

For the general second-order nonlinear ODE

$$y'' = f(x, y, y') \qquad (a < x < b) \tag{A.26}$$

with boundary conditions

$$y(a) = A, \qquad y(b) = B, \tag{A.13}$$

we may, in principle, solve (A.26) with the initial conditions

$$y(a) = A, \qquad y'(a) = V \tag{A.27}$$

for some unknown parameter $V$. We denote this solution by

$$y(x) = \varphi(x; V) \tag{A.28}$$

to indicate that $y(x)$ varies with different initial slope values. For simplicity, we assume that $\varphi(x; V)$ remains well defined and bounded in $a \le x \le b$ for the range of values of $V$ of interest. In that case, the problem again becomes one of choosing $V$ so that

$$y(b) = \varphi(b; V) = B\,. \tag{A.29}$$

The nonlinear equation (A.29) for $V$ may again be solved by an iterative method.

In general, a simple analytic form of the solution $\varphi(x; V)$ is not possible for (A.26). However, for the solution process outlined above, $\varphi(x; V)$ may be obtained by any suitable numerical method for IVP for any specific value of $V$. We merely have to recalculate $\varphi(x; V)$ for each iteration of (A.29) as $V$ changes from iteration to iteration. If a numerical solution of $\varphi(x; V)$ is to be used, we may consider a secant type method for (A.29) in which the derivative of $\varphi(b; V)$ with respect to $V$ is not needed:

$$V_{k+1} = V_k - \frac{\varphi(b; V_k)}{\varphi_{,V}(b; V_k)} \tag{A.30a}$$

$$\simeq V_k - \frac{\varphi(b; V_k)(V_k - V_{k-1})}{\varphi(b; V_k) - \varphi(b; V_{k-1})}\,. \tag{A.30b}$$

We have then the following approximate method of solution for the BVP (A.26) and (A.13):

**[A.3]** ***Solve the IVP (A.26) and (A.27) for two approximate values of*** $\mathbf{V}$***, say*** $\mathbf{V}_k$ ***and*** $\mathbf{V}_{k-1}$***. Then use the secant method (A.30b) to calculate an improved approximation*** $\mathbf{V}_{k+1}$ ***for*** $\mathbf{V}$***. Repeat the process (with*** $\mathbf{V}_{k+1}$ ***and*** $\mathbf{V}_k$ ***replacing*** $\mathbf{V}_k$ ***and*** $\mathbf{V}_{k-1}$***, respectively) until*** $\varphi(\mathbf{x}; \mathbf{V}_m)$ ***attains the desired accuracy as an approximate solution for*** $\mathbf{y}(\mathbf{x})$***.***

The rate of convergence of the secant method is known to be slower than the Newton's method (given that we have replaced the derivative $\varphi_{,V}$ by its difference quotient) and we now need two initial guesses $V_0$ and $V_1$ to get started.

If we are willing to do the additional computing needed for Newton's method to get faster convergence, we may calculate $\varphi_{,V}(b; V_k) \equiv Z_k(b)$ required by (A.30a) for each iteration by solving the associated BVP for $\varphi_{,V}(x; V_k) \equiv Z_k(x)$. With all terms in (A.26) and (A.27) involving $y$ dependent parametrically on $V$, we differentiate all these terms partially with respect to $V$ to get

$$Z_k'' = f_{,y}(x, \varphi_{,k}, \varphi_{,k}')Z_k + f_{,y'}(x, \varphi_{,k}, \varphi_{,k}')Z_k'\,, \tag{A.31}$$

$$Z_k(a) = 0\,, \qquad Z_k'(a) = 1\,, \tag{A.32}$$

where $\varphi_k = \varphi(x; V_k)$ and a prime continues to indicate differentiation with respect to $x$. The solution of the *linear* BVP (A.31) and (A.32) enables us to use the following alternative method for (A.26) and (a.13):

**[A.4]** ***Starting with some approximate value for*** **V*****, say*** $V_k$***, solve the IVP (A.26) and (A.27) and then the IVP (A.31) and (A.32). Use (A.30a) with*** $\varphi_{,V}(b; V_k) = Z_k(b)$ ***to calculate an improved approximation*** $V_{k+1}$***. Repeat the process (with*** $V_{k+1}$ ***replacing*** $V_k$***) until*** $\varphi_m(x) \equiv \varphi(x; V_m)$ ***attains the desired accuracy as an approximate solution for*** $y(x)$***.***

The solution method described above can be modified to accommodate other types of boundary conditions, higher-order equations and systems. Some examples of these new classes of problems can be found in the exercises. The number of unknown parameters to be determined iteratively by Newton's method in each pass of the solution process for the relevant IVP increases with the number of boundary conditions at $x = b$. If the number of conditions at $x = b$ exceeds the number of conditions at $x = a$, it may be more efficient to numerically solve the terminal-value problem (TVP) by satisfying all conditions at $x = b$ exactly and to determine the unknown terminal parameters by iterating on the conditions at $x = a$.

It should be noted that the ***shooting method*** described above is not the only method available for BVP; nor is it the most efficient. For example, it is often the case that the solution of the associated IVP grows too large too soon to cause machine overflow before we reach $x = b$. The method of ***multiple shooting*** or some other alternative is needed for this type of situation. In the next few sections, we will discuss another class of solution methods for admissible extremals which does not involve the solution of IVP and, thus, avoids the complication of fast growth rate and the attendant machine overflow problem.

## 5. Finite Difference Analogue

A different approach to obtaining an admissible extremal of the variational problem is to work with a finite difference analogue of the relevant two-point BVP. In this section, we discuss this solution process and the mathematical issues that arise in this approach for the linear problem

$$\mathscr{L}[y] \equiv -y'' + p(x)y' + q(x)y = r(x), \quad (a < x < b) \tag{A.33a}$$

$$y(a) = A, \qquad y(b) = B. \tag{A.33b}$$

Note that we have written the linear ODE here in a form slightly different from (A.12) and will continue to do so in the subsequent development to be consistent with the conventional practice.

To construct a finite difference analogue of the two-point BVP, we again introduce a set of mesh points $\{x_n = a + nh\}$ with the equal spacing $h = (b - a)/(N + 1)$ so that $x_0 = a$ and $x_{N+1} = b$. From the boundary conditions

(A.33b), we have $y_0 \equiv y(x_0) = y(a) = A$ and $y_{N+1} = y(x_{N+1}) = y(b) = B$. As in Section 2, we seek an approximate solution $y_n$ for $y(x_n)$, $n = 1, 2, \dots, N$, with the help of truncated Taylor's series for $y(x)$ and its derivatives. The truncated series (A.5) for $y'(x_n)$ has a local truncation error proportional to $h^2$, leading to an accumulation error proportional to $h$. For the present development, we need finite difference approximations for derivatives of higher-order accuracy and in a more symmetric form.

To get the desired finite difference formulas, we start with the Taylor series

$$y(x_{n+1}) = y(x_n) + hy'(x_n) + \frac{1}{2}h^2y''(x_n) + \frac{1}{3!}h^3y'''(x_n) + \frac{1}{4!}h^4y^{(iv)}(x_n) + \dots,$$

$$y(x_{n-1}) = y(x_n) - hy'(x_n) + \frac{1}{2}h^2y''(x_n) - \frac{1}{3!}h^3y'''(x_n) + \frac{1}{4!}h^4y^{(iv)}(x_n) + \dots,$$

and form the difference and sum of these two expressions to get

$$y'(x_n) = \frac{1}{2h}[y(x_{n+1}) - y(x_{n-1})] - \frac{1}{3!}h^2y'''(\xi_n),$$

$$y''(x_n) = \frac{1}{h^2}[y(x_{n+1}) - 2y(x_n) + y(x_{n-1})] - \frac{2}{4!}h^2y^{(iv)}(\zeta_n),$$

where we have assumed that $y$ is at least $C^4$ in $[a,b]$ so that the appropriate Taylor's formulas apply with $x_n \le \xi_n, \zeta_n \le x_{n+1}$. We obtain an approximation of the left-hand side of the ODE in (A.33) at a point $x_n$ if we replace $y'(x_n)$ by the *central difference quotient* $[y(x_{n+1}) - y(x_{n-1})]/2h$ and $y''(x_n)$ by $[y(x_{n+1}) - 2y(x_n) + y(x_{n-1})]/h^2$. We denote the resulting approximate (finite difference) solution of the problem by $\{y_n\}$ and the finite difference analogue of $\mathcal{L}[y]$ by $\mathcal{L}_h[y]$ with

$$\mathcal{L}_h[y] \equiv -\frac{y_{n+1} - 2y_n + y_{n-1}}{h^2} + p_n\frac{y_{n+1} - y_{n-1}}{2h} + q_ny_n = r_n, \quad \text{(A.34)}$$

where $p_n = p(x_n)$, $q_n = q(x_n)$, and $r_n = r(x_n)$ are the exact values of these functions at $x_n$. The linear equation (A.34) applies at each of the $N$ interior mesh points $x_1, \dots, x_N$. These $N$ linear equations contain only $N$ unknowns $\{y_1, \dots, y_N\}$ for we know $y(x_0) = y(a) = A$ and $y(x_{N+1}) = y(b) = B$ and can set $y_0 = A$ and $y_{N+1} = B$ in (A.34).

The linear system (A.34) with $1 \le n \le N$ may be rewritten in a more conventional form:

$$\frac{h^2}{2}\mathcal{L}_h[y] \equiv a_ny_{n-1} + b_ny_n + c_ny_{n+1} = \frac{h^2}{2}r_n, \quad \text{(A.35)}$$

where

$$\begin{pmatrix} a_n \\ c_n \end{pmatrix} = -\frac{1}{2}\left[1 \pm \frac{h}{2} p_n\right], \quad b_n = 1 + \frac{h^2}{2} q_n, \quad \bar{r}_n = \frac{h^2}{2} r_n \tag{A.36}$$

or, with $\mathbf{y} = (y_1, \dots, y_N)^T$,

$$C\mathbf{y} = \mathbf{z} \equiv (\bar{r}_1 - a_1 A, \bar{r}_2, \dots, \bar{r}_{N-1}, \bar{r}_N - c_N B)^T, \tag{A.37}$$

$$C = \begin{bmatrix} b_1 & c_1 & 0 & 0 & \cdot & \cdot & \cdot & 0 & 0 \\ a_2 & b_2 & c_2 & 0 & & & & \cdot & 0 \\ 0 & a_3 & b_3 & c_3 & & & & \cdot & \cdot \\ \cdot & \cdot & \cdot & \cdot & & & & & \cdot \\ \cdot & \cdot & \cdot & & & & & & \cdot \\ \cdot & \cdot & & & & & \cdot & \cdot & \cdot \\ \cdot & & & & & & b_{N-2} & c_{N-2} & 0 \\ 0 & & & & & & a_{N-1} & b_{N-1} & c_{N-1} \\ 0 & 0 & \cdot & \cdot & \cdot & \cdot & 0 & a_N & b_N \end{bmatrix} \tag{A.38}$$

The matrix $C$ is tridiagonal; a solution by Gaussian elimination is particularly effective. We implement the solution process by performing the $LU$ decomposition for $C$ to get $C = LU$ with

$$L = \begin{bmatrix} \beta_1 & 0 & \cdot & \cdot & \cdot & \cdot & 0 \\ a_2 & \beta_2 & 0 & \cdot & & & \cdot \\ 0 & a_3 & \cdot & & & & \cdot \\ \cdot & \cdot & & & & & \cdot \\ \cdot & \cdot & & & & & \cdot \\ \cdot & & & & & & 0 \\ 0 & \cdot & \cdot & \cdot & 0 & a_N & \beta_N \end{bmatrix} \tag{A.39a}$$

$$U = \begin{bmatrix} 1 & \gamma_1 & 0 & \cdot & \cdot & \cdot & 0 \\ 0 & 1 & \gamma_2 & \cdot & & & \cdot \\ \cdot & 0 & 1 & & & & \cdot \\ \cdot & \cdot & 0 & & & & \cdot \\ \cdot & & & & & & 0 \\ \cdot & & & & & 1 & \gamma_{N-1} \\ 0 & \cdot & \cdot & \cdot & \cdot & 0 & 1 \end{bmatrix} \tag{A.39b}$$

where

$$\begin{aligned} &\beta_1 = b_1, \qquad \gamma_1 = c_1/\beta_1, \\ &\beta_j = b_j - a_j\gamma_{j-1} \qquad (j = 2, 3, \ldots, N), \\ &\gamma_j = c_j/\beta_j, \qquad (j = 2, 3, \ldots, N-1). \end{aligned} \tag{A.40}$$

We then set $U\mathbf{y} = \mathbf{u}$ so that (A.37) becomes $L\mathbf{u} = \mathbf{z}$. The solution of the latter is easily obtained to be

$$u_1 = z_1/\beta_1, \qquad u_j = (z_j - a_j u_{j-1})/\beta_j, \qquad (j = 2, 3, \ldots, N). \tag{A.41a}$$

We can then solve $U\mathbf{y} = \mathbf{u}$ to get

$$y_N = u_N, \qquad y_k = u_k - \gamma_k y_{k+1}, \qquad (k = N-1, N-2, \ldots, 1). \tag{A.41b}$$

If Gausian elimination yields a solution for the difference analogue (A.37), we naturally expect this solution to provide an approximate solution of the original BVP. The following result establishes when the solution can be so obtained:

**[A.5]** *Let* $|\mathrm{p}(\mathrm{x})| \le \mathrm{P}$ *and* $0 < \mathrm{Q}_0 \le \mathrm{q}(\mathrm{x}) \le \mathrm{Q}_1$ *in* $[\mathrm{a}, \mathrm{b}]$ *for some positive constants* P, $\mathrm{Q}_0$ *and* $\mathrm{Q}_1$. *The system* $\mathrm{C}\mathbf{y} = \mathbf{z}$ *with the coefficient matrix* C *given by* (A.38) *and* (A.36) *has a unique solution given by* (A.41) *if the mesh spacing* h *is less than* $2/\mathrm{P}$.

The proof of this result is left as an exercise at the end of the chapter.

Although the requirements imposed on $p$ and $q$ in [A.5] guarantee the existence and uniqueness of the solution of (A.37), they are not necessary. The simple harmonic oscillator $-y'' - y = 0$ with $y(a) = 0$ and $y(b) = B$ is an instructive example. The exact solution of the ODE which satisfies $y(a) = 0$ is $y = c\sin(x - a)$. If $b - a \neq n\pi$, the exact solution of the BVP is $y = B\sin(x - a)/\sin(b - a)$. There exists a unique solution of the BVP even when $q = -1$ does not satisfy the hypothesis of [A.5]. On the other hand, there is no solution for the BVP if $b - a = n\pi$ unless $B = 0$, in which case there are infinitely many solutions (for different values of $c$). The corresponding situation for the finite difference analogue will be left as an exercise.

## 6. Accuracy of the Finite Difference Solution

Suppose $p$, $q$ and $h$ all meet the requirements of [A.5] so that we get a unique solution for the difference analogue of the original BVP. Naturally, we are interested in knowing how good an approximation we have obtained for the exact solution. How large is $|y_k - y(x_k)|$ at each mesh point? To answer this question, we note the following results for any set of values $\{v_0, v_1, \ldots, v_{N+1}\}$:

**[A.6]** *If* p, q, *and* h *satisfied the hypotheses of* [A.5], *we have*

$$|v_j| \leq M\{\max(|v_0|, |v_{N+1}|) + \max_{1 \leq i \leq N} |\mathscr{L}_h[v_i]|\}, \qquad (A.42)$$

*where* $M = \max(1, 1/Q_0)$.

Inequality (A.42) follows from the definition of $\mathscr{L}_h[v_i]$ (see p. 77 of H.B. Keller, 1968, for a proof).

We now introduce for any $C^2$ function $u(x)$ on $[a, b]$ the local truncation error $T_j[u]$ of the approximation of $\mathscr{L}[u]$ by $\mathscr{L}_h[u]$:

$$T_j[u] \equiv \mathscr{L}[u(x_j)] - \mathscr{L}_h[u(x_j)]$$

$$= -\left[u''(x_j) - \frac{u_{j+1} - 2u_j + u_{j-1}}{h^2}\right] + p(x_j)\left[u'(x_j) - \frac{u_{j+1} - u_{j-1}}{2h}\right]$$

for $1 \leq j \leq N$. If $u$ is at least $C^4$, we have from a previous observation

$$T_j[u] = \frac{h^2}{12}[u^{(iv)}(\xi_j) - 2p(x_j)u'''(\zeta_j)] \quad (1 \leq j \leq N), \qquad (A.43)$$

where $x_{j-1} \leq \xi_j, \zeta_j \leq x_{j+1}$. In terms of $T_j[u]$, the question we posed concerning the accuracy of the finite difference solution is answered by the following result.

**[A.7]** *Let* p(x), q(x) *and* h *satisfy the hypotheses of* [A.5]. *Then the bound*

$$|y(x_k) - y_k| \leq M\left\{\max_{1 \leq i \leq N} |T_i[y]|\right\} \qquad (A.44)$$

*holds for* $0 < k < N + 1$ *with* $M = \max[1, 1/Q_0]$ *where* $0 < Q_0 \le q(x)$. *If* $y$ *is at least* $C^4$ *in* $[a, b]$, *then (A.44) may be written as*

$$|y(x_k) - y_k| \le Mh^2(M_4 + 2PM_3)/12, \tag{A.45}$$

*where* $M_n \equiv \max|d^n y/dx^n|$ *in* $[a, b]$.

The error bound (A.45) is an immediate consequence of (A.44) and (A.43). To get (A.44), we note that

$$\mathcal{L}_h[y(x_k) - y_k] = \mathcal{L}_h[y(x_k)] - r_k = \mathcal{L}_h[y(x_k)] - \mathcal{L}[y(x_k)] = -T_k[y].$$

Now, for $e_n = y(x_n) - y_n$, we have from [A.6] (with $e_n$ taking the place of $v_n$) the inequality

$$|e_n| \le M\left\{\max_k |\mathcal{L}_h[e_k]|\right\}$$

given that $e_0 = e_{N+1} = 0$. Equivalently, we have

$$|y(x_n) - y_n| \le M\{\max_k |\mathcal{L}_h[y(x_k) - y_k]|\} \le M\{\max_k |T_k[y]|\}$$

for $1 \le n \le N$, which is just (A.44).

The error bound (A.45) tells us not only the accuracy of the finite difference approximation for a given $h(\le 2/P)$, but also the rate of decrease of the error with decreasing $h$. For instance, halving $h$ reduces the error by a factor of 4. With $|y(x_n) - y_n|$ proportional to $h^2$, the method is said to *converge quadratically with h*.

Although more refined results than [A.7] are available, including difference analogues with higher-order accuracy than (A.45), i.e., with error bounds proportional to $h^4$ or higher powers, the discussion of this section suffices to provide an introduction to the typical issues and results of another class of methods for obtaining approximate admissible extremals of variational problems.

## 7. Fixed Point Iteration

Consider now the more general case where the Euler DE takes the form

$$D[y] \equiv -y'' + f(x, y, y') = 0 \quad (a < x < b),$$
$$y(a) = A, \qquad y(b) = B. \tag{A.46}$$

For a set of evenly spaced mesh points $\{x_0 = a, x_1, \ldots, x_N, x_{N+1} = b\}$ in $[a, b]$, we take the finite difference analogue of (A.46) to be

$$D_k[y_k] = -\frac{y_{k+1} - 2y_k + y_{k-1}}{h^2} + f\left(x_k, y_k, \frac{y_{k+1} - y_{k-1}}{2h}\right) = 0,$$

$$y_0 = A, \qquad y_{N+1} = B. \tag{A.47}$$

Suppose we have succeeded in solving (A.47) to get $\{y_1, y_2, \ldots, y_N\}$ and the truncation error $T_i[y]$ is now understood to be

$$T_i[u] \equiv D[u(x_i)] - D_k[u(x_i)]$$

$$= \frac{h^2}{12}[u^{(iv)}(\xi_i) - 2f_{,y'}(x_i, u(x_i), u'(t_i))u'''(\zeta_i)] \tag{A.48a}$$

for some $\xi_i$, $t_i$ and $\zeta_i$ inside the interval $[x_i, x_{i+1}]$, end points included. It is not difficult to obtain the following result, which is a version of [A.7] for the nonlinear problem (A.46):

**[A.8]** *If f is at least $C^1$ on $[a,b]$ with $|f_{,y'}| \le P$ and $0 < Q_0 \le f_{,y} \le Q_1$, then (A.44) again applies so that*

$$|y(x_k) - y_k| \le M\{\max_i |T_i[y]|\} \quad (1 \le k \le N) \tag{A.48b}$$

*with $M = \max(1, 1/Q_0)$, where $Q_0$ is the (positive) lower bound of $f_{,y}$. If y is at least $C^4$ on $[a,b]$, then (A.45) again holds.*

The proof of this result is not difficult and can be found in Keller (1968).

With the error bound (A.45) also applicable to the nonlinear problem, we are assured that the finite difference solution does provide a useful approximation for the exact admissible extremal. We can then turn to the more important business of obtaining the solution of the system of nonlinear equations (A.47). For the simplest iterative method of solution for (A.47), we rewrite the system in the form

$$y_k = g_k(\mathbf{y}) \quad (k = 1, 2, \ldots, N) \tag{A.49a}$$

or in vector form

$$\mathbf{y} = \mathbf{g}(\mathbf{y}). \tag{A.49b}$$

This may be accomplished in many ways. One way would be to multiply (A.47) through by $h^2/2$ and add $\omega y_k$ to both sides for any real $\omega \ne -1$. The result may be written as

$$y_k = (1+\omega)^{-1}\{\frac{1}{2}(y_{k+1}+y_{k-1}) + \omega y_k - \frac{1}{2}h^2 f(x_k, y_k, z_k)\}$$

$$= y_k - (1+\omega)^{-1}\frac{h^2}{2}D_h[y_k] \equiv g_k(\mathbf{y}) \quad (k = 1, 2, \ldots, N) \tag{A.50}$$

with $y_0 = A$ and $y_{N+1} = B$. Relation (A.50) suggests the following iterative solution scheme for the finite difference solution $\mathbf{y} = (y_1, \ldots, y_N)^T$ of (A.47).

Take any initial estimate $\{y_k^{(0)}\}$ of the finite difference solution and calculate the corresponding right-hand side of (A.50) to get

$$y_k^{(1)} = y_k^{(0)} - (1+\omega)^{-1}\frac{h^2}{2}D_h[y_k^{(0)}] \qquad (k = 1, 2, \ldots, N).$$

If $y_k^{(1)} = y_k^{(0)}$ for $k = 1, 2, \ldots, N$, then $\{y_k^{(0)}\}$ is the solution of (A.47). Generally, this is not the case and we continue the iterative process with

$$y_k^{(m+1)} = g_k(\mathbf{y}^{(m)}) = y_k^{(m)} - \frac{h^2}{2(1+\omega)}D_h[y_k^{(m)}] \quad (k = 1, 2, \ldots, N) \tag{A.51}$$

for $m = 0, 1, 2, \ldots$

If $y_k^{(m)}$ tends to a limit $\bar{y}_k$ for all $k\,(1 \le k \le N)$ as $m$ increases, we say the *fixed point iteration* algorithm converges. More precisely, the sequence $\{\mathbf{y}^{(m)}\}$ converges (to the vector $\bar{\mathbf{y}}$) if for any given $\varepsilon > 0$ there is an $M_0(\varepsilon) > 0$ such that

$$|y_k^{(m)} - \bar{y}_k| < \varepsilon \tag{A.52}$$

for all $m > M_0$ and $1 \le k \le N$. The main result of interest here is

**[A.9]** ***The sequence $\{\mathbf{y}^{(m)}\}$ defined recursively by (A.51) converges for any initial estimate $\{\mathbf{y}^{(0)}\}$. The rate of convergence is most rapid with $\omega = h^2 Q_1/2$.***

To see this, we note that

$$y_k^{(m+1)} - \bar{y}_k = g_k(\mathbf{y}^{(m)}) - g_k(\bar{\mathbf{y}})$$

$$= (1+\omega)^{-1}[-a_k(y_{k-1}^{(m)} - \bar{y}_{k-1}) \tag{A.53}$$

$$+ (1+\omega-b_k)(y_k^{(m)} - \bar{y}_k) - c_k(y_{k+1}^{(m)} - \bar{y}_{k+1})]$$

for $1 \le k \le N$, where

$$a_k = -\frac{1}{2}\left\{1 + \frac{h^2}{2} f_{,y'}(x_k, \tilde{y}_k, \tilde{z}_k)\right\},$$

$$b_k = 1 + \frac{h^2}{2} f_{,y}(x_k, \tilde{y}_k, \tilde{z}_k), \quad c_k = -1 - a_k, \tag{A.54}$$

with

$$\tilde{y}_k = \bar{y}_k + \theta_k [y_k^{(m)} - \bar{y}_k],$$

$$\tilde{z}_k = \frac{\bar{y}_{k+1} - \bar{y}_{k-1}}{2h} + \theta_k \frac{(\bar{y}_{k+1}^{(m)} - \bar{y}_{k+1}) - (\bar{y}_{k-1}^{(m)} - \bar{y}_{k-1})}{2h} \tag{A.55}$$

for some $\theta_k$ in the interval $(0, 1)$, $0 < \theta_k < 1$. Result [A.5] for the linear case suggests that we take $h \le 2/P$. In that case $-a_k \ge 0$, $-c_k \ge 0$, and $1 + \omega - b_k \ge \omega - h^2 Q_1/2$. If we now take $\omega \ge h^2 Q_1/2$, then (A.53) implies

$$|y_k^{(m+1)} - \bar{y}_k| \le \frac{1 + \omega - a_k - b_k - c_k}{1 + \omega} \max_n |y_n^{(m)} - \bar{y}_n|. \tag{A.56}$$

Given the relation $a_k + b_k + c_k = h^2 f_{,y}(x_k, \tilde{y}_k, \tilde{z}_k)/2 \le h^2 Q_1/2$, we have

$$0 < \lambda_k(\omega) \equiv (1 + \omega - a_k - b_k - c_k)/(1 + \omega) < 1$$

with $\lambda_k(\omega)$ attaining its smallest value at $\omega^* = h^2 Q_1/2$. For this choice of $\omega$, we have from (A.53)

$$|y_k^{(m+1)} - \bar{y}_k| \le \lambda(\omega) \max_n |y_n^{(m)} - \bar{y}_n|, \tag{A.57}$$

where $0 < \lambda(\omega^*) < [1 + h^2(Q_1 - Q_0)/2]/[1 + h^2 Q_1/2] < 1$ (with $|f_{,y}| \le Q_0$ as in [A.8]). Thus, $\{|y_k^{(m)} - \bar{y}_k|\}$ is a monotone decreasing sequence in $m$ for any $\omega \ge h^2 Q_1/2$. It is bounded below by zero and, hence, converges. The rate of convergence is most rapid with $\omega = h^2 Q_1/2 \equiv \omega^*$.

If we set

$$\|\mathbf{u}\| = \|(u_1, \ldots, u_N)\| = \max_{1 \le i \le N} |u_i|, \tag{A.58}$$

then we may write the result (A.57) as

$$\|\mathbf{y}^{(m+1)} - \bar{\mathbf{y}}\| \le \lambda(\omega) \|\mathbf{y}^{(m)} - \bar{\mathbf{y}}\|. \tag{A.59}$$

The mapping of (A.49) with this property is known as a *contraction map*. The iterative scheme converges to a *fixed point* of the map [because $\bar{\mathbf{y}} = \mathbf{g}(\bar{\mathbf{y}})$] and is known as *fixed point iteration*.

## 8. Newton's Iteration

Even under the most favorable circumstances, $\lambda_k$ is close to 1 because

$$\lambda_k(\omega) = [1 + \omega - h^2 f_{,y}(x_k, \bar{y}_k, \bar{z}_k)/2]/(1+\omega)$$
$$\geq (1+\omega)^{-1} = 1 + 0(h^2)$$

given $\omega = 0(h^2)$ in (A.57) and $h$ expected to be small for a reasonable approximation of the exact solution [see (A.48a) and (A.48b)]. As a method for obtaining the finite difference solution for an admissible extremal, the fixed point iteration scheme is sure (as it converges for any initial guess) but slow. Once we get close to the actual solution after a few iterations, we may want to speed up the solution process by switching over to Newtons's iteration.

For this purpose, we write (A.47) as a vector equation

$$\mathbf{D}(\mathbf{y}) = \mathbf{0}, \tag{A.60}$$

where $\mathbf{y} = (y_1, \ldots, y_N)^T$ and $\mathbf{D} = (D_1, D_2, \ldots, D_N)^T$ with $D_n(\mathbf{y}) = h^2 D_h[y_n]/2$ and where the end conditions $y(a) = A$ and $y(b) = B$ have been used to eliminate $y_0$ and $y_{N+1}$ from (A.47). An approximate solution $\mathbf{y}^{(k)}$ obtained from the fixed point iteration scheme (or any other suitable method) generally does not satisfy (A.60). By Taylor's theorem, we have

$$\mathbf{0} = \mathbf{D}(\mathbf{y}) = \mathbf{D}(\mathbf{y}^{(k)}) + J(\mathbf{y}^{(k)})(\mathbf{y} - \mathbf{y}^{(k)}) + O(\|\mathbf{y} - \mathbf{y}^{(k)}\|^2)$$

or

$$\mathbf{y} = \mathbf{y}^{(k)} - J^{-1}(\mathbf{y}^{(k)})\mathbf{D}(\mathbf{y}^{(k)}) + O(\|\mathbf{y} - \mathbf{y}^{(k)}\|^2), \tag{A.61}$$

where the Jacobian $J(\mathbf{u})$ is given by

$$J(\mathbf{u}) = \begin{bmatrix} b_1(\mathbf{u}) & c_1(\mathbf{u}) & 0 & \cdot & \cdot & \cdot & \cdot & 0 \\ a_2(\mathbf{u}) & b_2(\mathbf{u}) & c_2(\mathbf{u}) & 0 & & & & \cdot \\ 0 & \cdot & & & & & & \cdot \\ \cdot & \cdot & & & & & & 0 \\ \cdot & \cdot & & & & a_{N-1}(\mathbf{u}) & b_{N-1}(\mathbf{u}) & c_{N-1}(\mathbf{u}) \\ 0 & \cdot & \cdot & \cdot & \cdot & 0 & a_N(\mathbf{u}) & b_N(\mathbf{u}) \end{bmatrix}$$

with

$$a_m(\mathbf{u}) = -\frac{1}{2}\left[1 + \frac{h}{2} f_{,y'}(x_m, u_m, v_m)\right],$$

$$b_m(\mathbf{u}) = 1 + \frac{h^2}{2} f_{,y}(x_m, u_m, v_m),$$

$$c_m(\mathbf{u}) = -\frac{1}{2}\left[1 - \frac{h}{2} f_{,y'}(x_m, u_m, v_m)\right],$$

and $v_m = (u_{m+1} - u_{m-1})/2h$. The iterative scheme

$$\mathbf{y}^{(k+1)} = \mathbf{y}^{(k)} + \Delta\mathbf{y}^{(k)}, \tag{A.62}$$

with $\Delta\mathbf{y}^{(k)}$ being the solution of

$$J(\mathbf{y}^{(k)})\Delta\mathbf{y}^{(k)} = -\mathbf{D}(\mathbf{y}^{(k)}), \tag{A.63}$$

follows from (A.61) and will be used to generate successively improved approximations of the solution of (A.47). The linear system (A.63) has a unique solution for any $\mathbf{y}^{(k)}$ if $h \leq 2/P$ and can be solved by Gaussian elimination or any appropriate method. It is well known that the Newton iteration scheme (A.62) and (A.63) converges quadratically if $\mathbf{y}^{(k)}$ is sufficiently close to the exact solution. We can get such a close approximation by performing a sufficiently large number of fixed point iterations (A.51).

## 9. Exercises

1. Solve $y'' + p(x)y' + q(x)y = f(x)$, $a < x < b$, for each of the following sets of boundary conditions by way of the numerical solutions $\varphi_1(x), \varphi_2(x)$, and $\varphi_p(x)$ for three suitable initial-value problems:

   (a) $y'(a) = V$, $y(b) = B$;

   (b) $\alpha y(a) + \beta y'(a) = V$, $y(b) = B$ $\quad(\alpha\beta \neq 0)$;

   (c) $y(a) = A$, $\alpha y(b) + \beta y'(b) = V$ $\quad(\alpha\beta \neq 0)$.

2. Solve by an appropriate fundamental set of solutions for each of the following boundary-value problems on the interval $[a, b]$:

   (a) $y''' + p(x)y'' + q(x)y = 0$ with $y(a) = A$, $y'(a) + y''(a) = 0$, $y'''(a) = 0$, and $y(b) - 3y''(b) = 0$.

   (b) $y_k' = p_k(x)y_1 + q_k(x)y_2 + r_k(x)y_3 + s_k(x)y_4 + f_k(x)$ $\quad(k = 1, 2, 3, 4)$ with $y_1(a) = 0$, $y_3(a) = 0$, $y_2(b) = 0$, and $y_4(b) = 0$.

   Note that $p, q, p_k, q_k, r_k, s_k$, and $f_k$ are known functions and $A$ is a known constant.

3. $\quad \varepsilon^2 y'' = y - 1, \quad y(0) = 0, \quad y(1) = 0.$

To find the solution of this BVP by the method of IVP, we need $\varphi_2(x)$ and $\varphi_p(x)$ for the expression $y = v_0\varphi_2(x) + \varphi_p(x)$, where $\varphi_2$ and $\varphi_p$ are a complementary solution and a particular solution of the ODE, respectively, with

$$\varphi_2(0) = \varphi_p(0) = \varphi_p'(0) = 0, \qquad \varphi_2'(0) = 1.$$

(a) For $\varepsilon = 10$, generate $\varphi_2(x)$ numerically by a method of your choice with $h = 0.1$.

(b) In your opinion, how small must $h$ be if we want two-digit accuracy by the *Simple Euler* method for $\varepsilon = 10^{-1}$? Give some justification for your answer.

4. Sketch a solution process for the BVP

$y_k' = f_k(x, y_1, \ldots, y_4), \quad (k = 1,2,3,4;\ a < x < b);$

$y_j(a) = Y_j \quad (j = 1,2,3), \quad y_4(b) = Y_4.$

How would you do the problem if the boundary conditions are replaced by $y_1(a) = y_1$ and $y_j(b) = Y_j \quad (j = 2,3,4)$?

5. Solve the following *nonlinear* BVP by iterating on the boundary condition at $x = 2$:

$$y'' + \frac{1}{x}y' = 1 \quad (1 < x < 2), \qquad y(1) = 0, \qquad y'(2) + e^{y(2)} = 0.$$

[Hint: The general solution of this linear ODE is

$$y(x) = c_0 + c_1 \ell n\, x + \frac{1}{4}x^2.]$$

6. Solve the following BVP by either iterating on the boundary condition at $x = z$:

$$y'' + x(y')^2 = 0 \quad (1 < x < 2), \qquad y(1) = 0, \qquad y'(2) = \ell n(81/25).$$

[Hint: The general solution of the nonlinear ODE is

$$y(x) = \frac{1}{c_1}\ell n\,\frac{x - c_1}{x + c_1} + c_2 \quad (c_1 > 0).]$$

7. $\quad y' = 1 + y^2 \quad (0 < x < 1), \qquad y(1) = 2.$

Find $y(0) \equiv \bar{y}_0$ so that $y(1) = 2$ as follows:

(a) Use Euler's method with $h = -0.1$ ($N = 10$) to integrate the ODE backward from $x = 1$ to $x = 0$ and get $y(0) = y_{10} \equiv \bar{y}_0$.

(b) As a check for accuracy, use $\bar{y}_0$ as obtained in part (a) to integrate the ODE numerically forward (using Euler's method with $h = 0.1$ again) from $x = 0$ to $x = 1$. Compare $y_{10}$ obtained in this way with the prescribed value which is $y(1) = 2$.

8. Find $y(0) = \bar{y}_0$ of Exercise 7, by the *shooting method*, i.e., generate $\varphi(x;y_0)$ by solving the ODE with $y(0) = y_0$ from $\varphi(1;y_0) = 2$ iteratively.

9. $$-y'' + ay = f(x) \quad (0 < x < 1), \quad y(0) = y(1) = 0.$$

Use the finite difference approximations $y'(x_i) \cong (y_{i+1} - y_{i-1})/2h$ and $y''(x_i) \cong (y_{i+1} - 2y_i + y_{i-1})/h^2$.

(a) Write down the $3 \times 3$ finite difference analogue of the BVP by taking $h = 1/4$.

(b) For $f(x) = 4\pi^2 \sin(2\pi x)$ and $a = 0$, solve for $\{y_i\}$ and find the difference between $y_i$ and the exact solution $y(x) = \sin(2\pi x)$ at $x_i = i/4$.

(c) Write down the $5 \times 5$ finite difference analogue of the BVP for the given ODE and $y'(0) = y'(1) = 0$ (instead of $y(0) = y(1) = 0$) by taking $h = 1/4$.

(d) For $a = 0$, show that the coefficient matrix in $\mathbf{Cy} = \mathbf{f}$ for part (c) is singular.

(e) For the continuous problem of part (c), show that the solution is not unique.

10. For the linear BVP $-y'' + p(x)y' + q(x)y = f(x)$ ($a < x < b$), $y(a) = A$, and $y(b) = B$, we have, with the difference approximations for $y'$ and $y''$ in Exercise 9, the finite difference analogue

$$L_h[y_k] \equiv \frac{2}{h^2}\left[a_k y_{k-1} + b_k y_k + c_k y_{k+1}\right] = f(x_k) \quad (1 \le k \le N),$$

where $h = (b - a)/(N + 1)$, $x_k = a + kh$, $y_0 = A$, $y_{N+1} = B$,

$$a_k = -\frac{1}{2}\left[1 + \frac{1}{2}hp(x_k)\right], \qquad b_k = 1 + \frac{1}{2}h^2 q(x_k),$$

$$c_k = -\frac{1}{2}\left[1 - \frac{1}{2}hp(x_k)\right]$$

for $1 \le k \le N$. The system may be written as $C\mathbf{y} = \mathbf{f}$ where

$$\mathbf{y} = (y_1, \ldots, y_N)^T$$

$$\mathbf{f} = \frac{1}{2} h^2 (f(x_1), \ldots, f(x_N))^T - (a_1 A, 0, \ldots, 0, c_N B)^T.$$

(a) Obtain the coefficient matrix $C$.

(b) Let $C = LU$, where $U \equiv [u_{ij}]$ and $L \equiv [\ell_{ij}]$ with $u_{kk} = 1$, $u_{k(k+1)} = \gamma_k, \ell_{kk} = \beta_k, \ell_{k(k-1)} = \alpha_k$, and $u_{kj} = 0$ otherwise. Show that $\alpha_k = a_k$, $\beta_1 = b_1, \gamma_1 = c_1/\beta_1 = b_k - a_k \gamma_{k-1}$, and $\gamma_k = c_k/\beta_k, k = 2, 3, \ldots, (N-1)$.

(c) Let $U\mathbf{y} = \mathbf{z}$. Then $C\mathbf{y} = \mathbf{f}$ can be written as $L\mathbf{z} = \mathbf{f}$. Show that $z_1 = f_1/\beta_1, z_k = (f_k - \alpha_k z_{k-1})/\beta_k$ for $k = 2, 3, \ldots N$, and $y_N = z_N, y_k = z_k - \gamma_k y_{k+1}$ for $k = N-1, N-2, \ldots, 1$.

11. Suppose $|b_k| > |a_k| + |c_k|$ in Exercise 10 ($a_1 = c_N = 0$). Prove the following:

(a) The coefficient matrix is nonsingular.

(b) $|\gamma_k| < 1$.

(c) $|b_k| - |a_k| \le |\beta_k| \le |b_k| + |a_k|$.

12. Suppose $|p(x)| \le P$ and $0 < Q_0 \le q(x) \le Q_1$ for $a \le x \le b$. Prove that the system $C\mathbf{y} = \mathbf{f}$ of Exercise 10 has a unique solution as given in part (c) of that problem if $h \le 2/P$.

# Bibliography


Barber, J.R., *Elasticity*, Kluwer Academic Publishers. Dordrecht/Boston/London, 1992.

Berkovitz, L.D., *Optimal Control Theory*, Springer-Verlag, New York, 1974.

Bliss, G.A., *Lectures on the Calculus of Variations*. The University of Chicago Press, Chicago, 1963.

Boyce, W.E., and DiPrima, R.C., *Elementary Differential Equations and Boundary Value Problems*, 3rd Ed., John Wiley & Sons, New York, 1976.

Brechtken–Manderscheid, U., *Introduction to the Calculus of Variations*, Chapman & Hall, 1991, London/New York.

Bryson, A., and Ho, Y.C., *Applied Optimal Control*, Ginn, Lexington, MA, 1969.

Bucciarelli, L.L., and Dworsky, N., *Sophie Germain*, D. Reidel, Dordrecht, 1980.

Burmeister, E., and Dobell, A.R., *Mathematical Theories of Economic Growth*, Macmillan, New York, 1970.

Caratheodory, C., *Variationsrechnung und partielle Differentialgleichungen erster Ordnung*, Teubner, Berlin, 1935.

Churchill, R.V., and Brown, J.W., *Fourier Series and Boundary Value Problems*, 3d ed., McGraw-Hill, New York, 1978.

Clark, C.W., *Mathematical Bio-Economics*, 2nd, Ed., John Wiley & Sons, New York, 1990.

Clarke, F.H., *Calculus of Variations and Optimal Control*, Lecture Notes for Math. 426, University of British Columbia, Vancouver, B.C., Canada, 1980.

Cole, J.D., and Cook, L.P., *Transonic Aerodynamics*, North Holland, New York, 1986.

Courant, R., and Hilbert, D., *Methods of Mathematical Physics*, Vol. 1, Intersicence New York, 1953.

Ewing, G.M., *Calculus of Variations with Applications*, Dover, New York, 1985.

Finzi, B., "Integrazione delle equazioni indefinite della meccanica dei sistemi continui," *Rend. Lincei* (6), 19 (1934), 578-584, 620-623.

Flanders, H., *Differential Forms*, Academic Press, New York, 1963.

Fleming, W.H., and Rishel, R.W., *Deterministic and Stochastic Optimal Control*, Springer-Verlag, Berlin, 1975.

Forsyth, A.R., *Calculus of Variations*, Dover, New York, 1960.

Friedrichs, K.O., and Dressler, R.F., "A boundary layer theory for elastic bending of plates," *Commun. Pure Appl. Math.*, 14 (1961), 1-33.

Fung, Y.C., *Foundations of Solid Mechanics*, Prentice-Hall, Englewood Cliffs, NJ, 1967

Gelfand, I.M., and Fomin, S.V., *Calculus of Variations*, Prentice-Hall, Englewood Cliffs, NJ, 1963.

Goldstine, H.H., *A History of the Calculus of Variations from the 17th Through 19th Century*, Springer-Verlag, Berlin, 1980.

Greenspan, H.P., and Benney, D.J., *Calculus: An Introduction to Applied Mathematics*, McGraw-Hill, New York, 1973.

Gregory, J., and Lin, C., *Constrained Optimization in the Calculus of Variations and Optimal Control Theory*, Van Nostrand Reinhold, New York, 1992.

Gregory, R.D., and Wan, F.Y.M., "On plate theories and Saint-Venant's principle", *Internat. J. Solids Structures*, 21 (1985), 1005-1024.

Gregory, R.D., and Wan, F.Y.M., "The interior solution for linear problems of elastic plates," *J. Appl. Mech.*, 55 (1988), 551-559.

Henrici, P., *Discrete Variable Methods in Ordinary Differential Equations*, John Wiley & Sons, Inc., New York, 1962.

Hestenes, M.R., *Calculus of Variations and Optimal Control Theory*, John Wiley & Sons, New York, 1966.

Hillier, F.S., and Lieberman, G.J., *Operations Research*, 2nd ed., Holden-Day, Inc., San Francisco, 1974.

Horgan, C.O., and Knowles, J.K., "Recent developments concerning Saint-Venant's principle," *Adv. Appl. Mech.*, 23 (1983), 179-269.

Kamien, M.I., and Schwartz, N.L., *Dynamic Optimization: The Calculus of Variations and Optimal Control in Economics and Management*, North-Holland, New York, 1981.

Kanemoto, Y., *Theories of Urban Externalities*, North Holland, New York, 1980.

Keller, H.B., *Numerical Methods for Two-Point Boundary-Value Problems*, Blaisdell, Waltham, MA, 1968.

Kevorkian, J., *Partial Differential Equations*, Wadsworth & Brooks/Cole, Pacific Grove, CA, 1990.

Kevorkian, J., and Cole, J.D., *Perturbation Methods in Applied Mathematics*, Springer-Verlag, Berlin, 1981.

Lamb, H., *Hydrodynamics*, 6th ed., Dover, New York, 1945.

Lee, E.B., and Markus, L., *Foundations of Optimal Control Theory*, John Wiley & Sons, New York, 1967.

Leitmann, G., *The Calculus of Variations and Optimal Control*, Plenum Press, New York, 1981.

Lin, C.C., "Hydrodynamics of helium II," Proc. Enrico Fermi Intern'l School of Physics, Varenna, Italy (July 3-5, 1961), ed. C. Careri, Italian Phys. Soc. Ministry of Educ., 1963, 93-146 (see p. 95).

Love, A.E.H., *A Treatise on the Mathematical Theory of Elasticity*, 4th Ed., Dover, New York, 1944.

Luke, J.C., "A variational principle for a fluid with a free surface," J. Fluid Mech. 27 (1967), 395-397.

Lutzen, J., *Joseph Liouville, 1809-1882: Master of Pure and Applied Mathematics*, Springer-Verlag, Berlin, 1990, p. 633.

Macki, J., and Strauss, A., *Introduction to Optimal Control Theory*, Springer-Verlag, Berlin, 1982.

Pars, L.A., *An Introduction to the Calculus of Variations*, Heinemann, London, 1962.

Pearson, C.E., *Theoretical Elasticity*, Harvard University Press, Cambridge, MA, 1959

Pontryagin, L.S., Boltyanskii, V.G., Gamkrelidze, R.V., and Mishchenko, E.F., *The Mathematical Theory of Optimal Processes*, Interscience, New York, 1962.

Reissner, E., "On the theory of bending of elastic plates," *J. Math. Phys.*, 23 (1944), 184-191.

Reissner, E., "The effect of transverse shear deformation on the bending of elastic plates," *J. Appl. Mech.*, 12 (1945), A69-A77; 13 (1946), A252.

Reissner, E., "On bending of elastic plates," *Quart. Appl. Math.*, 5 (1947), 55-68.

Reissner, E., "On a variational theorem in elasticity," J. Math. and Phys. 29, 1950, 90-95.

Reissner, E., "On a variational theorem for finite elastic deformations," J. Math. and Phys. 32, 1953, 129-135.

Reissner, E., "Finite twisting and bending of thin rectangular elastic plates," *J. Appl. Mech.*, 24 (1957), 391-396.

Reissner, E., "Variational considerations for elastic beams and shells," *Proc. Amer. Soc. Civil Engrs. (EM)*, 8 (1962), 23-57.

Reissner, E., "On the foundations of generalized linear shell theory," *Theory of Thin Shells*, Proc. 1967 IUTAM Symp., Copenhagen, IUTAM Pub., 1969, 15-30.

Reissner, E., "On the derivation of two-dimensional shell equations from three dimensional elasticity theory," *Studies Appl. Math.*, 49 (1970), 205-224.

Reissner, E., "On consistent first approximations in the general linear theory of thin elastic shells," *Ingenieur-Archiv*, 40 (1971), 402-419.

Reissner, E., "Some aspects of the variational-principles problem in elasticity," *Comput. Mech.*, 1 (1986), 3-9.

Reissner, E., "Variational principles in elasticity," *Finite Element Handbook*, ed. S. Atluri, McGraw-Hill, 1987, 2.3–2.19.

Reissner, E., and Wan, F.Y.M., "A note on Gunther's analysis of Couple stress," *Mechanics of Generalized Continua*, IUTAM Symposium, Freudenstadt-Stuttgart, 1967, 83-86.

Rockefellar, R.T., *Convex Analysis*, Princeton Univ. Press, Princeton, NJ. 1970.

Sagan, H., *Introduction to the Calculus of Variations*, Dover, New York, 1985.

Sedgewick, R., *Algorithms*, 2nd ed., Addison-Wesley Publishing Company, Reading, Mass., 1988.

Serrin, J., "Mathematical principles of classical fluid mechanics," *Encycl. of phys.* Vol. III/1, ed. S. Flugge, Springer-Verlag, Berlin, 1959, 148-149.

Smith, D.R., *Variational Methods in Optimization*, Prentice-Hall, Inc., Englewood Cliffs, NJ, 1974.

Sokolnikoff, I.S., *Mathematical Theory of Elasticity*, McGraw-Hill, New York, 1956.

Solow, R.M., "Congestion, density and the use of land in transportation," *Swedish J. Econ.*, 74 (1972), 161-173.

Solow, R.M., "Congestion cost and the use of land for streets," *Bell J. Econ. Management Sci.*, 4 (2) (1973), 602-618.

Solow, R.M., "Intergenerational equity and exhaustible resources," *Rev. Econ. Studies (Symp. Issue)* (1974), 29-46.

Solow, R.M., and Vickery, W.S., "Land use in a long narrow city," *J. Econ. Theory*, 3 (1971), 430-447.

Stoker, J.J., *Water Waves*, Wiley-Interscience, New York, 1957.

Strang, G., *Linear Algebra and Its Applications*, 2nd Ed., Academic Press, New York, 1980.

Strang, G., and Fix, G., *An Analysis of the Finite Element Method*, Prentice-Hall, Inc., Englewood Cliffs, N.J. 1973.

Struik, D., *Differential Geometry*, 2nd ed., Addison-Wesley, Reading, MA, 1961.

Timoshenko, S., and Goodier, J.N., *Theory of Elasticity*, 2nd Ed., McGraw-Hill, New York, 1951.

Timoshenko, S., and Woinowsky-Krieger, S., *Theory of Plates and Shelles*, 2nd Ed., McGraw-Hill, New York, 1959.

Tonelli, L., *Fondamenti di Calcolo delle Variazioni*, Zanichelli, Bologna, Vol. I, 1921, and II, 1923.

Troutman, J.L., *Variational Calculus with Elementary Convexity*. Springer-Verlag, Berlin, 1983.

Wan, F.Y.M., "Perturbation and asymptotic solution for problems in the theory of urban land rent," *Studies Appl. Math.*, 56 (1977), 219-239.

Wan, F.Y.M., "Constant sustainable consumption rate in optimal growth with a multi-grade exhaustible resource," *Studies Appl. Math.*, 63 (1980), 47-66.

Wan, F.Y.M., *Mathematical Models and Their Analysis*, Harper & Row Publishers, New York, 1989.

Wan, F.Y.M., "Perturbation solutions for the second-best land use problem," *Can. Appl. Math. Quart.*, 1, (1993), 1-33.

Washizu, K., *Variational Methods in Elasticity and Plasticity*, 2nd ed., Pergamon Press, London, 1975.

Weinstock, R., *Calculus of Variations with Applications to Physics and Engineering*, Dover, New York, 1974.

Whitham, G.B., "A general approach to linear and non-linear dispersive weves usuig a Lagrangian," J. Fluid Mech. 22 (1965), 273-283.

Whitham, G.B., "Non-linear dispersion of water waves," J. Fluid Mech. 27 (1967), 399-412.

Young, L.C., *Lectures on Calculus of Varoiations and Optimal Control Theory*, W.B. Saunders Co., Philadelphia, 1969.

# Index

Printed in the United States
by Baker & Taylor Publisher Services

Printed in the United States
by Baker & Taylor Publisher Services